OPERATION AND MAINTENANCE OF WASTEWATER COLLECTION SYSTEMS

VOLUME I

Sixth Edition

A Field Study Training Program

prepared by

Office of Water Programs
College of Engineering and Computer Science
California State University, Sacramento

in cooperation with the
California Water Environment Association

★★★★★★★★★★★★★★★★★★★★★★★★★★★★★★★★★

Kenneth D. Kerri, Project Director
John Brady, Co-Director

★★★★★★★★★★★★★★★★★★★★★★★★★★★★★★★★★

for the

U.S. Environmental Protection Agency
Office of Water Program Operations
Municipal Permits and Operations Division
First Edition, Grant No. T900494 (1976)
Spanish Edition, NETA Grant No. X-999658-01-1 (1999)

2003

NOTICE

This manual is revised and updated before each printing based on comments from persons using the manual.

FIRST EDITION, First Printing, 1976	5,000
SECOND EDITION, First Printing, 1983	5,000
THIRD EDITION, First Printing, 1987	8,000
Second Printing, 1991	8,000
FOURTH EDITION, First Printing, 1993	8,000
FIFTH EDITION, First Printing, 1996	8,000
Second Printing, 1999	10,000
Third Printing, 2001	5,000
SIXTH EDITION, First Printing, 2003	15,000
Second Printing, 2008	12,600
Third Printing, 2013	3,000

ISBN 1-884701-43-4 (Volume I)
ISBN 1-884701-45-0 (Volumes I and II)

OPERATOR TRAINING MANUALS

OPERATOR TRAINING MANUALS AND VIDEOS IN THIS SERIES are available from the Office of Water Programs, California State University, Sacramento, 6000 J Street, Sacramento, CA 95819-6025, phone: (916) 278-6142, e-mail: wateroffice@csus.edu or FAX (916) 278-5959, website: www.owp.csus.edu.

1. *OPERATION AND MAINTENANCE OF WASTEWATER COLLECTION SYSTEMS*, 2 Volumes,*
2. *OPERATION OF WASTEWATER TREATMENT PLANTS*, 2 Volumes,
3. *SMALL WASTEWATER SYSTEM OPERATION AND MAINTENANCE*, 2 Volumes,
4. *ADVANCED WASTE TREATMENT*,
5. *INDUSTRIAL WASTE TREATMENT*, 2 Volumes,
6. *TREATMENT OF METAL WASTESTREAMS*,
7. *PRETREATMENT FACILITY INSPECTION*,**
8. *WATER TREATMENT PLANT OPERATION*, 2 Volumes,
9. *SMALL WATER SYSTEM OPERATION AND MAINTENANCE*,
10. *WATER DISTRIBUTION SYSTEM OPERATION AND MAINTENANCE*, and
11. *UTILITY MANAGEMENT*.

* Other training materials and training aids developed by the Office of Water Programs to assist operators in improving collection system operation and maintenance and overall performance of their systems include:

1. *COLLECTION SYSTEMS: METHODS FOR EVALUATING AND IMPROVING PERFORMANCE*. This handbook presents detailed benchmarking procedures and worksheets for using performance indicators to evaluate the adequacy and effectiveness of existing O & M programs. It also describes how to identify problems and suggests many methods for improving the performance of a collection system.
2. *OPERATION AND MAINTENANCE TRAINING VIDEOS*. This series of six 30-minute videos demonstrates the equipment and procedures collection system crews use to safely and effectively operate and maintain their collection systems. These videos complement and reinforce the information presented in Volumes I and II of *OPERATION AND MAINTENANCE OF WASTEWATER COLLECTION SYSTEMS*.

** *PRETREATMENT FACILITY INSPECTION TRAINING VIDEOS*. This series of five 30-minute videos demonstrates the procedures to effectively inspect an industry, measure flows, and collect samples. These videos complement and reinforce the information presented in *PRETREATMENT FACILITY INSPECTION*.

The Office of Water Programs at California State University, Sacramento, has been designated by the U.S. Environmental Protection Agency as a *SMALL PUBLIC WATER SYSTEMS TECHNOLOGY ASSISTANCE CENTER*. This recognition will provide funding for the development of training videos for the operators and managers of small public water systems. Additional training materials will be produced to assist the operators and managers of small systems.

PREFACE TO THE FIRST AND SECOND EDITIONS

The purposes of this wastewater collection system home study course are:

1. To develop new qualified wastewater collection system operators,
2. To expand the abilities of existing operators, permitting better service to both their employers and the public, and
3. To prepare operators for civil service and *CERTIFICATION EXAMINATIONS.*[1]

To provide you with the information needed to operate and to maintain collection systems as efficiently and effectively as possible, experienced wastewater collection system operators prepared the material in each chapter of this manual.

For many years the Water Pollution Control Federation, its member associations, state and local utility agencies, and colleges and universities have sensed a need for improvement in the dissemination of information on the operation and maintenance of wastewater collection systems. They have also felt a need for better training opportunities in this field. However, because of the lack of communication between the practical people doing the work in the field and the professional people in charge of publishing and training activities, the dissemination of information and training in the collection system field has been, until a few years ago, almost negligible compared to that in the wastewater treatment plant field.

Following completion of and the successful results from EPA's national field study training program, *OPERATION OF WASTEWATER TREATMENT PLANTS,*[2] a similar effort appeared desirable for wastewater collection system operators. In cooperation with the California Water Pollution Control Association, the project directors prepared and submitted a proposal to the Environmental Protection Agency for financial support through the Foundation of California State University, Sacramento. CWPCA President Robert Bernicchi requested the Presidents of the Water Pollution Control Federation's member associations to urge EPA to support this national effort to develop a training manual for wastewater collection system operators. Response from the member associations was quick, the need was documented, and EPA approval received. Chapters were written, presented at small informal seminars, reviewed by consultants and reviewers from throughout the United States, field tested by potential and experienced collection system operators, reviewed by the Environmental Protection Agency, and revised after each step in accordance with the suggestions and experiences gained from these sources.

The California Water Pollution Control Association would like to acknowledge the cooperation and assistance in review that it received from the Personnel Advancement Committee and the Wastewater Collection Systems Committee of the Water Pollution Control Federation during the development of this publication.

The project directors are indebted to the authors of each chapter and the many potential and experienced collection system operators who field tested and reviewed the material. F.J. Ludzack, National Training Center, Office of Water Program Operations, Environmental Protection Agency, offered many technical improvements. Dr. Elie Namour, author of EPA's "Manpower Manuals for Wastewater Collection Systems," reviewed the manpower aspects of the manual. Our Program Manager, Robert Rose, Chief, State and Local Training Activities Section, EPA, served as a valuable source of information and guidance. A special note of thanks is due to our typists, Charlene Arora and Elaine Saika, who somehow managed to cope with illegible handwriting and unmeetable deadlines with smiles and perfect manuscripts. Illustrations were drawn by Martin Garrity. Some sketches and line drawings were prepared by George Gardin and Sue Hashimoto.

KENNETH D. KERRI
JOHN BRADY

1976

[1] *Certification Examination. An examination administered by a state agency that wastewater collection system operators take to indicate a level of professional competence. In many states certification of wastewater collection system operators is voluntary. Current trends indicate that more states, provinces, and employers will require wastewater collection system operators to be "certified" in the future.*

[2] *OPERATION OF WASTEWATER TREATMENT PLANTS, by Kenneth D. Kerri and Bill B. Dendy, California State University, Sacramento, 6000 J Street, Sacramento, CA 95819-6025.*

PREFACE TO THE THIRD EDITION

Collection system operators have always survived by their wits and ingenuity. Today operators are applying the most advanced technologies available to the operation and maintenance of wastewater collection systems. The TV technology used to inspect and record the status of collection systems is continually advancing. New instruments whose precision and reliability are improving over previous models are becoming available to make work in and around collection systems much safer. Personal computers are being used to inventory collection systems, maintain a history of the system, and schedule preventive maintenance when needed and before problems develop. Technology is allowing collection system operators to perform more and more tasks without having to enter manholes or sewers, thus reducing the risk of exposure to safety hazards.

Persons selected to work on this revision and expansion were recognized experts throughout the United States. Many of the collection system operators who successfully completed this course made helpful suggestions for improvement as they worked their way through the manual. Gay Kornweibel prepared the manuscript and took care of the administrative details. The efforts of everyone who contributed to the successful revision of this manual are sincerely appreciated.

1987 KENNETH D. KERRI

PREFACE TO THE SIXTH EDITION

Collection system operators and the technology they use to safely perform their jobs are continually becoming more sophisticated and efficient. The Office of Water Programs at California State University, Sacramento, continues to rely on dedicated collection system operators to help us keep our operator training manuals current with the rapidly advancing collection system technology and workforce.

This manual was first published in 1976 as a single volume. Since then, the manual has been expanded to two volumes and over 92,000 copies of Volumes I and II have been distributed all over the world. The current two volumes are the most widely used collection system O & M training references in the United States. O & M staff, engineers, regulators, and others involved in collection systems use these manuals extensively as references when performing their daily tasks.

Rick Arbour helped us revise the Fourth and Fifth Editions and provided some of the graphics used in the Sixth Edition. Gary Batis has provided us information regarding the newest technologies for the Sixth Edition. He also made many helpful suggestions to improve the usefulness of the Sixth Edition for collection system operators. David Jurgins and his collection system crews with the City of Fayetteville, Arkansas, and Victor Coles of Municipal Service Company, Inc., reviewed the drafts of the Sixth Edition and made helpful suggestions. Peg Hannah edited the material and prepared the manuscript for printing, Jan Weeks corrected printer's proofs and coordinated production of the manual, and Gay Kornweibel managed the administrative details. The efforts of everyone who contributed to the successful revision of the manual are sincerely appreciated.

2003 KENNETH D. KERRI

OBJECTIVES OF THIS MANUAL

Proper installations, inspections, operations, maintenance, and repairs of wastewater collection, conveyance, and treatment facilities have a significant impact on the operation and maintenance costs and effectiveness of these facilities. The objective of this manual is to provide wastewater collection system operators with the knowledge and skills where possible to effectively operate and to maintain collection systems, thus eliminating or reducing the following problems:

1. Health hazards caused by untreated wastewater flowing down streets and watercourses during stoppages and storms;
2. System failures that result from the lack of proper installation, inspection, preventive maintenance, surveillance, and repair programs designed to protect the public's investment in these facilities;
3. Odors from the collection system, lift stations, and treatment plants caused by collection system problems;
4. Shock loads from the clearing of stoppages that affect wastewater treatment processes;
5. Corrosion damages to equipment and structures in collection systems and treatment plants;
6. Inflow and infiltration that use a valuable portion of the capacity of the facilities;
7. Noise pollution from collection systems (noisy manhole lids) and lift stations; and
8. Complaints from the public or local officials due to the unreliability of the collection system. These complaints result from constant system failures and create a poor public image of the agency that is supposed to be managing the system.

SCOPE OF THIS MANUAL

This manual contains information on:

1. What collection systems are expected to achieve,
2. Why wastewater collection systems must be properly operated and maintained,
3. What the collection system operator is expected to do to keep the collection system functioning as intended,
4. How to inspect and test newly constructed sewers and new and old wastewater collection systems,
5. How to locate and evaluate problems such as stoppages, leaks, odors, and lift station failures,
6. Selection of procedures and equipment to correct identified problems and to minimize recurrence of problems and failures,
7. How to operate and maintain collection system inspection and cleaning equipment,
8. Methods of selection and application of chemicals,
9. Procedures for operating, maintaining, and repairing collection systems,
10. Instructions for operating, maintaining, troubleshooting, and repairing collection system equipment and facilities such as lift stations,
11. How to evaluate the status of a collection system and to select the appropriate rehabilitation measures when necessary,
12. Techniques for recognizing hazards and developing safe procedures, and
13. How to organize and administer the operation and maintenance of wastewater collection systems.

Material contained in this manual furnishes you with information concerning situations encountered by most collection system operators regardless of whether they are located in the North, South, East, or West, at sea level or in the mountains, in either humid or arid regions. These materials also provide you with an understanding of the basic operational and maintenance concepts for wastewater collection systems and with an ability to analyze and solve problems when they occur. Operation and maintenance programs for wastewater collection systems will vary with the age of the system, the extent and effectiveness of previous programs, and local conditions. You will have to adapt the information and procedures in this manual to your particular situation.

Technology is advancing very rapidly in the field of the operation and maintenance of wastewater collection systems. To keep pace with scientific advances, the material in this program must be periodically revised and updated. This means that you, the wastewater collection system operator, must be aware of new advances and recognize the need for continuous personal training reaching beyond this program. Training opportunities exist in your daily work experience, from your associates, and from attending meetings, workshops, conferences, and classes.

USES OF THIS MANUAL

This manual was developed to serve as a home-study course for collection system operators in remote areas or persons unable to attend formal classes either due to shift work, personal reasons, or the unavailability of suitable classes. This home-study training program uses the concepts of self-paced instruction where you are your own instructor and work at your own speed. In order to certify that a person has successfully completed this program, objective tests and special answer sheets for each chapter are provided when a person enrolls in this course.

Once collection system operators started using this manual for home study, they realized that it could serve effectively as a textbook in the classroom. Colleges and universities have used the manual as a text in formal classes often taught by foremen and supervisors. In areas where colleges were not available or were unable to offer classes in the operation and maintenance of wastewater collection systems, collection system operators and utility agencies joined together to offer their own courses using the manual.

Utility agencies have enrolled from three to over 70 of their collection system operators in this training program. A manual is purchased for each operator. A supervisor, foreman, or group of foremen are designated as instructors. These instructors help answer questions when the persons in the training program have questions or need assistance. The instructors grade the objective tests at the end of each chapter, record scores, and notify California State University, Sacramento, of the scores when a person successfully completes this program. This approach avoids the long wait while papers are being graded and returned by CSUS.

This manual was prepared to help collection system operators operate and maintain their collection systems. Please feel free to use it in the manner which best fits your training needs and the needs of other collection system operators. We will be happy to work with you to assist you in developing your training program. Please feel free to contact:

Office of Water Programs
California State University, Sacramento
6000 J Street
Sacramento, California 95819-6025

Phone (916) 278-6142

INSTRUCTIONS TO PARTICIPANTS IN HOME-STUDY COURSE

Procedures for reading the lessons and answering the questions are contained in this section.

To progress steadily through this program, you should establish a regular study schedule. For example, many operators in the past have set aside two hours during two evenings a week for study.

The study material is contained in two volumes with a total of thirteen chapters. Some chapters are longer and more difficult than others. For this reason, many of the chapters are divided into two or more lessons. The time required to complete a lesson will depend on your background, experience, and study habits. Some people may require an hour to complete a lesson, and some might require three hours, but that is perfectly all right. *THE IMPORTANT THING IS THAT YOU UNDERSTAND THE MATERIAL IN THE LESSON!*

To complete this program you will have to work all of the chapters. You may proceed in numerical order, or you may wish to work some lessons sooner. If you are working on a specific job, such as pipeline cleaning, you may wish to read Chapter 6 after Chapter 2.

SAFETY IS A VERY IMPORTANT CHAPTER. Everyone working in and around wastewater collection systems must always be safety conscious. You must take extreme care with your personal hygiene to prevent the spread of disease to yourself and your family. Collection system operators may have to work in and near excavations, embankments (fills), washouts, below-grade structures, traffic, and other dangerous situations and equipment that can cause a serious, disabling injury if the operator is not aware of the potential danger and does not exercise adequate precautions. Confined spaces such as manholes and wet wells may have harmful atmospheres such as toxic gases, explosive mixtures, or oxygen deficiencies. For these reasons, you may decide to work on Chapter 4, "Safe Procedures," and Chapter 11, "Safety/Survival Programs for Collection System Operators," before completing the other chapters.

Begin your study of each chapter by reading the list of objectives to find out what you are expected to learn. In most chapters, the objectives are followed by one or more pages of collection system words and their definitions. These words are referred to in the chapter and the definitions are provided as a reference in case some of the words are unfamiliar to you.

Each lesson is arranged for you to read a short section, write the answers to the questions at the end of the section, compare your answers against suggested answers; and then *YOU* decide if you understand the material sufficiently to continue or whether you should reread the section again. You will find that this procedure is slower than reading a typical textbook, but you will know and be able to use the information much better when you have finished the lesson.

Some discussion and review questions are provided following each lesson in some of the chapters. These questions review the important points you have covered in the lesson. Writing the answers to these questions in your notebook will help you remember the information.

After you have completed the last chapter in each volume, you will find a final examination. This exam is provided for you to review how well you remember the material. You may wish to review the entire manual before you take the final exam. Some of the questions are essay-type questions, which are used by some states and professional associations for higher-level certification examinations. After you have completed the final examination, grade your own paper and determine the areas in which you might need additional review before your next examination.

You are your own teacher in this program. You could merely look up the suggested answers at the end of each chapter or copy the answers from someone else, but you would not have given yourself a chance to understand the material. Consequently, you would not be able to apply the material to the operation and maintenance of your collection system nor recall it during an examination for certification or a civil service position.

YOU WILL GET OUT OF THIS PROGRAM WHAT YOU PUT INTO IT.

SPECIAL NOTICE

All job titles or descriptions of duties of wastewater collection system operators in this manual are identified with "Standard Occupational Titles." We have tried to describe job titles and duties with words that reflect or describe actual tasks performed. These words were used to help you understand the jobs and duties of each person on a crew.

Describing our job titles according to "Standard Occupational Titles" will help you and your employer realize the knowledge, skills, and traits you need to do your job. Use of "Standard Occupational Titles" helps operators prepare for and accept jobs throughout the United States because all of us are talking about the same job. If you work for a small agency, you may be expected to be a "jack of all trades" and do many different types of jobs, regardless of your job title. Operators for a large agency may become a specialist in one or two tasks, but often are expected to be knowledgeable about and skilled at several jobs. The U.S. Employment Service prepared the "Standard Occupational Titles" and emphasizes that these titles *"CANNOT BE CONSIDERED STANDARDS FOR SETTING WAGES OR HOURS, OR SETTLING JURISDICTIONAL MATTERS."*

ENROLLMENT FOR CREDIT AND CERTIFICATE

Students wishing to earn credits and a certificate for completing this course may enroll by contacting the Office of Water Programs, California State University, Sacramento, 6000 J Street, Sacramento, CA 95819-6025, (916) 278-6142. If you have already enrolled, the enrollment packet you were sent contains detailed instructions for completing and returning the objective tests. Please read these important instructions carefully before marking your answer sheets.

Following successful completion of each volume in this program, a Certificate of Completion will be sent to you. If you wish, the Certificate can be sent to your supervisor, the mayor of your town, or any other official you think appropriate. Some operators have been presented their Certificate at a City Council meeting, got their picture in the newspaper, and received a pay raise.

TECHNICAL CONSULTANTS

First and Second Editions	John Carvoretto, Ira Cotton, George Gardner, William Garber, James Kenmir, and Warren Prentice
Third Edition	John Brady and Russ Armstrong
Fourth and Fifth Editions	Rick Arbour
Sixth Edition	Gary Batis and Victor Coles

OPERATION AND MAINTENANCE OF WASTEWATER COLLECTION SYSTEMS

VOLUME I — COURSE OUTLINE

VOLUME II — COURSE OUTLINE

Other operator training aids that may be of interest to you are the collection system operation and maintenance training videos described below and our handbook for evaluating and improving the performance of a collection system.

COLLECTION SYSTEM OPERATION AND MAINTENANCE TRAINING VIDEOS

COURSE OUTLINE

Video	**Title**	**Topic**
1	Guardians of Health	Importance of Operators, Inspection, and Testing
2	TV Stars	Closed-Circuit Television Inspection
3	Pipe Detectives	Pipeline Cleaning and Maintenance Methods
4	Way Makers	Pipeline Cleaning and Chemical Control
5	Flow Movers	Operation of Wastewater Lift Stations
6	Motor Specialists	Maintenance of Wastewater Lift Stations

COLLECTION SYSTEMS: METHODS FOR EVALUATING AND IMPROVING PERFORMANCE

COURSE OUTLINE

Chapter	**Title**
1	Wastewater Collection System Problems/Needs
2	Research Conducted for This Project
3	Benchmark Data
4	Developing, Analyzing, and Interpreting O & M Performance Indicators
5	Suggested Methods for Improving Collection System Performance
6	Case Histories
Appendix A	Summary of Pertinent Literature Reviews
Appendix B	Data Collection Form
Appendix C	Benchmarking Worksheets

CHAPTER 1

THE WASTEWATER COLLECTION SYSTEM OPERATOR

by

Walt Driggs

TABLE OF CONTENTS

Chapter 1. THE WASTEWATER COLLECTION SYSTEM OPERATOR

OBJECTIVES

Chapter 1. THE WASTEWATER COLLECTION SYSTEM OPERATOR

At the beginning of each chapter in this manual you will find a list of *OBJECTIVES*. The purpose of this list is to stress those topics in the chapter that are most important. Contained in the list will be items you need to know and skills you must develop to operate, maintain, and repair your wastewater collection system as efficiently and safely as possible.

Following completion of Chapter 1, you should be able to:

1. Explain the type of work done by collection system operators,
2. Describe where to look for jobs in this profession, and
3. Outline how to learn or determine procedures necessary to perform the collection system operator's job.

CHAPTER 1. THE WASTEWATER COLLECTION SYSTEM OPERATOR

Chapter 1 is prepared especially for new or potential wastewater collection system operators. If you are an experienced sewer maintenance person or wastewater collection system operator, you may find some new viewpoints in this chapter.

1.0 LET'S START LEARNING

I know that there are some of you working in the wastewater collection profession who feel you are the forgotten person due to lack of available books to help you improve yourself. Admittedly many books have been written by and for engineers on wastewater collection systems, but very few have been written by and for collection system operators. There are also those of you who are just starting out in this field. We hope this manual will help all of you in your attempts to improve your knowledge of the operation and maintenance of wastewater collection systems. We, the authors, dedicate ourselves to opening the eyes of those responsible for the collection and conveying of wastewater to make them more aware of the need for adequate equipment and good operation and maintenance practices.

Personnel working in the operation and maintenance of wastewater collection systems may have a variety of different job classifications depending on their particular organization. These classifications can be general such as Operator 1, 2, or 3, Maintenance Person, or they can be specific such as mechanic, electrician, TV operator, or other similar titles. Throughout this manual people working with wastewater collection systems are called *COLLECTION SYSTEM OPERATORS*. Similarly, the term wastewater is gradually replacing the term sewage to describe the liquid and liquid-carried wastes from homes and businesses, and the collection system itself is now commonly referred to as a wastewater collection system rather than a sewerage system. By using words that better describe our jobs, we can improve our public image and upgrade our profession. The Environmental Protection Agency considers us to be in the business of operating and maintaining wastewater collection systems. The wastewater is usually collected and conveyed in sewer pipes. If you find words or terms you don't understand as you progress through this manual, try looking up their meaning in the Appendix, "Collection System Words," at the end of this manual. New words will be defined at the beginning of Chapter 2 and all the remaining chapters.

1.1 WHAT IS A WASTEWATER COLLECTION SYSTEM?

A wastewater collection system gathers the used water from our homes and businesses and conveys it to a wastewater treatment plant (Figure 1.1). This water may come from the kitchens or processes of businesses and industries. Sometimes groundwater, surface water, and storm water may be present, too. The collection system includes the gravity sewers, force mains, manholes, pumping equipment, and other facilities that collect and convey the water to a wastewater treatment plant. Collection systems designed solely for storm waters usually convey these waters to streams, rivers, lakes, or the ocean. Details on the parts of wastewater collection systems are described in Chapter 3.

1.2 WHAT IS MEANT BY OPERATION AND MAINTENANCE OF WASTEWATER COLLECTION SYSTEMS?

Operation and maintenance (O & M) of wastewater collection systems is similar to the O & M program of any other utility. The facilities should be kept in good operating condition. We would not be satisfied if the water company that serves

our homes did not deliver water every time we turn the handle of a faucet. The same goes for the gas company that supplies gas to heat our homes. We don't want excuses why the gas or water delivery system has failed. We expect to have all the gas and water we want when we need it. Recall how annoyed we become if our phone doesn't work? We wonder why the phone company can't keep the phone service working all the time.

What happens when wastewater collection systems fail due to lack of or improper operation and maintenance? Here are some of the problems that can develop.

1. Blockages occurring in the main sewer line that result in backups into homes, businesses, and other customer facilities that are connected to the wastewater collection system. At a minimum, backups usually result in the agency having to pay for cleanup and repair/replacement costs of the damage. Backups can also result in lawsuits that potentially may cost your agency tens of thousands of dollars.
2. Bypassing raw wastewater from the collection system or pump station as a result of system failures. If the raw wastewater is discharged into surface waters or groundwaters that are used for recreation and/or water supply, the consequences can range from water contamination and beach closings to violations of the Clean Water Act. In many cases, state and federal regulatory agencies will assess fines for such violations and the fines may easily add up to tens of thousands or hundreds of thousands or even millions of dollars.

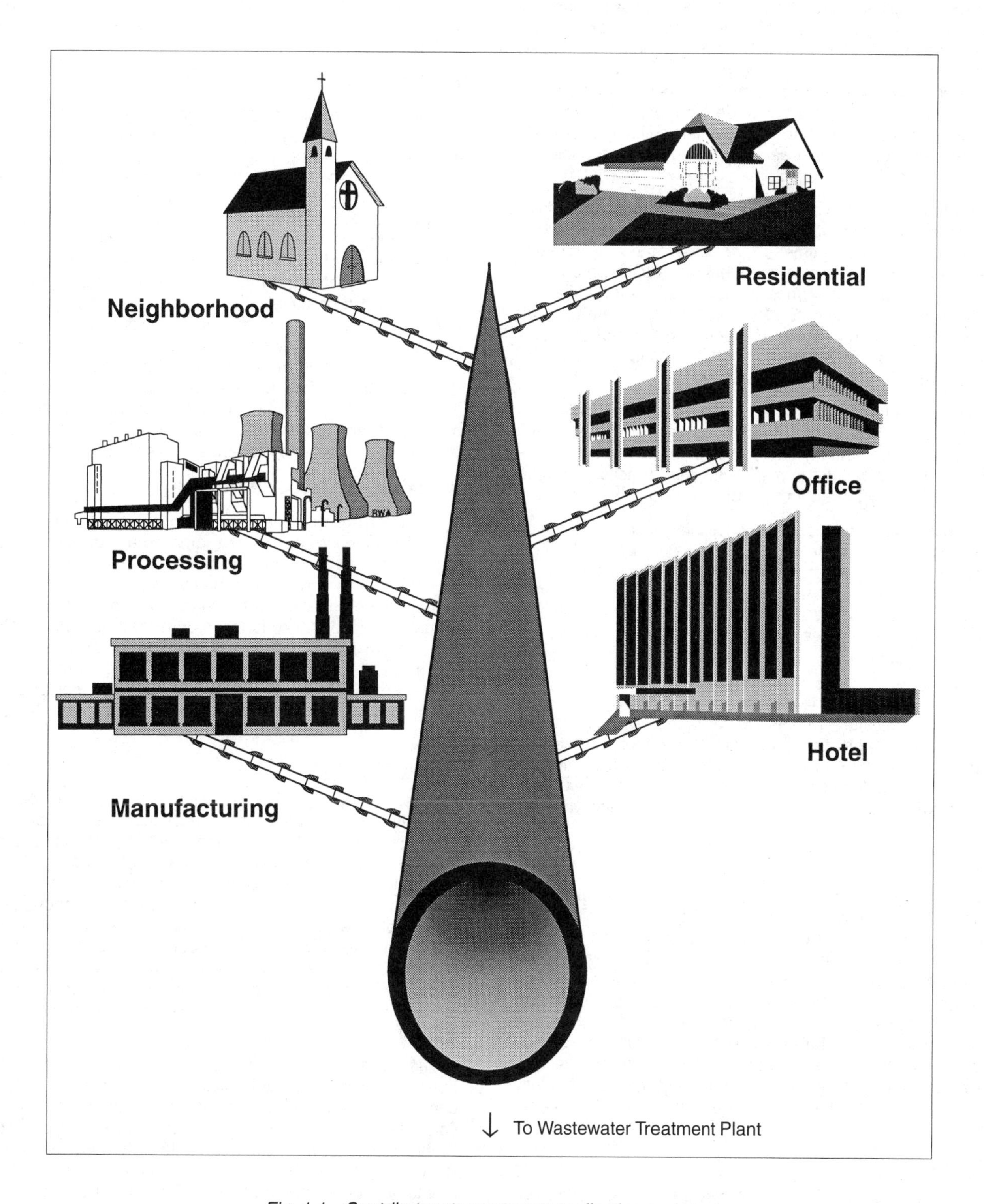

Fig. 1.1 Contributors to wastewater collection systems

3. Street collapse. Inadequate operation and maintenance affects other utilities and frequently the transportation system when the collection system fails. Leaking sewer mains saturate the ground surrounding the leaking pipe and gradually wash away some of the pipe bedding material and soil creating large cavities or voids under paved streets. Without a stable foundation, the street may eventually collapse causing injury and death to the public as well as extensive property damage to streets and adjacent buildings and foundations. Because other underground utilities are normally installed near the wastewater collection system, water service, electrical service, telephone, and cable TV service can also be disrupted.

Experience has shown that inadequate operation and maintenance of the wastewater collection system over the long term is actually several times more costly than a systematic, scheduled program of system operation and maintenance. Not only is the performance of the wastewater collection system affected, but the life of the collection system network can be reduced significantly. Expensive rehabilitation or replacement will be required sooner if the system is not adequately maintained on an ongoing basis.

Operation and maintenance of the wastewater collection system can be thought of, therefore, as protecting the capital investment the community has made in the wastewater collection system so that it performs its intended function and can be used efficiently throughout the planned life of the system.

Let's compare implications of service interruptions of common public services such as telephone, gas, water supply, electrical supply, and wastewater collection systems. We know each of these systems occasionally fails. Whenever any one of these services fails, a nuisance condition develops that tends to produce a highly emotional response depending upon the individual and the situation.

Telephone

A telephone failure means going to the nearest working telephone to report the problem. The telephone service people may find the cause of the problem in the central office or they may send repair personnel into the area. You may miss a few phone calls or be inconvenienced. Usually you can reach a nearby working phone.

Drinking Water

The water supply may stop as a result of a line break in your home or beyond your property line. If the failure occurs on your property, you shut off the water supply at the meter or curb and call a plumber. If the failure occurs beyond your meter or curb, you call the water supply service agency. They will correct their problem as soon as possible, depending upon the nature and extent of the service interruption and what is required to restore service. In the meantime you may have to depend on a friendly neighbor for an alternate temporary water supply. There may be water damage, but hopefully a serious washout will not be caused by a major line breakage.

Natural Gas

Natural gas service interruption may be the result of a stoppage or leaks in the delivery system. Gas is odorized to provide a warning if a leak occurs. Proceed with caution at any time there is any suspicion of gas leakage. Explosive limits for natural gas in air are from about 1 to 33 percent gas. All this mixture needs is a spark or source of ignition to cause a fire or an explosion. A defective heater, heater controls, flame blowout, or leaking pipes all can provide pockets of gas. When a gas leak is suspected or discovered, eliminate ignition sources (absolutely no smoking), ventilate the area, and call the municipal gas service immediately. Explosion and/or fire are serious hazards. Even a small one causes a messy cleanup job. A major explosion or fire may result in an injury or loss of life and extensive property damage or destruction. Preventive maintenance is the only safe and sane course of action.

Electric Power

Electric power interruption affects lighting, power, automatic controllers, alarms, and safety devices. The householder must use candles (if there are no gas leaks), flashlights, manual controls, and do-it-yourself ingenuity. Refrigerators, freezers, and air conditioners don't work. For a short time the refrigerator and freezer will protect their contents if they are kept closed. The electric service utility will take action as soon as they are notified, but they are not magicians. They must have time to locate the problem, gather staff and equipment, and repair the problem. The failure may be due to a fuse or thermal breaker in the house caused by defective wiring or appliances. A major brownout or blackout in the area could be caused by central or substation damage or failure. Line damage can be repaired relatively rapidly if only one or two lines are involved. A major ice storm or windstorm can affect many lines and a long time will be required to restore service even with imported maintenance crews and equipment.

Wastewater Collection System

The interruption of wastewater collection service, like drinking water service, generally promotes a more emotional response than failures of other utilities. Possibly this is because the customers have been underfunding collection system maintenance and repairs for years and want the municipality to make it disappear or function on a "free-ride" basis. Collection system maintenance and repair is an expensive operation and is usually a low priority item on the municipal budget when compared with streets, public buildings, parks, schools, police and fire protection that can be seen every day by the public.

What happens?

When a stoppage develops or flow in excess of the sewer capacity cannot escape, the wastewater starts to back up in the sewers and flows out manholes into the streets or through the lowest elevation basement drains or house drains connected to the sewer. Households at higher elevations may experience no problems because their drains are above the elevation of the stoppage and the backed-up wastewater. Actually these homes are contributing to the basement backflow in the lower elevation drains. The lowest households upstream from the stoppage receive the collected backflow. A stinking mess develops that creates a health, odor, and filth problem.

The low section households can shut off their water supply, but the sewer backflow continues from the higher elevation connections. Flooded basements may cause cross connections between the water supply and the sewer by allowing wastewater to reach water supply faucets, thus possibly contaminating the public water supply. When flooding of basements occurs, electric and gas utilities must be shut off immediately. Telephone lines become jammed because many of the users in the area are trying to complain and get the situation corrected. Any utility interruption is a serious inconvenience and a potential hazard. A collection system failure creates a health and sanitation hazard and possibly interrupts all of the utilities serving the affected area. Residents of affected buildings usually must find alternate shelter and services somewhere else until service can be restored to the affected area and the individual household can be cleaned, repaired, or replaced.

The wastewater collection system operator is a key individual with a major effect upon the health and general well-being of the community. The primary objectives for collection system operators are direct responsibility for public health and safety and protection of the environment. Recognition of this status is being improved. Better trained and motivated individuals working for collection system agencies are essential to present a better case for improved priorities and public recognition.

1.3 WHAT DOES A COLLECTION SYSTEM OPERATOR DO?

Simply described, collection system operators keep the system working. They inspect the system to keep the wastewater flowing today and in the future. Physically they have manual and power-operated equipment to help them with the job. Some of the jobs they do include inspecting and/or installing new service connections. The level of professionalism has increased and certification programs have been implemented. Formal training and certification play a big role in developing the knowledge and skills needed for one of the most important jobs in the community. The role of the collection system operator has changed drastically in the last several years due to the development of modern technology, using closed-circuit computerized television systems to inspect sewers where there is no other way to observe the system, sewer cleaning, removing stoppages in sewers, replacing or repairing lines, and maintaining equipment and facilities. From an overall viewpoint, collection system operators are responsible for the collection of wastewater and the conveying of the wastewater to a treatment plant. The better the job collection system operators do, the more effectively a wastewater treatment plant can do its intended job.

1.4 WHO HIRES COLLECTION SYSTEM OPERATORS?

Collection system operators are employed by a wide variety of organizations. Depending on how your utility is structured, you may be working for the city, the county, a sanitation district or a special district created specifically for wastewater collection systems, or any private utility company that owns and operates a collection system. Collection system operators also may be employed by consulting engineers who not only design wastewater collection systems but offer operation and maintenance services as well. There are also a large number of private companies that perform contract operation and maintenance of wastewater collection systems.

The need for qualified collection system operators continues to grow and this field offers a rewarding career regardless of the type of employer that hires you. Operators are playing an important part in maintaining the good health of the community. People living in the community who discharge their wastewaters to the collection system are the people for whom the collection system operators are really working.

QUESTIONS

Place an X by the *CORRECT ANSWER OR ANSWERS* (more than one answer may be correct). After you have answered all the questions, compare your answers with those given at the end of this chapter on page 10. Reread any sections you do not understand and then proceed to the next section. You are your own teacher in this training program, and YOU should decide when you understand the material and are ready to continue with new material.

EXAMPLE

This is a training manual on:

_______ 1. Arithmetic

_______ 2. Engineering

___X___ 3. Wastewater Collection Systems

_______ 4. Truck Driving

1.1A The main purpose of a wastewater collection system is to gather

_______ 1. Spent or used water from our homes and businesses.

_______ 2. Irrigation water from fields.

_______ 3. Waste material from junk yards.

_______ 4. Drinking water from lakes.

1.2A Why must wastewater collection systems be properly operated and maintained?

_______ 1. So wastewater won't flood our homes.

_______ 2. To keep your employer from being sued for damages caused by wastewater.

_______ 3. So collection system operators will have extra time to repair streets.

_______ 4. To keep the collection system's office phones from being busy all of the time.

Using TV to inspect sewer pipes?

1.3A What does a collection system operator do?

_________ 1. Keeps the collection system working.

_________ 2. Repairs underground pipes and manholes.

_________ 3. Removes stoppages.

_________ 4. Cleans sewers.

_________ 5. Maintains collection system equipment.

1.4A Who hires collection system operators?

_________ 1. City.

_________ 2. Sanitation district.

_________ 3. Pipeline operation and maintenance contractor.

1.5 WHERE DO COLLECTION SYSTEM OPERATORS WORK?

1.50 On the Job

You may be surprised, but if you are working for an agency or contractor that is up to date, you will spend very little time in a sewer or manhole. Modern technology has developed equipment and procedures to reduce risky and unpleasant tasks as much as possible. Most of the work is done from "topside." But if you must work underground (below the ground surface), learn to do it safely.

Throughout this manual, you will learn how to protect yourself and your fellow operators whether you are working underground or topside. If you must work underground or in confined spaces in the collection system, be sure to follow your supervisor's instruction. If you or your supervisor have questions or doubts, your supervisor should consult with an experienced associate or your agency or local safety officer for rules and regulations for your safety. If you work for a small contractor or agency that does not have a safety officer, consult with your state or federal OSHA (Occupational Safety and Health Act) safety people *BEFORE* you take a chance.

Your employer wants you to be a hit, but not a smash (*WORK SAFE*).

1.51 In the City and In the Country

Jobs are available for wastewater collection system operators wherever people live and discharge the wastewater from their homes into a common collection system. From the mountains to the sea and everywhere in between are found wastewater collection systems. You have the opportunity to work for either a small or large agency. The choice is yours if you can convince an employer that you have the knowledge and skills necessary to do the job.

1.6 WHAT PAY CAN A COLLECTION SYSTEM OPERATOR EXPECT?

In dollars? Prestige? Job satisfaction? Community service? In opportunities for advancement? By whatever scale you use, returns are what you make them. If you choose a large municipality, the pay is good and advancement prospects are tops. Choose a small town or contractor and the pay may not be as good, but job satisfaction, freedom from time-clock hours, community service, and prestige may well add up to outstanding personal achievement. If you have the ability and the opportunity, you can make this field your career and advance to the top. Many of these positions are or will be represented by a union that will try to obtain higher pay and other benefits for you. Total reward depends on you and how *YOU APPLY YOURSELF.*

1.7 WHAT DOES IT TAKE TO BE A COLLECTION SYSTEM OPERATOR?

DESIRE. First you must choose to enter this profession. You can do it with a grammar school, a high school, or a college degree. While some jobs will always exist for manual labor, the real and expanding need is for *QUALIFIED OPERATORS.* You must be willing to study and take an active role in upgrading your capabilities. The high-velocity cleaners and closed-circuit television inspection units are examples of the costly and sophisticated equipment used by our profession today. Employers will not risk their investment by allowing unqualified people to operate this complex equipment.

What will be your job title? This will depend on how your employer classifies job titles. Today a sewage treatment plant operator may be called a water quality control plant operator, a janitor may be called a custodian, and a garbage man a refuse collector. All of these new job titles were developed to make the job more acceptable to the general public. Whether you are called a wastewater collection system operator or a sewer maintenance operator, you should be proud of where you work and of your job. You can become a better qualified professional operator ready for advancement by increasing your knowledge today.

Obviously this manual cannot train you to be an expert equipment operator, familiar with each handle on all machines, but it will tell you how to select equipment, what it can and can't do, and how to keep records of your activities. All of these factors are essential to proper operation and maintenance of your collection system.

This training manual can be your start toward a future filled with a different challenge each day and job satisfaction that most people never achieve and enjoy.

1.8 PREPARING YOURSELF FOR THE FUTURE

1.80 Your Qualifications

What do you know about your job or the job you'd like to obtain? Perhaps a little, and perhaps a lot. You must evaluate the knowledge, skills, and experience you already have and what you will need to achieve future jobs and advancement.

The knowledge, skills, abilities, and judgment required for your job depend to a large degree on the size and type of collection system where you work. You may work on a large, complex system serving several hundred thousand persons and employing a hundred or more operators. In this case, you are probably a specialist in one or more phases of the collection system or lifting/pumping stations (such as being responsible for the wet wells, valves, pumps, and motors).

On the other hand, you may operate and maintain a small collection system serving only a thousand people or even fewer. You may be the only operator for the entire system or, at best, have only one or two helpers. If this is the case, you must be a "jack of all trades" because of the diversity of your tasks.

1.9 YOUR PERSONAL TRAINING COURSE

Beginning on this page you are starting on a training course which has been carefully prepared to allow you to improve your knowledge and ability to operate and to maintain wastewater collection systems.

You will be able to proceed at your own pace; you will have an opportunity to learn a little or a lot about each topic. The manual has been prepared this way to fit the various needs of collection system operators, depending on the size and type of collection system for which you are responsible. To study for certification and civil service exams you may have to cover most of the material in this manual. You will never know everything about wastewater collection systems and the equipment and procedures available for operation and maintenance, but you can begin to answer some very important questions about how, why, and when certain things happen in collection systems. You can also learn how to manage your collection system to minimize stoppages, odors, failures, and costs to taxpayers in the long run.

1.10 CERTIFICATION

Certification examinations are usually administered by professional associations. Collection system operators take these exams in order to obtain certificates that indicate a level of professional competence. You should continually strive to achieve higher levels of certification. Successful completion of this operator training program will help you achieve your certification goals. An excellent book to study when preparing for a certification exam is the "Wastewater Collection System Operator Certification Study Book" published by the Water Environment Federation (WEF) (for ordering information, please see item 4 in the list of publications at the end of this section).

This training course is not the only one available to help you improve your abilities. Some state water pollution control associations, community colleges, and universities offer training courses on both a short- and long-term basis. Many federal, state, local, and private agencies have conducted training programs and informative seminars.

Libraries can be an excellent source of useful journals and books. Many very helpful publications have been developed and distributed by the:

WATER ENVIRONMENT FEDERATION (WEF)
Publications Order Department
601 Wythe Street
Alexandria, VA 22314-1994 U.S.A.

When ordering publications from the Water Environment Federation, include your Member Identification Number in order to obtain the member discount.

WEF publications listed below will provide interesting reading and additional information. All prices listed in this manual are the most recent prices available when the manual was printed.

	WEF Members	Non-members
1. Manual of Practice (MOP) 1. *SAFETY AND HEALTH IN WASTEWATER SYSTEMS.* Order No. MO2001.	$19.99	$29.99
2. MOP 7. *WASTEWATER COLLECTION SYSTEMS MANAGEMENT.* Order No. M05000.	50.00	70.00
3. MOP FD-5. *GRAVITY SANITARY SEWER DESIGN AND CONSTRUCTION.* Order No. MFD5.	25.00	35.00
4. *WASTEWATER COLLECTION SYSTEM OPERATOR CERTIFICATION STUDY BOOK.* Order No. E20012.	38.00	54.00

Note that Water Environment Federation members can buy publications at lower prices. You can join the WEF by writing to the Water Environment Federation. The WEF can help you contact your own state and local water pollution control associations. These professional organizations can offer you many helpful training opportunities and educational materials when you join and actively participate with your associates in the field.

QUESTIONS

Write your answers in a notebook. When you *WRITE* your answers, you are helping yourself *LEARN AND REMEMBER* the answer. Compare your answers with the suggested answers on page 10.

1.5A Where do collection system people work on the job?

1.5B Who can help you determine safe procedures for working underground?

1.7A What does it take to be a collection system operator?

1.8A Where else besides this manual could you look for information and help to learn more about wastewater collection systems?

SUGGESTED ANSWERS

Chapter 1. THE WASTEWATER COLLECTION SYSTEM OPERATOR

You are not expected to have the exact answer suggested for questions requiring written answers, but you should have the correct idea.

Answers to questions on pages 7 and 8.

1.1A 1. Spent or used water from our homes and businesses.

1.2A 1. So wastewater won't flood our homes.
2. To keep your employer from being sued for damages caused by wastewater.

1.3A 1. Keeps the collection system working.
2. Repairs underground pipes and manholes.
3. Removes stoppages.
4. Cleans sewers.
5. Maintains collection system equipment.

1.4A 1. City.
2. Sanitation district.
3. Pipeline operation and maintenance contractor.

Answers to questions on page 9.

1.5A On the job collection system people work both topside and underground in manholes and sewers. Modern technology is continually developing equipment and procedures to make our job more pleasant and easier, as well as help us do a better job.

1.5B Your supervisor, an experienced associate, or your agency or local safety officer or a state or federal safety officer can help you determine safe procedures for working underground. This manual also outlines safe procedures.

1.7A *DESIRE* is the main requirement to be a collection system operator. Also, you have to be willing to study and upgrade your capabilities.

1.8A To learn more information and gain help regarding wastewater collection systems, contact the Water Environment Federation, state and local associations, state pollution control training agencies, colleges, and universities. Libraries can be an excellent source of useful journals and books.

CHAPTER 2

WHY COLLECTION SYSTEM OPERATION AND MAINTENANCE?

by

Walt Driggs

TABLE OF CONTENTS

Chapter 2. WHY COLLECTION SYSTEM OPERATION AND MAINTENANCE?

OBJECTIVES

Chapter 2. WHY COLLECTION SYSTEM OPERATION AND MAINTENANCE?

Following completion of Chapter 2, you should be able to:

1. Describe the problems of operating and maintaining a wastewater collection system,
2. Justify the need to operate and maintain the system, and
3. Tell what collection systems are expected to achieve.

OPERATOR'S PROJECT PRONUNCIATION KEY

by Warren L. Prentice

The Operator's Project Pronunciation Key is designed to aid you in the pronunciation of new words. While this key is based primarily on familiar sounds, it does not attempt to follow any particular pronunciation guide. This key is designed solely to aid operators in this program.

You may find it helpful to refer to other available sources for pronunciation help. Each current standard dictionary contains a guide to its own pronunciation key. Each key will be different from each other and from this key. Examples of the differences between the key used in this program and the *WEBSTER'S NEW WORLD COLLEGE DICTIONARY*[1] "Key" are shown below:

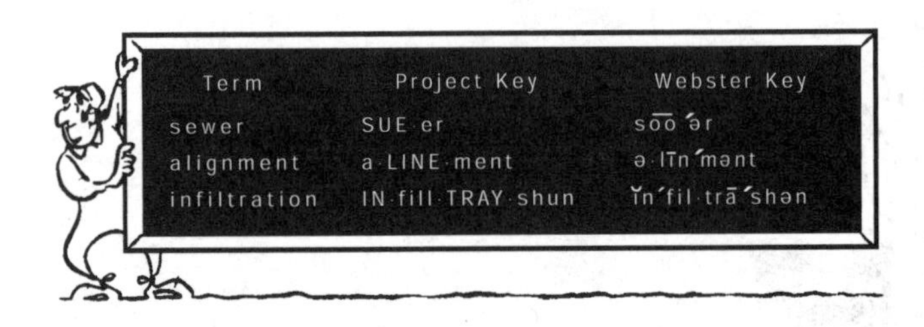

In using this key, you should accent (say louder) the syllable that appears in capital letters. The following chart is presented to give examples of how to pronounce words using the Operator's Project Pronunciation Key.

WORD	SYLLABLE			
	1st	2nd	3rd	4th
sewer	SUE	er		
alignment	a	LINE	ment	
infiltration	IN	fill	TRAY	shun

The first word, *SEWER*, has its first syllable accented. The second word, *ALIGNMENT*, has its second syllable accented. The third word, *INFILTRATION*, has its first and third syllables accented.

We hope you will find the key useful in unlocking the pronunciation of any new word.

EXPLANATION OF WORDS

The meanings of words in the glossary of this manual are based on current usage by the wastewater collection profession and definitions given in *GLOSSARY—WATER AND WASTEWATER CONTROL ENGINEERING*, prepared by the Joint Editorial Board representing APHA, ASCE, AWWA, and WEF,[2] 1981. Certain words used by wastewater collection system operators tend to have slightly different meanings in some regions of the United States. We have tried to standardize word meanings as much as possible.

The Appendix, "Collection System Words", is a complete listing of all the words defined at the beginning of each chapter.

[1] *The WEBSTER'S NEW WORLD COLLEGE DICTIONARY, Fourth Edition, 1999, was chosen rather than an unabridged dictionary because of its availability to most collection system operators. Other editions may be slightly different.*

[2] *APHA. American Public Health Association.*
ASCE. American Society of Civil Engineers.
AWWA. American Water Works Association.
WEF. Water Environment Federation.

WORDS

Chapter 2. WHY COLLECTION SYSTEM OPERATION AND MAINTENANCE?

APPURTENANCE (uh-PURR-ten-nans) APPURTENANCE

Machinery, appliances, structures and other parts of the main structure necessary to allow it to operate as intended, but not considered part of the main structure.

CMOM CMOM

Capacity Assurance, **M**anagement, **O**peration and **M**aintenance. A program developed by collection system agencies to ensure adequate capacity and also proper management and operation and maintenance of the collection system to prevent SSOs.

EXFILTRATION (EX-fill-TRAY-shun) EXFILTRATION

Liquid wastes and liquid-carried wastes which unintentionally leak out of a sewer pipe system and into the environment.

GRAVITY FLOW GRAVITY FLOW

Water or wastewater flowing from a higher elevation to a lower elevation due to the force of gravity. The water does not flow due to energy provided by a pump. Wherever possible, wastewater collection systems are designed to use the force of gravity to convey waste liquids and solids.

INFILTRATION (IN-fill-TRAY-shun) INFILTRATION

The seepage of groundwater into a sewer system, including service connections. Seepage frequently occurs through defective or cracked pipes, pipe joints, connections or manhole walls.

INFLOW INFLOW

Water discharged into a sewer system and service connections from such sources as, but not limited to, roof leaders, cellars, yard and area drains, foundation drains, cooling water discharges, drains from springs and swampy areas, around manhole covers or through holes in the covers, cross connections from storm and combined sewer systems, catch basins, storm waters, surface runoff, street wash waters or drainage. Inflow differs from infiltration in that it is a direct discharge into the sewer rather than a leak in the sewer itself. See INTERNAL INFLOW.

INTERNAL INFLOW INTERNAL INFLOW

Nonsanitary or industrial wastewaters generated inside of a domestic, commercial or industrial facility and being discharged into the sewer system. Examples are cooling tower waters, basement sump pump discharge waters, continuous-flow drinking fountains, and defective or leaking plumbing fixtures.

NPDES PERMIT NPDES PERMIT

National **P**ollutant **D**ischarge **E**limination **S**ystem permit is the regulatory agency document issued by either a federal or state agency which is designed to control all discharges of pollutants from point sources and storm water runoff into U.S. waterways. NPDES permits regulate discharges into navigable waters from all point sources of pollution, including industries, municipal wastewater treatment plants, sanitary landfills, large agricultural feedlots and return irrigation flows.

SSO SSO

Sanitary **S**ewer **O**verflow.

CHAPTER 2. WHY COLLECTION SYSTEM OPERATION AND MAINTENANCE?

2.0 STATEMENT OF THE PROBLEMS

Eric Hoffer has said, "One's capacity to endure is directly related to the quality of his maintenance." Certainly this is true of wastewater collection systems. Preventive maintenance is talked about but seldom practiced by some agencies. Decades of neglect, or grossly inadequate maintenance of some systems, are two reasons why wastewater collection systems now require over 65.8 billion dollars worth of rehabilitation and upgrading between now and 2016 and at a time when the economy can ill afford the cost.[1]

How did this come about?

1. Wastewater collection systems were designed as *GRAVITY FLOW*[2] sewers with inadequate flow capacities for the area served or for the unexpected population growth.

2. Collection systems were not installed as designed. Problems are caused by faulty construction, poor inspection, and low-bid shortcuts.

3. Little thought was given to the fact that sewers, although made of "permanent" material, could be considered only as permanent as the weakest joints. Earth movement, vibration from traffic, settling of structures, and construction disturbance (all occur from time to time) require a flexible pipe material or joint that can maintain tightness, yet joints were made rigid.

4. Corrosion of sewer pipes from either the trench bedding and backfill or the wastewater being transported by the collection system was a factor neglected during design. A major cause of corrosion in wastewater collection systems and treatment plants is hydrogen sulfide gas.

5. Not enough scientific knowledge existed or was available to designers about potential damage to pipe joints by plant roots. Although root intrusion into sewers was age-old, it was assumed that if the joint was watertight, it would be roottight. People did not realize that roots would be attracted by moisture and nutrient vapor unless the joints were vaportight (which means airtight). Roots can enter a pipe joint or walls microscopically (through extremely small holes or cracks); thus, open or leaking joints are not necessary for root intrusion in collection systems.

   Collection system environments are ideal for root growth. In this environment, roots enter, expand, and open joints and cracks. Root growth is a principal cause of pipe damage that allows *INFILTRATION*[3] and *EXFILTRATION*.[4] This creates a major concern for health and pollution control authorities because of wastewater treatment plant overload and groundwater pollution. Plant overload is also caused by *INFLOW*.[5]

6. The out of sight, out of mind nature of the wastewater collection system. Local taxpayers have invested more money in underground sewers than in all the structures above ground owned by their local government. Why has this great taxpayer investment been so grossly neglected? Because it is out of sight, and so, out of mind. Some cities spend more money repairing and replacing sidewalk sections because of earth movement and tree root damage than on preventive maintenance of wastewater collection systems.

7. Many collection systems are maintained by a department charged with street, sidewalk, storm drain, and sometimes water utility maintenance. You can easily guess where the money is spent. Money is usually spent where the taxpayer can see it, especially when the budget is inadequate for the total need. How often do you see a sign, "Your tax dollar is at work cleaning and repairing your sewer"?

[1] *"1996 Clean Water Needs Survey Report To Congress, Assessment of Needs for Publicly Owned Wastewater Treatment Facilities, Correction of Combined Sewer Overflows, and Management of Storm Water and Nonpoint Source Pollution in the United States," United States Environmental Protection Agency, Office of Water, Washington, DC 20460, 1997. EPA 832-R-97-003.*

[2] *Gravity Flow. Water or wastewater flowing from a higher elevation to a lower elevation due to the force of gravity. The water does not flow due to energy provided by a pump. Wherever possible, wastewater collection systems are designed to use the force of gravity to convey waste liquids and solids.*

[3] *Infiltration (IN-fill-TRAY-shun). The seepage of groundwater into a sewer system, including service connections. Seepage frequently occurs through defective or cracked pipes, pipe joints, connections or manhole walls.*

[4] *Exfiltration (EX-fill-TRAY-shun). Liquid wastes and liquid-carried wastes which unintentionally leak out of a sewer pipe system and into the environment.*

[5] *Inflow. Water discharged into a sewer system and service connections from such sources as, but not limited to, roof leaders, cellars, yard and area drains, foundation drains, cooling water discharges, drains from springs and swampy areas, around manhole covers or through holes in the covers, cross connections from storm and combined sewer systems, catch basins, storm waters, surface runoff, street wash waters or drainage. Inflow differs from infiltration in that it is a direct discharge into the sewer rather than a leak in the sewer itself.*

8. The ditches in which sewers are installed have the bottoms sloping downhill to produce gravity flow. Gravel and sand placed in the ditch for pipe bedding and cover create ideal conditions to move water for great distances down the trench. Water that enters this backfilled ditch does not easily percolate (seep) out of the ditch because the usual silt and clay soils have been compacted both on the ditch bottom and sidewalls by the heavy excavation equipment.

   Many of these ditches start at the toe (bottom) of foothill slopes where the natural soil conditions produce springs or lateral water flow into the ditch. Leaking pipe joints and exfiltration can further increase water flow. This water can cause flow down the ditch and create serious problems along a downstream sewer where no one would expect water in the ditch. Problems that can develop include groundwater infiltration into the sewer, possible flotation of the sewer, or structural failure of the sewer or joint.

9. The Administrator of the Environmental Protection Agency was not able to authorize grants for the construction of wastewater treatment plants if the wastewater collection system discharging into a plant was subject to excessive infiltration.[6] Consequently municipalities had to reduce infiltration to the extent that was cost effective and properly maintain the wastewater collection system to qualify for federal construction grants. Reducing infiltration/inflow is good practice because it allows the collection system to handle greater quantities of wastewater.

10. Negligence and vandalism can be the source of collection system problems. Vandals may place objects such as rocks and tree branches in manholes. Contractors repaving roads have accidentally allowed old pavement, soil and/or road base material to enter manholes. Any material in a sewer will slow the flow and allow other solids to settle.

11. Poor records regarding complaints from the public or the date and location of stoppages that had to be cleared can result in an ineffective maintenance program. Good records, regular analysis of the records, and use of this information can produce a cost-effective preventive maintenance program.

The discussion has explained why many wastewater collection systems now present such a mammoth problem to the taxpayer and governmental agencies. Fortunately there exist today properly designed, constructed, operated, and maintained collection systems that can serve as examples of how the job should be done. Leakage should not be a problem in properly constructed and maintained sewers.

Operation and maintenance of wastewater collection systems on a trouble or emergency basis has been the usual procedure and policy in many communities and districts. Planned operation and preventive maintenance of the collection system has been delayed or omitted, in spite of desires by collection system operators. Municipal officials tended to neglect collection systems as long as complaints were not excessive. To please constituents, officials often demanded street and sidewalk repair be done by collection system crews, but seldom have they ever demanded preventive work on collection systems.

2.1 WHY HAVE COLLECTION SYSTEMS?

First we should consider why we have wastewater collection systems; then it will be clear why we should take care of them. Collection systems were built to remove diseases that people come in contact with when they live surrounded by other people and their wastes. Roman and other early civilizations built drains and aqueducts to carry away their wastes. Wherever large population centers developed, a means of waste handling and removal was needed. In 1888, Dr. McVail reviewed the Methods and Objectives of Preventive Medicine. Three basic lines of defense were outlined as constituting the preventive measures. They were (1) sanitation (cleanliness), (2) inoculation (vaccination for protection against diseases), and (3) isolation (keeping sick people away from healthy people). Sanitation was considered the most important of the three. Also emphasized was the need for a good environment suitable for humans. If you wish more information, read Waring's book on *SEWERAGE AND LAND DRAINAGE*.[7] This was written in 1889, but many of the reasons for building collection systems are true today.

2.2 COMPARISONS OF COLLECTION SYSTEMS

You will find city and county officials who will know the cost of each building and their total building budget. Bridges and parks are obvious and everyone can see neglect, but you will find few city and county officials who know for sure how many feet of sewers they have and the number of manholes in their system. Few cities or agencies have correct maps of their wastewater collection system or of the connections to the system. Officials become aware of sewers mainly when the service is inadequate, a stoppage occurs, the system fails, or odors are produced.

---

[6] *Public Law 92-500, Section 201. "Federal Water Pollution Control Act Amendments of 1972." 92nd Congress, S.2770, Washington, DC, October 18, 1972.*

[7] *Waring, George Edwin, SEWERAGE AND LAND DRAINAGE, D. Van Nostrand, New York, NY, 1889.*

We know how many pumping plants or lift stations are in the collection system, where the wastewater treatment plants we have are located, and the number of people we think it takes to operate and maintain them. Numbers of operators will vary from city to city, but you will find some agreement. Unfortunately officials often forget or neglect adequate inspection during construction, and fail to budget adequately for personnel and equipment requirements for the proper operation and maintenance of collection systems. The following sections contain comparisons of collection systems with highways, traffic signals, and railroad crossings.

2.20 Highway and Street Comparison

Drive down a road. If it is too rough or bumpy you can call the highway department and complain. You and every other driver and passenger who have passed over that section of road know how bad it is. You become inspectors for the highway department. The highway patrol and local police will call in any unsafe section they see. Who notices a sewer if it isn't overflowing or stinking?

2.21 Traffic Signal Comparison

When the red lights stay on a few seconds longer, you start to stew, but let a traffic signal fail altogether and quite often you will spend your quarter to call and report its failure. What is going on underground? Who is looking out for the collection system?

2.22 Railroad Crossing Comparison

What happens when the tracks are too rough and every low-slung car hits bottom? Motorists call in and complain. The same thing happens when you have a manhole in the roadway that causes a bump, or the manhole cover rattles and keeps someone awake. They're interested in your collection system for the first time? Not quite so; just in the manhole that caused the bump or rattle.

You will find that the underground collection system won't draw much attention from anyone until the flow starts coming out of a manhole or some other low overflow point, such as a basement drain or shower drain in a home.

By now you are probably getting the point. One reason a staff is needed to operate and maintain a collection system is that you don't have all of the outside help to let you know when or where there is a problem. Most people tend to neglect devices and equipment until there is a serious squeak, rattle, or bump that makes the problem obvious and usually serious.

QUESTIONS

Write your answers in a notebook and then compare your answers with those on page 22.

2.0A List five problems or reasons why many wastewater collection systems are in need of repair or replacement.

2.1A What is the purpose of wastewater collection systems?

2.2A Why aren't people more concerned about collection systems?

2.3 WHY OPERATION AND MAINTENANCE OF YOUR WASTEWATER COLLECTION SYSTEM?

2.30 *SSOs*[8] and *CMOM*[9] Program Regulatory Compliance

The Clean Water Act prohibits the discharge of pollutants to waters of the United States, unless a discharge is authorized by a *NATIONAL POLLUTANT DISCHARGE ELIMINATION SYSTEM (NPDES) PERMIT.*[10] An unpermitted discharge from a sanitary sewer system into waters of the United States therefore constitutes a violation of the Clean Water Act. In some cases, the NPDES permit may incorporate requirements as part of the wastewater treatment plant permit that prohibit SSOs (Sanitary Sewer Overflows). The Clean Water Act requires that permittees shall at all times properly operate and maintain all treatment facilities and related *APPURTENANCES*[11] which are installed or used by the permittee to achieve compliance with permit conditions. The NPDES permits for municipal treatment plants require recordkeeping and reporting of overflows that result in a discharge. The U.S. Environmental Protection Agency (EPA) realizes that even very well-operated collection systems may experience unauthorized discharges under exceptional circumstances, but poor collection system operation and maintenance practices that result in SSOs would violate permit provisions.

In November 2001, the U.S. Environmental Protection Agency's Assistant Administrator for Water instructed the Office of Wastewater Management (OWM) to propose a new SSO Rule and CMOM Program. In response, the EPA is proposing revisions to the NPDES permit regulations to improve the operation of sanitary sewer collection systems, reduce the occurrence of sewer overflows, and provide more effective public notification when overflows do occur. This will provide communities with a framework for reducing health and environmental risks associated with overflowing sewers. The results will be fewer overflows, better information for local communities, and an extended lifetime for the nation's infrastructure. You can find additional information on SSOs and

---

[8] *SSO. **S**anitary **S**ewer **O**verflow.*

[9] *CMOM. **C**apacity Assurance, **M**anagement, **O**peration and **M**aintenance. A program developed by collection system agencies to ensure adequate capacity and also proper management and operation and maintenance of the collection system to prevent SSOs.*

[10] *NPDES Permit. **N**ational **P**ollutant **D**ischarge **E**limination **S**ystem permit is the regulatory agency document issued by either a federal or state agency which is designed to control all discharges of pollutants from point sources and storm water runoff into U.S. waterways. NPDES permits regulate discharges into navigable waters from all point sources of pollution, including industries, municipal wastewater treatment plants, sanitary landfills, large agricultural feedlots and return irrigation flows.*

[11] *Appurtenance (uh-PURR-ten-nans). Machinery, appliances, structures and other parts of the main structure necessary to allow it to operate as intended, but not considered part of the main structure.*

CMOM programs in *OPERATION AND MAINTENANCE OF WASTEWATER COLLECTION SYSTEMS*, Volume II,[12] in this series of operator training manuals.

2.31 What Is Operation and Maintenance?

Collection systems consist of sewers, manholes, pipelines, pumping or lift stations, and force mains. Operation and maintenance of a wastewater collection system can be simply defined as the O & M activities that result in conveying wastewater safely and efficiently to a treatment plant. The results of these activities can be measured by the performance of the collection system and the protection of public health and safety, the environment, and your community's capital investment.

Let's talk about something all of us understand—our car. You have to drive or operate your car to make it take you someplace. If you drive your car improperly or recklessly, you could have an accident and seriously injure yourself and damage your car. Also you have to consider such things as oil, water, grease, tires, and washing. This is the maintenance of your auto. Don't add or change oil and soon you will be walking. After this happens, any repairs you need could be caused by your lack of maintenance. So operation is driving your car and maintenance is keeping it in good running order so that you can depend upon it when you need it.

Operation and maintenance of your wastewater collection system is not letting wastewater overflow into the street or into someone's home. After it has overflowed, you must take emergency action because your operation and maintenance program did not prevent the stoppage.

Public safety also is affected by collection system O & M. Injuries have occurred to the public when vehicles were involved in cave-ins from collapsed pipes, pump station explosions, and accidents caused by loose or missing manhole covers.

Raw wastewater can contain high levels of pathogenic organisms, suspended solids, toxic pollutants, floatables, nutrients, oxygen-demanding organic components, oil, grease, and other pollutants. To protect the public health, discharge of these pollutants to areas where they present high risk of human exposure (such as streets, private property, basements, and receiving waters used as a drinking source, for fish as shell fishing, or for contact recreation such as swimming) should be prevented. The five major groups of disease-causing organisms or agents associated with raw wastewater are:

- Bacteria,
- Viruses,
- Protozoa (including *Giardia* and *Cryptosporidium*),
- Helminths (intestinal worms), and
- Bio-aerosols (which include molds, fungi, biotoxins, and some of the organisms named above).

The health problems caused by these organisms and agents range in severity from mild stomach cramps and diarrhea to diseases that could be life threatening, such as cholera, infectious hepatitis, giardiasis, or dysentery caused by *Cryptosporidium.*

Don't think the agency you work for is a failure if it has main line stoppages. These will occur in even a well-operated and well-maintained system. You will sometimes wonder how the material causing the stoppage got into the line, but everything from bikes, tree limbs, blocks of concrete, and even VW fenders have been found in our sewers and manholes.

2.32 How To Tell When You Need To Do a Better Job

If the stoppage is roots, for example, in the same place each time and your records show it has happened ten times, you should have corrected this problem long ago.

Let's go back to your car that had *NO OIL* and caused you to walk and pay a big repair bill. You would never let this happen ten times. You would soon find the cause for *NO OIL* and stop it from occurring again.

We need to have the same type of concern when a sewer stoppage occurs, the collection system produces obnoxious odors, or a lift station failure occurs.

2.4 THE COST OF YOUR WASTEWATER COLLECTION SYSTEM

When collection systems are installed, they sometimes cost significantly more than the cost of the wastewater treatment plant. If for some reason the collection system fails in, let's say, 20 years, the cost per foot for replacement could be as much as ten times its original cost. In some locations the replacement is practically impossible and extremely expensive because of the crowding of other utilities. The fact that most utilities are buried underground now makes this even more of a problem.

The cost effectiveness of a collection system is measured on how efficiently a city or district manages its O & M program to achieve regulatory compliance while consistently providing the desired level of service to the community. One of the characteristics of well-managed agencies is their recognition of the need to continuously monitor the use of human and material resources in their O & M program. Labor and equipment costs represent major components of the O & M budget, and substantial cost savings can be achieved while maintaining high levels of service by taking the following actions:

- Accurately tracking all labor costs, including direct labor, indirect labor, and overhead,
- Evaluating how work is planned, scheduled, and executed to identify obstacles to efficiency,
- Identifying work activities that can result in savings through outside contracting. This process more closely matches skills and equipment resources to the task,
- Emphasizing training to match skills to tasks,
- Shifting from highly specialized job classifications to increased cross training resulting in broad-based skills, and
- Streamlining organizational structures that contain many classifications and layers of management and supervision.

[12] *OPERATION AND MAINTENANCE OF WASTEWATER COLLECTION SYSTEMS, Volume II, available from the Office of Water Programs, California State University, Sacramento, 6000 J Street, Sacramento, CA 95819-6025. Price, $22.00.*

The U.S. Environmental Protection Agency (EPA) estimates that the proposed SSO Rule and CMOM Program would impose an additional total cost for municipalities of $93.5 million to $126.5 million each year, including costs associated with both planning and permitting. A collection system serving 7,500 people may need to spend an average of $6,000 each year to comply with this Rule. These cost estimates do not include the cost of repairing or upgrading existing sewers to meet existing Clean Water Act requirements.

Additional information about ways to improve the cost effectiveness of a collection system can be found in *COLLECTION SYSTEMS: METHODS FOR EVALUATING AND IMPROVING PERFORMANCE*[13] in this series of operator training manuals.

Public agencies are having a difficult time stretching their resources to meet all the demands they face from both internal and external sources. Some examples of these pressures are:

- Environmental Regulations—more federal, state, and local regulations
- Budget—reduction of federal, state, and local funds
- O & M Costs—rising costs of O & M
- Civil and Criminal Liability—fines and jail time for O & M staff
- Social Policy Laws—Americans with Disabilities Act (ADA), Equal Employment Opportunity (EEO)
- Safety—more regulations and enforcement
- Technology—increased use of technology in collection systems
- Skills—reduction of skills in the labor pool

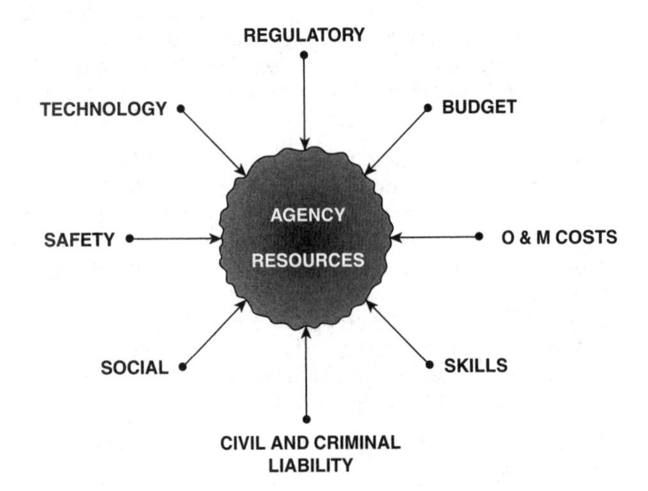

Because of these competing demands, those of us responsible for collection system O & M must a better job of managing our resources effectively. Those responsible for collection system O & M are in a better position than anyone else in an agency to manage resources.

Good collection system inspection, operation, and maintenance are by far the best ways to protect a large investment that the taxpayers have placed in your hands. An effective inspection program will alert you to small problems that should be corrected before they become big problems.

2.5 THE WASTEWATER COLLECTION SYSTEM DOCTOR

This manual will cover many ways to help you properly operate and diagnose the ills of wastewater collection systems. Even more important, it will teach you the cure for these ills.

Picture yourself now as a doctor of wastewater collection systems and seeking cures for the ills a collection system could acquire. This manual will equip you with many different ways to solve the same or similar problems. As you become more acquainted with these methods and know how to evaluate each one, you will become a better employee and your feeling of job accomplishment will grow.

You are not always going to diagnose the problem correctly the first time, and may work on a symptom and not the cause of the problem. One way to avoid this is to review your information carefully for gaps and the method you have selected to correct the problem. Keep good records to help you next time. If you don't know about mistakes, you can't learn from them. Experience is trying and learning what works and doesn't work, and then avoiding use of the "didn't work" procedures again for these conditions. Try to learn from the experience of others in your field.

EXAMPLE

If possible and appropriate, use closed-circuit television (CCTV) to inspect the section in question. See, before you attempt to diagnose the problem. Go back and televise the section after corrective action to see if your selection of a corrective measure was effective. If you don't have a TV unit, maintain careful and accurate records of a problem section. Attempt to evaluate corrective action with other methods such as visual inspection mirrors and cleaning devices that can provide data regarding the condition of the section and the effectiveness of the action taken.

[13] *COLLECTION SYSTEMS: METHODS FOR EVALUATING AND IMPROVING PERFORMANCE, available from the Office of Water Programs, California State University, Sacramento, 6000 J Street, Sacramento, CA 95819-6025. Price, $22.00.*

2.6 SUMMARY

Collection system operators are responsible for:

- Preserving the community's capital investment,
- Protecting public health and safety,
- Protecting the environment, and
- Controlling O & M expenses.

We operate a wide range of equipment:

- Jet or high-velocity cleaning machines,
- Rodders,
- Vacuum equipment,
- Bucket machines,
- Pickup trucks,
- Dump trucks,
- Backhoes,
- Generators,
- Compressors,
- Pump stations,
- TV inspection equipment, and
- Grouting equipment.

We need a broad range of knowledge and skills:

- Safety/survival,
- Electrical,
- Mechanical,
- Hydraulics,
- Construction, and
- Design.

Our jobs require:

- Planning,
- Teamwork,
- Thinking on our feet, and
- Decision making.

Because of the importance of collection systems, opportunities for anyone choosing to work in this field will continue to expand. The work will become even more interesting, challenging and rewarding for those who are prepared professionally to take advantage of the opportunities.

In this chapter you have learned why it is necessary to operate and maintain wastewater collection systems. We have discussed some of the problems and how or why they occur. This is intended to introduce you to these topics. You will find more detail in later chapters.

You are now ready to go on to Chapter 3, which describes collection systems and deals with design and construction. Many problems actually originate when the collection system is first designed and built. Chapter 2 has told you why collection systems need to be properly operated and maintained. Chapter 3 actually begins the discussions on how to do it.

QUESTIONS

Write your answers in a notebook and then compare your answers with those on page 22.

2.3A What is wastewater collection system operation and maintenance?

2.4A How can the taxpayers' investment in your wastewater collection system be protected?

2.5A What should you do after you have selected a method of solving a collection system problem and you think you have corrected the problem?

SUGGESTED ANSWERS

Chapter 2. WHY COLLECTION SYSTEM OPERATION AND MAINTENANCE?

Answers to questions on page 18.

2.0A Many wastewater collection systems are in need of repair or replacement because of

1. Inadequate capacities,
2. Improper construction, inspection, and/or low-bid shortcuts,
3. Joints that cannot flex and maintain tightness,
4. Corrosion of materials due to conditions found in collection systems,
5. Insufficient concern for potential damage to pipe joints by plant roots,
6. Lack of preventive maintenance,
7. Priorities among other jobs a collection system maintenance department may have to do such as maintain streets, sidewalks, and the drinking water system,
8. The characteristic of ditches that allows surface or groundwater to flow down ditches,
9. Restrictions on use of EPA grants,
10. Negligence and vandalism, and
11. Poor recordkeeping.

2.1A The purpose of wastewater collection systems is to carry our wastewater away from our homes and industries to the wastewater treatment plant.

2.2A People aren't very concerned about collection systems because they are out of sight and out of mind.

Answers to questions on page 21.

2.3A Operation and maintenance of your wastewater collection system is not letting it overflow into the street or into someone's home. This means keeping the system in such a condition that stoppages rarely occur and when they do, they are quickly relieved.

2.4A Good collection system operation and maintenance is the only protection of the large investment the taxpayers have in the system.

2.5A After a problem has been corrected, the effectiveness of the corrective action should be evaluated and good records should be kept for reference.

CHAPTER 3

WASTEWATER COLLECTION SYSTEMS
(Purpose, Components, and Design)

by

Steve Goodman

TABLE OF CONTENTS

Chapter 3. WASTEWATER COLLECTION SYSTEMS
(Purpose, Components, and Design)

OBJECTIVES

Chapter 3. WASTEWATER COLLECTION SYSTEMS

Following completion of Chapter 3, you should be able to:

1. List the parts of a wastewater collection system and explain the purpose of each part,
2. Communicate to design engineers the need to consider preventing operation and maintenance problems when designing collection systems,
3. Identify sources and calculate quantities of wastewater flow,
4. Estimate the velocity of water flowing in a sewer, and
5. Review plans and specifications for wastewater collection systems from the viewpoint of effective operation and maintenance of collection systems.

WORDS

Chapter 3. WASTEWATER COLLECTION SYSTEMS

ALIGNMENT (a-LINE-ment) ALIGNMENT

The proper positioning of parts in a system. The alignment of a pipeline or other line refers to its location and direction.

ANNULAR (AN-you-ler) SPACE ANNULAR SPACE

A ring-shaped space located between two circular objects. For example, the space between the outside of a pipe liner and the inside of a pipe.

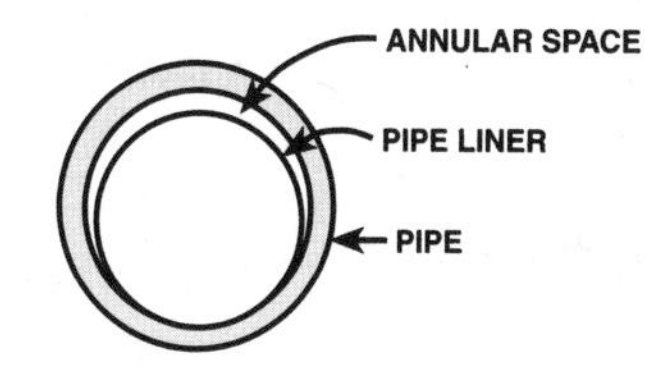

APPURTENANCE (uh-PURR-ten-nans) APPURTENANCE

Machinery, appliances, structures and other parts of the main structure necessary to allow it to operate as intended, but not considered part of the main structure.

BACKWATER GATE BACKWATER GATE

A gate installed at the end of a drain or outlet pipe to prevent the backward flow of water or wastewater. Generally used on storm sewer outlets into streams to prevent backward flow during times of flood or high tide. Also called a TIDE GATE.

BARREL BARREL

(1) The cylindrical part of a pipe that may have a bell on one end.

(2) The cylindrical part of a manhole between the cone at the top and the shelf at the bottom.

BEDDING BEDDING

The prepared base or bottom of a trench or excavation on which a pipe or other underground structure is supported.

BLOCKAGE BLOCKAGE

(1) Partial or complete interruption of flow as a result of some obstruction in a sewer.

(2) When a collection system becomes plugged and the flow backs up, it is said to have a "blockage." Commonly called a STOPPAGE.

BRANCH SEWER BRANCH SEWER

A sewer that receives wastewater from a relatively small area and discharges into a main sewer serving more than one branch sewer area.

BUILDING SEWER BUILDING SEWER

A gravity-flow pipeline connecting a building wastewater collection system to a lateral or branch sewer. The building sewer may begin at the outside of the building's foundation wall or some distance (such as 2 to 10 feet) from the wall, depending on local sewer ordinances. Also called a "house connection" or a "service connection."

CATCH BASIN CATCH BASIN

A chamber or well used with storm or combined sewers as a means of removing grit which might otherwise enter and be deposited in sewers. Also see STORM WATER INLET and CURB INLET.

CAULK (KAWK) CAULK

To stop up and make watertight the joints of a pipe by filling the joints with a waterproof compound or material.

CAULKING (KAWK-ing) CAULKING

(1) A waterproof compound or material used to fill a pipe joint.

(2) The act of using a waterproof compound or material to fill a pipe joint.

CHECK VALVE — CHECK VALVE

A special valve with a hinged disc or flap that opens in the direction of normal flow and is forced shut when flows attempt to go in the reverse or opposite direction of normal flows. Also see FLAP GATE and TIDE GATE.

CLEANOUT — CLEANOUT

An opening (usually covered or capped) in a wastewater collection system used for inserting tools, rods or snakes while cleaning a pipeline or clearing a stoppage.

COHESIVE (co-HE-sive) — COHESIVE

Tending to stick together.

COMBINED SEWER — COMBINED SEWER

A sewer designed to carry both sanitary wastewaters and storm or surface water runoff.

COMBINED WASTEWATER — COMBINED WASTEWATER

A mixture of storm or surface runoff and other wastewater such as domestic or industrial wastewater.

CONCENTRIC MANHOLE CONE — CONCENTRIC MANHOLE CONE

Cone tapers uniformly from barrel to manhole cover.

CURB INLET — CURB INLET

A chamber or well built at the curbline of a street to admit gutter flow to the storm water drainage system. Also see STORM WATER INLET and CATCH BASIN.

DEBRIS (de-BREE) — DEBRIS

Any material in wastewater found floating, suspended, settled or moving along the bottom of a sewer. This material may cause stoppages by getting hung up on roots or settling out in a sewer. Debris includes grit, paper, plastic, rubber, silt, and all materials except liquids.

ECCENTRIC MANHOLE CONE — ECCENTRIC MANHOLE CONE

Cone tapers nonuniformly from barrel to manhole cover with one side usually vertical.

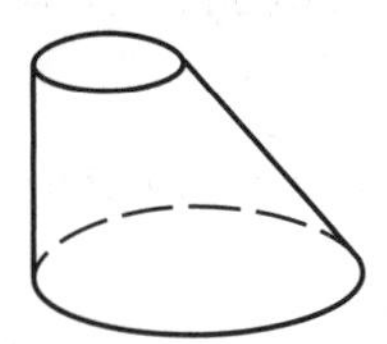

EXFILTRATION (EX-fill-TRAY-shun) — EXFILTRATION

Liquid wastes and liquid-carried wastes which unintentionally leak out of a sewer pipe system and into the environment.

FLAP GATE — FLAP GATE

A hinged gate that is mounted at the top of a pipe or channel to allow flow in only one direction. Flow in the wrong direction closes the gate. Also see CHECK VALVE and TIDE GATE.

FLOW LINE — FLOW LINE

(1) The top of the wetted line, the water surface or the hydraulic grade line of water flowing in an open channel or partially full conduit.

(2) The lowest point of the channel inside a pipe or manhole. See INVERT. *NOTE:* (2) is an improper definition, although used by some contractors.

FLUSHER BRANCH — FLUSHER BRANCH

A line built specifically to allow the introduction of large quantities of water to the collection system so the lines can be "flushed out" with water. Also installed to provide access for equipment to clear stoppages in a sewer.

FORCE MAIN — FORCE MAIN

A pipe that carries wastewater under pressure from the discharge side of a pump to a point of gravity flow downstream.

GRADE RING — GRADE RING

A precast concrete ring 4 to 12 inches high which is placed on top of a manhole cone to raise the manhole cover frame flush with the surface grade. Sometimes called a "spacer."

GRADIENT — GRADIENT

The upward or downward slope of a pipeline.

GRIT — GRIT

The heavy mineral material present in wastewater such as sand, coffee grounds, eggshells, gravel and cinders. Grit tends to settle out at flow velocities below 2 ft/sec and accumulate in the invert or bottoms of the pipelines. Also called "detritus."

HYDRAULIC GRADE LINE (HGL) — HYDRAULIC GRADE LINE (HGL)

The surface or profile of water flowing in an open channel or a pipe flowing partially full. If a pipe is under pressure, the hydraulic grade line is at the level water would rise to in a small tube connected to the pipe. To reduce the release of odors from sewers, the water surface or hydraulic grade line should be kept as smooth as possible.

HYDROGEN SULFIDE GAS (H_2S) — HYDROGEN SULFIDE GAS (H_2S)

Hydrogen sulfide is a gas with a rotten egg odor. This gas is produced under anaerobic conditions. Hydrogen sulfide gas is particularly dangerous because it dulls the sense of smell so that you don't notice it after you have been around it for a while. In high concentrations, hydrogen sulfide gas is only noticeable for a very short time before it dulls the sense of smell. The gas is very poisonous to the respiratory system, explosive, flammable, colorless and heavier than air.

INFILTRATION (IN-fill-TRAY-shun) — INFILTRATION

The seepage of groundwater into a sewer system, including service connections. Seepage frequently occurs through defective or cracked pipes, pipe joints, connections or manhole walls.

INFILTRATION/INFLOW — INFILTRATION/INFLOW

The total quantity of water from both infiltration and inflow without distinguishing the source. Abbreviated I & I or I/I.

INFLOW — INFLOW

Water discharged into a sewer system and service connections from such sources as, but not limited to, roof leaders, cellars, yard and area drains, foundation drains, cooling water discharges, drains from springs and swampy areas, around manhole covers or through holes in the covers, cross connections from storm and combined sewer systems, catch basins, storm waters, surface runoff, street wash waters or drainage. Inflow differs from infiltration in that it is a direct discharge into the sewer rather than a leak in the sewer itself. See INTERNAL INFLOW.

INLET — INLET

(1) A surface connection to a drain pipe.

(2) A chamber for collecting storm water with no well below the outlet pipe for collecting grit. Often connected to a CATCH BASIN or a "basin manhole" ("cleanout manhole") with a grit chamber.

INTERCEPTING SEWER — INTERCEPTING SEWER

A sewer that receives flow from a number of other large sewers or outlets and conducts the waters to a point for treatment or disposal. Often called an "interceptor."

INTERNAL INFLOW — INTERNAL INFLOW

Nonsanitary or industrial wastewaters generated inside of a domestic, commercial or industrial facility and being discharged into the sewer system. Examples are cooling tower waters, basement sump pump discharge waters, continuous-flow drinking fountains, and defective or leaking plumbing fixtures.

INVERT (IN-vert) — INVERT

The lowest point of the channel inside a pipe or manhole. See FLOW LINE.

INVERTED SIPHON — INVERTED SIPHON

A pressure pipeline used to carry wastewater flowing in a gravity collection system under a depression such as a valley or roadway or under a structure such as a building. Also called a "depressed sewer."

LAMP HOLE — LAMP HOLE

A small vertical pipe or shaft extending from the surface of the ground to a sewer. A light (or lamp) may be lowered down the pipe for the purpose of inspecting the sewer. Rarely constructed today.

LATERAL SEWER — LATERAL SEWER

A sewer that discharges into a branch or other sewer and has no other common sewer tributary to it. Sometimes called a "street sewer" because it collects wastewater from individual homes.

LIFT STATION — LIFT STATION

A wastewater pumping station that lifts the wastewater to a higher elevation when continuing the sewer at reasonable slopes would involve excessive depths of trench. Also, an installation of pumps that raise wastewater from areas too low to drain into available sewers. These stations may be equipped with air-operated ejectors or centrifugal pumps. Sometimes called a PUMP STATION, but this term is usually reserved for a similar type of facility that is discharging into a long FORCE MAIN, while a lift station has a discharge line or force main only up to the downstream gravity sewer. Throughout this manual when we refer to lift stations, we intend to include pump stations.

MAIN LINE — MAIN LINE

Branch or lateral sewers that collect wastewater from building sewers and service lines.

MAIN SEWER — MAIN SEWER

A sewer line that receives wastewater from many tributary branches and sewer lines and serves as an outlet for a large territory or is used to feed an intercepting sewer.

MANHOLE — MANHOLE

An opening in a sewer provided for the purpose of permitting operators or equipment to enter or leave a sewer. Sometimes called an "access hole" or a "maintenance hole."

NOMINAL DIAMETER — NOMINAL DIAMETER

An approximate measurement of the diameter of a pipe. Although the nominal diameter is used to describe the size or diameter of a pipe, it is usually not the exact inside diameter of the pipe.

NOMOGRAPH (NOME-o-graph) — NOMOGRAPH

A graphic representation or means of solving an equation or mathematical relationship. Results are obtained with the aid of a straightedge placed over important known values.

OUTFALL — OUTFALL

(1) The point, location or structure where wastewater or drainage discharges from a sewer, drain, or other conduit.

(2) The conduit leading to the final disposal point or area. See OUTFALL SEWER.

OUTFALL SEWER — OUTFALL SEWER

A sewer that receives wastewater from a collection system or from a wastewater treatment plant and carries it to a point of ultimate or final discharge in the environment. See OUTFALL.

OUTLET — OUTLET

Downstream opening or discharge end of a pipe, culvert, or canal.

PEAKING FACTOR — PEAKING FACTOR

Ratio of a maximum flow to the average flow, such as maximum hourly flow or maximum daily flow to the average daily flow.

PIG — PIG

Refers to a poly pig which is a bullet-shaped device made of hard rubber or similar material. This device is used to clean pipes. It is inserted in one end of a pipe, moves through the pipe under pressure, and is removed from the other end of the pipe.

PLAN — PLAN

A drawing showing the *TOP* view of sewers, manholes and streets.

PNEUMATIC EJECTOR (new-MAT-tik ee-JECK-tor) — PNEUMATIC EJECTOR

A device for raising wastewater, sludge or other liquid by compressed air. The liquid is alternately admitted through an inward-swinging check valve into the bottom of an airtight pot. When the pot is filled compressed air is applied to the top of the liquid. The compressed air forces the inlet valve closed and forces the liquid in the pot through an outward-swinging check valve, thus emptying the pot.

PROFILE — PROFILE

A drawing showing the *SIDE* view of sewers and manholes.

SANITARY COLLECTION SYSTEM — SANITARY COLLECTION SYSTEM

The pipe system for collecting and carrying liquid and liquid-carried wastes from domestic sources to a wastewater treatment plant. Also see WASTEWATER COLLECTION SYSTEM.

SANITARY SEWER — SANITARY SEWER

A pipe or conduit (sewer) intended to carry wastewater or waterborne wastes from homes, businesses, and industries to the POTW (**P**ublicly **O**wned **T**reatment **W**orks). Storm water runoff or unpolluted water should be collected and transported in a separate system of pipes or conduits (storm sewers) to natural watercourses.

SEDIMENTATION (SED-uh-men-TAY-shun) — SEDIMENTATION

The process of settling and depositing of suspended matter carried by wastewater. Sedimentation usually occurs by gravity when the velocity of the wastewater is reduced below the point at which it can transport the suspended material.

SEPTIC (SEP-tick) — SEPTIC

A condition produced by anaerobic bacteria. If severe, the wastewater produces hydrogen sulfide, turns black, gives off foul odors, contains little or no dissolved oxygen, and the wastewater has a high oxygen demand.

SEWAGE — SEWAGE

The used household water and water-carried solids that flow in sewers to a wastewater treatment plant. The preferred term is WASTEWATER.

SEWER — SEWER

A pipe or conduit that carries wastewater or drainage water. The term "collection line" is often used also.

SEWER GAS — SEWER GAS

(1) Gas in collection lines (sewers) that results from the decomposition of organic matter in the wastewater. When testing for gases found in sewers, test for lack of oxygen and also for explosive and toxic gases.

(2) Any gas present in the wastewater collection system, even though it is from such sources as gas mains, gasoline, and cleaning fluid.

SEWERAGE — SEWERAGE

System of piping with appurtenances for collecting, moving and treating wastewater from source to discharge.

SIPHON — SIPHON

A pipe or conduit through which water will flow above the hydraulic grade line (HGL) under certain conditions. Water (or other liquid) is first forced to flow or is sucked or drawn through the pipe by creation of a vacuum. As long as no air enters the pipe to interrupt flow, atmospheric pressure on the liquid at the elevated (higher) end of the siphon will cause the flow to continue.

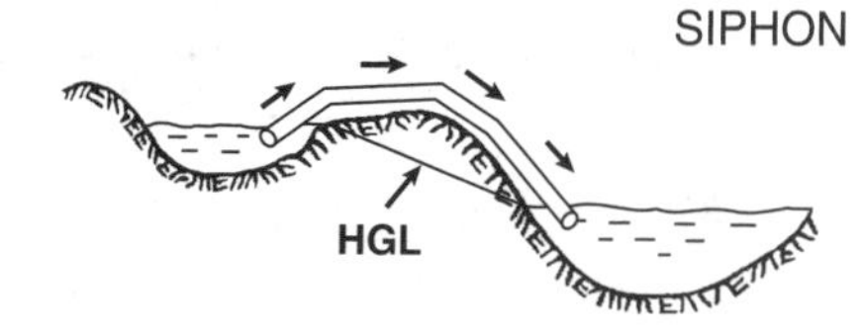

SLOPE — SLOPE

The slope or inclination of a sewer trench excavation is the ratio of the vertical distance to the horizontal distance or "rise over run."

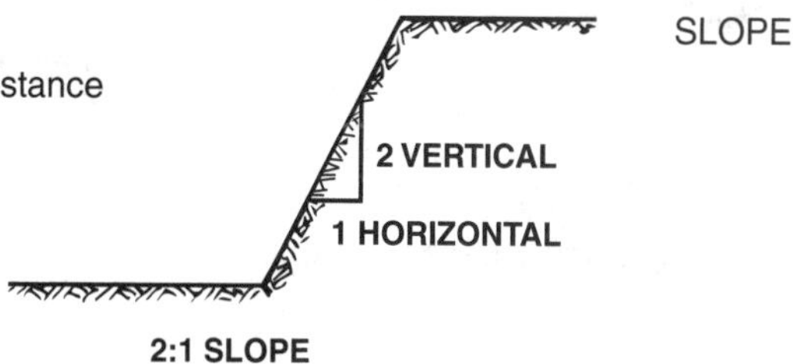

SNAKE — SNAKE

A stiff but flexible cable that is inserted into sewers to clear stoppages; also known as a "sewer cable."

SPRING LINE — SPRING LINE

Theoretical center of a pipeline. Also, the guideline for laying a course of bricks.

STOPPAGE — STOPPAGE

(1) Partial or complete interruption of flow as a result of some obstruction in a sewer.

(2) When a sewer system becomes plugged and the flow backs up, it is said to have a "stoppage." Also see BLOCKAGE.

STORM RUNOFF — STORM RUNOFF

The amount of runoff that reaches the point of measurement within a relatively short period of time after the occurrence of a storm or other form of precipitation. Also called "direct runoff."

STORM SEWER — STORM SEWER

A separate pipe, conduit or open channel (sewer) that carries runoff from storms, surface drainage, and street wash, but does not include domestic and industrial wastes. Storm sewers are often the recipients of hazardous or toxic substances due to the illegal dumping of hazardous wastes or spills created by accidents involving vehicles and trains transporting these substances. Also see SANITARY SEWER.

STORM WATER — STORM WATER

The excess water running off from the surface of a drainage area during and immediately after a period of rain. See STORM RUNOFF.

STORM WATER INLET — STORM WATER INLET

A device that admits surface waters to the storm water drainage system. Also see CURB INLET and CATCH BASIN.

SURFACE RUNOFF — SURFACE RUNOFF

(1) The precipitation that cannot be absorbed by the soil and flows across the surface by gravity.

(2) The water that reaches a stream by traveling over the soil surface or falls directly into the stream channels, including not only the large permanent streams but also the tiny rills and rivulets.

(3) Water that remains after infiltration, interception, and surface storage have been deducted from total precipitation.

TAP — TAP

A small hole in a sewer where a wastewater service line from a building is connected (tapped) into a lateral or branch sewer.

THRUST BLOCK — THRUST BLOCK

A mass of concrete or similar material appropriately placed around a pipe to prevent movement when the pipe is carrying water. Usually placed at bends and valve structures.

TIDE GATE — TIDE GATE

A gate with a flap suspended from a free-swinging horizontal hinge, usually placed at the end of a conduit discharging into a body of water having a fluctuating surface elevation. The gate is usually closed because of outside water pressure, but will open when the water head inside the pipe is great enough to overcome the outside pressure, the weight of the flap, and the friction of the hinge. Also called a BACKWATER GATE. Also see CHECK VALVE and FLAP GATE.

TOPOGRAPHY (toe-PAH-gruh-fee) — TOPOGRAPHY

The arrangement of hills and valleys in a geographic area.

TRUNK SEWER — TRUNK SEWER

A sewer that receives wastewater from many tributary branches or sewers and serves a large territory and contributing population. Also see MAIN SEWER.

TRUNK SYSTEM — TRUNK SYSTEM

A system of major sewers serving as transporting lines and not as local or lateral sewers.

U-TUBE — U-TUBE

(1) A pipe shaped like a U that is constructed in a force main to raise the dissolved oxygen concentration in the wastewater.

(2) U-tube manometers are used to indicate the pressure of a gas or liquid in a contained area, such as a pipeline or storage vessel.

WASTEWATER — WASTEWATER

A community's used water and water-carried solids that flow to a treatment plant. Storm water, surface water, and groundwater infiltration also may be included in the wastewater that enters a wastewater treatment plant. The term "sewage" usually refers to household wastes, but this word is being replaced by the term "wastewater."

WASTEWATER COLLECTION SYSTEM — WASTEWATER COLLECTION SYSTEM

The pipe system for collecting and carrying water and water-carried wastes from domestic and industrial sources to a wastewater treatment plant.

WASTEWATER FACILITIES — WASTEWATER FACILITIES

The pipes, conduits, structures, equipment, and processes required to collect, convey, and treat domestic and industrial wastes, and dispose of the effluent and sludge.

WASTEWATER TREATMENT PLANT WASTEWATER TREATMENT PLANT

(1) An arrangement of pipes, equipment, devices, tanks and structures for treating wastewater and industrial wastes.

(2) A water pollution control plant.

WEIR (weer) WEIR

(1) A wall or plate placed in an open channel and used to measure the flow of water. The depth of the flow over the weir can be used to calculate the flow rate, or a chart or conversion table may be used to convert depth to flow.

(2) A wall or obstruction used to control flow (from settling tanks and clarifiers) to ensure a uniform flow rate and avoid short-circuiting.

WETTED PERIMETER WETTED PERIMETER

The length of the wetted portion of a pipe covered by flowing wastewater.

PIPE

WETTED PERIMETER = DISTANCE FROM A to B

A

B

WATER

TOPOGRAPHY toe-POG-graphy

CHAPTER 3. WASTEWATER COLLECTION SYSTEMS

3.0 NEED TO KNOW

Persons responsible for the operation and maintenance of wastewater collection systems usually are not the ones responsible for the design, construction, and inspection[1] of these systems. However, a basic knowledge of the components of collection systems, their purpose, and how engineers design the systems will help you do a better job. At the same time, such knowledge will enable you to discuss design, construction, and inspection with engineers.

Collection system operators who know how and why wastewater collection systems work should and will be consulted by design engineers for the purpose of improving the design and construction of future systems.

Understanding the design and construction of the system you will be responsible for operating and maintaining also will assist you in determining the correct action to take when the system fails to operate properly. In some emergency situations you may be called upon to design and to construct a small, temporary portion of the wastewater system you are maintaining.

3.1 PURPOSE

The purpose of a wastewater collection system is to collect and convey the wastewater from a community's homes and industries. The water carries the wastes in the form of either dissolved or suspended solids. Wastewater collection systems must be properly designed and constructed to provide a water velocity of greater than 2.0 feet per second. Velocities that are too high or too low can interfere with the operation and maintenance of a collection system. The collection system conveys the wastewater and solids to a treatment plant where the pollutants are removed before the treated wastewater is discharged to a body of water or onto land.

The wastewater in collection systems is usually conveyed by gravity using the natural slope of the land. Wastewater pumps are used when the slope of the land requires lifting the wastewater to a higher elevation for a return to gravity flow. Pumps and other mechanical equipment require extensive maintenance and costly energy to operate so their use is avoided whenever possible. Thus, the location and design of the components of a collection system are strongly influenced by the topography (surface contour) of its service area. In some instances, where the topography is unfavorable for a gravity collection system, a pressure or vacuum collection system may be used.

Collection system operators need to recognize the differences between the sanitary wastewater collection systems described above, a similar storm water system, and a combined storm and wastewater collection system. A storm water collection system conveys water resulting from runoff of rainfall and snow melt from buildings and surrounding unpaved and paved areas to a natural watercourse or body of water, usually without treatment. This system also conveys excess water supplied to urban areas for irrigation and surface cleaning.

A combined wastewater and storm water collection system conveys a combination of these waters to a treatment plant. Occasionally an excessively heavy rain or cloudburst may produce flows that exceed the wastewater treatment plant's capacity. The excess flow may be diverted to specially designed holding tanks or lagoons for storage until the high flows recede. The stored water is then returned to the plant for proper treatment. In a few older communities with combined wastewater collection systems, provisions are usually made to relieve the collection system and treatment plant by diverting the excess flow and providing partial treatment plus disinfection before discharging the flow to water or onto land.

---

[1] *Collection system operators in small agencies are sometimes required to inspect sewer construction.*

QUESTIONS

Write your answers in a notebook and then compare your answers with those on page 75.

3.0A Why should a collection system operator understand how the system was designed and constructed?

3.1A What is the purpose of a wastewater collection system?

3.1B In areas where the topography is unfavorable for a gravity wastewater collection system, what types of collection systems may be used?

3.2 QUANTITY OF WASTEWATER

The amount of wastewater that a collection system will convey is determined by careful analysis of the present and probable future quantities of domestic, commercial, and industrial wastewaters, as well as anticipated groundwater *INFILTRATION*[2] and surface water *INFLOW*[3] produced in the service area. The sanitary sewers that make up a collection system are usually designed to convey the peak flow from all of these sources.

3.20 Design Period

Ideally a wastewater collection system should be designed to convey the estimated peak flow from its service area when the area has reached its maximum population and has been fully developed commercially and industrially. However, it is not always economically feasible to construct some of the major sewers in a system large enough to convey the ultimate peak flow from its service area. It also could be difficult to maintain system velocity (±2 ft/sec) if the design capacity is much greater than the actual flow. This situation does not allow the wet area of the pipe to scour away solids, which can cause anaerobic conditions that generate hydrogen sulfide gas. Under these circumstances the sewers are designed to convey peak flows that are estimated to occur within an appropriate design period, generally ranging from 10 to 30 years. Provisions should be made when planning a collection system to provide for expansion of the system before the actual flows become greater than the design flows.

3.21 Design Flows

Once the design period has been determined, design flows that will occur during the period are calculated based on the estimated population, per capita (per person) wastewater discharge and industrial and commercial wastewater discharges that will occur at the end of the period. The information in this section and Section 3.22 is provided to show you how a design engineer calculates needed flow capacities. It will also be useful to you, as the system operator, in predicting and managing the increasing flows that accompany commercial and residential development in your service area.

Population estimates can sometimes be obtained from a planning agency that has data covering the collection system's service area. If such information is not available, population estimates are made by using the most appropriate method.[4]

Suitable methods for making the estimates include an arithmetic increase per year, uniform percentage growth rate based on recent census periods, graphical comparison with the growth of other similar but larger cities, and graphical extension of past growth.

Domestic wastewater flow is calculated by multiplying the estimated population in the collection system service area by the per capita flow factors. Typical per capita flows from residential sources range from 70 to 100 gallons per day per person. Wastewater flows from commercial and industrial areas are usually estimated in terms of gallons per day per acre, based on the type of development and measured flows from similar existing developments. Estimation of flows from a proposed unsewered industrial area can be very difficult. Industrial discharges vary widely due to the types of manufacturing processes used by the industries. Information from local zoning commissions regarding the types and sizes of industries expected and permitted to use the area can be very helpful. Typical sewer capacity allowances for commercial and industrial areas are outlined in Table 3.1.

Some infiltration of groundwater and inflow of surface water occurs in all wastewater collection systems and capacity must be provided for this additional flow. The amount of infiltration entering through cracks in pipes, joints, and manholes will vary with the age and condition of the collection system and the portion of the system that is submerged in groundwater. The amount of inflow will vary with the number of manholes in a collection system that become submerged in surface water and the number of surface water drainage sources, such as roof drains, that are illegally connected to the collection system. Inflow can be reduced by sealing manhole covers that are subject to flooding. Table 3.2 outlines the infiltration design allowances for several cities. Design allowances for infiltration normally include allowances for inflow. Combined collection systems are designed similar to sanitary collection systems with the addition of storm water flows.

---

[2] *Infiltration (IN-fill-TRAY-shun). The seepage of groundwater into a sewer system, including service connections. Seepage frequently occurs through defective or cracked pipes, pipe joints, connections or manhole walls.*

[3] *Inflow. Water discharged into a sewer system and service connections from such sources as, but not limited to, roof leaders, cellars, yard and area drains, foundation drains, cooling water discharges, drains from springs and swampy areas, around manhole covers or through holes in the covers, cross connections from storm and combined sewer systems, catch basins, storm waters, surface runoff, street wash waters or drainage. Inflow differs from infiltration in that it is a direct discharge into the sewer rather than a leak in the sewer itself.*

[4] *Refer to GRAVITY SANITARY SEWER DESIGN AND CONSTRUCTION (MOP FD-5), pages 23 and 24. Obtain from Water Environment Federation (WEF), Publications Order Department, 601 Wythe Street, Alexandria, VA 22314-1994. Order No. MFD5. Price to members, $25.00; nonmembers, $35.00; plus shipping and handling.*

TABLE 3.1 SEWER CAPACITY ALLOWANCES FOR COMMERCIAL AND INDUSTRIAL AREAS [a]

City	Year Data Published Or Obtained	Commercial Allowances (GPD/acre)	Industrial Allowance (GPD/acre)
Cincinnati, OH	1980	Case-by-case determination after consultation with the Directors of Sewers.	
Dallas, TX	1960	30,000 added to domestic rate for downtown; 60,000 for tunnel relief sewers.	
Grand Rapids, MI	1980	Offices, 40 – 50 GPD/cap [b]; hotels, 400 – 500 GPD/room; hospitals, 200 GPD/bed; schools, 200 – 300 GPD/room.	
Hagerstown, MD	—	Hotels, 180 – 250 GPD/room; hospitals, 150 GPD/bed; schools, 120 – 150 GPD/room.	
Houston, TX	1960	Peak flows: offices 0.36 GPD/sq ft; retail 0 – 20 GPD/sq ft; hotels 0.93 GPD/sq ft.	
Las Vegas, NV	—	Resort hotels, 310 – 525 GPD/room; schools, 15 GPD/cap.	
Los Angeles, CA	1980	Commercial, 100 GPD/1,000 sq ft gross floor area; hospitals, 500 GPD/bed (surgical), 85 GPD/bed (convalescent); schools, elementary or junior high schools, 10 GPD/student; high schools, 15 GPD/student; universities, 20 GPD/student. The above values give peak flow rates. Divide by 3.0 to obtain average flow rates.	15,500
Los Angeles Co. San. District, CA	1980	4,000 – 6,000	
Lincoln, NE	1962	7,000	
Milwaukee, WI	1980	240,000 (max); 25,800 (min)	
St. Joseph, MO	1962	64,000 (downtown) 25,800 (neighborhood)	
St. Louis, MO	1960	90,000 average, 165,000 peak	
Santa Monica, CA	1980	Commercial, 9,700; hotels, 7,750	13,600
Toronto, Ont., Canada	1980	Analysis of actual water consumption in the commercial and industrial downtown area is approximately 20,000 GPD/acre.	

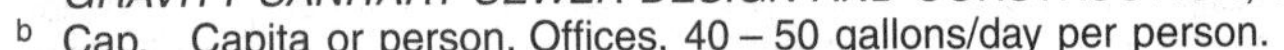

[a] *GRAVITY SANITARY SEWER DESIGN AND CONSTRUCTION*, Water Environment Federation, Alexandria, VA.
[b] Cap. Capita or person. Offices, 40 – 50 gallons/day per person.

TABLE 3.2 INFILTRATION DESIGN ALLOWANCE FOR SEVERAL CITIES[a]

City	Allowance (GPD/acre)[b]	Remarks
Seattle, WA	1,100	
Bay City, TX	1,000	
Lorain, OH	1,000	
Marion, OH	750	Calculations based on a proposed density, with a 100 GPD/cap average flow and peak of 400 GPD/cap also often used.
Ottumwa, IA	600	Infiltration and exfiltration shall not exceed 200 gal per inch of pipe diameter per mile of pipe per 24 hr period.[c]
West Springfield, MA	2,000	
Alma, MI	140	

[a] *GRAVITY SANITARY SEWER DESIGN AND CONSTRUCTION,* Water Environment Federation, Alexandria, VA.
[b] Higher allowances are used in areas where the cities have high groundwater tables and the sewer lines may be submerged for a portion of the year.
[c] Typical construction specification for the acceptance test for a new sewer line.

The flow of wastewater in a collection system (exclusive of infiltration and inflow) will vary during a 24-hour period with the minimum flow occurring during the early morning and the maximum occurring around noontime. Table 3.3 lists the ratio ranges of maximum and minimum flows to average daily wastewater flow recommended for use in design by various authorities. The ratios in Table 3.3 are often called *PEAKING FACTORS.*[5]

TABLE 3.3 DESIGN RATIOS OF MAXIMUM AND MINIMUM TO AVERAGE DAILY FLOW[a]

Population, in thousands	Ratio Ranges[b] Minimum	Maximum
1	0.20	3.9 – 5.5
5	0.26 – 0.27	3.3 – 4.5
10	0.29 – 0.32	2.9 – 4.0
50	0.39 – 0.44	2.3 – 2.9
100	0.43 – 0.50	2.0 – 2.6
200	0.48 – 0.57	1.8 – 2.3
500	0.55 – 0.69	1.5 – 1.9

[a] Adapted from *GRAVITY SANITARY SEWER DESIGN AND CONSTRUCTION,* Water Environment Federation, Alexandria, VA.
[b] *NOTE:* As the population increases the ratio ranges become closer to one, thus there is less fluctuation in flows.

3.22 Sample Design Flow Calculation

EXAMPLE 1

The estimated design flow for a typical wastewater collection system can be calculated given the following information on the system's service area and the assumed flow data.

Given Information:

Service area	10,000 acres
Present population	30,000 people
Estimated design period population	50,000 people
Estimated commercial area	500 acres
Estimated industrial area	700 acres

Assumed Flow Data:

Per capita flow	85 gallons per day per person (GPCD)
Commercial flow allowance	5,000 GPD/acre
Industrial flow allowance	13,000 GPD/acre
Infiltration flow allowance	600 GPD/acre

Design Flow:

Residential flow,

$$\frac{(50{,}000 \text{ people})(85 \text{ GPD/capita})}{1{,}000{,}000/\text{M}^{**}} = 4.25 \text{ MGD}^{*}$$

Peak residential flow,

$$4.25 \text{ MGD} \times 2.6 \text{ peaking factor} = 11.05 \text{ MGD}$$

Commercial capacity allowance,

$$\frac{(500 \text{ acres})(5{,}000 \text{ GPD/acre})}{1{,}000{,}000/\text{M}} = 2.50 \text{ MGD}$$

Industrial capacity allowance,

$$\frac{(700 \text{ acres})(13{,}000 \text{ GPD/acre})}{1{,}000{,}000/\text{M}} = 9.10 \text{ MGD}$$

Infiltration and inflow capacity allowance,

$$\frac{(10{,}000 \text{ acres})(600 \text{ GPD/acre})}{1{,}000{,}000/\text{M}} = 6.00 \text{ MGD}$$

Total design flow,

$$(11.05 + 2.50 + 9.10 + 6.00) = 28.65 \text{ MGD}$$

* MGD. Million Gallons per Day.
** 1,000,000/M. This means 1,000,000/Million which is similar to saying 12 in/ft.

While Example 1 provides a method to estimate design flow for a collection system, it should be noted that the data used for the calculation can vary widely from geographical area to geographical area for a wide variety of reasons. For example, in the southwest where drought conditions are experienced, very aggressive water conservation programs have been implemented in many areas. The required use of low-flow toilets and showerheads will significantly reduce the estimated flows used in the above example. These variations or considerations need to be taken into account when developing design flow data in your area.

[5] *Peaking Factor. Ratio of a maximum flow to the average flow, such as maximum hourly flow or maximum daily flow to the average daily flow.*

3.23 Flow Measurement

Collection system operators will need to know the fundamentals of wastewater flow measurement in a sewer pipe. There are many devices available for flow measurement. All of these flowmeters are based on the simple principle that the flow rate equals the velocity of flow multiplied by the cross-sectional area of the flow. This principle is expressed by the following formula:

Q, cubic feet per second = (Area, sq ft)(Velocity, ft/sec)

Calculation of the cross-sectional area of flow in a sewer line can be made by using a factor found in Table 3.4. This procedure is explained in Example 2.

EXAMPLE 2

The depth of flow in a 12-inch diameter sewer is 5 inches. Determine the cross-sectional area of the flow.

Known		Unknown
D or Diameter, in	= 12 in	Cross-sectional Area, sq ft
d or Depth, in	= 5 in	

To determine the cross-sectional area for a sewer pipe flowing partially full use the following steps:

1. Find the value for the depth, d, divided by the diameter, D.

$$\frac{\text{d, in}}{\text{D, in}} = \frac{5\text{ in}}{12\text{ in}}$$

$$= 0.42$$

2. Find the correct factor for 0.42 in Table 3.4.

$$\frac{d}{D} = 0.42 \qquad \text{Factor} = 0.3130$$

3. Calculate the cross-sectional area.

$$\text{Pipe Cross-sectional Area, sq ft} = \frac{(\text{Factor})(\text{Diameter, in})^2}{144\text{ sq in/sq ft}}$$

$$= \frac{(0.3130)(12\text{ in})^2}{144\text{ sq in/sq ft}}$$

$$= 0.313\text{ sq ft}$$

Velocities in sewers should be measured in the field to be sure *ACTUAL* velocities are high enough to prevent the deposition of solids. If roots or other obstructions get into a sewer, design velocities can be reduced and problems develop. Actual velocities in sewers can be measured by using dyes or floats. Realize that velocities vary during the day depending on the activities of the tributary population. A scouring velocity should be reached or exceeded during peak flows, but probably will not be reached during low-flow periods. Actual velocities should be measured in the field, rather than estimating the velocities by the use of hydraulic formulas which give the theoretical velocity.

DYES. Tracer dyes are very effective for estimating velocities. Put into the upstream manhole the smallest amount of dye that can be seen easily in the downstream manhole. Fluorescent tracer dyes can be obtained from chemical suppliers. The supplier of chemicals for a wastewater treatment plant is a good source to recommend the proper type of dye. The amount of dye used will depend on the type of dye, diameter of sewer, and estimated flow in the sewer.

TABLE 3.4 AREA OF PARTLY FILLED CIRCULAR PIPES

d/D	Factor	d/D	Factor
0.01	0.0013	0.51	0.4027
0.02	0.0037	0.52	0.4127
0.03	0.0069	0.53	0.4227
0.04	0.0105	0.54	0.4327
0.05	0.0174	0.55	0.4426
0.06	0.0192	0.56	0.4526
0.07	0.0242	0.57	0.4625
0.08	0.0294	0.58	0.4724
0.09	0.0350	0.59	0.4822
0.10	0.0409	0.60	0.4920
0.11	0.0470	0.61	0.5018
0.12	0.0534	0.62	0.5115
0.13	0.0600	0.63	0.5212
0.14	0.0668	0.64	0.5308
0.15	0.0739	0.65	0.5404
0.16	0.0811	0.66	0.5499
0.17	0.0885	0.67	0.5594
0.18	0.0961	0.68	0.5687
0.19	0.1039	0.69	0.5780
0.20	0.1118	0.70	0.5872
0.21	0.1199	0.71	0.5964
0.22	0.1281	0.72	0.6054
0.23	0.1365	0.73	0.6143
0.24	0.1449	0.74	0.6231
0.25	0.1535	0.75	0.6319
0.26	0.1623	0.76	0.6405
0.27	0.1711	0.77	0.6489
0.28	0.1800	0.78	0.6573
0.29	0.1890	0.79	0.6655
0.30	0.1982	0.80	0.6736
0.31	0.2074	0.81	0.6815
0.32	0.2167	0.82	0.6893
0.33	0.2260	0.83	0.6969
0.34	0.2355	0.84	0.7043
0.35	0.2450	0.85	0.7115
0.36	0.2545	0.86	0.7186
0.37	0.2642	0.87	0.7254
0.38	0.2739	0.88	0.7320
0.39	0.2836	0.89	0.7384
0.40	0.2934	0.90	0.7445
0.41	0.3032	0.91	0.7504
0.42	0.3130	0.92	0.7560
0.43	0.3229	0.93	0.7612
0.44	0.3328	0.94	0.7662
0.45	0.3428	0.95	0.7707
0.46	0.3527	0.96	0.7749
0.47	0.3627	0.97	0.7785
0.48	0.3727	0.98	0.7816
0.49	0.3827	0.99	0.7841
0.50	0.3927	1.00	0.7854

After the dye has been inserted in the flow at the upstream manhole, measure and record two readings:

1. Total travel time from insertion of dye until the dye is first seen at the downstream manhole (t_1); and

2. Total travel time from insertion of dye until the dye is no longer visible at the downstream manhole (t_2).

The average travel time is $(\frac{1}{2})(t_1 + t_2)$. The average velocity is calculated by dividing the distance (between manholes) by the average time.

$$\text{Velocity, ft/sec} = \frac{\text{Distance, ft}}{\text{Average Time, sec}}$$

FLOATS. Floats are used to estimate velocities in existing sewers flowing partially full. A float could be a stick, an orange, or any other object that can be easily identified and that travels down the sewer at a velocity similar to the wastewater. Velocity is determined by recording the time the float takes to travel a known distance, such as the distance between manholes. Most surface floats such as sticks or oranges travel from 10 to 15 percent faster than the average velocity due to the velocity differences between the surface and bottom sections of the flowing water. Float travel down a sewer line may be hindered by obstructions in the line such as protruding taps, roots, grease deposits, or the crown (top) of the pipe. If the float travel time is slower than normal, try double checking the time using a dye.

EXAMPLE 3

A stick travels the 400 feet between two manholes in 3 minutes and 20 seconds (200 seconds). Estimate the velocity in the sewer.

Known	**Unknown**
Distance, ft = 400 ft	Velocity, ft/sec
Time, sec = 200 sec	

$$\text{Velocity, ft/sec} = \frac{\text{Distance, ft}}{\text{Time, sec}}$$

$$= \frac{400 \text{ ft}}{200 \text{ sec}}$$

$$= 2 \text{ ft/sec}$$

NOTE: The velocity of the stick (2 ft/sec) is 10 to 15 percent faster than the average velocity of the wastewater in the sewer.

$$\text{Average Velocity, ft/sec} = 2 \text{ ft/sec} - 2 \text{ ft/sec } (0.10 \text{ to } 0.15)$$

$$= 2 \text{ ft/sec} - 0.2 \text{ to } 0.3 \text{ ft/sec}$$

$$= 1.7 \text{ to } 1.8 \text{ ft/sec}$$

Assuming the channel in the manhole at the downstream end of the section of sewer line in which the velocity was determined is similar in cross section to the sewer line, the cross section of the area of flow can be determined by measuring the depth of flow at the manhole.

Knowing the velocity and cross-sectional area of the flow in a sewer line, the quantity of flow or flow rate can then be calculated.

EXAMPLE 4

The velocity in a 12-inch sewer line is 1.7 ft/sec and the flow cross-sectional area is 0.313 sq ft. Determine the rate of flow.

Known	**Unknown**
Velocity, ft/sec = 1.7 ft/sec	Flow, cu ft/sec
Area, sq ft = 0.313 sq ft	

Calculate the rate of flow in cubic feet per second.

$$\text{Flow, cu ft/sec} = (\text{Velocity, ft/sec})(\text{Area, sq ft})$$

$$= (1.7 \text{ ft/sec})(0.313 \text{ sq ft})$$

$$= 0.53 \text{ cu ft/sec}$$

This calculation produces the flow rate in the sewer line in cubic feet per second. However, flow rates of wastewater are generally expressed in million gallons per day rather than cubic feet per second. Example 5 illustrates the method of converting cubic feet per second (CFS) to million gallons per day (MGD).

EXAMPLE 5

Convert a flow of 0.53 cubic feet per second to million gallons per day.

Known	**Unknown**
Flow, CFS = 0.53 CFS	Flow, MGD
1 cu ft = 7.48 gal	
1 day = 86,400 sec	

1. Develop a conversion factor for CFS to MGD. This conversion factor is very useful.

$$1 \text{ MGD} = \frac{(1 \text{ M Gal/day})(1{,}000{,}000/\text{M})}{(7.48 \text{ gal/cu ft})(86{,}400 \text{ sec/day})}$$

$$= 1.55 \text{ cu ft/sec}$$

$$= 1.55 \text{ CFS}$$

OR $$1 \text{ CFS} = \frac{1 \text{ CFS}}{1.55 \text{ CFS/MGD}}$$

$$= 0.646 \text{ MGD}$$

2. Calculate the flow in million gallons per day using the known flow in cubic feet per second.

$$\text{Flow, MGD} = \frac{\text{Flow, CFS}}{1.55 \text{ CFS/MGD}}$$

$$= \frac{0.53 \text{ CFS}}{1.55 \text{ CFS/MGD}}$$

$$= 0.34 \text{ MGD}$$

OR $$\text{Flow, MGD} = (\text{Flow, CFS})(0.646 \text{ MGD/CFS})$$

$$= (0.53 \text{ CFS})(0.646 \text{ MGD/CFS})$$

$$= 0.34 \text{ MGD}$$

For further information on how to insert flowmeters in sewers and open channels to measure flows, see Chapter 7, "Wastewater Flow Monitoring," in *PRETREATMENT FACILITY INSPECTION* in this series of training manuals.

3.24 Flow Variations

Knowledge of the daily and annual variations or fluctuations in a collection system's wastewater flow is useful in planning maintenance programs for the system. For example, collection system hydraulic cleaning operations using the flow in the system should be scheduled during the hours of sufficient flows. On the other hand, cleaning with bucket equipment should be scheduled during the hours of minimum flow in the system. Of course, upstream diversion of flow, if possible, should always be considered as a means of developing an optimum maintenance schedule.

Figure 3.1 illustrates daily fluctuations of dry weather wastewater flow from a typical residential community. The amount of flow is shown on the vertical axis of the chart and the time of day is shown on the horizontal axis of the chart. Beginning at 12:00 p.m. we see that flow is at 1.0 MGD and begins to drop off to a minimum of 0.5 MGD just after 4:00 a.m. in the morning. This is typical of residential flow when people go to bed at night and are no longer using water for showers, washing clothes, or doing dishes. A local industry that operates 24 hours a day and discharges process wastewater would have an effect on the minimum flow. The residential flow increases from the minimum at just after 4:00 a.m. as domestic and commercial water consumption and discharges increase and peak at 1.3 MGD at 10:00 a.m. Flow then begins to diminish; however, a second peak occurs just before 10:00 p.m. and the flow then drops off once again to 1.0 MGD at 12:00 p.m. The average daily flow is 0.9 MGD. The ratio of maximum flow to average flow is 1.3 MGD ÷ 0.9 MGD or a dry weather peaking factor of 1.4.

A diurnal (day/night) graph like the one in Figure 3.1 is a good diagnostic tool that can help the operator determine the condition of the collection system and the amount of inflow or infiltration occurring. If plotted on a frequent basis and analyzed regularly, diurnal flow graphs also alert the operator to problems, such as a broken or collapsed line. When the operator notices significant variations from historical flow patterns, it can be assumed that some change has occurred and the cause of the flow variations should be promptly investigated.

An industry with a significant wastewater discharge would impose a different flow fluctuation on a collection system, depending on its pattern of wastewater discharge. This can be determined by flow measurement. You could then construct a combined fluctuation chart, similar to Figure 3.1, for the flows from your combined residential and industrial sources, or the fluctuation in total flows could be measured in the collection system downstream of the point of combined flows.

To construct a chart showing your system's annual flow fluctuations, calculate the typical monthly residential flows. Add to these amounts any industrial flows, including known or estimated fluctuations. Next, add in your estimates of infiltration and inflow amounts, on a monthly basis. Infiltration and

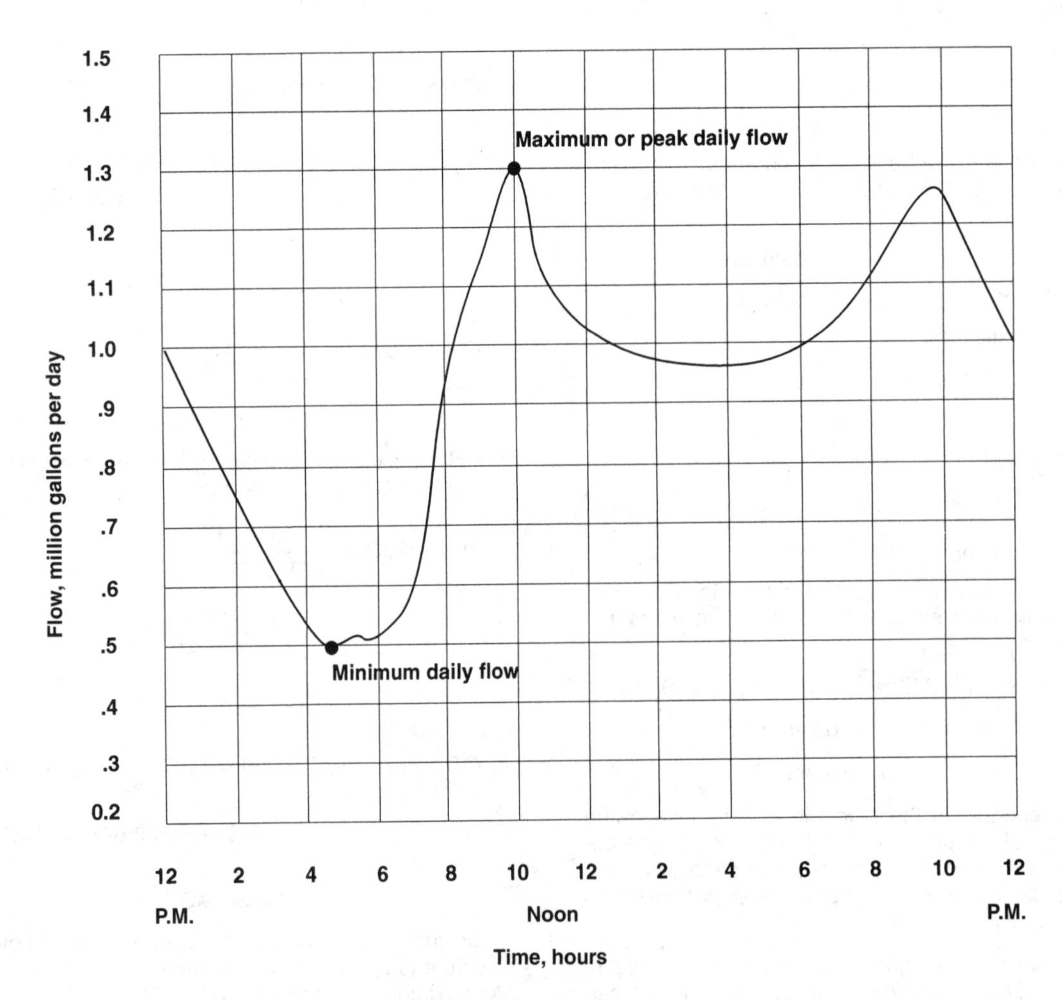

Fig. 3.1 Variations of dry weather wastewater flow from a typical community

inflow[6] will vary proportionally to the seasonal variation in groundwater levels and rainfall intensity. Graph the monthly amounts as you did for daily totals. The resulting chart can be used to help you plan long-term periodic maintenance activities. High levels of infiltration and inflow can be reduced, especially in older collection systems, by an effective sewer renewal or rehabilitation program (Chapter 10, Volume II).

QUESTIONS

Write your answers in a notebook and then compare your answers with those on page 75.

3.2A List the sources or types of wastewater conveyed by wastewater collection systems.

3.2B What factors are considered in estimating peak flow from an area?

3.2C What measurement aids are commonly used to measure actual flow velocity in sewer mains?

3.3 COMPONENTS

Let's take a wastewater collection system apart and identify the parts and their purposes. In doing this we will start at the upper end (highest point) and follow the flow path of wastewater through the system.

3.30 Gravity Collection System

Wastewater from residences, commercial facilities and industrial developments is collected and conveyed to wastewater treatment plants through the following parts of a gravity wastewater collection system:

BUILDING SEWERS. A building sewer connects a building's internal wastewater drainage system (plumbing) to the larger street sewer. Technically, a building sewer may connect with a lateral sewer, a main sewer, or a trunk sewer, depending on the layout of the system. (All of these sewers are parts of the sewer in the street.) The building sewer may begin immediately at the outside of a building or some distance (such as 2 to 10 feet) from the foundation, depending on local building codes. Where the sewer officially begins marks where the building plumber's responsibility ends and the collection system operator's responsibility starts for maintenance and repair of the system. Jurisdiction usually changes at the property line. Figure 3.2 illustrates a typical building sewer and *APPURTENANCES*.[7]

LATERAL AND BRANCH SEWERS are the upper ends of the street sewer components of a collection system. In some instances lateral and branch sewers may be located in easements. This location should be avoided where possible due to maintenance problems created by locations with difficult

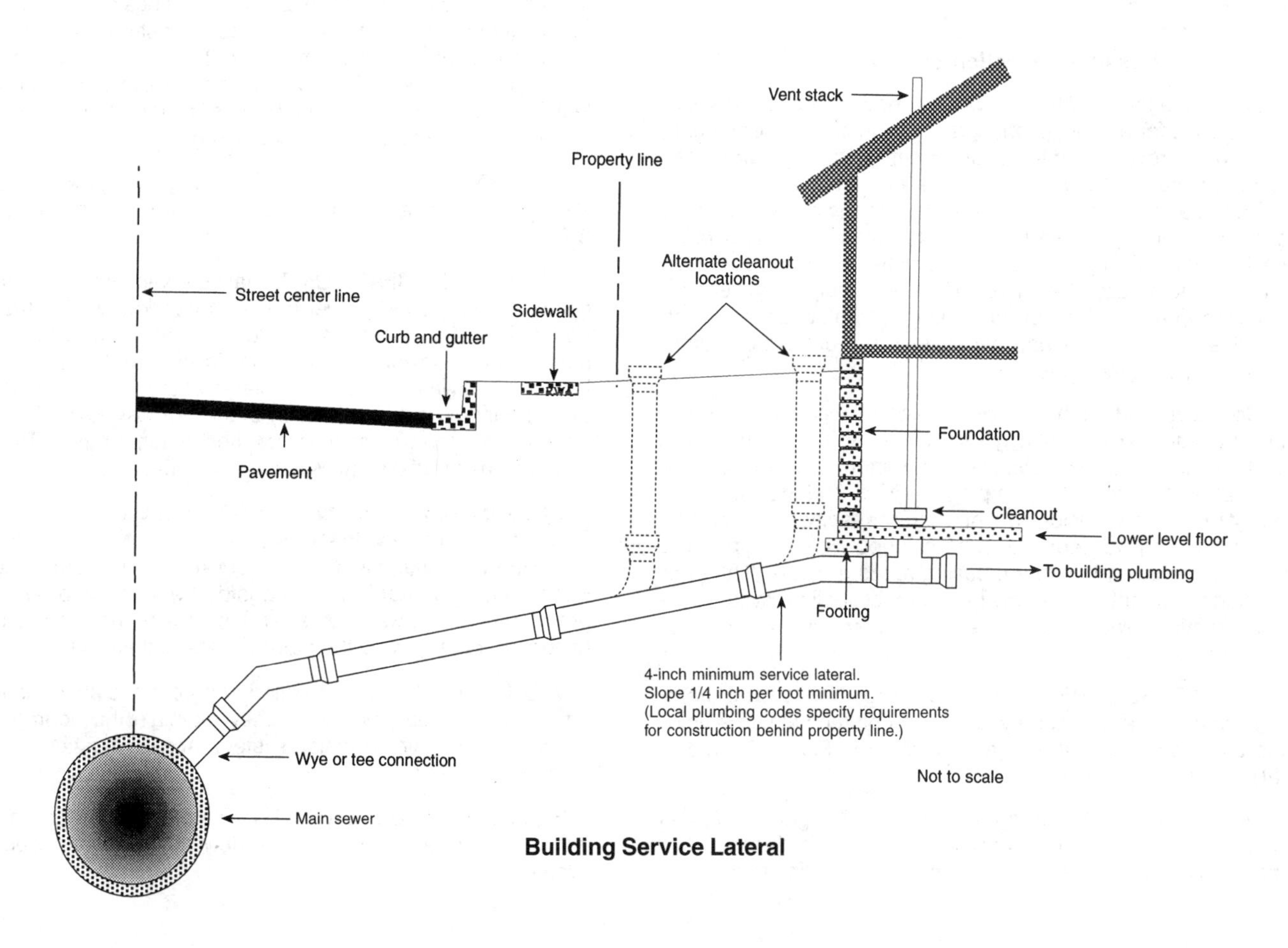

Fig. 3.2 Typical building service lateral

[6] *Infiltration and inflow may also be referred to as I & I or I/I.*

[7] *Appurtenance (uh-PURR-ten-nans). Machinery, appliances, structures and other parts of the main structure necessary to allow it to operate as intended, but not considered part of the main structure.*

access and limited work space. Figure 3.3 (page 43) illustrates the typical locations of lateral and branch sewers.

MAIN SEWERS collect the wastewater of several lateral and branch sewers from an area of several hundred acres and convey it (the wastewater) to larger trunk sewers. Figure 3.3 illustrates the typical locations of main sewers in a collection system.

TRUNK SEWERS are the main "arteries" of a wastewater collection system; they convey the wastewater from numerous main sewers either to a treatment plant or an interceptor sewer. Figure 3.3 illustrates the locations of trunk sewers in a typical collection system.

INTERCEPTING SEWERS receive the wastewater from trunk sewers and convey it to the treatment plant. Some of the trunk sewers may have previously discharged directly into bodies of water or may have routed wastewater to now-abandoned treatment plants. The location of an intercepting sewer in a typical collection system is illustrated in Figure 3.3.

LIFT STATIONS are used in a gravity collection system to lift (pump) wastewater to a higher elevation when the slope of the route followed by a gravity sewer would cause the sewer to be laid at an insufficient slope or at an impractical depth. Lift stations vary in size and type depending on the quantity of wastewater and the height the wastewater is to be lifted. Chapter 8 in Volume II of this manual describes the sizes and types of lift stations.

3.31 Low-Pressure Collection System

Where the topography and ground conditions of an area are not suitable for a conventional gravity collection system due to flat terrain, rocky conditions, or extremely high groundwater, low-pressure collection systems are now becoming a practical alternative. Pressure sewers may be installed instead of gravity sewers in an area because (1) a pipe slope is not practical to maintain gravity flow, (2) smaller pipe sizes can be used due to pressurization, and (3) reduced pipe sizes can be installed due to a lack of infiltration and inflow because the pipeline has no leaks and water does not enter the system through access manholes.

Operation and maintenance considerations when comparing pressure sewers with gravity systems include the facts that pressure systems have (1) higher energy costs for pumping, (2) greater costs for pumping facilities, (3) fewer stoppages, (4) no root intrusion, (5) no extra capacity for infiltration and inflow, (6) no deep trenches or deeply buried pipe, and (7) no inverted siphons for crossing roads or rivers. The principal components of a low-pressure collection system include gravity sewers, holding tanks, grinder pumps, and pressure mains.

GRAVITY SEWERS connect a building's wastewater drainage system to a buried pressurization unit (containing a holding tank) located on the lot as illustrated in Figure 3.4 (page 44).

HOLDING TANKS serve as a reservoir for grinder pumps and have a capacity of approximately 50 gallons. Figure 3.4 illustrates a typical pressurization unit with a holding tank.

GRINDER PUMPS serve both as a unit to grind the solids in the wastewater (solids could plug the downstream small-diameter pressure sewers and valves) and to pressurize the wastewater to help move it through the collection system. Figure 3.4 illustrates the location of the submersible grinder pump in the holding tank.

PRESSURE MAINS are the "arteries" of the low-pressure collection system; they convey the pressurized wastewater to a treatment plant. Since the wastewater is "pushed" by pressure, the mains are not dependent on a slope to create a gravity flow and can be laid at a uniform depth following the natural slope of the land along their routes. Low-pressure collection systems must have access for maintenance. This means line access where a *PIG*[8] (see Chapter 6) can be inserted into a line for cleaning and also removed from the line. Manholes or access boxes must have valves and pipe spools (two- to three-foot long flanged sections of pipe) that can be removed for cleaning the pipe or for pumping into or out of the system with a portable pump. Figure 3.5 illustrates the profile of a typical low-pressure main and Figure 3.6 illustrates a typical low-pressure collection system.

3.32 Vacuum Collection System

Vacuum collection systems are also being used as an alternative to a gravity collection system. These vacuum systems are installed for the same reasons that low-pressure collection systems are used. Operation and maintenance of vacuum systems are more difficult than low pressure systems because it is harder to maintain a vacuum due to the number of inlets and valves to the system. Principal components of a vacuum collection system include gravity sewers, holding tanks, vacuum valves, vacuum mains, and vacuum pumps.

GRAVITY SEWERS connect a building's wastewater drainage system to a vacuum interface unit, as illustrated in Figure 3.7.

VACUUM INTERFACE UNIT seals the vacuum service and connects the vacuum sewer to the gravity flow system from a home. This unit maintains the required vacuum level in the main vacuum sewer system. When approximately three gallons of wastewater have accumulated at the vacuum interface unit, a valve is activated which allows atmospheric air pressure to force the wastewater into the vacuum branch. The location of the interface unit is shown in Figure 3.7.

VACUUM MAINS are the "arteries" of a vacuum collection system; they convey the wastewater to a treatment plant. Since the wastewater is drawn or "sucked" by vacuum, the collection mains do not have to be laid at a slope to produce a gravity pull on the wastewater and the mains may be laid at a uniform depth following the natural slope of their routes.

VACUUM PUMPS are installed in a central station, usually adjacent to a treatment plant; they maintain the appropriate vacuum in the main collection system. Access must be provided for maintenance.

Figure 3.8 illustrates a typical street portion of a vacuum collection system and gives an overview of the entire collection system.

---

[8] *Pig. Refers to a poly pig which is a bullet-shaped device made of hard rubber or similar material. This device is used to clean pipes. It is inserted in one end of a pipe, moves through the pipe under pressure, and is removed from the other end of the pipe.*

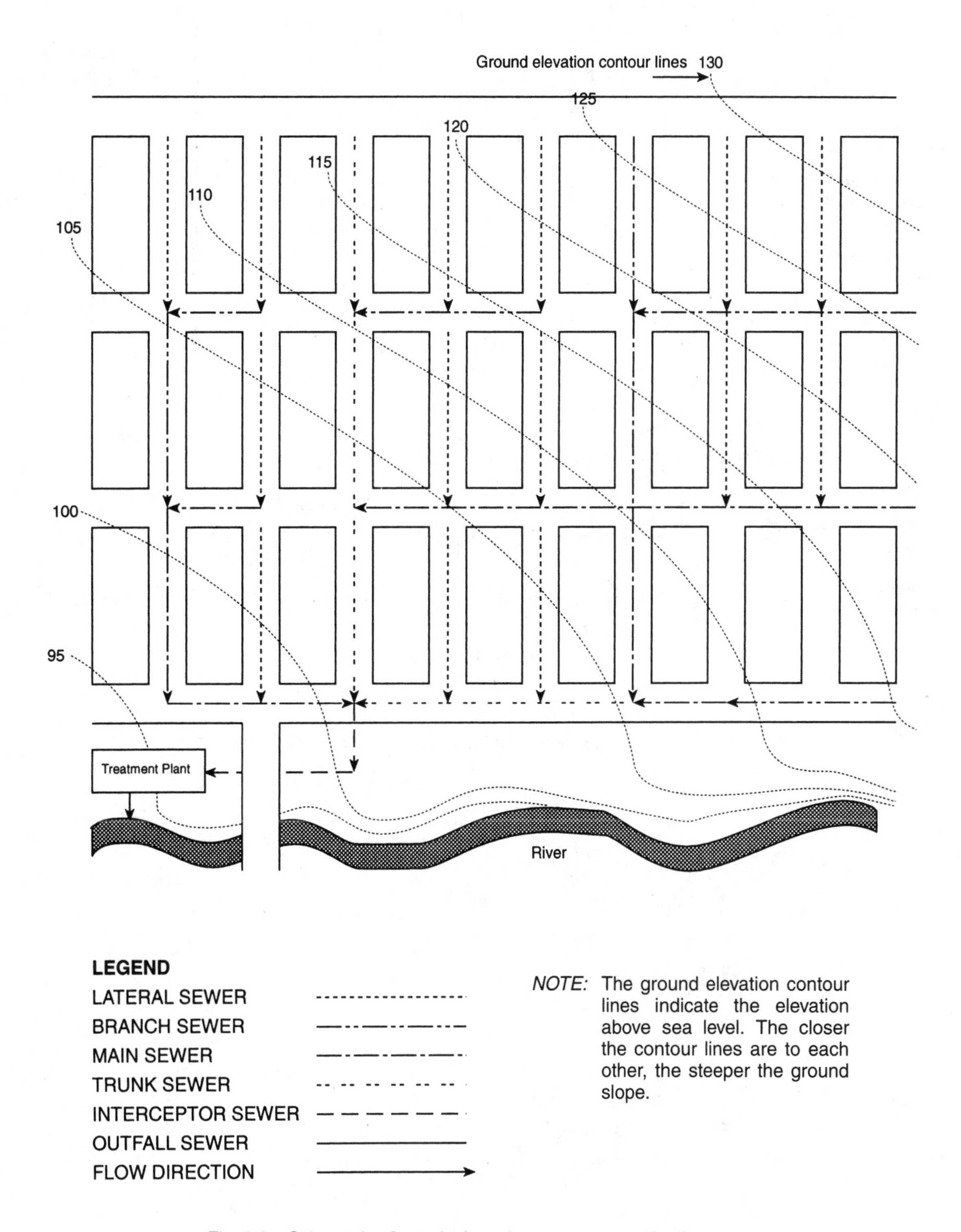

Fig. 3.3 Schematic of a typical gravity wastewater collection system

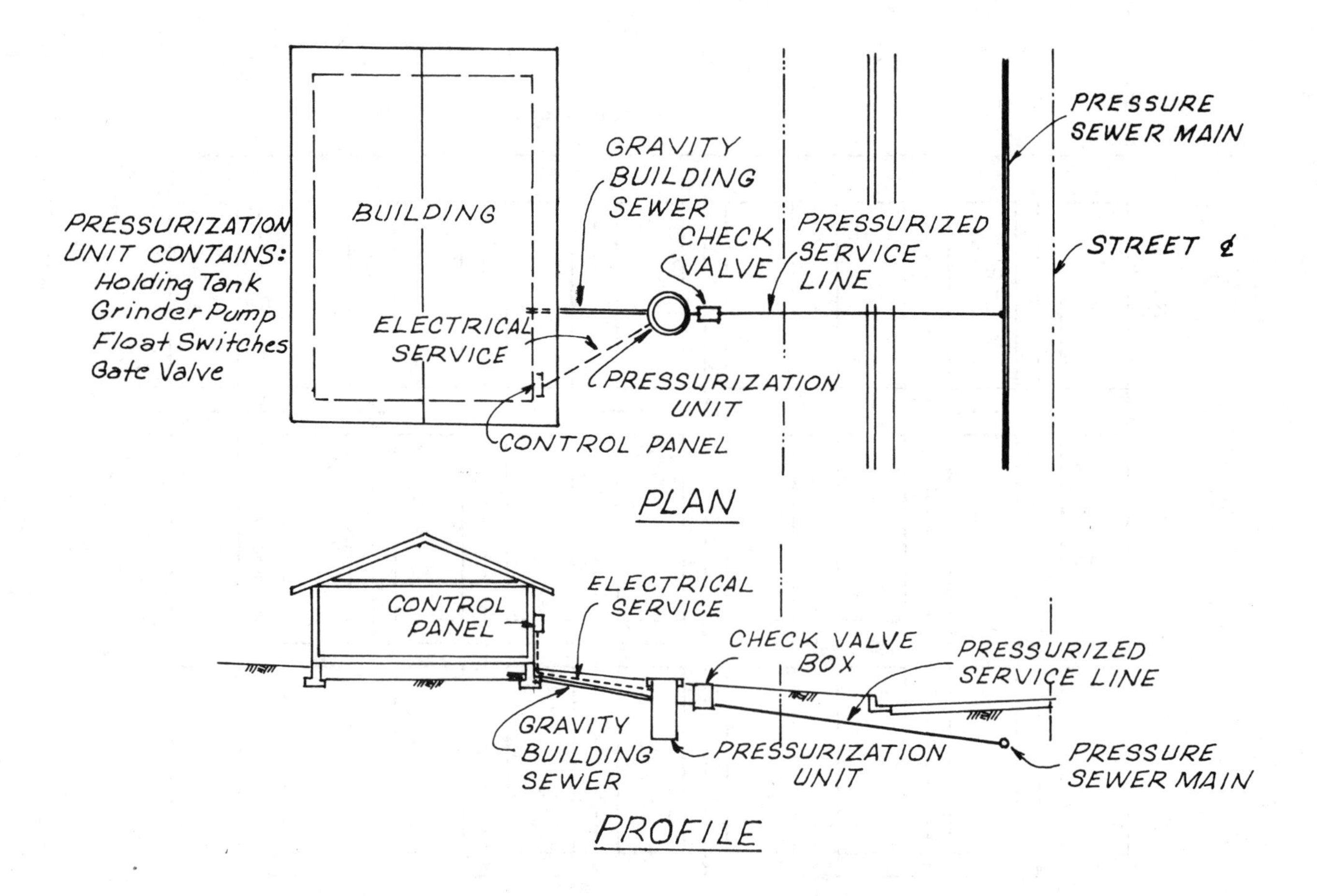

Fig. 3.4 Principal components of a typical low-pressure collection system

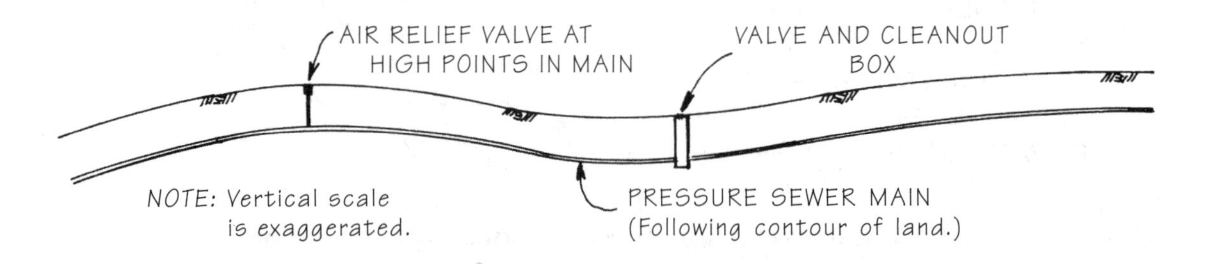

Fig. 3.5 Profile of a typical low-pressure main

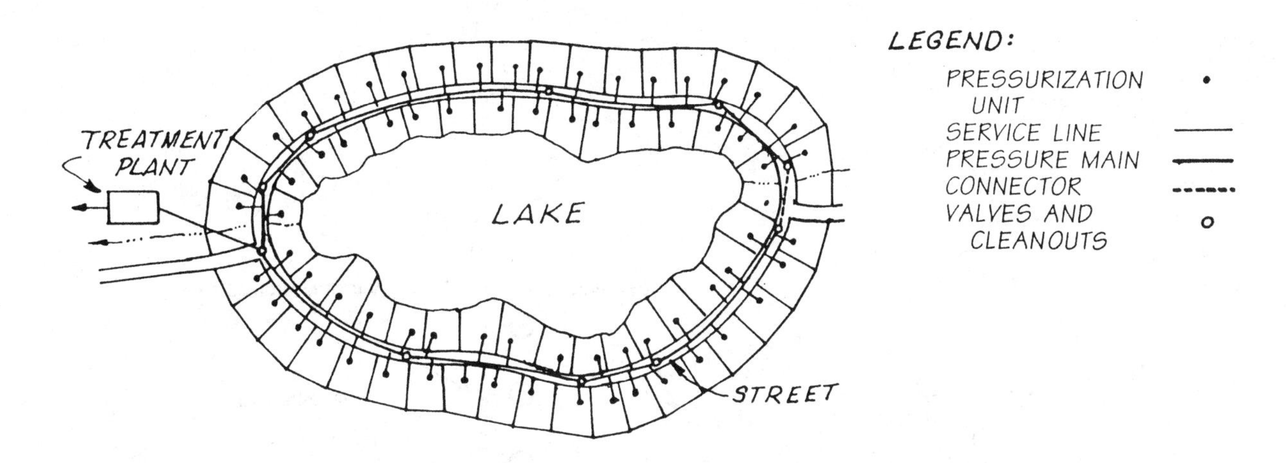

Fig. 3.6 Schematic of a typical low-pressure collection system

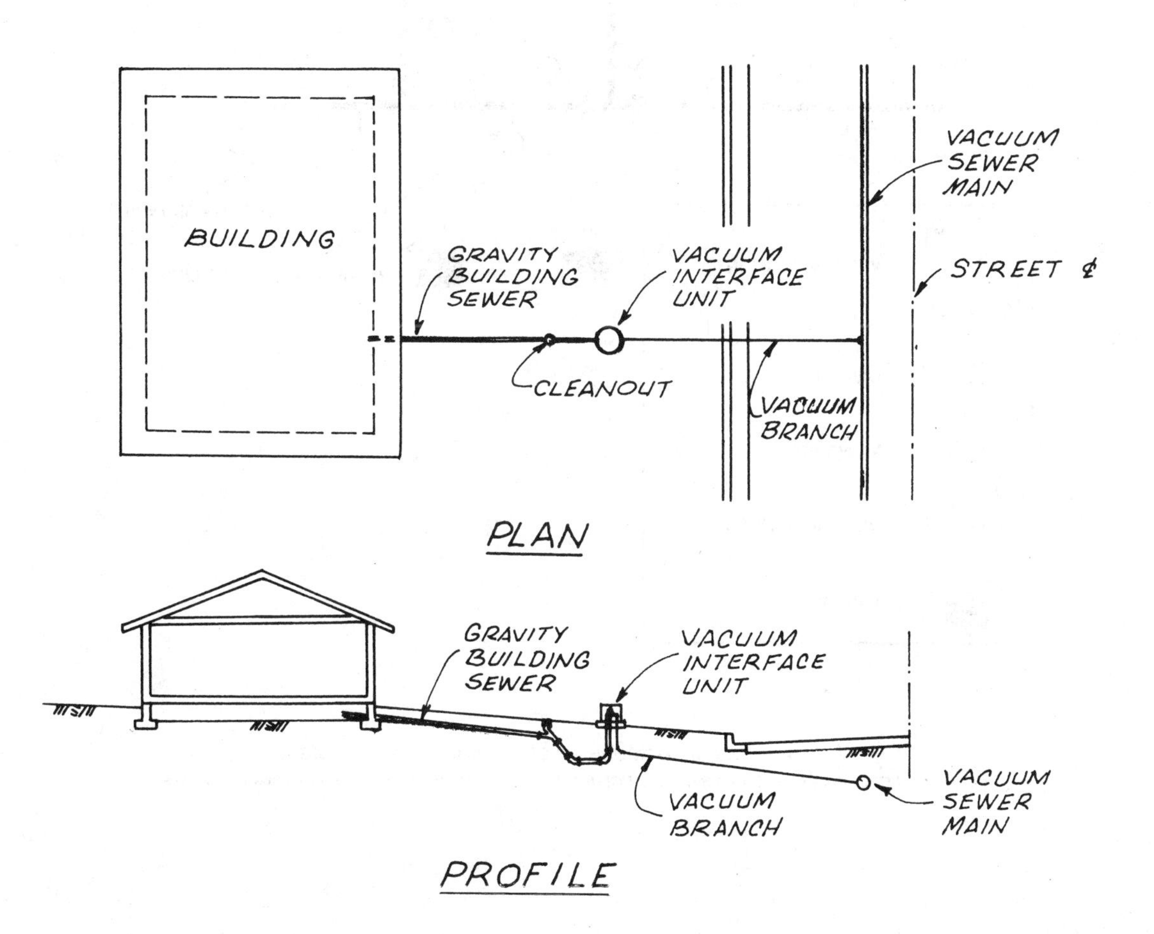

Fig. 3.7 Principal components of a typical vacuum collection system

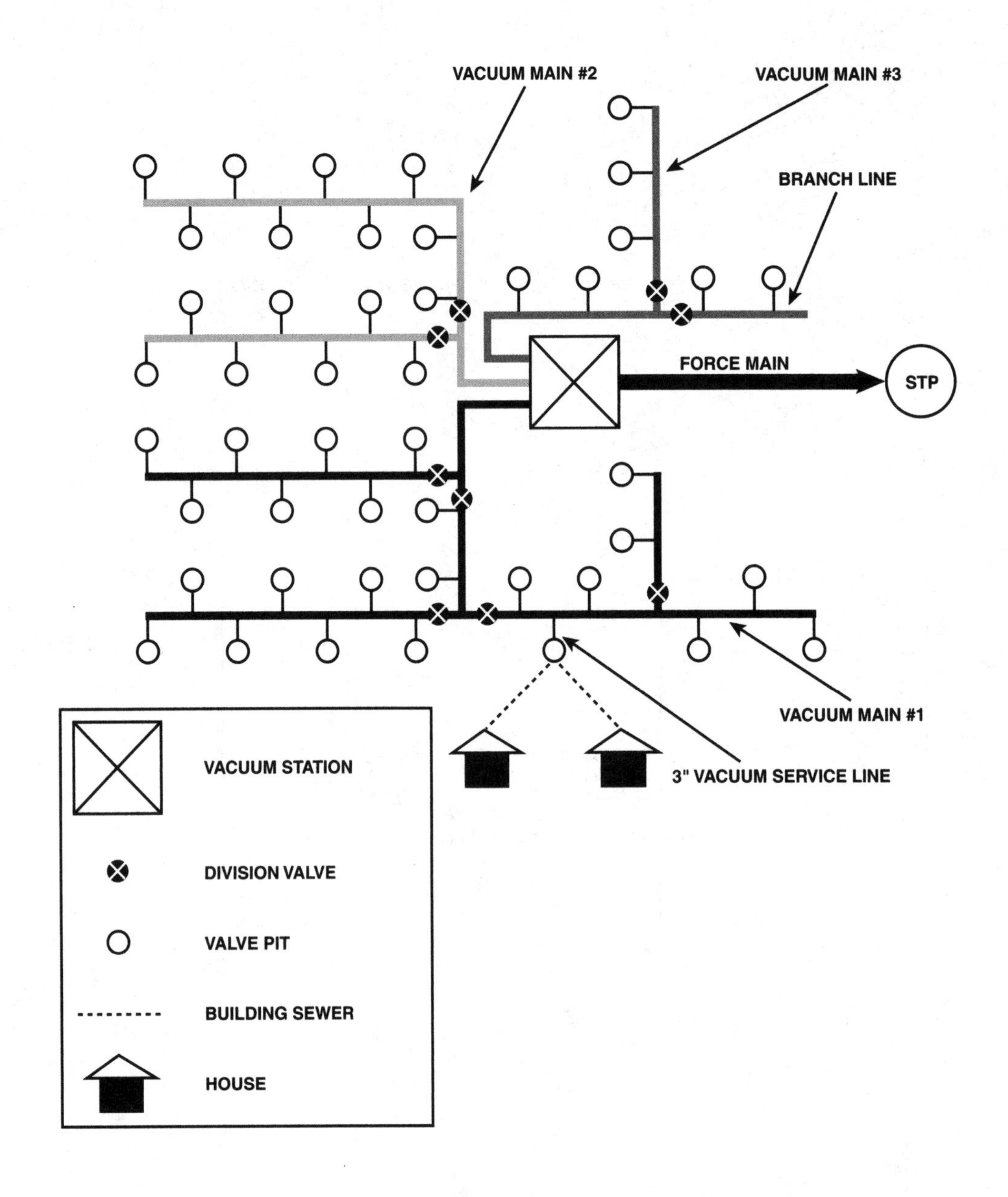

Fig. 3.8 Major components of a vacuum sewer system
(Source: *ALTERNATIVE WASTEWATER COLLECTION SYSTEMS*, U.S. EPA Technology Transfer Manual, October 1991)

QUESTIONS

Write your answers in a notebook and then compare your answers with those on page 75.

3.3A List the parts or components of a gravity wastewater collection system.

3.3B List the principal components of a low pressure collection system.

3.3C List the principal components of a vacuum collection system.

3.33 Appurtenances

The immediately preceding sections of this chapter described the principal components of a wastewater collection system consisting primarily of sewer piping and pumps. This section will describe the appurtenances to a collection system that facilitate and control the flow of wastewater and provide for the access to maintain the collection systems.

BACKFLOW PREVENTERS are used to stop the accidental backflow or reverse flow of wastewater into buildings. The type "A" backflow preventer shown in Figure 3.9 is used when the elevation at the top of the device is lower than the floor elevation of the building plumbing system. This device acts simply as a relief so when the water elevation in the main sewer line and manholes reaches the same elevation as the bottom of the ball, the ball is lifted and raw wastewater then flows out of the backflow preventer. The obvious disadvantage of this device is that it results in an overflow of raw wastewater. Depending on the severity of the main line sewer backup, a type "A" backflow preventer may not be able to release enough of the excess wastewater to protect the building from a backup.

A second type of backflow preventer is the type "B," which uses either a traditional swing check valve or ball check valve. As the raw wastewater in the main line backs up into the building service lateral, the check valve closes and prevents flow into the building plumbing system. This type of valve also has some disadvantages in that it requires maintenance to make sure that the check valve arm is operating properly and that debris does not build up in the bottom of the valve and prevent positive seating of the valve.

A third type of backflow prevention device is the automated system shown in Figure 3.10 (page 49). This valve uses a flexible diaphragm and a level sensing device in the building sewer. When the level sensor senses that a backflow is occurring in the building sewer, it automatically energizes an air control valve closing the diaphragm and closing the building lateral. Air is supplied by a small compressor and a storage tank. In addition, an alarm is activated to signal the homeowner that a backup is occurring and that some action is necessary. This system can be installed either inside or outside the building.

BUILDING SEWER VENT TRAPS[9] are used to prevent the passage of gases and odors from the main sewer system into a building's wastewater plumbing (drainage) system. They are not used extensively today since plumbing codes require adequate trapping and venting of a building's plumbing fixtures and because the traps tend to collect solids, thus causing stoppages in the building sewer. Also, removal of stoppages in a building sewer is made more difficult since cleaning equipment must negotiate the bends in the trap as illustrated in Figure 3.11. If vent traps are not properly screened, they can allow rodents access to building sewers and the collection system. Also, if traps are located in a low spot in a yard, this will serve as a source of inflow from surface runoff. Vent traps in low spots should be removed and adequate venting should be installed in the home or building.

BUILDING SEWER CLEANOUTS are located as required by local building codes and sewer ordinances. They are used to allow access to sewers to remove solids that cause stoppages. As a minimum, a cleanout should be located in the building sewer approximately three feet from the building foundation and another at the building's property line if the part of the sewer in the street is maintained by a sewer agency. In very cold climates building sewers are usually placed below the frost line and cleanouts are located in the inside wall of basements to provide convenient access in the winter. Figure 3.12 illustrates the types and locations of cleanouts in building sewers. Figure 3.2 (page 41) illustrates their typical location with respect to the residence and the property line.

BUILDING SEWER CONNECTIONS (also called TAPS) to street sewers are either provided during construction of these sewers, if the appropriate location is known, or after their construction. In either instance they must be constructed with care since these connections can be a source of groundwater infiltration and sewer stoppages if not constructed properly. The taps must not protrude into the street sewer. Figure 3.13 illustrates the types of building sewer connections.

LATERAL AND BRANCH SEWER CLEANOUTS are sometimes placed at the terminals of these sewers, in place of manholes, to provide a means of inserting cleaning equipment and water for flushing the line. Typical cleanouts are illustrated in Figures 3.14 and 3.15. The slant riser provides easier access to the lateral or branch sewer, but the impact of street traffic on the frame and cover tend to shear the cleanout pipe below the frame.

FLUSHER BRANCHES were used in the past for the introduction of large volumes of water in the upper ends of lateral and branch sewers to attempt to move deposited solids downstream to self-cleaning sewers. Those that still exist are similar to the vertical and slant lateral sewer cleanouts illustrated in Figures 3.14 and 3.15. However, other sewer cleaning methods are more effective than flushing and flusher branches are not used very often today.

[9] *Also called handhole trap, vent trap, main trap, or house trap.*

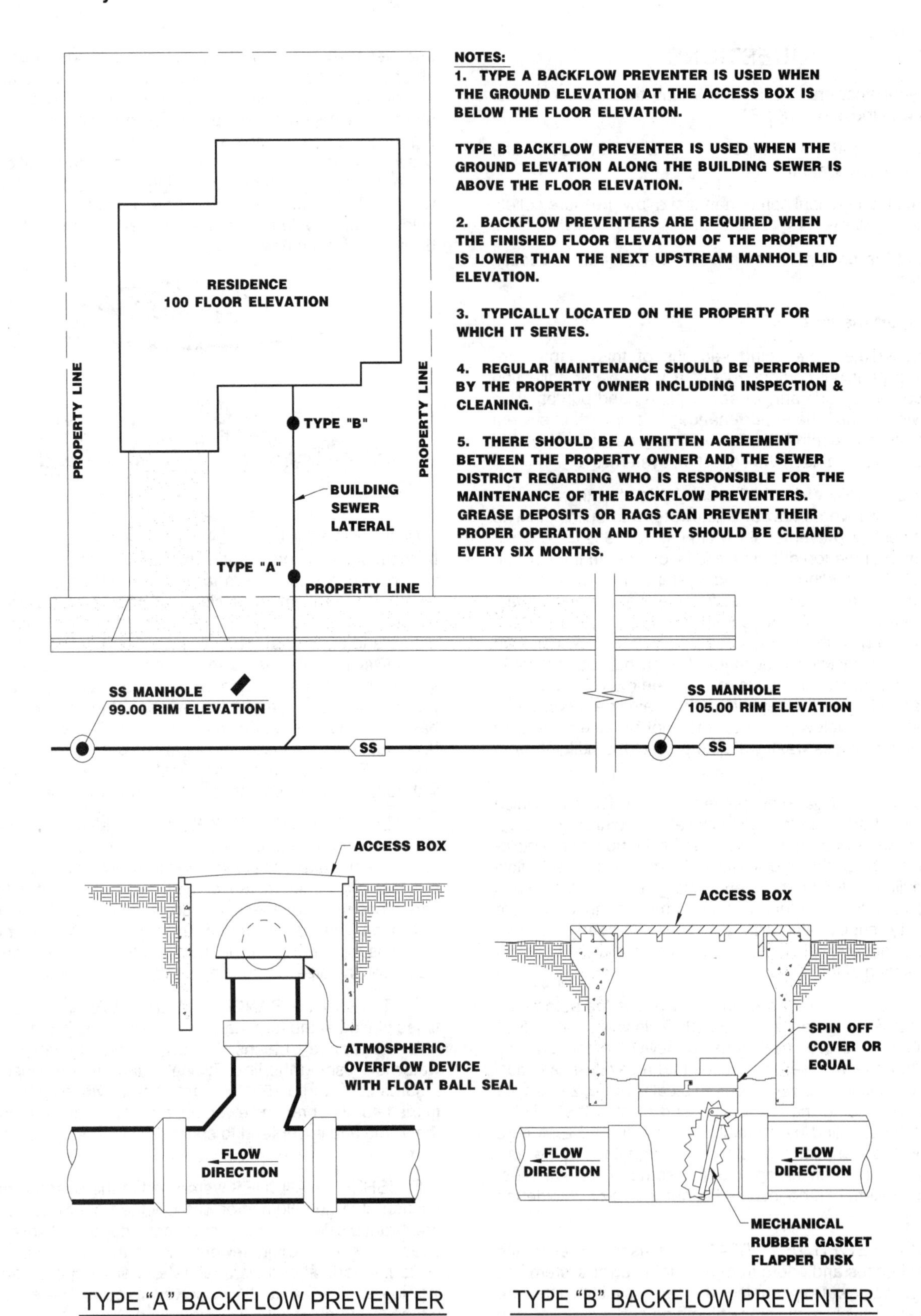

Fig. 3.9 Types and locations of wastewater backflow preventers

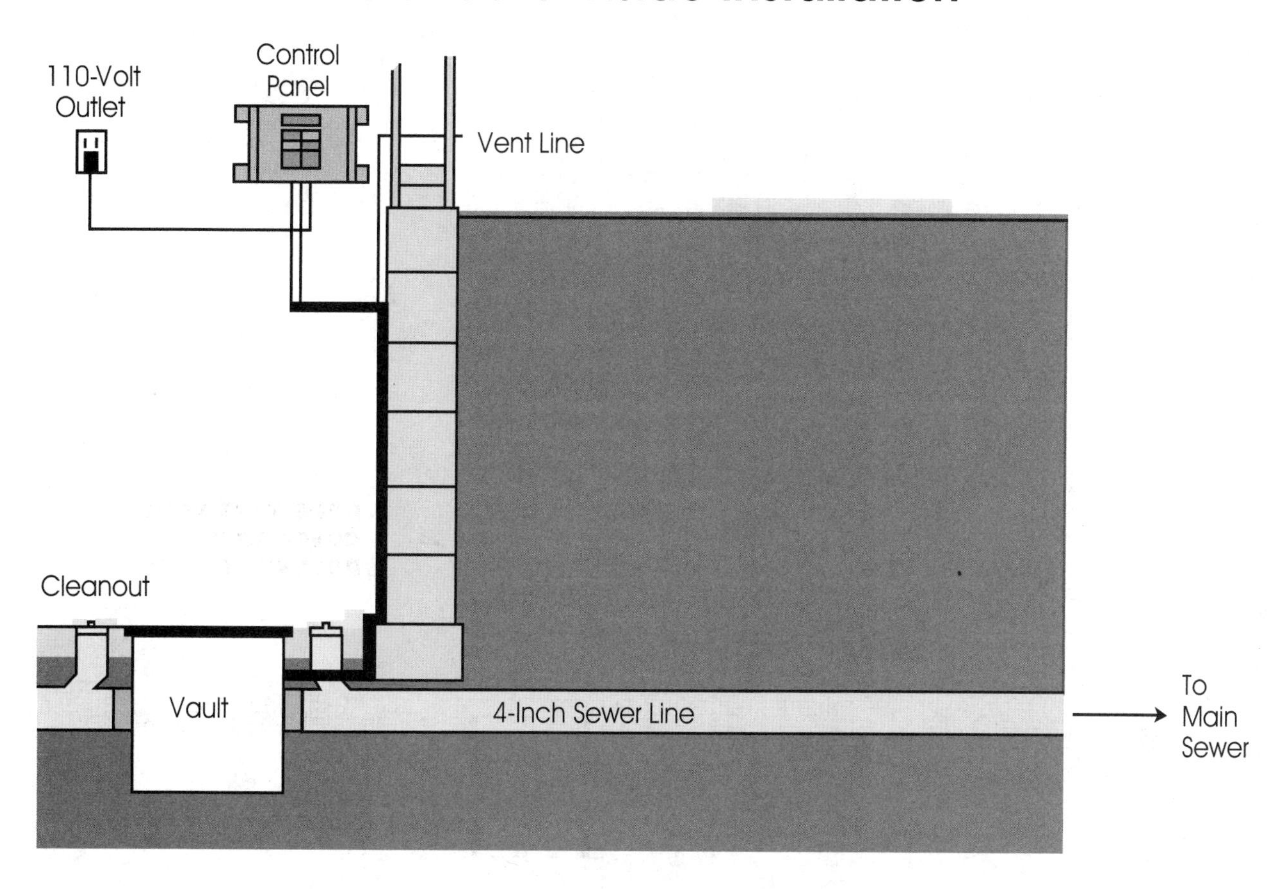

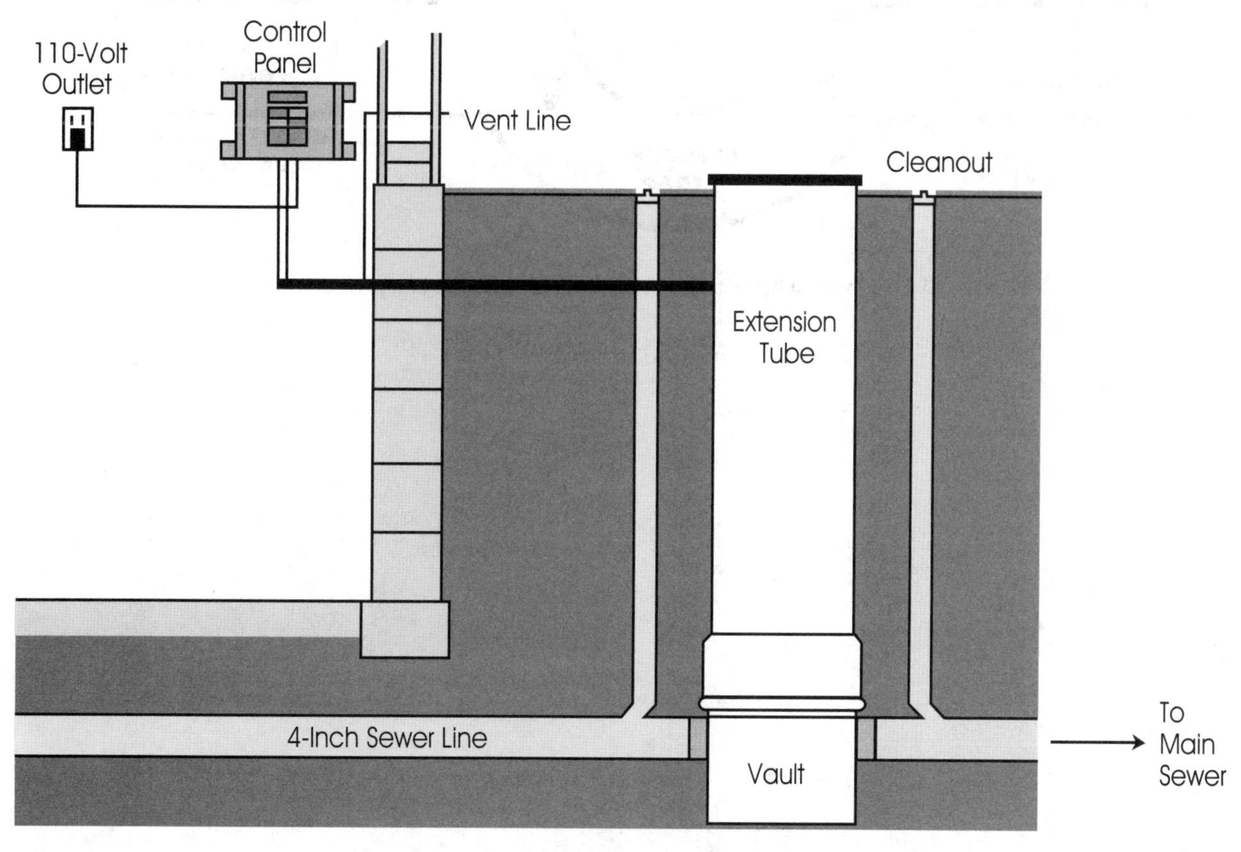

Fig. 3.10 Automated backflow prevention devices

(Permission of Peterson Valve Company)

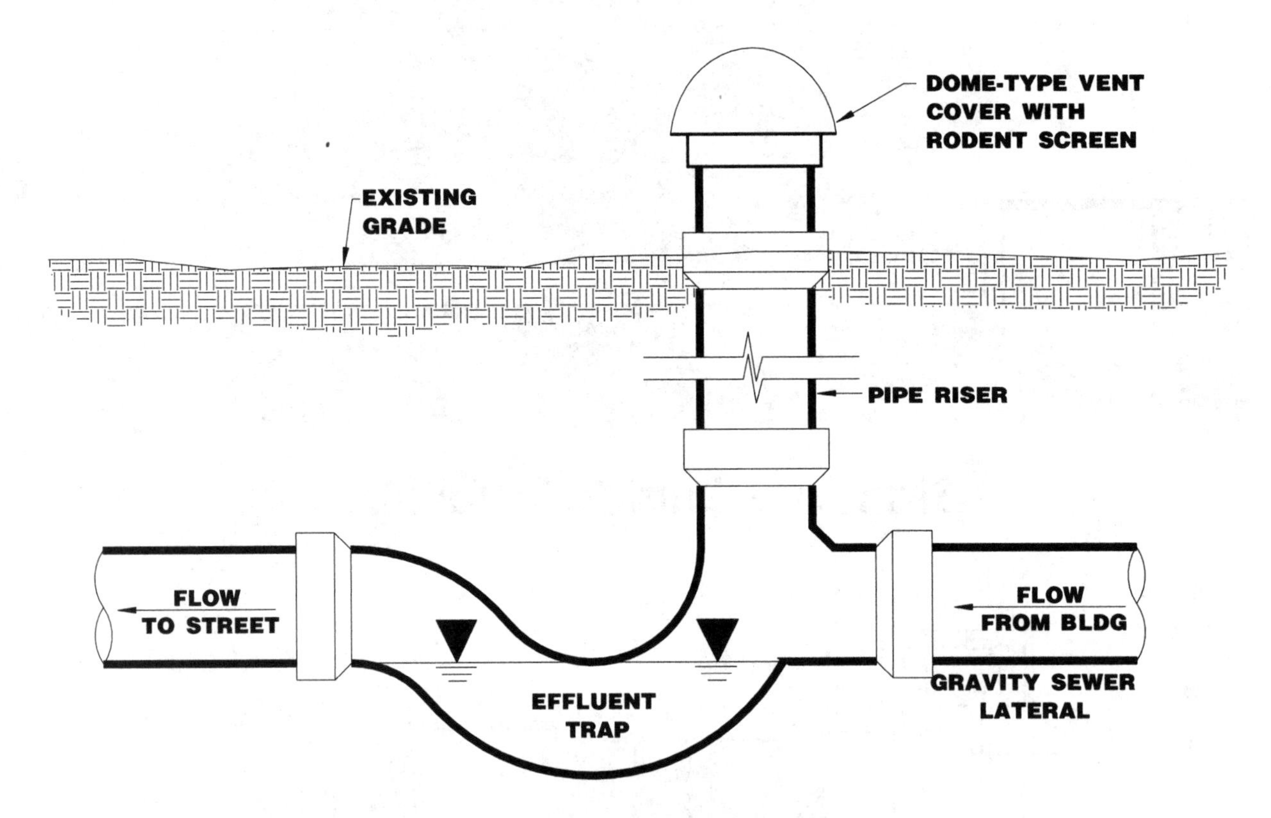

NOTE: A screen must be installed in the dome to prevent entry by rodents.

Fig. 3.11 Building sewer vent trap

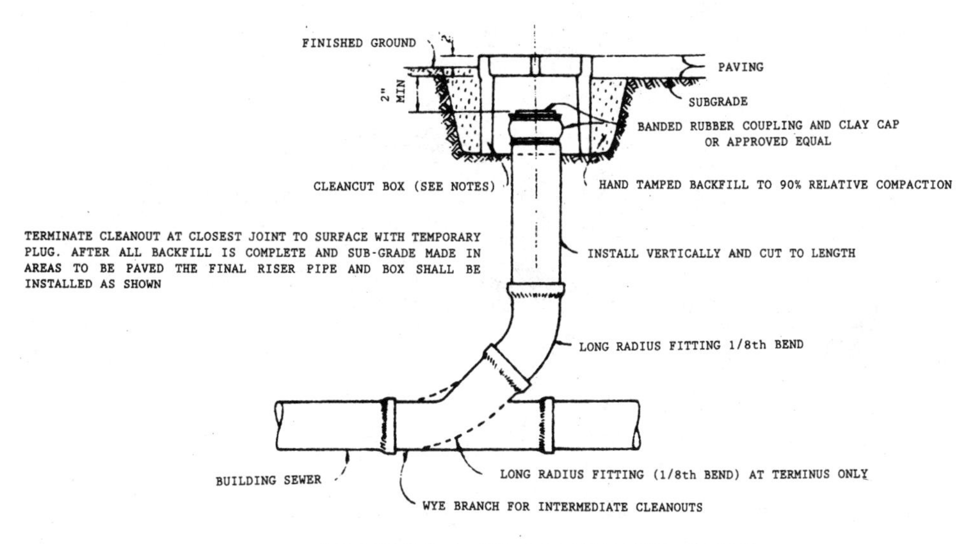

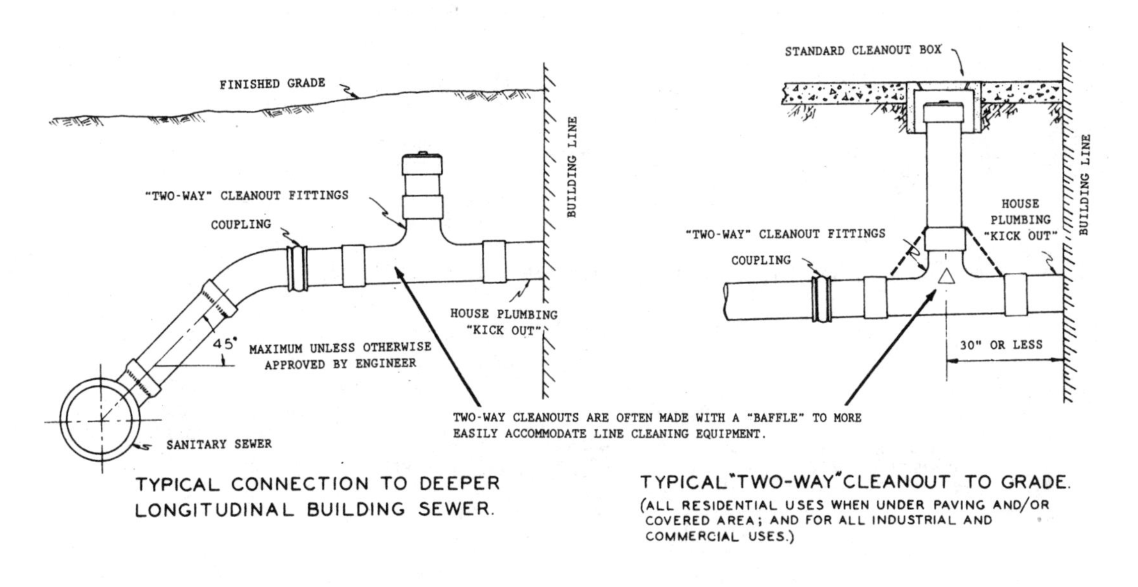

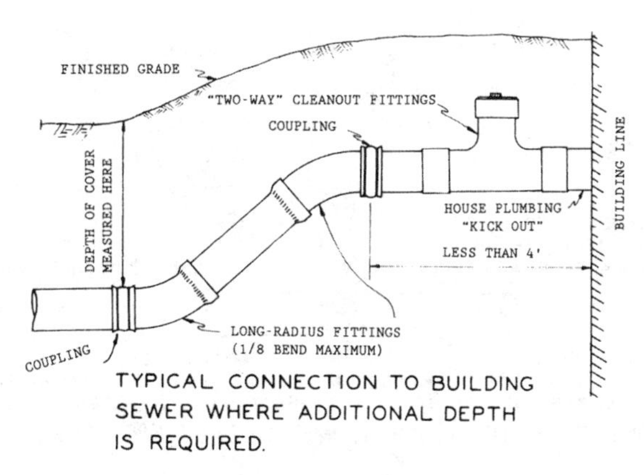

NOTES: 1. Cleanouts should be extended to surface so they are accessible without excavation in order to reduce maintenance costs and customer complaints regarding operators disturbing their yards.

2. "Two-Way" cleanout fittings may be difficult to push equipment through because of the right-angle entrance instead of a gradual entrance.

Fig. 3.12 Types and locations of building sewer cleanouts

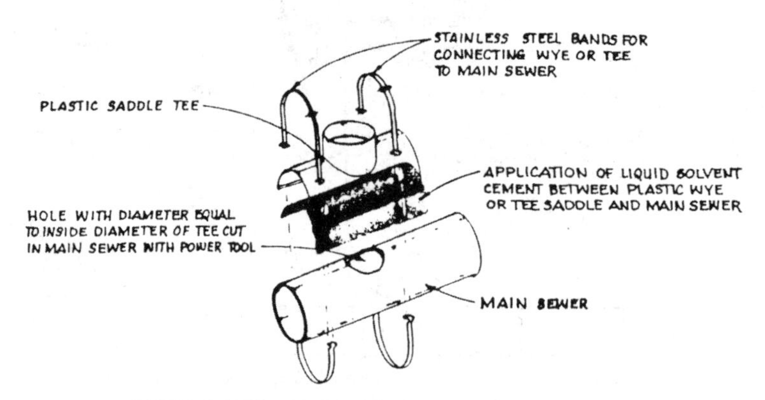

SADDLE TEE CEMENTED TO PLASTIC PIPE

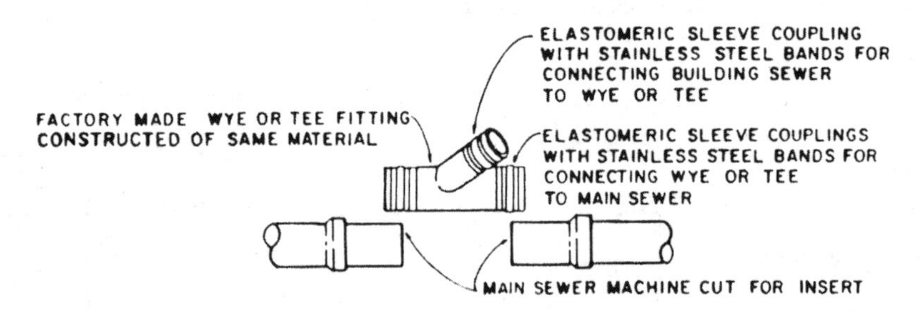

INSERTION OF FACTORY MADE WYE OR TEE

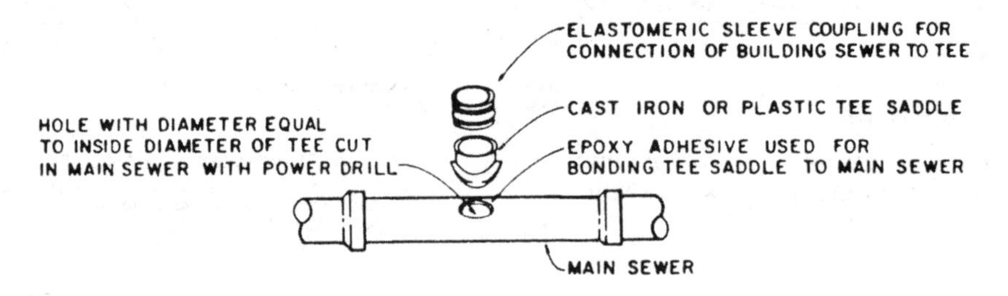

EPOXY BONDED SADDLE TEE

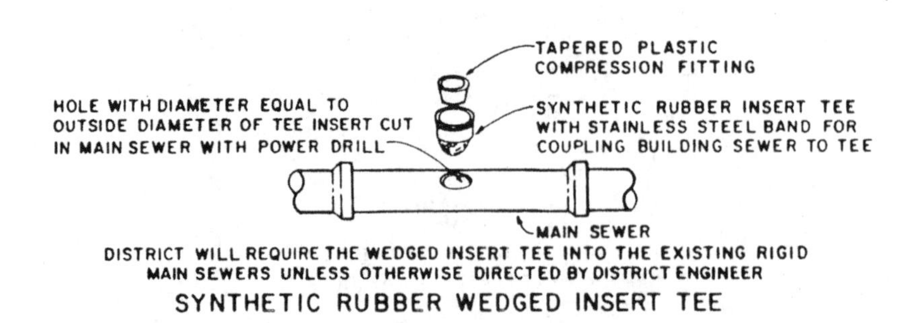

Fig. 3.13 Building sewer connections (taps)

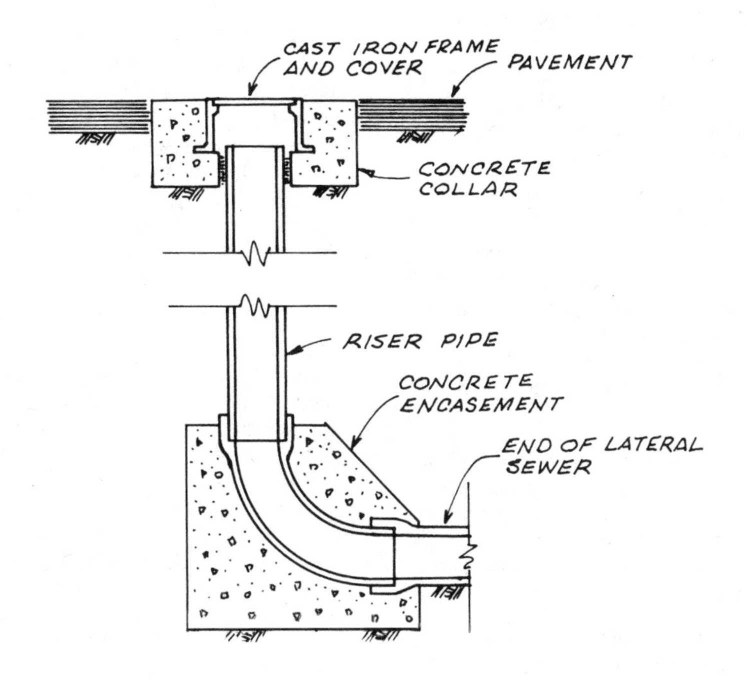

Fig. 3.14 Vertical lateral cleanout

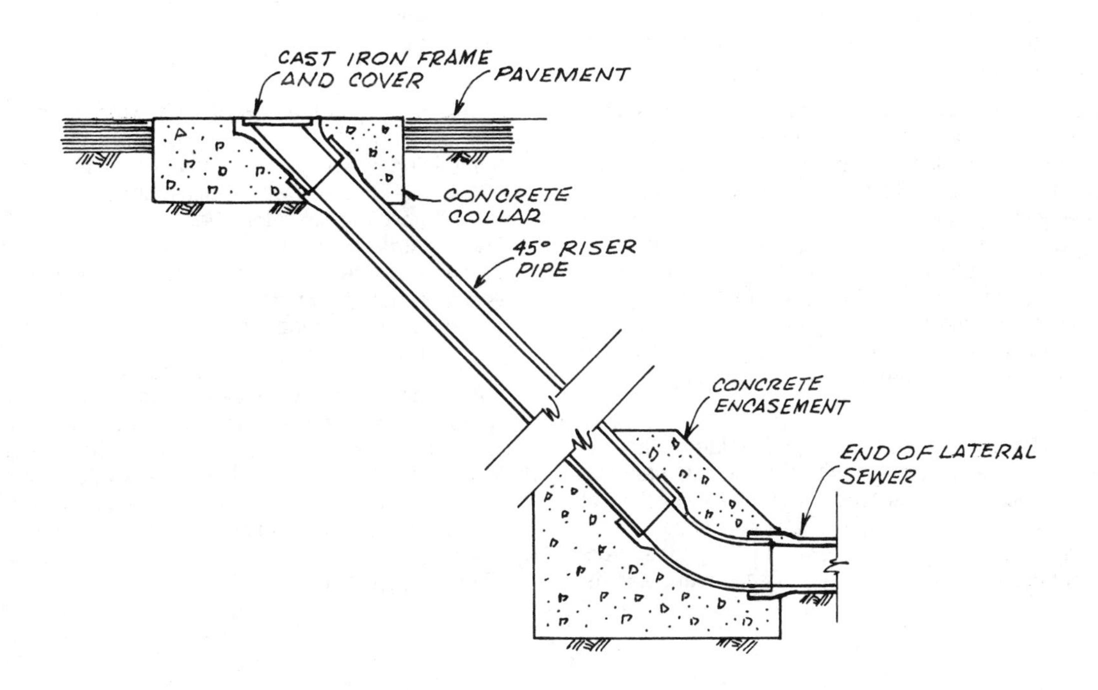

Fig. 3.15 Slant lateral cleanout

LAMP HOLES are rarely found today. They are similar to a vertical lateral sewer cleanout. Lamp holes were placed midway between adjacent manholes for the introduction of a lamp that would be viewed through the sewer line from the manholes to determine if the sewer line was free of solids or breaks that could cause a stoppage of the wastewater flow. The use of modern sewer line viewing methods, such as closed-circuit television, have made lamp holes obsolete. However, they may still exist in older collection systems where they should be sealed to prevent surface inflow.

MANHOLES are installed in lateral, main, trunk, and intercepting sewers for the purpose of placing persons, equipment, and materials into these sewers for inspection, maintenance, and the removal of solids from cleaning operations. Manholes in straight runs of sewer lines should be spaced no farther apart than the distance that can be cleaned by the appropriate type of equipment (usually 300 to 500 feet). Manholes are also placed at changes in sewer line direction, slope, elevation, pipe sizes, and junctions.

Drop manholes are used when the difference in the elevation of an incoming and outgoing sewer line cannot be accommodated by a drop in the manhole channel without creating excessive turbulence and splashing of wastewater. Steeper slopes for incoming sewers to manholes are highly recommended to avoid drop manholes.

Manholes may be furnished with or without steps, depending upon the practice of a sewer agency. Some agencies prefer to use ladders for entry and exit of manholes for safety reasons. With the confined space regulations, ladders or steps should never be used without a proper harness and person-rated winch, and proper confined space entry procedures must always be followed whenever anyone enters a confined space. Deterioration caused by corrosion could cause the steps or rungs to fail when the operator tries to use them.

Figure 3.16 (page 55) illustrates a typical manhole, including a drop structure, for sewers up to 24 inches in diameter. Larger diameter sewers require modification of the base and channel. However, the barrel, frame, and cover usually remain similar to the illustration.

The use of outside drop manholes (see Figure 3.16) should be considered very carefully. Under normal flow conditions, there is no problem since flow drops down the vertical section and then into the manhole. However, when an outside drop plugs, the overflow from the pipe then comes out the horizontal section and flows straight into the upper part of the manhole and falls to the bottom. While in some cases it may be possible to get a cleaning tool into the lower elbow and up into the vertical section to clear the blockage, in other cases an operator may have to enter the manhole and unplug the drop connection from inside the manhole. Obviously this is not a desirable job.

JUNCTION STRUCTURES are used to join large-diameter trunk sewers when the junction cannot be made in a manhole. The structure usually contains facilities for the regulation of wastewater flows in the sewers running out of the junction as illustrated in Figure 3.17. Flow of wastewater through the structure should not produce undesirable turbulence that could release hydrogen sulfide gas or create a partial blockage of the incoming wastewater. Hydrogen sulfide smells like rotten eggs, is a toxic gas, and forms corrosive sulfuric acid when combined with moisture in sewers.

INTERCONNECTOR SEWERS are short sewer lines that run between adjacent manholes to allow the regulation of wastewater flows in parallel sewers or the automatic diversion of wastewater to an alternate sewer if the primary sewer becomes blocked. Although the automatic diversion of wastewater from a blocked sewer may protect buildings from wastewater overflow, this type of diversion requires periodic inspection to determine if the primary sewer is blocked. If the wastewater in the primary sewer is blocked, it may become *SEPTIC*,[10] produce hydrogen sulfide, and cause odor problems.

INVERTED SIPHONS are sewer lines installed lower than the normal *GRADIENT*[11] of the sewer line to pass under obstructions such as watercourses and depressed roadways, as illustrated in Figure 3.18. Wastewater is "pushed" through the siphon by the pressure resulting from the upstream sewer (at the junction manhole) being higher than the downstream sewer. Siphons generally require more maintenance than gravity sewers since solids in the conveyed wastewater must be pushed up the downstream leg of the siphon by the velocity of the wastewater. Low velocities will leave heavy solids at the bottom of the siphon and these must be removed by one of several cleaning methods. For this reason, siphons are sometimes constructed with two parallel barrels (pipes) to provide an alternate route for the wastewater if one barrel becomes blocked.

AIR JUMPERS are sometimes constructed as part of an inverted siphon, as illustrated in Figure 3.18. Since the siphon is completely filled with wastewater, a blockage in the flow of air in the sewer line occurs without an air jumper. Continuation of the air flow is important if its stoppage would cause a corrosive, odorous, or toxic release of hydrogen sulfide at the upstream siphon manhole. Installation of air jumpers prevents this from happening by providing the downstream flow of air that usually occurs above the wastewater in a partially filled sewer line.

FLOW REGULATORS in a wastewater collection system are used to divert the flow of wastewater from one sewer line to another for the efficient use of the system and to prevent overloading one portion of the system. The regulators are usually simple in nature with a minimum of moving parts due to the nature of wastewater. Weirs are used in manholes and junction structure channels, as illustrated in Figures 3.17 and 3.19, to regulate the flow of wastewater in the multiple sewers as appropriate for operating conditions. The weirs are simply boards of an appropriate height that are inserted in preformed slots in the channels.

Another simple flow regulator is the Hydro-Brake developed by Hydro Research & Development; it is illustrated in Figure 3.20. The Brake imparts a centrifugal (moving outward from center) motion to the entering wastewater when a predetermined depth has been reached. This action effectively reduces the rate of discharge. The Brake can also be used as an energy dissipator to reduce turbulence at the lower end of a steep section of sewer line.

---

[10] *Septic (SEP-tick). A condition produced by anaerobic bacteria. If severe, the wastewater produces hydrogen sulfide, turns black, gives off foul odors, contains little or no dissolved oxygen, and the wastewater has a high oxygen demand.*

[11] *Gradient. The upward or downward slope of a pipeline.*

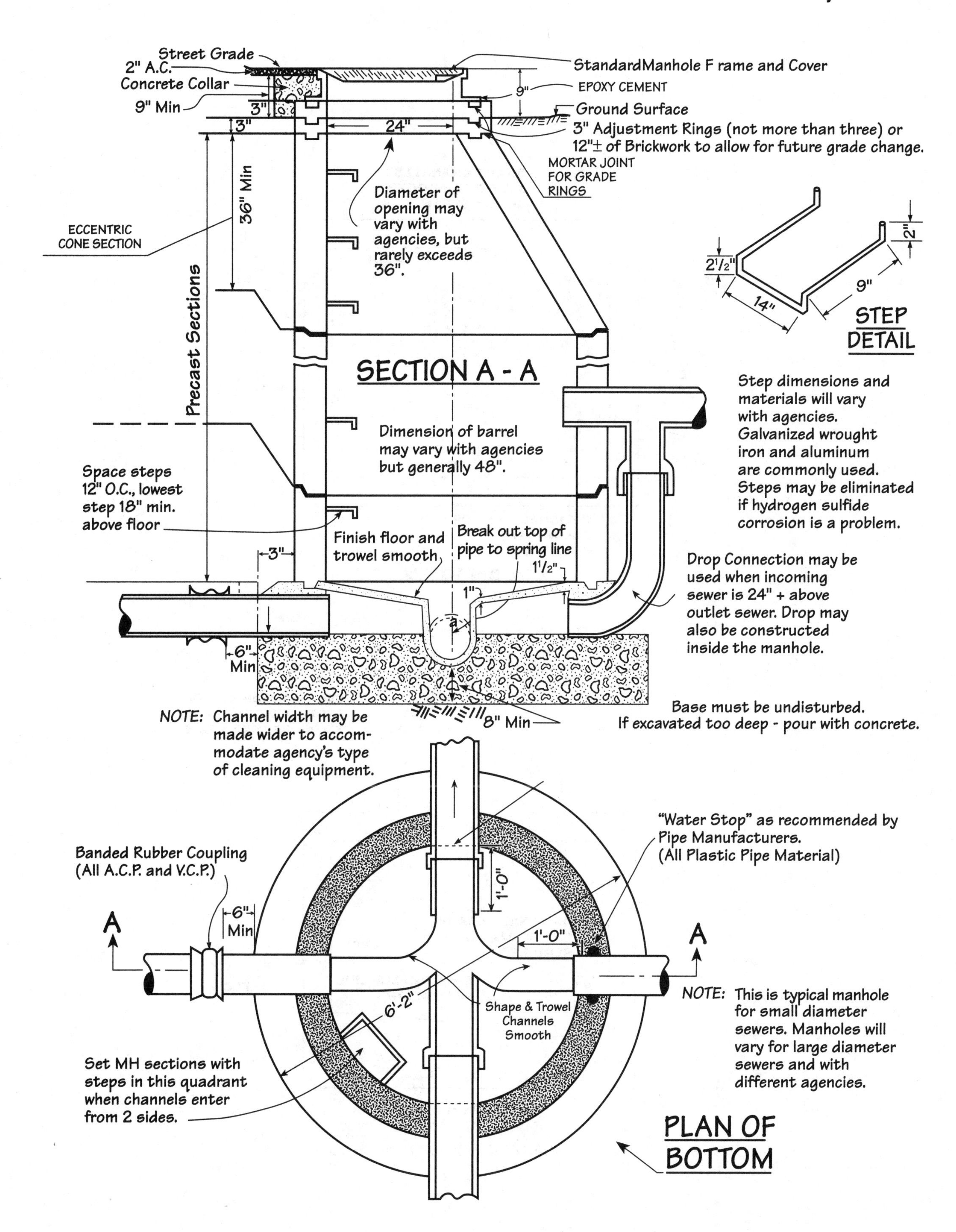

Fig. 3.16 Precast concrete manhole

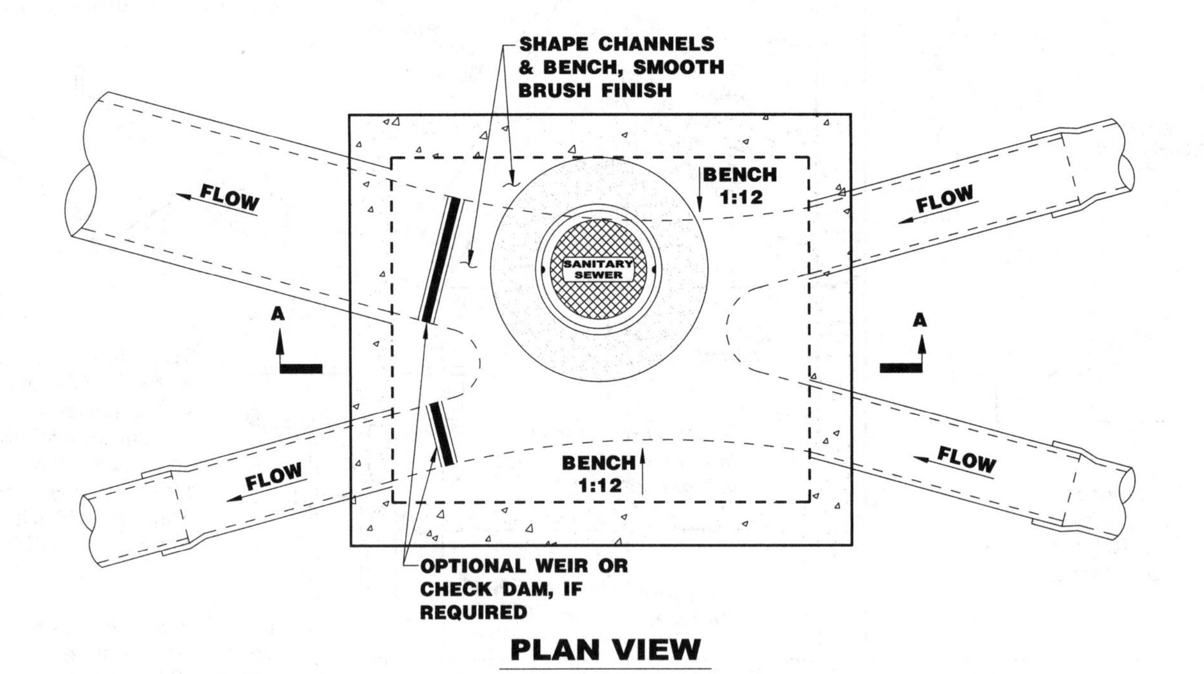

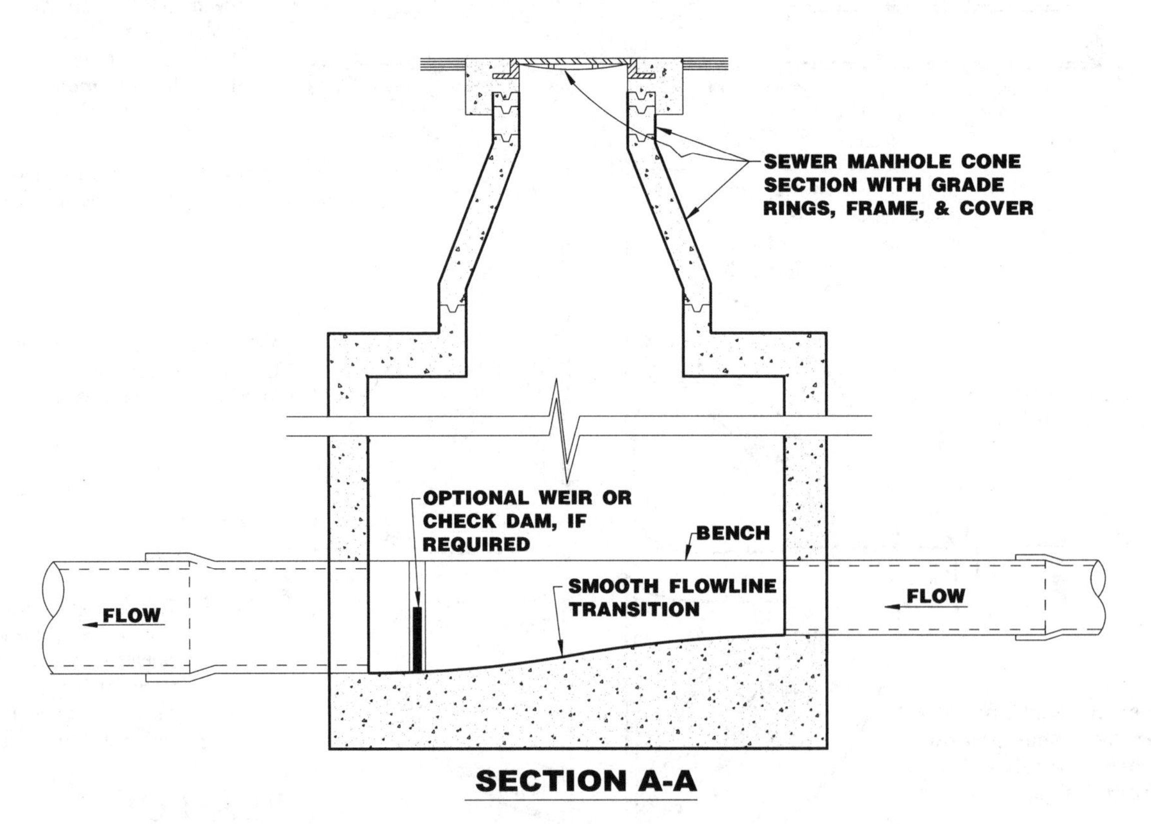

Fig. 3.17 Junction structure

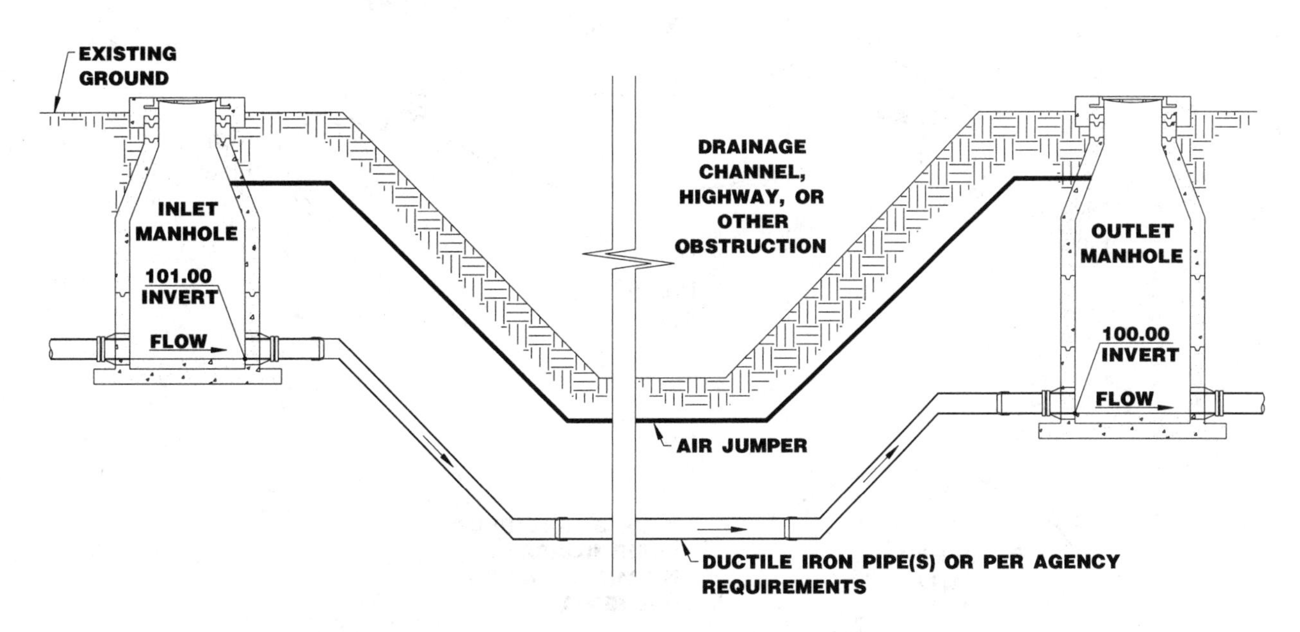

NOTE: Air jumpers are sometimes constructed as part of an inverted siphon. Since the siphon is completely filled with wastewater, a blockage in the flow of air in the sewer line occurs without an air jumper. This blockage may cause a continuous release of toxic, odorous, and corrosive hydrogen sulfide at the upstream siphon manhole. Installation of air jumpers prevents this from happening by providing the downstream flow of air that usually occurs above the wastewater in a partially filled sewer line.

Fig. 3.18 Inverted siphon with air jumper

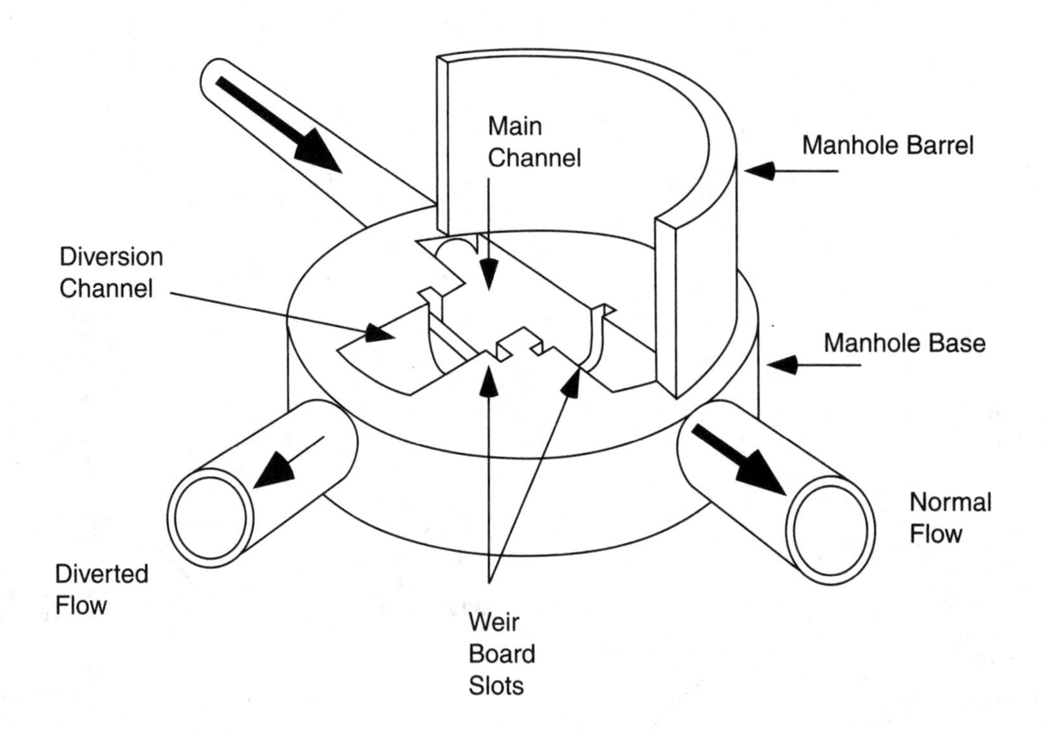

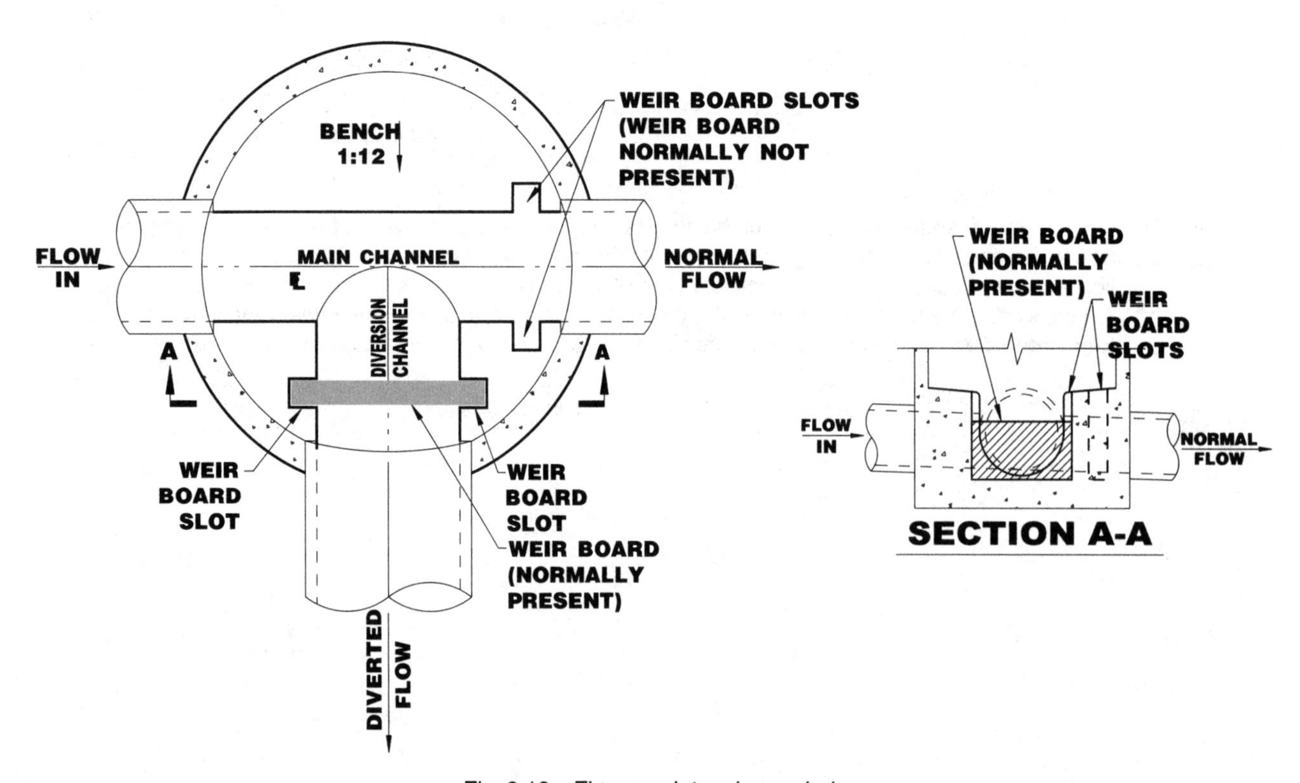

Fig. 3.19 Flow regulators in manhole

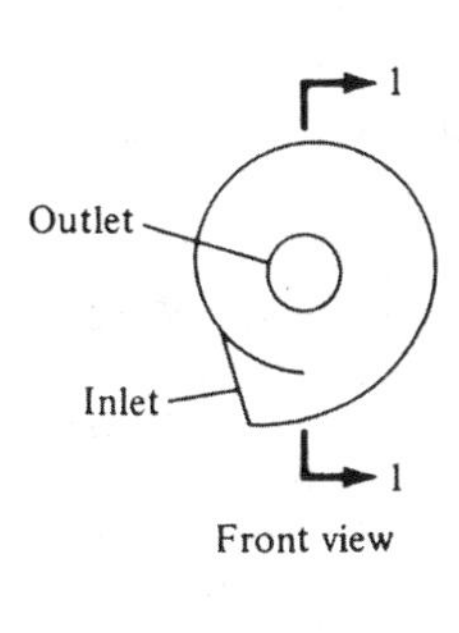

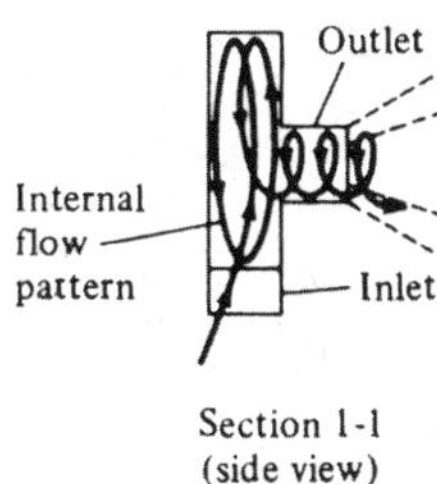

Fig. 3.20 Hydro-Brake flow regulator
(Permission of Hydro Research & Development)

QUESTIONS

Write your answers in a notebook and then compare your answers with those on page 75.

3.3D What is the purpose of a backflow preventer?

3.3E List the three main types of wastewater backflow preventers.

3.3F What are inverted siphons?

3.3G What is the purpose of flow regulators in wastewater collection systems?

3.4 DESIGN

Collection system operators need to be familiar with the elements of a wastewater collection system design that have an influence on the operation and maintenance (O & M) of the constructed system. This familiarity will allow an operator to effectively review the plans and specifications for newly designed collection systems to minimize later maintenance and to understand problems in existing collection systems caused by improper or unnecessary design factors. When operators review designs they should, for example, make sure that manholes are located in accessible locations and that overhead power lines and highway overpasses will not interfere with cranes used to lift pumps or other activities. Also, manholes should not be located in creek beds with intermittent flows or in the middle of major streets or highways.

3.40 Layout

A layout of a wastewater collection system shows the general location of laterals, mains, and trunk sewers within the service area, as illustrated by Figure 3.3 on page 43. The layout shows the direction of flow in the sewer system based on the contours of the area, but does not indicate the size or slope of sewer lines which are unknown at the time a layout is prepared.

3.41 Field Surveys

Field surveys are conducted by engineers to record the location of the physical features and the elevation of the ground surface along the planned route of the sewer lines in the collection system. Physical information includes locations and elevations of buildings to be served, surface improvements (such as sidewalks and street pavement), location of surface and underground utilities (gas, electricity, phone), and other items that will influence the location of a sewer line. Field surveys include both route surveys and level surveys.

ROUTE SURVEYS pinpoint the location of physical features that will influence the horizontal location of a sewer line. A surveyor's transit and tape are used to locate these features. By using measured angles and taped distances, the exact locations of physical features are referenced to the known location of the transit. Very accurate aerial photographs are sometimes used to determine the location of visible features, with field surveys being limited to the determination of invisible features (such as underground utilities).

LEVEL SURVEYS determine elevations of the ground surface along the route of the sewer line, buildings to be served, and (when possible) the elevation of underground obstructions and pipelines.

Since a collection system operator may conduct a level survey for construction or repair of a short section of sewer line, the following section describes the equipment, personnel, and procedures used for a level survey.

Personnel

One SURVEYOR and one ASSISTANT.

Equipment

1. Surveyor's level. The level establishes a level line of sight through a 360-degree angle using a horizontal cross-hair in a telescope.
2. Rod. The rod is marked in 0.01-foot (hundredths of a foot) intervals. It is about 7.5 feet long and extendable to 14 feet. The rod is used to measure vertical distances or elevations.

Procedure

1. SURVEYOR: Set the level at a convenient location where back (backward or behind) and fore (forward or ahead) sights can be made. Level the instrument.
2. ASSISTANT: Hold the rod plumb (use the level bubble) on a point of known elevation (a bench mark).
3. SURVEYOR: Back sight (BS) the telescope of the surveyor's level on the rod. Read the distance on the rod that the cross-hair is above the bench mark (5.65 feet in Figure 3.21).
4. SURVEYOR: To determine the height of the instrument (HI) above the bench mark (BM), add the BS distance to the bench mark elevation (HI = BM + BS = 263.47 + 5.65 + 8.20 = 277.32). The 8.20 is the measured distance from the *INVERT*[12] elevation of 263.47 to the top of the manhole as shown in Figure 3.21.
5. To determine the elevation of any other point, hold the rod plumb on the point. Fore Sight (FS) the level telescope on the rod and read the distance of the horizontal cross-hair above the point (5.38 feet at station 1 + 00 in Figure 3.21).
6. To calculate the elevation of the point, subtract the FS distance from the HI (EL. 1 + 00 = HI – FS = 277.32 – 5.38 = 271.94).
7. If you (the instrument surveyor) need to move the level due to (1) an obstruction in the line of sight, (2) the FS point being above the HI, or (3) the HI being above the upper end of the rod, establish a secondary bench mark (turning point) and carefully calculate its elevation. Move the level to a new location and determine the HI by back sighting on a

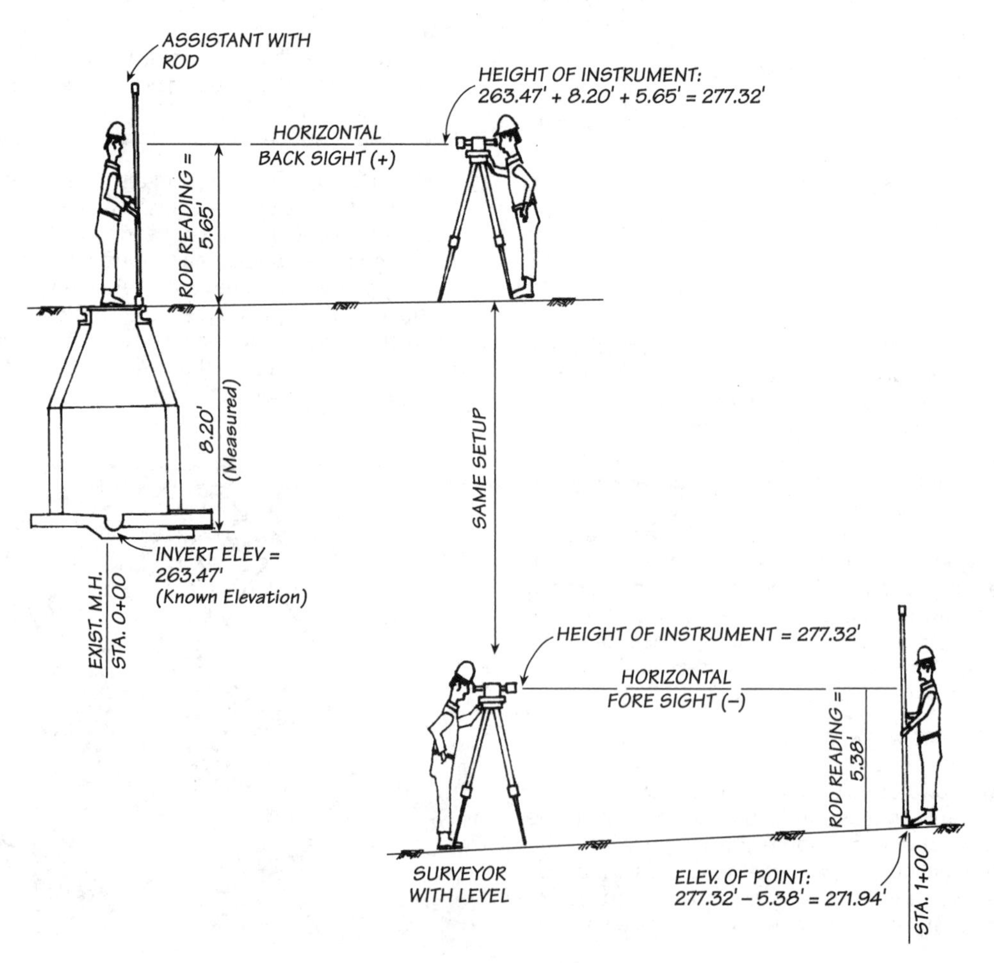

Fig. 3.21 Level survey

[12] *Invert (IN-vert). The lowest point of the channel inside a pipe or manhole.*

turning point. For example, in the level survey made to develop the field notes shown in Figure 3.21, the nail designating 0 + 00 with a previously determined elevation of 271.63 was used as a turning point for the purpose of setting the level for observation of a storm sewer that had to be considered in the design of the sanitary sewer. The height of the instrument was determined by back sighting on a rod held on the turning point and adding the reading to the elevation of the point (HI = TP + BS = 271.63 + 5.41 = 277.04). The elevation of the storm sewer manhole rim was determined by holding a rod on the rim and subtracting the fore sight rod reading from the HI (EL. MH Rim = HI – FS = 277.04 – 5.62 = 271.42).

8. The invert elevation of the storm sewer at the manhole was determined by using the surveyor's rod to measure the distance from the rim to the invert and then subtracting this distance from the rim elevation (EL. Invert = EL. Rim – Dip = 271.42 – 6.46 = 264.96).

As route and level surveys are conducted, field notes similar to those illustrated in Figure 3.22 (page 62) are usually made by the person in charge (surveyor or crew leader).

3.42 Slope and Size

The following factors are considered in determining the slope and size of a sewer line.

SLOPE OF SEWER should follow the slope of the land along the route of the sewer as closely as practical provided the slope is adequate to produce gravity flow and the minimum scouring velocity. Deviations from ground slope could cause the sewer to become too shallow for adequate cover on the pipe or too deep for safe and economical construction.

The slope of the sewer, also known as the grade, is the ratio in the change of the vertical distance to the horizontal distance of the pipe. The slope of the sewer will determine the velocity of the flow within the sewer; the steeper the slope the higher the velocity. Gravity sewers are normally designed to maintain minimum velocities to keep the solids suspended. However in some cases it may be necessary for certain sections of the line to be laid on a flatter slope. This could create problems with insufficient scouring velocities and the resultant buildup of solids in the bottom of the sewer. Minimum slopes also are related to the diameter of the pipe. For example, the recommended minimum slope that will maintain a scouring velocity in an 8-inch pipe is 0.40 foot for every 100 feet of pipe. Minimum slope for a 16-inch pipe would be 0.14 foot for every 100 feet of pipe, and for a 36-inch pipe it would be 0.046 foot for every 100 feet of pipe.

DESIGN FLOW is the estimated amount of wastewater to be conveyed by the sewer line being designed. Section 3.21 describes a method of determining the design flow of a collection system.

VELOCITY. The wastewater in a sewer line should move at a speed which will prevent the deposition and buildup of solids in the sewer; this is called a "scouring velocity." Experience has shown that wastewater velocities of two feet per second or greater will keep the solids usually contained in wastewater moving in a sewer line. A scouring or self-cleaning velocity may not be attainable during periods of minimum daily flows. However, a sewer should be designed to produce a scouring velocity during the average daily flow or, at the very least, during the maximum daily flow. Collection systems in flat regions with low flow velocities tend to have the solids settle out of the wastewater, thus causing sewer stoppages and the generation of toxic hydrogen sulfide gas. On the other hand, steep slopes on hilly terrain can produce high wastewater velocities in sewer lines. Velocities higher than 10 feet per second can separate solids from the wastewater, cause excessive turbulence at junctions, and erode the sewer line.

A velocity of 2.0 feet per second or more will keep normal solids in suspension in the flow, so very little material (either organic or inorganic) will build up in the invert of the pipe. When the velocity is 1.4 to 2.0 feet per second, the heavier inorganic grit, such as gravel and sand, begins to accumulate in the invert. At a velocity of 1.0 to 1.4 feet per second, inorganic grit and organic solids will accumulate. When the velocity drops below 1.0 foot per second, significant amounts of organic and inorganic solids accumulate in the sewer. It is therefore critical to your maintenance program to have flat grades or reverse grades corrected prior to acceptance on new lines.

SIZE. A sewer line should be large enough for the use of cleaning equipment. A properly sized sewer line will be approximately one-half full when conveying the peak daily dry weather design flow and just filled when it is conveying the peak wet weather design flow. Designing for a half-full sewer line allows for a beneficial flow of air above the wastewater and allows for some error in determining the design flow. The flow of air helps prevent the buildup of toxic or explosive gases, an oxygen deficiency, and moisture condensing on the pipe walls and crown above the waterline, which can contribute to sulfide corrosion. Most wastewater agencies require at least a four-inch diameter residential building sewer and a six-inch diameter lateral and branch sewer. Many agencies require an eight-inch diameter lateral and branch sewer to facilitate maintenance and to convey unexpected wastewater flows. Commercial buildings require a six-inch or larger sewer line and industrial sewer lines are six inches or larger and sized on the basis of expected flows.

QUESTIONS

Write your answers in a notebook and then compare your answers with those on page 75.

3.4A Why are field surveys conducted?

3.4B What is the difference between a route survey and a level survey?

3.4C Why must a scouring velocity be maintained in a sewer line?

3.4D What will happen in a sewer if a scouring velocity is not maintained?

3.43 Velocity and Flow

The Manning Equation is commonly used by engineers to interrelate slope, quantity of wastewater (design flow), and velocity of flow in a sewer line. Although you will probably not be designing a complete system, this equation will be useful if you need to design an extension of an existing sewer or com-

6" LATERAL AND 4" BUILDING SEWER
EXTENSION TO 19105 BROOKVIEW DRIVE

OBJ.	B.S.	H.I.	F.S.	ELEV.	B.M.
INV. 8"					263.47
	13.85	277.32			
SANITARY M.H. RIM			5.65	271.67	
0+00			5.69	271.63	
+50			5.57	271.75	
1+00			5.38	271.94	
1+19			5.31	272.01	
+50			5.20	272.12	
+75			5.06	272.26	
2+75			5.34	271.98	
3+05			5.43	271.89	
M.H. RIM			5.64	271.68	
0+00				271.63	
	5.41	277.04			
STORM M.H. RIM			5.62	271.42	
(DIP)				6.46	
INV.				264.96	
PALMTAG M.H. RIM			5.43	271.61	
(DIP)				4.56	
				267.05	

NOTE: CONSTRUCTION CONTROLS SET 5' RIGHT OF SEWERLINE.

6/28/02 DRY & WARM
L.T. ø
H.G. ⊼⊞

BROOKGLEN DRIVE
STORM M.H.
EXIST. MANHOLE
EXIST. 8" SANITARY SEWER
5'
12"
19105
75'
24" STORM
15'
BROOKVIEW DRIVE
STORM M.H.
PALMTAG DRIVE

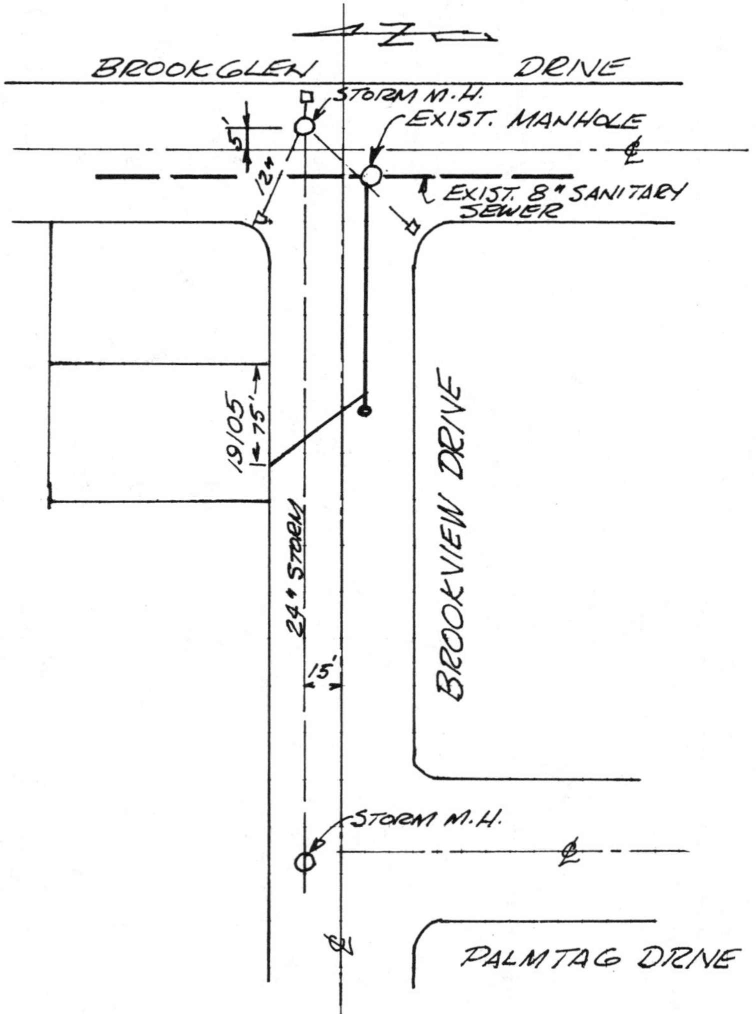

Fig. 3.22 Field notes

pare expected design velocities with actual velocities in sewers. The Manning Equation is as follows:

$$V = \frac{1.486}{n} R^{2/3} S^{1/2}$$

where V = Velocity, ft/sec,

n = Manning's coefficient[13] of pipe roughness,

R = Hydraulic Radius, ft,

$$= \frac{\text{Cross-sectional Area of Flow, sq ft}}{\text{Wetted Pipe Perimeter, ft}}$$

and S = Slope of Sewer, ft/ft.

Knowing the available slope and desired depth of flow in a sewer line and assuming a coefficient of pipe roughness, the velocity of flow can be calculated. Then, knowing the velocity and the cross-sectional area of flow, the flow rate can be calculated using the formula:

$$Q = AV$$

where Q = Flow, cubic feet per second,

A = Area, square feet, and

V = Velocity, ft/sec.

EXAMPLE 6

The slope of an 8-inch diameter sewer is 0.004 ft per ft, the pipe is flowing one-half full and the coefficient of pipe roughness is assumed to be 0.013. Determine the velocity and quantity (flow rate) of wastewater conveyed by the sewer.

Known	Unknown
Slope, S, ft/ft = 0.004 ft/ft	1. Velocity, V, ft/sec
Diameter, D, in = 8 in	2. Quantity, Q, MGD
Roughness, n = 0.013	

1. Calculate the cross-sectional area of the pipe when it is flowing half full (refer to Example 2, page 38).

$$\frac{\text{depth}}{\text{Diameter}} = \frac{(1/2)(8\text{ in})}{8\text{ in}}$$

$$= \frac{4\text{ in}}{8\text{ in}}$$

$$= 0.50$$

From Table 3.4, if d/D = 0.50, factor = 0.3927.

2. Calculate the pipe cross-sectional flow area.

$$\text{Pipe Flow Area, sq ft} = \frac{(\text{Factor})(\text{Diameter, in})^2}{144\text{ sq in/sq ft}}$$

$$= \frac{(0.3927)(8\text{ in})^2}{144\text{ sq in/sq ft}}$$

$$= 0.175\text{ sq ft}$$

3. Calculate the *WETTED PERIMETER*[14] for a pipe flowing half full.

$$\text{Wetted Perimeter, ft} = \frac{(1/2)(3.14)(\text{Diameter, in})}{12\text{ in/ft}}$$

$$= \frac{(0.5)(3.14)(8\text{ in})}{12\text{ in/ft}}$$

$$= 1.05\text{ ft}$$

4. Calculate the hydraulic radius, R, in feet.

$$\text{Hydraulic Radius, R, ft} = \frac{\text{Flow Area, A, sq ft}}{\text{Wetted Perimeter, P, ft}}$$

$$= \frac{0.175\text{ sq ft}}{1.05\text{ ft}}$$

$$= 0.167\text{ ft}$$

5. Calculate the flow velocity in feet per second.

$$\text{Velocity, V, ft/sec} = \frac{1.486}{n} R^{2/3} S^{1/2}$$

$$= \frac{1.486}{0.013} (0.167\text{ ft})^{2/3} (0.004)^{1/2}$$

$$= (114.3)(0.303)(0.063)$$

$$= 2.18\text{ ft/sec}$$

6. Calculate the quantity of flow in cubic feet per second.

$$\text{Quantity, Q, CFS} = (\text{Area, sq ft})(\text{Velocity, ft/sec})$$

$$= (0.175\text{ sq ft})(2.18\text{ ft/sec})$$

$$= 0.382\text{ cu ft/sec}$$

[13] *Values for pipe roughness coefficients may be found in hydraulics textbooks which contain values for Manning's n. The value of n increases with the roughness of the pipe which depends on the pipe material and age of the pipe.*

[14] *Wetted Perimeter. The length of the wetted portion of a pipe covered by flowing wastewater.*

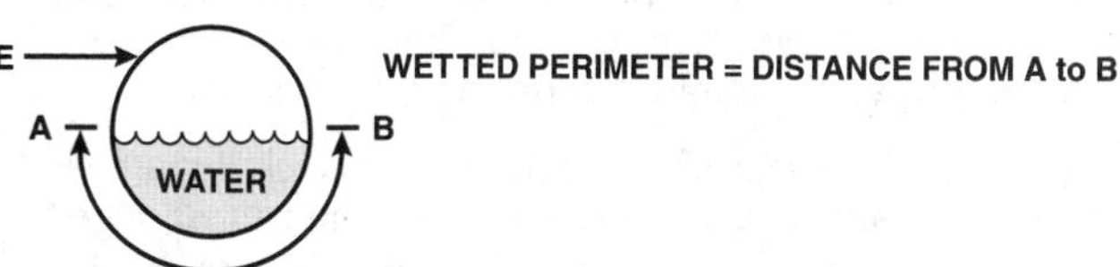

7. Convert the quantity of flow from cubic feet per second to million gallons per day, MGD.

$$\text{Quantity, Q, MGD} = \frac{(\text{Quantity, Q, cu ft/sec})(7.48\ \text{gal/cu ft})(60\ \text{sec/min})(60\ \text{min/hr})(24\ \text{hr/day})}{1{,}000{,}000/\text{M}}$$

$$= \frac{\text{Quantity, Q, cu ft/sec}}{1.55\ \text{cu ft/sec/MGD}}$$

$$= \frac{0.38\ \text{cu ft/sec}}{1.55\ \text{cu ft/sec/MGD}}$$

$$= 0.245\ \text{MGD}$$

The 8-inch diameter pipe would be acceptable for the proposed sewer if the peak daily dry weather design flow is approximately 0.245 MGD and the peak wet weather design flow is approximately 0.490 MGD. If not, either the design slope or pipe diameter or both would have to be increased until the appropriate combination of sewer line slope and diameter would produce the desired capacity and velocity.

3.44 Location and Alignment

Lateral, main, and trunk sewers are usually constructed at or near the center of public streets to equalize the length of building sewers and to provide convenient access to the sewer line manholes. In extremely wide and heavily traveled streets, lateral sewers are sometimes placed on each side of the street to reduce the length of building sewers and to lessen interference with traffic when the sewer is being maintained. Sewer lines are usually separated from water mains by a distance prescribed by the appropriate health agency. Sewers are sometimes placed in easements when the topography of the land or street layout requires their use. Use of easements should be minimized since sewers in easements are more difficult to gain access to and to maintain than sewers in streets. Emergency plans for access to manholes in an easement should be prepared prior to an emergency and should be part of the easement agreement.

With the greater use of curved streets today and the development of sewer cleaning equipment that is more versatile, sewers are sometimes constructed along curved paths. The radius of a curved sewer will be determined by the radius of the street in which the sewer is being constructed, allowable deflection (departure from a straight line) in the sewer pipe joints and, of major importance, the effect of radius upon cleaning procedures. Some state and local regulatory agencies prohibit the use of curved sewers because of maintenance problems. For the efficient use of modern sewer cleaning equipment, the radius of curvature must be greater than 200 feet. Also, most pipe manufacturers will specify the maximum allowable deflection in their pipe joints; this amount is usually sufficient for modern sewer cleaning equipment.

3.45 Depth

Lateral and main sewers should generally be buried approximately six feet deep. This depth will allow connecting building sewers to cross under most underground utilities and will serve the drainage systems of most buildings without basements. If most buildings in a collection system's service area include basements, the lateral and main sewers should be deep enough to serve drainage systems in the basements by gravity. The depth of trunk and interceptor sewers will depend mostly on the depth of connecting main sewers.

The depth and width of a trench, type of backfill material, and method of compaction determine the load placed on the sewer pipe (Figure 3.23). The amount of support provided the sewer pipe by the sides of the trench decreases as the width of the trench increases. The width of a sewer trench (controlled by the diameter of the pipe, clearance for pipe laying, soil stability for safety during excavation, and the excavation procedures) should be kept to a minimum.

Most sewer pipe manufacturers will provide information on the allowable loading on their pipe based on depth, trench width, and bedding conditions.

QUESTIONS

Write your answers in a notebook and then compare your answers with those on page 75.

3.4E If a 10-inch diameter sewer on a 0.005 ft/ft slope will not carry the peak wet weather design flow, what could be done to provide adequate capacity?

3.4F Why are sewers usually not placed in easements?

3.4G What factors control the width of a sewer trench?

3.46 Pipe Materials

Construction materials for sewers are selected for their resistance to deterioration by the wastewaters they convey, strength to withstand earth and surface loads, resistance to root intrusion, their ability to minimize infiltration and exfiltration, and the cost of the materials and their installation. Care must be taken in the selection of materials since sewers are expected to have an economical life of at least 30 years and most are in service for a much longer period.

Pipe for construction of sanitary sewers is manufactured from several types of material producing either flexible or rigid pipe with specific characteristics that will perform satisfactorily in a particular situation. A flexible pipe will bend or bulge from its usual shape when subjected to a load without being adequately supported. However, it will not crack under most loads and will return nearly to its original shape when a load is removed. A rigid pipe will require less support; however, it will crack when the combination of a load on the pipe and its support causes a stress that exceeds the strength of the pipe. It will not return to its original shape when the load is removed. Most of the commonly used sanitary sewer materials are described in the following subsections.

A wide variety of pipe materials are used for sanitary sewer construction. Each type is manufactured in accordance with nationally recognized material specifications. The type of pipe used depends on various factors which can be identified by answering the following questions:

- Will it be used for gravity or force main (pressure)?
- Is the wastewater abrasive?
- Are there any special installation requirements such as loading?
- Is there a potential for internal and/or external corrosion conditions from soil, hydrogen sulfide, or industrial wastes?
- What are the flow and capacity requirements based on design for pipe size, velocity, slope, and friction coefficient?
- Is infiltration or exfiltration a potential problem?
- What physical characteristics such as pipe size, fittings and connection requirements, and laying length need to be considered?
- How cost effective is the pipe, installation, maintenance, and what is the life expectancy?

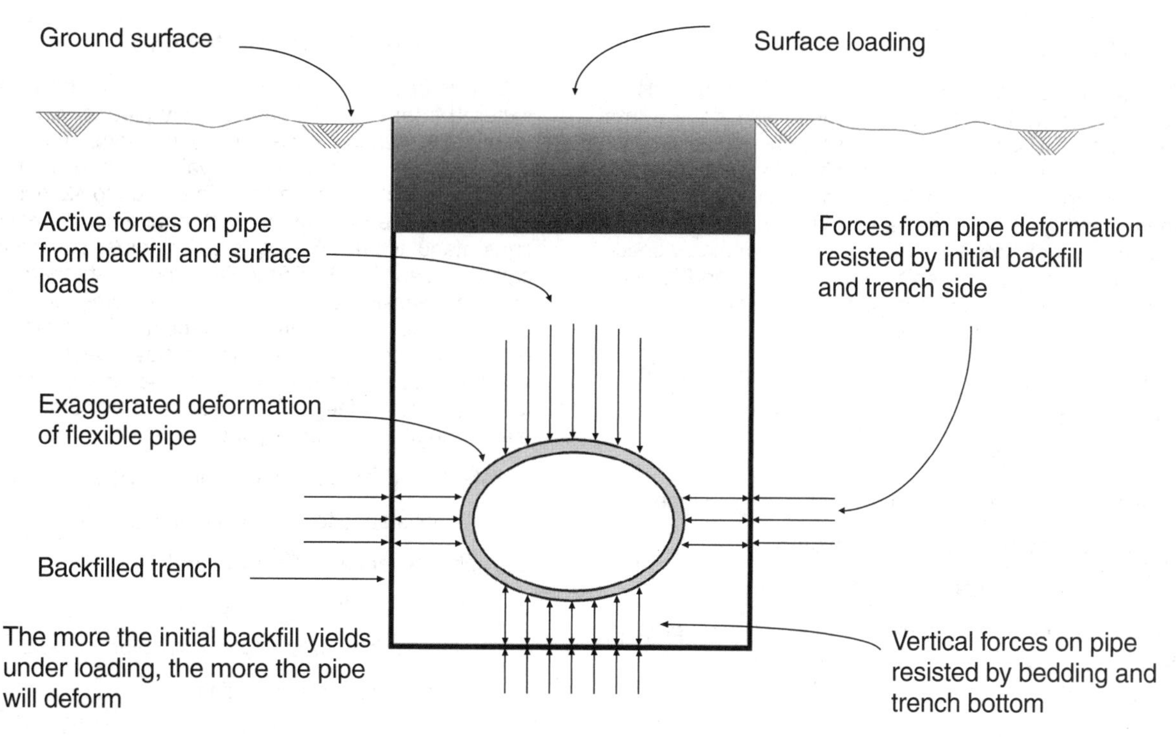

Fig. 3.23 Pipe loading

- What physical properties such as crush strength for rigid pipe, pipe stiffness or stiffness factor for flexible pipe, soil conditions, pipe loading strength, hoop strength for force main pipe, pipe shear loading strength, and pipe flexural strength need to be considered?
- What about handling requirements including weight, impact resistance, and ease of installation?

New pipe materials for use in sanitary sewers and improved installation methods are continually being developed. Operators need to be aware of these changes in order to take advantage of them.

The commonly used types of rigid pipe are:

- **A**sbestos-**C**ement **P**ipe (ACP) (no longer manufactured),
- **C**ast-**I**ron **P**ipe (CIP),
- **R**einforced **C**oncrete **P**ipe (RCP),
- **P**restressed **C**oncrete **P**ressure **P**ipe (PCPP), and
- **V**itrified **C**lay **P**ipe (VCP).

Types of flexible pipe most often used in sewer construction include the following:

- **D**uctile **I**ron **P**ipe (DIP)
- Steel Pipe
- Thermoplastic Pipe

  Acrylonitrile **B**utadiene **S**tyrene (ABS)

  ABS Composite

  Poly**e**thylene (PE)

  Poly**v**inyl **C**hloride (PVC)
- Thermoset Plastic Pipe

  Reinforced **P**lastic **M**ortar (RPM)

  Reinforced **T**hermosetting **R**esin (RTR)

Each of these types of rigid and flexible pipe material is discussed in more detail in the following sections.

3.460 Rigid Pipe

3.4600 ASBESTOS-CEMENT PIPE (ACP)

Although it is no longer manufactured, asbestos-cement pipe was frequently used for both gravity and pressure sanitary sewers in nominal diameters from 4 inches through 42 inches (for metric pipe sizes, please refer to page 583). A full range of fittings were manufactured to be compatible with the class of pipe being used. Joints were made by compressing elastomeric (rubberlike) rings between the pipe ends and sleeves or couplings. ACP manufactured for gravity sanitary sewer applications was available in seven strength classifications and was specified by pipe diameter and class or strength. (The class designation represents the minimum crushing strength of the pipe expressed in pounds per linear foot of pipe.) ACP is still found in many sanitary systems throughout the United States.

Potential advantages of asbestos-cement pipe include:

1. Long laying lengths,
2. Availability of a wide range of strength classifications,
3. Availability of a wide range of fittings, and
4. Resistance to abrasion.

Potential limitations of asbestos-cement pipe include:

1. Subject to corrosion where acids are present,
2. Subject to shear and beam breakage when improperly bedded,
3. Low beam strength, and
4. Restrictions by OSHA due to asbestos content (precautions need to be taken when repairing asbestos-cement pipe).

3.4601 CAST-IRON PIPE (CIP)

CIP (gray iron) is used for both gravity and pressure sanitary sewers in diameters from 2 inches through 48 inches with a variety of joints. CIP is manufactured in a number of thicknesses, classes, and strengths. CIP is specified by nominal diameter, class, lining, and type of joint. A cement mortar lining with an asphaltic seal coating may be specified on the interior of the pipe. An exterior asphaltic coating is also commonly specified. Cast-iron pipe is very resistant to crushing and is often used for creek and river crossings, shallow trench locations, and under heavy traffic load areas.

Potential advantages of cast-iron pipe (gray iron) include:

1. Long laying lengths (in some situations), and
2. High pressure and load bearing capacity.

Potential limitations of cast-iron pipe (gray iron) include:

1. Subject to corrosion where acids are present,
2. Subject to chemical attack in corrosive soils,
3. Subject to shear and beam breakage when improperly bedded,
4. High weight, and
5. High cost.

3.4602 CONCRETE PIPE

Reinforced (RCP) and non-reinforced concrete pipe are both used for gravity sanitary sewers. Reinforced concrete pressure pipe and prestressed concrete pressure pipe (PCPP) are used for pressure as well as gravity flow sanitary sewers. Non-reinforced concrete pipe is available in nominal diameters from 4 inches through 36 inches. Reinforced concrete pipe is available in nominal diameters from 12 inches through 200 inches. Pressure pipe is available in diameters from 12 through 120 inches.

Concrete fittings and appurtenances such as wyes, tees, and manhole sections are widely available. A number of jointing methods may be used, depending on the tightness required and the operating pressure. Various linings and coatings are also available. Gravity and pressure concrete pipe may be manufactured to any reasonable strength requirement by varying the wall thickness, concrete strength, and quantity and configuration of reinforcing steel or prestressing elements. Concrete pipe is normally specified by nominal diameter, class or D-load strength, and type of joint.

Potential advantages of concrete pipe include:

1. Wide range of structural and pressure strengths,
2. Wide range of nominal diameters, and
3. Wide range of laying lengths (4 to 24 ft).

Potential limitations of concrete pipe include:

1. High weight,
2. Subject to corrosion where acids are present,
3. Subject to shear and beam breakage when improperly bedded, and
4. Special field repair methods and special fittings may be required for PCPP.

3.4603 VITRIFIED CLAY PIPE (VCP)

VCP is used for gravity sanitary sewers. This pipe is manufactured from clay and shales. Clay pipe is made nonporous (vitrified) by heating it at a high enough temperature to fuse the clay mineral particles. It is available in diameters from 3 through 36 inches and in some areas up to 42 inches. Clay fittings are available to meet most requirements, with special fittings manufactured on request. A number of joining methods can be used. VCP is manufactured in standard and extra-strength classifications, although in some areas standard-strength pipe is not commonly manufactured in sizes 12 inches or smaller. The strength of vitrified clay pipe varies with the diameter and strength classification. The pipe is manufactured in lengths up to 10 feet. VCP is specified by nominal pipe diameter, strength, and type of joint.

Potential advantages of vitrified clay pipe include:

1. High resistance to chemical corrosion,
2. High resistance to abrasion, and
3. Wide range of fittings available.

Potential limitations of vitrified clay pipe include:

1. Limited range of sizes available,
2. High weight, and
3. Subject to shear and beam breakage when improperly bedded.

3.461 Flexible Pipe

3.4610 DUCTILE IRON PIPE (DIP)

DIP is used for both gravity and pressure sanitary sewers. DIP is manufactured by adding cerium or magnesium to cast (gray) iron just prior to the pipe casting process. The product is available in nominal diameters from 3 through 54 inches and in lengths to 20 feet. Cast-iron (gray iron) or ductile iron fittings are used with ductile iron pipe. Various jointing methods for the product are available.

DIP is manufactured in various thicknesses, classes, and strengths. Linings for the interior of the pipe (for example, cement mortar lining with asphaltic coating, coal tar epoxies, epoxies, or polyethylene) may be specified. An exterior asphaltic coating and polyethylene exterior wrapping are also commonly specified. DIP is specified by nominal diameter, class, lining, and type of joint.

Potential advantages of DIP include:

1. Long laying lengths (in some situations),
2. High pressure and load bearing capacity,
3. High impact strength, and
4. High beam strength.

Potential limitations of DIP include:

1. Subject to corrosion where acids are present,
2. Subject to chemical attack in corrosive soils,
3. High weight, and
4. Cost.

3.4611 STEEL PIPE

Steel pipe is rarely used for gravity sanitary sewers but is common for force mains. When used, it usually is specified with interior protective coatings or linings (polymeric, bituminous, asbestos). Steel pipe is fabricated in diameters from 8 through 120 inches and in lengths up to 40 feet. Pipe fittings include tees, wyes, and elbows fabricated from steel. Steel pipe is specified by size, shape, wall profile, gauge or wall thickness, and protective coating or lining.

Potential advantages of steel pipe include:

1. Light weight, and
2. Long laying lengths (in some situations).

Potential limitations of steel pipe include:

1. Subject to corrosion where acids are present,
2. Subject to chemical attack in corrosive soils,
3. Subject to excessive deflection when improperly bedded, and
4. Subject to turbulence abrasion.

3.4612 THERMOPLASTIC PIPE

Thermoplastic materials include a broad variety of plastics that can be repeatedly softened by heating and hardened by cooling through a temperature range characteristic for each specific plastic. Generally, thermoplastic materials used in sanitary sewers are limited to acrylonitrile butadiene styrene (ABS), polyethylene (PE), and polyvinyl chloride (PVC).

- Acrylonitrile Butadiene Styrene (ABS) Pipe

ABS pipe is used for both gravity and pressure sanitary sewers. Non-pressure rated ABS sewer pipe is available in nominal diameters from 3 through 12 inches and in lengths up to 35 feet. A variety of ABS fittings and several joint systems are available. ABS pipe is manufactured by extrusion of ABS plastic material. ABS gravity sanitary sewer pipe is available in three Dimension Ratio (DR) classifications: 23.5, 35, and 42. (The DR is the ratio of the average outside diameter to the minimum wall thickness of the pipe. When adopted as a standard, it is referred to as Standard Dimension Ratio or SDR.) ABS pipe is specified by nominal diameter, dimension ratio, pipe stiffness, and type of joint.

Potential advantages of ABS pipe include:

1. Light weight,
2. Long laying lengths,
3. High impact strength, and
4. Ease in field cutting and tapping.

Potential limitations of ABS pipe include:

1. Limited range of sizes available,
2. Subject to environmental stress cracking,
3. Subject to excessive deflection when improperly bedded,
4. Subject to attack by certain organic chemicals, and
5. Subject to surface change from long-term ultraviolet exposure.

- Acrylonitrile Butadiene Styrene (ABS) Composite Pipe

ABS composite pipe is used for gravity sanitary sewers. It is available in nominal diameters from 8 through 15 inches and in lengths from 6.25 to 12.5 feet. The joint systems available include elastomeric gasket joints and solvent cemented joints. ABS composite pipe is manufactured by extrusion of ABS plastic material with a series of truss beams which are filled with filler material such as lightweight Portland cement concrete. ABS composite pipe is specified by nominal diameter and type of joint.

Potential advantages of ABS composite pipe include:

1. Light weight,
2. Long laying lengths, and
3. Ease in field cutting.

Potential limitations of ABS composite pipe include:

1. Limited range of sizes available,
2. Subject to environmental stress cracking,
3. Subject to rupture when improperly bedded,
4. Subject to attack by certain organic chemicals, and
5. Subject to surface change due to long-term ultraviolet exposure.

- Polyethylene (PE) Pipe

PE pipe is used for both gravity and pressure sanitary sewers. Non-pressure PE pipe, primarily used for sewer relining, is available in nominal diameters from 4 through 48 inches. Jointing is primarily accomplished by butt fusion or flanged adapters. PE pipe is manufactured by extrusion of PE plastic material. PE pipe is specified by material designation, nominal diameter (inside or outside), standard dimension ratios, and type of joint.

Potential advantages of PE pipe include:

1. Long laying lengths,
2. Light weight,
3. High impact strength, and
4. Ease in field cutting.

Potential limitations of PE pipe include:

1. Relatively low tensile strength and pipe stiffness,
2. A limited range of sizes available,
3. Subject to environmental stress cracking,
4. Subject to excessive deflection when improperly bedded,
5. Subject to attack by certain organic chemicals,
6. Subject to surface change due to long-term ultraviolet exposure, and
7. Special tools are required for fusing joints.

- Polyvinyl Chloride (PVC) Pipe

  PVC pipe is used for both gravity and pressure sanitary sewers. Non-pressure PVC sewer pipe is available in nominal diameters from 4 through 36 inches. PVC pressure and non-pressure fittings are available. PVC pipe is generally available in lengths up to 20 feet. Jointing is primarily accomplished with elastomeric seal gasket joints, although solvent cement joints are available for special applications. PVC pipe is manufactured by extrusion of the plastic material.

  Non-pressure PVC sanitary sewer pipe is provided in two standard dimension ratios, SDR 35 and SDR 41. PVC pipe is specified by nominal diameter, dimension ratio, pipe stiffness, and type of joint.

  Potential advantages of PVC pipe include:

  1. Light weight,
  2. Long laying lengths,
  3. High impact strength, and
  4. Ease in field cutting and tapping.

  Potential limitations of PVC pipe include:

  1. Subject to attack by certain organic chemicals,
  2. Subject to excessive deflection when improperly bedded,
  3. Limited range of sizes available, and
  4. Subject to surface changes from long-term ultraviolet exposure.

3.4613 THERMOSET PLASTIC PIPE

Thermoset plastic materials include a broad variety of plastics. These plastics, after having been cured by heat or other means, are substantially infusible and insoluble. Generally, thermoset plastic pipe materials used in sanitary sewers are provided in two categories, reinforced thermosetting resin (RTR) and reinforced plastic mortar (RPM).

- Reinforced Thermosetting Resin (RTR) Pipe

  RTR pipe is used for both gravity and pressure sanitary sewers. RTR pipe is generally available in nominal diameters from 1 through 12 inches and is manufactured in accordance with ASTM (American Society for Testing and Materials) standard specifications. The product is available in nominal diameters from 12 through 144 inches. RTR fittings are available for small-diameter pipes but for larger diameters, RTR fittings are manufactured as required. A number of joining methods are available. Various methods of interior protection (for example, thermoplastic or thermosetting liners or coatings) are available. RTR pipe is manufactured using a number of methods including centrifugal casting, pressure laminating, and filament winding. In general, RTR pipe contains fibrous reinforcement materials, such as fiberglass, embedded in or surrounded by cured thermosetting resin. RTR pipe is specified by nominal diameter, pipe stiffness, lining and coating, method of manufacture, thermoset plastic material, and type of joints.

  Potential advantages of RTR pipe include:

  1. Light weight, and
  2. Long laying lengths.

  Potential limitations of RTR pipe include:

  1. Subject to strain corrosion in some environments,
  2. Subject to excessive deflection when improperly bedded,
  3. Subject to attack by certain organic chemicals, and
  4. Subject to surface change from long-term ultraviolet exposure.

- Reinforced Plastic Mortar (RPM) Pipe

  RPM pipe is used for both gravity and pressure sewers. RPM pipe is available in nominal diameters from 8 through 144 inches. In smaller diameters, RPM fittings are generally available, but in larger diameters the fittings are manufactured as required. A number of joining methods are available. Various methods of interior protection (for example, thermoplastic or thermosetting liners or coatings) are available. RPM pipe is manufactured containing fibrous reinforcements such as fiberglass and aggregates such as sand embedded in or surrounded by cured thermosetting resin. RPM pipe is specified by nominal diameter, pipe stiffness, stiffness factor, beam strength, hoop tensile strength, lining or coating, thermoset plastic material, and type of joint.

  Potential advantages of RPM pipe include:

  1. Light weight, and
  2. Long laying lengths.

  Potential limitations of RPM pipe include:

  1. Subject to strain corrosion in some environments,
  2. Subject to excessive deflection when improperly bedded,
  3. Subject to attack by certain organic chemicals, and
  4. Subject to surface change from long-term ultraviolet exposure.

3.47 Pipe Joints

Pipe joints are one of the most critical components of the piping system because of the need to control groundwater infiltration, wastewater exfiltration, and root intrusion in sanitary sewer systems. Many types of pipe joints are available for the different pipe materials used in sanitary sewer construction. Regardless of the type of sewer pipe used, reliable, tight pipe joints are essential. A good pipe joint must be watertight, root resistant, flexible, and durable. Various forms of gasket (elastomeric seal) pipe joints are now commonly used in sanitary sewer construction. In general, this type of joint can be assembled easily in a broad range of weather conditions and environments with good assurance of a reliable, tight seal that prevents leakage and root intrusion. Several of the most widely used types of sanitary sewer pipe joints are described in the following sections.

3.470 Gasketed Pipe Joints

Gasket joints create a seal against leakage through compression of an elastomeric seal or ring. The two main designs for gasket-type pipe joints are the push-on pipe joint and the mechanical compression pipe joint.

- Push-on Pipe Joint (Figure 3.24)

  This type of pipe joint uses a continuous elastomeric ring gasket which is compressed into the *ANNULAR SPACE*[15] formed by the pipe, fitting, or coupler socket and the spigot end of the pipe. The gasket provides a positive seal when the pipe spigot is pushed into the socket. When using this type of pipe joint in pressure sanitary sewers, thrust restraint (*THRUST BLOCKS*[16]) may be required to prevent joint separation under pressure. Push-on pipe joints (fittings, couplers, or integral bells) are available for nearly all types of pipe used in sewer systems today.

- Mechanical Compression Pipe Joint (Figure 3.24)

  This type of pipe joint uses a continuous elastomeric ring gasket which provides a positive seal when the gasket is compressed by means of a mechanical device. When using this type of pipe joint in pressure sanitary sewers, thrust restraint may be required to prevent joint separation under pressure. This type of pipe joint may be provided as an integral part of cast-iron or ductile iron pipe. When incorporated into a coupler, this type of pipe joint may be used to join two similarly sized plain spigot ends of any commonly used sewer pipe materials.

3.471 Flanged Pipe Joint (Figure 3.24)

Flanged connections are primarily used in pump station piping systems where cast-iron or ductile iron piping is common.

3.472 Cement Mortar Pipe Joint (Figure 3.24)

This type of pipe joint involves use of shrink-compensating cement mortar caulking placed into a bell and spigot pipe joint to provide a seal. The use of this joint is discouraged because reliable, watertight joints are not assured. Cement mortar joints are not flexible and may crack if there is any pipe movement.

3.473 Solvent Cement Pipe Joint (Figure 3.24)

Solvent cement pipe joints may be used in joining thermoplastic pipe materials such as ABS, ABS composite, and PVC pipe. This type of pipe joint involves bonding a sewer pipe spigot into a sewer pipe bell or coupler using a solvent cement. Solvent cement joints can provide a positive seal provided the proper cement is applied under suitable conditions with proper techniques. Precautions should be taken to ensure adequate trench ventilation and protection for operators installing the pipe. Solvent cement pipe joints may be needed in special situations and with some plastic fittings.

3.474 Rubber Coupling With Compression Bands (Figure 3.24)

External sealing bands of rubber are used on plain end pipe. The elastomeric band is slipped over each pipe end and then compression bands are tightened on the joint.

3.475 Welded Joint

Welded joints are typically found on steel pipe used for force mains. This type of joint is very strong and leaktight. The most common welded joint is the butt weld or a butt weld with the pipe ends beveled. A sleeve may also be used as shown in Figure 3.24.

3.476 Heat Fusion Pipe Joint

Heat fusion pipe joints are commonly used for polyethylene (PE) sanitary sewer pipe. The general method of jointing PE sanitary sewer pipe involves butt fusion of the pipe lengths, end to end. After the ends of two lengths of PE pipe are trimmed and softened to a melted state with heated metal plates, the pipe ends are forced together to the point of butt fusion, providing a positive seal. The pipe joint does not require thrust restraint in pressure applications. Trained technicians with special equipment are required to achieve reliable watertight pipe joints.

3.477 Bituminous Pipe Joint

This type of pipe joint involves use of hot-poured or cold-packed bituminous material forced into a bell and spigot pipe joint to provide a seal. While this type of joint can be found in older systems, it is not used in new construction.

3.478 Elastomeric Sealing Compound Pipe Joint

Elastomeric sealing compound may be used in joining properly prepared concrete sewer pipe for a gravity sanitary sewer. Pipe ends must be sandblasted and primed for elastomeric sealant application. The sealant is mixed on the job site and applied with a caulking gun and spatula. The pipe joint, when assembled with proper materials and procedures, provides a positive seal against leakage in gravity sewer pipe.

3.479 Mastic Pipe Joint

Mastic pipe joints are frequently used for special non-round shapes of concrete gravity sewer pipe which are not adaptable for use with gasket pipe joints. The mastic material is placed into the annular space to provide a positive seal. Application

---

[15] *Annular (AN-you-ler) Space. A ring-shaped space located between two circular objects. For example, the space between the outside of a pipe liner and the inside of a pipe.*

[16] *Thrust Block. A mass of concrete or similar material appropriately placed around a pipe to prevent movement when the pipe is carrying water. Usually placed at bends and valve structures.*

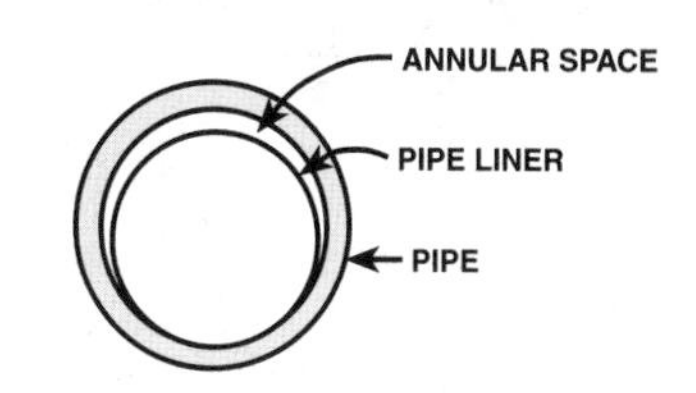

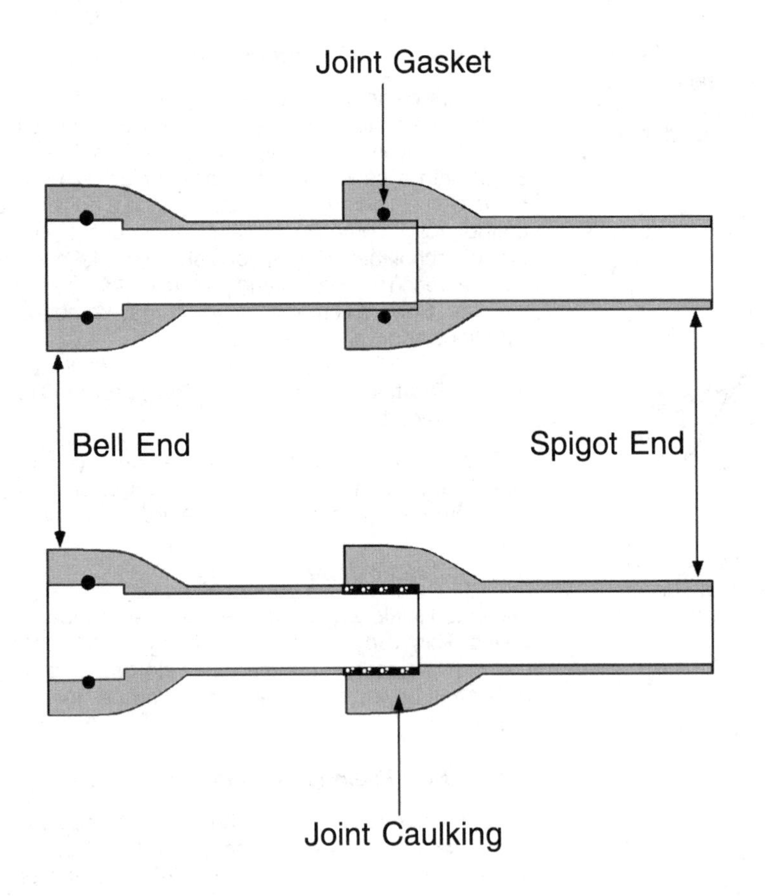

Push-on bell and spigot with gasketed pipe joint

Push-on bell and spigot with caulked pipe joint

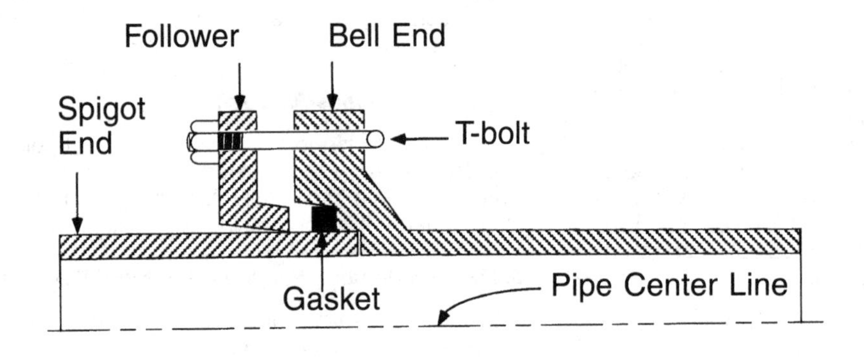

Push-on bell and spigot, mechanical joint with gland-type follower

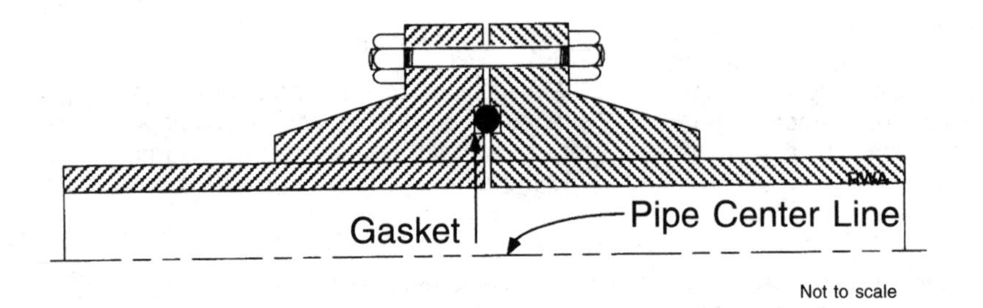

Flanged end pipe with gasket

Fig. 3.24 Pipe joints

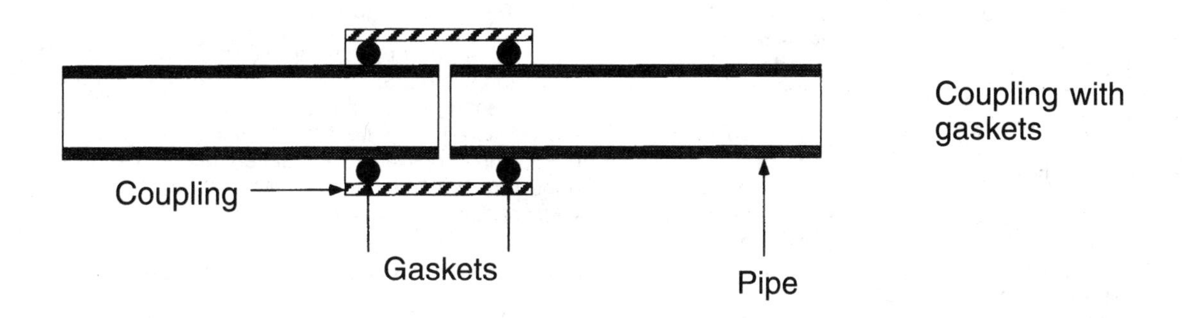

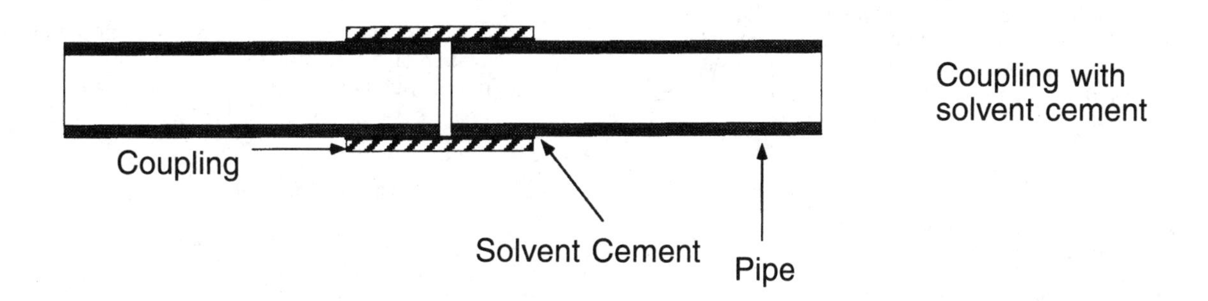

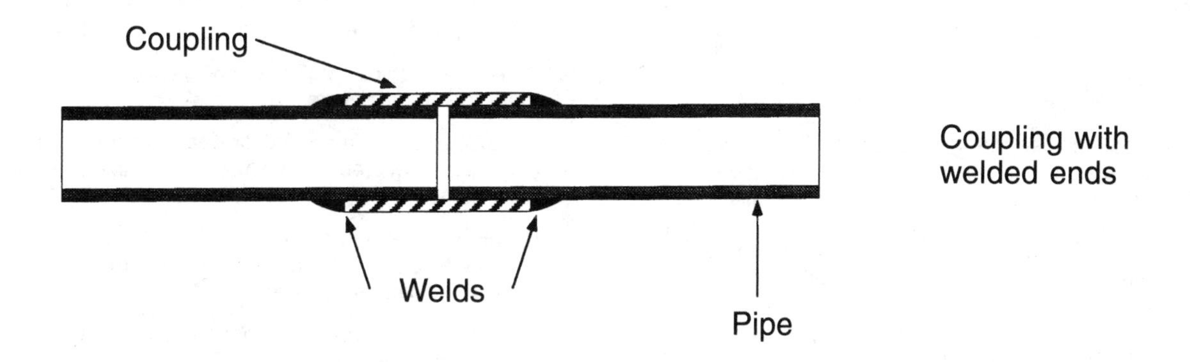

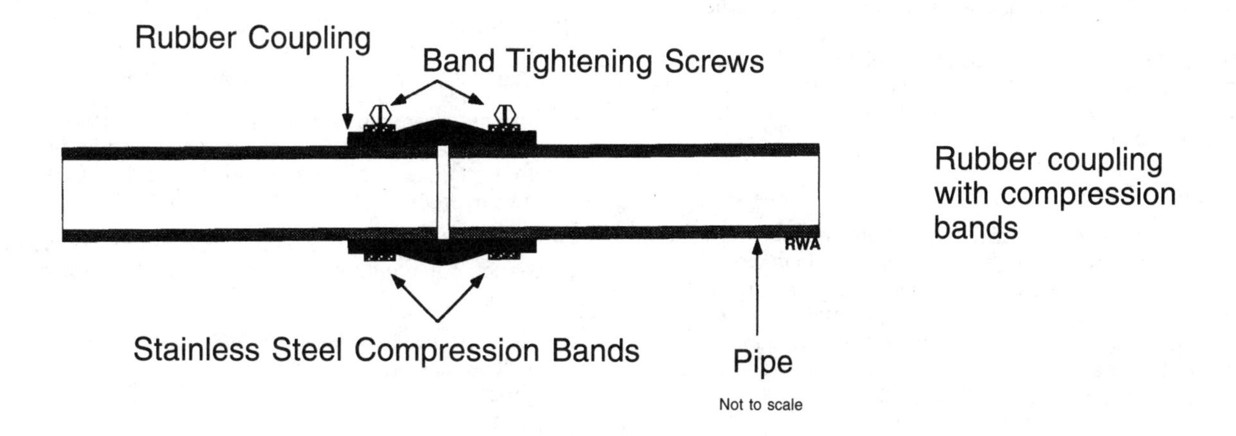

Fig. 3.24 Pipe joints (continued)

may be by troweling, caulking, or by the use of preformed segments of mastic material in a manner similar to gaskets. Satisfactory performance of the pipe joints depends upon the proper selection of primer and mastic material as well as good workmanship in application.

3.48 Manhole Materials

Manhole bases are made of either precast or cast-in-place unreinforced concrete with wastewater channels formed by hand before the concrete hardens. Manhole barrels are usually made of precast concrete sections with cement mortar or bitumastic joints. Manholes in older collection systems may have been made with clay bricks and cement mortar joints. Some manhole barrels are now being made with fiberglass for use in areas with high groundwater and corrosive material in the wastewater conveyed. Manhole frames and covers are made of cast iron with the mating surface between the frame and cover machined to ensure a firm seating. Manhole steps or rungs should be made of wrought iron or reinforced plastic to resist corrosion. In some jurisdictions steps are prohibited and ladders are used to eliminate the hazard of corroded steps.

QUESTIONS

Write your answers in a notebook and then compare your answers with those on pages 75 and 76.

3.4H What is the difference between flexible and rigid pipe?

3.4I List as many of the most commonly used sanitary sewer pipe materials as you can recall.

3.4J List the problems that can be caused by poor sewer-pipe joints.

3.4K Why are some manhole barrels made with fiberglass?

3.5 PLAN AND PROFILE

A civil engineer will use the field data described in Section 3.41 and the design criteria described in Sections 3.42 through 3.48 to prepare plans and specifications for the construction of a wastewater collection system. A typical plan (top view) and profile (side view) for a short sewer extension based on the field notes shown in Figure 3.22 are illustrated in Figure 3.25.

3.6 DESIGN REVIEW

Wastewater collection system operators should request the opportunity to review the design of collection systems that they will be responsible for maintaining after construction. To make an effective review, an operator must be familiar with the elements of design that can create problems during the life of a collection system. The following sections give some suggestions for the review of plans and profiles for construction of a collection system.

3.60 Office Study

When you receive a set of plans and profiles for review, look them over for an overview of how the system is to work. Then make a detailed review by asking yourself how the design may be improved for the elimination of future maintenance problems. Make a list of items that are not clear to you and ask the designer for a clarification. Some items that you should consider in your review are set forth in Section 3.62.

3.61 Field Investigation

Now that you are familiar with the plans and profiles for a collection system and have developed some ideas and questions concerning the effect of the design on maintenance, you are prepared for a field investigation. As you walk the proposed route, envision the maintenance of the sewer lines and appurtenances as planned and ask yourself how the actual field conditions will influence maintenance.

3.62 Design Items Influencing Maintenance

During both the office and field review of plans and profiles for a wastewater collection system, ask yourself the following questions concerning the design:

1. Is the route or alignment satisfactory?
2. Could there be a better route?
   a. Consider the effect of pipe diameters and depth of excavation on future maintenance.
   b. Think about maintenance equipment access.
3. Is there sufficient overhead clearance for maintenance and repair equipment?
4. How could repairs be made in case of failures?
   a. Availability of large-diameter pipe?
   b. Depth of excavation?
   c. Shoring requirements?
5. Can the system be easily and properly operated and maintained?
6. Will there be any industrial waste dischargers? If so, what will they discharge and will the discharge cause any problems? How can these problems be controlled?
7. Where can vehicles be parked when operators and equipment require access to a manhole?
8. Is traffic a problem around or near a manhole and if so, can it be relocated to minimize the problem? Will night work be necessary?
9. Can the atmosphere in the manhole be tested safely for oxygen deficiency, explosive gases and toxic gases? Can large ventilation blowers be used?
10. Can an operator reach the bottom of the manhole with a ladder if steps are not provided?
11. If there are steps in the manhole wall for access, will they be damaged by corrosion and become unsafe?
12. Is the shelf in the bottom of the manhole properly sloped? If the shelf is too flat, it will get dirty; if the shelf is too steep, it can be a slippery hazard to operators.
13. Are the channels properly shaped to provide smooth flow through a manhole to minimize splashing and turbulence that can cause the release of toxic, odorous, and corrosive hydrogen sulfide?
14. Are the channels at least as large as the pipe so maintenance equipment and plugs can be fitted or inserted into the pipes?
15. Can the available sewer maintenance equipment be effective between the manholes or is the distance too great?

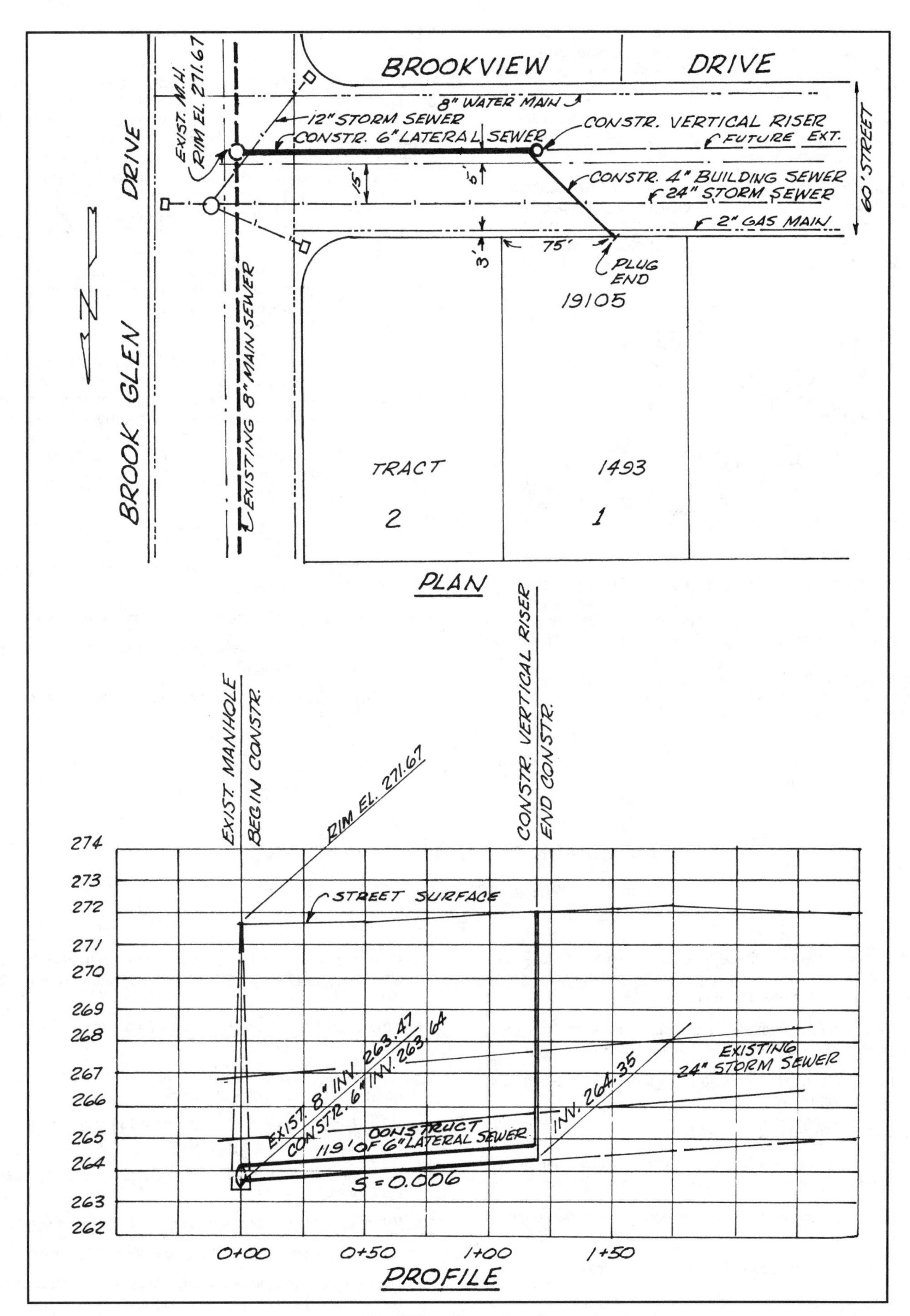

Fig. 3.25 Plan and profile

16. Is the manhole opening large enough to allow access for tools and equipment used to maintain a large sewer line? Is the opening sufficient to handle equipment and personnel during routine maintenance or inspection?

17. Does the area flood? How often? Can flows be diverted? Stored? How? For how long? What tools, equipment and procedures are needed?

In addition to the above questions, use your experience to raise other questions on how the design of a collection system can affect your future operation and maintenance of the system. Try to anticipate what information you will need to make repairs or build temporary facilities in emergency situations.

3.63 Communication

Your questions and ideas concerning the improvement of the design of a collection system should be given to the designer in a written and understandable form with an offer to meet to discuss them. The designer should be capable of explaining and justifying a design in terms of economics and installation. You should be able to justify your concerns on the basis of operation and maintenance requirements.

3.64 Construction, Testing, and Inspection

Designers must consider how the wastewater collection facilities they design will be constructed. Some facilities have been properly designed, but have not functioned as intended due to poor or improper construction materials and techniques. Good construction, inspection, and testing procedures are very important in order to eliminate operation and maintenance problems that could develop from poor construction practices. Whenever problems (stoppages, failures, or odors) develop in your collection system, try to identify the cause so it can be corrected now and in the future design of new systems. Did the problem start with the design, construction, or inspection of the facility; or was it caused by a waste discharger, by aging of the system, or by your maintenance and repair program? Whatever the cause, identify it and try to correct it.

In Chapter 7, "Underground Repair," details are provided regarding the construction, inspection, and testing of wastewater collection systems. Knowledge of this material will help you communicate with design engineers and also help you identify the causes of problems in your collection system. If your job today is not directly related to construction or construction inspection and testing, this information may be very helpful to you in the future.

QUESTIONS

Write your answers in a notebook and then compare your answers with those on page 76.

3.5A What is meant by the plan and profile of a sewer?

3.6A When reviewing plans for a new sewer, where could traffic be a problem?

3.6B Why is the slope of the shelf in a manhole important?

DISCUSSION AND REVIEW QUESTIONS

Chapter 3. WASTEWATER COLLECTION SYSTEMS

Write the answers to these questions in your notebook. The purpose of these questions is to indicate to you how well you understand the material in the chapter.

1. Why should collection system operators know how engineers design wastewater collection systems?
2. How does inflow and infiltration get into wastewater collection systems?
3. Why are lift stations installed?
4. Why are low-pressure collection systems sometimes used as an alternative to gravity collection systems?
5. Why are vacuum collection systems sometimes used as an alternative to gravity collection systems?
6. Why are manholes installed in sewer lines?
7. What is the purpose of interconnector sewers?
8. How can collection system operators communicate with engineers regarding wastewater collection system design?
9. Why should the slope of sewers follow the slope of the land?
10. Why are lateral, main and trunk sewers usually constructed at or near the center of a street?
11. What factors need to be considered in the selection of materials for sewer construction?
12. Why are good construction, inspection, and testing procedures very important?
13. When is a wastewater pump used in a collection system?
14. How would you measure the flow velocity in a sewer?
15. How can wastewater collection system operators (you) help engineers design better systems?

SUGGESTED ANSWERS

Chapter 3. WASTEWATER COLLECTION SYSTEMS

Answers to questions on page 35.

3.0A Collection system operators should understand how the systems they are responsible for operating and maintaining were designed and constructed so they can determine the correct action to take when the system fails to operate properly. Also they will be able to discuss design, construction, and inspection with engineers.

3.1A The purpose of a wastewater collection system is to collect the wastewater from a community's homes and industries and convey it at an appropriate velocity to a wastewater treatment plant.

3.1B In areas where the topography is unfavorable for a gravity wastewater collection system, a pressure or vacuum collection system may be used.

Answers to questions on page 41.

3.2A Sources or types of wastewater conveyed by wastewater collection systems include wastewater from domestic, commercial, and industrial sources, as well as groundwater infiltration and surface water inflows produced in the service area.

3.2B The peak flow from a service area is estimated on the basis of maximum population and full commercial and industrial development of the area, with additional allowances for infiltration and inflow.

3.2C The flow velocity in a sewer main can be measured using either dyes or floats.

Answers to questions on page 47.

3.3A The parts or components of a gravity wastewater collection system include building sewers, lateral and branch sewers, main sewers, trunk sewers, intercepting sewers, and lift stations.

3.3B The principal components of a low-pressure collection system include gravity sewers, holding tanks, grinder pumps, and pressure mains.

3.3C The principal components of a vacuum collection system include gravity sewers, holding tanks, vacuum valves, vacuum mains, and vacuum pumps.

Answers to questions on page 59.

3.3D Backflow preventers are used to stop the accidental backflow or reverse flow of wastewater into buildings from the sewer.

3.3E The three main types of wastewater backflow preventers are the (1) overflow type, (2) check valve type, and (3) automated system that uses a flexible diaphragm and level sensor.

3.3F Inverted siphons are sewer lines depressed under obstructions to the normal gradient of the sewer line such as watercourses and depressed roadways.

3.3G Flow regulators in wastewater collection systems are used to divert the flow of wastewater from one sewer line to another for the efficient use of the system and to prevent overloading one portion of the system. They are also used when performing maintenance on a main line or intercepting sewer.

Answers to questions on page 61.

3.4A Field surveys are conducted to determine the locations of the physical features and the elevations of the ground surface along the planned route of the sewer lines in the collection system. Field surveys include both route and level surveys.

3.4B Route surveys determine the locations of physical features that influence the horizontal location of a sewer line. Level surveys determine elevations of the ground surface along the route of the sewer line, buildings to be served, and the elevations of underground obstructions and pipelines.

3.4C A scouring velocity must be achieved at some time during the day to prevent the deposition and buildup of solids in the sewer line.

3.4D If a scouring velocity is not maintained in a sewer, the solids will settle out. These solids will cause sewer stoppages and the generation of hydrogen sulfide gas which is corrosive, toxic, and odorous.

Answers to questions on page 64.

3.4E If a 10-inch diameter sewer on a 0.005 ft/ft slope will not carry the peak wet weather design flow, either the diameter or the slope or both could be increased to provide sufficient capacity.

3.4F Sewers are usually not placed in easements because they are more difficult to gain access to and to maintain than sewers in streets.

3.4G The width of a sewer trench is controlled by the diameter of the sewer pipe, clearance for pipe laying, soil stability for safety during excavation and the excavation procedures.

Answers to questions on page 72.

3.4H A flexible pipe will bend or bulge from its usual shape when subjected to a load without being adequately supported. However, it will not crack under most loads and will return nearly to its original shape when a load is removed. A rigid pipe will require less support; however, it will crack when the combination of a load on the pipe and its support causes a stress exceeding the strength of the pipe. It will not return to its original shape when the load is removed.

3.4I The materials most commonly used for sanitary sewer pipelines include asbestos-cement pipe (ACP); cast-iron pipe (CIP); reinforced concrete pipe (RCP); non-reinforced concrete pipe; prestressed concrete pressure pipe (PCPP); vitrified clay pipe (VCP); ductile iron pipe (DIP); thermoplastic pipe such as acrylonitrile butadiene styrene (ABS), ABS composite pipe, polyethylene (PE) pipe, and polyvinyl chloride (PVC) pipe; and thermoset plastic pipe such as reinforced thermosetting resin (RTR) pipe and reinforced plastic mortar (RPM) pipe.

3.4J Poor sewer pipe joints can lead to problems caused by roots, infiltration, and exfiltration.

3.4K Some manhole barrels are made with fiberglass for use in areas with high groundwater and corrosive material in the wastewater conveyed.

Answers to questions on page 74.

3.5A The plan view is the top view of a sewer and shows the pipe, manholes and street. The profile is the side view of a sewer and shows the pipe, manholes, invert elevations, street surface, and underground pipes such as storm sewers, gas mains, water mains, and electrical cables.

3.6A When reviewing plans for a new sewer, traffic could be a problem wherever crews must work, such as around manholes. Also consider where vehicles will be parked when operators and equipment require access to a manhole.

3.6B The slope of the shelf of a manhole is important because if the slope is too flat, the shelf will get dirty; and, if too steep, the shelf can be a slippery hazard to operators.

CHAPTER 4

SAFE PROCEDURES

by

George Freeland

Revised by

Rick Arbour

Russ Armstrong

Gary Batis

TABLE OF CONTENTS

Chapter 4. SAFE PROCEDURES

OBJECTIVES

Chapter 4. SAFE PROCEDURES

Following completion of Chapter 4, you should be able to:

1. Perform duties of a collection system operator by following safe procedures,
2. Inspect safety features of vehicles and equipment,
3. Drive vehicles defensively and safely,
4. Route traffic around a job site,
5. Identify manhole hazards,
6. Assemble necessary equipment to safely enter a confined space,
7. Calibrate and use gas detection equipment,
8. Take necessary precautions prior to entering a confined space,
9. Follow safe procedures during confined space entry, work, and exit,
10. Work in excavations safely,
11. Avoid electrical hazards,
12. Protect yourself from excessive noise,
13. Extinguish fires, and
14. Interpret and comply with Worker Right-To-Know laws.

WORDS

Chapter 4. SAFE PROCEDURES

AEROBIC (AIR-O-bick) AEROBIC

A condition in which atmospheric or dissolved molecular oxygen is present in the aquatic (water) environment.

AMBIENT (AM-bee-ent) AMBIENT

Surrounding. Ambient or surrounding atmosphere.

ANAEROBIC (AN-air-O-bick) ANAEROBIC

A condition in which atmospheric or dissolved molecular oxygen is *NOT* present in the aquatic (water) environment.

ANAEROBIC (AN-air-O-bick) DECOMPOSITION ANAEROBIC DECOMPOSITION

The decay or breaking down of organic material in an environment containing no "free" or dissolved oxygen.

CFR CFR

Code of **F**ederal **R**egulations. A publication of the United States Government which contains all of the proposed and finalized federal regulations, including safety and environmental regulations.

COMPETENT PERSON COMPETENT PERSON

A competent person is defined by OSHA as a person capable of identifying existing and predictable hazards in the surroundings, or working conditions which are unsanitary, hazardous or dangerous to employees, and who has authorization to take prompt corrective measures to eliminate the hazards.

CONFINED SPACE CONFINED SPACE

Confined space means a space that:

A. Is large enough and so configured that an employee can bodily enter and perform assigned work; and

B. Has limited or restricted means for entry or exit (for example, tanks, vessels, silos, storage bins, hoppers, vaults, and pits are spaces that may have limited means of entry); and

C. Is not designed for continuous employee occupancy.

(Definition from the Code of Federal Regulations (CFR) Title 29 Part 1910.146.)

CONFINED SPACE, NON-PERMIT CONFINED SPACE, NON-PERMIT

A non-permit confined space is a confined space that does not contain or, with respect to atmospheric hazards, have the potential to contain any hazard capable of causing death or serious physical harm.

CONFINED SPACE, PERMIT-REQUIRED (PERMIT SPACE) CONFINED SPACE, PERMIT-REQUIRED (PERMIT SPACE)

A confined space that has one or more of the following characteristics:

- Contains or has a potential to contain a hazardous atmosphere,
- Contains a material that has the potential for engulfing an entrant,
- Has an internal configuration such that an entrant could be trapped or asphyxiated by inwardly converging walls or by a floor which slopes downward and tapers to a smaller cross section, or
- Contains any other recognized serious safety or health hazard.

(Definition from the Code of Federal Regulations (CFR) Title 29 Part 1910.146.)

DANGEROUS AIR CONTAMINATION DANGEROUS AIR CONTAMINATION

An atmosphere presenting a threat of causing death, injury, acute illness, or disablement due to the presence of flammable and/ or explosive, toxic or otherwise injurious or incapacitating substances.

A. Dangerous air contamination due to the flammability of a gas or vapor is defined as an atmosphere containing the gas or vapor at a concentration greater than 10 percent of its lower explosive (lower flammable) limit.

B. Dangerous air contamination due to a combustible particulate is defined as a concentration greater than 10 percent of the minimum explosive concentration of the particulate.

C. Dangerous air contamination due to the toxicity of a substance is defined as the atmospheric concentration immediately hazardous to life or health.

DECIBEL (DES-uh-bull) DECIBEL

A unit for expressing the relative intensity of sounds on a scale from zero for the average least perceptible sound to about 130 for the average level at which sound causes pain to humans. Abbreviated dB.

ENGULFMENT ENGULFMENT

Engulfment means the surrounding and effective capture of a person by a liquid or finely divided (flowable) solid substance that can be aspirated to cause death by filling or plugging the respiratory system or that can exert enough force on the body to cause death by strangulation, constriction, or crushing.

FIT TEST FIT TEST

The use of a procedure to qualitatively or quantitatively evaluate the fit of a respirator on an individual.

GROSS VEHICLE WEIGHT (GVW) GROSS VEHICLE WEIGHT (GVW)

The total weight of a single vehicle including its load.

GROSS VEHICLE WEIGHT RATING (GVWR) GROSS VEHICLE WEIGHT RATING (GVWR)

The maximum weight rating specified by the manufacturer for a single vehicle including its load.

IDLH IDLH

Immediately **D**angerous to **L**ife or **H**ealth. The atmospheric concentration of any toxic, corrosive, or asphyxiant substance that poses an immediate threat to life or would cause irreversible or delayed adverse health effects or would interfere with an individual's ability to escape from a dangerous atmosphere.

MATERIAL SAFETY DATA SHEET (MSDS) MATERIAL SAFETY DATA SHEET (MSDS)

A document which provides pertinent information and a profile of a particular hazardous substance or mixture. An MSDS is normally developed by the manufacturer or formulator of the hazardous substance or mixture. The MSDS is required to be made available to employees and operators whenever there is the likelihood of the hazardous substance or mixture being introduced into the workplace. Some manufacturers are preparing MSDSs for products that are not considered to be hazardous to show that the product or substance is *NOT* hazardous.

MERCAPTANS (mer-CAP-tans) MERCAPTANS

Compounds containing sulfur which have an extremely offensive skunk-like odor; also sometimes described as smelling like garlic or onions.

NIOSH (NYE-osh) NIOSH

The **N**ational **I**nstitute of **O**ccupational **S**afety and **H**ealth is an organization that tests and approves safety equipment for particular applications. NIOSH is the primary federal agency engaged in research in the national effort to eliminate on-the-job hazards to the health and safety of working people. The NIOSH Publications Catalog, Sixth Edition, NIOSH Pub. No. 84-118, lists the NIOSH publications concerning industrial hygiene and occupational health. To obtain a copy of the catalog, write to National Technical Information Service (NTIS), 5285 Port Royal Road, Springfield, VA 22161. NTIS Stock No. PB86-116787, price, $103.50, plus $5.00 shipping and handling per order.

NONSPARKING TOOLS NONSPARKING TOOLS

These tools will not produce a spark during use. They are made of a nonferrous material, usually a copper-beryllium alloy.

OSHA (O-shuh) OSHA

The Williams-Steiger **O**ccupational **S**afety and **H**ealth **A**ct of 1970 (OSHA) is a federal law designed to protect the health and safety of industrial workers and collection system operators. The Act regulates the design, construction, operation and maintenance of industrial plants and wastewater collection and treatment facilities. The Act does not apply directly to municipalities, *EXCEPT* in those states that have approved plans and have asserted jurisdiction under Section 18 of the OSHA Act. *HOWEVER, CONTRACT OPERATORS AND PRIVATE FACILITIES DO HAVE TO COMPLY WITH OSHA REQUIREMENTS.* Wastewater collection systems have come under stricter regulation in all phases of activity as a result of OSHA standards. OSHA also refers to the federal and state agencies which administer the OSHA regulations.

OLFACTORY (ol-FAK-tore-ee) FATIGUE — OLFACTORY FATIGUE

A condition in which a person's nose, after exposure to certain odors, is no longer able to detect the odor.

OXYGEN DEFICIENCY — OXYGEN DEFICIENCY

An atmosphere containing oxygen at a concentration of less than 19.5 percent by volume.

OXYGEN ENRICHMENT — OXYGEN ENRICHMENT

An atmosphere containing oxygen at a concentration of more than 23.5 percent by volume.

QUALITATIVE FIT TEST (QLFT) — QUALITATIVE FIT TEST (QLFT)

A pass/fail fit test to assess the adequacy of respirator fit that relies on the individual's response to the test agent.

QUANTITATIVE FIT TEST (QNFT) — QUANTITATIVE FIT TEST (QNFT)

An assessment of the adequacy of respirator fit by numerically measuring the amount of leakage into the respirator.

SEPTIC (SEP-tick) — SEPTIC

A condition produced by anaerobic bacteria. If severe, the wastewater produces hydrogen sulfide, turns black, gives off foul odors, contains little or no dissolved oxygen, and the wastewater has a high oxygen demand.

SEWER GAS — SEWER GAS

(1) Gas in collection lines (sewers) that results from the decomposition of organic matter in the wastewater. When testing for gases found in sewers, test for lack of oxygen and also for explosive and toxic gases.

(2) Any gas present in the wastewater collection system, even though it is from such sources as gas mains, gasoline, and cleaning fluid.

SPECIFIC GRAVITY — SPECIFIC GRAVITY

(1) Weight of a particle, substance or chemical solution in relation to the weight of an equal volume of water. Water has a specific gravity of 1.000 at 4°C (39°F). Wastewater particles or substances usually have a specific gravity of 0.5 to 2.5.

(2) Weight of a particular gas in relation to the weight of an equal volume of air at the same temperature and pressure (air has a specific gravity of 1.0). Chlorine has a specific gravity of 2.5 as a gas.

SPOIL — SPOIL

Excavated material such as soil from the trench of a sewer.

TIME WEIGHTED AVERAGE (TWA) — TIME WEIGHTED AVERAGE (TWA)

The average concentration of a pollutant (or sound) based on the times and levels of concentrations of the pollutant. The time weighted average is equal to the sum of the portion of each time period (as a decimal, such as 0.25 hour) multiplied by the pollutant concentration during the time period divided by the hours in the workday (usually 8 hours).

CHAPTER 4. SAFE PROCEDURES

(Lesson 1 of 3 Lessons)

4.0 PERFORM YOUR DUTIES SAFELY

Collection system operators have a unique and very challenging job because it includes a wide range of responsibilities and requires more than an average level of knowledge in many different areas. Clearly it requires a trained professional who is capable of exercising common sense under many different circumstances. In this job, more than in many others, an operator's knowledge, skills, and common sense relate directly to *SAFETY/SURVIVAL*. A trained professional operator demonstrates an awareness of hazards and a commitment to accomplish every task in a safe manner.

After receiving job assignments and discussing equipment needs, safety considerations, and other aspects of the tasks, the first thing an operator generally does is get into a utility vehicle of some sort to get to the job site. This vehicle may range from a pickup truck to a very large, cumbersome vehicle, or a combination of vehicle and towed equipment, for example:

1. Tankers,
2. Rodding machines,
3. Boom trucks,
4. Standby power generators,
5. Foaming trailers,
6. TV/grout rigs,
7. Compressors,
8. Trailer-mounted engine-driven pumps, and
9. Dump trucks and backhoes.

Not only is the type of vehicle varied, but also the driving conditions we frequently work in are varied. Most of our emergency situations occur when conditions are less than ideal so darkness, rain, cold, wind, fog, snow, ice, or other weather conditions affect our ability to drive safely. Even when conditions are ideal, the equipment and traffic conditions frequently require extraordinary driving skills just to get to the job site.

Once at the site, the crew may be assigned one or more of the following tasks:

1. Working in a confined space such as a manhole, a metering vault, or a pump station wet well;
2. Working on or around electrical equipment found in pump station motor control centers;
3. Working on mechanical equipment such as pumps, bar screens, or compressors;
4. Working with hydraulic or pneumatic systems when installing a flow measurement weir or a bag plug to stop or control the flow;
5. Sampling industrial waste downstream from an industrial discharger;
6. Cleaning sewers;
7. Unplugging stopped up sewers;
8. Repairing broken or damaged sewers;
9. Working in a roadway where traffic control is required; and
10. Doing construction work involving trenching and shoring in or around other utilities (natural gas lines, phone, and electrical utilities).

These are just a few examples that illustrate the range of equipment and jobs we must be familiar with.

OBJECTIVE

The objective of this chapter on *SAFE PROCEDURES* is to provide the collection system operator with insight into many of the actual hazards that exist in performing our everyday routine tasks. As always, awareness, knowledge, and a commitment to safety/survival, in addition to the technical skills required, will assist you in performing your assignments in a safe and professional manner. Specific sections in this chapter deal with the following topics:

1. Vehicles,
2. Driving,
3. Traffic control,
4. Confined spaces,
5. Electrical safety,
6. Noise,
7. Fire, and
8. Worker "Right-To-Know" laws.

In some cases, detailed procedures are included. Keep in mind, however, that these procedures may not cover all situations and, therefore, your own common sense is required. If state and local regulations and standards in your area require different procedures, you and your agency have a responsibility to develop and use other appropriate procedures. Confirm specific requirements for your area with the local safety agency. Carefully analyze each problem or regulation and develop appropriate procedures to meet those specific needs. Plan the way each job can be performed safely and you will not become an injury/fatality statistic to be included in the distressing statistics on the safety record of our profession as shown in Volume II, Chapter 11.

REMEMBER

SAFE OPERATIONS ALWAYS TAKE PRECEDENCE OVER EXPEDIENCY OR SHORTCUTS.

4.1 VEHICLE AND EQUIPMENT INSPECTION PROCEDURES

Most of the tasks that we do in our everyday life require preparation, regardless of how routine or mundane the task. For example, going to the grocery store requires that your car will start, that it will have enough gas, that you have the shopping list and the checkbook, and that you allocate a certain amount of time to accomplish the task of going to the store. Human nature being what it is, we typically will try to substitute a modified plan when we have used poor planning.

In our collection system business, the modifications and shortcuts that result from not following established procedures can have serious consequences. Because of the hazardous nature of our jobs, poor planning increases the risk of an injury and/or fatality to ourselves and co-workers. Therefore, one of the most critical aspects of your ability to perform your job in a professional manner occurs before you even leave the maintenance yard or shop. You and your agency should develop and follow established procedures for every collection system operation or activity.

Let us assume that today you are going to be working in a manhole and that access to the manhole is in the street. The collection agency's standard procedures require completion of the following activities before the crew leaves the maintenance yard:

1. Discuss the work assignments for the day with your colleagues, co-workers, and supervisors.
2. Determine equipment needed; for example, atmospheric testing devices, traffic control devices, material handling devices such as slings, and hand tools not normally carried on your utility vehicles. Do you need any specialized equipment?
3. Inspect each piece of equipment you will be using to ensure it is working. It doesn't make sense to have atmospheric testing devices available if they are not calibrated or they are inoperative. Your inclination may be to short-cut procedures once in the field. Don't do this; it will compromise you and your co-workers' safety. If there is any doubt about the functional capacity of an item, use replacement equipment while the defective or suspected equipment is repaired or further tested.
4. Inspect vehicles and towed equipment:
   a. Remember that most of our equipment is large/oversized/specialized equipment; therefore, procedures must be established to ensure safety and driveability once on the road. Driving equipment day in and day out may cause us to lose sight of the fact that this equipment has the following characteristics:
      1. Limited visibility,
      2. Reduced braking effectiveness,
      3. Top heavy (tankers),
      4. It may not be suitable for high-speed highway driving,
      5. Reduced maneuverability, such as the inability to perform quick maneuvers to avoid an accident,
      6. May have special requirements to prevent or accommodate load shifting,
      7. Difficult to handle under unpredictable events such as a tire blowout, and
      8. Susceptible to reduced control under icy, wet, or high-wind conditions.

   b. As part of the procedure before leaving the yard, the vehicle and any towed equipment should undergo a thorough mechanical/safety item inspection including:
      1. Mirrors and windows,
      2. Lighting system,
      3. Brakes,
      4. Tires,
      5. Trailer hitch/safety chain,
      6. Towed vehicle lighting system, and
      7. Auxiliary equipment such as winches or hoists.

Once your equipment and vehicle have been checked out, you and your crew are ready to depart the maintenance area. Ironically, many vehicle accidents occur before the crew is even out of the yard, due to the unusual nature of the equipment they are driving. Vehicles used in limited visibility situations should be equipped with a backup alarm that sounds when the vehicle is placed in reverse to alert personnel in the area that the vehicle is backing up. Congestion in the yard, as well as the amount of activity with other vehicles and crews, increases the hazard of an accident occurring at this time. An even better system, since most crews are two-person crews, is to have one person behind the vehicle to guide and warn the driver.

It is a good idea to weigh your loaded vehicle. You are responsible for making sure that the vehicle is not overloaded. Here are some definitions you should know:

Gross Vehicle Weight (GVW) The total weight of a single vehicle including its load.

Gross Vehicle Weight Rating (GVWR) The maximum weight rating specified by the manufacturer for a single vehicle including its load.

Depending on how vehicles are loaded and equipped, they will have varying payload allowances. Weighing is easily accomplished for a very nominal fee at a large commercial truck scale. In the event that you are involved in an accident with an overweight vehicle, you could be subject to criminal or civil charges.

Your state vehicle code and/or Department of Motor Vehicles may also require a special operator's license with endorsements to drive some of your maintenance vehicles. Confirm the requirements for your area.

4.2 DEFENSIVE DRIVING

Having made it safely out of the maintenance yard, the next hazard we face is getting to the job site safely, usually during rush hour and with heavy vehicle traffic conditions. Don't forget to *FASTEN YOUR SEAT BELTS.* Many states now have seat belt laws, and most agencies have seat belt policies. (See Volume II, Chapter 11, Section 11.8, "Safety/Survival Program Policies," for a typical Seat Belt Policy.)

Because collection systems are spread out, operators drive more miles annually on the job than the average person's total miles driven during a year. Therefore, defensive driving is an important factor in our daily routine.

Defensive driving is not only a mature attitude toward driving, but a strategy as well. You must always be on the alert for the large number of new drivers, inattentive drivers, sleepy or sleeping drivers, people driving under the influence of alcohol and/or drugs, and those drivers who are simply incompetent. In addition, weather, construction, and the type of equipment you are driving can also present hazards. Some of the elements of defensive driving are:

1. Always be aware of what is going on ahead of you, behind you, and on both sides of you;
2. Always have your vehicle under control;
3. Be willing to surrender your legal right-of-way if it might prevent an accident;
4. Know the limitations of the vehicle and towed equipment you are operating (stopping distance, impaired road vision, distance to change lanes safely, acceleration rate);
5. Take into consideration weather or other unusual conditions;
6. Develop a defensive driving attitude, that is, be aware of the situations that you are driving in and develop options to avoid accidents. For example, in residential areas you should assume that a child could run out into the street at any time, so always be thinking of what type of evasive action you could take. Be especially alert driving through construction areas during rush hour;
7. Rear end collisions, either due to driver inattentiveness or tailgating, are one of the most frequent types of vehicle accidents. Figure 4.1 illustrates the distance required for a vehicle to stop with ideal conditions and good brakes at various speeds, including the amount of time it takes for your brain to react. These figures are for passenger cars and, obviously, increase significantly with most of the equipment that we operate on the road; and
8. Keep in mind that you are driving a vehicle that is highly visible to the public. It may be necessary to swallow your pride when confronted by discourteous or unsafe drivers. Even when in the wrong, those types of drivers are quick to alert your agency, and thus your supervisor, to real or imagined violations you committed while driving an agency vehicle.

Formal defensive driving programs are offered by many public and private agencies and are one of the special programs that should be incorporated into your Collection System Safety/Survival Program (see Volume II, Chapter 11). These programs range from formal classroom training to actual defensive driving on a closed course using the vehicle you drive on your job every day. These courses are not only interesting and informative, but can be fun as well. Suggested contacts in your area for defensive driving opportunities include:

1. State highway patrol,
2. Fire departments and local public safety (police) agencies,
3. Private trucking companies,
4. Private bus companies, and
5. Public transportation companies.

For a nominal fee, these organizations frequently will allow you and your co-workers to participate in their defensive driving programs.

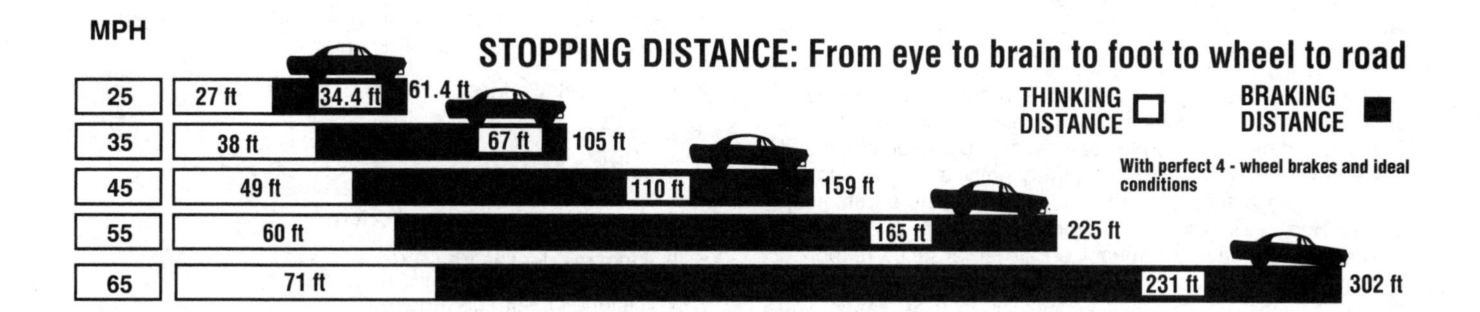

Fig. 4.1 Always maintain a safe stopping distance behind the vehicle in front of you

(Courtesy of California Highway Patrol)

QUESTIONS

Write your answers in a notebook and then compare your answers with those on page 176.

4.0A List the different types of vehicles, or combinations of vehicles and towed equipment, that a collection system operator may be expected to drive.

4.1A Why must collection system operators follow established procedures?

4.1B What tasks should be performed *BEFORE* leaving the maintenance yard or shop?

4.1C What should an operator look for when checking out an atmospheric testing device before taking it out to a job site?

4.2A What is defensive driving?

4.3 ROUTING TRAFFIC AROUND THE JOB SITE

4.30 Need for Traffic Control

The primary function of streets and highways is to provide for the movement of traffic. A common secondary use within the right-of-way of streets or highways is for the placement of public and private utilities such as sanitary sewers. While the movement of traffic is very important, streets need to be constructed, reconstructed, or maintained, and utility facilities need to be repaired, modified, or expanded. Consequently, traffic movements and street or utility repair work must be regulated to provide optimum safety and convenience for all.

Working in a roadway represents a significant hazard to a collection system operator as well as pedestrians and drivers. Motor vehicle drivers can be observed reading books, files, and newspapers, shaving and applying makeup, talking on cellular phones, and changing tapes, CDs, or radio stations (and using headsets to listen to them) rather than concentrating on driving. At any given time of the night or day, a certain percentage of drivers can be expected to be driving while under the influence of drugs or alcohol. Given the amount of time collection system operators work in traffic while performing inspection, cleaning, rehabilitation, and repairs, the control of traffic is necessary if we want to reduce the risk of injury or death while working in this hazardous area. The purpose of traffic control is to provide safe and effective work areas and to warn, control, protect, and expedite vehicular and pedestrian traffic. This can be accomplished by appropriate and prudent use of traffic control devices.

Most states, counties, and cities have adopted regulations to control traffic and reduce the risk under different circumstances. This section illustrates examples of traffic control which may or may not meet the specific requirements of the laws in your geographical area, but should serve to make you aware of various aspects of traffic control.

Any time traffic may be affected, appropriate authorities in your area must be notified before leaving for the job site. These could be state, county, or local depending on whether it is a state, county, or local street. Frequently, a permit must be issued by the authority that has jurisdiction before traffic can be diverted or disrupted. In some cases, traffic diversion or disruption may have an impact on the emergency response system in your area, such as access by fire or police, and so these agencies may be involved as well. In most cases, you will need to plan ahead to secure permits and notify authorities. This may mean only a phone call or two or it could mean several days' or weeks' advance planning if you need to make extensive traffic control arrangements.

Upon arrival at the job site, look for a safe place to park vehicles. If they must be parked in the street to do the job, route traffic around the job site *BEFORE* parking vehicles in the street. If practical, park vehicles between oncoming traffic and the job site to serve as a warning barricade and to discourage reckless drivers from plowing into operators. Remember—you need protection from the drivers as well!

Traffic must be warned of your presence in the street. "CREW WORKING IN STREET" and "CAUTION, CONSTRUCTION WORK" signs are effective. Signs with flags or flashers and vehicles with rotating flashing lights are used to warn other motorists (Figures 4.2, 4.3, and 4.4). Vehicle-mounted traffic guides are also helpful (Figure 4.5). Use flaggers to alert drivers and to direct traffic around the work site (Figure 4.6). Warning signs and flaggers must be located far enough in advance of the work area to allow motorists time to realize they must slow down, be alert for activity, and safely change lanes or follow a detour. Exact distances and the nature of the advance warning depend on traffic speed, congestion, roadway conditions, and local regulations.

Once motorists have been warned, they must be safely routed around the job site. Traffic routing includes the use of flaggers and directional signs. Properly placed barricades and traffic cones can effectively channel traffic from one lane to another around the job site. Retain a traffic officer to direct traffic if streets are narrow or if there is considerable traffic or congestion.

When work is not in progress or the hazard no longer exists, promptly remove, fold, or cover the traffic control devices. Drivers quickly begin to ignore traffic control signs when it becomes obvious no work is in progress, and they assume construction hazards exist only when crews are on the job.

Answers to several questions will determine how traffic control should be accomplished.

1. Is traffic moving at a low speed (0 – 35 MPH or 0 – 56 km/hr) or a high speed (40 – 55 MPH or 64 – 88 km/hr)?
2. Is the street two-lane, one-way, or two-way?
3. Is it undivided four-lane?
4. Is it multi-lane one-way?
5. Are pedestrian walkways affected?
6. Is it in a residential area?
7. Will a lane closure be required?
8. Will more than one lane be closed?
9. Will traffic control be required during peak traffic periods or at night?

Fig. 4.2 Signs warning traffic (advance warning area)

Fig. 4.3 Barricades and traffic cones (transition area)

Fig. 4.4 High flags, barricades, and cones routing traffic
(Courtesy of Water Quality Division, Sacramento County)

Fig. 4.5 Vehicle-mounted traffic guide
(Permission of Traffic Guide)

Fig. 4.6 Alerting and directing traffic
(Courtesy of the Bureau of Contract Administration, City of Los Angeles)

Because of the need for consistency when traveling from one city, town, or state to another, most states have developed work zone traffic controls based on the U.S. Department of Transportation's *MANUAL ON UNIFORM TRAFFIC CONTROL DEVICES* (see Section 4.39, "Additional Reading," reference 1). The information that follows in Section 4.31 through Section 4.35 on work zone traffic control was adapted from a state handbook. Check your local and state laws and regulations for traffic control guidelines in specific circumstances and incorporate these control requirements into your agency's procedures.

4.31 Definitions Of Terms

The following definitions apply to work zone traffic control:

Short-term Work Any work on a street or highway where it is anticipated the activity will take one work shift (typically 6 to 12 hours or less) to complete. This time period may be modified by written approval of the governing road authority.

Intermediate-term Work Any work on a street or highway that occupies a location from overnight to three days.

Street or Highway The entire width between boundary lines of any way or place that is open to the use of the public, as a matter of right, for the purpose of vehicular traffic.

Low-speed Street or Highway Any street or highway where the speed limit is 35 miles per hour or less.

High-speed Street or Highway Any street or highway where the speed limit is 40 miles per hour (64 km/hr) or greater.

Low-speed Residential Street or Highway Any low-speed street or highway that serves only residential areas where through traffic is deliberately discouraged.

Low-speed Collector or Arterial Street or Highway Any low-speed street or highway that serves as residential and business access and is primarily used to collect traffic from residential streets and channel it into the arterial system, or any low-speed street or highway that permits travel between communities and serves as a through facility.

Traffic Control Plan (TCP) A TCP describes traffic controls to be used for facilitating all traffic through a temporary traffic control zone. The degree of detail in the TCP depends entirely on the complexity of the job site. Persons knowledgeable about the principles of temporary traffic controls should prepare TCPs.

Traffic Control Zone The area from the point where traffic is first affected by the activity and/or traffic controls to a point where traffic is no longer affected. Typically this is the distance between the first advance warning sign and the point beyond the work area where traffic is no longer affected.

Stationary Traffic Control Zone Any traffic control zone that remains in one place for longer than 15 minutes.

Mobile Traffic Control Zone Any traffic control zone where the traffic control zone remains in one place for less than 15 minutes.

Moving Traffic Control Zone Any traffic control zone that is continuously moving.

Special Traffic Control Zone Any traffic control zone where the workers are performing tasks with little or no interference to traffic. Generally the presence of the vehicle and worker should not "surprise" the driver or cause any erratic maneuvers.

Low-volume Street or Highway Any street or highway where the ADT (Average Daily Traffic) is less than 1,500 vehicles.

Decision Sight Distance The distance required by a driver to properly react to hazardous, unusual, or unexpected situations even where an evasive maneuver is more desirable than a hurried stop.

Good Visibility Location Any location where the sight distance to the work area is sufficient to meet decision sight distance.

QUESTIONS

Write your answers in a notebook and then compare your answers with those on page 177.

4.3A What is the purpose of traffic control around work areas in streets and highways?

4.3B Who should be contacted before setting up a work site in a street?

4.3C How can traffic be warned of your presence in a street?

4.3D How can motorists be safely routed around a job site?

4.3E What area at a work site is considered a "traffic control zone"?

4.3F What is the meaning of the term "decision sight distance"?

4.32 Individuals Qualified to Control Traffic

Each person whose actions affect traffic zone safety during maintenance, construction, utility, and incident management should receive training appropriate to the job decisions they will be required to make. Individuals become qualified to control traffic by gaining the following knowledge and experience:

- A basic understanding of the principles of traffic control in work zones,
- Knowledge of the standards and regulations governing traffic control,
- Adequate training in safe traffic control practices, and
- Experience in applying traffic control in work zones.

Only qualified individuals should supervise the selection, placement, and maintenance of traffic control devices in work and incident management areas.

4.33 Permission to Work Within the Right-of-Way of Streets or Highways

Any work in public streets or highways must be regulated to ensure proper coordination of the work, thus protecting the public's interest. To accomplish this, any person, firm, corporation, or agency must obtain permission from the governing road authority before starting work within the right-of-way of any street or highway. Regulatory traffic control devices or signs must be approved by the governing road authority before they are installed on any street or highway.

When working in or near an intersection where traffic is controlled by a signal light, the owner of the signal must be notified before work begins. This is necessary to ensure the proper operation of the signal while the work is in progress. Also, the signals may be used to control traffic in the vicinity of the work area.

4.34 General Responsibilities

In most areas of the country, any agency performing work within the right-of-way of streets or highways is responsible for:

1. Supplying, installing, and maintaining all necessary traffic control devices as required by the governing road authority to protect the work area and safely direct traffic around the work zone,
2. Supplying their own flagger(s) when required,
3. Informing occupants of abutting properties, either orally or by written notice, of parking prohibitions or access limitations,
4. Notifying the governing road authority when existing traffic signs need to be removed or relocated or any regulatory sign must be installed for construction or maintenance work,
5. Replacing or reimbursing the governing road authority for any damage to or loss of existing traffic signs,
6. Keeping all traffic control devices clean and in proper position to ensure optimum effectiveness,
7. Removing traffic control equipment when it is no longer required or appropriate, and
8. Keeping proper records of traffic control that contain starting and ending times, location, names of personnel, and traffic controls used.

4.35 Regulations Concerning All Street or Highway Work

4.350 Time of Work

The governing road authority may determine or define the times when work may be performed. During peak traffic periods, construction work may be restricted or not permitted. Peak periods of traffic movement may vary in different areas.

The governing road authority may require work to be performed at night or on weekends when the location of the work within the roadway is considered critical from a traffic standpoint. Some examples of these locations include the vicinity of major signalized intersections, shopping centers, and downtown areas.

When work is planned for nighttime hours, the traffic controls should be modified to adequately control traffic through the work zone. Some items that should be considered for nighttime activities are the addition of warning lights to traffic control devices and signs, use of larger, more reflective channelizing devices such as drums, use of floodlighting for any flagger stations and for the work area, and addition of reflective devices to work vehicles that are typically used for nighttime work.

4.351 Specific Situations

Situations may arise where the typical traffic control guidelines do not apply and a special traffic control plan is needed. This special control plan should be developed by people with traffic engineering expertise.

Although it is usually desirable to set up traffic controls as shown in the example layouts on pages 100 through 105, situations sometimes arise where this becomes impractical and modifications to the typical layouts are needed. When modifications are made, factors such as traffic volume, speed, sight distance, and type of work must be considered.

Situations may occur where consideration should be given to using a stationary zone where a mobile zone normally would be indicated. If the tasks to be performed require that the work area remain in one place less than 15 minutes, but several individual work areas are concentrated within a relatively short section of roadway, it would be desirable to handle this situation as a stationary zone. An example of this type of work would be the repair of many potholes on a long section of roadway.

4.352 Partial Street or Highway Closures

Unless a section of street is to be completely closed to vehicular traffic, the work needs to be done so that as few traffic lanes as practical are blocked. Where a traffic lane is blocked, parking prohibition(s) may be necessary to permit continued traffic movement.

The governing road authority may require that excavations and cuts be properly backfilled, bridged, or plated to allow the passage of vehicles. The governing road authority may allow an exception to these requirements when traffic volumes or other special conditions warrant such action. All *SPOIL*[1] material from the excavation is to be removed from the pavement surface when bridges or plates are used. All bridging or plating needs to be of sufficient strength to accommodate all legal traffic loads. The plates need to be secured in place so they will not slip out of position during use.

[1] *Spoil. Excavated material such as soil from the trench of a sewer.*

4.353 Complete Street or Highway Closures and Detours

Permission needs to be obtained from the governing road authority for all street and highway closures. When detours are necessary and authorized by the governing road authority, they must be installed in accordance with the local regulations.

4.354 Emergency Situations

An emergency situation may arise where immediate action to protect the safety of the general public requires work to be done on a street or highway even though full compliance with the local regulations cannot be immediately provided. In this case, proper traffic control must be provided as soon as possible.

QUESTIONS

Write your answers in a notebook and then compare your answers with those on page 177.

4.3G How could an operator become qualified to control traffic?

4.3H Records of traffic control activities should contain what types of information?

4.3I When collection system work is scheduled for nighttime hours, what additional traffic control safety measures should be considered?

4.36 Traffic Control Zones

When traffic is affected by construction, maintenance, or utility activities, traffic control is needed to safely guide and protect motorists, pedestrians, and operators in a traffic control zone. The traffic control zone is the area between the first advance warning sign and the point beyond the work area where traffic is no longer affected.

Most traffic control zones can be divided into these specific areas:

1. Advance warning area,
2. Transition area,
3. Buffer space,
4. Work area, and
5. Termination area.

Figure 4.7 illustrates these five parts of a traffic control zone. The following paragraphs discuss each of the five parts for one direction of travel. If the work activity affects more than one direction of travel, the same principles apply to traffic in all directions.

4.360 Advance Warning Area

An advance warning area is necessary for all traffic control zones because drivers need to know what to expect. Before reaching the work area, drivers should have enough time to alter their driving patterns. The advance warning area may vary from a series of signs starting a mile in advance of the work area to a single sign or flashing lights on a vehicle.

When the work area, including access to the work area, is entirely off the shoulder and the work does not interfere with traffic, an advance warning sign may not be needed. However, an advance warning sign should be used when any problems or conflicts with the flow of traffic are anticipated.

The advance warning area, from the first sign to the start of the next area, should be long enough to give the motorists adequate time to respond to the conditions. For most activities the length can be:

- One-half mile (0.8 km) to one mile (1.6 km) for freeways or expressways,
- 1,500 feet (450 m) for most other roadways or highway conditions, and
- At least one block for urban streets.

Available decision time should allow a driver time to detect a hazard in a cluttered environment, recognize it and its potential hazard, and select the appropriate speed and path. If there is only enough time to stop the vehicle after seeing a hazard, the driver usually will not have time to perform an evasive maneuver, which is often a safer action. Therefore the recommended decision sight distance should be provided when the following conditions exist:

- When the driver needs to process relatively complex information,
- When the hazard is difficult to perceive,
- When unexpected or unusual maneuvers are required, and
- Where an evasive maneuver is more desirable than a hurried stop.

This decision sight distance is the sum of perception time, reaction time, and vehicle maneuver time. In other words, it is the time it takes a driver to recognize the hazard plus the time it takes to react plus the time for the vehicle to actually respond to the driver's reaction. Typical stopping distances are shown in Table 4.1 and typical decision sight distances for use in short-term work zones are shown in Table 4.2.

TABLE 4.1 STOPPING DISTANCES[a]

Speed (MPH)[b]	Driver Reaction Distance (feet)[c]	Braking Distance (feet)[c]	Total Stopping Distance (feet)[c]
20	22 (20 + 2)	18 – 22	40 – 44
40	44 (40 + 4)	64 – 80	108 – 124
55	60 (55 + 5)	132 – 165	192 – 225
65	71 (65 + 6)	160 – 224	231 – 295

[a] Source: State of California, Department of Motor Vehicles Hand Book
[b] Multiply MPH x 1.6 to obtain km/hr.
[c] Multiply ft x 0.3 to obtain meters.

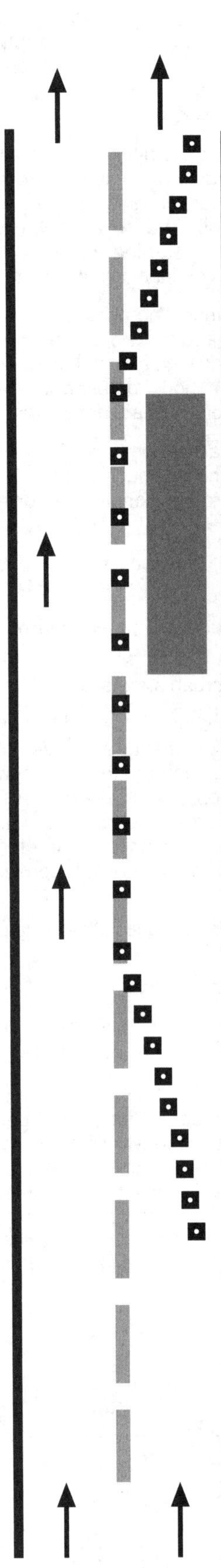

TERMINATION AREA

The termination area provides a short distance for traffic to clear the work area and to return to the normal traffic lanes. A downstream taper may be placed in the termination area. A downstream taper is used at the downstream end of the work area to indicate to drivers that they can move back into the lane that was closed.

WORK AREA

The work area is that portion of the roadway which contains the work activity and is closed to traffic and set aside for exclusive use by workers, equipment, and construction materials. Work areas may remain in fixed locations or may move as work progresses. The work area is usually delineated by channelizing devices or shielded by barriers.

BUFFER SPACE

The buffer space is the open or unoccupied space between the transition and work areas. With a mobile traffic control zone, the buffer space is the space between the shadow vehicle, if one is used, and the work vehicle. The buffer space provides recovery space for an out-of-control vehicle. Neither work activity nor storage of equipment, vehicles, or material should occur in this space.

TRANSITION AREA

When work is performed within one or more traveled lanes, a lane closure(s) is required. In the transition area, traffic is channelized from the normal highway lanes to the path required to move traffic around the work area. The transition area contains the tapers which are used to close lanes.

ADVANCE WARNING AREA

An advance warning area is necessary for all traffic control zones because drivers need to know what to expect. Before reaching the work area, drivers should have enough time to alter their driving patterns. The advance warning area may vary from a series of signs starting a mile in advance of the work area to a single sign or flashing lights on a vehicle. The true test of whether sign spacing is adequate is to evaluate how much time the driver has to perceive and to react to the conditions ahead.

Fig. 4.7 Parts of a traffic control zone

TABLE 4.2 SUGGESTED DECISION SIGHT DISTANCE

Posted Speed Limit (MPH)[a]	Suggested Decision Sight Distance (feet)[b]
0 – 35	750
40 – 45	950
50 – 55	1,200
60 – 65	1,400

[a] Multiply MPH x 1.6 to obtain km/hr.
[b] Multiply ft x 0.3 to obtain meters.

4.361 Transition Area (Figures 4.3 and 4.8)

When work is performed within one or more traveled lanes, a lane closure(s) is required. (If no lane or shoulder closure is involved, the transition area is not used.) In the transition area, traffic is channelized from the normal highway lanes to the path required to move traffic around the work area. The transition area contains the tapers that are used to close lanes. A taper is a series of channelizing devices and/or pavement markings placed on an angle to move traffic out of its normal path.

Four general types of tapers used in traffic control zones are:

- Lane closure tapers necessary for closing lanes of moving traffic (sometimes referred to as channelizing or merging tapers);
- Two-way traffic tapers needed to control two-way traffic where traffic is required to alternately use a single lane (commonly used when flaggers are present);
- Shoulder closure tapers needed to close shoulder areas; and
- Downstream tapers installed to direct traffic back into its normal path.

Lane Closure Taper. The length of taper used to close a lane is determined by the speed of traffic and, in some locations, the width of the lane to be closed. Table 4.3 lists typical tapering guidelines.

Fig. 4.8 Portable arrows reduce driver confusion in construction area

(Permission of Safety Tech, Inc.)

If sight distance is restricted by obstacles such as sharp vertical or horizontal curves, the taper should begin well in advance of the view obstruction. The beginning of tapers should not be hidden behind curves.

Generally, tapers should be lengthened, not shortened, to increase their effectiveness. Observe traffic to see if the taper is working correctly. Frequent use of brakes and evidence of skid marks may be indications that either the taper is too short or the advance warning is inadequate.

Two-Way Traffic Taper. The two-way traffic taper is used in advance of a work area that occupies part of a two-way road. The taper is placed in such a way that the remainder of the road can be used alternately by vehicles traveling in either direction. A short taper is used to cause traffic to slow down by giving the appearance of restricted alignment and a flagger is usually stationed at the taper to assign the right of way. Table 4.4 illustrates details on these tapers.

Shoulder Closure Taper. When an improved shoulder is closed on a high-speed roadway, it should be treated as a closure of a portion of the roadway which the motorist expects to use in an emergency. The work area on the shoulder should be preceded by a taper that is one-fourth of the length of a lane closure taper, provided the shoulder is not used as a travel lane.

Downstream Taper. The downstream taper is used in terminating areas to provide a visual cue to drivers that access is available to the original lane. The taper should have a maximum length of about 100 feet (30 meters) per lane, with channelizing devices spaced about 20 feet (6 meters) apart.

4.362 Buffer Space

The buffer space is the open or unoccupied space between the transition and work areas. With a mobile traffic control zone, the buffer space is the space between the shadow vehicle, if one is used, and the work vehicle.

The buffer space provides a margin of safety for both traffic and operators. If a driver does not see the advance warning or fails to negotiate the transition, a buffer space provides room

TABLE 4.3 LANE CLOSURE TAPER SPECIFICATIONS[a]

Approach Speed (MPH)[b]	Taper Length (feet)[c]	Number of Cones (for taper)	Cone Spacing Along Taper (feet)[c]
25	125	6	25
30	180	7	30
35	245	8	35
40	320	9	40
45	540	13	45
50	600	13	50
+50	1,000	21	50

[a] Source: State of California, Manual of Traffic Control
[b] Multiply MPH x 1.6 to obtain km/hr.
[c] Multiply ft x 0.3 to obtain meters.

TABLE 4.4 MINIMUM RECOMMENDED DELINEATOR AND SIGN PLACEMENT FOR TRAFFIC CONTROL

Traffic Speed (MPH)[a]	Taper Length Each Lane (feet)[b]	Delineator Spacing		Sign Spacing in Advance of Taper and Between Signs (feet)[b]
		Taper (feet)[b]	Work Area (feet)[b]	
25	150	25	50	150
30	200	30	60	200
35	250	35	70	250
40	350	40	80	350
45	550	45	90	550
50	600	50	100	600
55	1,000	50	100	1,000

[a] Multiply MPH x 1.6 to obtain km/hr.
[b] Multiply ft x 0.3 to obtain meters.

to stop before entering the work area. It is important that the buffer space be free of equipment, operators, materials, and operators' vehicles.

Channelizing devices should be placed along the edge of the buffer space. In general, a spacing of one-half to one times the posted speed limit is recommended on low-speed streets or highways. A spacing of two to four times the posted speed limit in feet is recommended on high-speed streets or highways.

Situations occur where opposing streams of traffic on multi-lane streets or highways are transitioned so one lane of traffic uses a lane that normally flows in the opposite direction. To help prevent head-on collisions, a buffer space should be used to separate the two tapers for opposing directions of traffic.

4.363 Work Area

The work area is that portion of the roadway which contains the work activity and is closed to traffic and set aside for exclusive use by operators, equipment, and construction materials. Work areas may remain in fixed locations or they may move as work progresses. An empty buffer space may be included at the upstream end. The work area is usually delineated by channelizing devices (Table 4.4) or shielded by barriers to exclude traffic and pedestrians.

Conflicts between traffic and the work activity or potential hazards increase when the following conditions occur:

- The work area is close to the traveled lanes,
- Physical deterrents to normal operation exist, such as uneven pavements or vehicle loading and unloading,
- Speed and volume of traffic increase, and
- The change in travel path gets more complex, shifting traffic across the median and into lanes normally used by opposing traffic rather than shifting traffic a few feet.

Work areas that remain set up overnight have a greater need for delineation than daytime activities. Consideration should be given to selecting larger, more reflective devices and use of additional lighting devices for short-term work zones that are in place during nighttime hours.

Traffic control measures for work areas should meet the following guidelines:

- Use traffic control devices to make the work area clearly visible to traffic;
- Place channelizing devices along the edge of the work area. To accomplish this, a spacing of one-half to one times the posted speed limit in feet is recommended on low-speed streets or highways and a spacing of two to four times the posted speed limit in feet is recommended on high-speed streets or highways. This spacing should be reduced or larger channelizing devices should be used if needed to keep vehicles entirely outside the work area;
- Provide a safe entrance and exit for work vehicles;
- Protect mobile traffic control zones with adequate warning devices on the work and/or shadow vehicles; and
- Use flashing lights or flags on work vehicles exposed to traffic.

Avoid gaps in the traffic control that may falsely lead drivers to think they have passed through the work area. For example, if the work area includes intermittent activity throughout a one-mile (1.6-km) section, the drivers should be reminded periodically that they are still in the work area. The primary purpose of the guide sign ROAD CONSTRUCTION NEXT ______ MILES is to inform drivers of the length of the work area. The sign should not be set up until work begins.

4.364 Termination Area

The termination area provides a short distance for traffic to clear the work area and return to the normal traffic lanes. A downstream taper may be placed in the termination area to indicate to drivers that they can move back into the lane that was closed. Closing tapers are optional and may not be advisable when material trucks move into the work area by backing up from the downstream end of the work area. Closing tapers are similar in length and spacing to two-way traffic tapers.

There are occasions where the termination area could include a transition. For example, if a taper was used to shift traffic into opposing lanes around the work area, then the termination area should have a taper to shift traffic back to its normal path. This taper would then be in the transition area for the opposing direction of traffic. It is advisable to use a buffer space between the tapers for opposing traffic.

Pages 100 through 105 show how to position high flags, signs, and cones for flagger control and work under five typical conditions:

1. Work area in center of road,
2. Work beyond intersection,
3. Closing left lane,
4. Closing half of roadway, and
5. Closing right lane.

QUESTIONS

Write your answers in a notebook and then compare your answers with those on page 177.

4.3J List the five different areas of a traffic control zone.

4.3K The decision sight distance should give motorists time to do what?

4.3L What is a traffic control taper?

4.3M What should be done to prevent head-on collisions on a multi-lane street when one lane of traffic uses a lane that normally flows in the opposite direction?

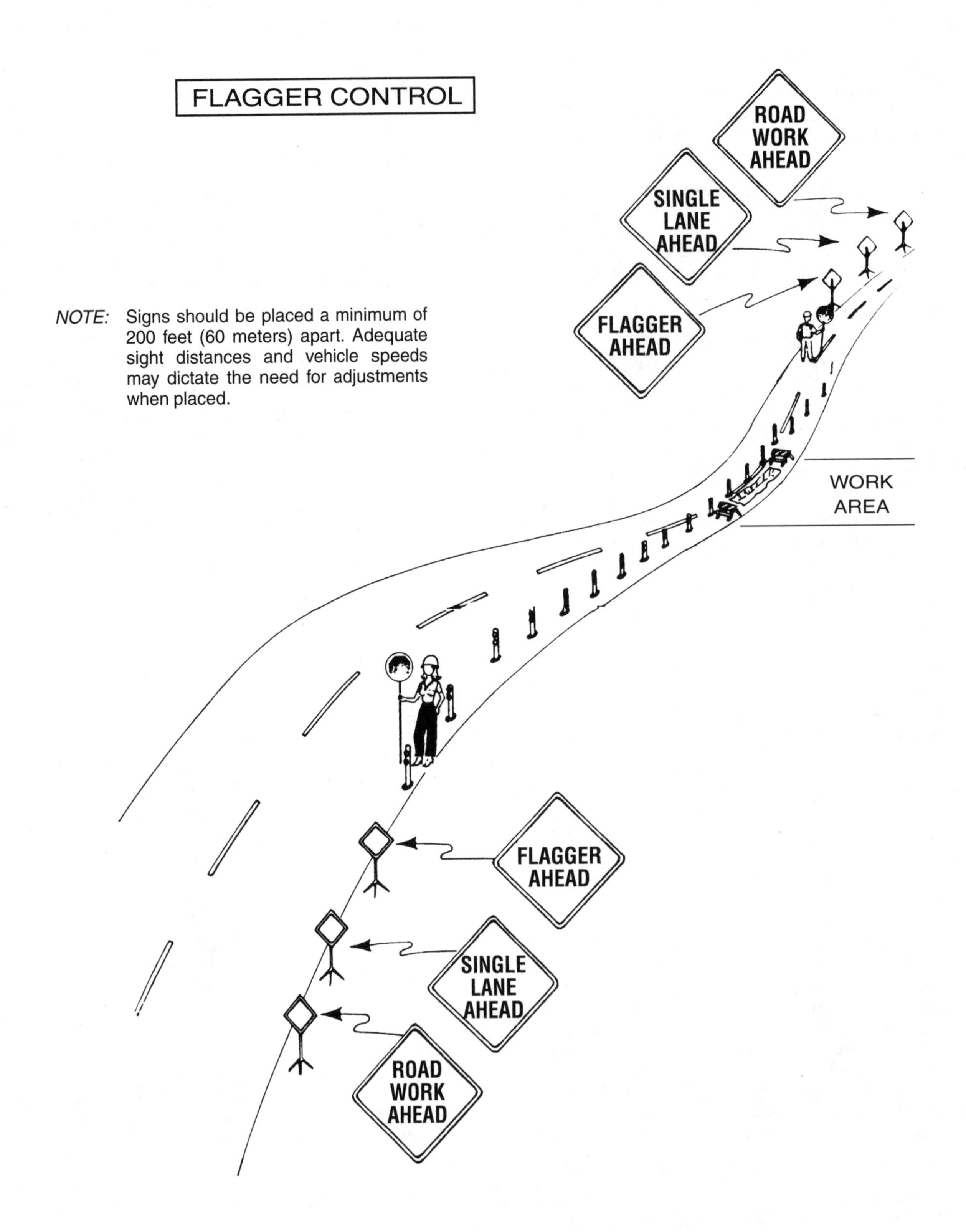

FLAGGER CONTROL

(Courtesy San Diego Chapter, American Public Works Association, illustration modified 1999)

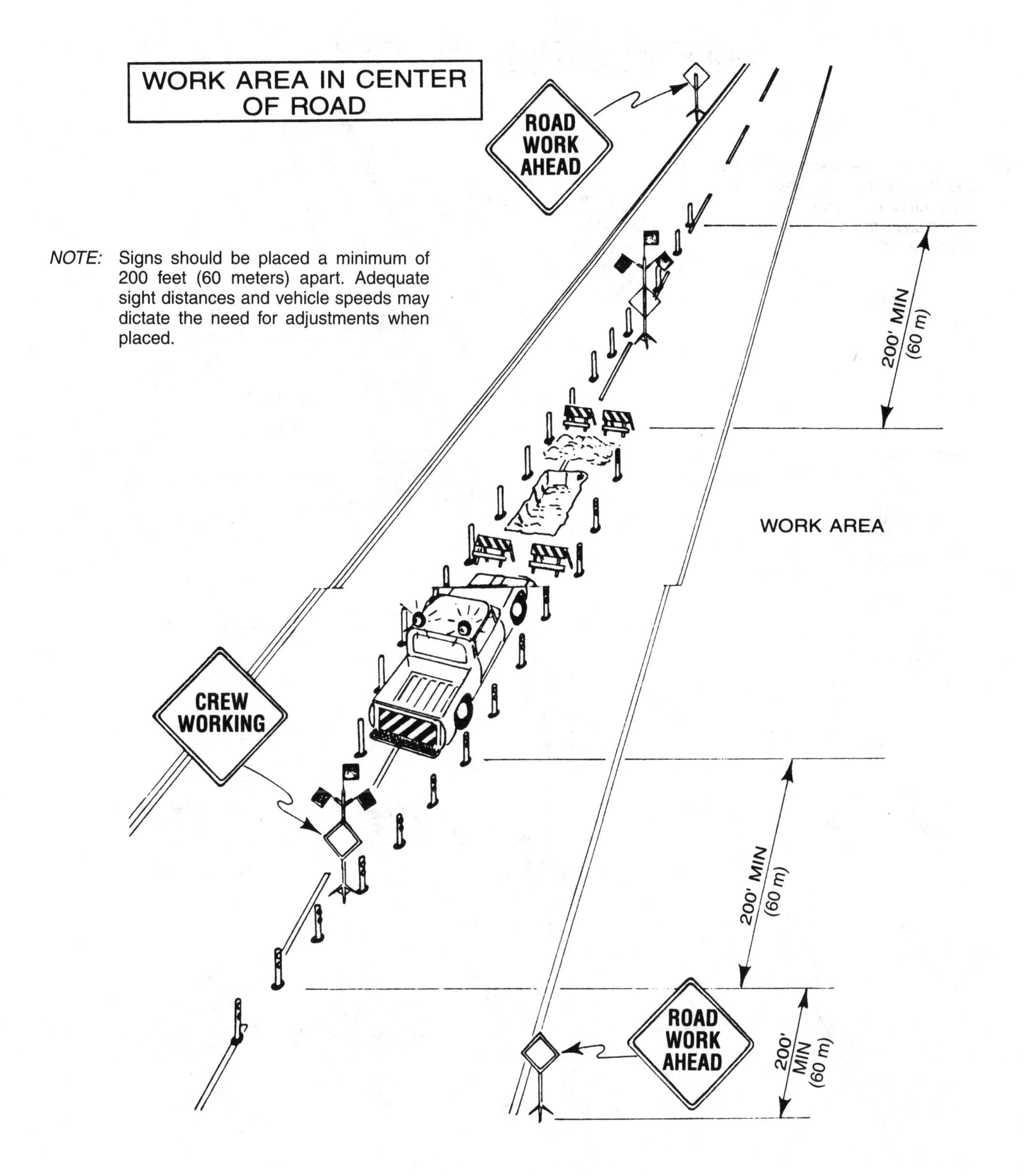

WORK AREA IN CENTER OF ROAD

(Courtesy San Diego Chapter, American Public Works Association, illustration modified 1999)

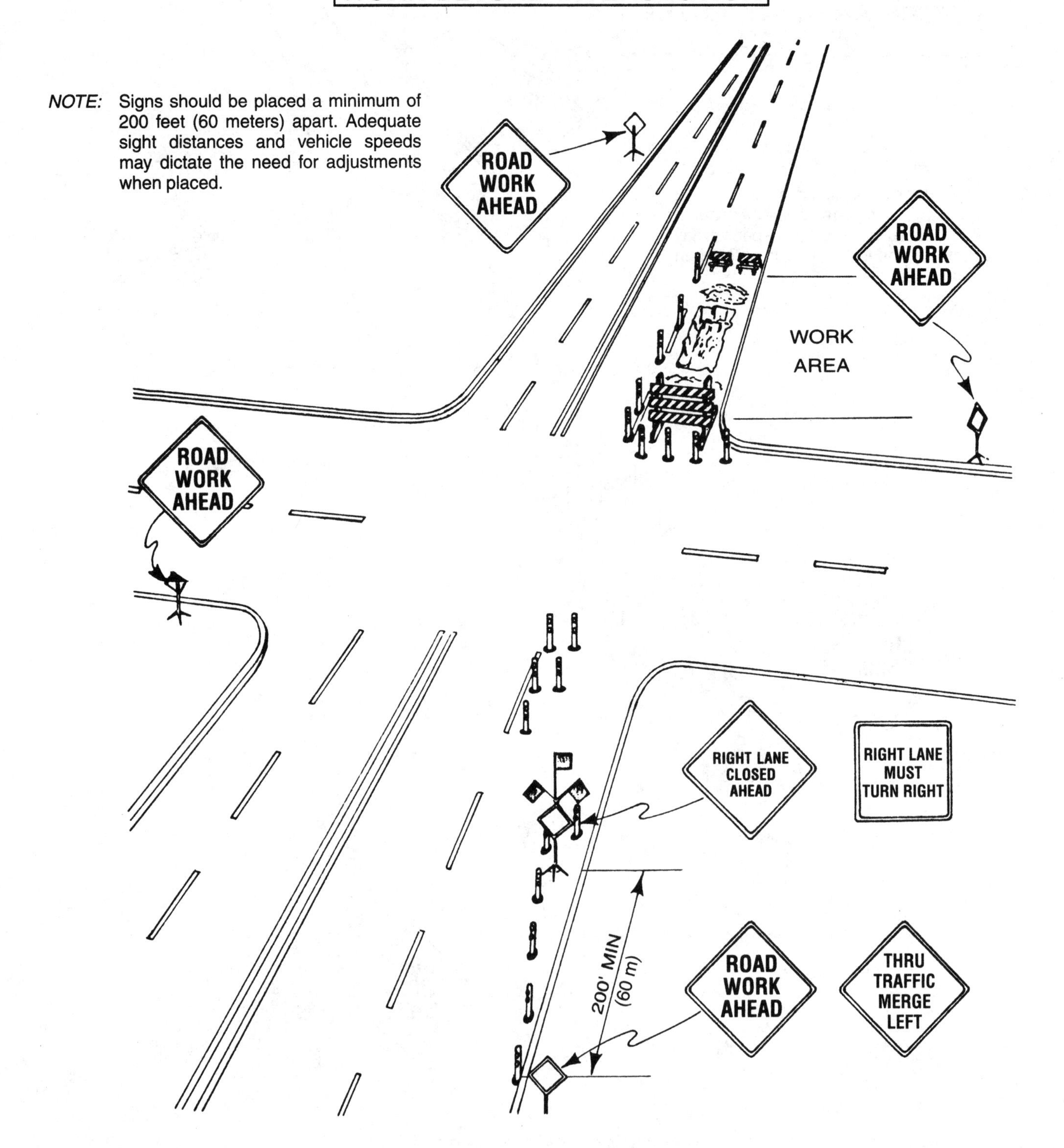

WORK BEYOND INTERSECTION

(Courtesy San Diego Chapter, American Public Works Association, illustration modified 1999)

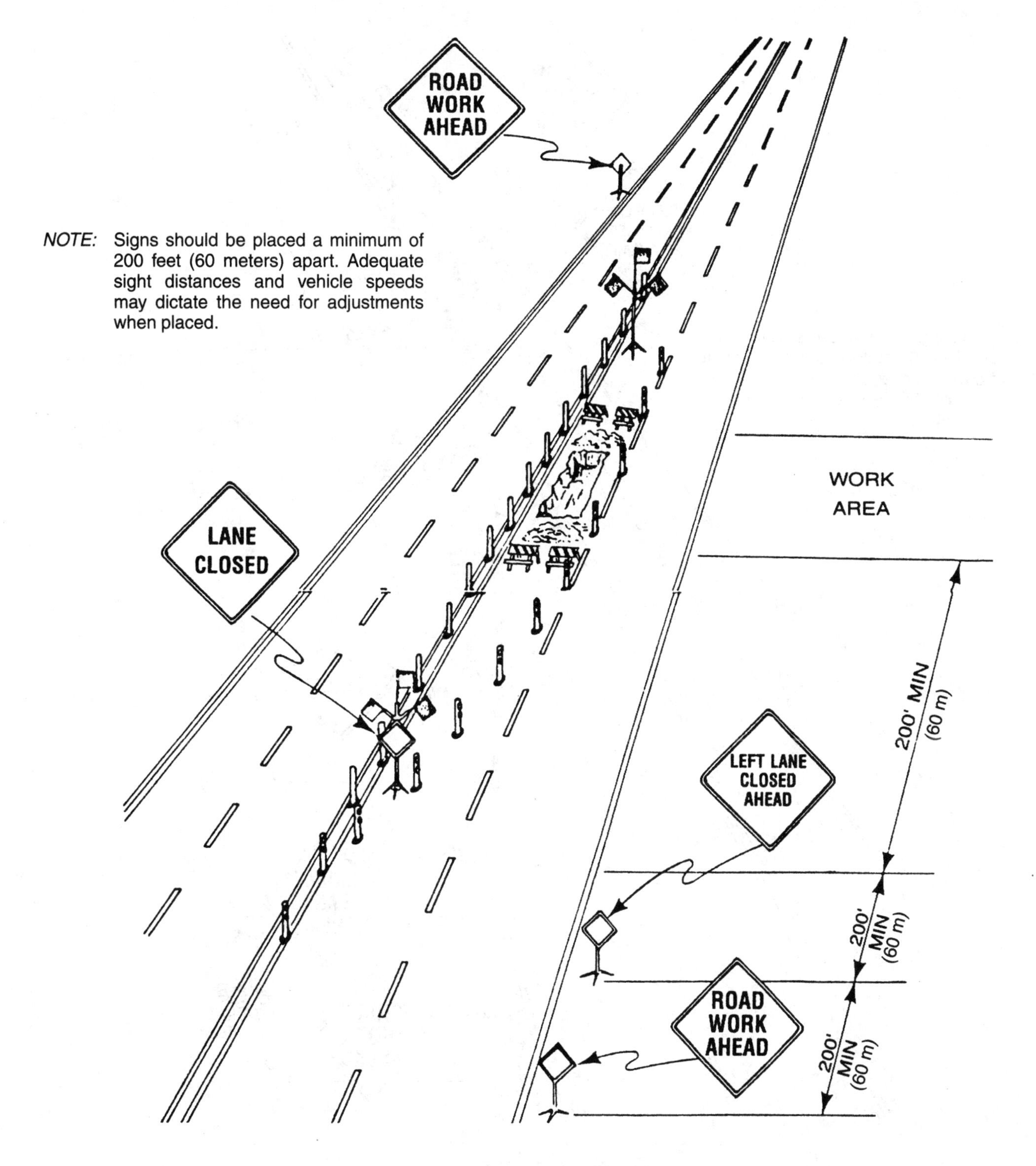

CLOSING LEFT LANE

(Courtesy San Diego Chapter, American Public Works Association, illustration modified 1999)

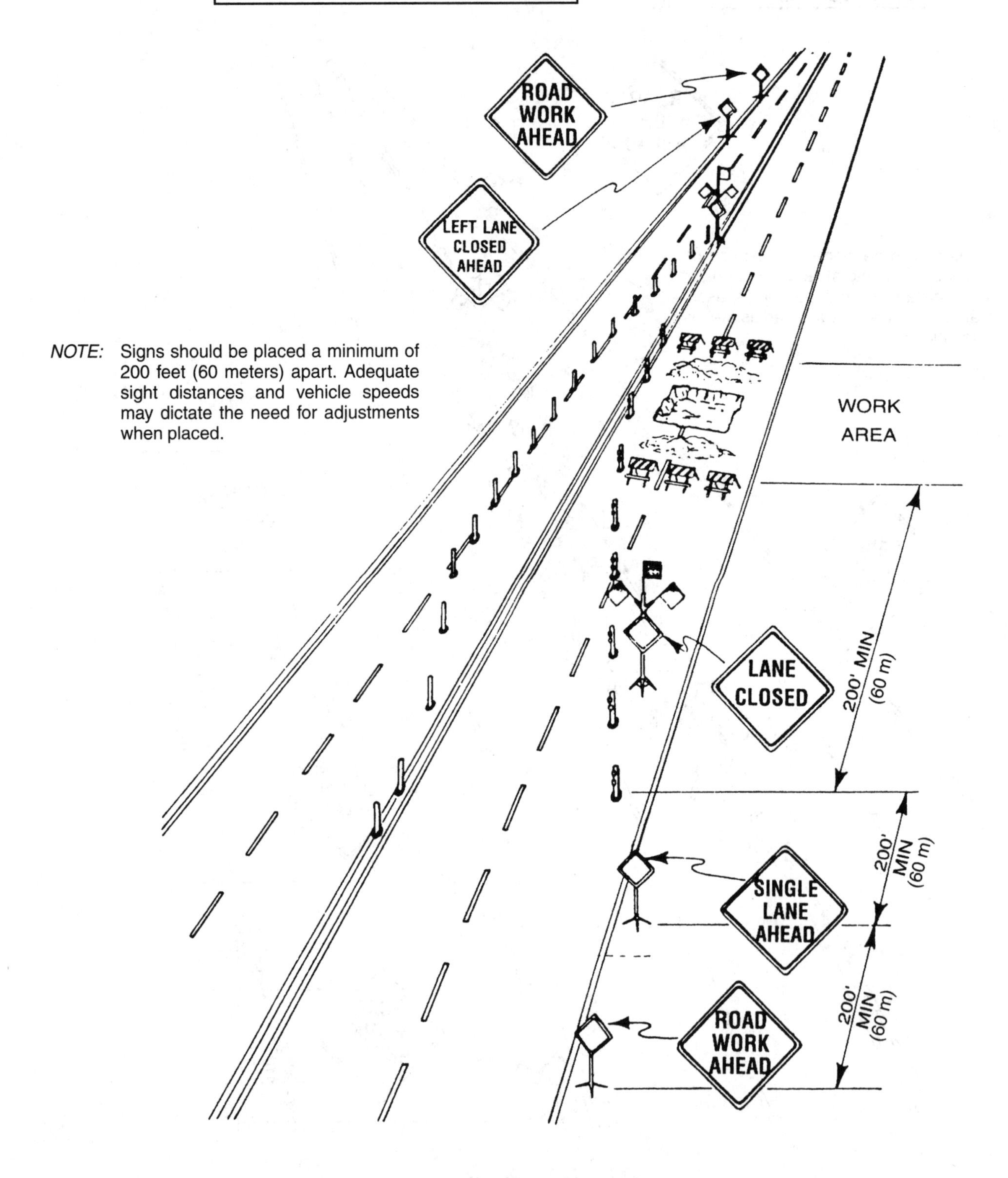

CLOSING HALF-ROADWAY

(Courtesy San Diego Chapter, American Public Works Association, illustration modified 1999)

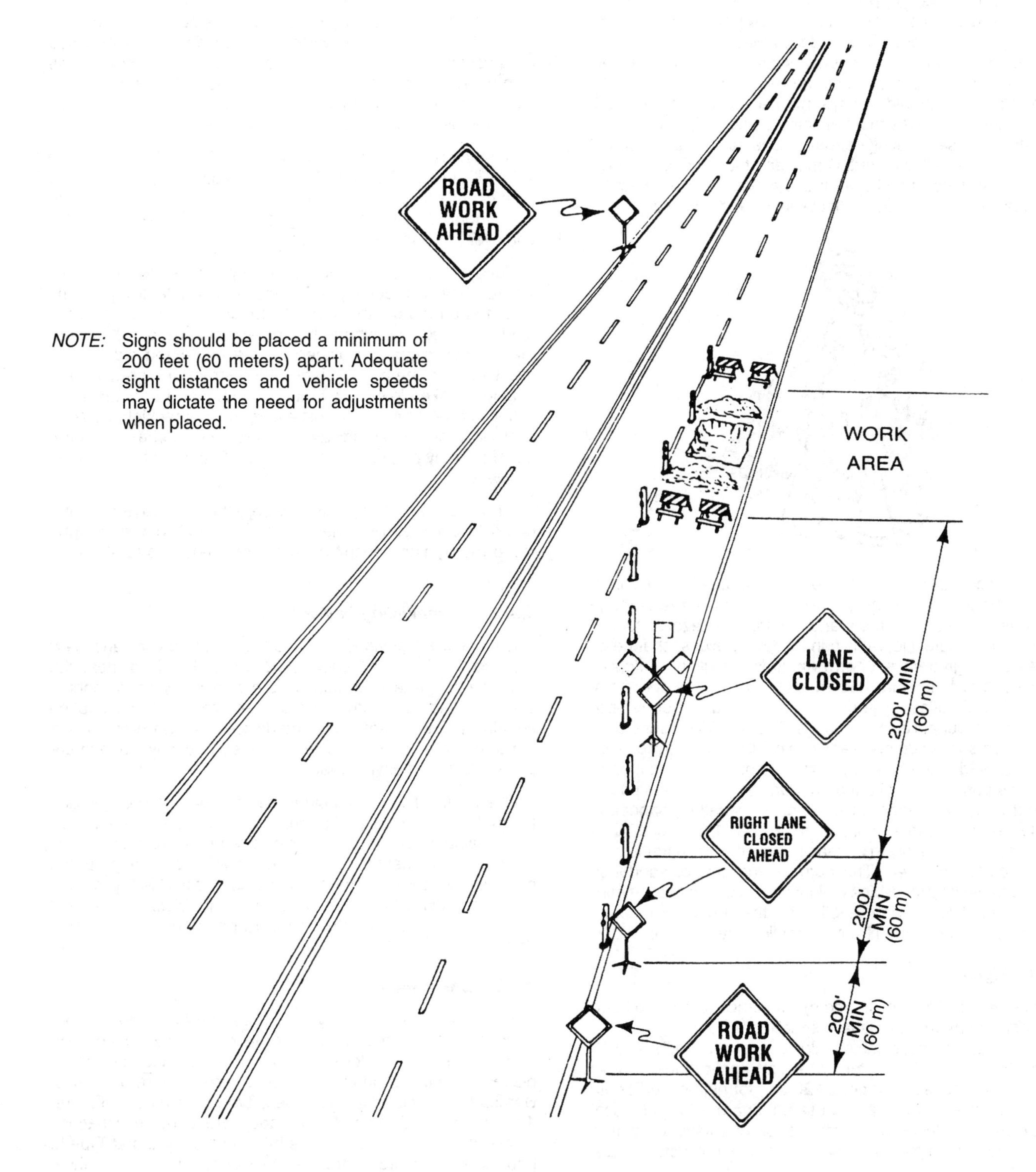

CLOSING RIGHT LANE

(Courtesy San Diego Chapter, American Public Works Association, illustration modified 1999)

4.37 How To Use Traffic Control Devices

4.370 General Requirements

One of the best ways to increase the safety of both drivers and operators at construction sites is by using traffic control devices in a consistent, predictable manner. Drivers learn by experience what to expect when they see traffic control devices. If the devices themselves and their placement in the roadway are consistent over time, drivers can more quickly understand and respond to the changed road conditions. Traffic control devices should always be placed where they will convey their messages most effectively and give drivers adequate time to react. All signs, barricades, drums, and vertical panels must be reflectorized. Cones and tubes need to be reflectorized if used at night. All traffic control devices must be kept clean to ensure proper effectiveness and reflectivity.

All traffic control devices should be constructed to yield upon impact. The purpose of this precaution is to minimize damage to a vehicle that strikes them and to minimize hazards to motorists and workers. No traffic control device (signs, channelizing devices, sign trailers, or arrow boards) should be weighted so heavily that it becomes hazardous to motorists or operators. The approved ballast system for devices mounted on temporary portable supports is sandbags. The sandbags should be constructed so they do not readily rot or allow the sand to leach when exposed to the highway environment. Also, the sandbag should be constructed of a material which will allow the bag to break if struck by a vehicle. During cold weather, sandbags should be filled with a mixture of sand and a deicer. They should not be too heavy to be readily moved when a traffic control device is relocated. The number and size of sandbags used as traffic control device ballast should be kept to the minimum needed to provide stability for the device. Sandbags should not be suspended from the traffic control device.

4.371 Signs

The sizes of construction and maintenance signs also need to be consistent and are usually specified by local regulations. For example, warning signs should be at least 36 inches by 36 inches (90 cm x 90 cm) for low-speed applications and 48 inches by 48 inches (120 cm x 120 cm) for high-speed applications. Smaller signs may be used, if approved by the local regulations, on low-volume streets or highways or where larger signs become an additional hazard to motorists and pedestrians.

Construction and maintenance signs must not be mounted on existing traffic signs, posts, or other utility structures without permission from the proper authority. The minimum mounting height on fixed supports should be seven feet (2.1 m) from the ground to the bottom of the sign in urban districts and five feet (1.5 m) from pavement elevation to the bottom of the sign in rural districts. Signs mounted on barricades or other temporary supports may be installed at lower heights, but the bottom of the sign should not be less than one foot (0.3 m) above the pavement elevation.

For maximum mobility, which may be needed on certain types of maintenance activities, a large sign may be mounted on a vehicle stationed in advance of the work. These mobile sign displays may be mounted on a trailer with self-contained electric power units for flashers and lights, or they may be mounted on a regular maintenance vehicle.

Regulatory signs, such as STOP, YIELD, NO PARKING, NO TURNS, and SPEED LIMIT, impose legal obligations and/or restrictions on all traffic. It is essential, therefore, that their use be authorized by the governing road authority having jurisdiction over traffic in the work area.

Construction and maintenance signs need to be placed where they will convey their messages effectively. Signs mounted on temporary supports should not be placed in the open, traveled lane where they pose a hazard to traffic. Generally these signs are placed on the right-hand side of the roadway or in the parking lane of the street or highway. Care should be taken to ensure that the motorists' view of the signs is not blocked by parked vehicles, trees, or other sight obstructions on or near the roadway. When special emphasis is needed, signs may be placed on both the left and right sides of the roadway.

All signs used at night should be either reflective or illuminated to show similar shape and color both day and night. Roadway lighting doesn't meet the requirements for illumination.

4.372 Channelizing Devices

Channelizing devices are used to warn drivers and alert them to conditions created by work activities in or near the roadway, to protect workers in the temporary traffic control zone, and to guide drivers and pedestrians. Consideration should be given to equipping the devices with warning lights in fog or rain areas, along severely curved roadways, and in unusually cluttered environments.

Channelizing devices include barricades, traffic cones and tubes, drums, and vertical panels. These devices are not interchangeable because they have different effects on traffic. Therefore, judgment must be used in selecting the appropriate channelizing device(s) for a specific work zone. Many factors, such as traffic volumes, speeds, and sight distances, should be considered when selecting the appropriate channelizing device(s).

4.373 Barricades

Barricades are commonly used to outline excavation or construction areas, close or restrict the right-of-way, channelize traffic, mark hazards, or mount signs. Barricades need to be placed so that the diagonal stripes slope down toward traffic. Barricades are classified as Type I, Type II, or Type III (Figure 4.9). Type I and Type II barricades are used in situations where traffic continues to move through the area, and Type III barricades are used where the street or highway is partially or totally closed. Type III barricades should be set up at the point of closure. Where provision is made for access of authorized equipment and vehicles, the responsibility for Type III barricades should be assigned to a person to ensure proper closure at the end of each work period.

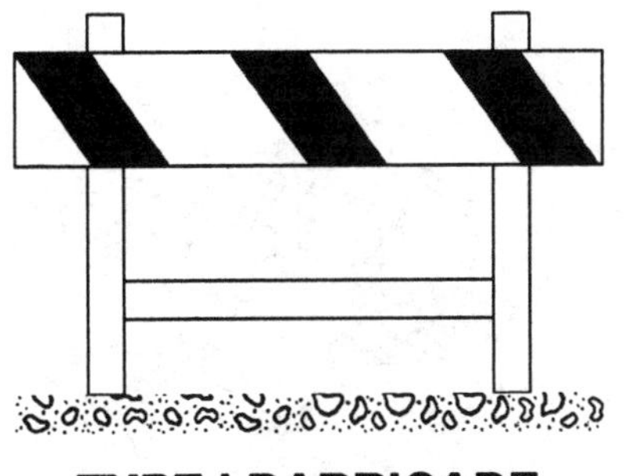

TYPE I BARRICADE

TYPE II BARRICADE

TYPE III BARRICADE

Not To Scale

Fig. 4.9 Types of barricades

4.374 *Traffic Cones and Tubes* (Figure 4.10)

Traffic cones and tubes are an effective method of channelizing traffic along a specified route during daylight hours. Use of cones or tubes in transition areas during nighttime activities is discouraged. Traffic cones and tubes are easily moved by passing vehicles so monitoring is necessary to keep them in place. Steps should be taken to minimize cone displacement. Cones can be doubled up to increase their weight. Some cones are constructed with bases that can be filled with ballast while others have special weighted bases.

4.375 *Drums* (Figure 4.10)

Drums are commonly used to channelize or delineate traffic flow. Drums should not be weighted to the extent that they become hazardous to motorists. They can be designed with a base that will separate from the drum if struck by a vehicle. If ballast is used, it should be sand or water. Drums should also be constructed of material that will be flexible or collapsible if struck by a vehicle.

4.376 *Vertical Panels* (Figure 4.10)

Vertical panels are most commonly used for traffic separation or shoulder barricading where only limited space is available.

4.377 *Lighting Devices*

Construction and maintenance activities often create conditions that become particularly hazardous at night. Therefore, lighting devices should be used in addition to the traffic control devices. Warning lights are classified as Type A, B, or C as follows:

- Type A lights are low-intensity, flashing yellow lights most commonly mounted on barricades, drums, or advance warning signs, and are intended to warn drivers that they are approaching, or in, an area of potential danger.
- Type B lights are high-intensity, flashing yellow lights normally mounted on advance warning signs or on independent supports. They are designed to operate 24 hours per day.
- Type C lights are steady-burn lights used to delineate (define) the edge of the desired path, such as on detour curves, on lane changes, and on lane closures.

Flashing lights (Types A and B) must not be used for delineation because a series of flashers may obscure the desired path and would be confusing to the drivers. When used, flashing warning lights are normally installed above the first advance warning sign. Flashing warning lights also may be used to supplement barricades at hazardous locations, particularly where there is a sudden change in alignment, such as T-intersections.

4.378 *Advance Warning Arrow Boards* (Figures 4.3, 4.4, and 4.11)

Advance warning arrow boards (or flashing arrow signs, FAS) are rectangular black sign panels covered with yellow lights capable of displaying a flashing or sequential arrow. These arrow panels are intended to supplement other traffic control devices, not replace them. Advance warning arrow boards will not solve difficult traffic problems by themselves, but they can be very effective when properly used to reinforce signs, barricades, cones, and other traffic control devices. All required traffic control devices should be used in combination with the advance warning arrow panel.

Arrow boards are vehicle- or trailer-mounted for easy transport to a job site. The displays are operated from a remote control panel and powered by a self-contained power source (batteries or an electric generator) mounted on the vehicle or trailer.

4.379 *Flashing Vehicle Lights*

Work vehicles in or near the traffic areas should be equipped with flashing lights. The vehicle warning lights may be emergency flashers, flashing strobe lights, or rotating beacons. High-intensity lights are effective for both day and nighttime work. The governing road authority should be contacted concerning requirements for flashing vehicle lights.

4.3710 *High-Level Warning Devices* (Figures 4.12 and 4.13)

High-level warning devices are most commonly used in urban high-density traffic situations to supplement other traffic control devices. The high-level warning device should be placed behind channelization and near the center of the blocked traffic lane. High-level warning devices may be attached to a service vehicle when such a vehicle is placed in advance of a work area.

4.3711 *Portable Changeable Message Sign (PCMS)* (Figures 4.11 and 4.12)

A portable changeable message sign (PCMS) is a lighted, electronic traffic control device with the flexibility to display a variety of messages. The PCMS should always be used in conjunction with conventional signs, pavement markings, and lighting. They are used most frequently on high-density urban freeways, but have applications on all types of streets and highways where alignment and traffic routing problems require advance warning and information. PCMSs serve a wide variety of functions in work zones including advising motorists of road closures, accident management instructions, narrow lanes, construction schedules, traffic management and diversion activities, and adverse conditions.

PCMSs should be placed in advance of a temporary traffic control zone and should not replace any required sign. Each PCMS display should convey a single thought in as brief a message as possible. The entire message cycle should be readable at least twice at the posted speed.

4.3712 *Flagging* (Figures 4.6 and 4.14)

Flagging procedures, when used, can provide positive guidance to motorists driving through the work area. Specific methods, procedures, and specifications for flagging must be used. In general, flaggers may be needed when the following conditions exist:

1. Operators or equipment intermittently block a traffic lane,
2. One lane must be used for two directions of traffic (a flagger is required for each direction of traffic), and
3. The safety of the public and/or operators requires it.

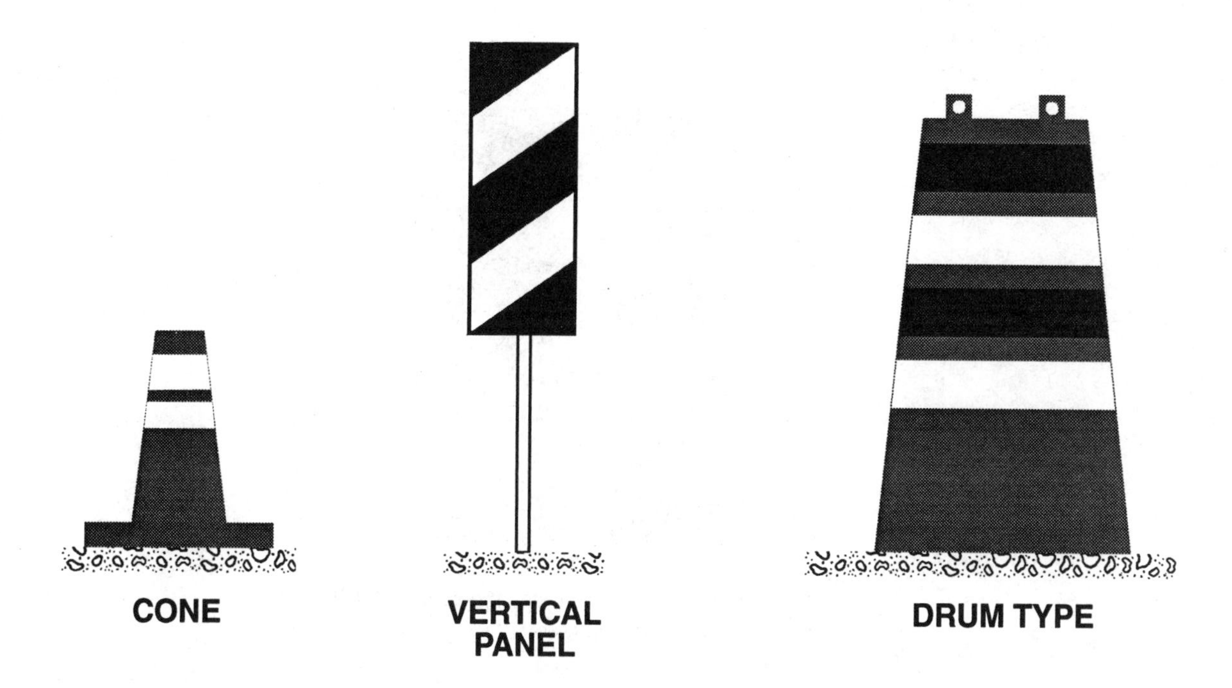

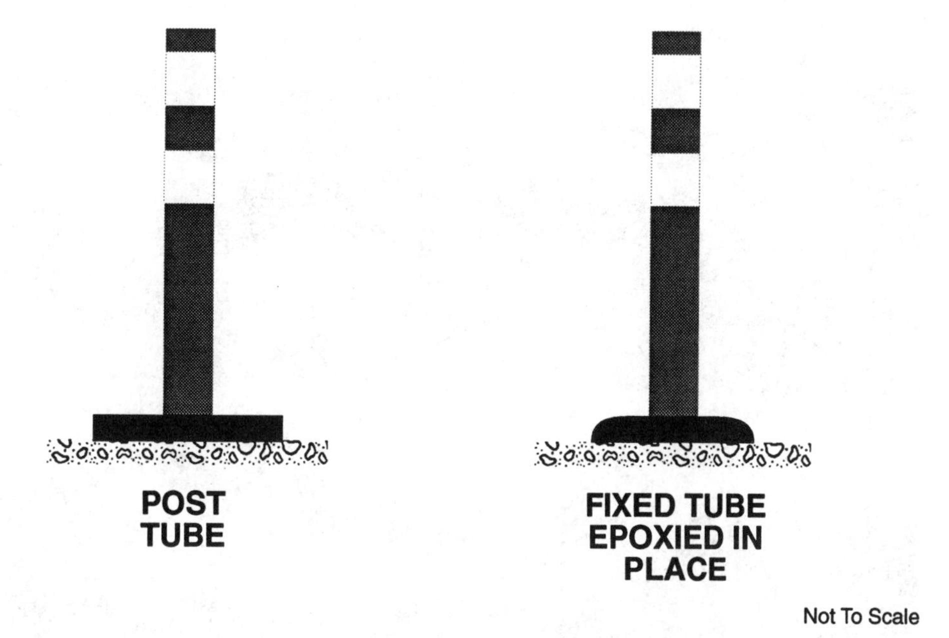

Fig. 4.10 Delineating and channelizing devices

Fig. 4.11 Portable changeable message sign (PCMS)

(Permission of Wanco, Inc.)

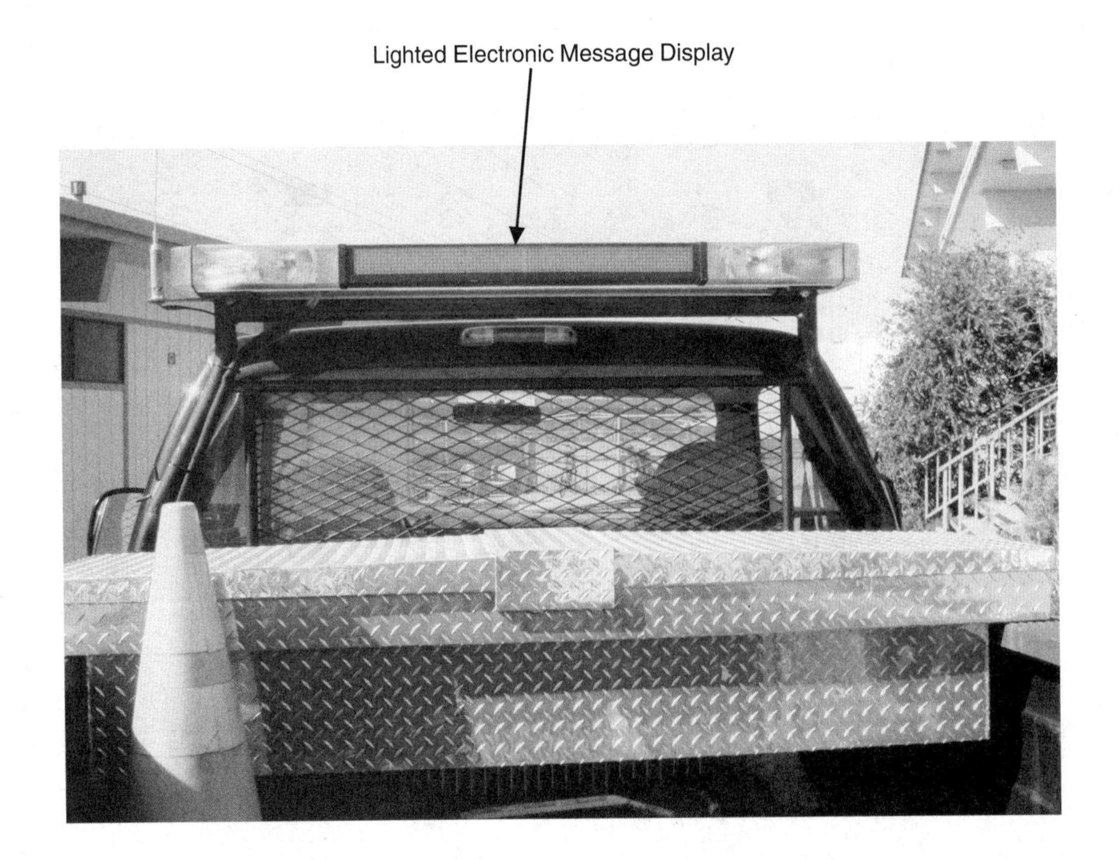

Fig. 4.12 Vehicle-mounted changeable message sign

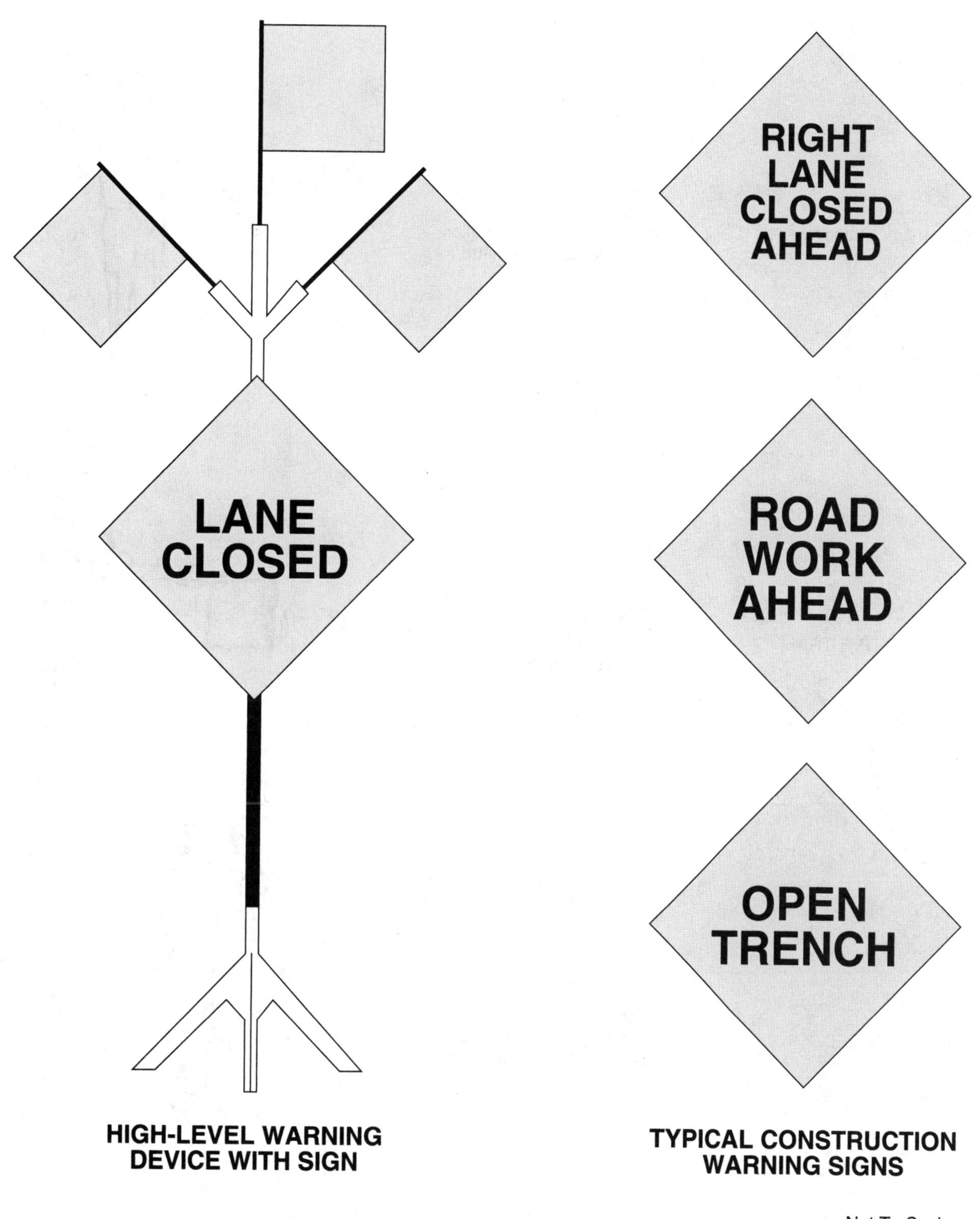

Fig. 4.13 Warning devices and signs

TO *SLOW* DAYTIME TRAFFIC

To Slow Traffic
(Short up and down motion with extended right hand.)

TO *STOP* DAYTIME TRAFFIC

To Stop Traffic
("Stop" paddle high enough to be above garment.)

TRAFFIC PROCEED

Traffic Proceed
(Never use a flag or paddle as a signal to move traffic.)

Fig. 4.14 Flaggers and hand signaling motions and devices

The flaggers should be positioned at least 100 feet (30 m) in front of the work space and a sign reading FLAG CONTROL AHEAD should be placed as far ahead of the flaggers as practical (500 feet or 150 meters minimum). (Be sure to check with your governing road authority for local regulations concerning the use of flaggers to control traffic.) Figure 4.14 shows flaggers and hand signaling motions and devices. Flaggers are expected to regulate and control the flow of traffic in a safe and orderly fashion. Because flaggers are responsible for employee and public safety, they must have a sense of responsibility and they must receive appropriate training in traffic control practices. Flaggers and utilities have been sued because improper actions by flaggers caused an accident. Once again, consistency is the key to traffic control safety. Use standard devices and consistent signal motions to avoid confusing motorists.

REMEMBER, YOU ARE SPEAKING A SIGN LANGUAGE... MAKE YOUR SIGNS SPEAK LOUD AND CLEAR!

4.38 Other Traffic Control Concerns

4.380 Pedestrian and Bicyclist Needs

If the work will block pedestrian walkways, bicycle lanes, or bicycle pathways, the governing road authority may require you to provide alternative walkways or bikeways. Alternative walkways or bikeways must provide protection from hazardous work areas and vehicular traffic.

4.381 Worker Visibility

All workers not separated from traffic by a positive barrier must be required to wear high-visibility clothing such as a vest, shirt, or jacket as approved by the governing road authority. In addition, any time it is raining, snowing, sleeting, or hailing, and any other time visibility is reduced by weather, smoke, fog, or other conditions, reflectorized garments should be worn. The reflective material may be orange, yellow, white, or silver and fluoresced versions and should be visible from at least 1,000 feet (300 m). The shape of the reflective garment should be visible through the full range of body motions. Details on the approved types of high-visibility garments should be obtained from the governing road authority.

4.382 Speed Limits in Work Zones

Proper uniform application of speed limits in work zones will help to improve the safety of operators working in and near roadways as well as protecting the motoring public. In high-traffic, high-hazard areas, consideration should be given to reducing the speed of traffic through regulatory speed zoning, use of police, lane reduction, or flaggers.

4.383 Worker Considerations

The safety of workers in a temporary traffic control zone is as important as the safety of the public. In addition to worker visibility and speed controls mentioned earlier, the following elements should also be included to ensure worker safety:

- Training—all workers must be trained in how to safely work next to traffic. Workers with traffic control responsibilities must be trained in traffic control techniques and device usage and placement.
- Barriers—barriers along the work area should be considered if the work area is close to traffic, if traffic speed is excessive, or if traffic volume is high.
- Lighting—if work is required at night, lighting the area and approaches will provide drivers with a better understanding of the requirements being imposed.
- Public information—a public relations effort that tells what the nature of the work is, its duration, anticipated effects on traffic, and alternate routes may improve drivers' performance and reduce traffic exposure.
- Road closure—if an alternate route is available to handle detoured traffic, a temporary road closure would reduce traffic hazards greatly and would probably allow the job to be completed much faster.

4.39 Additional Reading

1. *MANUAL ON UNIFORM TRAFFIC CONTROL DEVICES*, Federal Highway Administration (FHWA). The 2000 Edition is currently available as an electronic version on the FHWA website at www.fhwa.dot.gov. Hard copies are available from the U.S. Government Printing Office, Superintendent of Documents, PO Box 371954, Pittsburgh, PA 15250-7954. Order No. 050-001-00332-5. Price, $110.00.
2. *WORK AREA TRAFFIC CONTROL HANDBOOK*. Obtain from BNi Building News, 1612 South Clementine Street, Anaheim, CA 92802. Price, $6.95.

QUESTIONS

Write your answers in a notebook and then compare your answers with those on page 177.

4.3N Why is it important to use traffic control devices in a consistent, predictable manner?

4.3O List four types of channelizing devices.

4.3P What is a portable changeable message sign (PCMS) and when is it used?

4.3Q Describe the motions a flagger would use to slow daytime traffic.

END OF LESSON 1 OF 3 LESSONS
on
SAFE PROCEDURES

Please answer the discussion and review questions next.

DISCUSSION AND REVIEW QUESTIONS

Chapter 4. SAFE PROCEDURES

(Lesson 1 of 3 Lessons)

At the end of each lesson in this chapter you will find some discussion and review questions. The purpose of these questions is to indicate to you how well you understand the material in the lesson. Write the answers to these questions in your notebook before continuing.

1. What types of hazardous tasks are performed by collection system operators?
2. Which items on a vehicle and towed equipment should be inspected before leaving the yard?
3. How could you obtain information on defensive driving programs in your area?
4. How can traffic be warned of your presence in the street or highway?
5. What factors should be considered when deciding where to station flaggers?
6. What are the definitions of a low-speed and high-speed street or highway?
7. What is the purpose of the advance warning area?
8. Why should traffic control devices be constructed to yield upon impact?
9. List six types of traffic control devices.
10. What types of clothing should operators wear to increase their visibility to motorists?

CHAPTER 4. SAFE PROCEDURES

(Lesson 2 of 3 Lessons)

4.4 CLASSIFICATION AND DESCRIPTION OF MANHOLE HAZARDS

Let's review what we have accomplished so far:

1. Received our work assignments;
2. Reviewed and discussed work assignments with colleagues and supervisors;
3. Arranged for the necessary permits (see Section 4.40);
4. Inventoried tools, equipment, and materials to ensure that we have the appropriate items to accomplish the work assignment;
5. Inspected all tools and equipment, to ensure they are in good, safe working order;
6. Inspected and checked all safety equipment and calibrated testing devices as required;
7. Inspected vehicles or other rolling stock that we will be using;
8. Driven safely to the first job site; and
9. Set up adequate traffic controls.

We are now ready to proceed with our first work assignment for the day. We are going to inspect a manhole in an interceptor that is upstream from one of our lift stations. It is suspected of surcharging due to accumulated grease, causing a partial stoppage of the line. This brings us to another area of routine assignments that can be particularly hazardous and requires extreme care and professionalism in order to accomplish the task safely and survive potential hazards. Let's look in more detail at what kinds of manhole hazards we might encounter.

There are six major categories of hazards an operator may encounter when entering a manhole. These hazards are discussed in order of known frequency of accidents and deaths to operators—atmospheric, physical injury (slips, falls, falling objects, sharp objects, bumps, and structural failures), infection and disease, insects and biting critters, toxic exposure, and drowning.

4.40 Confined Spaces

Special safety rules and regulations have been developed to protect the life and health of persons who must enter confined spaces. Manholes are confined spaces. Throughout this manual safe procedures for safely entering manholes and other confined spaces will be emphasized.

Confined spaces are regulated under the OSHA Code of Federal Regulations, CFR Title 29 Part 1910.146, "Permit-required Confined Spaces." Each state, and in most cases each agency, has either adopted the federal regulation or has developed confined space procedures based on this regulation. There are five Appendices to Part 1910.146:

1. Appendix A - a decision flow chart,
2. Appendix B - atmospheric testing procedures,
3. Appendix C - examples of confined space programs,
4. Appendix D - sample permits, and
5. Appendix E - sewer system entry.

Sewer system confined space entry differs in several vital aspects from other permit and non-permit entries. In some cases there rarely exists any way to completely isolate the space (due to a section of the system that is continuous) needed to make an entry to perform work. If the isolation is not complete, the atmosphere may suddenly and unpredictably become hazardous (toxic, flammable, or explosive) from a cause beyond the control of the entrant, attendant, or supervisor. Experienced collection system operators are especially knowledgeable in entry procedures and work tasks in their respective permit or non-permitted confined space entries because of their frequent entries. Unlike many other workers for whom permit space entry is a rare and exceptional event, collection system operators usually work in an environment that is a permitted space.

When entering into a large sewer system, you may be required to use special equipment. The type of equipment might include atmospheric monitoring devices with alarms. In the event of a sudden or unpredictable atmospheric change, an emergency escape breathing apparatus (EEBA) with at least a 10-minute air supply or other *NIOSH*[2] approved, self-contained breathing apparatus should be worn for escape purposes. If in a boat or raft, a rope may be needed for pulling operators around bends and corners. A waterproof flashlight, life jacket, and radio may also be needed during sewer maintenance operations.

Confined space regulations apply to a wide variety of industries so they are very broad in scope. Even within an agency, there frequently are variations between the entry procedures for treatment plants and collection systems. Consequently, a permit-required confined space procedure for one agency may differ significantly from another agency's procedure since it reflects the needs of the specific agency. Operators must always remember that confined spaces are dangerous and

---

[2] *NIOSH (NYE-osh). The **N**ational **I**nstitute of **O**ccupational **S**afety and **H**ealth is an organization that tests and approves safety equipment for particular applications. NIOSH is the primary federal agency engaged in research in the national effort to eliminate on-the-job hazards to the health and safety of working people. The NIOSH Publications Catalog, Sixth Edition, NIOSH Pub. No. 84-118, lists the NIOSH publications concerning industrial hygiene and occupational health. To obtain a copy of the catalog, write to National Technical Information Service (NTIS), 5285 Port Royal Road, Springfield, VA 22161. NTIS Stock No. PB86-116787, price, $103.50, plus $5.00 shipping and handling per order.*

are probably responsible for more deaths of collection system operators than any other cause.

DEFINITIONS

Confined Space[3] means a space that:

- Is large enough and so configured that an operator can bodily enter and perform assigned work,
- Has limited or restricted means for entry or exit, and
- Is not designed for continuous operator occupancy.

In addition, if the space has or may have one or more of the following conditions, it is a permit-required confined space.

- A hazardous atmosphere,
- The potential for engulfing an operator. This means the space could contain a material, such as wastewater or granular or powdered chemicals, that might flow down and around an operator thereby surrounding and trapping the operator,
- An internal configuration (shape) such that an operator could be trapped or asphyxiated (suffocated) by inwardly converging walls or by a floor that slopes downward and tapers to a smaller cross section, and/or
- Any other recognized serious safety or health hazard.

Let's apply a little common sense to simplify the definition of a confined space. First, we can get into the space. Second, even though we can get into the space, it has limited means of getting into it and getting back out, which is important in case rescue is required. Third, the space has the potential for the existence of a toxic or explosive atmosphere or an oxygen deficiency or enrichment that could kill us or injure us. Finally, there is a potential for engulfment by a material such as raw wastewater. Some examples of confined spaces in the collection system include manholes, meter vaults, wet wells, chemical storage tanks, storm water sewers, pipelines, and sumps.

Non-Permit Confined Space Non-permit confined space means a confined space that does not contain or, with respect to atmospheric hazards, have the potential to contain any hazard capable of causing death or serious physical harm. A permit is not required for entry into a non-permit space but conditions in the space should be evaluated prior to entry and a safe environment must be maintained while anyone is working in the space.

Confined Space Entry Permit A written checklist used for evaluation of a confined space for potential atmospheric and physical hazards.

Engulfment means the surrounding and effective capture of a person by a liquid or finely divided (flowable) solid substance. The likely result of engulfment is that the substance will be breathed in (aspirated) and cause death by filling or plugging the respiratory system, or that the engulfing material will exert enough force on the body to cause death by strangulation, constriction, or crushing.

Hazardous Atmosphere means an atmosphere that may expose operators to the risk of death, incapacitation, injury, or acute (sudden and severe) illness, or which may reduce the operator's ability to self-rescue or escape unaided from a permitted space. The following conditions are considered hazardous atmospheres:

- Flammable gas, vapor, or mist in excess of 10 percent of its **L**ower **E**xplosive **L**imit (LEL). The Lower Explosive Limit is a number, expressed in percent (%), that tells us at what concentration a flammable gas, vapor, or mist becomes explosive in the atmosphere (see Figure 4.15);

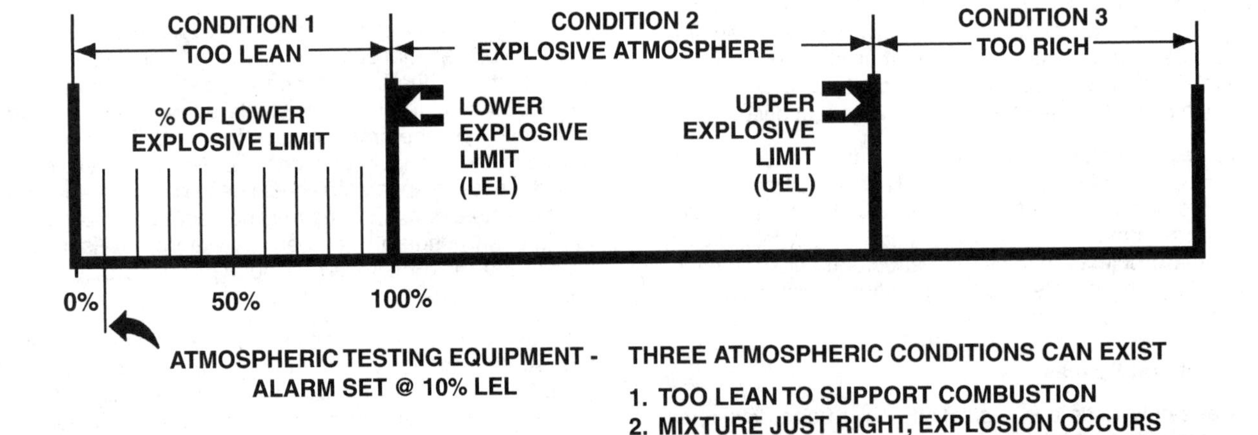

Fig. 4.15 Lower explosive limit or level (LEL)

[3] *Confined Space definition adapted from OSHA CFR Title 29 Part 1910.146. NOTE: Your state or province may have a DIFFERENT legal definition of a confined space.*

- Airborne combustible dust at a concentration that meets or exceeds its LEL. A rough approximation of this concentration is when the dust obscures vision (makes it hard to see an object) at a distance of five feet or less;
- Atmospheric oxygen concentration below 19.5 percent or above 23.5 percent;
- Atmospheric concentration of any substance for which a dose or a permissible exposure limit is published in Subpart G, "Occupational Health and Environmental Control," or in Subpart Z, "Toxic and Hazardous Substances," which could result in operator exposure in excess of its dose or permissible exposure limit; and
- Any other atmospheric condition that is immediately dangerous to life or health.

Hot Work Includes all work involving electric or gas welding, cutting, brazing, or similar flame- or spark-producing operations.

Hot Work Permit A written authorization to perform operations capable of providing a source of ignition.

Immediately Dangerous to Life or Health (IDLH) Any condition that poses an immediate or delayed threat to life or that would cause irreversible adverse health effects or that would interfere with an individual's ability to escape unaided from a permit space.

Lock Out/Tag Out (LOTO) The procedure by which an energy source (for example, electrical, mechanical, hydraulic, pneumatic, chemical, thermal, or liquid) is disabled by a positive means, such as a lock, to prevent accidental energy release during servicing and/or maintenance of equipment or machinery. A tag must also be placed at the point where the energy source is disabled to identify the person responsible for the lockout.

Upper Explosive Limit (UEL) The maximum concentration of flammable gas or vapor in air that will support combustion if an ignition source is present. This is the upper end of the flammable range. Concentrations greater than the UEL are "too rich"; there is too much flammable gas/vapor (fuel) and not enough oxygen to support combustion in the presence of an ignition source.

Lower Explosive Limit (LEL) The minimum concentration of flammable gas or vapor in air that will support combustion if an ignition source is present. This is the lower end of the flammable range. Concentrations of less than the LEL are "too lean"; there is not enough flammable gas or vapor (fuel) to support combustion in the presence of an ignition source.

4.41 Atmospheric Hazards

Atmospheric hazards consist of three major types: explosive or flammable, toxic atmospheres, and depletion or enrichment of breathable oxygen. Do not allow unhealthy odors to distract your attention from the three major types of hazards that could kill you.

1. Explosive or flammable atmospheres can develop at any time in the collection system. Flammable gases or vapors may enter a sewer or manhole from a variety of legal, illegal, or accidental sources. These conditions can be measured by the use of meters, calibrated for the expected gas, that indicate the explosive or flammable level of the atmosphere. The atmosphere in a manhole must be tested before anyone enters it and while the manhole is occupied.

   Methane (CH_4) gas is one of the products of *ANAEROBIC*[4] waste decomposition. This gas can be produced almost anywhere in a collection system. Methane is also the major flammable gas in the natural gas piped underground by utility companies. Leaks in these pipes will saturate the soil around a sewer pipe, and seepage will result in the gas entering the collection system and endangering operators in a manhole.

   Methane gas in its normal state is colorless, tasteless, and odorless, and therefore we cannot detect it through our normal senses. Natural gas that is supplied through the piping network of a natural gas utility has an odorant (*MERCAPTANS*[5]) added to it so that we are able to detect by smell the presence of natural gas leaks; for example, in our home heating system. When a leak occurs underground and the gas travels through the ground, the ground acts as an odor filter and removes the odor that was added. As a result, the underground natural gas leak is most likely going to be undetectable by smell because of the absence of the odorant. Natural gas is 85 percent methane, 10 percent ethane, and 5 percent other gases.

   Because methane and natural gases are lighter than air, a small portion of the gas will diffuse or escape from a manhole if there is natural ventilation. On the other hand, propane, gasoline, solvents, and other explosive fuel gases may be as much as two-and-one-half to four times heavier than air. They will tend to accumulate (if there is not ventilation) in the pockets of the lower portions of a collection system where they may form explosive mixtures or displace air.

   In summary, methane gas is 55 percent as heavy as air and will generally be found near the top of a structure. It is both colorless and odorless and acts as an asphyxiant. It also acts to mechanically deprive the body tissue of oxygen and does not support life. There is probably no limit to the amount of methane that can be tolerated, provided there is also sufficient oxygen to sustain life. It is important to remember that a space may contain methane in a concentration above its LEL (making the space explosive) and still have an oxygen content above 19.5 percent, allowing you to breathe normally. Methane is explosive at a concentration between 5 percent (50,000 ppm) and 15 percent (150,000 ppm). It is also an asphyxiant and incomplete combustion produces carbon monoxide (CO). The alarm set point (the point at which the alarm is activated) for methane is at 10 percent of the LEL.

METHANE (CH_4)

%	ppm	Hazard
85	850,000	Amount in natural gas
65	650,000	Amount in digester gas
15	150,000	Upper Explosive Limit (UEL)
5	50,000	Lower Explosive Limit (LEL)
0.5	5,000	Alarm set point (10 percent of LEL)

[4] *Anaerobic (AN-air-O-bick). A condition in which atmospheric or dissolved molecular oxygen is NOT present in the aquatic (water) environment.*
[5] *Mercaptans (mer-CAP-tans). Compounds containing sulfur which have an extremely offensive skunk-like odor; also sometimes described as smelling like garlic or onions.*

Underground sources of explosive or flammable gases include gasoline storage tanks, gas stations, and petroleum product pipelines that may be leaking. Accidental spills caused by traffic accidents involving tank trucks hauling flammable materials are very common and, more often than not, will find their way into the inlet basins connected to storm drains or combined sewers. Similarly, discharging of flammable materials, either accidentally or illegally, from industrial sources is a common source of flammables that end up in the collection system. Many of these compounds also have the ability to displace available oxygen, so even if they do not cause a fire or explosion, they may cause death or injury by asphyxiation. If a gas is heavier than air (*SPECIFIC GRAVITY*[6] of greater than 1, which is equal to air), it will tend to accumulate at the bottom of the manhole or underground structure and displace the oxygen in the lower section. Therefore, you can't assume that the bottom section of the structure will have a safe atmosphere, even if the top portion tests OK. Remember, air movement and temperature changes may disperse gases throughout a collection system, regardless of their specific gravity.

Gasoline vapor is three to four times heavier than air and the vapor is generally found at the bottom of a structure. Gasoline occurs frequently in storm and sanitary sewers as a result of accidents involving vehicles, leaking underground storage tanks, and/or illegal industrial discharges. It is colorless and has an odor that is noticeable at 0.03 percent. Gasoline vapor has an anesthetic effect when inhaled and is rapidly fatal at 2.4 percent. A short exposure to a concentration of 1.1 to 2.2 percent is dangerous. Gasoline vapor has little to no specific effect on the lungs, but acts after being absorbed into the blood and transported to the tissues of the body. Gasoline vapor is flammable and explosive in concentrations of 1.3 to 7.6 percent in air.

2. Toxic atmospheres (poisonous air) in wastewater collection systems or storm water collection systems are most likely to result from the presence of hydrogen sulfide (H_2S), a gas produced by the decomposition of certain materials containing sulfur compounds. Hydrogen sulfide gas quickly mixes with air and goes wherever the air goes. If there is no ventilation or air movement, hydrogen sulfide tends to accumulate in the lower sections of the collection system since it has a specific gravity of 1.19 and is therefore heavier than air.

Hydrogen sulfide can be detected by the smell of rotten eggs, although this is not a reliable method of testing because hydrogen sulfide gas also has the unique ability to affect your sense of smell. After a short exposure to hydrogen sulfide gas, you will lose your ability to detect hydrogen sulfide gas by smell; therefore, instrumentation must be used to detect the gas.

As with the flammable or explosive materials, toxics may also enter the collection system from accidental sources such as spills in industrial plants or accidents involving mobile tank transporters such as rail cars and tank trucks. Other toxic gases that may be encountered include carbon monoxide, chlorinated solvents, and industrial toxins (poisons), depending on wastes discharged to the collection system. As a professional collection system operator, you should be aware of the potential sources of these materials from industrial or commercial operations in your specific area.

In summary, hydrogen sulfide is 19 percent heavier than air and will generally be found near the bottom of a structure, unless the gas is heated and the air is very humid. It is a by-product of *ANAEROBIC DECOMPOSITION*[7] of organic material containing sulfur. This gas is colorless and has the odor of rotten eggs, even in small concentrations. This odor is not noticeable at a high concentration. Breathing hydrogen sulfide gas at a concentration of 0.2 percent can be *FATAL* in a few minutes. Exposure to concentrations of 0.07 to 0.1 percent will rapidly cause acute poisoning and paralyze the respiratory center of the body. Under these circumstances our nose suffers *OLFACTORY FATIGUE*[8] and can no longer smell the rotten egg odor. Many operators have died because their nose no longer detected hydrogen sulfide gas and they thought they were breathing "safe" air. The maximum amount that can be tolerated for a 60-minute period is 0.02 to 0.03 percent in air. At 700 to 1,000 ppm, rapid loss of consciousness occurs and breathing stops. The alarm set point for hydrogen sulfide is 10 ppm. The LEL is 4.3 percent in air and the UEL is 46 percent in air.

HYDROGEN SULFIDE (H_2S)

%	ppm	Hazard
46	460,000	Upper Explosive Limit (UEL)
4.3	43,000	Lower Explosive Limit (LEL)
0.1	1,000	**Dead**
0.07	700	Rapid loss of consciousness
0.06	100	IDLH
0.005	50	Eye tissue damage
0.002	20	Eye, nose irritant
0.001	10	Alarm set point

[6] *Specific Gravity. (1) Weight of a particle, substance or chemical solution in relation to the weight of an equal volume of water. Water has a specific gravity of 1.000 at 4°C (39°F). Wastewater particles or substances usually have a specific gravity of 0.5 to 2.5. (2) Weight of a particular gas in relation to the weight of an equal volume of air at the same temperature and pressure (air has a specific gravity of 1.0). Chlorine has a specific gravity of 2.5 as a gas.*

[7] *Anaerobic (AN-air-O-bick) Decomposition. The decay or breaking down of organic material in an environment containing no "free" or dissolved oxygen.*

[8] *Olfactory (ol-FAK-tore-ee) Fatigue. A condition in which a person's nose, after exposure to certain odors, is no longer able to detect the odor.*

Carbon monoxide (CO) is slightly lighter than air and is generally found near the top of structures. It is colorless, odorless, tasteless, flammable, and poisonous. It has a LEL of 12.5 and an UEL of 74. Carbon monoxide is a product of combustion; therefore, adequate ventilation is necessary when using engine-driven equipment. This gas acts as an asphyxiant. The maximum amount that can be tolerated for a 60-minute period is 0.04 percent in air. Carbon monoxide in concentrations of 0.2 percent to 0.25 percent would cause a person to become unconscious in 30 minutes. It is fatal in a concentration of 0.1 percent for a four-hour period and is immediately dangerous to life and health (IDLH) at 1,500 ppm. Carbon monoxide will cause headaches at a concentration of 0.02 percent in a two-hour period or less.

CARBON MONOXIDE (CO)

%	ppm	Hazard
74	740,000	Upper Explosive Limit (UEL)
12.5	125,000	Lower Explosive Limit (LEL)
0.2	2,000	Unconscious in 30 minutes
0.15	1,500	**IDLH**
0.05	500	Severe headache
0.02	200	Headache after 2 to 3 hours
0.0035	35	8-hour exposure limit
0.0035	35	Alarm set point

3. The amount of breathable oxygen present in a manhole or other underground structure can be dangerously decreased or even eliminated when the air is mixed with or replaced by another gas. Meters are available that measure the concentration of oxygen in the air. Do not work in confined spaces where the atmosphere contains less than 19.5 percent oxygen or more than 23.5 percent oxygen. *ALWAYS VENTILATE THE MANHOLE* before entry and continuously during occupancy. *TEST* all depths of the manhole for oxygen deficiency or enrichment and explosive and toxic conditions *BEFORE* anyone enters a manhole. Whenever anyone is in a manhole, *CONTINUOUSLY VENTILATE* the manhole and *CONTINUOUSLY TEST* the atmosphere near the working person.

Concentrations of oxygen in a structure may exceed the level in the air we normally breathe (20.9 percent) when pure oxygen (O_2) is used to prevent *SEPTIC*[9] conditions and the production of hydrogen sulfide in the wastewater collection system (see Chapter 6, Section 6.5, "Hydrogen Sulfide Control"). If the manhole is continuously ventilated and the oxygen level continuously measured, no problems should develop. A slightly higher-than-normal level of oxygen won't hurt you, but will increase the fire or explosive hazard. Personnel should leave a work area whenever the oxygen concentration approaches 23.5 percent due to the increase in fire/explosion hazard.

In summary, oxygen is 11 percent heavier than air and will be found at different levels in a structure or confined space. In an undisturbed confined space with an oxygen deficiency, the oxygen level decreases with increased depth or distance from the top or ground surface. The ambient (surrounding) temperature affects the natural ventilation of confined spaces. In cold weather, for example, natural ventilation is more efficient than in warm temperatures. Oxygen is both colorless and tasteless. Normal air contains about 21 percent oxygen. Breathing air with less than 10 percent oxygen is dangerous to life, and below 8 percent would be fatal. Oxygen in a concentration above 23.5 percent also can be dangerous because it speeds up combustion. Because of this, compressed oxygen must never be used to improve the oxygen content of the atmosphere of a confined space.

OXYGEN (O_2)

%	ppm	Hazard
23.5	230,500	Accelerates combustion
20.9	209,000	Oxygen content of normal air
19.5	195,000	Minimum permissible level
8	8,000	**DEAD** in 6 minutes
6	6,000	Coma in 40 seconds, then **DEAD**

QUESTIONS

Write your answers in a notebook and then compare your answers with those on pages 177 and 178.

4.4A List the major categories of hazards you may encounter when entering a manhole.

4.4B What is a confined space?

4.4C Why can the top portion of the atmosphere in a manhole test "safe" when the bottom portion could be dangerous?

4.4D What kinds of atmospheric hazards are encountered in manholes?

4.42 Mechanical and Electrical Hazards

Because of the shape, size and layout of many confined spaces, operators are often working close to electrical or me-

[9] *Septic (SEP-tick). A condition produced by anaerobic bacteria. If severe, the wastewater produces hydrogen sulfide, turns black, gives off foul odors, contains little or no dissolved oxygen, and the wastewater has a high oxygen demand.*

chanical hazards. The limited space may reduce our ability to avoid operating mechanical equipment or electrical circuits and we may accidentally get caught in equipment or be exposed to electrical hazards. Mechanical hazards include hot or cold surfaces, steam leaks, rotating shafts, and check valve arms. Electrical hazards include conduits, open control panels or live circuits, lights, portable power tools, and water in the confined space. Each confined space needs to be evaluated for these potential hazards prior to entering. If necessary, perform the appropriate lockout/tagout procedures. (See "Lockout/Tagout Procedure," page 159.)

4.43 Engulfment Hazards

Engulfment means being surrounded or covered and trapped by a liquid, powder, or granular material. Examples of materials that can engulf an operator in a confined space are raw wastewater, sludge, chemicals, and plant water. As with mechanical and electrical hazards, evaluate each confined space for potential engulfment hazards and take appropriate safety precautions before any operator enters the space.

4.44 Physical Injury

Physical injury during manhole entry can occur from several causes. Operators in restricted spaces with uneven footing often have poor balance and decreased coordination. The handling of tools in restricted spaces often means an operator is working in awkward positions, which can cause strained muscles or tendons, bruises, or torn skin if the operator is not careful.

The use of heavy ladders for manhole entry and exit has been prohibited in many communities after these ladders were dropped by accident while an operator was in a manhole. Portable, lightweight, fiberglass or aluminum alloy ladders and rope ladders have proven to be satisfactory substitutes. Similarly, the practice of installing metal rungs in manhole walls also has been discontinued by some agencies because the rungs or the concrete holding them was being eaten away and destroyed by the highly corrosive atmospheres of the collection system. Manhole rungs should always be suspected of being unsafe. *NEVER* use them without some type of fall protection device.

Dropping tools to operators in a manhole and tossing the tools back out causes many physical injuries. In order for you to see to catch a tool being dropped, you must look upward and into the brighter light, causing temporary loss of visual capacity. In addition to this problem, dust and debris from the street or manhole ring can fall into your eyes, again causing vision problems as well as possible eye infection if you attempt to wipe your eyes with a hand or glove that has been exposed to the wastewater environment.

Tools should be lowered into and pulled out of manholes in a bucket or with a sling, preferably before the operator enters and after the operator exits the manhole. In the event that it is necessary to lower or raise tools and equipment while in the manhole, they must be secured in such a manner that if the bucket tips or the bottom of the bucket gets hung up on a manhole rung and tips, the tools cannot fall out of the pouch or bucket and hit the operator working in the manhole.

The use of glasses or safety goggles in a manhole may be difficult because of their tendency to fog. Auxiliary ventilation can help reduce a fog problem. Several commercial products are also available to prevent "fogging" of protective eye/face wear. Contact your local safety equipment distributor for additional information. Glasses also can become smeared by moisture in a manhole environment and thus reduce the operator's vision. If a chipping gun or other tool is being used to chip concrete or pipe, safety glasses or goggles *MUST* be worn. In some cases, a full-face safety shield may work better than safety glasses or goggles. Hard hats must also be worn when working in manholes.

When working in a manhole, beware of sharp objects that can cut or penetrate your skin and cause a serious infection. Typical sharp objects include razor blades, pins, hypodermic needles, and broken pieces of glass and metal.

4.45 Infections and Diseases

Infections are always potentially present when you enter a manhole. Every disease, parasite, infection, virus, and illness of a community can end up in the wastewater collection system. Persons required to enter manholes are thus automatically exposed to these infections and diseases. Leptospirosis can be transmitted to operators through the urine and feces of rats living in sewers. This disease causes fever, headaches, nausea, muscular pains, vomiting, and thirst. Contact your safety officer or doctor regarding inoculations for typhoid, paratyphoid, polio, Hepatitis B (HBV), and tetanus. As previously mentioned, the practice of dropping tools into a manhole can cause contaminated particles to fall into an operator's eyes. Other sources of infection include rubbing one's face or eyes with contaminated hands or clothing; cuts from sharp objects in a contaminated environment; and eating or smoking on the job without first thoroughly washing your hands. Personal cleanliness is your best means of protection against infection.

Boots, heavy gloves, disposable coveralls, and rain suits are examples of protective clothing suitable for working in manholes. Uniforms, coveralls, and other clothing that become soiled should be cleaned by a commercial laundry service rather than being washed with the family laundry. Keep your work clothes separated from your street clothes by using two lockers, one for uniforms and one for street clothes.

If you must work in a manhole that is located reasonably close to the discharge of a hospital, venereal disease treatment center, clinical laboratory, surgical facility, or a veterinarian's office and hospital, it may be advisable to disinfect the manhole and the upstream line one or two hours ahead of scheduled entry. Disinfection may be accomplished by placing HTH (High Test Hypochlorite) tablets in a cloth sack tied to a rope and lowered into the flowing wastewater. This action could result in other problems and strong odors in the manhole, but will significantly reduce the risk of exposure to the diseases that may possibly be present in such manholes. Because disinfectants are poisons, do not enter a manhole when the odor of a disinfectant is present. Test the atmosphere with a chlorine monitor if HTH is used.

Most state and local health departments have laws and regulations that prevent the discharge of infectious wastes into wastewater collection systems until after they have been disinfected (usually in an autoclave that disinfects by using steam). Contact the medical officer in charge of any medical facility immediately upstream from manholes and sewers in which you must work. This person can explain the disinfection procedures used by the medical facility, indicate if the threat to your health is greater than usual, and recommend any special precautions or disinfectants that may be necessary.

4.46 Insects, Bugs, and Rodents

In sect, bug, and rodent bites, though somewhat uncommon, are a double hazard to operators. The insects themselves may be poisonous to humans, such as the black widow spider and the violin spider. Many other types of insect, bug, and rodent bites can lead to infection or serious illness. Rat bites, for example, can transmit rabies and mosquito bites sometimes transmit malaria.

Wear protective clothing and be alert to the possible danger to your health. If you are bitten, gently wash the area with soapy water. Get prompt medical attention if you develop any signs of redness or swelling, or if you develop a fever or an allergic reaction. Following is a list of insects and bugs which have been found in manholes.

Stinging insects such as wasps, mud daubers, and bees
Ticks
Fleas
Lice
Mosquitos
Houseflies
Blowflies
Cockroaches
Spiders
Centipedes
Scorpions

Always inspect a manhole for insects, bugs, and rodents before entering.

Where insects have been a problem, and where rats or other vermin may occupy a collection system or where epidemics of insect-borne diseases may be present, spraying a manhole with an insecticide is suggested. The solution should be water-soluble and leave a toxic residue to be effective against the next hatch of any insects present and breeding in the collection system. Contact your local health agency to determine the appropriate insecticide if in doubt. Spraying should be conducted at the time of entry if insects or other problems are observed. Ventilate the manhole so insecticide will not be inhaled by operators. If a manhole is especially filthy or odorous from an insecticide or hydrogen sulfide, wash down the manhole with a high-velocity stream of water. This wash is best accomplished within an hour or so before manhole entry.

4.47 Toxicants

Toxicants include any substances that act as poisons. Acids, bases, and other hazardous liquid or solid chemicals are sometimes discharged into the wastewater collection system by either accidental spills or deliberate action by industry or the public. Exposure to such toxic substances is always a potential health hazard for the collection system operator. Proper boots and gloves are effective means of protection against these toxicants.

Contact your pretreatment inspector or industrial waste section when working in industrial/commercial areas for possible toxicants that may be encountered when working in the collection system in the area. If you discover a strong or unusual odor or color in the wastewater, request your supervisor to notify the pretreatment inspection or industrial waste section of your agency.

4.48 Drowning

With the trend toward larger regional wastewater treatment plants, intercepting sewers are being constructed to convey large flows. These large-diameter sewers and high flows increase the chances of an operator drowning from an accidental slip or fall into the flowing wastewater. To avoid drowning, watch your step at all times, especially when working near or in large flows. Wear life jackets or buoyant work vests and use lifelines if the flow in the sewer is capable of washing a fallen or unconscious operator away or if an operator will fit into the downstream pipe.

4.49 Summary

Manhole entry always must be considered a hazardous task due to the potential exposure to chemicals, toxic and explosive gases, physical injury, insects, and infectious conditions. Work in manholes can be done safely when the proper procedures are followed.

QUESTIONS

Write your answers in a notebook and then compare your answers with those on page 178.

4.4E What are some of the causes of physical injuries in manholes?

4.4F How can you protect yourself from diseases when working in a manhole?

4.4G How can you protect yourself from insects when working in a manhole?

4.4H How can you protect yourself from drowning when working in a manhole?

4.5 SAFETY EQUIPMENT AND PROCEDURES FOR CONFINED SPACE ENTRY

Based on the information in the previous sections, it should be clear that wastewater collection system operators are faced with a variety of hazards when working in confined spaces.

We can reduce these hazards to a minimum by using the appropriate equipment and procedures whenever we need to work in a confined space. This section describes different pieces of equipment that are used for entering and working in confined spaces.

As with most types of collection system activities, planning prior to making a confined space entry goes a long way toward a safe activity. Remember, a confined space can be a manhole, sewer, diversion structure, pump station, underground metering vault, or any other structure that meets the definition of a confined space. Each of these specific types of confined spaces requires specific planning prior to the entry. For example, before entering a manhole in a gravity sewer system, the crew should be knowledgeable about the gravity sewer itself. That is, how much flow should be expected, are there any upstream industrial dischargers to the system, do septic tank service trucks discharge upstream, are there any combined sewers that would contribute to high flow during wet weather or if a thunderstorm occurs, and where is the manhole physically located? With this type of information you will be able to plan to have appropriate safety equipment on hand, arrange for effective traffic control, follow any lockout/tagout procedures necessary, notify affected industries, or take any other necessary steps to ensure the safety of the operator in the manhole.

The federal regulations regarding confined spaces identify both permit-required confined spaces and non-permit-required confined spaces. All spaces must be considered permit-required confined spaces until a qualified person following prescribed pre-entry procedures determines otherwise. Clearly, any operator who is either making this type of evaluation or will be entering the confined space must have completed minimum confined space entry training requirements. A written copy of Operating and Rescue Procedures that are required by the confined space entry procedure must be at the work site during any work. The checklist includes the control of atmospheric and engulfment (see Section 4.43) hazards, surveillance of the surrounding area to avoid hazards such as drifting vapors from tanks, piping, or sewers, and atmospheric testing for oxygen deficiency, lower explosive limit, and hydrogen sulfide gas. Specific entry procedures must also be followed in a non-permitted confined space entry.

Confined spaces may be entered without the need for a written permit or an additional person standing by provided that the only hazard posed by the permit space is an actual or potential hazardous atmosphere and continuous forced-air ventilation alone can maintain the space safe for entry.

Live sewers (sewers containing raw wastewater), and also pump station wet wells, gravity sewer manholes, siphon headworks structures, or other structures that are connected directly to a live sewer are generally considered a permit-required confined space since there is no way to completely isolate them from the rest of the system. Sudden changes in the atmosphere may take place unpredictably from toxic, flammable, or explosive materials. This means that the procedures required by your regulations for permit-required confined spaces must be followed.

If you arrange to have a contractor perform work in confined spaces within your collection system, you must inform the contractor:

- That all personnel must comply with confined space regulations,
- Of hazards that you have identified and your experience with the space(s),
- Of precautions or procedures you have implemented for the protection of operators in or near the space where the contractor's personnel will be working, and
- That a "debriefing" must occur at the conclusion of the entry operations regarding the confined space program followed and any hazards encountered or created during the entry operations.

To enhance safety, communications, and coordination concerning confined space activities, the contractor is also required, prior to the job, to obtain available information from you and to inform you of the confined space program the contractor will follow.

NOTE: Almost every agency or company that does confined space entries has developed procedures based on their interpretation of federal and state regulations. The information in this manual is not intended to provide a confined space entry procedure based on interpretations of federal OSHA regulations, but rather is intended to provide an overall *AWARENESS* of the types of procedures and safety equipment necessary when making confined space entries in collection systems.

4.50 Respiratory Protection

Respirators are one of the most important pieces of equipment used in confined space entry. Your life may depend on a respirator and therefore operators need a clear understanding of how to select and use respirators. Respiratory protection is regulated under the OSHA Code of Federal Regulations, CFR Title 29 Part 1910.134.

The four most common respirators are:

1. Self-contained Breathing Apparatus (SCBA),
2. Supplied Air Respirators (SAR),
3. Powered Air Purifying Respirators (PAPR), and
4. Air Purifying Respirators (APR).

SARs and SCBAs, known as air supplying respirators, supply clean air to the wearer from an independent source and provide the highest levels of protection. The air source may either be remotely located tanks or a tank carried by the user.

APRs and PAPRs take air from the immediate surroundings and purify it by passing it through filters, cartridges, or canisters. APRs and PAPRs provide lower levels of protection than air supplying types of respirators. These respirators (APRs and PAPRs) **ARE NOT** suitable for confined space entry because the oxygen level may be too low.

Positive pressure respirators, in which the pressure inside the facepiece during inhalation (breathing) remains higher than the pressure outside, help prevent contaminants from entering and offer the greatest protection. Conversely, negative pressure respirators, which allow the interior facepiece pressure to drop below the outside pressure during the inhalation cycle, may not prevent contaminants from leaking into an improperly sealed facepiece.

Positive pressure SCBAs (Figure 4.16) and full facepiece positive pressure SARs with an escape SCBA can be used in oxygen-deficient atmospheres (containing less than 19.5 percent oxygen) and atmospheres that are considered immediately dangerous to life or health (IDLH) according to OSHA's regulations.

Negative pressure respirators (PAPRs and APRs) are not approved for use in oxygen-deficient atmospheres or in atmospheres that are immediately dangerous to life or health.

Selection of a respirator is based upon the type of hazard and the contaminant concentration. Each respirator type is given an assigned protection factor (APF) by OSHA indicating the maximum contaminant level for which the respirator can be used. Contaminants are rated by OSHA according to their permissible exposure limit (PEL). A contaminant's PEL is the legally established maximum time-weighted average level of contaminant to which an operator can be exposed during a work shift. The proper respirator is therefore chosen according to its APF and the PEL of the contaminant. For example, a respirator with an APF of 100, approved for a given contaminant, can be used in atmospheres containing 100 times the PEL of the contaminant.

Once a respirator has been chosen, it is necessary to conduct a *FIT TEST*[10] (29 CFR, Part 1910 General Industrial) for the use in a work area. The respirator must be fitted to the individual operator and training must be provided. During operator training special emphasis is placed on checking the respirator for proper fit each time it is worn. Operator comfort is an important consideration in making sure that respirators are used properly and working correctly. Silicone rubber facemasks are regarded by many operators as the most comfortable and offer the best sealing characteristics for a wide range of temperatures encountered in different working conditions.

A self-contained breathing apparatus (SCBA) allows the user to carry the air supply since the air source is a tank. Two types are commonly used, closed-circuit SCBAs and open-circuit SCBAs. Open-circuit SCBAs are also called "rebreathers" since a major part of the user's exhaled air is recirculated through a carbon dioxide scrubber, then supplemented by oxygen and rebreathed. These devices can provide an air supply for a period of 1 to 4 hours. Figure 4.16 illustrates a 60-minute rebreather-type respirator that is lightweight, portable, and allows the operator to enter and exit small-diameter manholes. When purchasing SCBAs, be aware that in some cases it may be impossible to enter or exit a manhole wearing this type of equipment and so it would be useless in a rescue operation.

Closed-circuit SCBAs vent the wearer's exhaled breath to the atmosphere. The air source is a lightweight aluminum composite cylinder, with air under 2,216 to 4,500 pounds per square inch (psi) of pressure. SCBA cylinders have air supplies ranging from 30 to 60 minutes. Operators must be trained to allow ample time for entry and escape and to monitor their air supply while working. The National Institute of Occupational Safety and Health (NIOSH) recommends that the minimum service time for SCBA units should be calculated by adding together the entry time, plus the maximum work period, plus twice the estimated escape time for a safety margin.

Combination SAR/SCBAs can be used when clean air is available for delivery by an air line hose from an air compressor, storage cylinder, or plant air system. Combination units of this type have both a positive pressure supplied air respirator and a small SCBA with 5 or 10 minutes of air. The SCBA portion of the combination unit is certified by NIOSH only for use as an escape device. SCBAs with air capacities of 30 and 60 minutes are also available with SAR attachments. Because of the possibility that wearing a respirator may create a tricky or awkward exit, provision must be made to ensure adequate air supply and ease of movement.

Compressed air, compressed oxygen, liquid air, and liquid oxygen used for respiration must be of a high purity. Oxygen must meet the requirements for medical or breathing oxygen. Breathing air must meet at least the requirements of the specification for Grade D breathing air (Compressed Gas Association Commodity Specification G-7.1-1966). Compressed oxygen should not be used in supplied air respirators or in open-circuit self-contained breathing apparatus that have previously used compressed air. Oxygen must never be used with respirators that have an air line.

Breathing air can be supplied to respirators from cylinders or from a specialized type of breathing air compressor. Cylinders must be tested and maintained in accordance with appropriate specifications (Shipping Container Specification Regulations of the Department of Transportation, 49 CFR Part 178). Compressors for supplying breathing air need to be equipped with a receiver of sufficient capacity to enable the respirator wearer to escape from a contaminated atmosphere in the event of compressor failure. Alarms to indicate compressor failure and overheating must also be installed in the system. If an oil-lubricated compressor is used, it must have a high-temperature or carbon monoxide alarm, or both. If a carbon monoxide alarm is not provided, frequent testing for carbon monoxide must be performed to ensure that the air meets the specifications for Grade D breathing air.

An established part of your confined space policy should specify that only under extreme circumstances should the collection system operator work in a confined space when the atmosphere presents an immediate danger to life or health. In other words, if you are getting an atmospheric alarm condition, get out and stay out until the condition is cleared. There may be rare occasions when an operator cannot avoid working in a confined space under these circumstances. In such cases, a

[10] *Fit Test. The use of a procedure to qualitatively[11] or quantitatively[12] evaluate the fit of a respirator on an individual.*

[11] *Qualitative Fit Test (QLFT). A pass/fail fit test to assess the adequacy of respirator fit that relies on the individual's response to the test agent.*

[12] *Quantitative Fit Test (QNFT). An assessment of the adequacy of respirator fit by numerically measuring the amount of leakage into the respirator.*

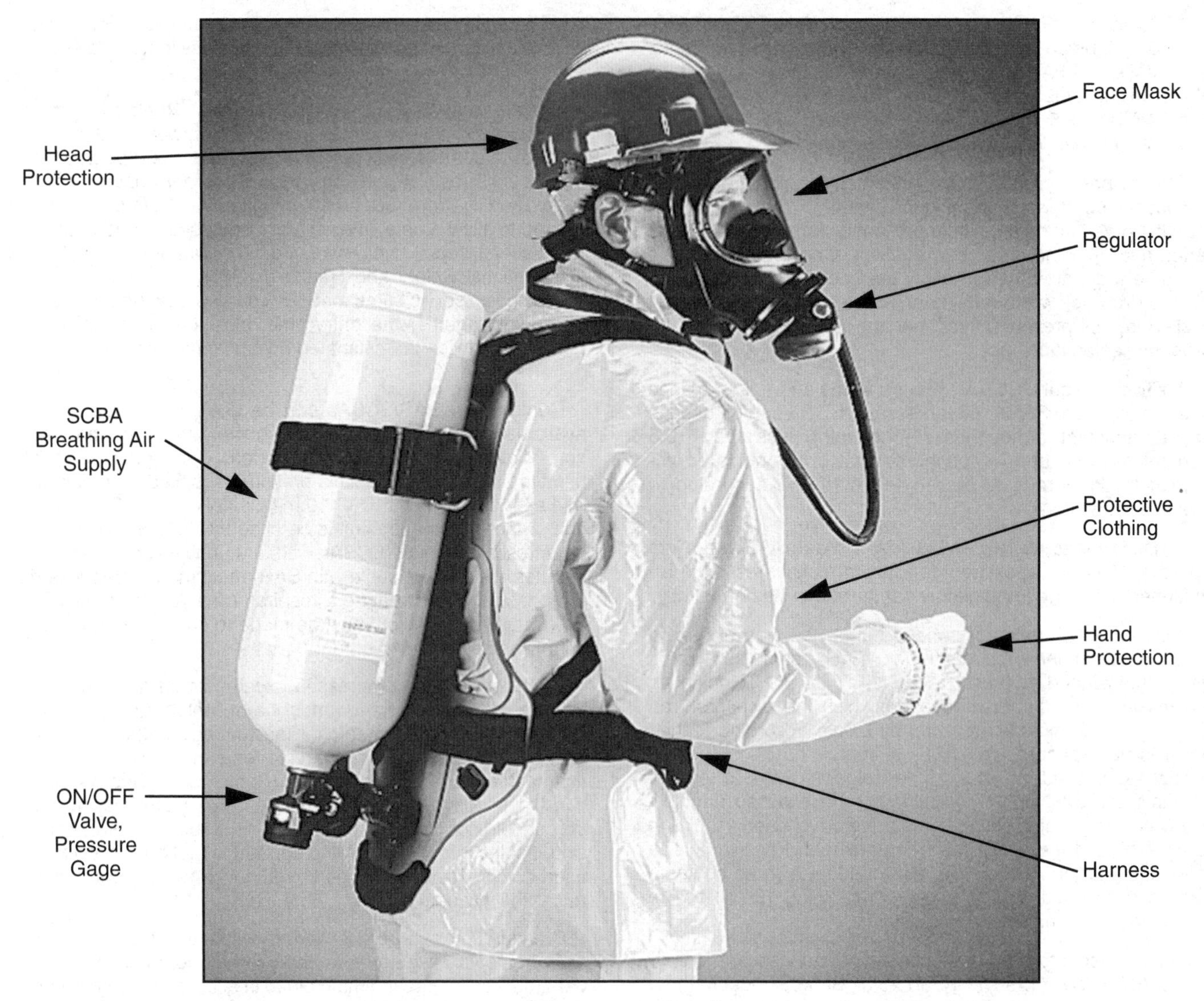

Fig. 4.16 Self-contained breathing apparatus with personal protective equipment
(Permission of Industrial Safety Supply)

self-contained breathing apparatus must be used. In general, though, the self-contained breathing apparatus (SCBA) will be used by rescue personnel; therefore, it must be readily available when working in a confined space.

Any person who may be expected to use respiratory protection must have a medical examination and be tested to determine their ability to safely use this equipment. If an operator must work in a confined space under hazardous conditions, the operator must wear a SCBA. The American National Standards Institute (ANSI) Z88.2 – 1980 "Practices for Respiratory Protection" states that "a respirator equipped with a facepiece shall not be worn if facial hair comes between the sealing periphery of the facepiece and the face or if facial hair interferes with valve function."

Rescue workers must be standing by wearing SCBAs so that any unexpected rescue operation can *BEGIN IMMEDIATELY* and *BEFORE THE STRICKEN PERSON IN THE CONFINED SPACE DIES.* If the operator(s) must enter confined spaces to perform rescue services, they must be trained specifically to perform the assigned rescue duty and to use required personal protective equipment (PPE) and rescue equipment. Rescue practice sessions must be held at least once every 12 months and at least one member of the rescue service must hold a current certification in first aid and cardiopulmonary resuscitation (CPR).

As a possible alternative to SCBAs, many agencies require the operator to have a 5-minute escape air pack in the confined space. This provides a 5-minute supply of compressed breathing air when activated to escape. Check the appropriateness of this with your local safety agency.

The atmosphere in large sewer systems may suddenly and unpredictably become hazardous and the collection system operator needs to exit the working area immediately. Training and specialized equipment may assist the operator in exiting the area safely. An emergency escape breathing apparatus (EEBA), such as the one shown in Figure 4.17, is a self-contained breathing apparatus that provides the operator with a constant flow of air for a short period of time. The purpose of the device is to provide respiratory protection long enough to enable the operator to get out of a hazardous area or oxygen-

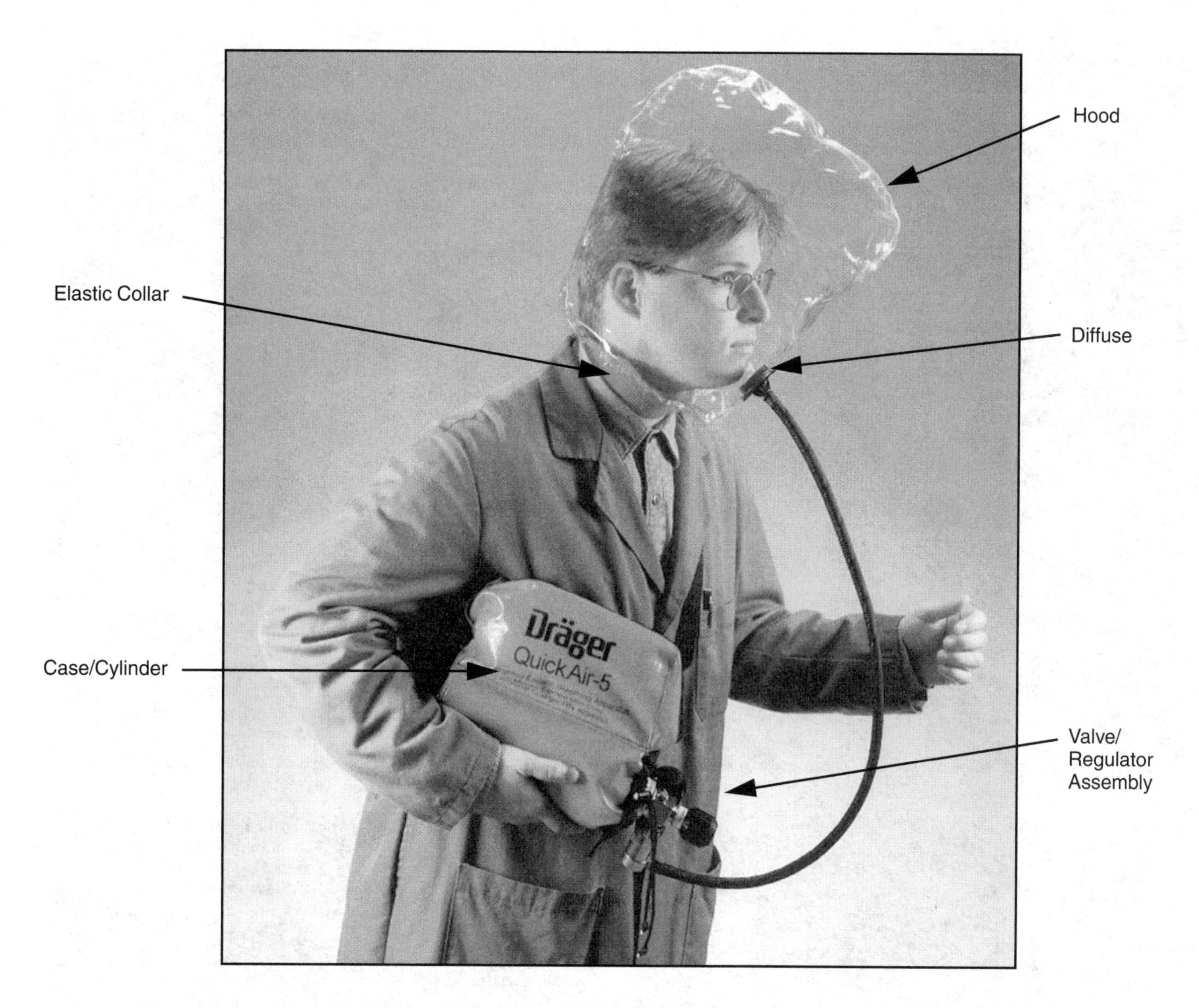

NOTE: The emergency escape breathing apparatus is intended for use in escape and emergency situations only.

Fig. 4.17 Emergency escape breathing apparatus (EEBA)
(Permission of Industrial Safety Supply)

deficient atmosphere. The devices will only provide air to the user for 5 to 10 minutes, depending on the model selected; therefore, a safe escape route should be planned in advance to ensure that the operator will have enough time to reach a safe area before the air supply runs out.

4.51 Safety Harness With Lifeline, Tripod, and Winch

Whenever anyone must enter and work in a manhole or other confined space, a safety harness with a lifeline must be worn. The harness and lifeline may be used to lower an operator into a confined space. Whenever an operator is underground (in a manhole), one operator must remain topside observing the actions of the operator in the harness. A third person must be nearby because two people will be required to pull an unconscious operator out of a manhole.

A chest or full-body harness such as the one shown in Figure 4.18 must be used. This type of harness will maintain the body in an erect position. The lifeline must be attached at the center of the entrant's back near shoulder level, or above the entrant's head. Not only is it difficult to maneuver a tilted or doubled-over body through the 24-inch manhole opening, but you could easily injure a person who is being removed in such a position. Some agencies use "wristlets" (Figure 4.19) instead of a harness. If you choose to use wristlets, you must be able to demonstrate that the use of a chest or full-body harness is not feasible or creates a greater hazard and that wristlets are the safest and most effective alternative.

A winch is usually used to raise or lower an operator wearing a safety harness. The use of person-rated, truck-mounted, davit-type winches is not generally recommended in collection systems operation. Because vehicles are parked in traffic, there is always the potential for the vehicle to be hit by another vehicle. If a disabled operator being rescued is still attached down in the manhole when a truck-mounted winch is struck by another vehicle, the result could be either very serious injury or death. For this same reason, vehicles are not normally used as tie off points for lifelines.

Lightweight, portable retrieval systems have been developed and are increasingly popular with operators. They are portable, easy to set up and transport from one area to an-

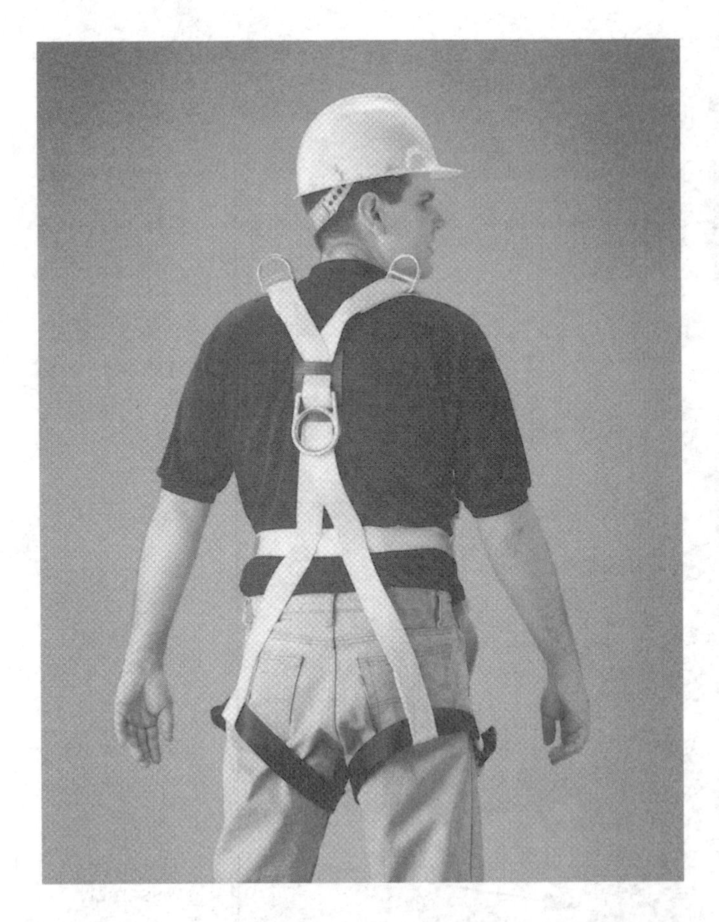

Fig. 4.18 Full-body harnesses

(Photos courtesy of Miller® Equipment)

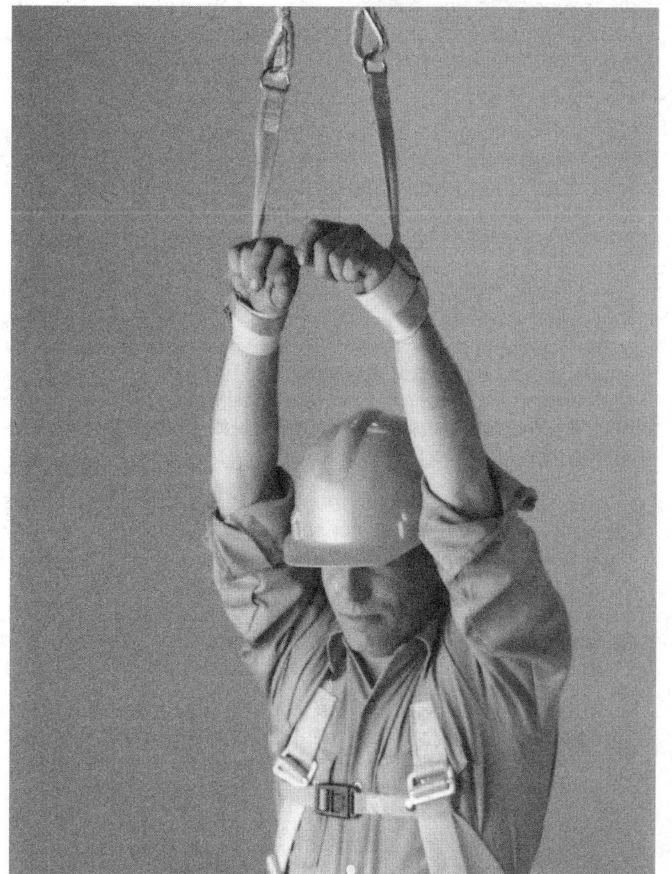

Fig. 4.19 Wristlets
(Photos courtesy of Miller® Equipment)

other, and they provide a method for retrieval of an operator from a manhole with only one person above. Figure 4.20 illustrates such a device. This one uses a triple braking system and a two-speed winch. It is corrosion resistant, has 120 feet of cable capacity, a level-wind cable device and a maximum capacity of 350 pounds. The tripod is adjustable with safety chains and rubber safety shoes, and weighs 110 pounds.

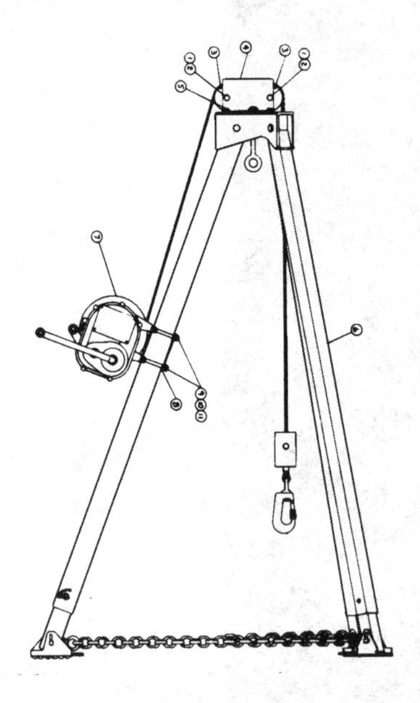

Fig. 4.20 Tripod with two-speed winch
(Permission of DBI/SALA)

Figures 4.21 and 4.22 illustrate some variations of a tripod combination with a retractable steel cable safety block and a manual retrieval winch (come-along). The safety block is used as a fall prevention device. When a fall occurs, inertia activates the block to prevent a free fall. The come-along enables a single crew member on the top to retrieve an unconscious person.

OSHA regulations require that a "mechanical device" be available to retrieve personnel from vertical-type permit spaces more than 5 feet (1.52 m) deep. No matter what specific type of tripod you use, remember that the tripod must be properly centered over the manhole opening so a person can be lifted up and out of the manhole without the person touching the rough surfaces of the manhole ring.

QUESTIONS

Write your answers in a notebook and then compare your answers with those on page 178.

4.5A Why is it necessary to have a confined space permit and follow confined space procedures to enter sewers containing raw wastewater, pump station wet wells, gravity sewer manholes, and siphon headworks structures?

4.5B What type of respirators provide the greatest protection for operators wearing them?

4.5C When an operator is working in a confined space and an atmospheric alarm condition develops, what should the operator do?

4.5D What types of situations require the use of a safety harness with a lifeline?

4.52 Portable Atmospheric Alarm Unit (Figure 4.23)

4.520 Purpose and Features

An atmospheric alarm unit should continuously sample the atmosphere from a manhole and test it for the presence of explosive or flammable gases, toxic gases (hydrogen sulfide) and the percentage of oxygen. If pick holes are available in the manhole cover, the atmosphere should be tested prior to removing the cover. In the event the cover does not have any pick holes, the cover should be carefully removed using *NON-SPARKING TOOLS*[13] so that a probe can be inserted or an air sample can be taken.

Oxygen content should be the first test performed with atmospheric analyzers because most combustible gas meters are oxygen dependent and will not provide reliable readings in an oxygen-deficient atmosphere. Combustible gases are tested for next because of the threat of fire or explosion, and then test for toxic gases.

One type of atmospheric analyzer withdraws air from the manhole for testing; another type is equipped with a sensor or probe that can be lowered into the manhole. The device should have an audible and visible alarm that will warn when flammable gases (such as methane, CH_4) exceed 10 percent of the lower explosive limit (LEL), when hydrogen sulfide concentration exceeds 10 ppm, or when the oxygen percentage drops below 19.5 percent or rises above 23.5 percent. Battery-operated units also should have an alarm to indicate low battery power before the power becomes inadequate for proper functioning. Gas testing devices must be maintained and in proper operating condition at all times. Be sure these devices receive regular calibration and preventive maintenance by a qualified person.

Atmospheric alarm units should have the following features (in addition to being safe for use in flammable/explosive atmospheres):

1. Portable and battery operated;
2. Continuously and simultaneously monitor for toxic and flammable gases as well as oxygen deficiency/enrichment;
3. Audible and visible *ALARMS* for each hazardous condition detected (operators in a manhole should not have to read meters nor decide if a hazardous condition exists—when an alarm sounds or flashes, *GET OUT NOW*);

[13] *Nonsparking Tools. These tools will not produce a spark during use. They are made of a nonferrous material, usually a copper-beryllium alloy.*

Fig. 4.21 Tripod rescue and retrieval systems
(Permission of DBI/SALA)

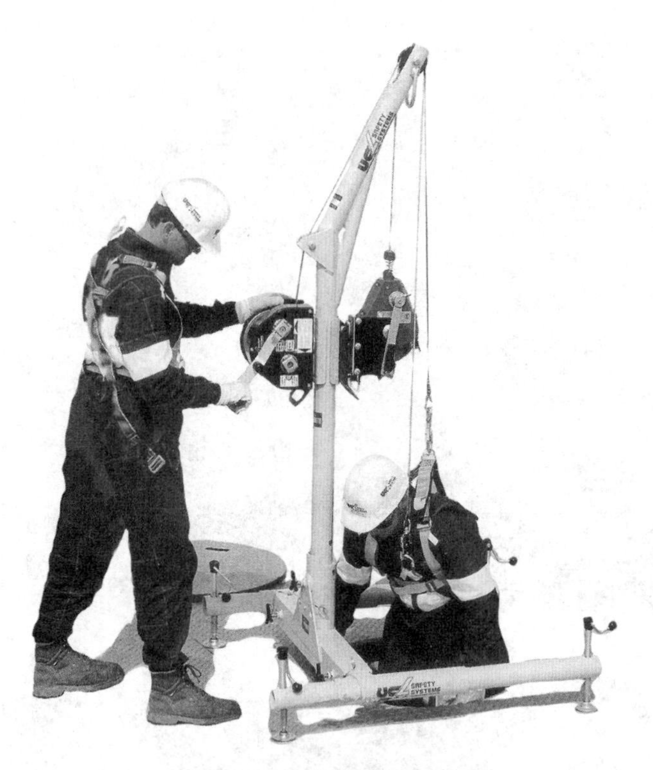

Five-Piece Safety System

(Breaks down into lightweight, manageable components for storage and is transported on a cart that moves easily over smooth or rough terrain.)

Vehicle Hitch Mount Sleeve

(Designed to install into a 2" hitch receiver attended vehicle to provide a portable anchor point for confined space entry, retrieval, rescue, and fall arrest.)

Fig 4.22 Manhole and confined space entry/retrieval equipment

(Permission of Unique Concepts Ltd.)

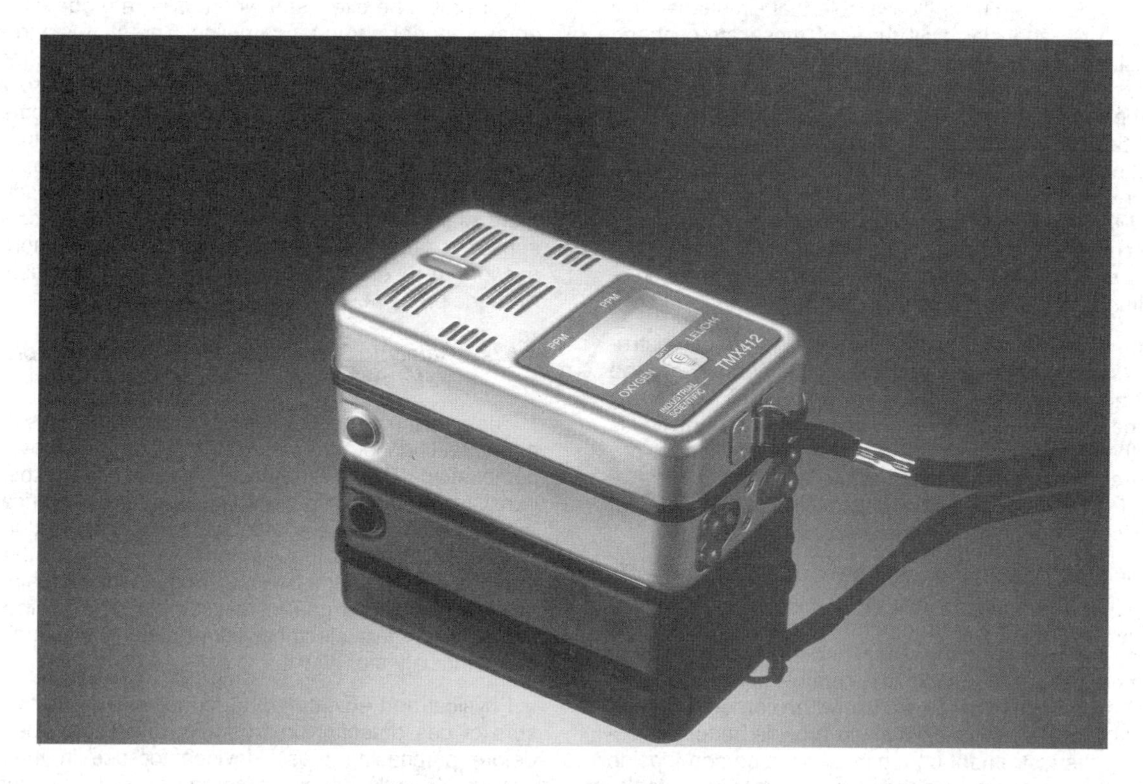

Fig. 4.23 Portable multi-gas monitor
(Permission of Industrial Scientific Corporation, 1-800-DETECTS)

4. Lighted indicator to confirm that the unit is operating;

5. Small enough and light enough to be worn by an operator in a manhole;

6. Over 10 hours of continuous operation on one full battery charge;

7. Automatic audible and visible low battery alarm;

8. Sample hose and aspirator or sample pump to test the atmosphere through the holes in a manhole cover and at various levels within the confined space; and

9. Gas test kit to calibrate atmospheric alarms. Set the alarm to sound when the following levels are reached:

TOXIC:	More than 10 ppm hydrogen sulfide (H_2S).
FLAMMABLE:	More than 10 percent of the Lower Explosive Limit (LEL) of methane (CH_4).
OXYGEN:	Less than 19.5 percent by volume.

Follow the manufacturer's recommendations regarding procedures and frequency of instrument calibration and preventive maintenance.

Advances in microelectronics and sensing elements have drastically improved testing instruments. Previously, instruments tended to be very large, bulky, cumbersome, sensitive to temperature changes or other ambient (surrounding) conditions, and frequently unreliable. Technological advances in instrumentation have produced a number of instruments without the limitations of the early equipment. Many such instruments are readily available to collection system operators.

The most important fact to keep in mind when using instruments in confined space work is the following:

ATMOSPHERIC TESTING EQUIPMENT IS JUST ONE TOOL YOU USE AND SHOULD NOT BE CONSIDERED THE ULTIMATE TEST. INSTRUMENTS ARE NOT 100 PERCENT RELIABLE; THEY CAN FAIL OR GIVE ERRONEOUS READINGS. THEY SHOULD NOT, REPEAT SHOULD NOT, PRECLUDE THE USE OF VENTILATION OF SOME SORT IN THE CONFINED SPACE, BUT RATHER SHOULD BE USED IN CONJUNCTION WITH VENTILATION.

This section will acquaint you with the basic features of testing/alarm instruments and some important concepts related to their use. Before buying testing/alarm instruments, carefully analyze their features and suitability for use in the collection system environment, since reliability and ease of maintenance will vary with the manufacturer.

As discussed earlier, operators are concerned with three specific guidelines in the working atmosphere:

1. Oxygen content,

2. Explosive or flammable conditions, and

3. Toxic gases (mainly hydrogen sulfide).

Test instruments are currently available that simultaneously sample, analyze, and alarm all three of these atmospheric conditions. Instruments are also available that test oxygen and explosive conditions only or a single gas instrument which would sample only one condition such as hydrogen sulfide. Figure 4.23 illustrates an instrument that monitors oxygen content, explosive or flammable conditions, and toxic gases, plus carbon monoxide (CO), which may be required by some agencies. Many instruments are also available in which a sensor(s) can be changed easily. This enables the operator to monitor for various other atmospheric hazards such as chlorine or sulfur dioxide with the same instrument.

Because of the limited access and working space, instruments that sample three or four items simultaneously are the most desirable since they reduce the amount of equipment that must be taken into the confined space. A good practice is to secure the instrument not only with the loop provided by the manufacturer, but with an additional line as well. The instrument could fall out of its case and into the flow and be lost unless this secondary safety line is used.

Gas detection instruments may be carried by operators working in manholes or left on top of the manhole with a person who monitors the instruments. If the instrument is left on top, then the person on top must keep the suction hose or wand near the person working in the manhole at all times. The instrument is exposed to greater abuse when carried by an operator working in a manhole, but could provide better protection. Leaving the instrument on top makes the person topside closely observe the operator in the manhole while keeping the suction hose or wand near the operator. Also the topside person may be able to hear alarms better than an operator working in a manhole, although some instrument manufacturers can provide accessory devices to enhance audible and visual alarms for work in high noise level areas.

The gas detection instruments available today are small, lightweight, and rugged, and are designed for a wide range of operating conditions. They consist of two major elements: (1) the sensing element, and (2) the electronic/alarm indicating section.

4.521 Diffusion-Type Sensing Element

Most of the portable personal-type instruments produced today have diffusion-type sensing heads, which simply means that the ambient air is allowed to contact the sensor through natural air circulation. An air pump is not required to pump the sample into the sensing area. Diffusion heads are used because they are low cost, rugged, and generally require only simple maintenance. The sensor head should be easily removable from the electronic section because they do wear out and require replacement.

Optional equipment available for most gas detection instruments includes an aspirator or pump which is used with a probe or tubing that can be lowered into a structure, such as a manhole. This feature will be useful if manhole covers in your system have a pick hole in the cover that will allow you to sample the atmosphere prior to removing the manhole cover. However, non-ventilated manhole covers are much more widely used today because of odor considerations and the introduction of storm water through the cover holes. Therefore, it is usually not possible to sample manhole atmospheres prior to lifting the cover.

Care must be exercised when using a probe or tubing with an aspirator or pump for sampling. The pumps and aspirators are capable of drawing water into the instrument. Instruments contaminated with water usually experience severe failure and require extensive and expensive repair work. Some types of sensors are also sensitive to humidity, temperature, or toxic conditions. The high concentrations of silicone, paint, solvents, and fluorocarbons that are sometimes present in treatment plants and lift stations can reduce the accuracy of a sensor or damage the sensor beyond repair. "Poison resistant" sensors, a recent development, are now being used to minimize the problem.

4.522 Catalytic Sensing Element (Used in Explosive Meters)

The sensing elements used in most of the personal atmospheric testing instruments operate on the principle of catalytic combustion. They are relatively simple and inexpensive and consist of a sensor filament that is heated electrically. When the filament comes in contact with an air/gas mixture, the gas burns and changes the electrical resistance of the filament. The amount of change that occurs depends on the percentage of the combustible gas in the sample. The microelectronics then convert this electrical information to an audible and a visual indication or alarm.

Physical and environmental conditions can affect the sensors of gas detection instruments and should be evaluated before purchasing these devices for use in the collection system. Caustic gases, temperature, dirty air, humidity, air velocity, and vibration are all factors which should be considered since they can cause inaccurate gas detection meter readings. Hydrogen sulfide, for example, will "poison" conventional catalytic-type sensors used in the typical explosive meter.

The catalytic sensor has a limited life and therefore must be replaced periodically.

4.523 Electrochemical Sensors (Hydrogen Sulfide and Oxygen Meters)

Electrochemical sensors use a chemical reaction to generate an electrical current when hydrogen sulfide comes in contact with the sensing element. The electrochemical reaction is proportional to the level of hydrogen sulfide in the air. The electrical current produced by this reaction is amplified in the electronic circuitry and is sent to the alarm and display indicator. The electrochemical sensor also has a limited life and must be replaced periodically, since it does lose its sensitivity to hydrogen sulfide.

4.524 Electronic Section

The electronic section contains the microelectronic circuitry including a microprocessor and the audio and visual alarms. Typically visual alarms are light emitting diode (LED) readouts in one or more colors, or a single LED display that lights up when the alarm set point is reached. The display may also indicate actual level in ppm using a liquid crystal display (LCD). Monitors should have an audible alarm feature, which is a signal you can hear. It is usually specified as a minimum sound level of 80 *DECIBELS*[14] at 10 inches to allow you to hear the alarm in a noisy working environment.

The electronics should be housed in a high-impact, rugged plastic case, classified as intrinsically safe for use in Class I, Division I areas.[15] A rating of "intrinsically safe" means any arcing or sparks generated in the electrical circuits will not have sufficient energy to initiate an explosion in a combustible or explosive atmosphere. It is important that the housing be waterproof or at least moisture-resistant, since that is the type of environment where you will use it. The instrument should not be left overnight in vehicles, particularly if you are in an area where it is cold, since temperature can affect the sensors and the battery. The instrument should be returned each night to the shop. If standard batteries are used, they should be checked and replaced. If rechargeable batteries are used, the instrument should be placed on the charger so it is ready for use again the following day.

4.525 Alarm Set Points

When a flammable gas is introduced into the atmosphere, fresh air is displaced until the area is filled with the flammable gas. During the process when the atmosphere goes from a normal atmosphere to one that is 100 percent saturated with flammable gas, three specific conditions occur (refer to Figure 4.15, page 116, illustrating Lower Explosive Limit). These conditions are:

1. The atmosphere is too lean to initiate an explosion if a spark or some other ignition source is introduced (not enough flammable gas in the mixture);
2. The air/gas mixture is in the range where an explosion will occur if an ignition source is present; and
3. The atmosphere is too rich to result in an explosion, even if an ignition source is present (too much flammable gas and not enough air in the mixture).

Explosive meters are calibrated in percent of the LEL as shown on Figure 4.15. Most instruments have adjustable alarm set points. That is, you can determine the level at which the alarm will be activated. Most agencies select 10 percent as the standard for explosive atmospheres. This setting allows the collection system operator time to evaluate the atmosphere of the area when an alarm condition occurs, rather than waiting until the atmosphere does become explosive. The instrument should also indicate other alarm conditions including low battery or other failure conditions.

Similarly, 10 parts per million is generally accepted as the alarm set point for hydrogen sulfide. The maximum concentration that can be inhaled for one hour is 200 to 300 ppm. Death may occur rapidly through exposure to hydrogen sulfide at concentrations above 700 ppm for 0.5 to 1 hour.

The typical alarm set points for oxygen content are 19.5 and 23.5 percent; 16 percent is the lowest oxygen level that will sustain consciousness. All alarm set points should match the values identified in the OSHA definition of a hazardous atmosphere and are selected to allow an operator time to evacuate before reaching the points at which levels become critical.

4.526 Calibration

It is extremely important to perform routine calibration before the use of atmospheric sensing instruments, since the electronics may drift off the desired alarm set points. In addition, the sensors have a limited "shelf life" and will only function for a certain period of time regardless of whether the device is being used or stored. This period may range from several months to a year, so even unused sensors may not be operable if stored for longer periods of time. Instruments should allow a sensor change without returning the entire unit to the manufacturer where it would be unavailable for several weeks. A satisfactory number of spare units should be available to compensate for instrument maintenance activities (calibration, sensor replacement).

Calibration kits are available and should be purchased along with the equipment. These kits contain a small gas cylinder with an accurately measured percentage of gas; for example, 2.5 percent methane and 17 percent oxygen. In addition, an adapter is provided that allows the sensor head(s) to be exposed to the gas.

Before working in any confined space environment, it is essential to have the right tools to ensure safe entry. Multi-gas utility monitors now can be calibrated automatically with an automated instrument management system, as shown on Figure 4.24. This system is designed to reduce the time and costs spent manually calibrating instruments while reducing the liabilities associated with instrument maintenance and recordkeeping. The system consists of a master control and PC interface station and up to five instrument docking modules for complete two-way instrument communication. The control software and database allow the system to automatically log instruments into a data management system, initiate function-

---

[14] *Decibel (DES-uh-bull). A unit for expressing the relative intensity of sounds on a scale from zero for the average least perceptible sound to about 130 for the average level at which sound causes pain to humans. Abbreviated dB.*

[15] *All sensing instruments should be classified for use in Class I, Division I areas, as specified by the National Electrical Code (NEC). For further information about the classification of areas, see the section on "EXPLOSIVE ATMOSPHERES" later in this chapter on page 157.*

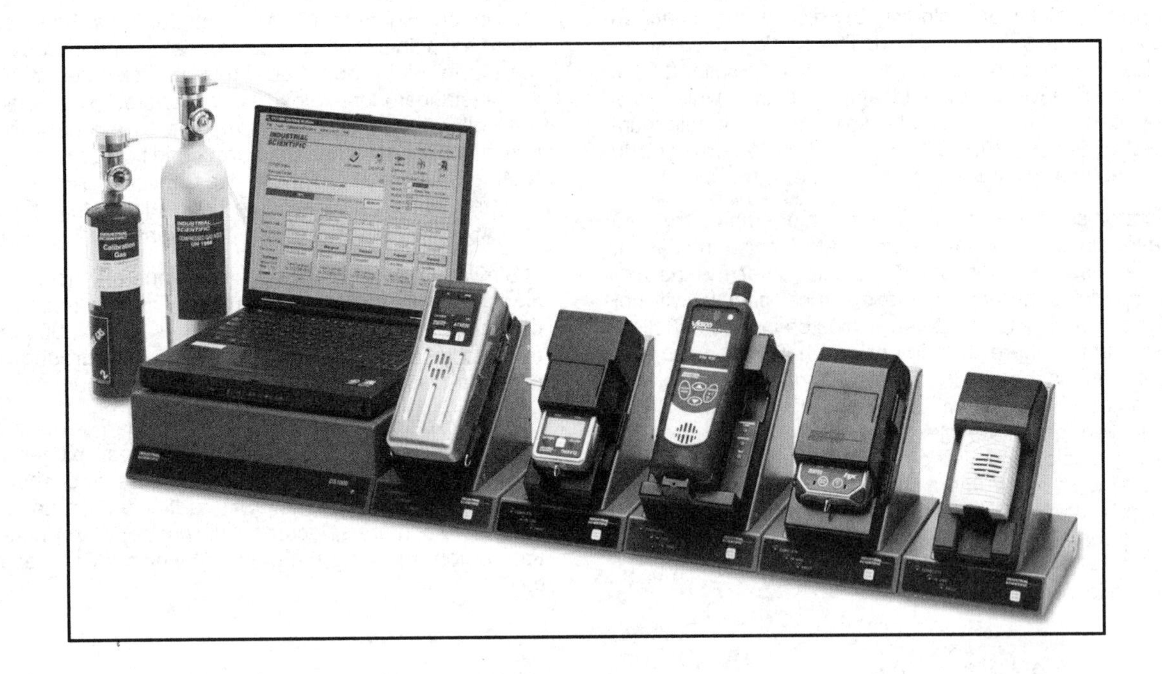

Fig. 4.24 Automated gas detection/instrument management system

(Permission of Industrial Scientific Corporation, 1-800-DETECTS)

al tests and full instrument calibration, download and retrieve calibration information and survey data, store all calibration and maintenance records, set and upload instrument alarm and calibration guidelines, and recharge batteries. The master control station is capable of accepting up to 12 gas inputs, which allows numerous instrument configurations to be managed in the same database.

QUESTIONS

Write your answers in a notebook and then compare your answers with those on page 178.

4.5E An atmospheric alarm unit is used to test for what hazardous conditions?

4.5F What is the sequence of testing for hazardous atmospheres?

4.5G What are the three types of sensing elements commonly used on atmospheric alarm units?

4.5H What are some physical and environmental conditions that could affect the accuracy of gas detection instruments?

4.5I What is the usual alarm set point for hydrogen sulfide?

4.53 Ventilation Blower With Hose (Figure 4.25)

Even after testing all levels of a manhole for toxic gases, explosive conditions, and oxygen deficiency/enrichment, run the ventilation blower for at least 10 to 15 minutes with the hose outlet positioned at the bottom of the manhole *BEFORE* entering the manhole. A fan-type blower driven by an electric motor or truck power should have between 750 and 850 CFM (cubic feet per minute) capacity to adequately ventilate a standard manhole with a 4-foot diameter at the bottom. The standard unit will have a 15-foot long hose, 8 inches in diameter, to conduct the blower air to the bottom of the manhole. Hose couplings and extensions must be available because some manholes are over 30 feet (10 m) deep. Gasoline engine-powered blowers are less desirable due to the high noise factor and the possibility of producing undesirable fumes and gasoline leaks. Some agencies use blowers powered by bottled gas or electric blowers powered by a vehicle-mounted generator. One problem encountered in using blowers is that blowing air into a manhole can create pockets of gas. Ventilation can be improved by removing the nearest upstream or downstream manhole cover. Ventilation can be further enhanced by using a combination of dilution ventilation (adding air to the entry manhole) and exhaust ventilation (pulling air from a downstream or upstream manhole). *CAUTION: Blowers used for exhaust ventilation* ***must be explosion proof*** *so that they do not provide a source of ignition in the event a flammable atmosphere is encountered.*

Gasoline and Electric Centrifugal Blower

(Permission of Wanco Inc.)

Blower Hose Canisters

(Permission of Wanco Inc.)

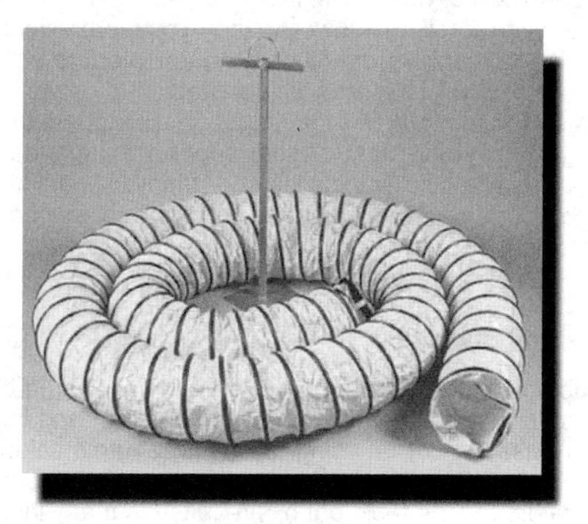

Ventilation Hose Carrier

(Permission of Wanco Inc.)

Ventilation of a Manhole

NOTE: Air inlet should be at least two feet above the street level so that street trash will not be picked up by the flowing air. Position the hose inlet in such a way that it will not pick up and blow exhaust gases from work vehicles or traffic into the manhole.

Fig. 4.25 Ventilation blower with hose

(Permission of Industrial Safety Supply)

4.54 Manhole Enclosure (Figure 4.26)

An enclosure is sometimes set up around an open manhole to prevent unauthorized persons from endangering the operator in the manhole or from falling in themselves. This structure should be of adequate strength to support the weight of an operator when required. Do not use an enclosure if it interferes with the work of a crew or their safety.

Some agencies use a "toe board" in addition to the enclosure. This is a metal ring that fits into the manhole frame and extends a minimum of four inches above the ground. These devices prevent objects from rolling into the manhole and provide protection against people walking into an open manhole.

4.55 Ladders and Tripods

Lightweight fiberglass or aluminum-alloy extension ladders are used in shallow manholes of sufficient diameter. They are easy to handle and to climb. However, they occupy precious space in a 24-inch manhole opening and often interfere with entry, especially if SCBA is used. Many operators use strap or rope ladders with rigid metal or rope rungs. These ladders are lightweight, do not require firm bottom support (suspended from the top), and are quickly installed and removed with little danger to the operator in the manhole. Tripods with winches are used to remove injured persons quickly and safely without the rescuer being exposed to hazardous conditions.

4.56 Ropes and Buckets

Tools for use by operators in a manhole must be lowered to them and recovered from them by the use of a bucket and tag line. Lower or raise the bucket using a hook rope with a safety clasp that will swivel. Dropping tools for operators to catch and allowing operators to toss tools out has resulted in many injuries and should result in immediate disciplining of the guilty operators.

4.57 Head, Hand, and Body Protection

Safety in a collection system begins in the design stage where the designer incorporates "engineering" solutions to safety problems. However, it is not practical to eliminate every hazard in the workplace simply by design. When it is impractical or impossible to eliminate the source of the hazard, then the operator must be protected in some other way. This is done by wearing approved personal protective gear such as hard hats, safety shoes, chemical-resistant clothing, gloves, toe guards, and other forms of equipment that will protect the operator's head, hands, and body.

Protective safety gear is sometimes less comfortable to wear than ordinary work clothes; therefore, some individuals are tempted to lay it aside when the "boss" isn't around. That operator becomes a gambler who is betting his/her life, or eyesight, or other physical well-being that "it won't happen to me." Losing that bet can easily mean a lifetime of discomfort rather than the minimal discomfort that would have been experienced by wearing the safety equipment for the duration of the job. Safety, in this instance, is a knowledge of the hazards, knowledge of the protection available, and a frame of mind that makes use of available protection a safe work habit.

Working in a wastewater collection system creates a potential hazard to health because wastewater is a carrier of disease-producing organisms and various chemical materials. Two important factors need to be considered when working in this environment: hygiene and protective equipment for the head, hands, and body.

The number one rule of hygiene is to prevent oral or skin contact with wastewater. This means no smoking, eating, or drinking until after washing thoroughly with a disinfectant soap and hot water. Waterless antiseptic hand cleaners also are effective and approved for washing purposes. Paper towels may be used for the wiping process. This is very important if the operator smokes or is going on a lunch or coffee break.

Clean clothes are also a must. Clothes worn on the job should be changed regularly and should be washed when dirty. Protective overalls or disposable coveralls will help maintain cleanliness. When the work day is over and the operators have returned to the maintenance yard, facilities should be provided for showering and changing into clean clothes before going home. As mentioned earlier, work clothes should remain at work and should never be washed with the family laundry. This will reduce the chances of infecting the operator's family.

Skin Protection

Cuts, bruises, blisters, or any breaks in the skin provide a pathway for the entry of disease-causing organisms that may be carried by either air or water. These skin conditions may permit infections to develop and possibly lead to very serious physical problems. The preventive measures that need to be taken include:

- Protect yourself from injury by wearing proper clothing such as gloves, chemical protective suits, hard hats, and safety shoes;
- Use barrier creams to prevent absorbing disease-causing organisms or substances through the skin, and to reduce dermatitis;
- Use first aid immediately—clean the affected area and apply approved antiseptic, cover with a waterproof bandage, then report; and
- For injuries that are more serious than minor cuts, abrasions, or blisters, follow approved first aid or rescue practices and see a physician as soon as possible.

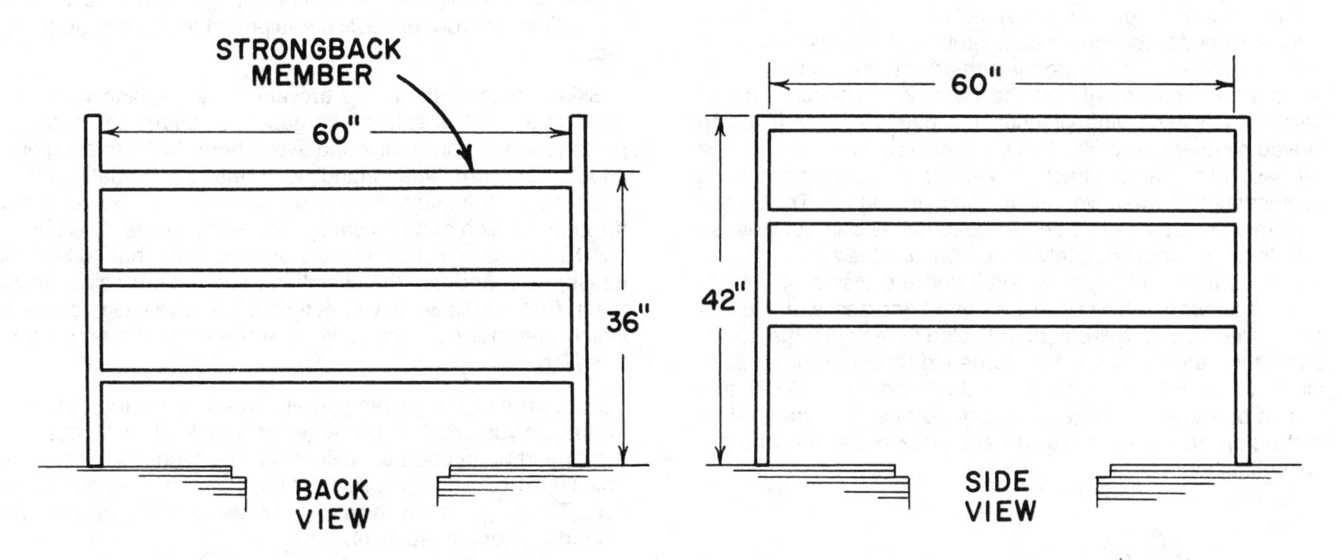

MATERIALS SCHEDULE

STRONGBACK 1 –1½" BLACK PIPE

REMAINDER – ¾" BLACK PIPE

(OR OTHER WITH EQUIVALENT STRENGTH)

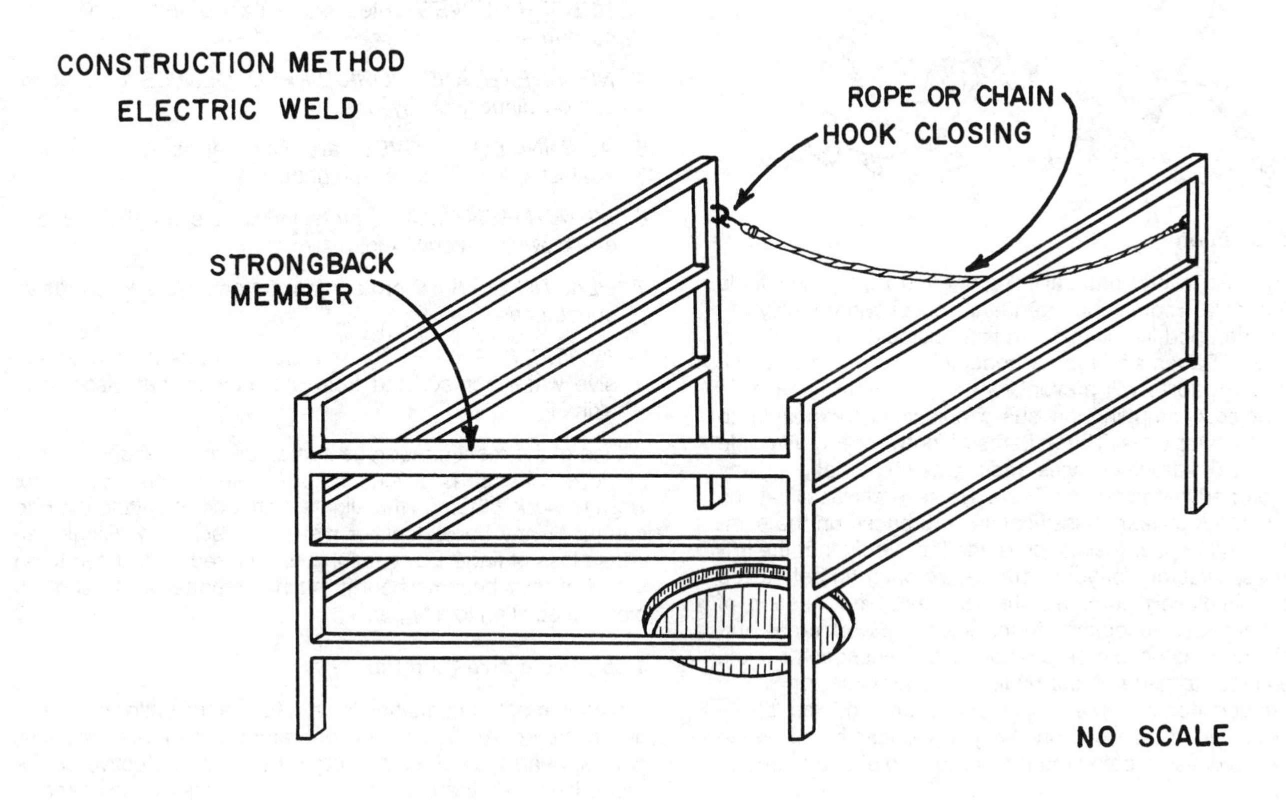

FINISH

SAFETY ORANGE OR YELLOW

NOTE: Strongback members and both sides should be coupled together so they can be stacked for transportation and quickly assembled if needed.

Fig. 4.26 Typical portable manhole safety enclosure

Head Protection

Many activities performed by collection system operators involve working above or below ground levels, movement of materials overhead, or working near construction and machinery. In such operations, the hazards constantly exist of being struck by falling objects, machinery, or loads being moved by machinery. Hard hats are provided to prevent head injuries from being struck by falling objects and bumps against objects when working in confined spaces. The proper protection is provided when the head harness is adjusted so that there is approximately one inch of clearance, plus or minus 1/8 inch, between the skull and the inside of the hat when it is worn. When the harness becomes worn to the extent that it no longer can be adjusted to maintain that clearance, the hard hat should be turned in for repair or replacement. Hard hats must meet ANSI Standards (Z89.1) and should be worn by all operators and visitors at all times. Note that bump hats do not provide adequate protection and are not suitable for use.

Foot Protection

Many tasks involve manual lifting or handling of heavy tools and materials. Foot injuries frequently occur when heavy objects are dropped, resulting in bruises, dislocations, fractures, or crushes. Shoes and rubber boots that are reinforced with steel toes or soles will prevent foot injuries from impacts of falling objects, stepping on sharp objects, or exposure to blades of power tools. These items of footwear are available in a variety of attractive styles and are as comfortable as any pair of properly fitted shoes. There are several classifications of safety toe footwear; classifications are based on the compression and impact resistance of the toe. Verify that the ratings are satisfactory for your potential exposure(s) before you (or your employer) purchase the footwear. The wearing of sandals or canvas sneakers (tennis shoes) in work areas outside office areas should be prohibited; safety shoes or boots are required. In general, sturdy leather shoes that provide a deep-lugged nonslip sole, ankle protection, and toe safety should be standard wear on the job. Rubber boots, when used, should also meet standards for toe, sole, and arch protection.

Hand Protection

Hands must be protected from rough surfaces, sharp edges, toxic or irritating materials, heat, and electrical equipment. Gloves are available to protect against any of these hazards and various types are available for the specific job application. Gloves with leather palms should be worn when handling rough edges or abrasive materials or when the work exposes your hands to possible lacerations, puncturing, or burns. When working around machinery that is revolving, wearing gloves or other hand protection can be dangerous. If a glove gets caught in the machinery, you could become injured. Don't let your protective equipment itself become a hazard.

Skin irritations should be prevented by washing with soap and water, not a solvent or gasoline. Learn to recognize poison ivy and poison oak and avoid them. Appropriate gloves should be worn when handling irritating materials and all chemicals. Compounds such as solvents can be absorbed through the skin and can cause long-term effects. Consult the MSDS for the specific chemical or substance that you will be dealing with and use the specified personal protective equipment (PPE). Many glove materials are available, some of which may not be compatible with the chemical you will be handling.

Be sure that the gloves you are wearing are the right type for the job you are doing. The gloves should allow for quick removal and be in good condition. Always check for cracks and holes, flexibility, and grip. Keep them clean and in good condition. There are many types of gloves and the proper type should be worn for each job.

1. *CLOTH GLOVES* protect from general wear, dirt, chafing, abrasions, and low heat.
2. *LEATHER GLOVES* protect from sparks, chips, rough material, wood slivers, and moderate heat.
3. *RUBBER GLOVES* protect against acids and some chemical burns.
4. *NEOPRENE AND CORK-DIPPED GLOVES* give better grip on slippery or oily jobs.
5. *ALUMINIZED GLOVES* are heat-resistant to protect against sparks, flames, and heat.
6. *METAL MESH GLOVES* protect from cuts, rough materials, and blows from edge tools.
7. *PLASTIC GLOVES* protect from chemicals and corrosive substances.
8. *INVISIBLE GLOVES* (barrier cream) protect from excessive water contact and from substances that dissolve in skin oil.

One of the most dangerous human ornamentations to wear in industrial work is a ring. Rings should be removed or not worn to work if there is the slightest chance of getting the ring caught in any hook, tool, or piece of machinery. Rings can cause loss of fingers or painful cuts and frequently have to be cut off if they become bent in such a manner as to shut off blood circulation to a finger.

4.58 Protective Clothing

Under most circumstances, a basic work uniform is adequate protective clothing for operators, along with goggles, gloves, head protection, and other personal protective equipment. However, there are times when we may be exposed to chemicals and we need additional protection. This exposure could occur from an industrial discharge, accident, or emergency situation, or routine application or use of chemicals in the collection system.

Protection against chemicals can be simple and cover a specific part of the body, or the protective gear can be complex and entirely cover an operator. An operator may be exposed to a chemical that is only mildly irritating. In such a case, it may be enough to wear safety goggles and the proper type of gloves to protect the eyes and hands. For example, when using a dilute acid in small amounts, we would need to shield the eyes and skin from the chemical.

The other extreme is a work environment that involves a highly toxic substance that can get into the air and be dangerous to breathe as well as hazardous if it touches the skin. Benzene and many pesticides are examples of such toxic substances. Full-body protection and a respiratory device may be needed in such a situation.

Some chemicals may be a problem only with long-term, repeated exposure. Others may be deadly from a single, brief contact. The Environmental Protection Agency's Office of Emergency and Remedial Response defines four levels of chemical hazards and states the type of protection required for each.

LEVEL A Highest level of respiratory and skin protection. A totally encapsulating, chemical-resistant suit is needed, as is a self-contained breathing apparatus worn inside the suit (so that chemicals won't damage it). At such a protection level, you must be sure the suit material specifically resists the substances involved.

LEVEL B High-level respiratory protection; lower level skin protection. Lightweight protective clothing is needed. Specific chemicals should be tested on the protective clothing material.

LEVEL C Respiratory protection; protection against chemical inhalation. This will usually involve an air purifying system with a full facepiece. Level C protection may also involve gloves, goggles, head protection, and face shields, depending on the situation.

LEVEL D Least hazardous. A basic work uniform provides adequate protection. Goggles, gloves, head protection, and other personal protective equipment are optional, depending on the work.

4.59 Other Equipment and Procedures

1. *CONES, BARRICADES, AND HIGH-LEVEL FLAGS:* When working in areas near moving traffic, always use cones, barricades, and high-level flags to warn motorists of your presence and to protect yourself. A utility truck is an effective barricade. This equipment and its proper application are discussed in Section 4.3, "Routing Traffic Around the Job Site."

2. *FIRST-AID KIT:* A suitably equipped first-aid kit should be on hand and immediately available for use where operators will be working in manholes. Cardiopulmonary resuscitation (CPR) devices are essential parts of this kit. At least one crew member, preferably more, must be trained in the use of first-aid equipment and procedures, including CPR (cardiopulmonary resuscitation), unless you arrange for outside rescue services. If your agency designates operators to provide first aid and/or CPR, those operators should be included in a Bloodborne Pathogens (BBP) program. The designated operators may be exposed to contact with blood or other potentially infectious materials from the performance of their duties. A body fluid cleanup kit (Figure 4.27) should be a part of the BBP program. The BBP program includes exposure potential determination, engineering and work practice controls, personal protective equipment, and the availability of the Hepatitis B vaccination series (29 CFR 1910.1030). Eye washing solutions as well as the irrigation or rinsing eye cups should be included. If crews must work in areas where they could come in contact with poison ivy or poison oak, solutions to relieve itching are essential.

QUESTIONS

Write your answers in a notebook and then compare your answers with those on page 178.

4.5J What types of ladders are often used in shallow manholes?

4.5K What is the number one rule of hygiene for collection system operators?

4.5L On a properly adjusted hard hat, how much clearance should there be between the harness and the person's head?

4.5M How can jewelry such as rings be a hazard to operators?

4.6 FINAL PRECAUTIONS BEFORE MANHOLE ENTRY

4.60 Health Conditions of Operators

Any operator entering a manhole should conform to the following rules at the time of entry:

1. Be in good health. If you are recovering from a recent injury, illness, or surgery, do not enter a manhole until fully recovered.
2. Be in sound physical condition.
3. Be completely free from the influence of alcoholic beverages or drugs and from the impairment of a hangover.
4. Have no open sores, skin irritations (including such problems as poison oak), fungus infection (athlete's foot, for example), or serious sunburn.
5. Have current immunizations. Operators in a wastewater collection system should have current immunizations against illnesses and diseases that might be encountered in the collection system. Many agencies require typhoid and tetanus shots with booster shots every five years as a matter of policy. Contact your safety officer or personal physician for recommendations.

Operators who do not require corrective glasses for manhole tasks are preferred for such work. Other operators on a manhole entry crew also should be in good health with necessary physical capacities to accomplish required work.

4.61 Required Tools, Materials, and Equipment

Before sending an operator into a manhole, carefully inventory and examine the condition of all required tools, materials, and equipment needed for the work. Exposure to injury is

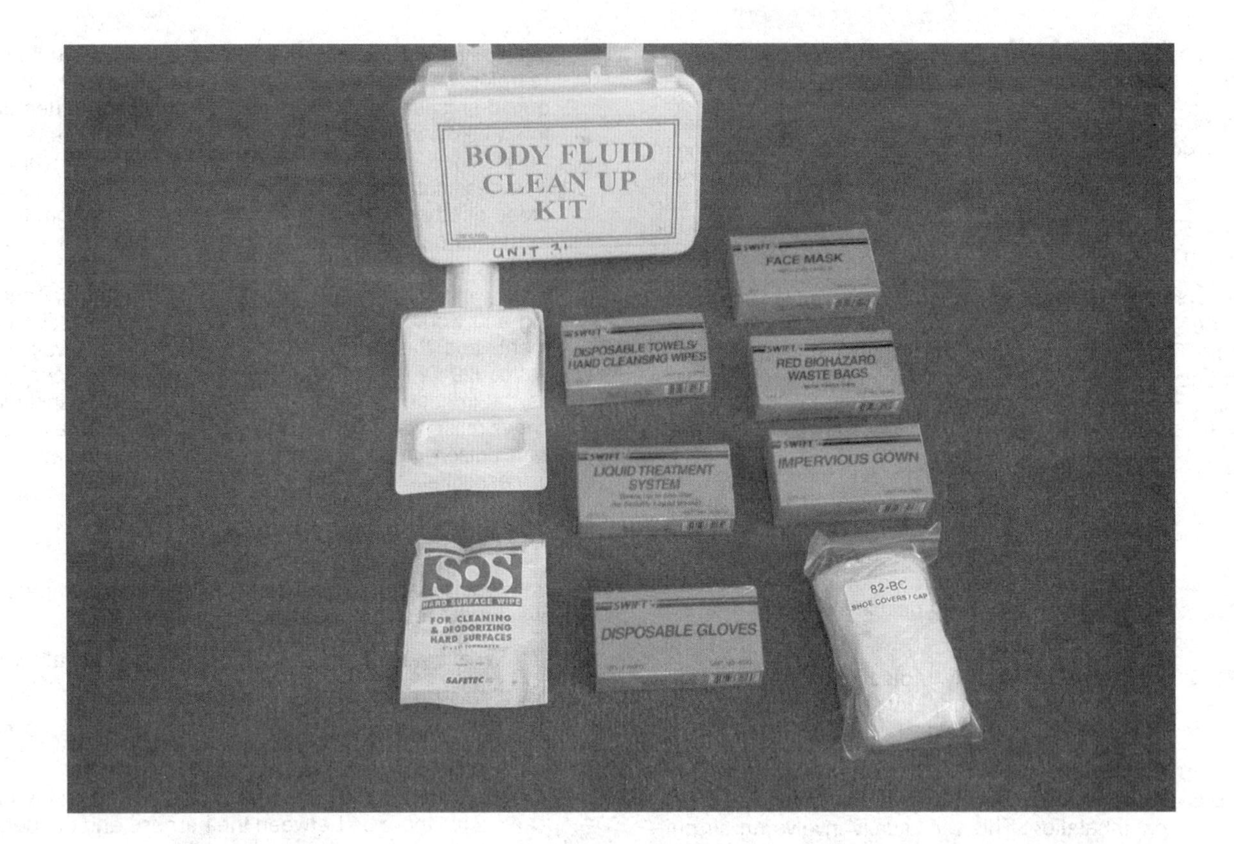

1. Scoop With Scraper
2. Surface Cleaning Wipes
3. Disposable Towels/Hand Wipes
4. Liquid Treatment System
5. Disposable Gloves
6. Face Shield
7. Red Biohazard Waste Bags
8. Gown
9. Shoe Covers/Cap

Fig. 4.27 Body fluid cleanup kit

greatest while an operator is descending into or climbing out of a manhole. Therefore, job organization and equipment should be arranged to permit all work to be accomplished with a single entry and exit of the manhole.

4.62 Briefing

Just prior to entry, the supervisor or crew leader should hold a short briefing on the project to explain the work sequence and the safety rules that must be observed. All members of the crew should be fully informed of all aspects of the project and procedures that will be followed in event of any emergency.

All agencies should require that a confined space entry form (Figures 4.28, 4.29, and 4.30) be completed and signed before manhole entry to ensure that all of the necessary safety checks have been performed.

QUESTIONS

Write your answers in a notebook and then compare your answers with those on page 178.

4.6A What are some of the health conditions of operators that should be considered before entering a manhole?

4.6B What topics should be discussed during the briefing before manhole entry?

4.7 PROCEDURES DURING MANHOLE ENTRY

There is no single procedure, instrument, or method that will provide full protection in a manhole entry situation, but rather a combination of factors. Many agencies have discovered new ways of doing collection system maintenance and minimizing entry (and therefore risk) into manhole and underground spaces. Collection system operators are among the most innovative people in the world. Special long-handled hooks, grapplers, and other tools have been fabricated to remove debris and install equipment without entering a manhole.

Confined Space Pre-Entry Checklist/Confined Space Entry Permit

Date and Time Issued: ________ Date and Time Expires: ________ Job Site/Space I.D.: ________

Job Supervisor: ________ Equipment to be worked on: ________ Work to be performed: ________

Standby personnel: ________ ________ ________

1. Atmospheric Checks: Time ________ Oxygen ________ % Toxic ________ ppm
   Explosive ________ % LEL Carbon Monoxide ________ ppm

2. Tester's signature: ________
3. Source isolation: (No Entry)

	N/A	Yes	No
Pumps or lines blinded disconnected, or blocked	( )	( )	( )

4. Ventilation Modification:

	N/A	Yes	No
Mechanical	( )	( )	( )
Natural ventilation only	( )	( )	( )

5. Atmospheric check after isolation and ventilation: Time ________
   Oxygen ________ % > 19.5% < 23.5% Toxic ________ ppm < 10 ppm H_2S
   Explosive ________ % LFL < 10% Carbon Monoxide ________ ppm < 35 ppm CO

Tester's signature: ________

6. Communication procedures: ________
7. Rescue procedures: ________
8. Entry, standby, and backup persons

	Yes	No
Successfully completed required training?	( )	( )
Is training current?	( )	( )

9. Equipment:

	N/A	Yes	No
Direct reading gas monitor tested	( )	( )	( )
Safety harnesses and lifelines for entry and standby persons	( )	( )	( )
Hoisting equipment	( )	( )	( )
Powered communications	( )	( )	( )
SCBAs for entry and standby persons	( )	( )	( )
Protective clothing	( )	( )	( )
All electric equipment listed for Class I, Division I, Groups A, B, C, and D and nonsparking tools	( )	( )	( )

10. Periodic atmospheric tests:

Oxygen:	____% Time ____;	____% Time ____;	____% Time ____;	____% Time ____;
Explosive:	____% Time ____;	____% Time ____;	____% Time ____;	____% Time ____;
Toxic:	____ppm Time ____;	____ppm Time ____;	____ppm Time ____;	____ppm Time ____;
Carbon Monoxide:	____ppm Time ____;	____ppm Time ____;	____ppm Time ____;	____ppm Time ____;

We have reviewed the work authorized by this permit and the information contained herein. Written instructions and safety procedures have been received and are understood. Entry cannot be approved if any brackets ( ) are marked in the "No" column. This permit is not valid unless all appropriate items are completed.

Permit Prepared By: (Supervisor) ________ Approved By: (Unit Supervisor) ________

Reviewed By: (CS Operations Personnel) ________
(Entrant) (Attendant) (Entry Supervisor)

This permit to be kept at job site. Return job site copy to Safety Office following job completion.

Fig. 4.28 Confined space pre-entry checklist/permit

MWCC CONFINED SPACE PERMIT

Date ____________ Location ______________________________

Structure Entering ______________________ Time in ________ Time out ________

Permit issued ______________ Permit expires ______________

Oxygen reading ____________ % (min. 19.5%, max. 22.5%) LEL reading ____________% (max. 10%) H2S reading ________ ppm (max. 10ppm)

Instrument No (s). ______________ Field Calibration ______________

1. Description of hazards ______________________________

	NA	COMPLETE	OTHER (#8)
2. Unit pumped out	☐	☐	☐
3. All valves off and/or hoses disconnected	☐	☐	☐
4. All valves tagged, dated & signed	☐	☐	☐
5. Drive mechanisms locked out & tried	☐	☐	☐
6. Hot work permit completed	☐	☐	☐
7. Protective Equipment & Rescue Devices:			
a. Forced ventilation used	☐	☐	☐
b. Harness on person(s) entering	☐	☐	☐
c. Lifeline on person(s) entering	☐	☐	☐
d. Five (5) minute escape capsule with person(s) entering	☐	☐	☐
e. SCBA on person(s) entering	☐	☐	☐
f. Harness on & lifeline near watcher	☐	☐	☐
g. Extra SCBA for watcher	☐	☐	☐
h. Emergency procedures and communication signals reviewed	☐	☐	☐

8. Other conditions (specify) ______________________________

9. Special procedures and/or equipment ______________________________

THIS UNIT HAS BEEN PROPERLY PREPARED. PERSONNEL INVOLVED KNOW THE SAFETY PROCEDURES AND HAVE BEEN DULY INFORMED.

Person (s) entering ______________________________

Watch person (s) ______________________________

Atmospheric tester ______________________________

Authorization (name & position) ______________________________
(signatures)

Fig. 4.29 Confined space permit

City of Columbia Sanitary Sewer Utility
PERMIT-REQUIRED CONFINED SPACE ENTRY PERMIT

NOTE: definitions of confined space and permit-required confined space on reverse side of this permit.

PERMIT VALID FOR DURATION OF THIS ENTRY ONLY

ENTRY SUPERVISOR: ______ DATE: ______

ENTRANT(S): ______ TIME PERMIT ISSUED: ______ am/pm

ATTENDANT(S): ______

PURPOSE FOR ENTRY: Repair ____ Cleaning ____ Inspection ____ Other ______

MH #______ DEPTH Top to Bottom ______ OUTLET PIPE SIZE ______

Description & location of confined space if other than MH: ______

PERMIT-REQUIRED CONFINED SPACE HAZARDS:

ATMOSPHERIC HAZARDS • PHYSICAL HAZARDS (DROWNING, FALLING, INSECTS/RODENTS, TRAFFIC) • TEMPERATURE EXTREMES • MECHANICAL HAZARDS • ELECTRICAL HAZARDS • BIOLOGICAL HAZARDS

ENTRY PROCEDURES	N/A	YES	NO
ATMOSPHERE TESTING Prior to Entry **MANDATORY**			
ATMOSPHERIC METER WORN BY ENTRANT DURING ENTRY			
VENTILATION for 5 Minutes Prior to Entry			
VENTILATION During Entry			
HARD HAT Worn by Entrant(s)			
SAFETY HARNESS Worn by Entrant(s)			
RETRIEVAL LINE Attached to Safety Harness			
MECHANICAL LIFT Available on Site			
MECHANICAL LIFT Used by Entrant(s)			
WORK ZONE Traffic Control			
COMMUNICATION MAINTAINED WITH ENTRANT **MANDATORY** ______ Direct Verbal Communication ______ Other, Specify ______			
MECHANICAL HAZARDS LOCKED/TAGGED-OUT			
ELECTRICAL HAZARDS LOCKED/TAGGED-OUT			
HOT WORK PERFORMED (If YES, Hot Work Permit Shall Be Attached.)			
OTHER:			

ATMOSPHERIC MEASUREMENTS by NEOTRONICS METER NO. ______

ACCEPTABLE ENTRY CONDITIONS	BEFORE VENTILATION	DURING ENTRY
% OXYGEN (O_2) *No less than 19.5%, No greater than 23.5%*		
% LEL as METHANE (CH_4) No greater than 10%		
HYDROGEN SULFIDE (H_2S) *No greater than 10 ppm*		

COMMENTS: ______

______

ENTRY SUPERVISOR: ______ *(Signature)*

For RESCUE or EMERGENCY ASSISTANCE Contact 911 via Radio or Telephone

TIME PERMIT CANCELLED: ______ am/pm *(PRCSEP.SEW 1/26/95)*

Fig. 4.30 Confined space entry permit

A well-planned maintenance program also helps reduce risk since it identifies properly maintained sewers where routine bucketing or cleaning operations may not be necessary and, if reduced, will lower the number of times operators must enter manholes.

In any case, it will not be possible to eliminate entry and work in confined spaces 100 percent of the time. Therefore, a confined space work procedure must be established. The following is a suggested procedure that one agency developed to meet the needs of their collection system operators. In some cases, this procedure may not meet your state or local requirements. The purpose of including these procedures here is to provide you with the essential elements of a workable procedure.

4.70 Confined Space Work Procedures

Figures 4.28, 4.29, and 4.30 are examples of permits and/or checklists for confined space entry. The first is from CFR 1910.146 Appendix D of the OSHA regulation, the second from a large regional agency, and the third from the City of Columbia (Missouri) Sanitary Sewer Utility. Remember that all confined spaces require a permit until a qualified person following prescribed pre-entry procedures determines otherwise (see page 121). A manhole with a flowmeter and force main, since it is isolated from a live sewer, can be classified as a non-permit space once a qualified person has verified (through the pre-entry checklist) that the manhole does not have any potential to contain any hazard capable of causing death or serious physical harm. Figure 4.31 is a Hot Work Permit, also part of Columbia's confined space procedures.

Whenever it is necessary to enter a manhole (confined space), the following procedure shall be adhered to:

1. A confined space entry form (Figure 4.28) shall be used to review the necessary precautions and conditions prior to entry. This form must be fully completed and understood by each employee involved with the job task. Additional equipment that may be required includes use of a tripod. The supervisor should specify the number of watch persons to be at the site of each confined space entry. Some agencies will add an oxygen content maximum of 23.5 percent. Prior to entry into any confined space, the supervisor should assemble the crew involved and review the sequence of activities to complete the job, each person's role, safety equipment needed, and safe procedures.
2. All traffic control measures shall be taken.
3. All valves or applicable power sources shall be locked out and properly tagged.
4. An initial test of the atmosphere must be performed to ensure that the oxygen content, explosive vapors, and toxic gases are at acceptable levels. Unacceptable levels will result in an alarm from the instrument. Important steps include:
   a. All instrumentation shall be properly calibrated before each use in accordance with manufacturers' instructions;
   b. Careful attention should be given to following the operating instructions that accompany every instrument;
   c. If an "alarm condition" develops, the area shall be evacuated, ventilated, and retested;
   d. Ventilation shall not be considered a substitute for testing a confined space;
   e. The fresh air intake on ventilation equipment shall not be placed in a contaminated area (for example, avoid an area with engine exhaust that will produce carbon monoxide);
   f. Fresh air shall be continuously circulated as long as the confined space is occupied.
5. All persons who enter or assist in the entry of a confined space shall be instructed (by their immediate supervisor) as to the nature of the hazards involved, the necessary precautions to be taken, and in the use of protective and emergency equipment required as approved by the Safety Officer.
6. All persons entering a confined space shall carry a five-minute or a ten-minute escape air capsule.
7. All persons entering a confined space shall wear a rescue harness with attached lifeline. Any variation of this requirement shall be approved by the area manager/plant manager prior to entry.
8. No smoking shall be permitted inside or within ten feet (3 meters) of a confined space.
9. At least one person shall remain outside the confined space while it is occupied. This person's job task shall be that of a watch person and to maintain communications with operator(s) inside the confined space. This person shall remain at the entrance of the confined space as long as it is occupied. A third person must be within calling distance to assist the topside individual with rescue efforts.
10. Atmospheric testing shall continue while the confined space is occupied.
11. All persons in a confined space shall vacate immediately if the warning alarm on any instrument is activated.
12. A hard hat shall be worn at all times in a confined space.
13. If a necessary piece of equipment is in need of repair or is missing, work shall be halted until the equipment is replaced or repaired.
14. Hot work shall not be permitted if the atmosphere contains explosive gases greater than 10 percent LEL (Lower Explosive Limit). The definition of hot work as used in this procedure includes, but is not necessarily limited to, burning, welding, brazing, soldering, cutting, heating, grinding, drilling, sanding, and chipping. (A separate permit (Figure 4.31) is required when it is necessary to perform hot work in a confined space.)
15. Any hazardous condition encountered shall be entered on the Confined Space Entry Form (Figure 4.29, item 8) for future reference (for example, steps in manhole are in poor condition).

The actual field form is completed each time a confined space is entered. Copies of the form are kept on file, with one copy sent to the Safety Department for their file.

Keep in mind that when you and other members of your crew sign this form, you agree to follow the listed procedures. If you then elect not to follow the established procedures and your action results in an injury or fatality, you have assumed the responsibility for your actions (and those of your crew, if you are a field supervisor). You could be held legally liable for the consequences of your actions or those of your crew.

The remainder of this section is an expansion of the "Confined Space Work Procedure."

CITY OF COLUMBIA PUBLIC WORKS DEPARTMENT

HOT WORK PERMIT

(ATTACH TO PERMIT-REQUIRED CONFINED SPACE ENTRY PERMIT)

NOTE: A Hot Work Permit must be completed for all operations performed within a confined space that require workers to weld, cut, or use other open-flame or a spark producing devices in a confined space. Only workers trained in confined space hot work and emergency procedures shall be allowed to perform hot work under this permit.

INSTRUCTIONS:

(1) Complete permit and eliminate or control all hazardous conditions before hot work begins.

(2) **Do not cut, weld, or use other open-flame or spark producing equipment until the proper precautions have been taken. Do not bring gas cylinders into the confined space.**

(3) Remove welding hoses, leads, etc. from confined space when not in use.

MH # or Location of Confined Space if other than MH:
PERMIT ISSUED: DATE ____________ TIME ________ AM/PM
TYPE OF HOT WORK: _____ CUT _____ WELD _____ GRIND _____ REPAIR **OTHER (DESCRIBE):**
EQUIPMENT:
FIRE WATCH: (NAME)
PERSONS PERFORMING HOT WORK:

PRECAUTIONS TAKEN (check YES or NO)	**YES**	**NO**
Are **ALL** atmospheric tests acceptable?		
Have flame/spark producing devices been inspected and in good condition?		
Have **ALL** flammable/combustible materials, combustible dust and ignitable residues been removed or purged from the confined space?		
Is a proper type fire extinguisher available & has it been inspected?		
Is a fire watch posted?		
Is all electrical equipment intrinsically safe (explosion proof)?		
Will ventilation remove fumes & smoke at the source?		
Are proper type respirators available and in use when required?		
Is welding and other equipment safely located, grounded & spark controlled?		

NOTE: If any of the above questions are answered "NO" do not perform hot work. Contact your immediate supervisor.

ENTRY SUPERVISOR:______________________________ (Signature)

TIME PERMIT CANCELLED:____________ **am/pm** WP60\FILES\HOTWORK.PER 12-21-94

Fig. 4.31 Hot work permit

4.71 Operations of Manhole Entry

The minimum crew required to be present for an operator to enter a collection system manhole is three operators: the operator who will go into the hole, the lifeline and winch attendant, and an assistant on the surface. The arrangement of the safety and other equipment is generally accomplished by the entire crew; however, a support crew may be necessary to direct traffic and provide other needed assistance.

1. Place the manhole safety enclosure around the manhole, if necessary.
2. Calibrate the portable atmospheric monitor *BEFORE* removing the manhole cover. Test the manhole from top to bottom for oxygen deficiency, explosive gases, and toxic (hydrogen sulfide) gases. If possible, test for explosive mixtures before removing the manhole cover because removal of the cover may produce a spark and cause an explosion. Also, it is helpful to know conditions in the manhole before any ventilation occurs.

IF AN EXPLOSIVE ATMOSPHERE IS DISCOVERED IN A MANHOLE:

a. If you have not removed the manhole cover, DO NOT do so. If you have already removed the manhole cover, DO NOT put it back on.
b. Immediately notify your supervisor that an explosive condition has been discovered and provide as many details as possible, including location. Request notification of police and fire departments and also the industrial pretreatment facility inspector.
c. Monitor the atmosphere in the immediate area around the outside of the manhole to determine the extent and degree of the hazard area.
d. Turn off any running engines within the hazard area that could cause a spark.
e. Route vehicles around the hazard area using cones, flags, and barricades.
f. Inspect upstream and downstream manholes for explosive conditions to determine the extent of the problem.
g. If necessary, route traffic off the street to reduce the potential for explosion.
h. Notify industrial waste and/or pretreatment facility inspectors and wastewater treatment plant operators, as appropriate.
i. Attempt to locate the source of the problem and correct the situation.
j. Cautiously ventilate the system with a large blower to eliminate the explosive hazard. Try ventilating from a safe upstream or downstream manhole. Ensure that exhaust air from your ventilation operation is discharged to a safe downwind location so that it does not create a hazard to operators or to the public.
k. Be sure there is *NO SMOKING* in the area.

If the atmosphere in the manhole tests OK, continue with entry procedures.

3. Never use hands to remove or replace the manhole cover. Always use manhole hooks or approved lifts.
4. Open manholes upstream and downstream from the work area to encourage natural ventilation of the sewer. Cover open manholes with grating and place barricades around manholes to warn traffic and pedestrians.
5. Sweep the area before removing the manhole cover. After the cover is removed, clean the area immediately around the manhole opening, including the manhole ring and lid ledge.
6. Before entering the manhole, start the ventilation blower and thoroughly ventilate the manhole. The blower should be located in an area upwind of the manhole and at least 10 feet (3 m) from the manhole opening. If the blower has a gas-driven engine, the exhaust must be downwind from the manhole. Place the air intake to the blower from 2 to 5 feet (0.6 to 1.5 m) above the ground surface, depending on conditions (higher for dusty ground surfaces). Some agencies prefer to exhaust or pull air from the downstream manhole if possible. Blowers used for exhaust ventilation should be explosion proof so that they won't provide a source of ignition in the event that a flammable or explosive atmosphere is encountered.
7. Once the operator going into the hole has put on the safety harness and the lifeline has been attached, the other operator on the crew or the supervisor should check it for proper fit and attachment.
8. Continue to use the atmospheric monitoring system to test for the presence of an oxygen deficiency/enrichment, explosive gases, and/or toxic gases in the manhole atmosphere the entire time the operator is in the manhole.

4.72 While Operator Is in the Manhole

1. The end of the lifeline must be secured outside of the confined space.
2. Whenever an operator is in a manhole, continuously test the atmosphere for oxygen deficiency and for explosive and toxic (hydrogen sulfide) gases. Proper ventilation generally will prevent any problems from developing in the manhole atmosphere unless a chemical spill or dump (discharge) up sewer from the manhole occurs. If atmospheric conditions begin to change, this is an indication that ventilation is ineffective.
3. The operator in the manhole must be observed continuously by a crew member within the safety enclosure. (Safety enclosures are not considered necessary by some agencies.) This person shall perform no other function, but keep constant watch over the operator in the manhole and call for help if needed. The operator observing the person in the manhole should be careful to secure any objects in shirt or jacket pockets so that they will not fall into the manhole

when bending over it. Also the operator must be careful not to accidentally kick any tools or objects over the edge of the manhole. As long as the operator is in the manhole, the operator topside should carefully watch and not distract the operator in the manhole. Always listen and respond to the needs and condition of the operator in the manhole.

4. If there are any indications of trouble, such as unusual behavior or alarms from the atmospheric monitor, immediately remove the operator from the manhole.

Fig. 4.32 Holding lifeline

How many safety hazards can you spot in Figure 4.32? No gloves, no hard hat, manhole cover balances on curb between operator's legs instead of out of the way, second operator topside missing, safety cones to warn pedestrians missing, and vehicle too close to manhole, thus not allowing adequate working room. A hoisting device should be used. The blower should be at least 10 feet (3 m) from the manhole. Is the atmosphere in the manhole being tested continuously?

4.73 Special Problems of Manhole Work

1. Manholes are frequently reservoirs for strong odors and many of these odors can cause olfactory fatigue. This is a condition where a sharp odor or prolonged breathing of an odorous atmosphere will cause the sense of smell to be temporarily lost.

   If a smell of gas or some dangerous substance is noticed when first opening a manhole lid, the lack of such a smell at a later time must never be taken as an indication that the source or danger has been eliminated. Continuously ventilate and test the manhole atmosphere for oxygen deficiency and for explosive and toxic gases whenever someone is in a manhole. Don't depend on odors to provide warnings; but when you do smell a potentially harmful odor (such as hydrogen sulfide), recheck the atmospheric testing system. If your lips tingle or become numb, these are signals to get out of the manhole *NOW.*

2. When the atmosphere in a manhole has been displaced by a gas that has no breathable oxygen in it, life expectancy for anyone entering the manhole is approximately 180 seconds with awareness of a problem lasting less than 30 seconds.

3. A lowering of the oxygen content in the air you breathe can rapidly result in a serious injury. *NEVER* allow anyone to enter a manhole until the oxygen content tests greater than 19.5 percent oxygen. The early warning signs that the topside operator should be looking for include:

   a. Labored breathing (shortness of breath),

   b. Chest heaving, and

   c. Change from usual responses.

   Victims suffering from a lack of oxygen may require cardiopulmonary resuscitation (CPR) if the case is severe. If the victim is still breathing without assistance, pure fresh air is safe and effective.

   WARNING: IF AN OPERATOR IS SUFFERING FROM LACK OF OXYGEN, ONLY AN EXPERT SHOULD ATTEMPT TO REVIVE OR ASSIST THE OPERATOR BY ADMINISTERING PURE OXYGEN. TOO MUCH OXYGEN IN AN OPERATOR'S BLOOD CAN BE AS DEADLY AS TOO LITTLE.

4. Any equipment manufactured and sold as safety equipment is required to pass rigid performance and quality tests. Such labels as Underwriters' Laboratories and Factory Mutual indicate that the equipment meets those qualifications. Always test equipment under conditions of use to learn how the equipment works and to be assured that it will work when needed.

4.74 After Leaving Manhole

A person who has been in a collection system manhole should take a hot shower and put on clean clothes before leaving work at the end of the day. Under no circumstances should the clothing worn in the manhole be worn home. Never expose your family to any contagious infection that might have come into contact with your clothing. *NO CLOTHING* worn in sewer maintenance or repair work should be worn home or washed with the family laundry. Store your work clothes and street clothes in separate lockers.

4.75 Responsibility

The crew leader or supervisor has the responsibility to be sure all work is conducted in a safe and prescribed manner. In some states, the person in charge can be held liable and subject to both a fine and imprisonment if damage or injury results from failure to use prescribed procedures. The employer may pay the fine, but the responsible person must serve the time in jail. Every agency must prepare, publish, and circulate to the collection system operators safety requirements and procedures for manhole entry. The agency must also provide required training and any necessary equipment. However, the operator(s) must consistently apply the training received and follow the agency's procedures if the job is to be done safely.

MANHOLE ENTRY SUMMARY

1. Test for oxygen deficiency/enrichment, flammable or explosive conditions, and toxic gases.
2. If flammable or explosive conditions are discovered, notify supervisors and experts. *DO NOT ENTER.*
3. If toxic gases and/or oxygen deficiency is found, *DO NOT ENTER* unless an emergency/rescue situation exists. Take all necessary safety precautions. Try ventilating the area to clear up the hazardous atmosphere and retest.
4. If atmospheric conditions test OK, ventilate the manhole and monitor conditions continuously while working in the manhole. Wear a safety harness and lifeline. Evacuate the manhole immediately if conditions change, alarms sound, or if you start to feel "different."

FOLLOW SAFE PRACTICES
USE GOOD JUDGMENT AT ALL TIMES

QUESTIONS

Write your answers in a notebook and then compare your answers with those on page 178.

4.7A What is the minimum size of a crew when someone enters a manhole?

4.7B What should the operator at the manhole entrance be doing while another operator is in the manhole?

4.7C What is the minimum level of oxygen in air for safe breathing?

4.7D What should an operator do after leaving a manhole?

END OF LESSON 2 OF 3 LESSONS
on
SAFE PROCEDURES

Please answer the discussion and review questions next.

DISCUSSION AND REVIEW QUESTIONS

Chapter 4. SAFE PROCEDURES

(Lesson 2 of 3 Lessons)

Write the answers to these questions in your notebook before continuing. The question numbering continues from Lesson 1.

11. What is the definition of a "confined space"?
12. What is the definition of the phrase "dangerous air contamination"?
13. How can explosive or flammable atmospheres develop in a collection system?
14. How can toxic acids, bases, and other hazardous liquid or solid chemicals enter wastewater collection systems?
15. What types of hazardous atmospheres should an atmospheric test unit be able to detect in confined spaces?
16. If operators are scheduled to work in a manhole, when should the atmosphere in the manhole be tested?
17. When a blower is used to ventilate a manhole, where should the blower be located?
18. List the safety equipment recommended for use when operators are required to enter a confined space.
19. What are some early signs that an operator working in a manhole or other confined space is not getting enough oxygen?

CHAPTER 4. SAFE PROCEDURES

(Lesson 3 of 3 Lessons)

4.8 EXCAVATIONS

Excavations can present a combination of hazards to the operator when access to the collection system is required. This section briefly outlines some of the requirements for work in excavations. More detailed information can be found in Chapter 7 of this manual.

If it becomes necessary for you to excavate a portion of your collection system, remember to contact utility companies to locate underground telephone, gas, fuel, electric, and water lines BEFORE opening the excavation. If you can't establish the exact locations of the underground utilities before starting work, you must proceed with extreme caution and use detection equipment, if possible. These lines can present a very significant hazard to operators and to the public during excavation activities. Also become familiar with the fundamentals of excavating and the various protective systems that can be used BEFORE becoming involved in excavation work.

Without a proper protective system, the bank (wall) of a trench or excavation can cave in and kill you. OSHA requires that you always use a protective system to protect operators in a trench five or more feet (1.5 m) in depth. *Excavations less than 5 feet (1.5 m) deep may also require protection if a competent person determines that there are indications of a potential cave-in.* A competent person is defined by OSHA as "a person capable of identifying existing and predictable hazards in the surroundings, or working conditions which are unsanitary, hazardous or dangerous to employees, and who has authorization to take prompt corrective measures to eliminate the hazards."

4.80 Shoring

Shoring is a complete framework of wood and/or metal that is designed to support the walls of a trench (see Figure 4.33). Sheeting is the solid material placed directly against the side of the trench. Either wooden sheets or metal plates might be used. Uprights are used to support the sheeting. They are usually placed vertically along the face of the trench wall. Spacing between the uprights varies depending on the stability of the soil. Stringers (or wales) are placed horizontally along the uprights. Trench braces are attached to the stringers and run across the excavation. The trench braces must be adequate to support the weight of the wall to prevent a cave-in. Examples of different types of trench braces include solid wood or steel, screw jacks, or hydraulic jacks.

4.81 Shielding

Shielding is accomplished by using a two-sided, braced steel box that is open on the top, bottom, and ends (see Figure 4.34). This "drag shield," as it is sometimes called, is pulled through the excavation as the trench is dug out in front and filled in behind. Operators using a drag shield must always work within the walls of the shield and are not allowed in the shield when it is being installed, removed, or moved.

4.82 Sloping/Benching

Sloping and benching are practices that simply remove the trench wall itself (see Figure 4.35). The amount of soil needed to be removed will vary depending on the stability of the soil. A good rule of thumb is to always slope or bench at least one foot (0.3 m) back for every one foot of depth on *both sides of the trench.*

Exact sloping angles and benching dimensions largely depend on the type of soil that is being excavated. OSHA has established three soil classifications, Type A, Type B, and Type C, in decreasing order of stability. The type of soil dictates sloping/benching requirements. A competent person must examine the work site and determine the soil classification in order to determine requirements for a specific site.

4.83 Inspections

Excavations and adjacent areas must be inspected on a daily basis by a competent person for evidence of potential cave-ins, protective system failures, hazardous atmospheres, or other hazardous conditions. The inspections are only required if operator exposure is anticipated.

4.84 Atmospheric Monitoring

Atmospheric monitoring prior to entering an excavation more than four feet (1.2 m) deep is required if a hazardous atmosphere can reasonably be expected to exist. Excavations in landfill areas, in areas where hazardous substances are stored nearby, or in which connections or openings are being made to in-service sewers or manholes are examples of situations where a hazardous atmosphere could exist or develop. Emergency rescue equipment, such as breathing apparatus and safety harness/lifeline, may also be required. If rescue equipment is required, it must also be attended.

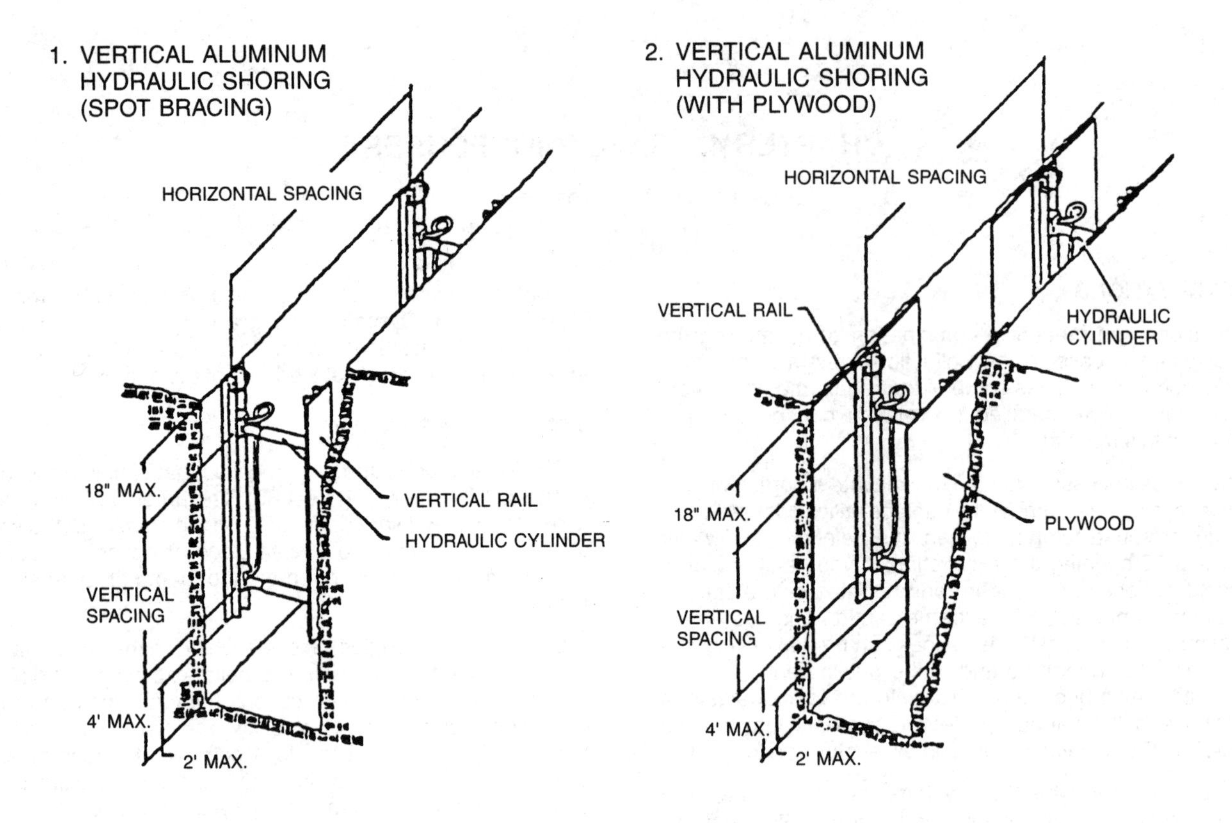

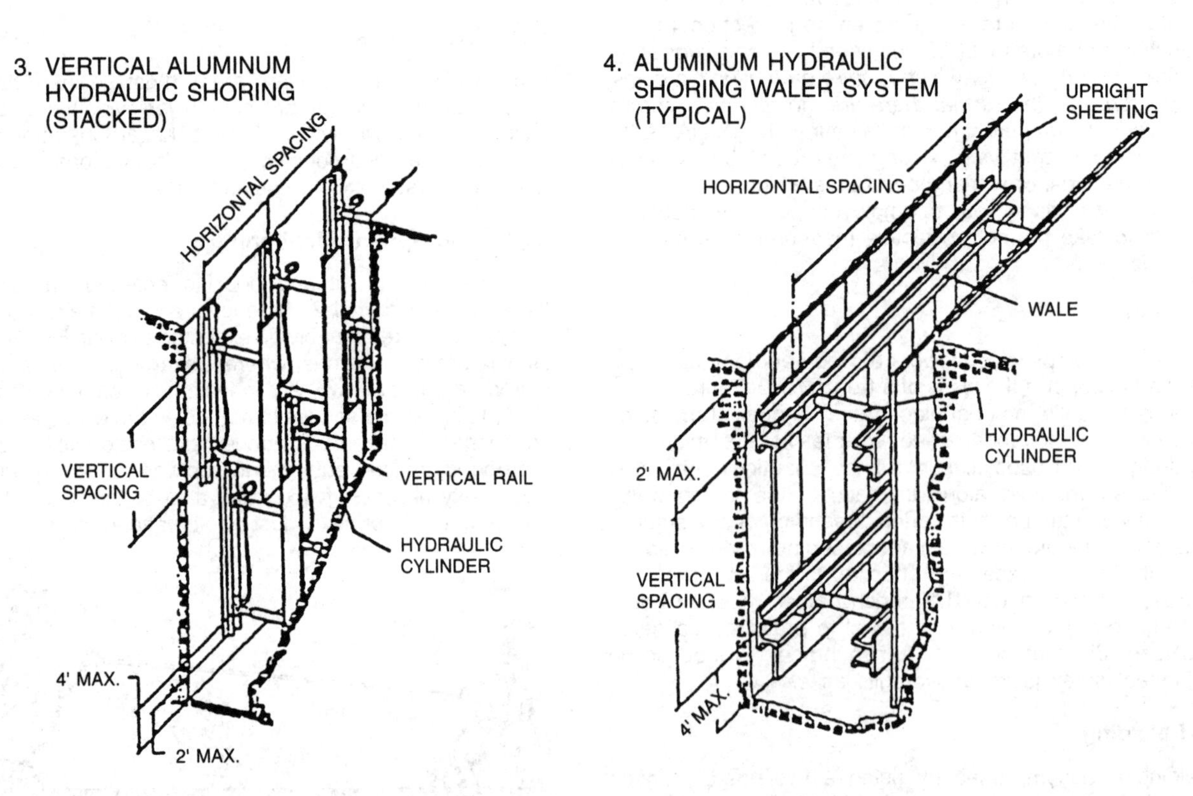

Fig. 4.33 Typical installations of aluminum hydraulic shoring

(Source: 29 Code of Federal Regulations 1926)

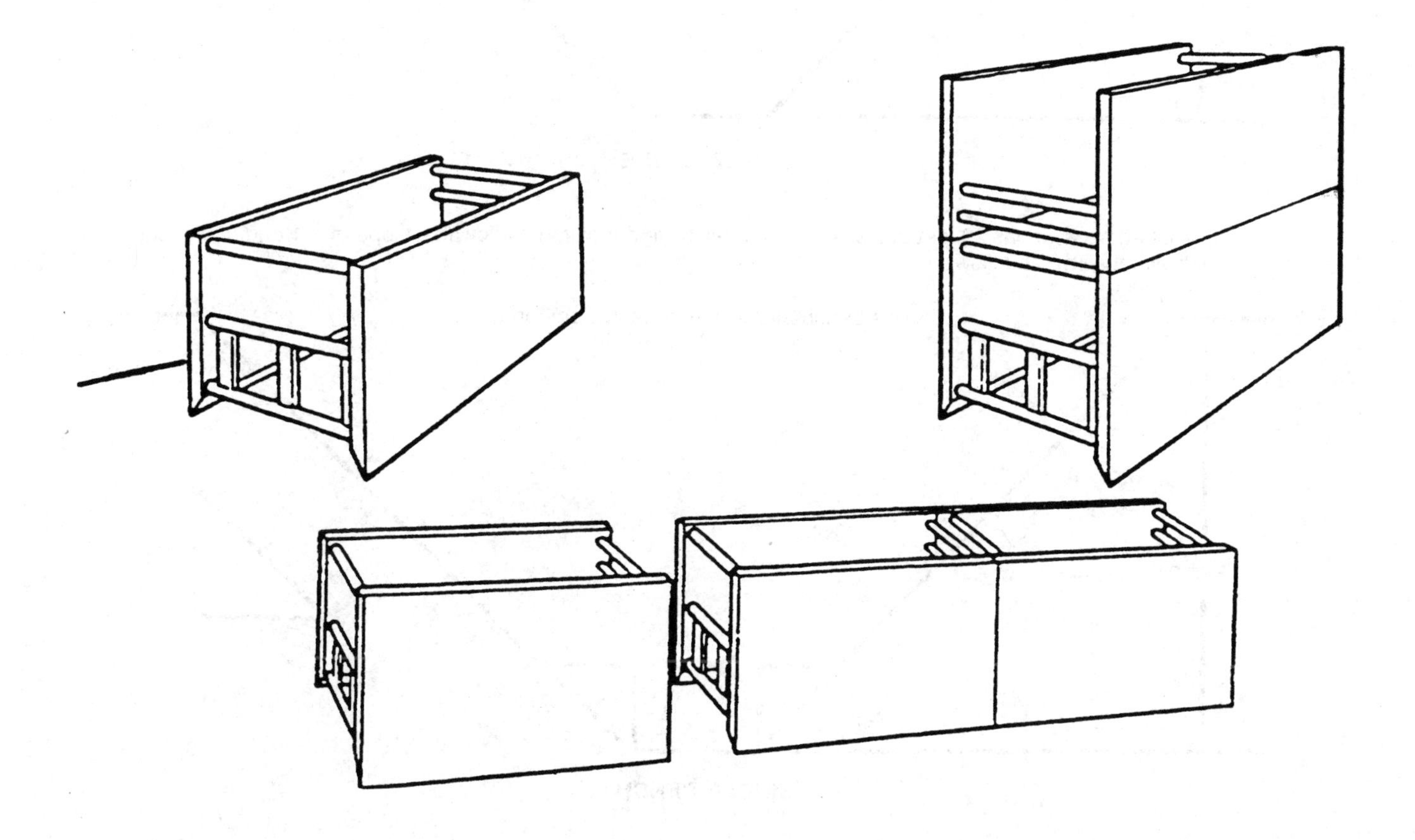

Fig. 4.34 Trench shields

(Source: 29 Code of Federal Regulations 1926)

EXCAVATIONS MADE IN TYPE B SOIL

1. All simple slope excavations 20 feet or less in depth shall have a maximum allowable slope of 1:1.

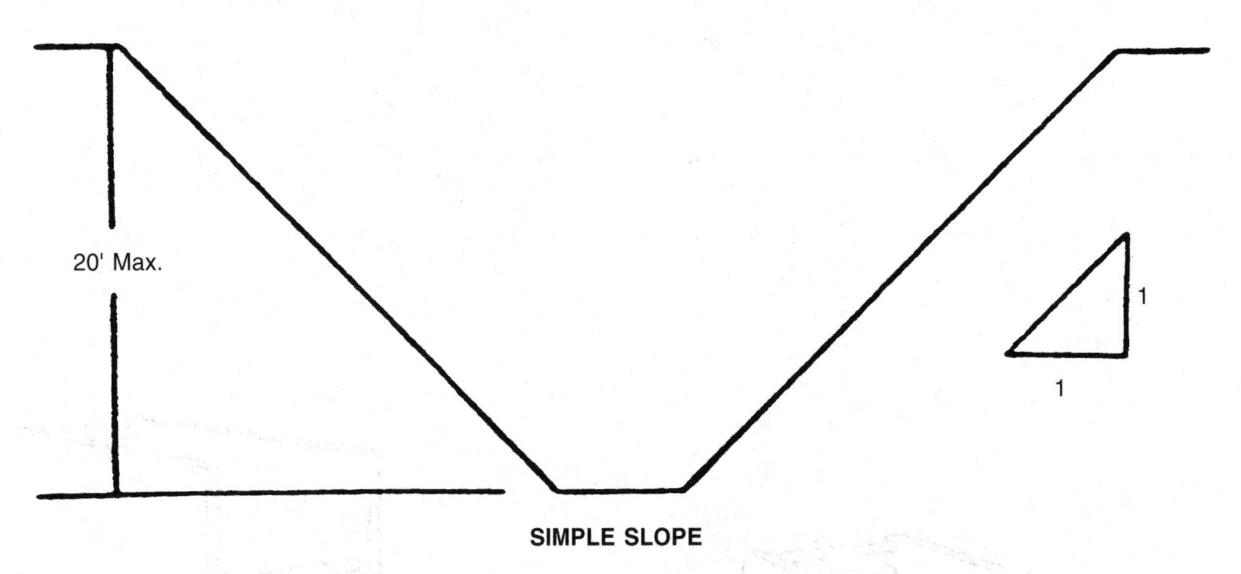

SIMPLE SLOPE

2. All benched excavations 20 feet or less in depth shall have a maximum allowable slope of 1:1 and maximum bench dimensions as follows:

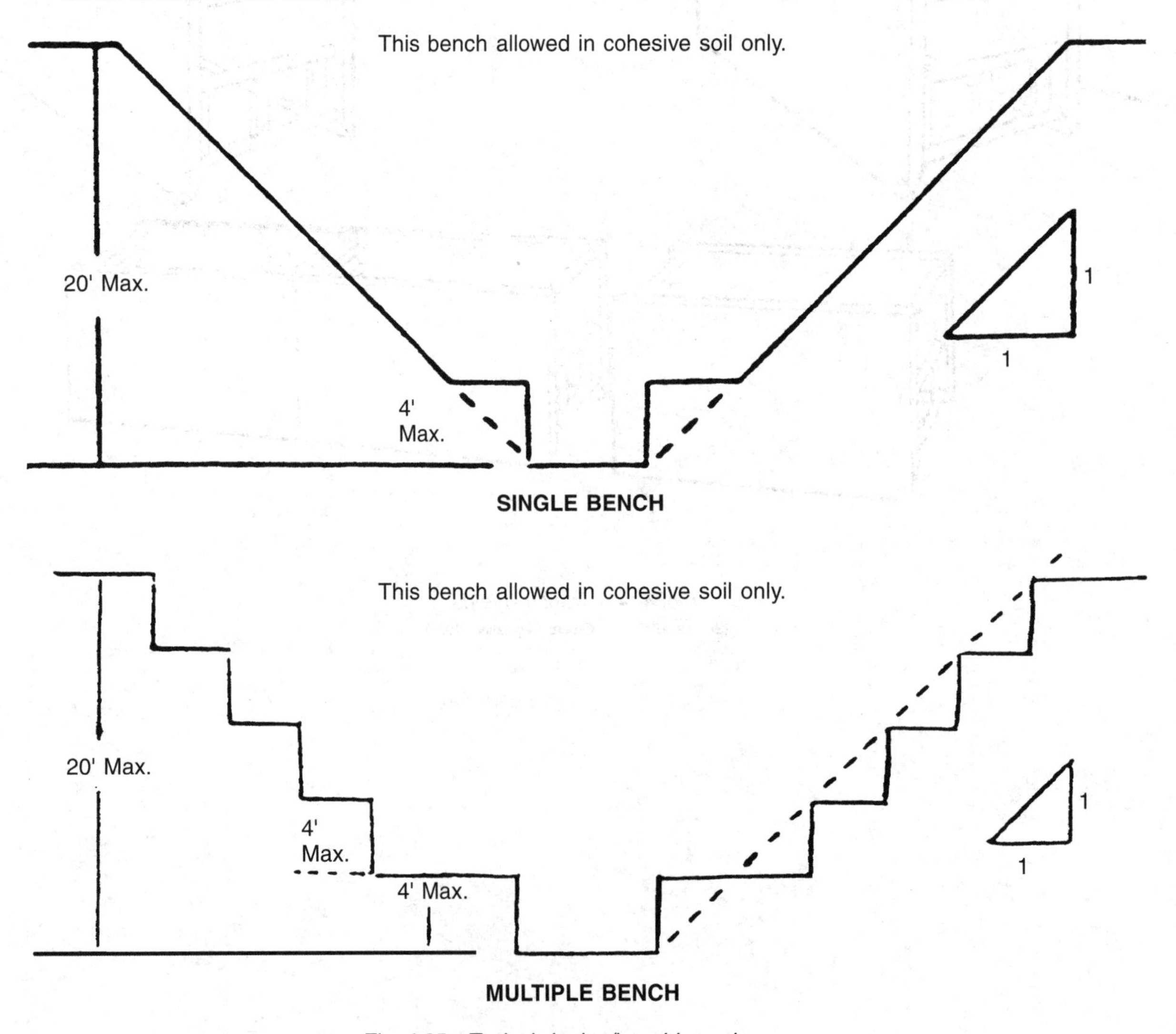

MULTIPLE BENCH

Fig. 4.35 Typical sloping/benching schemes
(Source: 29 Code of Federal Regulations 1926)

4.85 Other Requirements

The soil removed from the trench (spoil) should be placed at least two feet (0.6 m) from the trench and should be placed on one side of the trench only. A stairway, ramp, or ladder is required in the trench if it is four or more feet (1.2 m) deep. This means of egress (exiting) must be placed so that no more than 25 feet (7.5 m) of travel is required to exit the trench. Walkways must also be provided where operators must cross the trench, and if the trench is six feet (2 m) or more in depth at the crossing area, guardrails must be provided on the walkway.

QUESTIONS

Write your answers in a notebook and then compare your answers with those on page 179.

4.8A At what excavation depth does OSHA require a protective system?

4.8B List three protective systems for excavations.

4.8C How far from the trench should the "spoil" be placed?

4.8D A good rule of thumb is to slope or bench at least __________ foot/feet back for every one foot of trench depth.

4.9 ELECTRICAL SAFETY

Collection system operators are routinely exposed to a variety of field conditions related to electrical equipment and hazards. The objective of this section is *NOT* to make you an instant electrical expert qualified to work on the electrical equipment that we have in our systems, but it is to make you more *KNOWLEDGEABLE* so that you have an *AWARENESS* of the hazards concerned with electrical systems.

The first rule that applies to electrical systems is, *"IF YOU ARE NOT QUALIFIED, DO NOT, UNDER ANY CIRCUMSTANCES, ATTEMPT TO WORK ON ELECTRICAL SYSTEMS."*

Let's take a look at some of the types of equipment and work situations that increase our exposure to electrical hazards:

1. Lift stations have a variety of electrical systems, single phase, three phase, alternating current (A.C.), direct current (D.C.), low voltage and medium voltage;
2. Flow metering stations, usually wet locations, may require explosion-proof wiring methods;
3. Telemetry systems;
4. Standby/emergency power equipment;
5. Voltage ranges from low to high, voltages range from 24 volts to 4,160 volts, and as high as 13,800 volts (typical exposure is in the 120-volt to 480-volt range);
6. Much of our equipment is located in wet or damp locations;
7. We frequently have to work under adverse conditions such as thunderstorms, where downed power lines are frequent hazards;
8. Failures occur frequently at night when it is dark;
9. We may be required to provide standby power through the use of portable generators and temporary wiring;
10. When working under emergency conditions, time is usually very limited. In order to prevent bypassing of raw wastewater or backups into private homes, we are under a great deal of pressure to rig pumps, generators, or other equipment; and
11. Construction repair projects increase the hazard of encountering live underground power lines.

These hazards increase the possibility of:

- Electrocution resulting in slight to serious burns or even death;
- Physical injury due to failure to lock out or disconnect electrical systems; for example, someone starts a pump while you have your hand in the impeller cleaning out rags;
- Explosions resulting from electrical equipment operating in explosive atmospheres, such as wet wells and underground vaults;
- Protective equipment such as circuit breakers exploding due to high short-circuit currents; and
- Temporary or permanent blindness due to flashing or arcing.

To further increase your understanding of electrical systems, which many operators view as mysterious, we can compare them to a hydraulic system consisting of a pump, a pipeline, and a valve.

1. Refer to Figure 4.36, where water pressure is compared to voltage.
2. The water, which is the movement of water molecules, is equal to current, which is the movement of electrons. In either case, we are moving physical things or material.
3. The valve is equal to the resistance. If you close down the valve, you increase the resistance and reduce the ability of the water to move. Similarly, resistance in an electrical system is a blockage of the movement of electrons. In electrical systems this resistance is usually the "load," which may take the form of motor windings in an electric drill, a light filament, or even a human body.

In the case of the electrical system, the power company or utility can be compared to the pump. It supplies the electrical energy just as the pump supplies hydraulic energy to the pipeline.

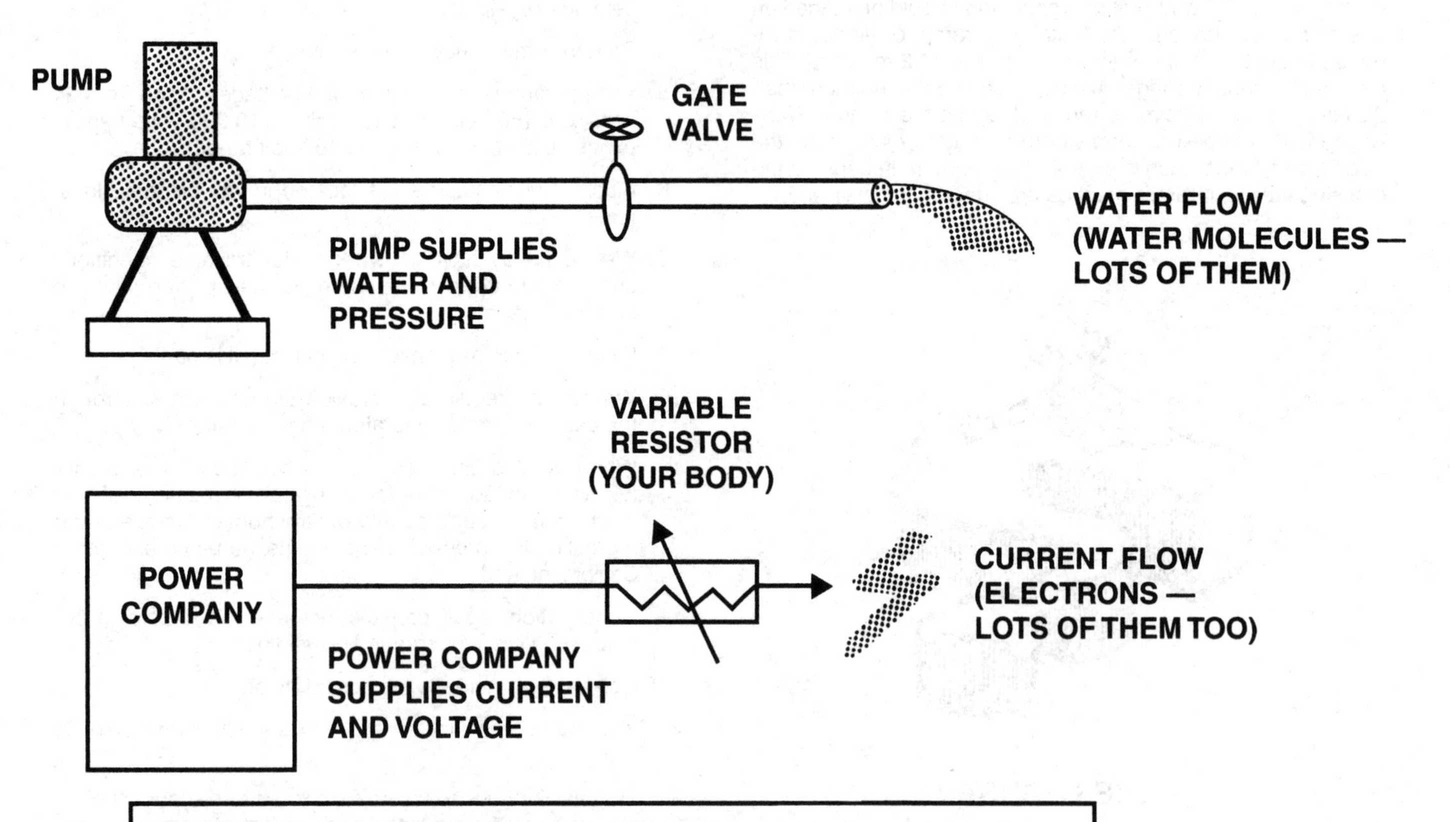

MAXIMUM RESISTANCE = VALVE CLOSED = NO WATER FLOW
MAXIMUM RESISTANCE = OPEN CIRCUIT = NO CURRENT FLOW

MINIMUM RESISTANCE = VALVE OPEN = MAXIMUM WATER FLOW
MINIMUM RESISTANCE = SHORT CIRCUIT = MAXIMUM CURRENT FLOW

	ENERGY (POTENTIAL)	MASS (PARTICLES)	RESTRICTION (TO FLOW)
HYDRAULIC SYSTEM	PRESSURE	WATER (MOLECULES)	VARIABLE VALVE
ELECTRICAL SYSTEM	VOLTAGE	CURRENT (ELECTRONS)	VARIABLE RESISTOR

ELECTRICAL SYSTEM AND HYDRAULIC SYSTEM SIMILARITIES

Fig. 4.36 Water pressure equal to voltage

Other similarities exist between hydraulic systems and electrical systems. Both of these systems want to become a "closed loop." In other words, in an electrical system the electrons (current measured in amps, milliamps, or microamps) want to return to the source, which is commonly referred to as ground (at a zero voltage state). In the case of the hydraulic system, the pressure in the pipe also wants to return to a zero pressure state. When you open a closed, pressurized pipeline, the pressure will want to drop to zero if the pump is off.

The significance of this, in collection system work, is that when we use the human body to become the "load," current is allowed to flow through the body to return to its ground state. The amount of current that flows through the human body influences the amount of damage to our body.

Many misconceptions exist about electricity. Among the more common is that low voltage (120 volts) is less dangerous than a higher voltage (240 volts, 3 phase or 480 volts, 3 phase). In fact, it is the current flow (electrons) through the body that causes the problems. The amount of current flowing through the body depends on the voltage and the resistance and can be expressed in the following formula:

Amps are equal to voltage divided by resistance.

$$\text{Current, amps} = \frac{\text{Voltage, volts}}{\text{Resistance, ohms}} \text{ or } \frac{\text{Watts}}{\text{Volts}} = \text{Amps}$$

For example, a 100-watt light bulb that is in a 120-volt (A.C.) circuit in your house has a resistance in the filament of approximately 144 ohms.

$$\text{Current, amps} = \frac{120 \text{ volts}}{144 \text{ ohms}} = 0.83 \text{ amp or } \frac{100 \text{ W}}{120 \text{ V}} = 0.83 \text{ amp}$$

Similarly, a 7.5-watt bulb in a 120-volt circuit results in a current flow of 0.06 amp. This appears to be a very small amount of current indeed, but let's take a look at the effects of current in the human body.

In this case we will use milliamps, which is a thousandth (0.001) of an amp; 0.06 amp becomes 60 milliamps and 0.83 amp becomes 830 milliamps.

- 1 milliamp or less, no sensation, not felt.
- More than 5 milliamps, painful shock. (This is 5 thousandths of an amp, 0.005 amp.)
- More than 10 milliamps, local muscle contractions sufficient to cause "freezing" of the muscles for 2.5 percent of the population.
- More than 15 milliamps, local muscle contractions sufficient to cause "freezing" of the muscles to 50 percent of the population.
- More than 30 milliamps, breathing is difficult, can cause unconsciousness.
- 50 to 100 milliamps, possible ventricular fibrillation of the heart (uncontrolled, rapid beating of the heart muscle).
- Over 200 milliamps, severe burns and muscular contractions, heart more apt to suffer stoppage rather than only fibrillation. (This is only two-tenths of an amp.)
- Over a few amperes, irreparable damage to the body tissue.

So the current flowing in the 120-volt circuit that lights a 7.5-watt light bulb is enough to cause severe muscular contractions of the heart. Again, the controlling factor that determines the extent of injury is the amount of resistance to the flow of current. Figure 4.37 illustrates some typical body resistances and the subsequent current flow.

- If your head happens to become the load in an electrical circuit, the current flows through the brain from ear to ear. At 100 volts that could allow one amp of current to flow, which is 1,000 milliamps. Chances for permanent damage or death are very high, since the head offers only 100 ohms of resistance.

$$\frac{\text{Volts}}{\text{Ohms}} = \text{Amps} \qquad \frac{100 \text{ Volts}}{100 \text{ Ohms}} = 1 \text{ Amp}$$

- If the current flows through the hand, down through the heart to your foot, this represents a resistance of about 500 ohms, which will allow 220 milliamps of current to flow. Again, potential death through severe muscular contraction of the heart is likely.
- Resistance to current flow is significantly altered if the skin is wet. Dry skin will have a resistance of 100,000 to 600,000 ohms while wet skin will have a resistance of only 1,000 ohms. When dealing with higher voltages such as 220 volts, or 480 volts, or even higher, wet skin allows more current to flow through the body.

ONCE AGAIN, IF YOU ARE NOT QUALIFIED YOU HAVE NO BUSINESS WORKING IN OR AROUND ELECTRICAL SYSTEMS. NOT ONLY ARE YOU CREATING A HAZARD TO YOURSELF, BUT TO YOUR COLLEAGUES AND THE PUBLIC AS WELL. ELECTRICAL MAINTENANCE/INSTALLATION IS CONTROLLED BY RIGID FEDERAL AND STATE REGULATIONS INCLUDING THE NATIONAL ELECTRICAL CODE (NEC) AND LOCAL BUILDING CODES.

If you are qualified and are working on electrical equipment, observe the basic rules listed below.

- Disconnect and lock out all equipment prior to working on it. (See *LOCKOUT/TAGOUT PROCEDURE* at the end of this section.)
- Observe common sense procedures that will prevent or reduce the possibility of electrocution through the accidental contact with a live circuit. Use rubber mats, rubber boots, leather gloves, or some other form of isolation between you and ground.
- Avoid wearing eyeglasses with metal frames, watches, jewelry, or metal belt buckles that can increase the possibility of severe burns and electrocution if they come in contact with electrical circuits.

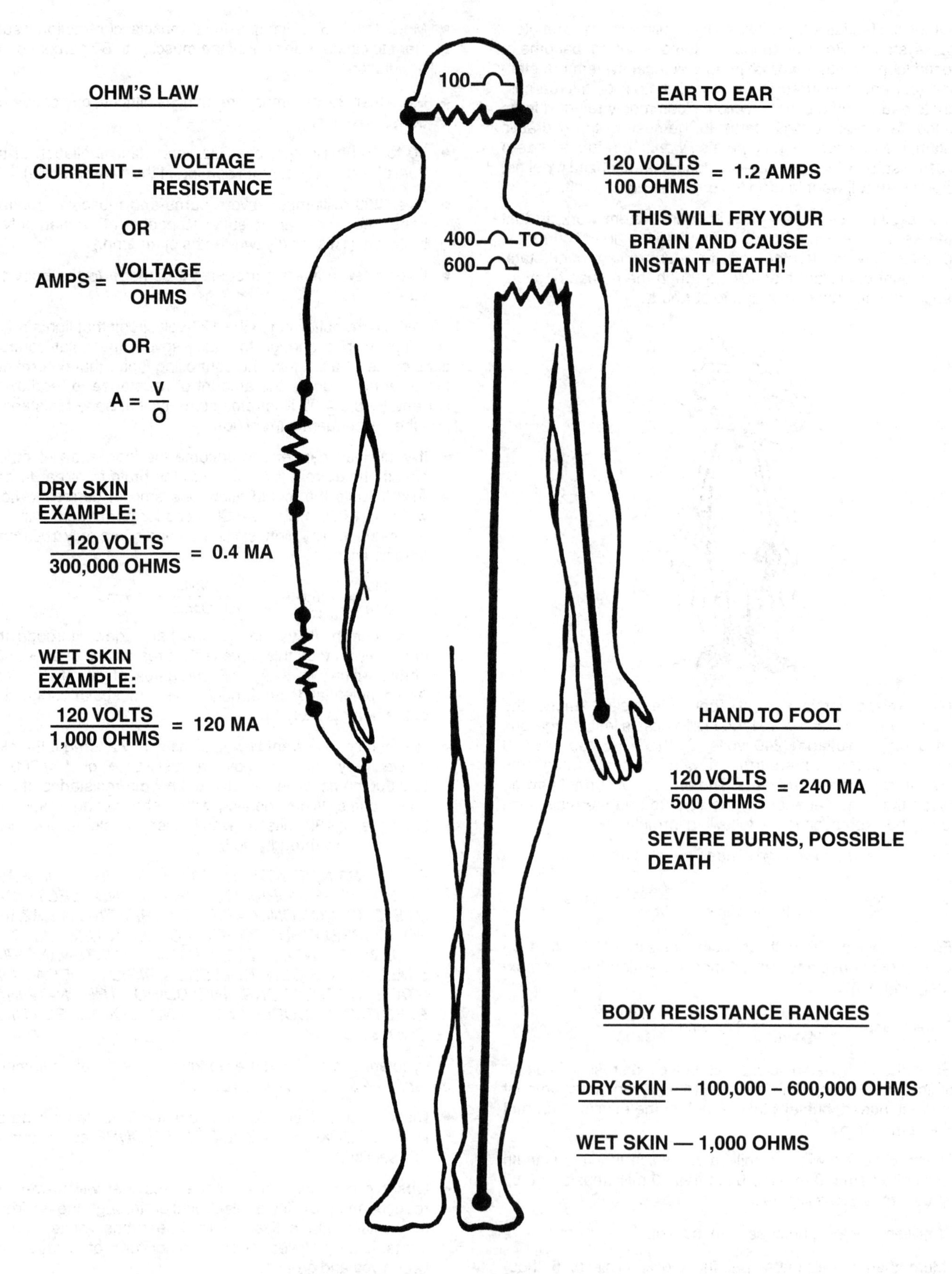

Fig. 4.37 Typical body resistances and currents

(Permission of General Electric)

- Always maintain adequate clearance of body parts with live circuits.
- Professional electricians will try to work with one hand when working around energized circuits. This, coupled with adequate isolation from the ground through rubber mats and gloves, minimizes the possibility of completing a circuit.
- Electrical codes in many states require that an "observer" be present when you work on energized circuits at or above a certain voltage level. The observer's job is to watch your work activity (a "second set of eyes") to prevent an accident and to render assistance in the event you are injured. The observer should be currently certified in cardiopulmonary resuscitation (CPR) and first aid in case of electrocution.
- Wear safety glasses when working on live equipment. Tinted glasses and a face shield are preferred. Flashing or arcing can damage the eyes through the instantaneous generation and explosion of molten metal and high-intensity arcing from short circuits.
- When resetting tripped circuit breakers, always investigate the cause of the circuit breaker tripping. Although circuit breakers are prone to nuisance-type trips, they are also protective devices indicating a possible short circuit or other fault condition downstream. There are cases where even qualified people have neglected to check out the cause of the tripping and have tried to reset the circuit breaker which then exploded. Enclosures should always be closed. For example, shut panel doors when re-energizing a circuit. The doors are a protective device designed to protect you against the explosion of circuit breakers or other electrical components within the enclosure.

Another possibility for electrocution occurs when we are working in damp and wet places and using electric hand tools (drills, saws) for routine maintenance work. Solid state electronics and the recognition that very small amounts of current flowing through the human body can cause fatal results prompted the development of a device called a "ground fault interrupter" (GFI) in the 1970s. Today most of us have these types of devices in our homes because the 1975 NEC electrical code requires them in bathrooms and for outdoor receptacles. Hotels and other commercial facilities have installed them as well.

A ground fault interrupter provides increased protection for the collection system operator who is almost always working in a damp environment. A portable ground fault interrupter receptacle, available for less than $50, should be used when operating electric hand tools.

A GFI (Figure 4.38) consists of three essential elements: a sensing element known as a differential transformer; an electronic amplifying section; and a circuit interrupting section. The differential transformer continuously monitors the current in the two conductors of a circuit, which are the phase, or "hot line," and the "neutral" line. Under normal conditions these two currents are always equal. If the differential transformer senses a difference between them of as little as 5 milliamps, it assumes the difference is being lost to ground through a fault of some sort. A partially shredded winding in your electric drill, for example, may cause some of the current to flow into the body of the drill and then into your hand. The electronic amplifying section of the GFI boosts the signal and trips the circuit in 1/40th of a second or less.

Ground fault receptacles should be installed in convenience receptacles where it is wet or damp, and should be carried on your vehicle to use when using portable electric tools. In addition, a good electrical preventive maintenance program significantly reduces the amount of electrical equipment failure, and therefore exposure to the collection system operator (see Chapter 8, "Lift Stations," and Volume II, Chapter 9, "Equipment Maintenance").

EXPLOSIVE ATMOSPHERES

The use of electrically powered equipment under certain conditions creates such a serious safety hazard that stringent regulations now govern installations of this kind. The National Electrical Code (NEC) defines different locations and conditions, which it calls Classes and Divisions, and specifies what equipment can be used in each class and how it is to be installed and repaired. The part of the Code that applies to collection system facilities is described below.

Class 1 location (gases). This is defined as an area where flammable gases or vapors may be present in the air in a quantity sufficient to produce explosive or ignitable mixtures.

Classes are further broken down into divisions. Division 1 is an area where concentrations of gases exist continuously, intermittently, or periodically under normal, everyday operating conditions, including hazardous materials that might be released during repair or maintenance operations or in the breakdown or faulty operation of equipment.

The National Electrical Code does not identify specific structures but it does describe the conditions that might exist. The National Fire Protection Association Standard 820 (NFPA 820), "Fire Protection in Wastewater Treatment Plants and Collection Facilities," describes the requirements for electrical classification, ventilation, gas detection, and fire control methods in wastewater and collection system areas. Comparing these requirements with your collection system's existing design and equipment may indicate deficiencies that should be fixed to minimize potential hazards.

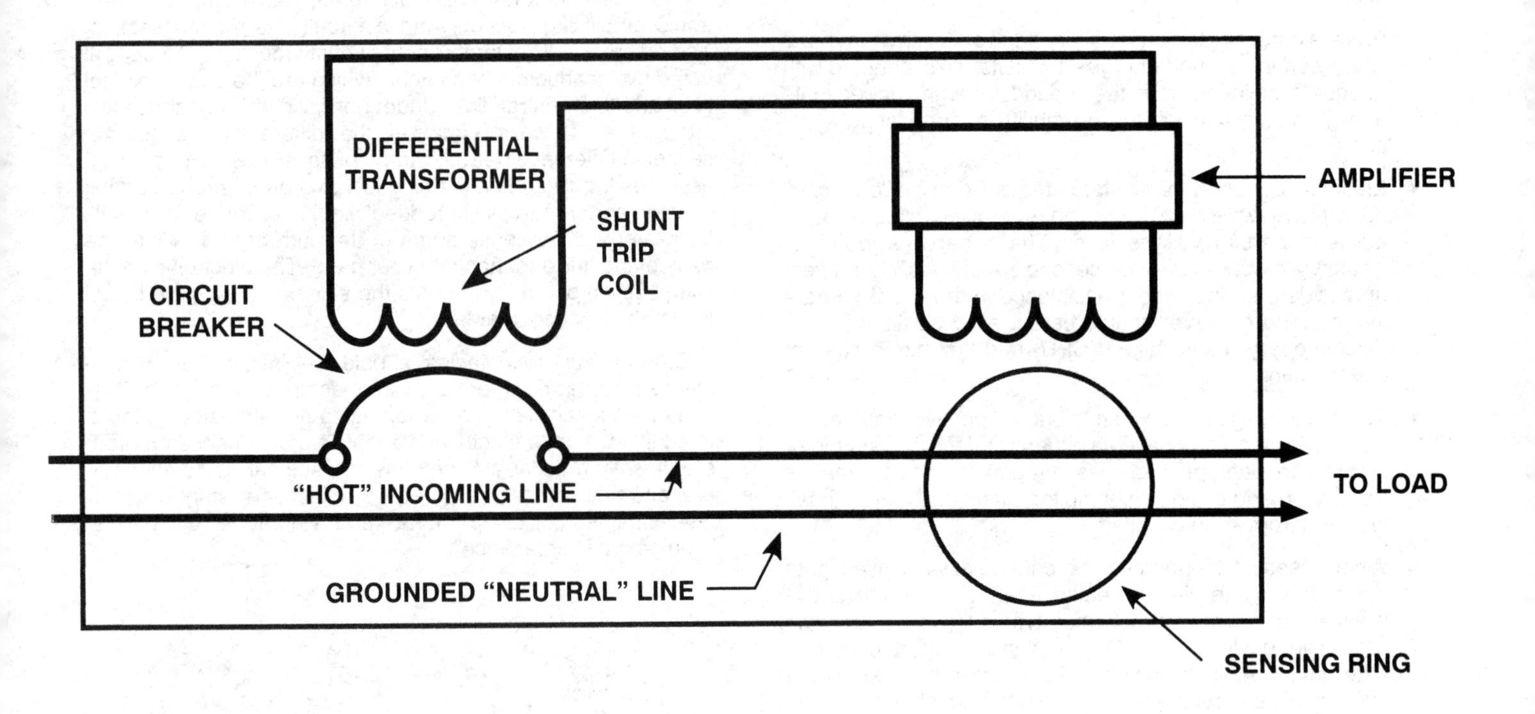

GROUND FAULT INTERRUPTER (GFI) ELEMENTS

THE DIFFERENTIAL TRANSFORMER CONTINUOUSLY MEASURES THE CURRENT FLOW IN THE "HOT" AND "NEUTRAL" LINES. UNDER NORMAL CONDITIONS THE CURRENT IS EQUAL IN EACH LINE. IF THERE IS A DIFFERENCE OF AS LITTLE AS 5 MILLIAMPERES (0.005 AMPS) THE AMPLIFIER ENERGIZES THE SHUNT TRIP COIL WHICH CAUSES THE CIRCUIT BREAKER TO TRIP IN 1/40TH OF A SECOND OR LESS.

EXAMPLE:

A HAND DRILL HAS A DEFECTIVE MOTOR WINDING ALLOWING A PORTION OF THE CURRENT TO FLOW TO THE METAL CASE AND THUS THROUGH YOUR BODY CAUSING A SHOCK AND POSSIBLE ELECTROCUTION.

ALWAYS USE A GFI WHEN USING ELECTRICAL EQUIPMENT OUTDOORS AND IN DAMP, WET LOCATIONS. ALWAYS MAKE SURE YOUR ELECTRICAL TOOLS ARE IN GOOD SHAPE.

Fig. 4.38 Ground fault interrupter

Explosive or flammable atmospheres frequently exist in collection system structures. Some examples are the confined spaces in gravity sewers, pump station wet wells, or underground vaults where natural gas leaks can accumulate and create explosive atmospheres. (One of the most significant examples of this occurred in the City of Louisville, Kentucky, which suffered $40,000,000 in damage to their sewer system as a result of an explosion.)

Wet wells frequently have blowers, lights, bar screens, or other types of systems that require electricity. The theory of explosion-proof construction is that an explosion can in fact occur within the device; for example, a light fixture. However, because of the construction (cast iron or cast aluminum) the explosion is contained within the fixture. In addition, all surfaces are machined to very close tolerances and are threaded so that as the ignited vapors or gases attempt to work their way out of the fixture, they encounter a long, tortuous route. By the time they do escape to the atmosphere, they have cooled enough to prevent the ignition of the wet well atmosphere.

Electrical equipment and wiring systems designed to operate in Class 1, Division 1 environments are specially designed to eliminate any source of ignition, such as a spark or arcing, that might cause a fire or explosion. Such equipment requires specialized knowledge for installation and maintenance. Under no circumstances should maintenance or repair of these devices, whether it be a motor, light fixture, or conduit system, be attempted by anyone but qualified personnel.

LOCKOUT/TAGOUT PROCEDURE
(OSHA 29 CFR 1910.147, "Control of Hazardous Energy")

Collection system operators are frequently required to work on mechanical, hydraulic, and electrical systems. All of these have the potential of causing serious injury unless some action is taken to prevent the accidental startup of a system while maintenance work is being performed.

Many forms of energy are present in the typical collection system; electrical energy in a pump station, mechanical energy in a surge-relief valve, thermal energy from high temperatures, and energy associated with hydraulic pressure, such as static head pressure in a pipeline. Typical equipment systems where these hazards exist are motor control centers, pumps, belts, and spring-loaded devices such as check valves.

It is essential for the collection system operator to take appropriate precautions to prevent the accidental discharge of energy from these systems because it could cause serious injury or death. Before performing any maintenance, be sure you thoroughly understand the systems you are working on and the precautions you need to take.

The following is a general lockout/tagout procedure that you can modify for your specific needs when performing maintenance where potential hazards exist.

Whenever it is necessary to work on a piece of equipment or machinery, the following procedure shall be adhered to:

1. The main power source shall be locked out, in the OFF position, with a multiple lockout device and padlock (see Figure 4.39). (*NOTE:* the pulling of fuses shall not be considered a substitute for locking out.)
2. All pneumatic, hydraulic, and other fluid lines shall be bled, drained, purged, or blanked off to prevent pressure and/or contents from causing movement and mechanisms under spring tension or compression shall be blocked, clamped, or chained in position. Blocks may also be needed on some machinery or equipment to prevent gravitational movement.
3. Never place locks where the disconnect can be bypassed at other locations.
4. Multiple lockout devices shall be used in all cases. Each employee who is engaged in working on machinery or equipment shall install a lock on the multiple lockout device. Each lock shall be individually keyed to prevent unauthorized removal from the lockout device. The operator using the padlock is the only person normally authorized to remove it (see step 9 below).
5. A tagging system shall be used which will advise other employees of the work being performed.
6. *CAUTION:* It is mandatory that the disconnect or valve be tried to make sure it cannot be moved to ON. In addition, the machine controls themselves shall be tried to make certain that the energy is OFF.
7. This procedure shall be followed at the start of each shift or workday.
8. The immediate supervisor shall check to determine that all requirements of this procedure have been complied with and shall give the personnel involved the clearance to proceed with the required task.
9. In case of emergency, supervisory personnel may authorize removal of the lock after all precautions have been taken. An attempt shall be made to contact the person who initially locked out the equipment (see step 5). A check shall be made to ensure that all personnel, tools, and other items will not be exposed to harm or injury before the lock is removed. The operator signing the tag shall be notified of the removal of the lock at the start of the operator's next scheduled workday.

Equipment Lockout and Tagout

(Courtesy of "We're Into Safety")

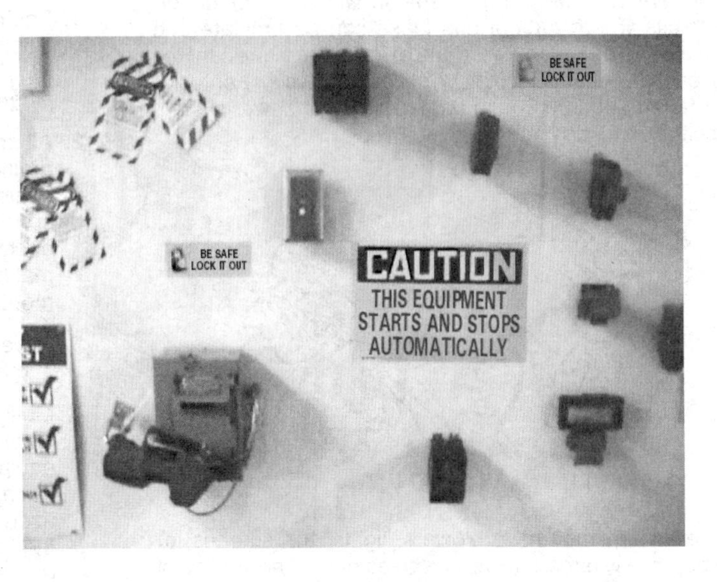

Lockout/Tagout Devices

(Courtesy of "We're Into Safety")

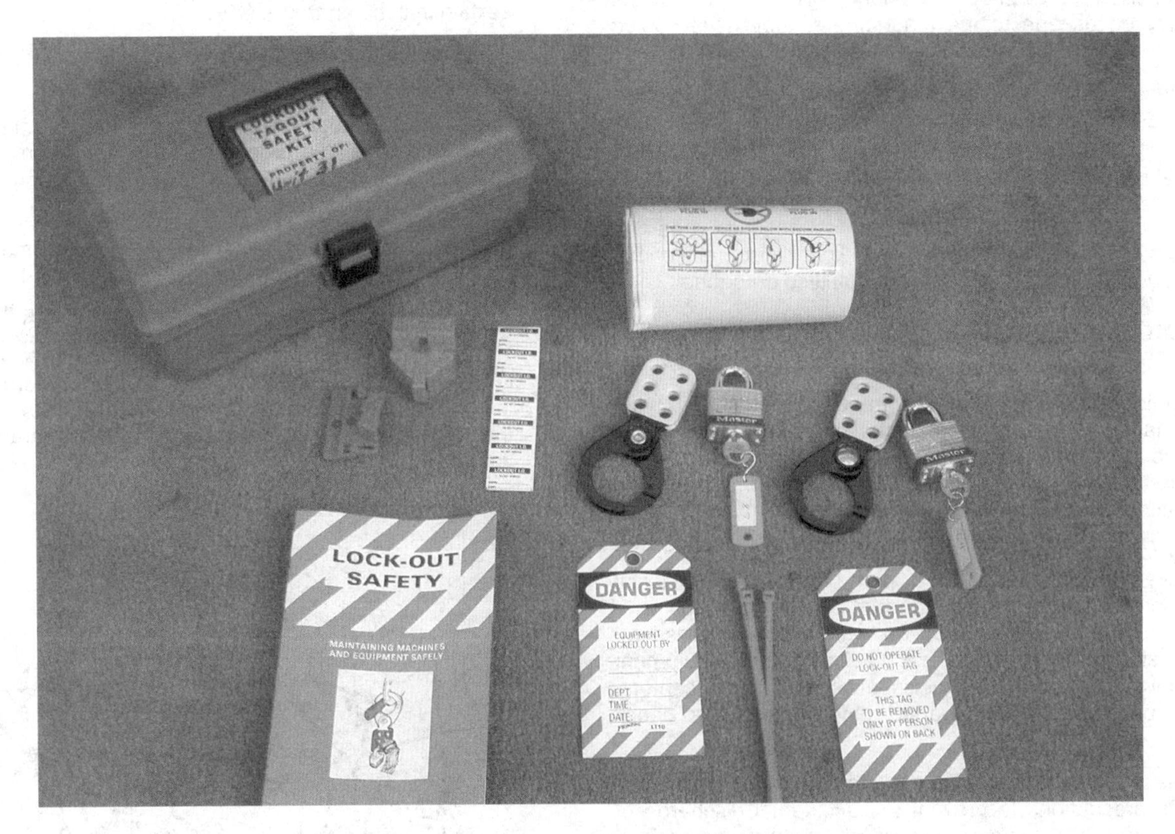

Fig. 4.39 Lockout/tagout kit

Your agency should develop a site-specific written lockout/tagout procedure and provide training to all of the operators who may use it. Each operator who must use lockout/tagout to safely do the job must be aware of applicable energy sources and the methods necessary for their effective isolation and control.

If you hire a contractor you must inform each other of your respective lockout/tagout procedures. You must ensure that you and your employees or co-workers comply with the restrictions and prohibitions of the contractor's program.

Periodic inspections of your lockout/tagout program are also required (at least annually) to ensure that the requirements of the program are being followed. The inspection(s) must be done by someone other than the one(s) using the procedure.

QUESTIONS

Write your answers in a notebook and then compare your answers with those on page 179.

4.9A What is the first rule that applies to electrical systems?

4.9B What is the "closed loop" similarity between hydraulic and electrical systems?

4.9C When working with electrical circuits, how can you provide isolation between you and the ground?

4.9D What is a ground fault interrupter (GFI)?

4.10 WORKING NEAR NOISE

Collection system equipment and facilities produce noise both inside and outside the immediate environment. Typical locations and equipment in a collection system that contribute to the operators' noise exposure are:

- Cleaning equipment such as rodding and flushing rigs,
- Air compressors,
- Pump stations,
- Engine generators, and
- Jackhammers and compactors.

Noise can cause fatigue on a short-term basis or permanent hearing loss on a long-term basis. When hearing loss occurs, the process can be very slow and usually goes unnoticed until it is too late. Hearing loss from noise exposure can be permanent.

Noise control measures include source control, path control, and receiver control. Source control usually takes place during design or when specifying equipment. Installation of a muffler on an engine generator is an example of source control.

Path control also is usually implemented during design by isolating high noise level sources in enclosures or sound treatment to limit the propagation of noise. In high noise level areas that are occupied for a full shift, an enclosure may be provided for operators.

OSHA requires that source control and path control measures (also known as engineering controls) be implemented whenever feasible, in preference to receiver controls (operator protection). Engineering noise out of the system *eliminates* the hazard whereas receiver controls only attempt to *control* the hazard.

Receiver control is a means of reducing the exposure to sound by limiting the time of exposure or providing personal protection in the way of hearing protection devices. Receiver controls (also known as administrative controls) are the least desirable method of controlling operator noise exposure, although they are frequently the only feasible option. Receiver controls are the least desirable because they rely on the operator to leave the high noise level area after a certain exposure time and/or to wear hearing protection devices conscientiously; neither of these actions may occur consistently.

The human ear is sensitive to different frequencies of sound. The frequencies we hear best are between 500 and 5,000 cycles per second. For frequencies above and below this range, the human ear is not as sensitive. (This is why stereo systems usually have some type of "bass" adjustment to make bass frequencies louder.) Meters used to measure noise levels have the ability to "filter" the sound pressure level so that the readings reflect the way a human ear perceives sound. The basic unit of measurement for sounds is the decibel (dB). When the meter adjusts or "filters" the sound, which is called "A-weighting," the unit of measurement is indicated as dBA. Noise surveys, using sound level meters, should be conducted not only to determine actual noise levels in different areas of the collection system facilities, but also noise levels of the tools and equipment used in maintenance activities.

Beginning at ages 25 to 30, hearing sensitivity gradually deteriorates with age, especially in the higher frequencies. This is part of the natural aging process that we cannot change; however, we do have the ability to prevent hearing loss from exposure to high noise levels in the workplace.

Noise exposure limits have been defined by OSHA (29 CFR Part 1910.95, "Occupational Noise Exposure") in terms of permitted exposure time at various noise levels. For example, if the noise level is 90 dBA, we can work in that area for 8 hours. If the noise level is 100 dBA, the maximum time limit is 2 hours. Limiting an operator's time of exposure to noise by shortening work hours in areas of high noise is one method of receiver control. The operator moves out of the high noise area to one of low noise on a schedule that will effectively reduce the operator's total exposure to noise.

The second method for protecting operators from the effects of noise is to provide them with protective equipment which will reduce sound levels below the levels listed in Table 4.5.

TABLE 4.5 PERMISSIBLE NOISE EXPOSURES[a]

1. When employees are exposed to loud or extended noise, earplugs and/or protective devices shall be provided and employees shall wear them accordingly.
2. Protection against the effects of noise exposure shall be provided when sound levels exceed those shown in rule number 7. These measurements shall be made on the A-scale of a standard sound level meter at a slow response.
3. When employees are subjected to sound levels exceeding those listed in rule number 7, feasible administrative and/or engineering controls shall be utilized. If such controls fail to reduce employees' exposure to within the permissible noise levels listed in this table, personal protective equipment shall be provided and used.
4. Plain cotton shall not be considered acceptable as an earplug.
5. Machinery creating excessive noise shall be equipped with mufflers.
6. Exposure to impulsive or impact noise should not exceed 140 dB peak sound pressure level.
7. Employees shall not be exposed to noise which exceeds the levels listed below for a time period not to exceed those listed below.

DURATION: HOURS PER DAY		SOUND LEVEL: dBA SLOW RESPONSE
8		90
6		92
4		95
3		97
2		100
1 1/2		102
1		105
1/2	(thirty minutes)	110
1/4	(one-fourth hour or less)	115

[a] Occupational Safety and Health Standards, Table G-16, Noise Exposure, 29 CFR 1910.95.

There are two basic types of ear protection devices, the full earmuff type (Figure 4.40) and the earplug insertion type (Figure 4.41). Each type has several different variations, for example, some muff types are designed for use with hard hats. The insertion types are manufactured with different materials; some are disposable and some reusable. Regardless of type, hearing protection devices are all given a Noise Reduction Rating (NRR) that is used to evaluate the effectiveness of the particular device. To estimate the adequacy of a hearing protector, use the NRR shown on the hearing protector package. Subtract 7 dB from the NRR and then subtract the remainder from the operator's A-weighted Time Weighted Average (TWA)[16] noise environment to obtain the estimated A-weighted TWA under the hearing protector. To provide adequate protection, the value under the hearing protector should be 85 dB or less; the lower the better. The selection of hearing protection depends on many things:

- Attenuation (sound reduction) required,
- Comfort and fit,
- Frequency of use, and
- Work area.

The employer is responsible for evaluating noise protection equipment for the specific noise environments in which the protective equipment will be used.

The employer is also required to identify and monitor operators whose normal noise exposure might equal or exceed an 8-hour TWA of 85 dBA. All operators whose normal exposure equals or exceeds the 8-hour TWA of 85 dBA must be included in a hearing conservation program. The primary elements of the program are monitoring to establish noise exposure, audiometric testing, hearing protection, training in the use of protective equipment and procedures, access to noise level information, and recordkeeping. The purpose of the program is to prevent hearing loss that might affect an operator's ability to hear and understand normal speech.

It is essential that operators be trained and use properly rated protective devices in high-noise areas or during high-noise activities, such as operating a jackhammer.

Employee training must include: (1) the effects of noise on hearing; (2) the purpose of hearing protectors, the advantages/disadvantages of each type, the attenuation (noise reduction) value of each type, and instruction on selection, fitting, use, and care; and (3) the purpose of the audiometric testing and an explanation of the test procedures. The training must be repeated annually. Audiometric test records must be retained for as long as the affected operator works for your agency.

4.11 FIREFIGHTING

Everyone must know what to do in case of a fire. If proper action is taken quickly, fires are less likely to cause damage and injury. In order to put a fire out promptly, the proper type of fire extinguisher must be used.

Fire classifications are important for determining the type of fire extinguisher needed to control the fire. Classifications also aid in recordkeeping. Fires are classified as A, B, C, or D fires based on the type of material being consumed: A, ordinary combustibles; B, flammable liquids and vapors; C, energized electrical equipment; and D, combustible metals. Fire extinguishers are also classified as A, B, C, or D to correspond with the class of fire each will extinguish.

Class A fires: ordinary combustibles such as wood, paper, cloth, rubber, many plastics, dried grass, hay, and stubble. Use foam, water, soda-acid, carbon dioxide gas, or almost any type of extinguisher.

Class B fires: flammable and combustible liquids such as gasoline, oil, grease, tar, oil-based paint, lacquer, and solvents, and also flammable gases. Use foam, carbon dioxide, or dry chemical extinguishers.

Class C fires: energized electrical equipment such as starters, breakers, and motors. Use carbon dioxide or dry chemical extinguishers to smother the fire; both types are nonconductors of electricity.

Class D fires: combustible metals such as magnesium, sodium, zinc, and potassium. Operators rarely encounter this type of fire. Use a Class D extinguisher or use fine dry soda ash, sand, or graphite to smother the fire. Consult with your

[16] *Time Weighted Average (TWA). The average concentration of a pollutant (or sound) based on the times and levels of concentrations of the pollutant. The time weighted average is equal to the sum of the portion of each time period (as a decimal, such as 0.25 hour) multiplied by the pollutant concentration during the time period divided by the hours in the workday (usually 8 hours).*

Fig. 4.40 Muff-type hearing protection
(Permission of Lyons Safety)

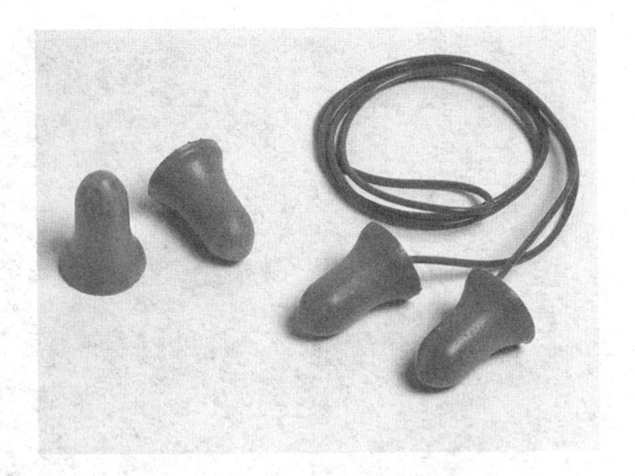

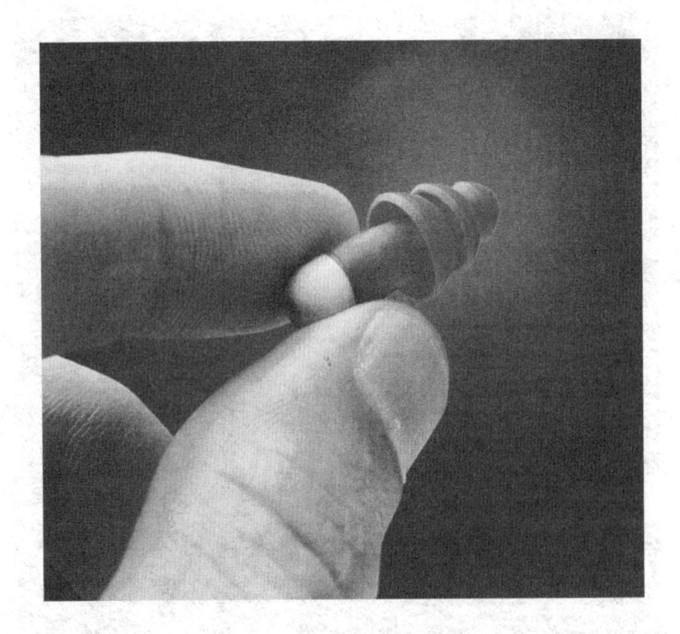

Fig. 4.41 Earplug insertion-type hearing protection
(Permission of Industrial Safety Supply)

local fire department about the best methods to use for specific hazards that exist at your facility.

Multipurpose extinguishers are also available, such as a Class BC carbon dioxide extinguisher that can be used to smother Class B and Class C fires. A multipurpose ABC carbon dioxide extinguisher will handle most laboratory fire situations. (When using carbon dioxide extinguishers, remember that the carbon dioxide can displace oxygen—take appropriate precautions.)

There is no single type of fire extinguisher that is effective for all fires so it is important that you understand the class of fire you are trying to control. You must be trained in the use of the different types of extinguishers, and the proper type should be located near the area where that class of fire may occur.

Mount fire extinguishers in conspicuous locations and install signs or lights nearby to assist people in finding them. Everyone should know how fires are classified and the appropriate extinguisher for each class. Post the phone number of the local fire department in an obvious location by every phone.

Portable extinguishers must be visually inspected monthly and must receive an annual maintenance check. Maintenance checks must be documented and the records must be retained for one year. Hydrostatic testing of extinguishers is also required every 5 to 12 years, depending on the type of extinguisher. Remember, always have extinguishers serviced promptly after use so they'll be available if needed. If your agency provides portable fire extinguishers for your use, it must also provide training in the principles of fire extinguisher use and the hazards involved. Refresher training must also be provided each year.

Clearly mark all exits from pump stations and other buildings. Post conspicuous signs identifying flammable storage areas and signs prohibiting smoking where smoking is not allowed.

The best way to fight fires is to prevent them from occurring. Fires can be prevented by careful extinguishing of cigarettes and good housekeeping of combustibles, flammable liquids, and electrical equipment and circuits.[17]

QUESTIONS

Write your answers in a notebook and then compare your answers with those on page 179.

4.10A What should collection system operators do if noise levels cannot be controlled within acceptable limits?

4.11A What different types of materials can cause different types or classes of fires?

4.11B How can each class or type of fire be extinguished?

4.11C How often should firefighting equipment and extinguishers be inspected?

4.12 HAZARD COMMUNICATION (WORKER RIGHT-TO-KNOW LAWS)
(OSHA 29 CFR 1910.1200)

In the past few years, there has been an increased emphasis nationally on hazardous materials and wastes. Much of this attention has focused on hazardous and toxic waste dumps and the efforts to clean them up after the long-term effects on human health were recognized. Each year thousands of new chemical compounds are produced for industrial, commercial, and household use. Frequently, the long-term effects of these chemicals are unknown. Exposure to the operator can occur from one or more of the following situations:

1. Use of the collection system as an intentional or accidental disposal method for hazardous materials, for example,
   a. Industrial solvents,
   b. Acids,
   c. Flammable/explosive compounds,
   d. Caustics, and
   e. Toxics.
2. Chemicals and chemical compounds that we use every day for maintenance in the collection system, for example,
   a. Solvents,
   b. Degreasers,
   c. Acids,
   d. Chlorine, and
   e. Industrial cleaners.

Federal and state laws have been enacted to control all aspects of hazardous materials handling and use. These laws are more commonly known as Worker Right-To-Know (R-T-K) laws. Every state is covered by one or more laws regarding Worker Right-To-Know. The Federal Occupational Safety and

---

[17] *A fire safety videotape is available from the National Fire Protection Association: "Fire Extinguishers, Fight or Flight," Order No. FL-80-VH. Price, $477.00, plus $6.95 shipping and handling. Order from National Fire Protection Association (NFPA), 11 Tracy Drive, Avon, MA 02322-9908.*

Health Administration (OSHA) Standard 29 CFR 1910.1200—Hazard Communication forms the basis of most laws. Although the federal standards were originally directed at the manufacturing sector, they now include all industries including the public sector and, therefore, collection systems.

In many cases, the individual states have the authority under the OSHA standard to develop their own state Worker Right-To-Know laws and most states have adopted their own laws. Unfortunately, state laws vary significantly from state to state. The state laws that have been passed are at least as stringent as the federal standard and, in most cases, are even more stringent and already apply to collection system operators.

State laws are also under continuous revision and, because a strong emphasis is being placed on hazardous materials and worker exposure, state laws can be expected to be amended in the near future to apply to virtually everybody in the workplace.

The purpose of this section is to familiarize you with general requirements of Worker RTK elements so that you are better prepared as a collection system operator to minimize risk to yourself and your co-workers from hazardous materials and to comply with state and federal laws as well. Because of the wide diversity of existing laws, these guidelines may or may not meet your state's requirements. This section will give you an overview of the basic elements of a hazard communication program, which can be particularly useful if your agency currently does not have one. An extremely effective program can be developed in house through your safety program and committee (see Volume II, Chapter 11). By cutting through the bureaucratic/legal language and applying some common sense to what the law is trying to accomplish (protection of the worker in the workplace from hazardous materials) an effective program can be developed.

The basic elements of a Worker Right-To-Know program are listed in Table 4.6 and described in the following paragraphs.

1. Identify Hazardous Materials

While there are thousands and thousands of chemical compounds that would fall under this definition in a technical sense, collection system operators should be concerned, first of all, with the materials they use in their everyday maintenance activities, and also with materials that can be introduced to the collection system through intentional or accidental spills. Information on materials that could be introduced into the collection system can be obtained from industrial pretreatment inspectors. (Also see the *PRETREATMENT FACILITY INSPECTION* manual in this series of manuals.)

In particular, you should be familiar with the industries in your area and be aware of the materials they routinely discharge into the collection system as well as the types of hazardous materials they might accidentally discharge. In some cases, special precautions may have to be taken. For example, when performing routine maintenance with a high-velocity cleaner (jet machine) in a section of line, you may wish to draw up an agreement with an industrial discharger upstream to halt discharges on the day the sewer maintenance is being performed. Hazardous materials can be broken down into general categories as follows:

a. Corrosives,
b. Toxics,
c. Flammables and explosives,
d. Asphyxiants,
e. Harmful physical agents, and
f. Infectious agents.

A complete inventory of materials in use will produce a list similar to the following:

a. Corrosives
   1. Sodium hydroxide
   2. Calcium oxide (lime)
   3. Hydrochloric acid
   4. Ferric chloride

b. Toxics
   1. Hydrogen sulfide
   2. Chlorine
   3. Carbon monoxide

c. Flammables and Explosives
   1. Methane
   2. Acetylene
   3. Gasoline
   4. Solvents

d. Asphyxiants
   1. Carbon dioxide
   2. Nitrogen

e. Harmful Physical Agents
   1. Noise
   2. Temperature
   3. Radiation

f. Infectious Agents

2. Obtain Chemical Information and Define Hazardous Conditions

Once the inventory is complete, the next step is to obtain specific information on each of the chemicals and hazardous conditions. This information is generally incorporated into a standard format form called the Material Safety Data Sheet (MSDS). Figure 4.42 (pages 169 and 170) is the Material Safety Data Sheet (OSHA 174, September, 1985) produced by the U.S. Department of Labor, Occupational Safety and Health Administration. This information is commonly available from manufacturers. Many agencies request an MSDS when the purchase order is generated and will refuse to accept delivery of the shipment if the MSDS is not included.

The purpose of the MSDS is to have a readily available reference document that includes complete information on common names, safe exposure level, effects of exposure, symptoms of exposure, flammability rating, type of first-aid

TABLE 4.6 ELEMENTS OF A HAZARD COMMUNICATION PROGRAM

I. *WRITTEN HAZARD COMMUNICATION PROGRAM*

A. Hazard Determination (Chemical manufacturers and importers only)

1. Person(s) responsible for evaluating the chemical(s)
2. List of sources to be consulted
3. Criteria to be used to evaluate studies
4. A plan for reviewing information to update MSDS if new and significant information is found

B. Labels and Other Forms of Warning

1. Person(s) responsible for labeling in-plant containers, if used
2. Person(s) responsible for labeling shipped containers
3. Description of labeling system
4. Description of written alternatives to labeling of in-plant containers, if used
5. Procedure to review and update label information

C. Material Safety Data Sheets

1. Person(s) responsible for obtaining/maintaining the MSDS
2. Description of how the MSDSs will be made available to employees
3. Procedure to follow when the MSDS is not received at time of first shipment
4. Procedure to review and update MSDS information
5. Description of alternatives to actual data sheets in the workplace, if used

D. Training

1. Person(s) responsible for conducting training
2. Format of the program to be used
3. Elements of the training program

a. Requirements of the OSHA standard
b. Operations, where hazardous chemicals are present in routine tasks, non-routine tasks, and foreseeable emergencies
c. Location and availability of

i. Written Hazard Communication Program (HCP)
ii. List of hazardous chemicals
iii. MSDS

d. Methods and observations that may be used to detect the presence or release of hazardous chemicals in the work area
e. Physical and health hazards of chemicals in the work area
f. Measures employees can take to protect themselves from hazards
g. Details for the Hazard Communication Program developed

4. Procedure to train new employees, as well as current employees, when a new chemical hazard is introduced into the workplace

E. List of hazardous chemicals

F. Procedure to inform employees of the hazards of chemicals in unlabeled pipes

G. Procedure to inform on-site contractors

II. *LABELS AND OTHER FORMS OF WARNING*

A. Labels or other markings on each container of hazardous chemicals

1. Process vessels
2. Storage tanks
3. Compressed gas cylinders
4. Product containers
5. Tank truck/tank car labels

III. *MATERIAL SAFETY DATA SHEETS*

A. MSDSs developed or obtained for all hazardous chemicals

B. Employees have access on each shift

C. MSDS completed appropriately

D. When no MSDS is available, the documentation requesting an MSDS from supplier is maintained

TABLE 4.6 ELEMENTS OF A HAZARD COMMUNICATION PROGRAM (continued)

IV. *TRAINING*

A. Employee training files

B. Employee questioning:

1. Awareness of the Hazard Communication Program (HCP) and its requirements
2. Have they received training
3. Ability to locate the MSDSs
4. General familiarity with the hazardous properties of the chemicals in their workplace

procedures, and other information about each hazardous substance.

Operators must be trained to read and understand the MSDS forms. The forms themselves must be stored in a convenient location where they are readily available for reference. Pages 171 and 172 contain an MSDS for hydrogen sulfide (H_2S). The pages are reproduced with the permission of the Genium Publishing Corporation.

3. Properly Label Hazards

Once the physical, chemical, and health hazards have been identified and listed, a labeling and training program must be implemented. To meet labeling requirements on hazardous materials, specialized labeling is available from a number of sources, including commercial label manufacturers. Exemptions to labeling requirements do exist, so consult your local safety regulatory agency for specific details.

Each container of a hazardous substance in the workplace must be labeled, tagged, or marked with the name of the hazardous substance and appropriate warning labels. You are not required to label portable containers that you transfer hazardous substances into for your personal use during your shift, but it is a good habit to develop. Then, if your work is interrupted before you complete the transfer, or if another person finishes the job, there will be no confusion about the contents of the portable container. A number of acceptable labeling systems are available, including one designed by The National Fire Protection Association (NFPA) (Figure 4.3). Private labeling systems such as the one illustrated in Figure 4.44 produced by J.T. Baker Chemical Company are also widely available. A third illustration of warning labels produced by a commercial label maker is shown in Figure 4.45.

These are standardized label formats that use a combination of pictographs, a numbering system, and colors to indicate various levels of conditions. In some cases, the MSDS can be incorporated into the labeling requirements by locating the appropriate MSDS in close proximity to drums or storage areas. The labeling requirements offer the collection system operator virtually instant recognition of the hazards in dealing with specific substances, protective equipment required, and other information. (The United States and Europe are working on an internationally acceptable labeling system, which ultimately could lead to standardized labeling for 575,000 chemical mixtures.)

4. Train Operators

The last element in the Hazard Communication Worker Right-To-Know Program is training and making information available to the collection system operator. A common sense approach eliminates the confusing issue of which of the thousands of substances operators should be trained for, and concentrates on those that they will be exposed to or use in everyday maintenance routines. Obviously, the protection from hazardous materials is tied in with the Confined Space policy, since average domestic wastewater found in the collection system does not contain sufficient concentrations of hazardous materials to require training on each of the compounds that may be found. The use of ventilating equipment, atmospheric testing instrumentation, and the other precautions defined in the Confined Space procedure assist in protecting the collection system operator from hazardous materials.

Industrial dischargers can and do discharge significant concentrations of hazardous materials, either intentionally or unintentionally. A collection system operator should be familiar with each industry that discharges to the collection system, since some special precautions may need to be taken to minimize exposure to specific hazardous materials.

The last element is a formal training program provided by qualified personnel designed to accomplish the following objectives:

1. Make operators and other employees aware of the hazard communication standard and what its requirements are;

2. Familiarize operators with potentially dangerous operations and hazardous substances that are present in their routine and nonroutine maintenance tasks and how to deal with emergency situations involving those materials and conditions;

3. Inform employees of the location and availability of the written hazard communication program;

4. Train employees in the methods and observations they may use to detect the presence or release of a hazardous substance in the work area, such as visual appearance, odor, and monitoring;

5. Inform employees of the measures they can take to protect themselves from the physical and health hazards of the chemicals in the work area;

Material Safety Data Sheet
May be used to comply with
OSHA's Hazard Communication Standard,
29 CFR 1910.1200 Standard must be
consulted for specific requirements.

U.S. Department of Labor
Occupational Safety and Health Administration
(Non-Mandatory Form)
Form Approved
OMB No. 1218-0072

IDENTITY *(As Used on Label and List)*	*Note: Blank spaces are not permitted. If any item is not applicable, or no information is available, the space must be marked to indicate that.*

Section I

Manufacturer's Name	Emergency Telephone Number
Address *(Number, Street, City, State, and ZIP Code)*	Telephone Number for Information
	Date Prepared
	Signature of Preparer *(optional)*

Section II — Hazardous Ingredients/Identity Information

Hazardous Components (Specific Chemical Identity: Common Name(s))	OSHA PEL	ACGIH TLV	Other Limits Recommended	%*(optional)*

Section III — Physical/Chemical Characteristics

Boiling Point		Specific Gravity (H_2O = 1)	
Vapor Pressure (mm Hg.)		Melting Point	
Vapor Density (AIR = 1)		Evaporation Rate (Butyl Acetate = 1)	
Solubility in Water			
Appearance and Odor			

Section IV — Fire and Explosion Hazard Data

Flash Point (Method Used)	Flammable Limits	LEL	UEL
Extinguishing Media			
Special Fire Fighting Procedures			
Unusual Fire and Explosion Hazards			

(Reproduce locally) OSHA 174, Sept. 1985

Fig. 4.42 Material Safety Data Sheet

Section V — Reactivity Data

Stability	Unstable		Conditions to Avoid
	Stable		

Incompatibility *(Materials to Avoid)*

Hazardous Decomposition or Byproducts

Hazardous Polymerization	May Occur		Condition to Avoid
	Will Not Occur		

Section VI — Health Hazard Data

Route(s) of Entry:	Inhalation?	Skin?	Ingestion?

Health Hazards *(Acute and Chronic)*

Carcinogenicity	NTP?	IARC Monographs?	OSHA Regulated?

Signs and Symptoms of Exposure

Medical Conditions
Generally Aggravated by Exposure

Emergency and First Aid Procedures

Section VII — Precautions for Safe Handling and Use

Steps to Be Taken in Case Material is Released or Spilled

Waste Disposal Method

Precautions to be Taken in Handling and Storing

Other Precautions

Section VIII — Control Measures

Respiratory Protection (Specify Type)

Ventilation	Local Exhaust	Special
	Mechanical *(General)*	Other

Protective Gloves	Eye Protection

Other Protective Clothing or Equipment

Work/Hygienic Practices

Page 2

* USGPO 1986-491-529/45775

Fig. 4.42 Material Safety Data Sheet (continued)

Genium Publishing Corporation
One Genium Plaza
Schenectady, NY 12304-4690 USA
(518) 377-8854

Material Safety Data Sheets Collection:

Sheet No. 52
Hydrogen Sulfide

Issued: 7/79 Revision: B, 9/92

Section 1. Material Identification

39

Hydrogen Sulfide (H_2S) Description: Formed as a byproduct of many industrial processes (breweries, tanneries, slaughter houses), around oil wells, where petroleum products are used, in decaying organic matter, and naturally occurring in coal, natural gas, oil, volcanic gases, and sulfur springs. Derived commercially by reacting iron sulfide with dilute sulfuric or hydrochloric acid, or by reacting hydrogen with vaporized sulfur. Used in the production of various inorganic sulfides and sulfuric acid, in agriculture as a disinfectant, in the manufacture of heavy water, in precipitating sulfides of metals; as a source of hydrogen and sulfur, and as an analytical reagent.

R 2
I 4
S 3
K 3

NFPA: 4 (top), 3 (left), 0 (right), – (bottom)

Other Designations: CAS No. 7783-06-4, dihydrogen monosulfide, hydrosulfuric acid, sewer gas, stink damp, sulfuretted hydrogen, sulfur hydride.

Manufacturer: Contact your supplier or distributor. Consult latest *Chemical Week Buyers' Guide*[73] for a suppliers list.

Cautions: Hydrogen sulfide is a highly flammable gas and reacts vigorously with oxidizing materials. It is highly toxic and can be instantly fatal if inhaled at concentrations of 1000 ppm or greater. Be aware that the sense of smell becomes rapidly fatigued at 50 to 150 ppm, and that its strong rotten-egg odor is not noticeable even at very high concentrations.

HMIS
H 3
F 4
R 0
PPE*
* Sec. 8

Section 2. Ingredients and Occupational Exposure Limits

Hydrogen sulfide: 98.5% *technical*, 99.5% *purified*, and CP *(chemically pure grade)*

1991 OSHA PELs
8-hr TWA: 10 ppm (14 mg/m^3)
15-min STEL: 15 ppm (21 mg/m^3)

1990 IDLH Level
300 ppm

1990 NIOSH REL
10-min Ceiling: 10 ppm (15 mg/m^3)

1992-93 ACGIH TLVs
TWA: 10 ppm (14 mg/m^3)
STEL: 15 ppm (21 mg/m^3)

1990 DFG (Germany) MAK
TWA: 10 ppm (15 mg/m^3)
Category V: Substances having intense odor
Peak exposure limit 20 ppm, 10 min momentary value, 4/shift

1985-86 Toxicity Data*
Human, inhalation, LC_{Lo}: 600 ppm/30 min; toxic effects not yet reviewed
Man, inhalation, LD_{Lo}: 5700 μg/kg caused coma and pulmonary edema or congestion.
Rat, intravenous, LD_{50}: 270 μg/kg; no toxic effect noted

* See NIOSH, *RTECS* (MX1225000), for additional toxicity data.

Section 3. Physical Data

Boiling Point: -76 °F (-60 °C)
Freezing Point: -122 °F (-86 °C)
Vapor Pressure: 18.5 atm at 68 °F (20 °C)
Vapor Density (Air = 1): 1.175
pH: 4.5 (freshly prepared saturated aqueous solution)
Viscosity: 0.01166 cP at 32 °F/0 °C and 1 atm
Liquid Surface Tension (est): 30 dyne/cm at -77.8 °F/-61 °C
Molecular Weight: 34.1
Density: 1.54 g/L at 32 °F (0 °C)
Water Solubility: Soluble*; 1g/187 mL (50 °F/10 °C), 1g/242 mL (68 °F/20 °C), 1g/ 314 mL (86 °F/30 °C)
Other Solubilities: Soluble in ethyl alcohol, gasoline, kerosine, crude oil, and ethylene glycol.
Odor threshold: 0.06 to 1.0 ppm†

Appearance and Odor: Colorless gas with a rotten-egg smell.

* H_2S solutions are not stable. Absorbed oxygen causes turbidity and precipitation of sulfur. In a 50:50 mixture of water and glycerol, H_2S is stable.
† Sense of smell becomes rapidly fatigued and can not be relied upon to warn of continuous H_2S presence.

Section 4. Fire and Explosion Data

Flash Point: None reported	Autoignition Temperature: 500 °F (260 °C)	LEL: 4.3% v/v	UEL: 46% v/v

Extinguishing Media: Let small fires burn unless leak can be stopped immediately. For large fires, use water spray, fog, or regular foam. **Unusual Fire or Explosion Hazards:** H_2S burns with a blue flame giving off sulfur dioxide. Its burning rate is 2.3 mm/min. Gas may travel to a source of ignition and flash back. **Special Fire-fighting Procedures:** Because fire may produce toxic thermal decomposition products, wear a self-contained breathing apparatus (SCBA) with a full facepiece operated in pressure-demand or positive-pressure mode. Structural firefighter's protective clothing is not effective for fires involving H_2S. If possible without risk, stop leak. Use unmanned device to cool containers until well after fire is out. Withdraw immediately if you hear a rising sound from venting safety device or notice any tank discoloration due to fire. Do not release runoff from fire control methods to sewers or waterways.

Section 5. Reactivity Data

Stability/Polymerization: H_2S is stable at room temperature in closed containers under normal storage and handling conditions. Hazardous polymerization cannot occur. **Chemical Incompatibilities:** Hydrogen sulfide attacks metals forming sulfides and is incompatible with 1,1-bis(2-azidoethoxy) ethane + ethanol, 4-bromobenzenediazonium chloride, powdered copper + oxygen, metal oxides, finely divided tungsten or copper, nitrogen trichloride, silver fulminate, rust, soda-lime, and all other oxidants. **Conditions to Avoid:** Exposure to heat and contact with incompatibles. **Hazardous Products of Decomposition:** Thermal oxidative decomposition of hydrogen sulfide can produce toxic sulfur dioxide .

Section 6. Health Hazard Data

Carcinogenicity: The IARC,[164] NTP,[169] and OSHA[164] do not list hydrogen sulfide as a carcinogen. **Summary of Risks**: H_2S combines with the alkali present in moist surface tissues to form caustic sodium sulfide, causing irritation of the eyes, nose, and throat at low levels (50 to 100 ppm). Immediate death due to respiratory paralysis occurs at levels greater than 1000 ppm. Heavy exposure has resulted in neurological problems, however recovery is usually complete. H_2S exerts most of it's toxicity on the respiratory system. It inhibits the respiratory enzyme cytochrome oxidase, by binding iron and blocking the necessary oxydo-reduction process. Electrocardiograph changes after over-exposure have suggested direct damage to the cardiac muscle, however some authorities debate this. **Medical Conditions Aggravated by Long-Term Exposure:** Eye and nervous system disorders. **Target Organs:** Eyes, respiratory system and central nervous system. **Primary Entry Routes:** Inhalation, eye and skin contact. **Acute Effects:** Inhalation of low levels can cause headache, dizziness, nausea, cramps, vomiting, diarrhea, sneezing, staggering, excitability, pale

Continued on next page

No. 52 Hydrogen Sulfide 9/92

Section 6. Health Hazard Data, *continued*

complexion, dry cough, muscular weakness, and drowsiness. Prolonged exposure to 50 ppm, can cause rhinitis, bronchitis, pharyngitis, and pneumonia. High level exposure leads to pulmonary edema (after prolonged exposure to 250 ppm), asphyxia, tremors, weakness and numbing of extremeties, convulsions, unconsciousness, and death due to respiratory paralysis. Concentrations near 100 ppm may be odorless due to olfactory fatigue, thus the victim may have no warning. Lactic acidosis may be noted in survivors. The gas does not affect the skin although the liquid (compressed gas) can cause frostbite. The eyes are very susceptible to H_2S keratoconjunctivitis known as 'gas eye' by sewer and sugar workers. This injury is characterized by palpebral edema, bulbar conjunctivitis, mucous-puss secretions, and possible reduction in visible capacity.
Chronic Effects: Chronic effects are not well established. Some authorities have reported repeated exposure to cause fatigue, headache, inflammation of the conjunctiva and eyelids, digestive disturbances, weight loss, dizziness, a grayish-green gum line, and irritability. Others say these symptoms result from recurring acute exposures. There is a report of encephalopathy in a 20 month old child after low-level chronic exposure.
FIRST AID Eyes: *Do not* allow victim to rub or keep eyes tightly shut. Gently lift eyelids and flush immediately and continuously with flooding amounts of water. Treat with boric acid or isotonic physiological solutions. Serious exposures may require adrenaline drops. Olive oil drops (3 to 4) provides immediate treatment until transported to an emergency medical facility. Consult a physician immediately. **Skin:** *Quickly* remove contaminated clothing and rinse with flooding amounts of water. For frostbite, rewarm in 107.6°F (42 °C) water until skin temperature is normal. *Do not* use dry heat. **Inhalation:** Remove exposed person to fresh air and administer 100% oxygen. Give hyperbaric oxygen if possible. **Ingestion:** Unlikely since H_2S is a gas above -60 °C. **Note to Physicians:** The efficacy of nitrite therapy is unproven. Normal blood contains < 0.05 mg/L H_2S; reliable tests need to be taken within 2 hr of exposure.

Section 7. Spill, Leak, and Disposal Procedures

Spill/Leak: Immediately notify safety personnel, isolate and ventilate area, deny entry, and stay upwind. Shut off all ignition sources. Use water spray to cool, dilute, and disperse vapors. Neutralize runoff with crushed limestone, agricultural (slaked) lime, or sodium bicarbonate. If leak can't be stopped in place, remove cylinder to safe, outside area and repair or let empty. Follow applicable OSHA regulations (29 CFR 1910.120).
Ecotoxicity Values: Bluegill sunfish, TLm = 0.0448 mg/L/96 hr at 71.6 °F/22 °C; fathead minnow, TLm = 0.0071 to 0.55 mg/L/96 hr at 6 to 24 °C.
Environmental Degradation: In air, hydrogen sulfides residency (1 to 40 days) is affected by temperature, humidity, sunshine, and the presence of other pollutants. It does not undergo photolysis but is oxidated by oxygen containing radicals to sulfur dioxide and sulfates. In water, H_2S converts to elemental sulfur. In soil, due to its low boiling point, much of H_2S evaporates quickly if spilled. Although, if soil is moist or precipitation occurs at time of spill, H_2S becomes slightly mobile due to its water solubility. H_2S does not bioaccumulate but is degraded rapidly by certain soil and water bacteria. **Disposal:** Aerate or oxygenate with compressor. For in situ amelioration, carbon removes some H_2S. Anion exchanges may also be effective. A potential candidate for rotary kiln incineration (1508 to 2912 °F/820 to 1600 °C) or fluidized bed incineration (842 to 1796 °F/450 to 980 °C). Contact your supplier or a licensed contractor for detailed recommendations. Follow applicable Federal, state, and local regulations.

EPA Designations
Listed as a RCRA Hazardous Waste (40 CFR 261.33): No. U135
SARA Toxic Chemical (40 CFR 372.65): Not listed
Listed as a SARA Extremely Hazardous Substance (40 CFR 355), TPQ: 500 lb
Listed as a CERCLA Hazardous Substance* (40 CFR 302.4): Final Reportable Quantity (RQ), 100 lb (45.4 kg) [* per RCRA, Sec. 3001 & CWA, Sec. 311 (b)(4)]

OSHA Designations
Listed as an Air Contaminant (29 CFR 1910.1000, Table Z-1-A & Z-2)
Listed as a Process Safety Hazardous Material (29 CFR 1910.119), TQ: 1500 lb

Section 8. Special Protection Data

Goggles: Wear protective eyeglasses or chemical safety goggles, per OSHA eye- and face-protection regulations (29 CFR 1910.133). Because contact lens use in industry is controversial, establish your own policy. **Respirator:** Seek professional advice prior to respirator selection and use. Follow OSHA respirator regulations (29 CFR 1910.134) and, if necessary, wear a MSHA/NIOSH-approved respirator. For < 100 ppm, use a supplied-air respirator (SAR) or SCBA. For < 250 ppm, use a SAR operated in continuous-flow mode. For < 300 ppm, use a SAR or SCBA with a full facepiece. For emergency or nonroutine operations (cleaning spills, reactor vessels, or storage tanks), wear an SCBA. *Warning! Air-purifying respirators do not protect workers in oxygen-deficient atmospheres.* If respirators are used, OSHA requires a respiratory protection program that includes at least: a written program, medical certification, training, fit-testing, periodic environmental monitoring, maintenance, inspection, cleaning, and convenient, sanitary storage areas. **Other:** Wear chemically protective gloves, boots, aprons, and gauntlets to prevent skin contact. Polycarbonate, butyl rubber, polyvinyl chloride, and neoprene are suitable materials for PPE. **Ventilation:** Provide general & local exhaust ventilation systems to maintain airborne concentrations below the OSHA PEL (Sec. 2). Local exhaust ventilation is preferred because it prevents contaminant dispersion into the work area by controlling it at its source.[103] **Safety Stations:** Make available in the work area emergency eyewash stations, safety/quick-drench showers, and washing facilities. **Contaminated Equipment:** Separate contaminated work clothes from street clothes and launder before reuse. Clean PPE. **Comments:** Never eat, drink, or smoke in work areas. Practice good personal hygiene after using this material, especially before eating, drinking, smoking, using the toilet, or applying cosmetics.

Section 9. Special Precautions and Comments

Storage Requirements: Prevent physical damage to containers. Store in steel cylinders in a cool, dry, well-ventilated area away from incompatibles (Sec. 5). Install electrical equipment of Class 1, Group C. Outside or detached storage is preferred. **Engineering Controls:** To reduce potential health hazards, use sufficient dilution or local exhaust ventilation to control airborne contaminants and to keep levels as low as possible. Enclose processes and continuously monitor H_2S levels in the plant air. Keep pipes clear of rust as H_2S can ignite if passed through rusty pipes. Purge and determine H_2S concentration before entering a confined area that may contain H_2S. The worker entering the confined space should have a safety belt and life line and be observed by a worker from the outside. Follow applicable OSHA regulations (1910.146) for confined spaces. H_2S can be trapped in sludge in sewers or process vessels and may be released during agitation. Calcium chloride or ferrous sulfate should be added to neutralize process wash water each time H_2S formation occurs. Control H_2S emissions with a wet flare stack/scrubbing tower. **Administrative Controls:** Consider preplacement and periodic medical exams of exposed workers emphasizing the eyes, nervous and respiratory system.

Transportation Data (49 CFR 172.101)

DOT Shipping Name: Hydrogen sulfide, liquefied
DOT Hazard Class: 2.3
ID No.: UN1053
DOT Packaging Group: --
DOT Label: Poison Gas, Flammable Gas
Special Provisions (172.102): 2, B9, B14

Packaging Authorizations
Exceptions: --
Non-bulk Packaging: 304
Bulk Packaging: 314, 315

Vessel Stowage Requirements
Vessel Stowage: D
Other: 40

Quantity Limitations
Passenger, Aircraft, or Railcar: Forbidden
Cargo Aircraft Only: Forbidden

MSDS Collection **References:** 26, 73, 89, 100, 101, 103, 124, 126, 127, 132, 136, 140, 148, 149, 153, 159, 163, 164, 168, 171, 180
Prepared by: M Gannon, BA; **Industrial Hygiene Review:** PA Roy, MPH, CIH; **Medical Review:** AC Darlington, MPH, MD

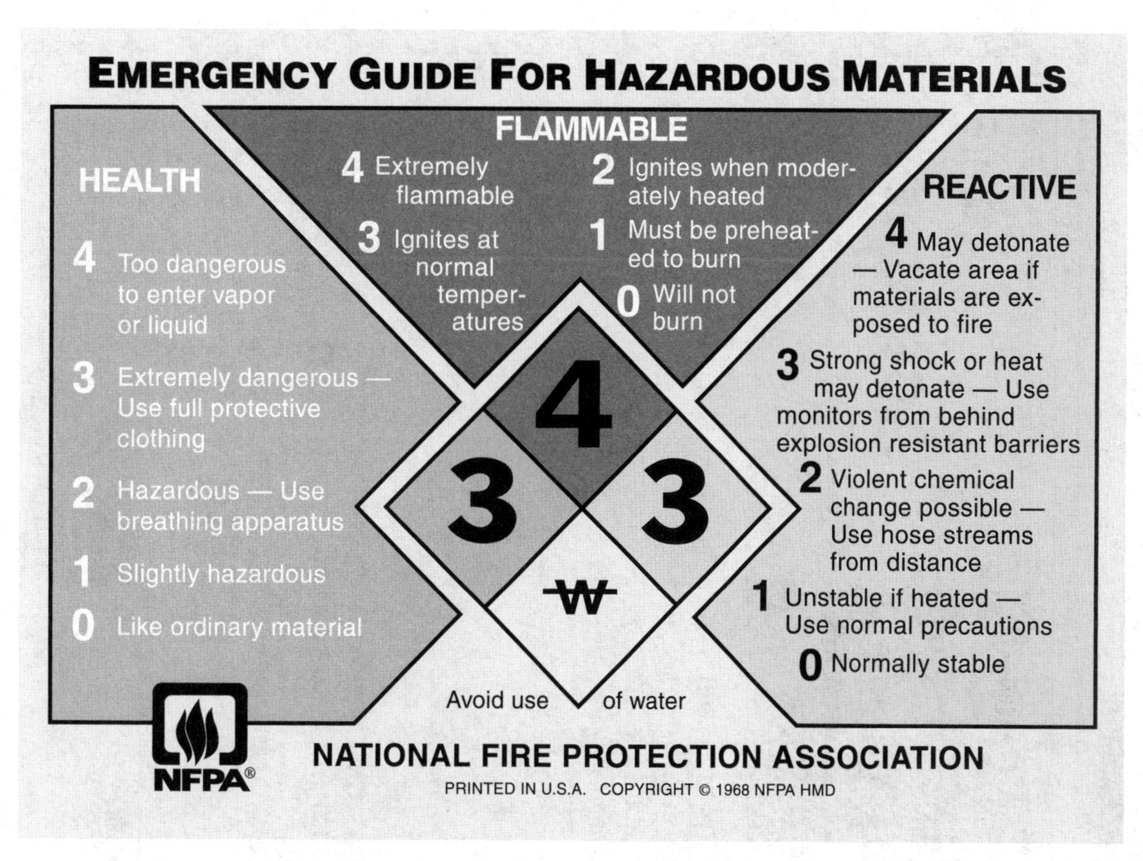

Fig. 4.43 NFPA hazard warning label

(Permission of National Fire Protection Association)

6. Explain how to read and interpret MSDS forms and have an MSDS file readily available for reference;
7. Train operators and other employees to use the labeling format; and
8. Document what training has been performed.

The hazard communication standard and the individual state requirements are a very complex set of regulations. Remember, however, the ultimate goal of these regulations and other collection system procedures is to provide additional operator protection. These standards and regulations, once the intent is understood, are relatively easy to implement and can certainly be accomplished in house by the safety structure as suggested in Volume II, Chapter 11.

When all is said and done, common sense, knowledge, awareness and commitment are the keys to complying with the hazardous material and the Worker Right-To-Know regulations.

Depending on what state you are in, your agency could be in violation of federal and state regulations if it does not currently have a Worker Right-To-Know policy. As more states which are covered by the federal standard revise their laws, it is anticipated that in the near future all employees will be covered by Worker Right-To-Know laws.

4.13 ADDITIONAL READING

1. *SAFETY AND HEALTH IN WASTEWATER SYSTEMS* (MOP 1). Obtain from Water Environment Federation (WEF), Publications Order Department, 601 Wythe Street, Alexandria, VA 22314-1994. Order No. MO2001. Price to members, $19.99; nonmembers, $29.99; plus shipping and handling.
2. *CONFINED SPACE ENTRY.* Obtain from Water Environment Federation (WEF), Publications Order Department, 601 Wythe Street, Alexandria, VA 22314-1994. Order No. P07115. Price to members, $19.99; nonmembers, $29.99; plus shipping and handling.
3. In this manual, Chapter 3, "Wastewater Collection Systems," Section 3.70, "Excavation and Shoring."
4. Chapter 11, "Safety/Survival Programs for Collection System Operators," in Volume II of this course.
5. Safety-related training products are available from Communication Arts Multimedia, Inc., RD 4, Box 22, Ligonier, PA 15658. The titles are:
   a. Confined Space Safety. Price for video, $295.00; CD-ROM, $395.00.
   b. Shoring and Trenching. Price for video, $245.00; CD-ROM, $345.00.

The cost of shipping and handling is $2.00 per item.

BAKER SAF-T-DATA™ Guide

An easy-to-understand hazard classification appears on J. T. Baker labels. It will help increase your awareness of vital occupational health and safety practices. Using the BAKER SAF-T-DATA™ System as a guide you can quickly learn the hazards each substance presents to your health and safety, personal laboratory protective equipment that should be used for handling, and the recommended storage of compatible products by color code.

A NUMERICAL HAZARD CODE

Substances are rated on a scale of 0 (non-hazardous) to 4 (extremely hazardous) in each of four hazard categories:

- **Health hazard** - the danger or toxic effect a substance presents if inhaled, ingested, or absorbed.
- **Flammable hazard** - the tendency of the substance to burn.
- **Reactivity hazard** - the potential of a substance to explode or react violently with air, water or other substances.
- **Contact hazard** - the danger a substance presents when exposed to skin, eyes, and mucous membranes.

Rating Scale

4	3	2	1	0
Extreme	Severe	Moderate	Slight	None*

*No scientific data in the standard references that suggests the substance is hazardous.

HAZARD SYMBOL

A substance rated 3 or 4 in any hazard category will also display a hazard symbol. These easy-to-understand pictograms emphasize the serious hazards related to a substance:

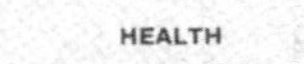

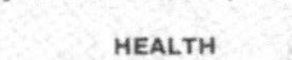

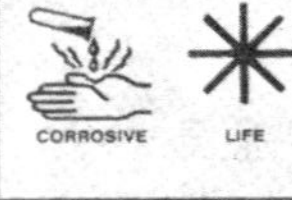

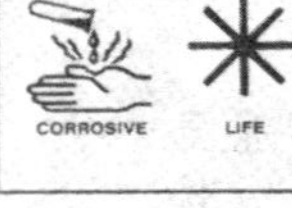

B LABORATORY PROTECTIVE EQUIPMENT

This series of pictograms suggests the personal protective clothing and equipment recommended for use when handling the substance in a laboratory situation. The pictograms relate to the combination of hazards presented by the substance.

The stop sign indicates the substance represents a special extreme hazard and the MSDS and other references should be consulted before handling.

C STORAGE COLOR CODING

The SAF-T-DATA label suggests a unique method for setting up your chemical storage area. Compatible products are labelled with the same color. Simply group these colors together and follow the recommendations for appropriate storage:

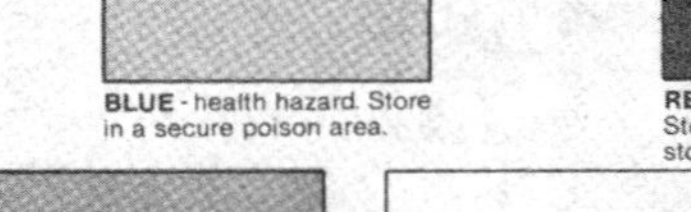

BLUE - health hazard. Store in a secure poison area.

RED – flammable hazard. Store in a flammable liquid storage area.

YELLOW - reactivity hazard. Store separately and away from flammable or combustible materials.

WHITE – contact hazard. Store in a corrosion-proof area.

ORANGE - substances with no rating higher than 2 in any hazard category. Store in a general chemical storage area.

STRIPED LABELS - incompatible materials of the same color class have striped labels. These approximately 40 products should not be stored adjacent to substances with the same colored labels. Proper storage must be individually assessed.

D SPILL CONTROL CODE

This statement indicates which J.T. Baker spill control kit is recommended for use with the substance.

E NFPA SYSTEM*

This system was adopted by the NFPA in 1975 to safeguard the lives of firefighters. It is based on the hazards created by a substance in a fire situation. For this reason, the hazard ratings in the Baker SAF-T-DATA™ System, which are based on substances in a laboratory situation, will not always correspond with the NFPA ratings.

*National Fire Protection Association

Fig. 4.44 J.T. Baker hazardous substance labeling system

(Permission of J.T. Baker Chemical Company)

CHECK

DANGER
EYE PROTECTION REQUIRED HERE

DANGER
HAND PROTECTION REQUIRED

DANGER
RESPIRATOR REQUIRED

DANGER
FOOT PROTECTION REQUIRED

CAUTION
USE APRON

CAUTION
USE PROTECTIVE CLOTHING

CAUTION
USE FACE SHIELD

CAUTION
USE CHEMICAL GOGGLES

DANGER
"NO SMOKING OR OPEN FLAME"

CAUTION
GROUND BEFORE POURING

DANGER

	CHECK
TOXIC	
ACID	
ALKALINE	
FLAMMABLE	
COMBUSTIBLE	
USE VENTILATION	

CHEMICAL NAME ____________________

USE ____________________

Fig. 4.45 Commercial warning labels

(Permission of the SIGNMARK Division, W.H. Brady Co.)

QUESTIONS

Write your answers in a notebook and then compare your answers with those on page 179.

4.12A List the four basic elements of a hazard communication program.

4.12B List the general categories of hazardous materials.

4.12C What is the purpose of the MSDS?

4.12D What information must be on the label of a hazardous material container?

END OF LESSON 3 OF 3 LESSONS
on
SAFE PROCEDURES

Please answer the discussion and review questions.

DISCUSSION AND REVIEW QUESTIONS

Chapter 4. SAFE PROCEDURES

(Lesson 3 of 3 Lessons)

Write the answers to these questions in your notebook. The question numbering continues from Lesson 2.

20. What types of equipment and situations increase collection system operators' exposure to electrical hazards?

21. What are the basic rules for qualified persons working with electricity?

22. What are some examples of places in a collection system where explosion-proof wiring should be used?

23. How can collection system operators be protected from injury by the accidental discharge of stored energy?

24. How can collection system operators protect their hearing from loud noises?

25. How would you extinguish a fire?

26. How can collection system operators be exposed to hazardous materials?

27. Why should collection system operators be familiar with worker "Right-To-Know" laws?

SUGGESTED ANSWERS

Chapter 4. SAFE PROCEDURES

ANSWERS TO QUESTIONS IN LESSON 1

Answers to questions on page 88.

4.0A Different types of vehicles, or combinations of vehicles and towed equipment, that a collection system operator may be expected to drive include tankers, rodding machines, boom trucks, standby power generators, foaming trailers, TV/grout rigs, compressors, trailer-mounted engine-driven pumps, and construction equipment such as dump trucks and backhoes.

4.1A Collection system operators must follow established procedures because of the serious consequences that could result due to the nature of the job and the increased risk of injury and/or fatality to the operator and co-workers.

4.1B Before leaving the maintenance yard or shop, discuss work assignments, determine equipment needed, inspect equipment that will be used, and inspect vehicles and towed equipment.

4.1C Before taking an atmospheric testing device out to a job site it should be calibrated and working properly.

4.2A Defensive driving is not only a mature attitude toward driving, but a strategy as well. You must always be on the alert for the large number of new drivers, inattentive, sleepy, or sleeping drivers, people driving under the influence of alcohol and/or drugs, and those drivers who are simply incompetent. In addition, weather, construction, and the type of equipment you are driving can also present hazards.

Answers to question on page 93.

4.3A The purpose of traffic control around work areas in streets and highways is to provide safe and effective work areas and to warn, control, protect, and expedite vehicular and pedestrian traffic.

4.3B Any time traffic may be affected, appropriate authorities must be notified; these could be state, county, or local. Frequently a permit must be issued by the authority that has jurisdiction before traffic can be diverted or disrupted.

4.3C Traffic can be warned of your presence in a street by "CREW WORKING IN STREET" and "CAUTION, CONSTRUCTION WORK" signs. Signs with flags or flashers and vehicles with rotating flashing lights are used to warn other motorists. Vehicle-mounted traffic guides are also helpful. Use flaggers to alert drivers and to direct traffic around the work site.

4.3D Once motorists have been warned, they must be safely routed around the job site. Traffic routing includes the use of flaggers and directional signs. Barricades and traffic cones can effectively channel traffic from one lane to another around the job site. Traffic officers can be used to direct traffic if streets are narrow and/or if there is considerable traffic or congestion.

4.3E The traffic control zone is the area from the point where traffic is first affected by the activity and/or traffic controls to a point where traffic is no longer affected. Typically this is the distance between the first advance warning sign and the point beyond the work area where traffic is no longer affected.

4.3F Decision sight distance is the distance required by a driver to properly react to hazardous, unusual, or unexpected situations even where an evasive maneuver is more desirable than a hurried stop.

Answers to questions on page 95.

4.3G An operator can become qualified to control traffic by gaining the following knowledge and experience:

1. A basic understanding of the principles of traffic control in work zones,
2. Knowledge of the standards and regulations governing traffic control,
3. Adequate training in safe traffic control practices, and
4. Experience in applying traffic control in work zones.

4.3H Records of traffic control activities should contain the starting and ending times, location, names of personnel involved, and the traffic controls used.

4.3I When work is planned for nighttime hours, consider adding warning lights to traffic control devices and signs, using larger, more reflective channelizing devices such as drums, using floodlights for any flagger stations and for the work area, and adding reflective devices to work vehicles that are typically used for nighttime work.

Answers to questions on page 99.

4.3J The five different areas of a traffic control zone are: (1) advance warning, (2) transition, (3) buffer, (4) work, and (5) termination areas.

4.3K The decision sight distance should give a motorist time to recognize a hazard, plus time to react to the hazard, and time for the vehicle to respond to the driver's reaction.

4.3L A taper is a series of channelizing devices and/or pavement markings placed on an angle to move traffic out of its normal path.

4.3M A buffer space should be used to separate the two tapers for opposing directions of traffic to prevent head-on collisions on multi-lane streets when one lane of traffic must use a lane that normally flows in the opposite direction.

Answers to questions on page 113.

4.3N Using traffic control devices in a consistent, predictable manner is one of the best ways to increase the safety of both drivers and operators at construction sites. Drivers learn by experience what to expect when they see traffic control devices. If the devices themselves and their placement in the roadway are consistent over time, drivers can more quickly understand and respond to the changed road conditions.

4.3O Types of channelizing devices include barricades, traffic cones and tubes, drums, and vertical panels.

4.3P A portable changeable message sign (PCMS) is a lighted, electronic traffic control device with the flexibility to display a variety of messages. PCMSs are used to give motorists advance warning of road closures, accident management instructions, narrow lanes, construction schedules, traffic management and diversion activities, and adverse conditions.

4.3Q To slow daytime traffic, the flagger would face oncoming vehicles, hold a paddle or sign marked SLOW in the left hand with the sign above shoulder height, and make short up and down motions with the extended right hand.

ANSWERS TO QUESTIONS IN LESSON 2

Answers to questions on page 119.

4.4A The major categories of hazards you may encounter when entering a manhole include atmospheric, physical injury, infection and disease, insects and biting critters, toxic exposure, and drowning.

4.4B Confined space means a space that:

A. Is large enough and so configured that an employee can bodily enter and perform assigned work; and
B. Has limited or restricted means for entry or exit (for example, tanks, vessels, silos, storage bins, hoppers, vaults, and pits are spaces that may have limited means of entry); and
C. Is not designed for continuous employee occupancy.

4.4C The top portion of the atmosphere in a manhole can test "safe" when the bottom portion is dangerous because dangerous gases that are heavier than air can accumulate in the bottom of a manhole. Also, gases may be found throughout the manhole due to air movement or temperature changes, regardless of their specific gravity.

4.4D Atmospheric hazards encountered in manholes may include:

1. Explosive or flammable atmospheres
   a. Methane gas
   b. Propane and explosive fuel gases
2. Toxic atmospheres, such as caused by hydrogen sulfide gas, and
3. Lack of oxygen or oxygen enrichment.

Answers to questions on page 121.

4.4E Physical injuries in manholes can be caused by:

1. Restricted space,
2. Uneven and slippery footing,
3. Defective metal rungs for entry and exit,
4. Dropping or tossing tools in and out of manholes, and
5. Sharp objects.

4.4F Personal cleanliness is the best means of protection from diseases when working in a manhole. Inoculations may also assist in reducing disease risk.

4.4G To protect yourself from insects when working in a manhole, always inspect the manhole before entry and wear protective clothing.

4.4H To protect yourself from drowning, watch your step at all times, especially when working near or in large flows. Wear life jackets or buoyant work vests and use lifelines if the flow in the sewer is capable of washing a fallen or unconscious operator away or if an operator will fit into the downstream pipe.

Answers to questions on page 128.

4.5A Sewers containing raw wastewater, pump station wet wells, gravity sewer manholes, and siphon headworks structures are considered permit-required confined spaces because there is no way to completely isolate them from the rest of the system. Sudden changes in the atmosphere may take place unpredictably from toxic, flammable, or explosive materials.

4.5B Air supplying respirators, such as the supplied air respirator (SAR) and the self-contained breathing apparatus (SCBA), supply clean air to the wearer from an independent source and provide the highest levels of protection.

4.5C If an atmospheric alarm condition develops while an operator is working in a confined space, the operator should get out and stay out until the alarm condition is cleared.

4.5D Whenever anyone must enter and work in a manhole or other confined space, a safety harness with a lifeline must be worn.

Answers to questions on page 134.

4.5E An atmospheric alarm unit is used to test for the presence of explosive or flammable gases, toxic gases (mainly hydrogen sulfide), and the percentage of oxygen in a confined space.

4.5F Oxygen content should be the first test performed with atmospheric analyzers, then test for combustible gases, and test for toxic gases last.

4.5G The three types of sensing elements used on atmospheric alarm units are diffusion-type sensors, catalytic sensing elements, and electrochemical sensors.

4.5H Physical and environmental conditions that can affect the sensors of gas detection instruments include caustic gases, temperature, dirty air, humidity, air velocity, and vibration.

4.5I The usual alarm set point for hydrogen sulfide is 10 parts per million.

Answers to questions on page 139.

4.5J Lightweight fiberglass or aluminum-alloy extension ladders are often used in shallow manholes. As an alternative, many operators use strap or rope ladders with rigid metal or rope rungs.

4.5K The number one rule of hygiene for collection system operators is to prevent oral or skin contact with wastewater. This means no smoking, eating, or drinking until after washing thoroughly with a disinfectant soap and water. Waterless antiseptic hand cleaners also are effective and paper towels may be used for the wiping process.

4.5L On a properly adjusted hard hat there will be approximately one inch of clearance, plus or minus $^1/_8$ inch, between the head harness on the person's head and the inside of the hard hat.

4.5M Rings can be a hazard if there is the slightest chance of them getting caught in any hook, tool, or piece of machinery. If the ring gets caught on equipment or becomes bent, the operator could suffer painful cuts or even loss of a finger.

Answers to questions on page 140.

4.6A Health conditions that should be considered before entering a manhole include:

1. Good health,
2. No hangover, not under influence of alcohol or drugs,
3. Good physical condition,
4. No open sores, infections, irritations, and
5. Current immunizations.

4.6B Topics discussed during the briefing before manhole entry should include:

1. Safety aspects of the job,
2. Order of doing tasks to accomplish the job, and
3. Emergency procedures.

Answers to questions on page 148.

4.7A The minimum crew when someone enters a manhole is three. One person enters the manhole and two must be standing by topside.

4.7B The person standing by in the safety enclosure should carefully watch the person working in the manhole. The manhole operator's responses and actions can help indicate if the operator is suffering from lack of oxygen or fatigue. If there are any questions regarding safety, the operator should be brought up immediately.

4.7C The minimum level of oxygen in air for safe breathing is 19.5 percent.

4.7D After leaving a manhole, an operator should change clothes and wash thoroughly *BEFORE* going home.

ANSWERS TO QUESTIONS IN LESSON 3

Answers to questions on page 153.

4.8A OSHA requires that excavations five feet or more in depth be provided with a protective system. Excavations less than five feet deep may also require protection if a competent person determines that there are indications of a potential cave-in.

4.8B Three protective systems for excavations include shoring, shielding, and sloping/benching.

4.8C Spoil should be placed at least two feet from the trench.

4.8D A good rule of thumb is to slope or bench at least one foot back for every one foot of trench depth.

Answers to questions on page 161.

4.9A The first rule that applies to electrical systems is, *"IF YOU ARE NOT QUALIFIED, DO NOT, UNDER ANY CIRCUMSTANCES, ATTEMPT TO WORK ON ELECTRICAL SYSTEMS."*

4.9B The "closed loop" similarity between hydraulic and electrical systems refers to the fact that in electrical systems the electrons want to return to the source, which is commonly referred to as ground. In hydraulic systems, the pressure in the pipe also wants to return to a zero pressure state.

4.9C When working with electrical circuits, provide isolation between you and the ground by using rubber mats, rubber boots, or leather gloves.

4.9D A ground fault interrupter (GFI) is a device that senses the amount of current flowing on the two conductors of an electric circuit. Under normal conditions these two currents are always equal. When a difference of as little as five milliamps occurs, indicating that some current is being lost to ground, the device trips the circuit in 1/40th of a second or less.

Answers to questions on page 165.

4.10A If noise levels cannot be controlled within acceptable limits, operators must be provided with and must use approved hearing protection devices.

4.11A The different types of materials that can cause fires are ordinary combustibles, flammable and combustible liquids, electricity, and metals.

4.11B The different types or classes of fires can be extinguished by the following methods:

CLASS	METHOD OF EXTINGUISHING
A	Foam, water, soda-acid, carbon dioxide gas, or almost any type of extinguisher
B	Foam, carbon dioxide, or dry chemical extinguishers
C	Carbon dioxide or dry chemical extinguishers to smother the fire
D	Class D extinguisher or use fine dry soda ash, sand, or graphite to smother the fire

4.11C Firefighting equipment and extinguishers must be visually inspected each month. They must receive an annual maintenance check also.

Answers to questions on page 176.

4.12A The four basic elements of a hazard communication program include:

1. Identify hazardous materials,
2. Obtain chemical information and define hazardous conditions,
3. Properly label hazards, and
4. Train operators.

4.12B The general categories of hazardous materials include:

1. Corrosives,
2. Toxics,
3. Flammables and explosives,
4. Asphyxiants,
5. Harmful physical agents, and
6. Infectious agents.

4.12C The purpose of the MSDS is to have a readily available document to be used for training and as an immediate reference, since it includes information on common names, safe exposure level, what the effects of exposure are, what the symptoms of exposure are, whether it is flammable, what type of first aid should be taken, and other information.

4.12D The label on a hazardous material container must indicate the name of the hazardous substance and appropriate hazard warnings.

CHAPTER 5

INSPECTING AND TESTING COLLECTION SYSTEMS

by

George Freeland

Revised by

Bob Rentfro

Rick Arbour

Gary Batis

TABLE OF CONTENTS

Chapter 5. INSPECTING AND TESTING COLLECTION SYSTEMS

APPENDIX

OBJECTIVES

Chapter 5. INSPECTING AND TESTING COLLECTION SYSTEMS

Following completion of Chapter 5, you should be able to:

1. Inspect existing sewers for operation and maintenance problems,
2. Inspect new sewers and replacement sewers for installation as planned by examining line and grade, joint and junction adequacy, and proper installation of manholes and appurtenances,
3. Test for leaks in joints, taps, sewers, and manholes of existing facilities,
4. Find legal, illegal, unauthorized, and improper connections,
5. Locate and determine the seriousness of inflow, infiltration, exfiltration, and diversion flow problems,
6. Identify and evaluate damage due to corrosion, cracking, crushing, subsidence (soil settling), root intrusion, stoppages, washouts, and improper connections,
7. Provide meaningful reports to supervisors so they can assign priorities in order to develop cost-effective maintenance or corrective action programs, and
8. Effectively use the inspecting and testing tools and procedures described in this chapter including closed-circuit television, smoke testing, dye testing, and pipeline lamping.

WORDS

Chapter 5. INSPECTING AND TESTING COLLECTION SYSTEMS

CABLE STRAIN RELIEF — CABLE STRAIN RELIEF

A mesh type of device that grips the power cable to prevent any strain on the cable from reaching the connections.

CORROSION — CORROSION

The gradual decomposition or destruction of a material due to chemical action, often due to an electrochemical reaction. Corrosion starts at the surface of a material and moves inward, such as the chemical action upon manholes and sewer pipe materials.

ELEVATION — ELEVATION

The height to which something is elevated, such as the height above sea level.

FAIR LEAD PULLEY (fair LEE-d pully) — FAIR LEAD PULLEY

A pulley that is placed in a manhole to guide TV camera electric cables and the pull cable into the sewer when inspecting pipelines.

FLOAT LINE — FLOAT LINE

A length of rope or heavy twine attached to a float, plastic jug or parachute to be carried by the flow in a sewer from one manhole to the next. This is called "stringing the line" and is used for pulling through winch cables, such as for bucket machine work or closed-circuit television work.

FOCAL LENGTH — FOCAL LENGTH

The distance of a focus from the surface of a lens (such as a camera lens) to the focal point.

HERTZ — HERTZ

The number of complete electromagnetic cycles or waves in one second of an electric or electronic circuit. Also called the frequency of the current. Abbreviated Hz.

LAMPING — LAMPING

Using reflected sunlight or a powerful light beam to inspect a sewer between two adjacent manholes. The light is directed down the pipe from one manhole. If it can be seen from the next manhole, it indicates that the line is open and straight.

OFFSET — OFFSET

(1) A combination of elbows or bends which brings one section of a line of pipe out of line with, but into a line parallel with, another section.

(2) A pipe fitting in the approximate form of a reverse curve, made to accomplish the same purpose.

(3) A pipe joint that has lost its bedding support and one of the pipe sections has dropped or slipped, thus creating a condition where the pipes no longer line up properly.

PRE-CLEANING — PRE-CLEANING

Sewer line cleaning, commonly done by high-velocity cleaners, that is done prior to the TV inspection of a pipeline to remove grease, slime, and grit to allow for a clearer and more accurate identification of defects and problems.

SADDLE CONNECTION — SADDLE CONNECTION

A building service connection made to a sewer main with a device called a saddle.

SEWER GAS — SEWER GAS

(1) Gas in collection lines (sewers) that results from the decomposition of organic matter in the wastewater. When testing for gases found in sewers, test for lack of oxygen and also for explosive and toxic gases.

(2) Any gas present in the wastewater collection system, even though it is from such sources as gas mains, gasoline, and cleaning fluid.

SUBSIDENCE (sub-SIDE-ence) SUBSIDENCE

The dropping or lowering of the ground surface as a result of removing excess water (overdraft or overpumping) from an aquifer. After excess water has been removed, the soil will settle, become compacted and the ground surface will drop and can cause the settling of underground utilities.

SURCHARGE SURCHARGE

Sewers are surcharged when the supply of water to be carried is greater than the capacity of the pipes to carry the flow. The surface of the wastewater in manholes rises above the top of the sewer pipe, and the sewer is under pressure or a head, rather than at atmospheric pressure.

TAG LINE TAG LINE

A line, rope or cable that follows equipment through a sewer so that equipment can be pulled back out if it encounters an obstruction or becomes stuck. Equipment is pulled forward with a pull line.

TV LOG TV LOG

A written record of the internal pipe conditions observed during a sewer line TV inspection.

WATER TABLE WATER TABLE

The upper surface of the zone of saturation of groundwater in an unconfined aquifer.

CHAPTER 5. INSPECTING AND TESTING COLLECTION SYSTEMS

(Lesson 1 of 5 Lessons)

5.0 REASONS FOR INSPECTING AND TESTING

Wastewater collection systems are intended to be a reliable method of conveying wastewater from individual dischargers to wastewater treatment plants. Inspection and testing are the techniques used to gather information to develop operation and maintenance programs to ensure that new and existing collection systems serve their intended purposes on a continuing basis. Inspection and testing are necessary to do the following:

1. Identify existing or potential problem areas in the collection system,
2. Evaluate the seriousness of detected problems,
3. Locate the position of problems, and
4. Provide clear, concise, and meaningful reports to supervisors regarding problems.

Two major purposes of inspecting and testing are to prevent leaks from developing in the wastewater collection system and to identify existing leaks so they can be corrected. The existence of leaks in a wastewater collection system is a serious and often expensive problem. When a sewer is under a water table, infiltration can take place and occupy valuable capacity in the sewer and the downstream treatment plant. Sewers located above a water table can exfiltrate, allowing raw wastewater to pollute soil and groundwater. A health hazard is created when the wastewater reaches a nearby well or open ditches where children and pets play. Leaks in sewers also invite root intrusion which eventually can cause stoppages, damage to the sewer and infiltration or exfiltration depending on the elevation of the water table. The location and elimination of leaks in a wastewater collection system is one of the major concerns of wastewater collection system operators and their supervisors.

QUESTIONS

Write your answers in a notebook and then compare your answers with those on page 253.

5.0A Why should wastewater collection systems be inspected and tested?

5.0B Why are leaks a problem in wastewater collection systems?

5.1 INSPECTING COLLECTION SYSTEMS

Inspection of wastewater collection systems is a very important phase of the operation and maintenance of collection systems. When inspecting, look for the causes of problems or the start of a potential problem. Once a problem has been identified, its exact location must be determined. The size and kind of problem must be identified so the problem can be solved. This section outlines the kinds of problems found in collection systems.

5.10 Types of Collection System Problems

Collection system problems may be created or caused by design, construction, sewer use, operation, maintenance or outside forces or events. Major sources of problems include:

1. Design-related deficiencies. Design engineers must consider and make provisions for special local conditions such as the ability of soil to support pipe and manhole weights, shifting soils, and vibrating or crushing forces from traffic. Other design-related problems include inadequate provisions for joint flexing and pipe bedding.

2. Improper installation. Plans and specifications must be followed. Improper line, grade and joint installation and shortcuts in bedding, connections and backfilling can cause future problems. Other examples of improper installation include use of inferior or damaged materials.
3. Inadequate sewer-use ordinances; poor communication between dischargers to sewers and enforcers of ordinances; and lack of enforcement.
4. Improper inspection and enforcement of tap-ins or service connections by individuals, plumbers or contractors that result in illegal or incorrect connections.
5. Changing patterns of activities and population shifts that result in *SURCHARGING*[1] some sections of the sewer system and excessive residence time of wastewater and solids in other sections.
6. Disaster or contingency situations such as explosions, earthquakes, *SUBSIDENCE*,[2] or wastewater flow shifts of major proportions.
7. Problems of a recurrent nature in the collection system such as accumulations of grease, debris and trash which result in stoppages or restrictions that reduce the capacity of the sewer. Also of great concern to many of the "sun belt" cities is the infestation of insects such as cockroaches. Although they do not affect the workings of a collection system, their presence does pose a public relations problem that must be dealt with daily.
8. Problems characteristic of the region which hasten the deterioration of sewers, for example, climate, high sulfate content in waters, high soil or wastewater temperatures, rapid root growth and mucky soils.
9. Poor coordination and cooperation between local agencies. Street construction and repair as well as activities of other utility agencies should be coordinated. Consideration must be given to locations of previous installations of utilities. Provisions must be made for early notification of accidents or a workable system for handling regular "accidental" dumps into the sewers.
10. Problems related to an old and frequently neglected wastewater collection system.

Once the source or cause of a problem has been identified, provisions can be made to correct the problem and also to prevent similar problems from occurring in the future.

QUESTIONS

Write your answers in a notebook and then compare your answers with those on page 253.

5.1A What is the purpose of inspecting wastewater collection systems?

5.1B List the major sources of wastewater collection system problems.

5.11 Inflow and Infiltration

Infiltration and inflow, while overlooked in many collection systems for decades, have now gained recognition as major defects that can cause failure of a collection system. In most cases, this failure results in hydraulic overloads (too much water) of the collection system or the wastewater treatment plant.

1. In the case of a collection system, hydraulic overloads result in surcharged manholes, manholes overflowing and exposure of a community to the diseases and pollutants carried by the wastewater in a collection system. This type of failure is also known as a sanitary sewer overflow (SSO). The U.S. Environmental Protection Agency is becoming increasingly concerned about SSOs. If the hydraulic overload in a gravity sewer is great enough, internal hydraulic pressure can build up and cause the pipe itself to fail since it is not designed as a pressure pipe. Also, when gravity sewers are pressurized, exfiltration can occur at joints or cracks in the pipe. The escaping water can wash away the bedding material and soil, causing voids (empty spaces) to develop around the outside of the pipe. If this problem goes undetected and uncorrected, subsidence may occur.

2. In the case of a wastewater treatment plant, infiltration and inflow can result in plant loads in excess of the plant capacity. Bypassing raw wastewater to the environment has been the only answer in the past, but this practice is no longer allowed.

5.110 Detection and Correction of Inflow

Inflow detection and correction depend upon the type and source of inflow causing the problem. Inflow is water that is not polluted and should not be in a wastewater collection system, but in a storm water drainage system. Inflow is water that enters a sewer as a result of a deliberate illegal connection or by deliberate drainage of flooded areas into a wastewater collection system.

Inflow detection requires a combination of flow studies and inspections by smoke or dye testing of buildings or private property if they appear to be the location of the source. The source of inflow is usually surface drainage, roof drains, yard drains, parking lot drains, basement sump pumps, heating and cooling tower water and air conditioning condensate drains.

---

[1] *Surcharge. Sewers are surcharged when the supply of water to be carried is greater than the capacity of the pipes to carry the flow. The surface of the wastewater in manholes rises above the top of the sewer pipe, and the sewer is under pressure or a head, rather than at atmospheric pressure.*

[2] *Subsidence (sub-SIDE-ence). The dropping or lowering of the ground surface as a result of removing excess water (overdraft or overpumping) from an aquifer. After excess water has been removed, the soil will settle, become compacted and the ground surface will drop and can cause the settling of underground utilities.*

Closed-circuit television inspections can reveal service connections with heavy discharges of clear water, or continuous discharges not typical of a domestic waste discharge, but TV inspection is a slow process. This procedure requires the camera to observe a service connection to determine the type and quantities of flow. If a location appears to be discharging greater than typical flows, contact residents of the home or building. Determine if water use is taking place and if the use is the only source of the flows observed.

Corrections to eliminate private sector inflow require a sewer-use ordinance rather than sewer maintenance projects. Local codes or ordinances governing discharges to the collection system should contain provisions giving enforcement authority to the agency's industrial waste section or legal department. A wastewater collection system maintenance crew should never be required to do more than report and provide reliable proof of inflow violations by a property owner.

Where illegal inflows are detected and certified, the maintenance crew may be required to evaluate the effectiveness of possible corrections or to accurately measure flows for legal and billing purposes.

Sources of surface drainage are best located by smoke testing (Section 5.4). These can partially be corrected when the collection system crew has the proper authority. On private property removal of improper connections, such as roof leaders connected to a wastewater collection system, should be the responsibility of some agency of a municipality or district other than the collection system crew.

Significant storm water inflow can also occur through manholes which are located at points lower than the surrounding area grade. Surface water can accumulate at manholes (ponding) and enter the manhole through pick holes in the cover, a poor seal between the cover and the casting, and through defects around the exterior of the casting.

5.111 Identification and Correction of Infiltration

Infiltration is created by high levels of groundwater and enters the wastewater collection system through deteriorated or broken pipes and joints. As with inflow rates which increase with the duration and intensity of rainfall, infiltration increases in a similar fashion in relation to groundwater level and water saturation of the soil or backfill in which the sewers are laid.

Infiltration is detected by metering flows in sections of the collection system to determine which area is providing the greatest flows, especially during the early morning hours when flows from residences are small. Typical flow metering devices don't work in areas where the sewers are surcharged (water in manholes over top of sewers). Under these conditions, measure[3] the depth of flow or surface of flow by measuring the distance from the manhole lid to the surface of the water to identify areas of high flow during periods of expected low flows.

After areas with high infiltration are located, conduct a field verification of the problem areas using visual inspection, smoke testing (Section 5.4), dye testing (Section 5.5) and/or TV inspection (Section 5.3). Visual inspection of manholes for infiltration should be performed after several storms have occurred and then during a heavy storm to confirm incoming flow from the suspected area. TV inspection can pinpoint the location and severity of an infiltration problem. TV inspection crews must be prepared to act quickly after heavy storms and during a period of time when the groundwater is above the elevation of the sanitary sewers in the area.

In many areas the main line portion of the collection system is relatively tight. A major source of infiltration in this situation can be the house service lines. They can be tested for leaks using smoke testing procedures and, with the development of small cameras and robotic equipment, they can be partially televised from the main line or points of cleanout access. Repair of service lines can be very expensive and complicated if the property owner is responsible for the service lines. For this reason some agencies have assumed the maintenance responsibilities of service lines where groundwater infiltration is a major problem.

Correction or elimination of inflow/infiltration depends on the type and location of the source of the problem. Typical solutions to inflow/infiltration problems include:

1. Manholes.
   a. Raise rim elevation by use of grade rings if not located in streets (inflow).
   b. Install watertight covers where needed (inflow).
   c. Install inflow protection covers (inflow).
   d. Seal covers (inflow).
   e. Seal or repair barrels (infiltration).
2. Cleanouts, vents and other appurtenances.
   a. Seal covers and/or divert surface waters away from possible areas of entry to a sewer (inflow).
   b. Repair damages (infiltration).
3. Sewer pipes (infiltration).

   (See Chapter 7, "Underground Repair.")

   a. Seal segment of damaged pipes and joints by pressure grouting.
   b. Dig up and replace damaged pipes and joints.
   c. Line sewer with a plastic liner and/or fiber liner materials.

[3] *Please see Chapter 7, "Wastewater Flow Monitoring," in PRETREATMENT FACILITY INSPECTION, in this series of training manuals for procedures on how to measure flows in sewers.*

5.112 Exfiltration

Exfiltration is the leakage of wastewater out of the collection system through broken or damaged pipes and manholes. All wastewater collection systems, except some constructed in recent years, have many leaks. These systems may exfiltrate wastewater through defective pipe joints and cracks. The wastewater that does exfiltrate may contaminate shallow wells or open ditches where children and pets play. To make an old collection system airtight would be extremely expensive and not very cost-effective. Major points of infiltration or exfiltration in a collection system can be identified by the use of television or smoke testing and can then be corrected.

5.113 Repair Costs

In order to qualify for grants of federal monies for collection system and wastewater treatment plant construction, a survey of the collection system is required. The purpose of this survey is to determine the presence or absence of infiltration and inflow and to evaluate the cost of correction in relation to the cost of constructing a treatment plant large enough to accommodate any additional flows.

There are a few factors which can make the methods of a system survey both unreliable and potentially very costly to a community that obtains a federal construction grant. If a collection system flow analysis shows the system conveying less than 100 gallons per person per day, the system is considered to be in good condition. It is possible, however, that the collection system was installed above a water table and is exfiltrating 100 gallons per person per day. This exfiltration could go on until a health authority traces the source of groundwater pollution to exfiltrating collection system pipes. At that point the real trouble falls on the shoulders of the community because the responsibility for maintenance and repair of a collection system will automatically fall to the community if a federal grant was used to construct a wastewater treatment plant. Required corrections to an exfiltrating collection system will have to be made by the community from its own financial resources.

To protect your community from high future repair costs, air testing of lines not deemed to be infiltrating is highly recommended. An air test procedure designed to use basic TV inspection and grouting equipment can be used to check each pipe joint with a low-pressure air test. Substandard joints are easily identified and repair methods evaluated. An air test in joint-by-joint increments (Chapter 7, Section 7.45) may be used to reveal possible sources of exfiltration in areas where infiltration and exfiltration are known to be a problem.

QUESTIONS

Write your answers in a notebook and then compare your answers with those on pages 253 and 254.

5.1C Why must inflow and infiltration be controlled?

5.1D What are the sources of inflow?

5.1E How can sources of infiltration be located?

5.1F How can infiltration problems be corrected or eliminated?

5.1G Why should exfiltration be controlled?

Please answer the discussion and review questions next.

DISCUSSION AND REVIEW QUESTIONS

Chapter 5. INSPECTING AND TESTING COLLECTION SYSTEMS

(Lesson 1 of 5 Lessons)

At the end of each lesson in this chapter you will find some discussion and review questions. The purpose of these questions is to indicate to you how well you understand the material in the lesson. Write the answers to these questions in your notebook before continuing.

1. What are the reasons for inspecting and testing wastewater collection systems?
2. What happens when there is inflow and infiltration in a collection system?
3. What can be done to correct or eliminate inflow and infiltration problems?
4. What types of design-related deficiencies cause problems in collection systems?
5. What are some examples of improper installation of sewers?
6. Why should exfiltration be controlled?

CHAPTER 5. INSPECTING AND TESTING COLLECTION SYSTEMS

(Lesson 2 of 5 Lessons)

5.2 INSPECTION OF MANHOLES

5.20 Objective of Manhole Inspection

Because they are part of the collection system, manholes will require the same inspection and attention as the rest of the sewer network. When located in streets, these structures are subject to the vibrations and pounding of vehicle traffic. Manholes may settle at a different rate than the connected sewer, thus creating cracks in joints. Easement locations on private property are subject to misuse and changes of ground surface due to construction or landscaping work. Your local sewer-use ordinances relating to manholes located in easements should clearly inform the public of the agency's unlimited right of access to perform collection system inspection and maintenance activities. The objectives of manhole inspection are, therefore, to determine the proper elevations or grades around the lid, to be sure the lid isn't buried, and to examine structural integrity (look for cracks) of the manhole and its functional capacity. An indication of the condition of the pipelines coming into a manhole may be gained merely by observing the content and volume of flows from a specific direction.

5.21 Safety

See Section 4.3, "Routing Traffic Around the Job Site," and Section 4.4, "Classification and Description of Manhole Hazards."

TRAFFIC ROUTING

MANHOLE HAZARDS

5.22 Equipment Required

1. An inspection report form which acts as both a record and a checklist to make certain all types of defects or maintenance requirements are adequately noted (see typical forms at the end of this section on pages 195 to 197).
2. A powerful flashlight which will permit the inspector to see down into the manhole. A mirror may be used if the sun is shining.

3. A map of the collection system to reference manhole locations on the report form.
4. A manhole hook or other type of lid removal device.
5. Scrapers and wire brushes for cleaning the manhole ring.
6. A rigid straightedge approximately 5 or 6 feet long. A board, piece of pipe, angle iron or similar device will do.
7. A pair of sturdy leather gloves.
8. Where the manhole is located in traffic, a vehicle with adequate warning devices, safety cones and other traffic safety devices.
9. Where required, a metal detector and a pick and shovel to expose manhole lids which may have been buried or paved over.
10. Gas detection devices (their use is explained in Section 5.24, "Procedure") capable of measuring and monitoring for oxygen deficiency/enrichment, explosive conditions, and toxic gases (hydrogen sulfide).
11. A telescoping measuring device with graduations to measure depth of flows, verify pipe diameters, and measure depth of manhole inverts.
12. A blower and hose for ventilating manhole.

5.23 Staffing Requirements

One operator is usually all that is essential for manhole inspection. In some locations where traffic is heavy, flaggers may be necessary. In rare situations, an operator may have to enter the manhole for inspection purposes and proper manhole safety procedures must be followed (see Section 4.3, "Routing Traffic Around the Job Site," and Section 4.4, "Classification and Description of Manhole Hazards"). As a safety precaution, it is advisable and common practice in the wastewater collection field for work crews to consist of a minimum of two workers per crew.

5.24 Procedure

The manhole inspection consists of five steps:

Step 1:

Locate the manhole and check the area around it for proper drainage away from the lid. The straightedge may be useful in this operation. If the manhole is located in a street, use the straightedge to verify that the manhole is installed at the proper elevation. In areas subject to snow, a high manhole ring or lid can cause the blades of snow removal equipment to catch on it, damaging both manhole and equipment. If the lid is excessively below or above finished street grade, the manhole structure will be subject to pounding when traffic crosses. These situations should be considered defects.

In easement areas, the manhole lid should be two or more inches above the soil level around it (except in driveways, walkways or parking lots where street elevation requirements apply).

Enter any grade or elevation defects on the report form.

Step 2:

Test atmospheric conditions in the manhole by inserting the probe of a gas detection device through an opening in the manhole cover.

Step 3:

Safety: Manhole Entry. (See Sections 4.3, "Routing Traffic Around the Job Site," and 4.4, "Classification and Description of Manhole Hazards.")

Remove the manhole lid (if atmospheric conditions are safe) and pull it to one side.

CAUTION

BECAUSE THE UNDERSIDE OF A MANHOLE LID CAN HARBOR DANGEROUS INSECTS SUCH AS THE BLACK WIDOW SPIDER, WASPS, HORNETS AND MUD DAUBERS, WEAR PROTECTIVE GLOVES AND EXERCISE CAUTION WHEN HANDLING MANHOLE LIDS.

Mark the inspection form in accordance with the configuration of the pipes entering the manhole. Be sure to write neatly and clearly so others can read your writing and understand what you observed.

With the flashlight, inspect all surfaces and joints inside the manhole. Write on the form the locations and types of any defects which may be observed such as:

1. Cracks or breaks in the walls or bottom.
2. Infiltration at any place. Estimate the flow in gallons per minute.
3. Joint security. Joints in a manhole should be tight. There should not be any visible cracks large enough to allow significant infiltration or exfiltration or to harbor insects or vermin.
4. Offsets or misalignment of any parts.
5. Root intrusions.
6. Grease accumulations around arch or inside of sewers entering and leaving manhole.
7. Gravel or debris accumulations in invert.
8. Concrete or grout in the invert or pipe causing flow turbulence.
9. Deterioration of the grout bed of frame.
10. Condition of steps or rungs (if provided).
11. Debris on shelf, steps or rungs.
12. Sluggish flow or wastewater backing up into the manhole.
13. Corrosion.

Where there is minimal background noise, the inspector can squat and listen for the possible noise of turbulence or squirting water. Where the pounding of traffic has driven the manhole structure as little as a quarter of an inch below constructed grade, pipes connected to it can be cracked or broken at or near the connection point to the manhole. Infiltration or turbulence in the flow will sometimes make a detectable noise at these breaks. Look into the sewer piping upstream and downstream of the manhole for cracks or breaks that could have been caused by settling. Using a light or reflected sunlight to inspect a sewer between two adjacent manholes is known as "lamping."

When inspecting a manhole in a climate subject to ground freezing, examine the grade rings for any separation because freezing is a major cause of separation. Leaks that develop at points of separation can introduce great quantities of infiltration during periods of snow melt.

Step 4:

With a wire brush and scraper, clean the ledge of the manhole ring. Inspect carefully for any cracks in the metal parts.

NOTE: ONE CRACK IN A MANHOLE RING WEAKENS THE REST OF THE METAL PART, THUS POSSIBLY RESULTING IN SUDDEN FAILURE UNDER A VEHICLE LOAD.

If a crack is observed, the manhole should be barricaded and the ring should be replaced immediately.

Step 5:

Replace the manhole lid. Look for evidence of a warped or misfit lid by standing at various positions around the lid. The lid should not rattle or "rock." Where traffic passes over a lid with a rattle, the noise and rattle will increase with time. Make certain that the ring is clean and does not have a pebble or other object preventing proper fit before recording a rattle as a defect. A rocking or improperly seated manhole lid can be flipped completely out of the ring by a vehicle driving over the cover.

Defective manhole lids and rings can be removed and machined, or milled in place with special equipment, or replaced. Gaskets may be used to reduce rattles, but gaskets have to be replaced periodically.

5.25 Frequency of Manhole Inspections

Any new manhole should be carefully inspected before acceptance in the same manner as the rest of the pipe system. Existing manholes should be inspected once every one to five years with those in areas of heavy vehicle traffic being inspected more often. If you are inspecting manholes and discovering many serious defects that need correcting, then the manholes should be inspected more frequently.

When a crew working in a manhole for other than manhole inspection (rodding, cleaning, televising) discovers a defect, a specially designed written report or "work request" should be prepared so the necessary manhole repairs are not overlooked.

Equally important as the properly filled out manhole inspection form is the method by which the report is entered into your system of paper work or computers. An orderly recordkeeping system helps to ensure that work requests receive the necessary attention from the appropriate work crews.

5.26 Sample Inspection Forms

Figures 5.1 and 5.3 show two examples of manhole inspection report forms and Figure 5.2 shows how to record data from a manhole inspection.

QUESTIONS

Write your answers in a notebook and then compare your answers with those on page 254.

5.2A Why should manholes be inspected?

5.2B How many operators are usually needed to inspect a manhole?

5.2C What items should be examined when inspecting a manhole?

5.2D How often should a manhole be inspected?

Please answer the discussion and review questions next.

DISCUSSION AND REVIEW QUESTIONS

Chapter 5. INSPECTING AND TESTING COLLECTION SYSTEMS

(Lesson 2 of 5 Lessons)

Write the answers to these questions in your notebook before continuing. The question numbering continues from Lesson 1.

7. Why should gloves be worn when handling a manhole lid?
8. How can a collection system operator see into a manhole during an inspection visit?
9. How would you determine if there is proper drainage away from a manhole lid?
10. Why should manhole lids be installed at the proper elevation?
11. How would you inspect a manhole?

MANHOLE INSPECTION REPORT

MH NO.__________ DATE__________ TIME__________ INSPECTOR__________

ELEVATION__________ DEPTH TO INVERT__________ CLEANLINESS__________

TYPE CONSTRUCTION__________ STREET REFERENCES__________

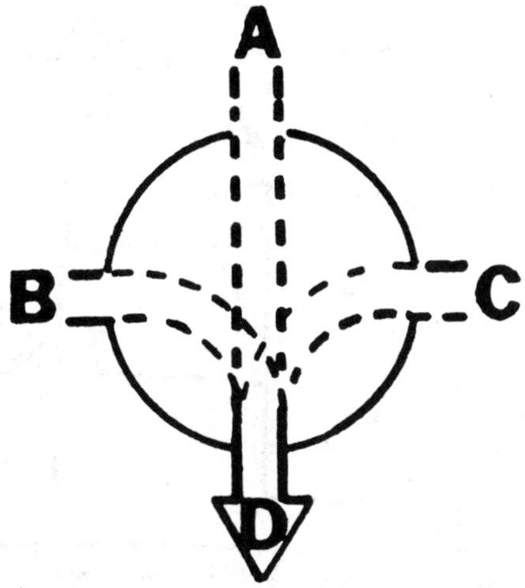

A

B

C

D

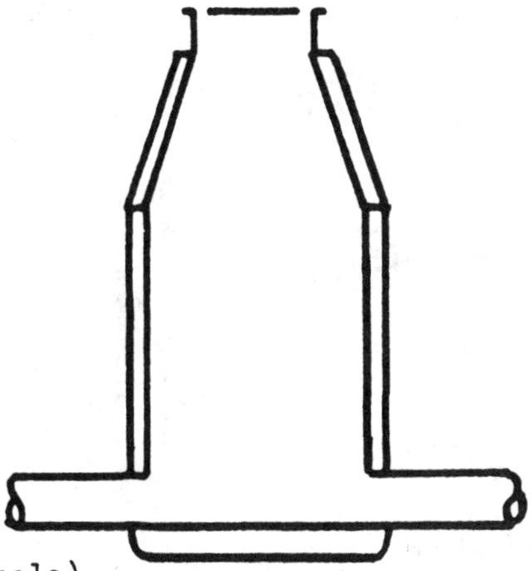

DEFECTS:
(Cover, frame, grout, steps, shelf, pipes, or channels)

1.__________

2.__________

3.__________

4.__________

5.__________

6.__________

7.__________

8.__________

(USE REVERSE SIDE FOR ADDITIONAL DEFECTS TO BE NOTED.)

	PIPE SIZE	LENGTH	TO MH#	EST. FLOW	TYPE FLOW
A-					
B-					
C-					
D-					

REMARKS:
(Include need for repairs)

____

____

MANHOLE INSPECTION REPORT

Fig. 5.1 Sample manhole inspection report form

MANHOLE INSPECTION REPORT

MH NO. 6822 DATE 7-20-02 TIME 10:15AM INSPECTOR J.S.

ELEVATION ______ DEPTH TO INVERT 9'7" CLEANLINESS OK

TYPE CONSTRUCTION CONC-CAST STREET REFERENCES 34 AVE & AITKIN

DEFECTS:
(Cover, frame, grout, walls, steps, shelf, pipes, or channels)

1. MH RING DEPRESSED 1 1/4"
2. DIAGONAL CRACK IN BARREL
3. SEAL FAILURE AT JOINT - 1-GPM LEAK
4. SEAL FAILURE & PIPE CRACK - 2-GPM LEAK
5. ______
6. ______
7. ______
8. ______

(USE REVERSE SIDE FOR ADDITIONAL DEFECTS TO BE NOTED.)

	PIPE SIZE	LENGTH	TO MH#	EST. FLOW	TYPE FLOW
A–	8	275	6823	2"	SOAPY
B–					
C–					
D–	8	320	6821	2"	

REMARKS:
(Include need for repairs)

AUDIBLE LEAK(S) IN PIPES NEAR MANHOLE. DEPRESSED RING MAY INDICATE DROP JOINTS OR BREAKS IN LINE.

MANHOLE INSPECTION FORM

Fig. 5.2 Completed manhole inspection report

CITY OF PHOENIX — WASTEWATER COLLECTION DIVISION

¼ SEC. #: SANITARY SEWER MANHOLE INSPECTION FORM

M.H. DEPTH: FORM: M.H. #: DATE:

I. INITIAL INSPECTION

A. LOCATION:
1. Roadway ☐
2. Gutter ☐
3. Paved Alley ☐
4. Unpaved Alley ☐
5. Easement ☐
6. Other ☐

B. MANHOLE COVER:
1. Serviceable ☐
2. Damaged ☐
3. Displaced ☐
4. Missing Grout ☐
5. Needs Raising ☐
6. Needs Lowering ☐

C. RING & FRAME:
1. Serviceable ☐
2. Loose ☐
3. Displaced ☐
4. Missing Grout ☐
5. Needs Raising ☐
6. Needs Lowering ☐

D. MANHOLE MATERIAL:
1. Brick ☐
2. Concrete ☐

E. SIZE M.H. COVER:
1. 24 inch ☐
2. 30 inch ☐

F. MANHOLE SIZE:
1. 4 foot ☐
2. 5 foot ☐

II. STRUCTURAL INSPECTION

A. STEPS:
1. Serviceable ☐
2. Unsafe ☐
3. Missing (No.) ☐
4. Corroded ☐

B. CONE:
1. Serviceable ☐
2. Broken ☐
3. Sulfided ☐
4. Misaligned ☐
5. Leaking/Bad Joints ☐

C. RISER:
1. Serviceable ☐
2. Broken ☐
3. Sulfided ☐
4. Misaligned ☐
5. Leaking/Bad Joints ☐

D. SHELF:
1. Serviceable ☐
2. Broken ☐
3. Dirty ☐
4. Sulfided ☐
5. Bad Base Joint ☐

E. CHANNEL:
1. Serviceable ☐
2. Obstructed ☐
3. Sulfided ☐
4. Bad Pipe Joint ☐
5. Silt ☐
6. Poor Struct. Cond. ☐

III. HYDRAULIC INSPECTION

A. INFLOW INDICATIONS:
1. Debris on Sides/Shelf ☐

B. SURCHARGE INDICATIONS:
1. Grease/Debris on Sides & Shelf ☐

C. CLARITY OF FLOW:
1. Turbid Appearance ☐
2. Clear Appearance ☐

D. FLOW:
1. Steady ☐
2. Pulsing ☐
3. Turbulent ☐
4. Surcharging ☐
5. Sluggish ☐

E. FLOW DEPTH COMPARED TO ADJACENT MANHOLES:
1. Same ☐
2. Lower ☐
3. Higher ☐

F. FLOW DEPTH:

______________ Inches

Time __________ AM/PM

IV. VERMIN
1. Roaches ☐
2. Rats ☐
3. Other ☐

OBSERVATION SUMMARY:

FOREMAN II RECOMMENDATIONS:

SUPERVISOR I APPROVAL & COMMENTS:

145-16OD

Fig. 5.3 Manhole inspection form

CHAPTER 5. INSPECTING AND TESTING COLLECTION SYSTEMS

(Lesson 3 of 5 Lessons)

5.3 CLOSED-CIRCUIT TELEVISION (CCTV) INSPECTIONS

Television inspection of a sewer provides positive and reliable answers to what, where, how bad, and how much in a way that is very informative for collection system operators. Closed-circuit television inspection overcomes the conjecture, guesswork, and wild estimates that were made without TV, when many thousands of dollars were wasted on pipeline replacement work or excavation repairs that were made at the wrong place or were not really necessary.

In the words of one operator who had made a careful study of operational and repair costs to a moderately sized system, comparing expenditures before and after acquisition of a closed-circuit television outfit,

> "Before television, we were like blind umpires trying to call a baseball game. We had an idea of what was going on and what was needed and did what we thought was the right thing to take care of the system. That we made some pretty costly mistakes is a matter of record, but we really didn't know how bad that record was until the television showed us the truth. If there is only as little as five miles of sewer system, today's replacement costs could exceed a million dollars. Inadequate installation, improper maintenance attempts, poor quality repairs and allowing problems to reach emergency status will, in addition to causing unreasonably increased costs over the years, hasten the date when full replacement is necessary. The engineers, superintendents, and other officials who attempt to maintain such a system without the positive knowledge from regular television inspections are like we were before we acquired our television equipment—blind, ignorant, and a heavy tax burden to our community."

Most cities and sanitation districts use closed-circuit television equipment for sewer line inspection. These agencies either own the TV equipment, rent the equipment or contract to have CCTV inspections made in their systems. Like the words of the operator in the previous paragraph, a great majority of people concerned with the design, construction and maintenance of the sewer lines are impressed by the facts revealed by CCTV inspection. As a result, engineering design concepts, construction methods and system maintenance procedures are in the process of revision and replacement.

Closed-circuit television can be used to inspect and evaluate public and private wastewater collection systems of all sizes. Because smaller agencies and collection systems will usually have limited financial capacities, the importance of the precise and accurate data on sewer conditions, performance of required work and the priority of such work can justify the use of CCTV to avoid the catastrophic burdens of many emergencies. Smaller sized collection systems usually have higher costs of contract emergency repair work, and this type of work results in losses that take away a heavier percentage of available financial assets. Larger systems with the opportunity for many problem areas have no problem in recognizing the advantages of and vast savings from the ability to schedule personnel, equipment and funds for maximum use of CCTV in determining system requirements.

Whether a CCTV system is owned, leased, or rented or the work is contracted out, operators employed to maintain wastewater collection systems need to know how the CCTV equipment works and the information it can provide. This section covers the operational phase of closed-circuit television inspections of wastewater collection system pipes. Details on the individual components of a television system can be found in the Appendix of this chapter, Section 5.8, "Closed-Circuit Television Equipment Details."

5.30 Use of Television

A closed-circuit television (CCTV) inspection consists of a special television camera passing through the sewer pipe with the signal (picture) being observed on a monitor similar to a home television screen. This picture allows a visual inspection of the interior of the sewer.

While older TV inspection units are limited to use in 4-inch through 48-inch (for metric pipe sizes, see page 583) diameter sewers, the more recent equipment is able to inspect virtually any size sewer because of advances in technology. Some of these advances include the use of solid state cameras with digital electronics that require much less light, more efficient light heads, the use of kevlar armored cable (which reduces the weight of the cable) and the advances in self-propelled or tractor cameras. These advances in equipment have resulted in reduced crew size, higher production rates for TV inspections, and improved quality. Other advances include the use of microprocessor-controlled systems that allow direct data entry of inspection information into a computer database where it will then be used in a CMMS (Computer Maintenance Management System). Reliability of equipment has also improved significantly through use of better materials and equipment design.

Camera improvements include wide-angle lenses with automatic focus and adjustable irises, pan-and-tilt cameras for inspecting joints and service laterals, and the ability to either push the camera into the sewer or pull it through with a winch. A self-propelled tractor or a sewer cleaner jetting hose attached to a camera may be used to push or pull a skid-mounted camera through pipes being televised. This technique is used for cameras with cables directly connected to CCTV mounted in vans, trucks, campers, or trailers (Figure 5.4) and also sewer cleaners without cables using wireless transmission of the video signal by radio.

Sewer scanner and evaluation technology (Figure 5.5) offers advanced digital imaging and pipeline assessment capability. During inspection operations, the camera is able to scan the complete interior of the pipeline (360 degrees) with no need to identify, stop, and isolate on laterals, connections, or defects.

5.300 Purposes of Televising Collection Systems

A television inspection of a collection system can be conducted for the following reasons:

1. Inspect conditions and determine location of problem areas such as pipe or joint separations, drops, ruptures, leaks, service connections, obstructions, corrosion, pipe alignment, and root intrusion;
2. Look for damage to sewers caused by excavation and construction on nearby utilities, paving and building construction;
3. Search for unrecorded connections that may reveal illegal taps, such as industrial, storm water, or surface drainage;
4. Locate inflow or infiltration sources and amounts;
5. Inspect taps or pipes installed and/or repaired by plumbers, contractors or maintenance crews;
6. Evaluate effectiveness of solutions and equipment in maintaining or correcting problem areas in sewers; and
7. Locate buried or lost manholes.

5.301 Actions Possible From CCTV Inspection Information

From the information obtained by a closed-circuit collection system inspection, all of the following actions are possible:

1. Require contractor or construction agency to correct defects observed in new sewers prior to acceptance;
2. Identify or verify exact locations of service connections and other pipe construction points and correct any as-built maps (record drawings) as needed;
3. Establish priorities for corrective work in old pipes on an "as needed" basis;
4. Provide for maximum effective and economical use of personnel and equipment for a preventive maintenance and repair program covering an entire collection system; and
5. Inspect construction of house service connections to the lateral and branch sewers. Determine by observation of these connections if infiltration, roots, debris accumulations and/or some types of internal inflow are adversely affecting the capacity of the sewers.

5.31 Equipment and Staffing

NOTICE

Technology is rapidly advancing throughout the wastewater collection system field. CCTV systems are taking a new approach in the application of wireless technology (Figure 5.6). Sewer inspection CCTV systems are equipped with computers, monitors, and wireless cameras that operate by a radio video signal from the camera. The wireless transmission is then received by the receiver, which is installed with a hose protector located at the manhole. The received signal is then displayed on a video monitor at the sewer inspection CCTV operator's station.

Industry is striving to help us do our job safer, faster, better, and at a lower cost. You must realize that equipment, staffing, and procedures described throughout this manual are presented to give you an idea of the types of equipment available, personnel needed, and the kind of procedures necessary to do the job properly and safely. When you read this material, recognize that there are other pieces of similar equipment that might have slightly different staffing requirements and/or operational procedures. For example, power winches are available that replace the "pull winch" at the far manhole. Also modern camera power cable contains strong pulling membranes which eliminate the need for the "camera return winch." These are controlled electrically by the TV operator giving that person full control of camera travel in both directions. Some TV inspection equipment is so light and compact that it can be contained easily in a van, station wagon, or pickup truck and removed from the vehicle when not in service. The industry also offers many other equipment variations and accessories too numerous to detail in the manual. The purpose of this section is to explain the basic steps in setting up and using CCTV equipment.

5.310 Equipment and Components Required for Televising

Essential components for televising sewers are listed in this section. Figures 5.7 and 5.8 show a typical setup. Figure 5.4 shows a computerized CCTV operator's station, and Figures 5.9 through 5.21 are photos of some of the components.

1. Television camera and remote control
2. Camera light
3. Power cable reel video unit (Figure 5.9)
4. Television picture monitor (screen)
5. System power control operator and data center
6. Internal or external portable power source
7. Camera skids (Figure 5.17) or camera crawler tractor unit (Figure 5.16)
8. Camera pulling winch
9. Camera downhole and cable guide system (Figure 5.15)
10. Footage counter (Figure 5.15)

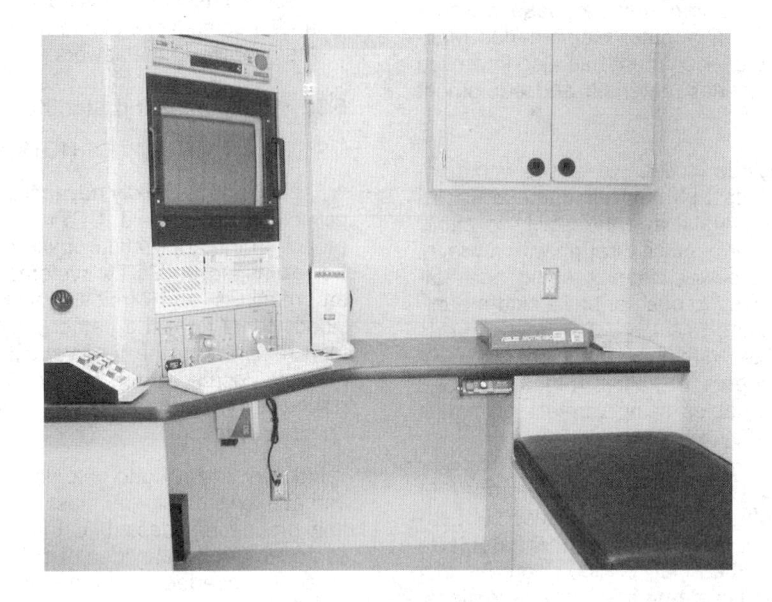

Step Van

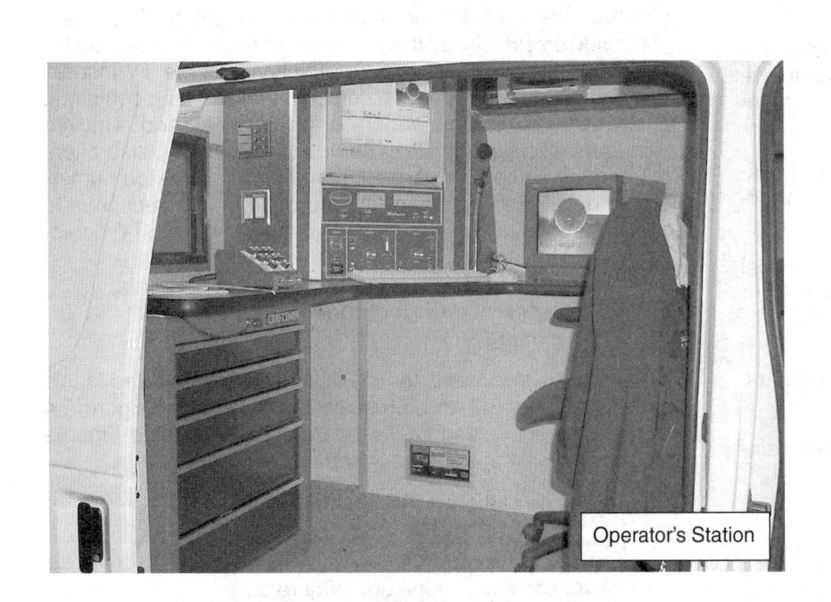

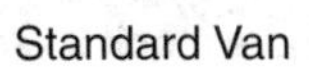

Standard Van

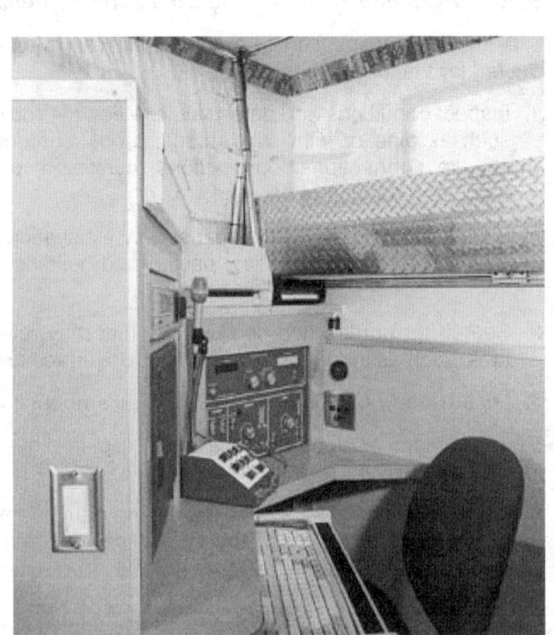

Camper Unit

Fig. 5.4 Computerized CCTV operator's station

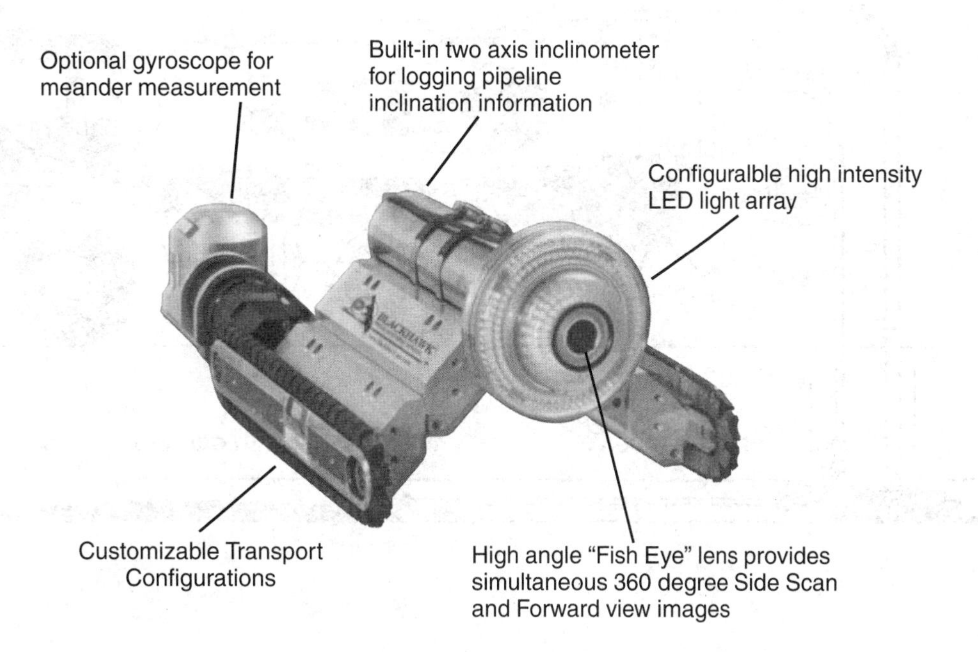

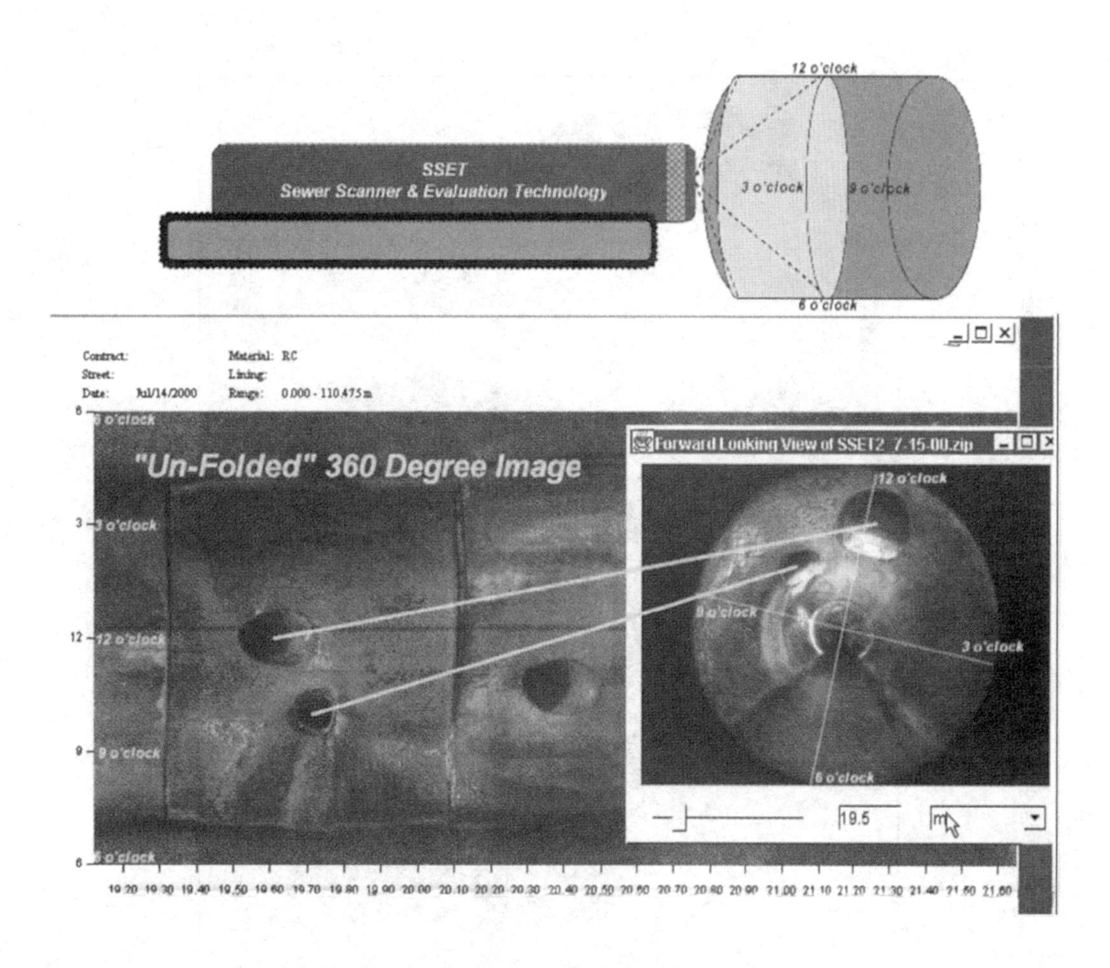

Fig. 5.5 Sewer scanner and evaluation technology with camera

(Permission of Blackhawk Pipeline Assessment Services)

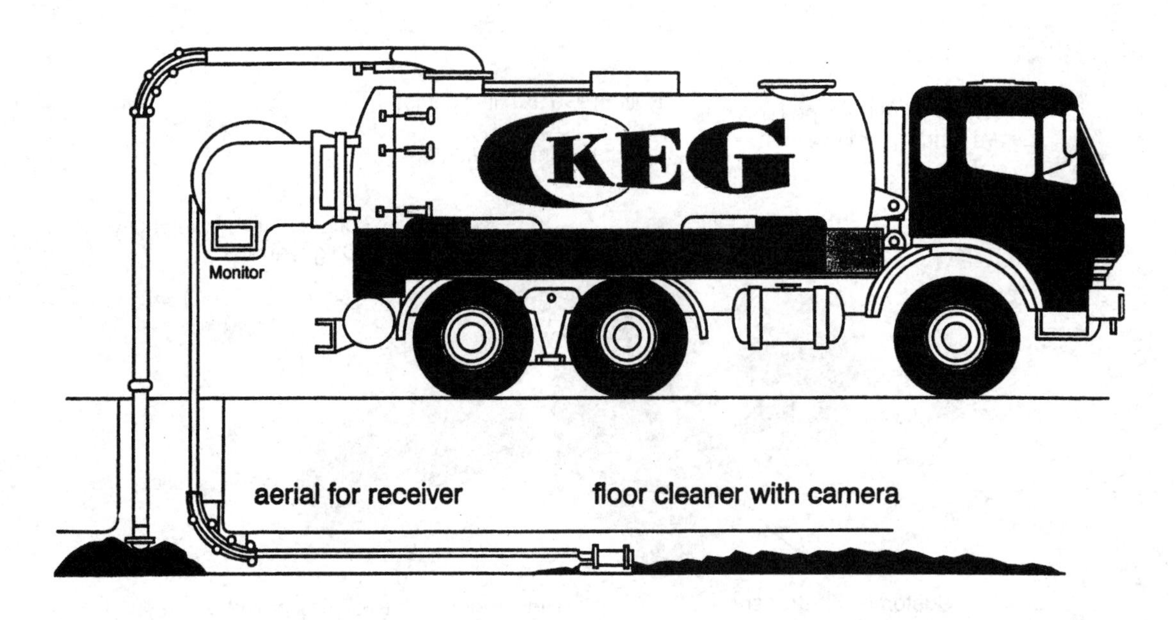

Fig 5.6 Combination sewer cleaner and wireless camera system

(Permission of KEG Technologies, Inc.)

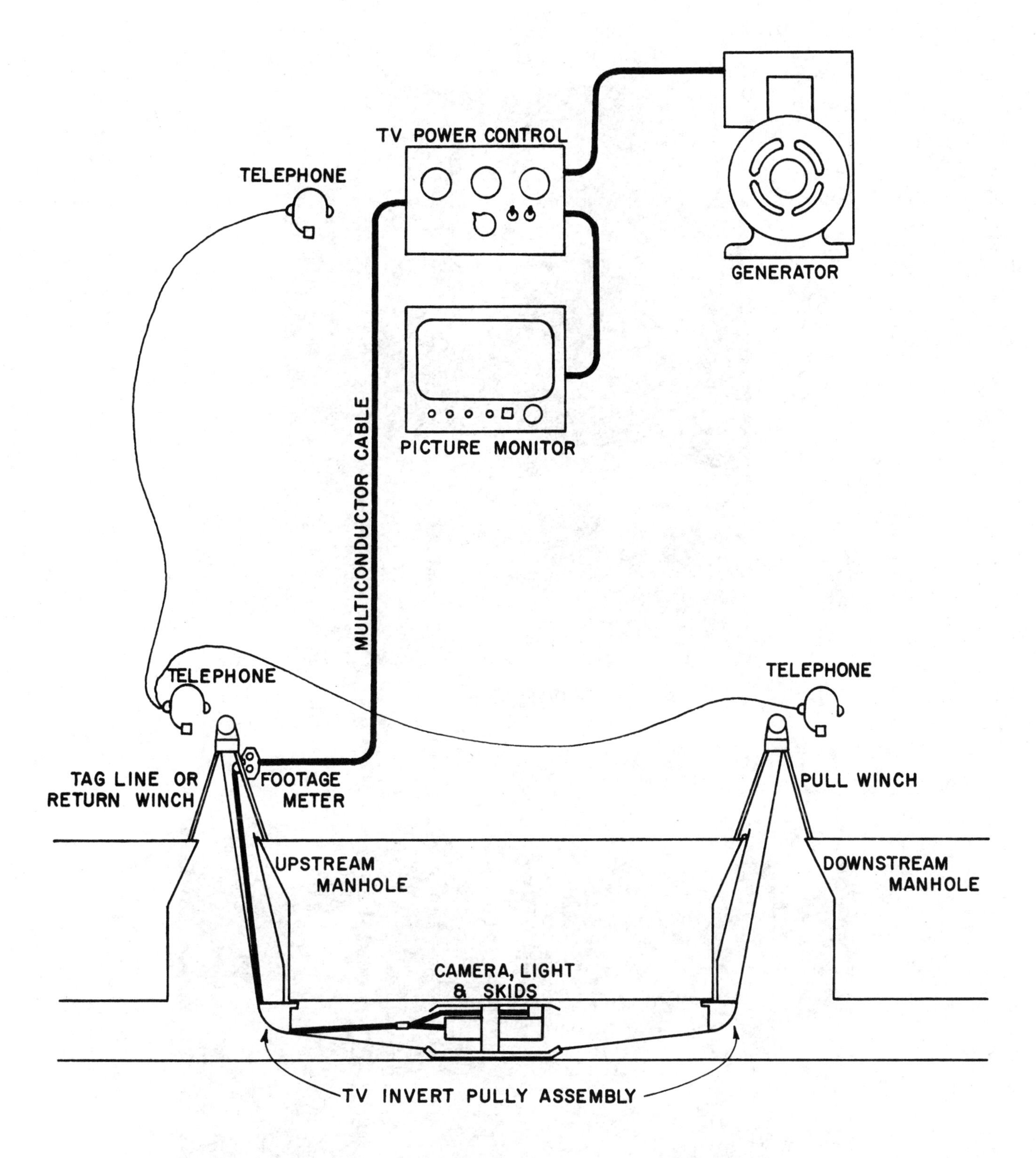

NOTE: There are three ways to control the far manhole winch pull:

1. Worker/sound-powered telephone,
2. Worker/walkie-talkie, or
3. No worker/remote control power winch.

NOTE: If a camera-carrying crawler tractor unit is used, no equipment is needed in the downstream manhole.

Fig. 5.7 Setup of basic TV system

TV Rig Setup

TV/Ground Rig Setup With Remote Power Winch

Fig. 5.8 Truck-mounted CCTV systems

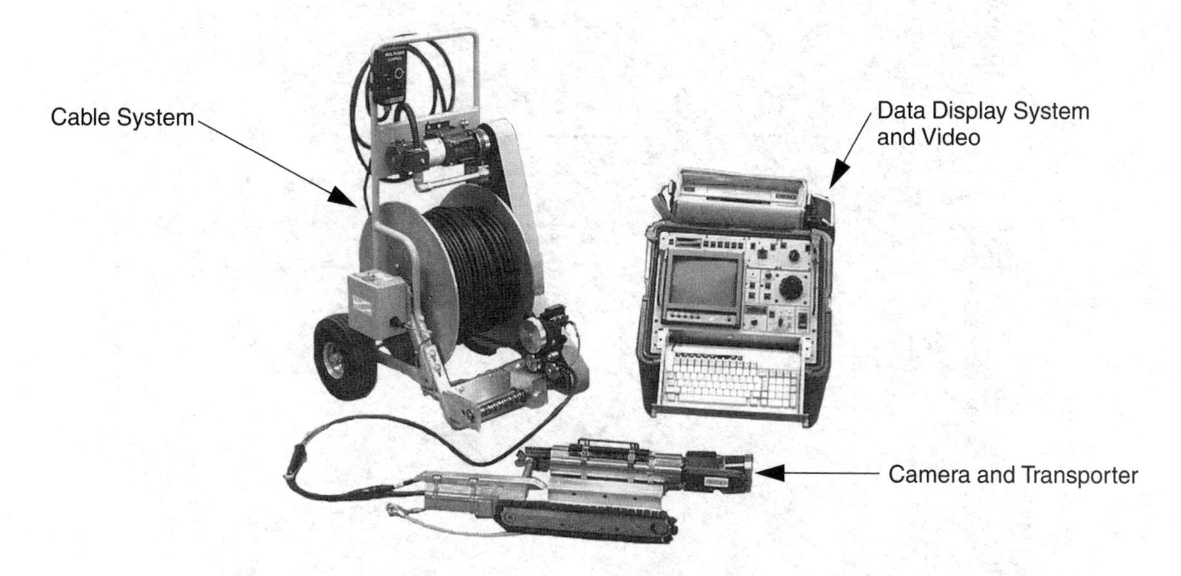

Mainline Inspection System

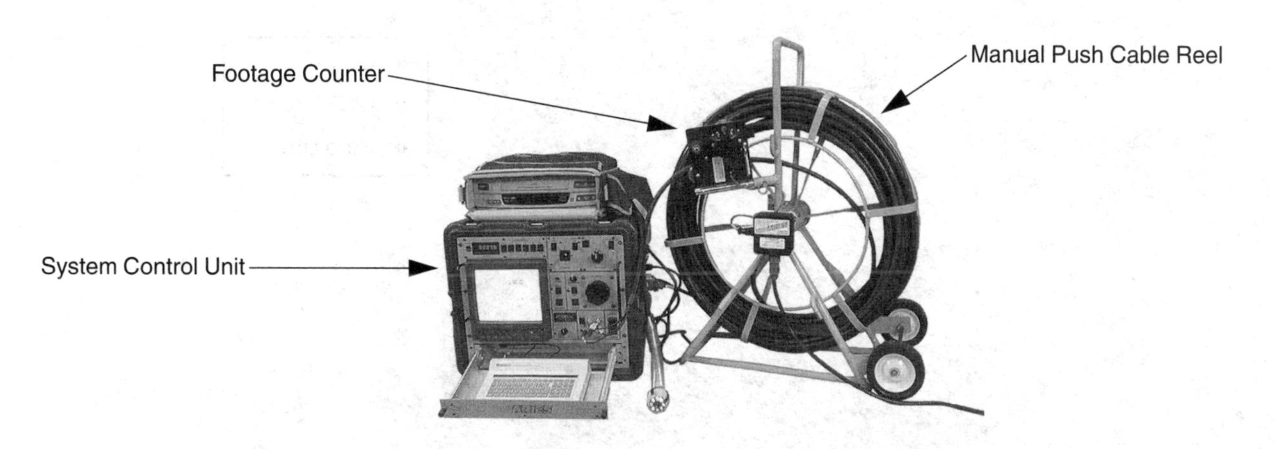

Push System

Fig. 5.9 Portable color TV inspection system for sewers

(Permission of Aries Industries, Inc.)

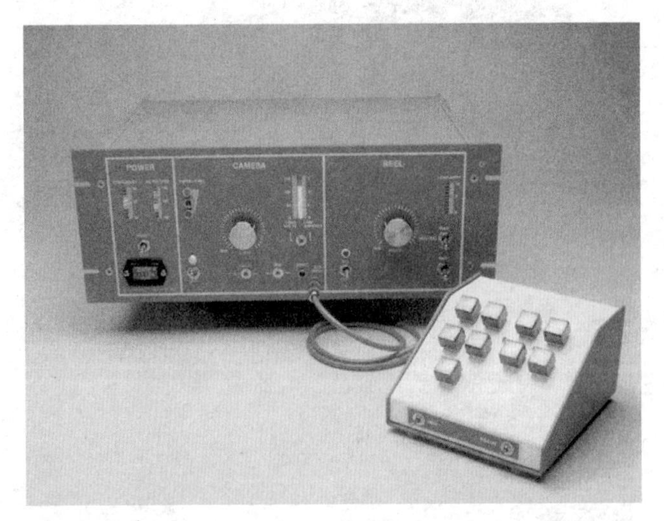

TV Camera Power Control Unit and Remote Camera Control

Fig. 5.10 System power control station

(Permission of RST Services, Inc.)

Side View With Bumper Hoist

Rear View With Camera and Accessories

Operator's Station

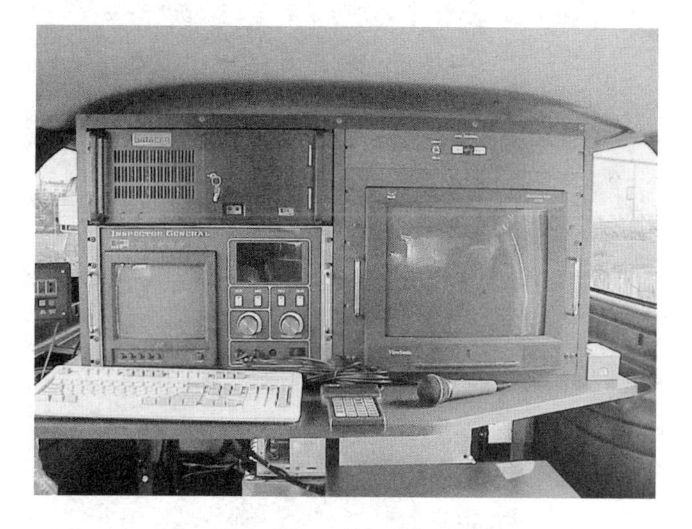

Mainline Camera Control Center

Fig. 5.11 Sports utility CCTV system

(Permission of Cues, Inc.)

Step Van

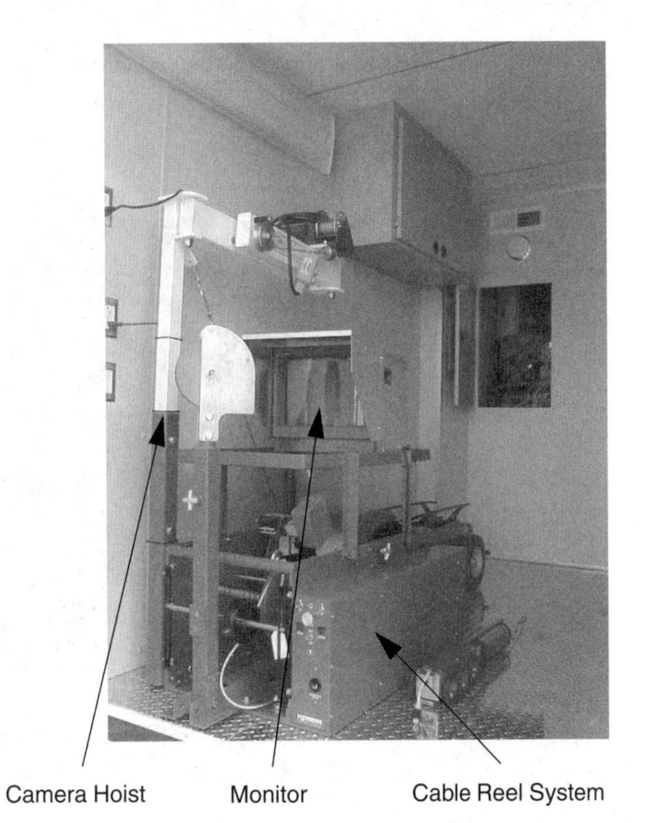

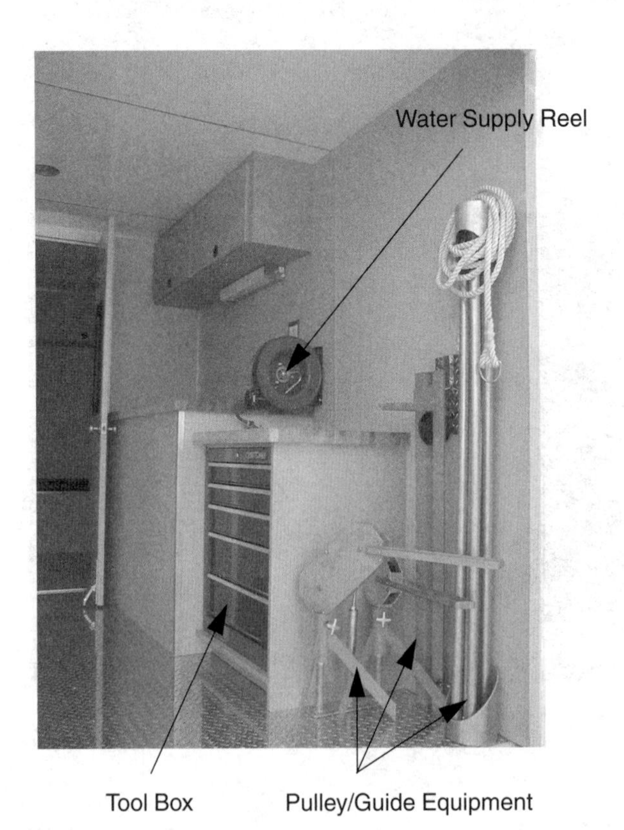

Fig. 5.12 Step van CCTV system
(Permission of RST Services, Inc.)

Trailer System

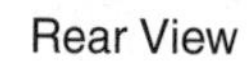

Rear View

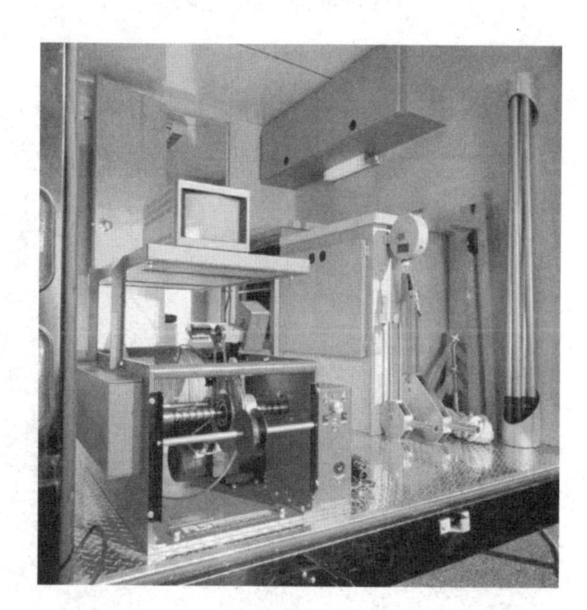

Cable Reel Unit

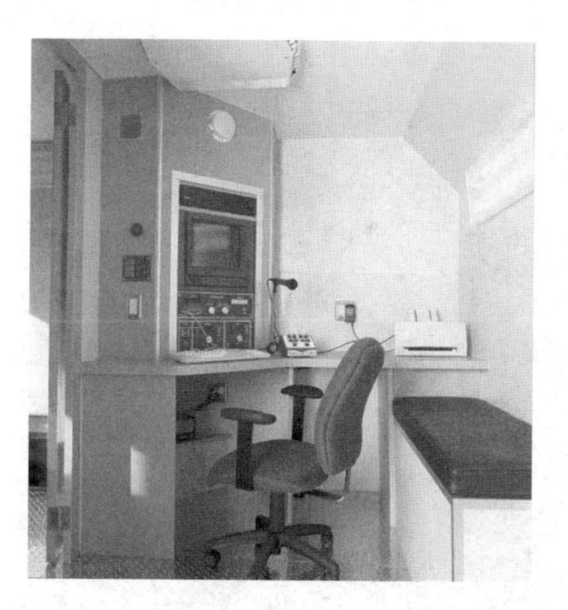

Main Control Station

Fig. 5.13 CCTV trailer system
(Permission of RST Services, Inc.)

Camera/Cable/Control System

Easement Camera System

Vehicle Transport Van

Portable Power Source

Fig. 5.14 Gator easement TV system

(Permission of Cues, Inc.)

Compact Portable Reel

Compact Reel System

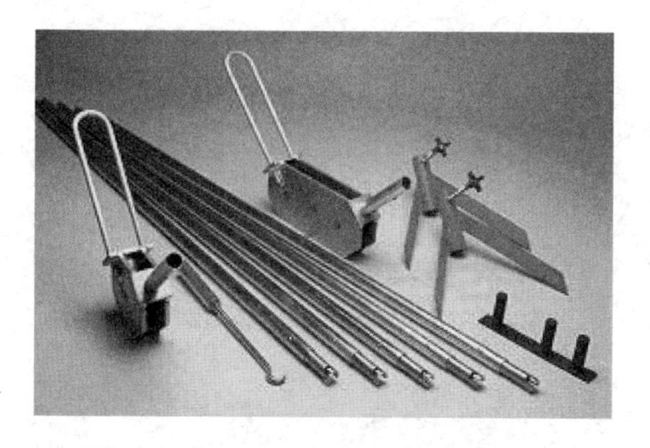

Roller Assemblies/Poles/Manhole Adapter Hooks

Tag Line

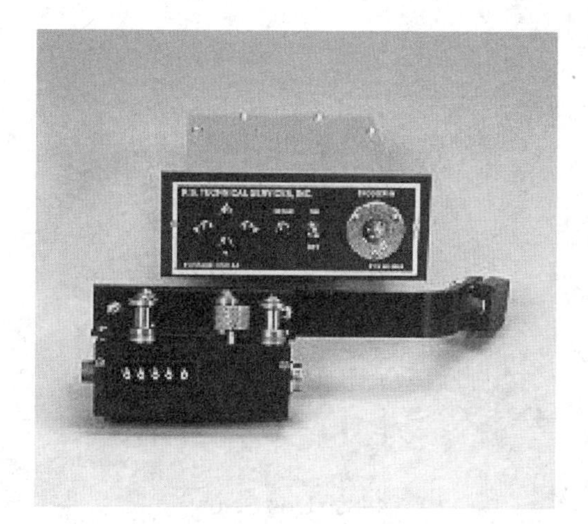

Footage Counter System

Fig. 5.15 TV equipment and accessories

(Permission of RST Services, Inc.)

Fig. 5.15 TV equipment and accessories
(Permission of RST Services, Inc.)

Steerable Storm Drain Tractor

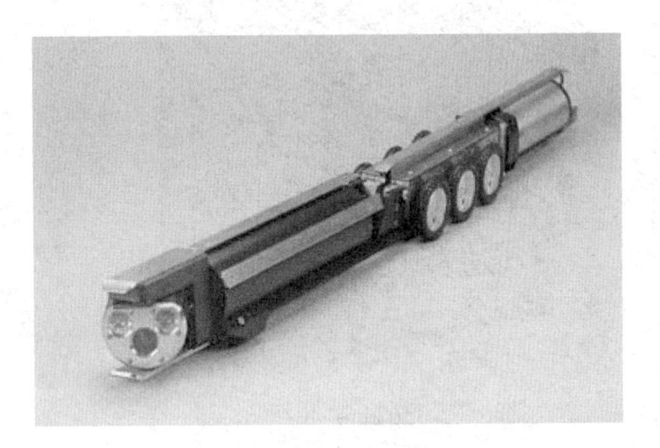

Tire and Wheel, Tractor Driven

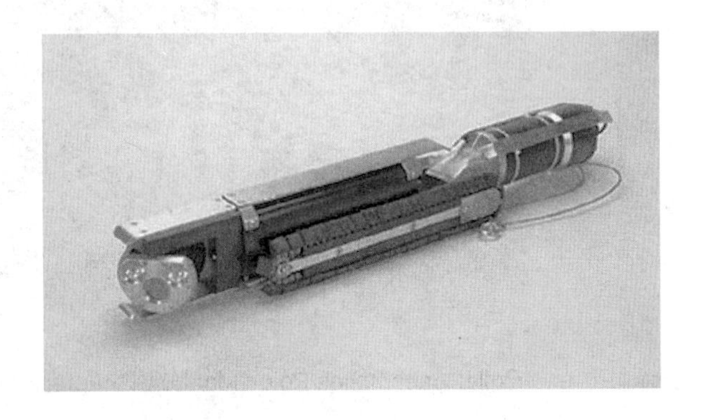

Crawler Driven

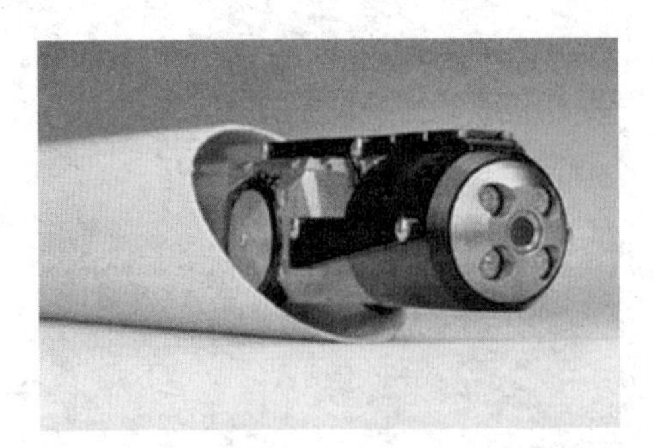

4-inch Line Setup

Fig. 5.16 Cameras and transport vehicles
(Permission of RST Services, Inc.)

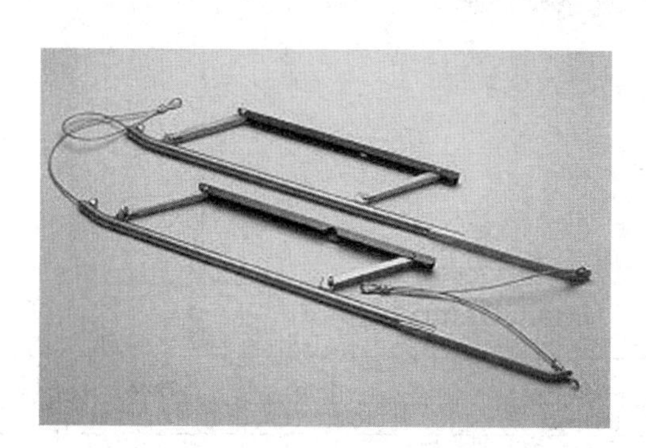

Camera Skid 6 Inches to 16 Inches

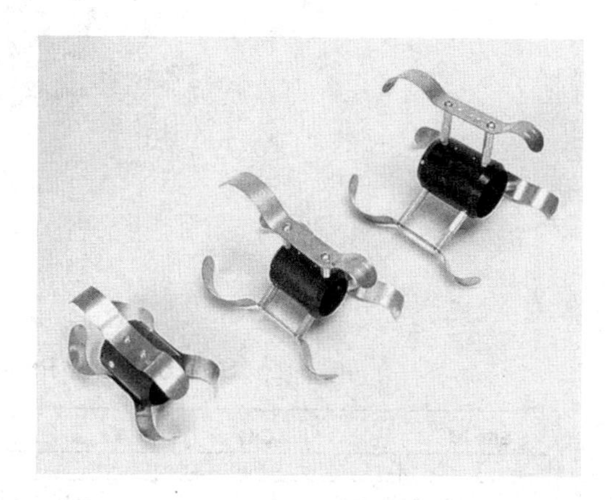

Push Camera Skids

Pontoon Floatable Skid With Camera

Camera Tractor Tires

Fig. 5.17 Transporting camera system components

(Permission of RST Services, Inc.)

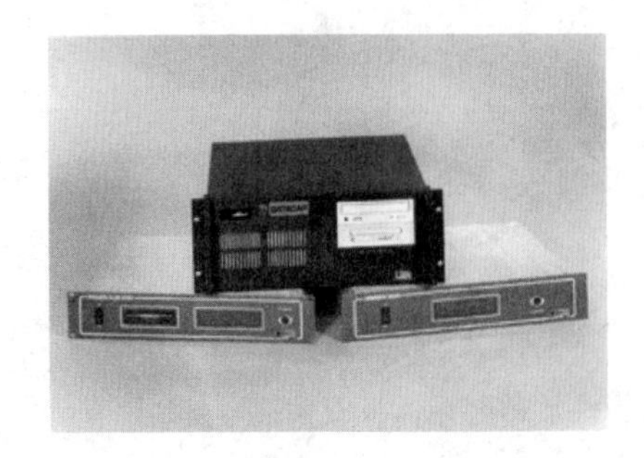

Site Data for Project: Ukiah, Demo

Site ID	City	Street	Date	Time
1	Ukiah	300-Hillcrest	10/02/2002	01:14:29 PM

M.H. Start	M.H. Stop	M.H. Depth	Starting Dist	Final Dist
344	end	4.0	8.0	390.8

Type of Pipe	Pipe Size (in)	Sec. lgth	Direction	Surface Condition	Operator
Concrete	6	4	Away-D	Paved Asphalt	Rich

Comment
Demo on 10-02-02 for the city of Ukiah

Observation Data

Obs ID	Ft	Lat Ft	Category	Category Details	Clock Pos	Sevr Lv	Ph 1 ID	Ph 2 ID	Vclip ID	Vid ID	Tape Cnt
1	8.0		Other	Start Run					34.68		
2	10.1		Pipe Problem	Roots	9 O'Clock	Level 2	1252.jpg	2252.jpg	153.41		
3	12.5		Joint Problem	Roots in Joint	360	Level 2			245.27		
23	12.6										
4	22.3		Joint Problem	Roots in Joint	360	Level 1	1254.jpg				
5	32.2		Joint Problem	Roots in Joint	7 O'Clock	Level 1					
6	42.7		Joint Problem	Roots in Joint		Level 1					
7	53.2		Joint Problem	Moderate Offset	360	Level 2	1257.jpg				
8	53.9		Service Conn.	Roots	9 O'Clock	Level 3	1258.jpg		294.15		
9	55.1		Service Conn.	Roots	9 O'Clock	Level 4			395.01		
10	66.7		Service Conn.	Roots	3 O'Clock	Level 4	12510.jpg		490.14		
11	107.2		Service Conn.	Roots	9 O'Clock	Level 3			587.34		
12	108.1		Service Conn.	Domestic Flow	12 O'Clock	Level 2			647.56		
13	130.9		Service Conn.	Roots	3 O'Clock	Level 2			747.57		
14	152.9		Service Conn.	Roots	9 O'Clock	Level 3			814.34		
15	175.2		Service Conn.	Roots	3 O'Clock	Level 4			891.64		
16	212.7		Service Conn.	Connection	9 O'Clock	Level 1			1010.09		
17	234.5		Service Conn.	Connection	9 O'Clock	Level 1			1073.80		
18	316.2		Joint Problem	Severe Offset	9 to 3	Level 2			1236.35		
19	346.6		Pipe Problem	Sag	bottom half	Level 2			1383.69		

Fig. 5.18 Data display module with video footage readout and observation report

(Permission of Cues, Inc.)

Site Data

Project Name	Site ID	City	Street	Starting Ftg.	Date	Time	Tape ID
Orlando	1	Orlando	3445 Gore	8.0	Jul. 30 1998	01:04:17 PM	12

M.H. Start	M.H. Stop	M.H.Depth	Pipe Size (in)	Type of Pipe	Direction	Surface Condition	Final Footage
55g	57h	3 (ft) 2 (in)	8	Clay	Away-DS	Wet	221.4

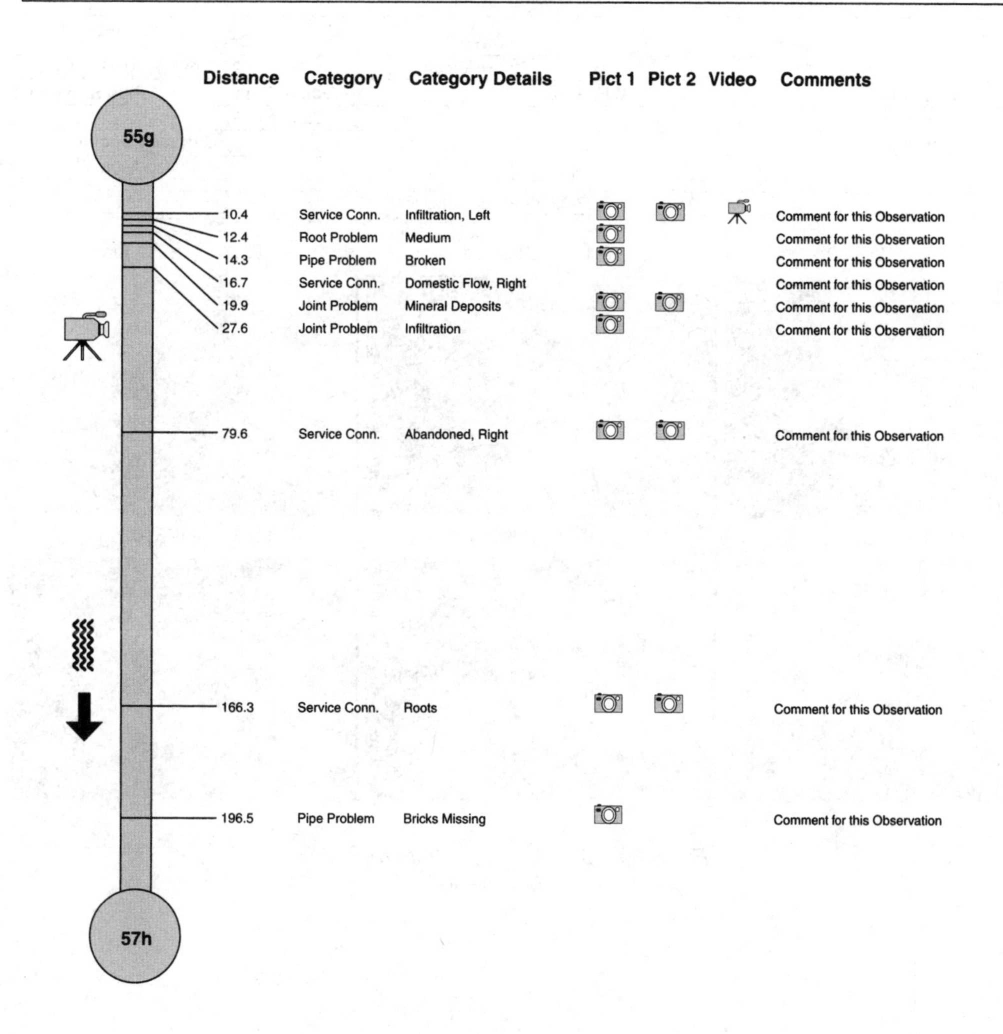

Fig. 5.19 Video footage readout and site report
(Permission of Cues, Inc.)

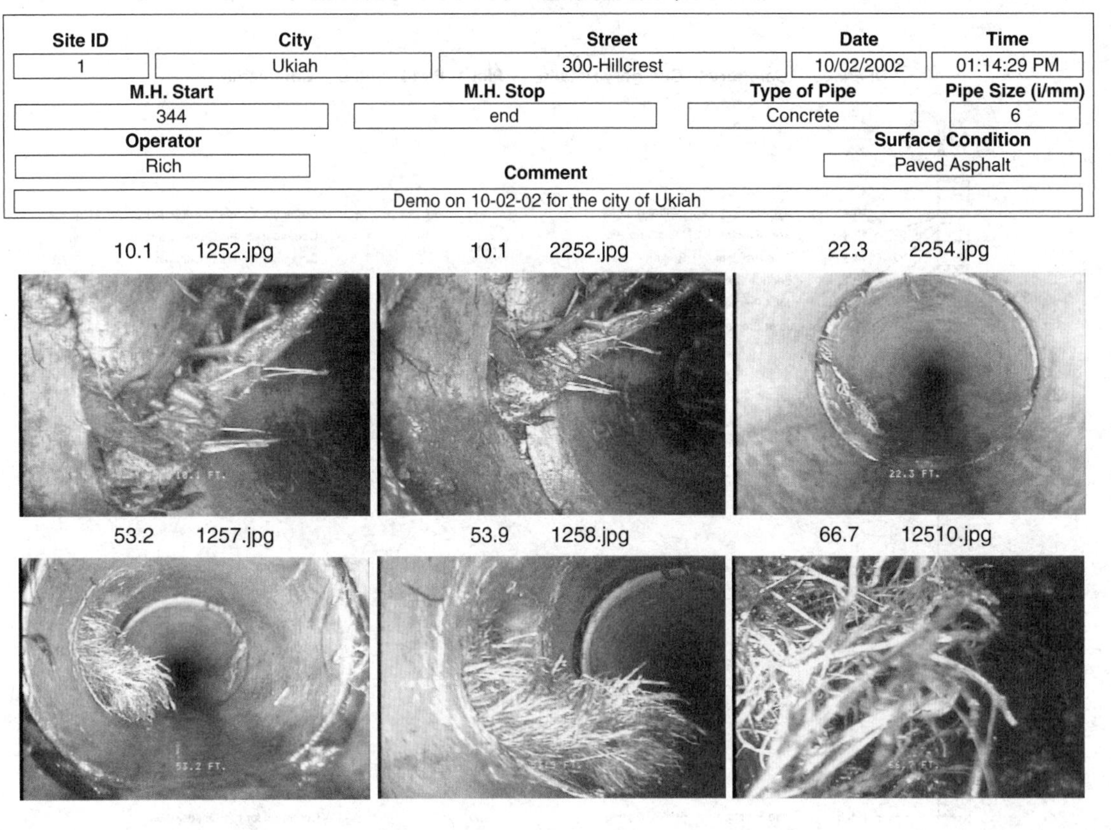

Site Data and Photos: Ukiah, Demo

Site ID	City	Street	Date	Time
1	Ukiah	300-Hillcrest	10/02/2002	01:14:29 PM

M.H. Start	M.H. Stop	Type of Pipe	Pipe Size (i/mm)
344	end	Concrete	6

Operator	Surface Condition
Rich	Paved Asphalt

Comment
Demo on 10-02-02 for the city of Ukiah

Fig. 5.20 Site data and photo report

(Permission of Cues, Inc.)

Photo #2 Data for Project: Ukiah, Demo

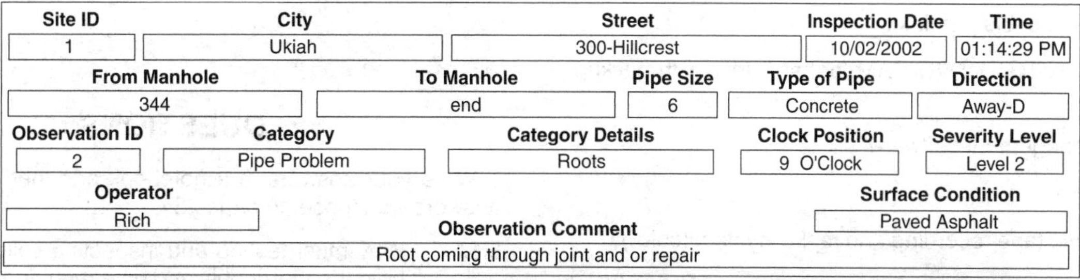

Site ID	City	Street	Inspection Date	Time
1	Ukiah	300-Hillcrest	10/02/2002	01:14:29 PM

From Manhole	To Manhole	Pipe Size	Type of Pipe	Direction
344	end	6	Concrete	Away-D

Observation ID	Category	Category Details	Clock Position	Severity Level
2	Pipe Problem	Roots	9 O'Clock	Level 2

Operator	Surface Condition
Rich	Paved Asphalt

Observation Comment
Root coming through joint and or repair

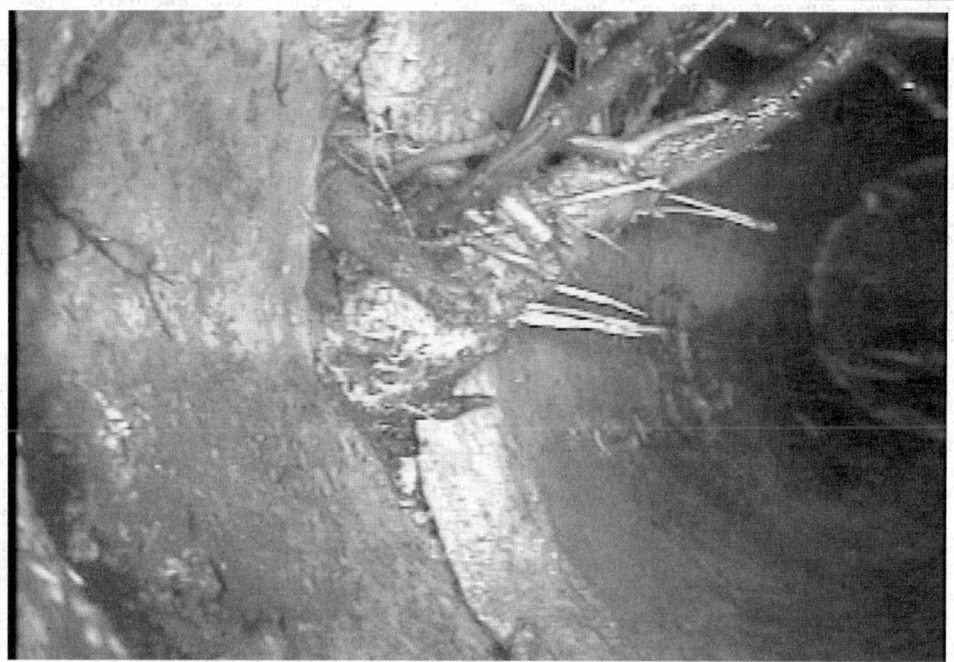

Photo #2 2252.jpg **Distance (ft/m): 10.1**

Fig. 5.21 Single site data and photo report

(Permission of Cues, Inc.)

11. Operator two-way communication system or devices
12. Camera extension poles (Figure 5.15)
13. Invert pulley assemblies
14. Manhole top roller assembly

Other items that are not essential to televising the sewer but are capable of saving time, useful in protecting the equipment, and helpful with the work to be done include:

1. Television cable reel
2. Water tank with pump, spring reel, and hose with nozzle
3. Tool box with voltage meter, electrical tape, and connectors
4. Camera and cable repair kit
5. Computer with CD, DVD-RAM, or hard drive with backup drive
6. CCTV management software
7. Color printer
8. Video/audio tape recording and replay system (color)
9. Usual tools and equipment required for safe manhole work (see Section 4.3, "Routing Traffic Around the Job Site" and Section 4.5, "Safety Equipment and Procedures for Confined Space Entry")
10. Articulating or adjustable cable guide arm
11. Rear TV monitor

5.311 Staffing Requirements

Usually the televising crew consists of three people: one TV operator and two collection system operators. Newer TV systems using self-propelled tractor drives may have smaller crews.

1. TV Operator

   The TV operator is the lead person of the three-party crew whose job it is to observe the TV monitor and control the progress of the TV camera from the operator's station.

   This operator must be thoroughly experienced in the inspection and repair of sanitary sewers. This person must be able to recognize and evaluate conditions found in the pipe. A knowledge of computers and TV electronics is not required but the ability to successfully troubleshoot the equipment is desirable.

2. Collection System Operator

   Helps to make initial setup; then is stationed at the far manhole to:

   a. Operate the pulling hand winch, or

   b. Observe the operation of the power winch, or

   c. Disconnect the self-propelled camera from the power supply so the power cable can be retracted.

3. Collection System Operator

   Helps to make initial setup; then is stationed at the rear of the TV unit or truck to:

   a. Operate the return hand winch, or

   b. Operate the electric controls on the reel return.

   c. Workers at both manholes must also observe "down hole" cable guides or protection devices used to protect the power and winch cables from sharp edges.

4. Additional help is required when working in heavy traffic and congested areas.

TRAFFIC ROUTING

PLEASE TURN BACK AND REVIEW SECTION 4.3 ON PAGE 88

MANHOLE HAZARDS

PLEASE TURN BACK AND REVIEW SECTION 4.4 ON PAGE 115

QUESTIONS

Write your answers in a notebook and then compare your answers with those on page 254.

5.3A Why must testing and inspecting collection systems be done thoroughly and on a regular basis?

5.30A What are the advantages of closed-circuit television inspection of wastewater collection systems?

5.30B List the kinds of information that can be obtained from a television inspection of a collection system.

5.31A How many people are usually on a collection system televising crew?

5.32 Procedure

5.320 Setting Up a CCTV Operation

In this example procedure, the TV camera will be used to inspect a sewer between two adjacent manholes. *THE CAMERA SHOULD ALWAYS BE PULLED IN THE DIRECTION OF FLOW OR FROM THE UPSTREAM MANHOLE TO THE DOWNSTREAM MANHOLE.* By moving a camera downstream, any debris in the pipe should be washed away in front of the camera and any buildup on the back of the camera should be removed as the camera moves along.

However, there are instances when the TV camera can be pulled upstream or against the flow, such as in a new line (not yet in service) or in a sewer line containing minimal flows. When doing this, it is important that the data be recorded on a properly designed "log sheet" (see example, Figure 5.24, page 224). The TV inspection report or "log sheet" must contain information and a diagram showing the direction of camera travel in the section being televised. A detailed explanation of the logging of information on the inspection report will be presented in Section 5.321.

Sewer lines usually should be cleaned prior to televising by using either high-velocity cleaners, water jet-vacuum (combination) machines, power rodders, bucket machines, or balling. Cleaning is generally necessary to remove sludge, grit, roots, and grease that could interfere with the operation of the TV camera and cover up cracks, leaks, and open joints. Whether or not a sewer should be cleaned before televising depends on the diameter and condition of the sewer. Cleaning also gets rid of obstructions, thus allowing better flow conditions, or identifies the location of obstructions that could cause problems during TV inspection.

Stringing of the pull cable into the sewer line can be accomplished by using a high-velocity cleaner or by using a parachute with an extractor fan. Lightweight string lines can also be floated from one manhole to another by using a small rubber ball and then attaching the cable to be inserted into the line. Crawler tractor units eliminate the need to string the pull cable.

If the sewer to be televised is flowing so full that the TV camera will be submerged, special considerations must be evaluated. Upstream lines can be plugged temporarily to reduce flows but should only be done by experienced personnel and only after careful study to ensure that property damage will not occur due to the flow restriction. If a plug is inserted in a sewer to reduce flows downstream in order to televise the sewer, have someone observe the level of the water in an upstream manhole above the installed plug. Be sure the water does not become so high that plumbing fixtures in nearby homes will overflow and flood the homes. An additional safeguard would be to set up "bypass pumping" of the flows during the time required for the TV inspection of the sewer line. A second alternative is to schedule the work to be done during normal low-flow times, such as between midnight and 5:00 a.m. Also, the plugging of sewers will cause downstream flows and conditions to change, which could interrupt downstream flow measurements, sampling, testing, and wastewater treatment processes; therefore, always notify the treatment plant operators and industrial waste inspectors whose activities could be affected by your actions.

The following sections contain typical steps that should be followed to inspect a sewer using closed-circuit TV. ALWAYS FOLLOW THE DIRECTIONS PROVIDED BY THE MANUFACTURER OF YOUR TV EQUIPMENT. Newer TV systems using self-propelled tractor drives will use slightly different procedures.

1. Safety. Refer to Sections 4.3, "Routing Traffic Around the Job Site," and 4.4, "Classification and Description of Manhole Hazards," for safe procedures to route traffic around the job site and enter a manhole.

2. String the sewer from manhole to manhole. The objective is to bring the pull winch cable from the downstream manhole, through the sewer, to the upstream manhole for the attachment of the camera. Any of the methods used to preclean sewer lines can be an effective means to string the line. The most positive way is with the use of a high-velocity cleaner or power rodder. "Floating" lines for stringing is possible but this method is the least desirable due to the effects of low spots, roots, and partial blockages in preventing the line from completing the procedure.

3. While the line is being strung through the sewer, the pull winch is moved to the downstream manhole and set up.

4. When used, the telephone cable and system are laid out from manhole to manhole and a worker at each manhole puts on the telephone headset. Voice communication is tested to make certain communications between manholes are adequate. In situations where traffic over the telephone cable is heavy, consider the use of a CB radio base station (at the operator's station) and a walkie-talkie for the sewer worker at the far hole.

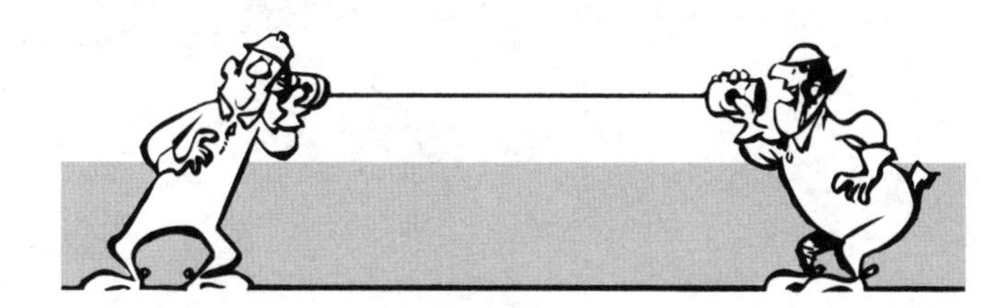

5. Also while the line is being strung, the generator is started and the system tested for adequate power in volts and for proper cycles (*HERTZ*[4]) of the power supply. Most systems operate on a voltage range of 110 to 125 volts and 60 cycles per second (Hertz) (59 to 61 range) of A.C. (alternating electrical current). Meters indicating both of these power values are generally included in the power control module.

   a. If voltage is too high or too low, the control module should have a multiple-tap transformer with marked settings to increase or decrease the voltage input to the system.

   b. Keep the generator tuned and set by experienced generator technicians to avoid problems.

6. As soon as the power generator has been started, inspected, and adjusted as necessary for proper power supply, the operator must turn off the master switch of the power control unit or such switches as are provided to the camera circuit and lighting circuit.

7. The crew makes necessary changes on the camera skid assembly to prepare the assembly for the size of pipe to be televised. The proper skid assembly is important because the TV camera lens must be as close as possible to the center of the pipe when the camera moves down the pipe. Connect the tow and tag line bridles at this time.

8. The operator brings the end of the television transmission cable into the location where both the front of the camera (Figure 5.22) and the television monitor can be seen. The lens cover is removed from the camera. Protection of the camera lens depends on the type of electronics. Modern electronic chips do not need to be protected from sunlight or bright light while the old vidicon tubes must be protected.

9. The operator connects the leads of the television transmission cable to the camera and turns on necessary switches to activate the camera power and video circuit,

---

[4] *Hertz. The number of complete electromagnetic cycles or waves in one second of an electric or electronic circuit. Also called the frequency of the current. Abbreviated Hz.*

Eagle Eye Camera With 450-Watt Triplex Lighthead
for Lines up to 20-Foot Diameter

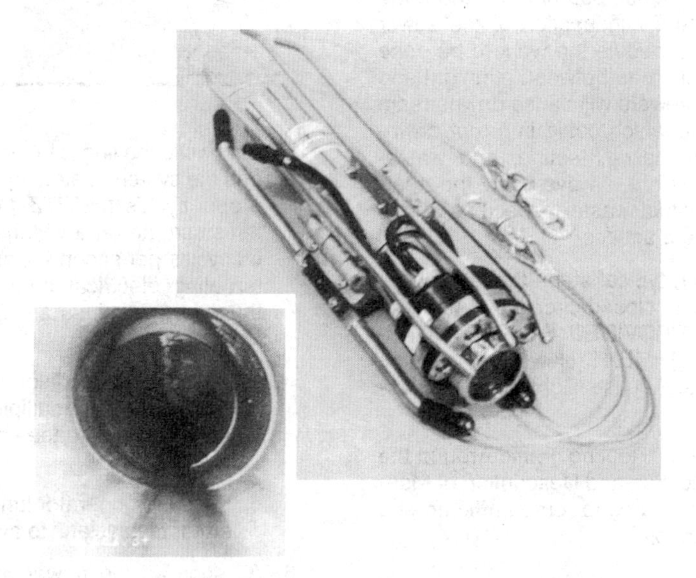

Eagle Eye Camera With 100-Watt Low Profile
Lighthead for 6" to 24" Lines

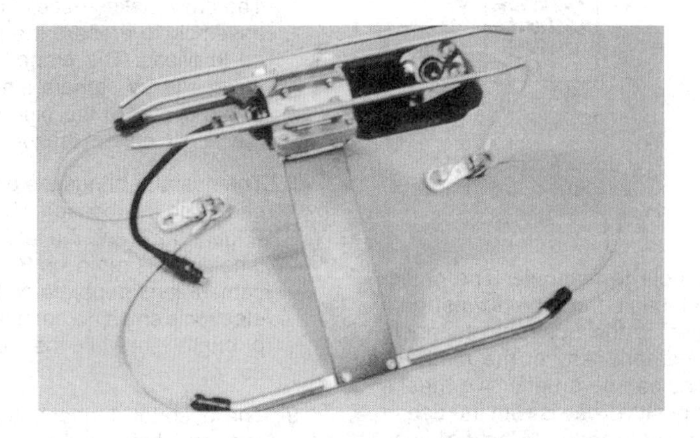

Pan and Tilt Camera With Directional Lightheads

Fig. 5.22 Sewer line TV cameras ready for operation

(Permission of Aries Industries, Inc.)

including turning on the television monitor. Do not activate the camera light head circuit, but make certain that it is off or set to zero on the intensity control.

10. As soon as a picture appears in the monitor, the operator focuses the camera for a preset distance from the camera, depending upon the diameter of the pipe being televised. Some of the newer cameras available are equipped with an automatic focus switch located on the operator's panel and can be adjusted automatically after the TV camera is placed inside the sewer line.

    a. Sewer inspection cameras are provided with various types of wide-angle lenses such as 33°, 53°, 64°, and 90°. Each one of these different wide-angle lenses works with a different focal distance in any given pipe size. A 90° lens has a picture edge only 8 1/2 inches away in a 12-inch pipe. In the same pipe, a 33° lens must look almost 2 feet down the pipe before a picture is available.

    b. Sewer pan and tilt cameras may rotate 360 degrees with full viewing capability and pan 240 to 270 degrees from centerline to centerline. Some cameras have night light enhancement features and may view obstructions and movements up to 240 feet away in a pipe.

    c. Where focal distances are unknown, consult the manufacturer's instructions.

11. Before placing the camera in the sewer line, check the camera light power supply to be sure all connections are secure. The lighting circuit should only be tested, and then should be turned off before any heat can develop in the light head. Once this has been done, turn off all power to the TV camera.

12. The operator inspects the lens cover on the camera.

13. Making certain that all power to the camera assembly is off, the operator takes the camera to the manhole where the television inspection is to start.

    Attach the *CABLE STRAIN RELIEF*[5] to the tag line bridle (when applicable) and keep the electrical connections intact.

14. At the close manhole, a worker attaches the pull winch cable that has been strung through the sewer line to the tow cable (swivel connection) of the TV camera. The camera is now ready to be placed into the sewer line. The TV rollers should be put in place so the line will not be cut or damaged. Crawler tractor units do not use a pull winch cable.

    The TV camera is inserted in the sewer while the crew remains topside. The camera is placed on the skid assembly and lowered into the manhole using the pull cable. Also the camera is positioned using pull cables. With the crawler tractor unit, a mechanical crane/hoist is used to lower the camera into the manhole. A pole with a hook is used to separate the camera from the lifting/lowering cradle and to properly position the camera in the sewer.

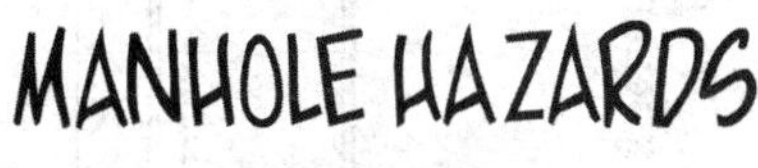

15. When the camera is properly positioned in the sewer, the footage meter is set for proper reading. All footage readings must be estimated as closely as possible and should always start at the CENTER of the manhole so repair crews will always measure from the same point. Since the TV camera has been inserted into the sewer and is ready to move down the line, the focal point could be 9 feet (3 m) from the center of the beginning manhole. The footage meter is set to read this number prior to beginning the inspection (Figure 5.23). This is extremely important so that all measurements are taken from the center of the starting (upstream) manhole.

16. When all is ready, the TV operator starts the camera moving down the sewer.

    CAUTION: WHENEVER PULLING A CAMERA BACK MORE THAN ONE FOOT, THE SLACK MUST BE TAKEN OUT OF THE POWER CABLE WHILE THE CAMERA IS MOVING BACKWARD. FAILURE TO DO THIS CAN PERMIT THE POWER CABLE TO BECOME TANGLED WITH THE CAMERA AND SKID ASSEMBLY AND CAUSE EXCESSIVE WEAR POINTS AND POSSIBLE SEVERE DAMAGE TO THE POWER CABLE ON CERTAIN TYPES OF EQUIPMENT.

17. There is one essential rule that must never be violated: while the camera is being moved through the line, the operator must never stop looking at the monitor picture. Any obstruction that could wedge or damage the camera or the light must be seen in time to stop the camera's forward progress. If the picture is lost for any reason, the forward movement of the camera must be stopped. If the picture cannot be reestablished, the camera must be pulled backward out of the line.

---

[5] *Cable Strain Relief. A mesh type of device that grips the power cable to prevent any strain on the cable from reaching the connections.*

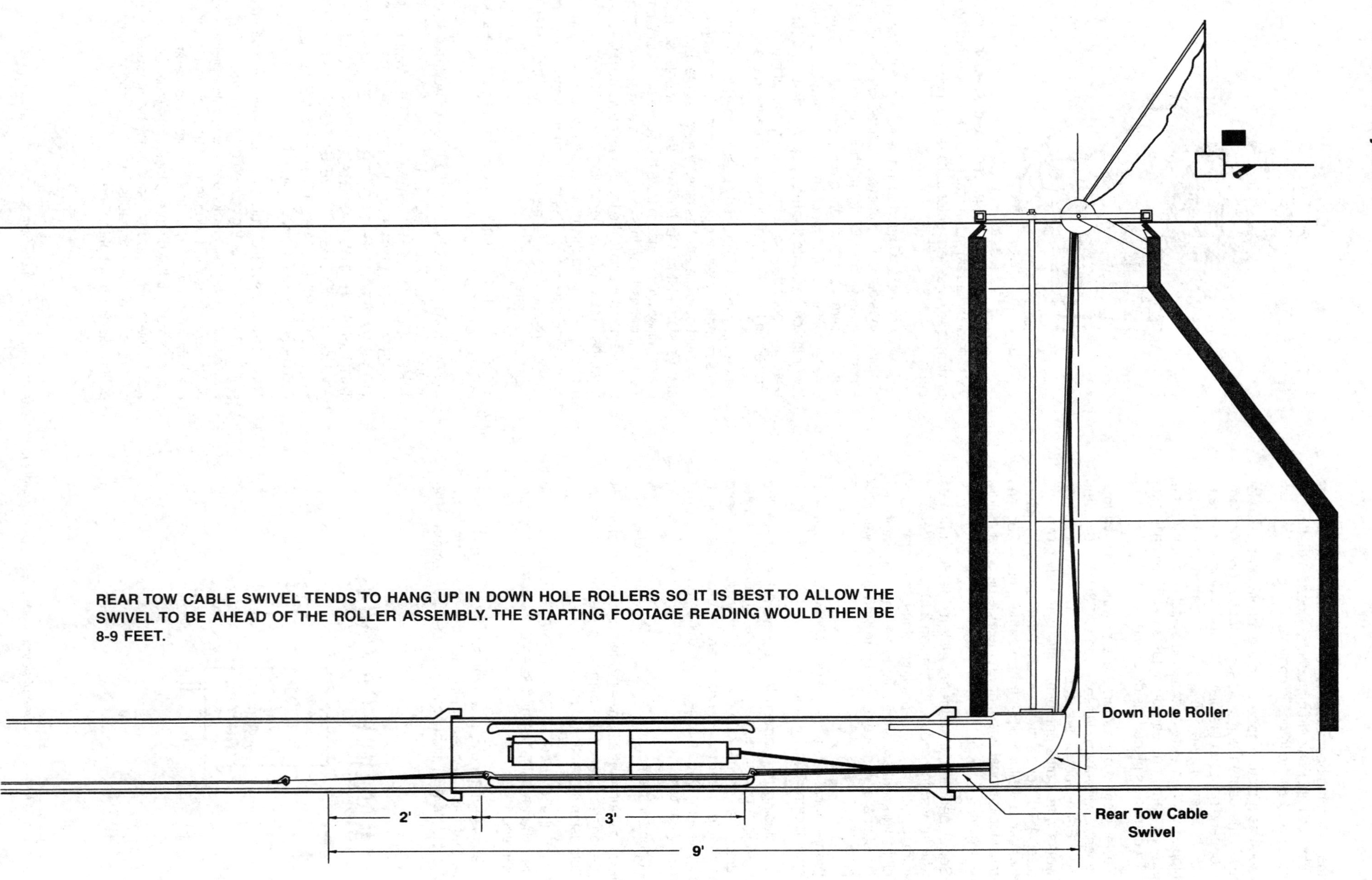

Fig. 5.23 Adjustment of footage meter

18. If the operator begins to experience excessive drag or wedging, this fact must be investigated immediately. Pull the camera back a few feet to see what could cause the drag or possible wedging. An out of round pipe can cause wedging. If a camera and skid assembly become locked or wedged in a pipe, the sewer must be dug up to recover the camera. Therefore, it is very important that the TV operator study the movement of the camera at all times.

QUESTIONS

Write your answers in a notebook and then compare your answers with those on page 254.

5.32A Why should the TV camera *ALWAYS* be pulled from the upstream manhole to the downstream manhole?

5.32B What should a TV inspection crew do first when they arrive at a sewer to be inspected with a TV camera?

5.32C Is the TV pull winch located at the upstream or downstream manhole?

5.32D Why must the proper skids be used with the TV camera?

5.32E Where in the pipeline should the footage meter be set to read zero?

5.32F What essential rule must never be violated while the TV camera is moving through the line?

5.32G Why must extreme care be taken to prevent the TV camera from becoming wedged in the pipe?

5.321 Logging and Recording Television Inspections

A television inspection of a pipeline requires that all data are observed and recorded. The operator of the TV monitor must determine what is to be recorded, photographed, or videotaped. As an operator you also should have the camera's progress halted every time you take your eyes off the monitor to record information.

1. Most television inspection records are written on a standard log (see example on Figure 5.24). Individual operators may desire to slightly redesign a log sheet especially for their particular needs. However, certain basic information should be included on all sewer line television report forms. This information includes agency name, date, location, map number, TV operator's name, type of pipe, and identification of section being televised (manhole numbers). Other information that could be helpful yet not critical includes a sketch showing the direction of flow in the sewer line, the direction of the TV inspection, a north arrow, identification number of equipment being used, videotape identification number, "new work" or "in use" sewer designation, and name of inspector or concerned individual present. Coded initials and symbols or numbers are used for simplicity and to save time. Some of the suggested codes are:

   - SC—Service connection. Quadrant number of position of entry is usually added. Some operators use a time clock notation rather than a quadrant number. (See Figure 5.24, sample log.)
   - R—Roots. Generally followed by number indicating growth. (See Figure 5.25, root rating chart.)
   - Infiltration. May be followed by number to indicate estimated gallons per minute and a quadrant number to show point of entry. To learn how to estimate infiltration flows, if possible, observe typical flows at locations or in a lab where actual flows can be measured. By comparing estimated flows with actual flows, accurate flow estimates can be made by experienced operators in the field.
   - DL-G—Dirty line, grease, or soap cake. For sand or silt deposits, DL-S; for rocks, DL-K.
   - C/B—Cracked or broken place. This can have a remark indicating possible collapse of the pipe for extensive damage of this type.
   - OJ—Offset joint. Can be followed by a number or fraction to indicate inches or fractions of an inch out of line.
   - GDS—Grade defect, starting point.
   - GDE—Grade defect, ending point.
   - CUW—Camera under water. Point where lens was more than half covered by water.
   - COW—Camera out of water. Point where lens was less than half covered by water.

   See the Appendix at the end of this chapter for a three-letter coding system for sewer inspection, defects, and actions/responses (Section 5.9, page 259).

   Combinations of codes are used to provide information about one location. Example: 263 SC-1-R2-I5 = "Station 263, Service Connection from Quadrant One, Roots to rating two, infiltration at approximately five gallons per minute, all from service lateral."

2. Polaroid still photographs, black and white, are often used to support log entries. Figure 5.26 shows some examples of backup photos of infiltration points, bad offsets, damaged laterals, root growths, and broken pipe locations. When each photo is taken, the footage and log entry are recorded on the back of the photo (if a video footage readout system (Figures 5.18 – 5.21) is not provided or used).

   a. Polaroid cameras have an automatic exposure device that automatically makes a good picture from the light in the television monitor. There must be a minimum of other light sources that can affect the camera's light adjustment process. In some situations, a special hood may be necessary to eliminate other light sources. This hood must permit the front face of the camera, as well as the electric eye port, to be exposed to the monitor picture.

   b. To be effective for focus, a Polaroid camera must have a close-up lens and sighting kit.

   c. Bright lighting behind a Polaroid camera can cause reflections in the glass face of the monitor. Adequate curtains or shrouding should be provided to block out background light.

Date 6-24-02 Viewer BECHTEL

Page No. 17 to M.H. No. 5

Book No. 268 Camera Direction M.H. No. 2

T.V. INSPECTION REPORT

SACRAMENTO COUNTY PUBLIC WORKS
WATER QUALITY DIVISION

District	Pipe size & type:	Street / Easement	Quadrant Code
SAC	6" VCP	Street ☒ / Easement ☐	4 1 / 3 2
Distance:	**Cleanliness of line:**	**M.H. Condition:**	**Grade of Line:**
260'	ROOTS	GOOD	POOR

Distance Reading	Quadrant 1	Quadrant 2	Quadrant 3	Quadrant 4	Photo No.	Remarks	Repairs	Root Rating
5.0						LIGHT OFFSET JOINT		
6.2						1" WATER		
9.5						2" WATER		
10.6						3" WATER		
11.6						ROOTS @ JOINT		3
18.3						" "		4
21.2						" "		3
24.3						" "		3
27.3						" "		3
30.3						" "		1
33.5						" "		8
36.2						" "		7
39.3						" "		6
45.5						" "		7
48.4	✓					HOUSE CONNECTION		
51.3						ROOTS @ JOINT		1
↓						1" WATER		
123.4						2" WATER		
126.7	✓			✓		HOUSE CONNECTION		
167.0				✓		HOUSE CONNECTION		
216.4						ROOTS @ JOINT		1
220.4						" "		1
222.3						" "		1
225.4						" "		1
228.4						" "		1
234.9						" "		2
240.9						" "		3
243.9						" "		1
249.0	✓			✓	#1	HOUSE CONNECTION W/ROOTS		
260.0						END OF LINE		

Fig. 5.24 TV inspection report

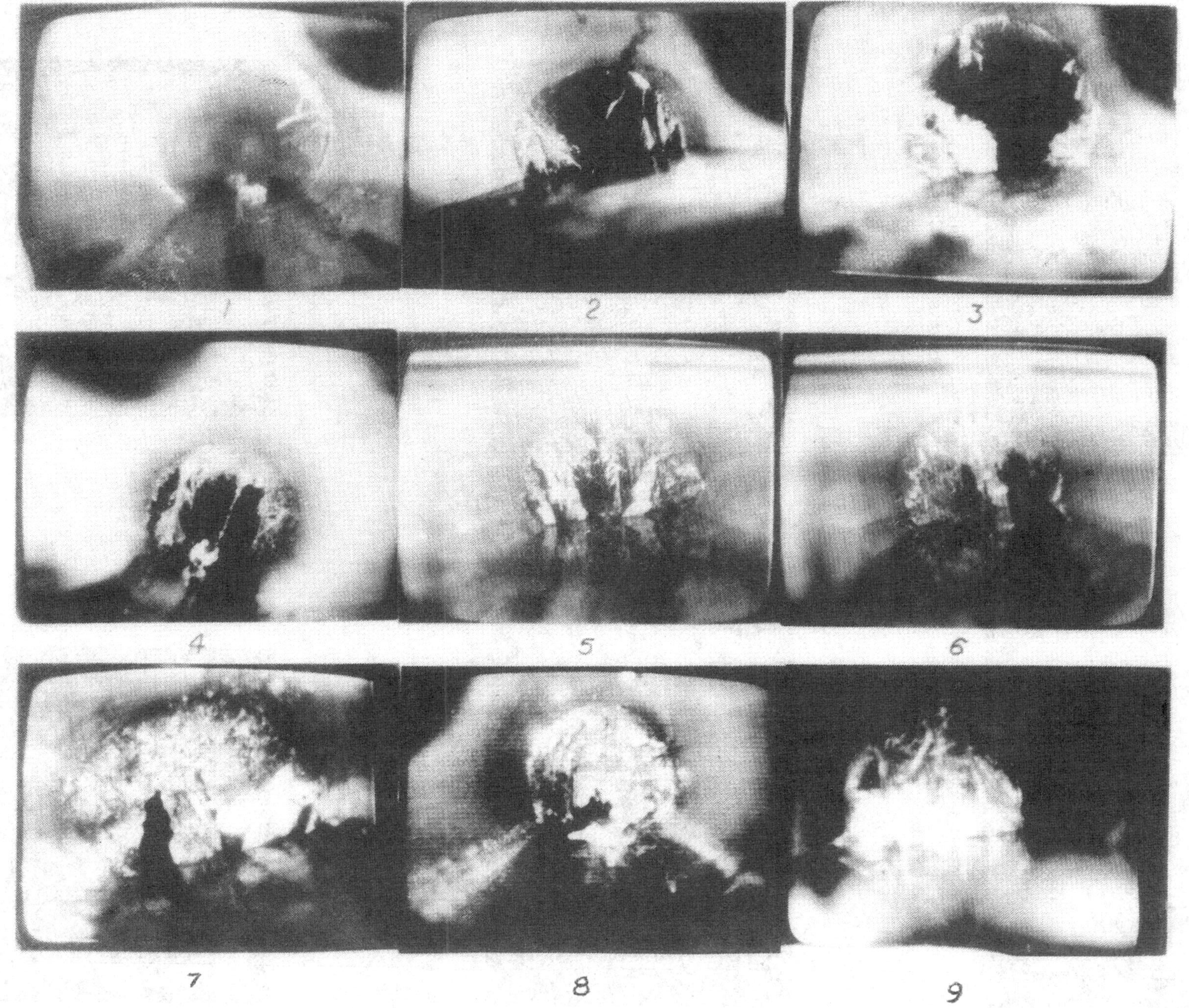

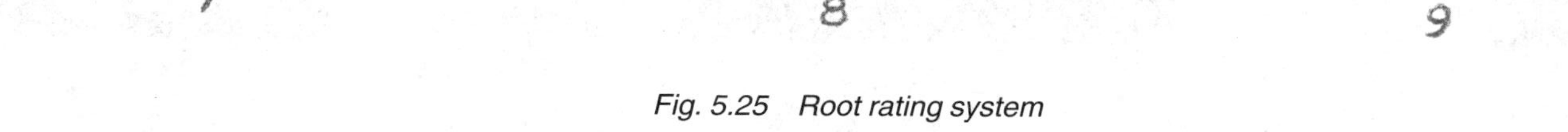

Fig. 5.25 Root rating system

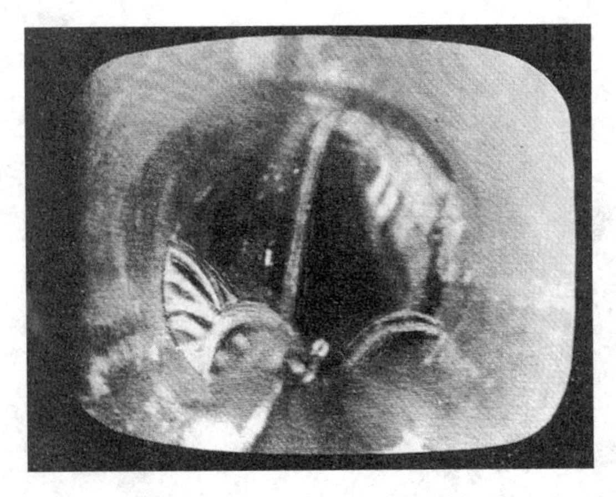

Infiltration at Defective Pipe Joint

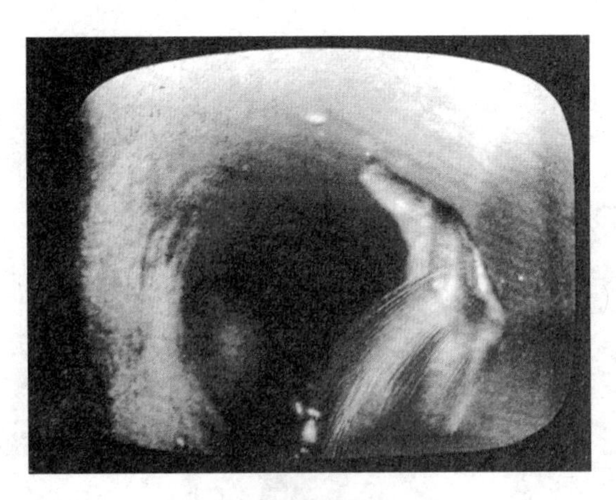

Infiltration or Inflow at
Defective Service Connection

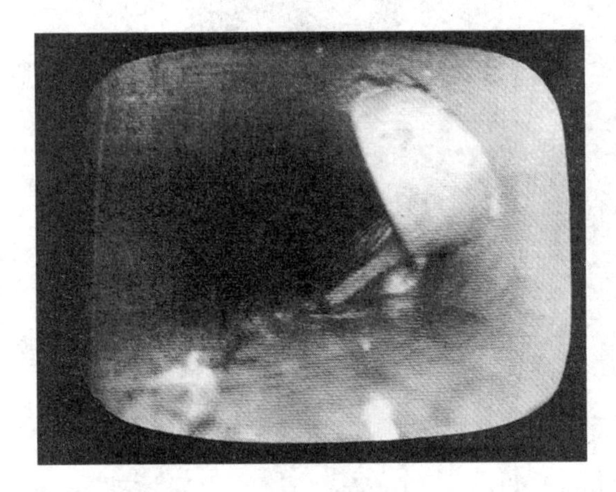

Protruding Service Tap

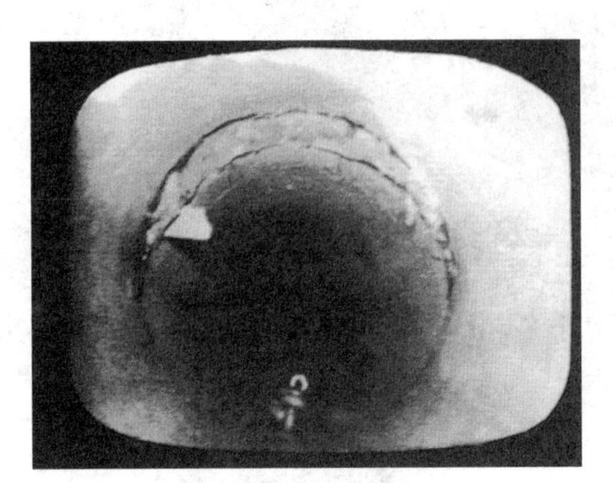

Drop Joint

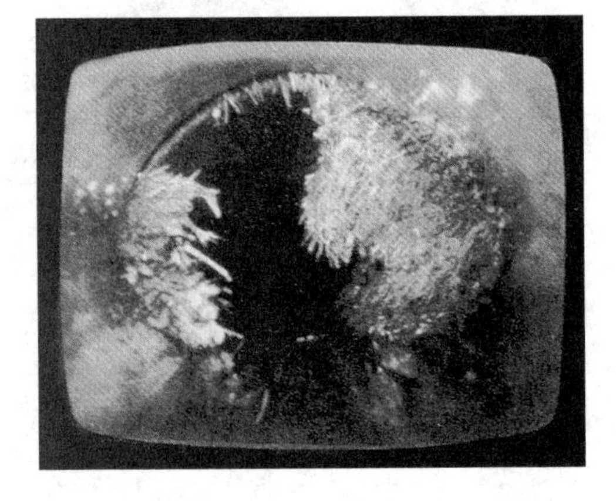

Root Intrusion

Dirty Line, Gravel, and Debris

Fig. 5.26 Typical data available from CCTV sewer inspections

Heavy Grease

Grease Buildup

Fig. 5.26 Typical data available from CCTV sewer inspections (continued)

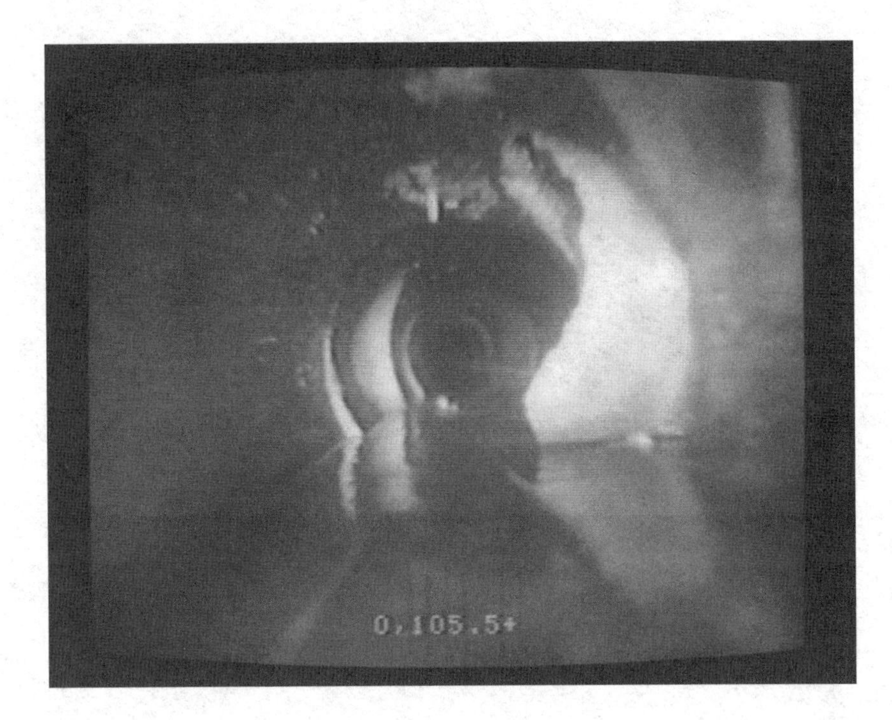

Protruding Tap

Detergent Buildup

Fig. 5.26 Typical data available from CCTV sewer inspections (continued)

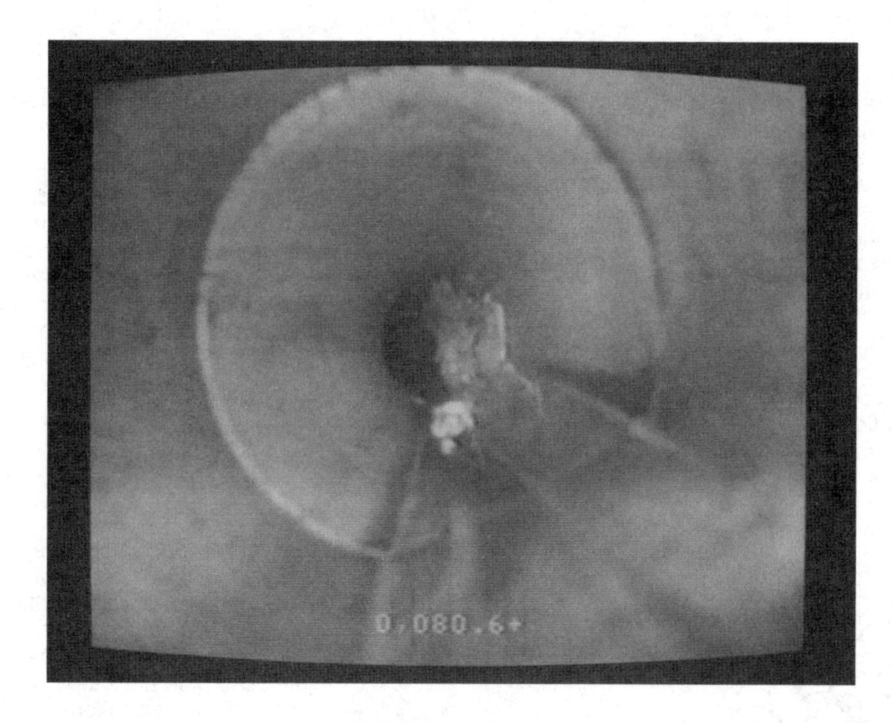

Roots in PVC Pipe

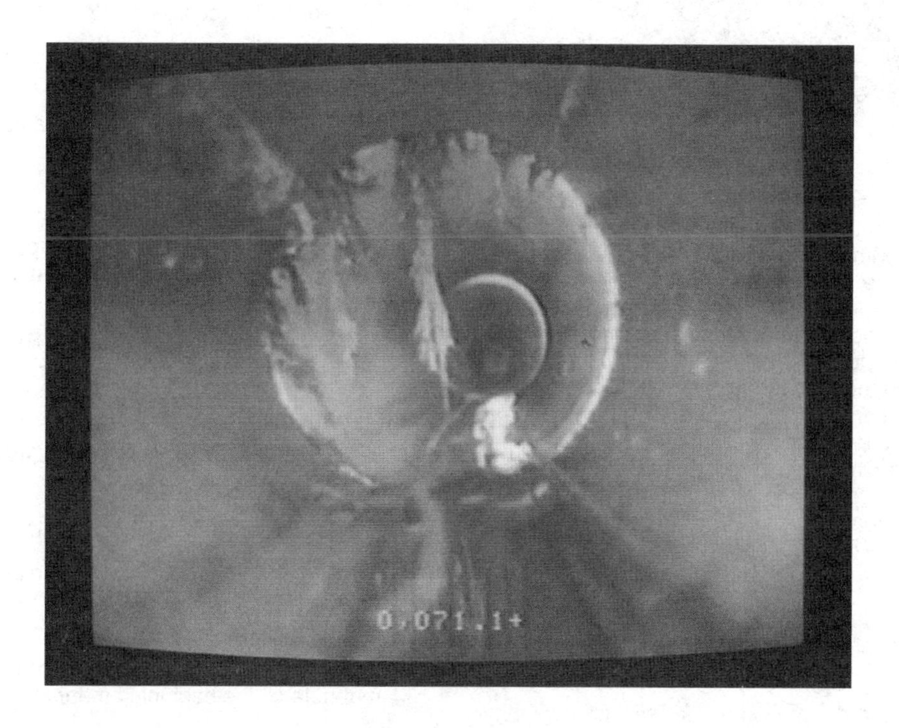

Root Intrusion

Fig. 5.26 Typical data available from CCTV sewer inspections (continued)

3. Videotape recordings can be used in place of or in addition to a Polaroid camera. Videotapes are able to provide a better record of some types of pipe situations. While most videotape recorders have the ability to record the voice of the TV operator when describing the condition of the sewer, defects observed, and the footage of the situation being viewed, voice records are not required when a video footage readout system is provided and used. Video footage readout systems are helpful, but a TV operator's actual description and comments are very useful when replaying the videotape in an office for someone who is not very familiar with the conditions in the collection system. These systems cannot replace the expertise of an experienced operator who can accurately describe the condition of the pipe in the audio portion of the videotape. In addition, when background noises from automobile traffic, industrial areas, and airport traffic areas make it impossible to use the audio at the time of the actual inspection, the TV operator can replay the tape later at a quieter site and "dub" in the voice comments at that time.

Another feature usually provided with video recording equipment is a selector switch. This three-position switch permits televising without recording, televising and recording at the same time, and instant replay of recorded tape after tape rewind. If the operator sees an object or situation to be videotaped, the operator starts the recorder, talks into the microphone to describe the situation, footages, or other remarks, and then stops the television camera progress after the TV camera has passed the problem area. The operator may then rewind the videotape to the place it was when the recording was started, switch to "VIR" (Video Instant Replay) and replay the recording in the monitor to make certain the desired picture and description are recorded correctly. The operator can then switch back to televise, switch off the recorder, and proceed on down the pipe.

Some of the advantages of a video recording system are:

a. The system records motion while still photography freezes motion. In an infiltration situation, the ability to record water flow in action for later viewing has definite advantages.

b. Where a pipe is off grade, operating the video recording during the full passage is desirable. When the supervisor reviews the tape, the supervisor can determine the seriousness of the situation and the necessary corrective action. If a video footage readout system is used, footages are automatically recorded on the tape. When no video footage readout system is provided, the operator should call out footages at frequent and regular intervals into the audio recording system of the tape recorder.

NOTE: In collection systems with dips or sags where water is allowed to stand, solids can settle out and create a dangerous nuisance by producing hydrogen sulfide as well as by causing high maintenance costs. Videotape documentation of the extent and severity of these locations will provide the necessary information so that economical and quick repair or correction of the situation can be made.

NOTE: If you stop the movement of the camera to examine a problem, turn off the video so a supervisor reviewing the video won't have to watch a still camera video while you are analyzing the problem.

NOTE: Many collection system agencies have libraries with shelves full of videotapes of televised sewer inspections. This information is difficult to retrieve and can only be shown to interested persons at one location. Today this information is being put on CD-ROMs. The CD-ROMs are easily reproduced. Several persons at different locations can easily and quickly observe and evaluate the same problem area.

QUESTIONS

Write your answers in a notebook and then compare your answers with those on page 254.

5.32I Why should the TV camera be stopped while the operator records data?

5.32J What items are photographed with a Polaroid camera?

5.32K What are some of the advantages of using a videotape over a Polaroid camera?

5.322 TV Log Sheet Explanation

Different agencies use different approaches to logging TV inspection data. This section illustrates the procedures used by the City of Phoenix.

The TV Inspection Report illustrated in Figure 5.27 shows a TV log completed by the TV operator.

Explanation:

- Flow direction arrows and TV direction arrows are drawn in as they occur in that specific run.
- Run is ✓'d as "routine," "complaint," or "new work."
- Starting manhole is placed in diagram on the left and ending manhole on the right with actual length of line recorded in center (432 LF).
- Starting stationing is always recorded as #0+00 (center of manhole) in the beginning manhole.
- "TN FACT" or "TN TAP" distinguishes between a factory- or field-tapped service connection into the main.
- City of Phoenix does not use "quadrant code." A "severity code" of 1-10 is used to describe the severity of the condition noted. For example, "roots @ joint 5" indicates that the root intrusion is obstructing approximately 1/2 of the diameter of the pipe. In addition, "pipe cracked 9" would indicate a severely cracked pipe, very near collapse.

CITY OF PHOENIX SANITARY SEWER REPORT

¼ SEC. #: 12-24 T.V. OPER: RENTFRO DATE: 10-15-02 ROUTINE: ☑

PIPE DIA: 8" PIPE TYPE: VCP PIPE LENGTH: 6' EQUIP. #: 481 COMPLAINT: ☐ NEW WORK: ☐

FLOW DIR. →

M.H.# 303 — 432 LF. — M.H.# 304

T.V. DIR. →

V.T. # 34 START: 000 STOP: 135

COMMENTS: SEVERE CRACKED PIPE AS NOTED
ROD LINE - ROOTS
CLEAN LINE - SLUDGE & ROCKS

STREET: W. CAMELBACK

STA. #:		VT
0+00	M.H. #303	
0+16.5	TN FACT.	
0+18.5	TS TAP	
0+42.0	ROOTS @ JOINT ④	000
0+54.0	" " ⑤	010
0+72.5	PIPE CRACKED ②	014
0+84.5	" " ⑥	020
0+90.5	TN TAP	
0+92.5	TS FACT.	
1+16.5	OFF-SET JOINT 1"+	028
1+22.0	2" WATER & SLUDGE	031
1+28.0	3" WATER	033
1+34.0	CUW	036
1+37.0	COW	039
1+52.0	PIPE CRACKED ⑨	042
1+64.5	TN FACT.	
1+66.5	TS TAP	
1+90.0	OFF-SET JOINT 1"+	047
2+10.0	SLUDGE - HVY.	051
2+14.0	ROCKS - SLUDGE	055
2+20.0	LOW JOINT - 1½" WATER	060
2+32.0	TN FACT.	
2+34.0	TS FACT.	
2+64.0	ROOTS @ JOINT ①	066
2+70.0	" " ④	069
2+90.0	TN FACT - CRACKED ⑥	072
2+92.0	TS TAP - CRACKED ⑨	078
3+10.0	OFF-SET JOINT 2"+	084
3+16.0	" " " 2"+	088

STA. #:		VT
3+28.0	PIPE CRACKED ⑤	092
3+40.0	" " ⑨	101
3+58.5	TN FACT.	
3+60.5	TS FACT.	
3+84.0	ROOTS @ JOINT ①	105
3+90.0	" " ①	↑
3+96.0	" " ④	│
4+02.0	" " ⑥	│
4+08.0	" " ⑧	│
4+14.0	" " ⑤	↓
4+20.0	" " ⑧	130
4+32.0	MH # 304	

INSPECTOR:

PERMIT #

Fig. 5.27 TV inspection report. Also see reviewed TV inspection report (Figure 5.28).

CITY OF PHOENIX SANITARY SEWER REPORT

¼ SEC. #: 12-24 T.V. OPER: RENTFRO DATE: 10-15-02 ROUTINE: ☑ COMPLAINT: ☐ NEW WORK: ☐

PIPE DIA: 8″ PIPE TYPE: VCP PIPE LENGTH: 6′ EQUIP. #: 481

FLOW DIR. → M.H. # 303 432 LF. M.H. # 304 T.V. DIR. →

V.T. # 34 START: 000 STOP: 135

COMMENTS:
SEVERE CRACKED PIPE AS NOTED
ROD LINE - ROOTS
CLEAN LINE - SLUDGE & ROCKS

STREET: W. CAMELBACK

STA. #:		VT
0+00	M.H. # 303	
0+16.5	TN FACT.	
0+18.5	TS TAP	
0+42.0	ROOTS @ JOINT ④	000
0+54.0	" " ⑤	010
0+72.5	PIPE CRACKED ②	014
0+84.5	" " ⑥	020
0+90.5	TN TAP	
0+92.5	TS FACT.	
1+16.5	OFF-SET JOINT 1″+	028
1+22.0	2″ WATER & SLUDGE	031
1+28.0	3″ WATER	033
1+34.0	CUW	036
1+37.0	COW	039
1+52.0	PIPE CRACKED ⑨	042
1+64.5	TN FACT.	
1+66.5	TS TAP	
1+90.0	OFF-SET JOINT 1″+	047
2+10.0	SLUDGE - HVY.	051
2+14.0	ROCKS - SLUDGE	055
2+20.0	LOW JOINT - 1½″ WATER	060
2+32.0	TN FACT.	
2+34.0	TS FACT.	
2+64.0	ROOTS @ JOINT ①	066
2+70.0	" " ④	069
2+90.0	TN FACT - CRACKED ⑥	072
2+92.0	TS TAP - CRACKED ⑨	078
3+10.0	OFF-SET JOINT 2″+	084
3+16.0	" " " 2″+	088

STA. #:		VT
3+28.0	PIPE CRACKED ⑤	092
3+40.0	" " ⑨	101
3+58.5	TN FACT.	
3+60.5	TS FACT.	
3+84.0	ROOTS @ JOINT ⑦	105
3+90.0	" " ⑦	↑
3+96.0	" " ④	
4+02.0	" " ⑥	
4+08.0	" " ⑧	
4+14.0	" " ⑤	↓
4+20.0	" " ⑧	130
4+32.0	MH # 304	

SERVICE REQUEST SUBMITTED
TYPE CONST. - ROD - CLEAN
DATE 10-17-02 [signature]

INSPECTOR:

PERMIT #

Fig. 5.28 Reviewed TV inspection report

- The columns headed by "VT" indicate the minute reading on the videotape where the identified defect can be viewed. This is very helpful because it allows a supervisor to "fast forward" to a particular location to quickly evaluate and establish repair priorities.

Explanation: (Figure 5.28)

- On the form illustrated in Figure 5.28, the addition of the "service request" stamp by the person in authority indicates that the TV log has been reviewed and the recommended action shown in the "comments" section has been approved. The service request calls for three activities:
   1. Construction crew to make needed repairs,
   2. Rodding crew to power rod entire line to remove roots, and
   3. Cleaning (high-velocity cleaner) crew to remove sludge and debris.
- Appropriate follow-up paper work is then activated and forwarded to the proper crew. Response reporting is encouraged through the form's design (see Figure 5.29 for forms activated by TV inspection report).

TV crews should realize that it is only necessary to videotape *DEFECTS* in active sewer lines. Taping the entire pipe run, unless it is for a specific presentation or purpose, would become quite costly and soon create tape storage problems for most agencies. On the other hand, taping the entire pipe run when inspecting sewer lines for acceptance of a new project is highly recommended. A comparison tape can then be made several months into the contractor's guarantee period to check for slowly developing faults that could seriously affect the long-term performance of a sewer line.

5.323 Special Precautions in Televising Collection Systems

Certain types of conditions encountered when televising sewers require special precautions. The operator and the crew should brief each other on the procedures to follow in the event these situations are encountered and there is a risk of damage or wedging of the camera in the sewer.

1. Offset joints restrict the pipe diameter and can cause a camera to wedge tightly in place rather than pass through. Two different types of common offsets require two different procedures.

   a. A horizontal offset occurs when the joint section has moved out of alignment to the right or left. As an operator gains experience in reading the severity of these offsets, the operator will know if the camera assembly can pass these or not. Where there is doubt, the TV operator must approach the offset very slowly and cautiously. If you find a sudden resistance to the progress of the camera, *STOP IMMEDIATELY.*

   Attempting to force the camera past such an obstruction can cause it to be wedged so tightly that the camera cannot return back out the sewer. When the operator finds that the camera cannot move ahead, the footage should then be recorded on the TV log with the appropriate explanation such as "TV camera would not pass." The camera should then be pulled back out of the line. The bad offset should be repaired before another attempt is made to televise that run of pipe.

   b. The two types of vertical offsets that may be encountered are illustrated in Figure 5.30. Each requires a different procedure. To pass a "DROP JOINT" the camera must drop down to get into the next section of pipe; for a "JUMP JOINT" the camera has to be lifted up in order to get into the next section of pipe.

   (1) Where the TV operator sees a drop joint ahead, the camera progress is reduced to an extremely slow speed. In most cases, and with a little experience, a TV operator can determine if the joint several inches ahead of the camera can be passed without damage or wedging. If there is any doubt, the attempt to pass a drop joint should not be made. If the camera comes to a sudden stop, the operator should not attempt to pull it farther.

   As with a tight horizontal offset, the footage must be recorded on the log, and the camera should be backed out of the line and the pipe joint repaired before televising that run again. In addition to possible damage to the camera in passing such a joint, or attempting to, the ability of the camera to return may be lost after the joint is passed. Should a point farther down the pipe prevent any more forward progress, the camera will have to be returned back out of the line and a drop joint will become a jump joint. If the camera becomes wedged tight in the pipe, excavation to recover the camera will be necessary.

   (2) Where the TV operator sees a jump joint ahead, the operator must be careful. The camera may be moved without danger until the lower skids butt into the ledge and progress comes to a halt. The TV operator must decide whether to try to 'jump' the camera or back it out. The extent or severity of the offset is the factor to be evaluated. To 'jump' the camera assembly, the operator tries to make the skids or crawler tractor move up over the ledge.

   In general, any jump joint that can be negotiated in the above manner can be pulled back through without problem in the event the camera has to be backed out of the line.

2. When the camera goes under water to a point where the top half of the pipe cannot be clearly seen, the operator is no longer able to see obstructions or offset joints adequately to prevent wedging or damage. The TV operator should note the circumstances by announcing "CAMERA UNDER WATER," AND TAKE IT SLOW AND EASY. If the operator finds the camera has stopped and cannot be moved, the TV operator should stop camera movement. To be safe, back the camera out of the sewer and have the sewer repaired at that point before trying to pull the camera through.

   If the submerged section of the sewer must be inspected, carefully assess the precautions necessary with respect to plugging the line for any length of time. Flow bypass pumping and working during low-flow times must be considered in high-volume lines to avoid the possibility of incurring property damage from wastewater backing up into residences and businesses. One method of televising through low spots or sags in the line is to use a high-velocity cleaner with closed-end nozzle to clean directly ahead of the TV camera. The jet of water from the nozzle will push the water clear of the pipe for a short distance. When the TV picture becomes submerged again, stop the high-velocity cleaner and allow the water to flow out the pipe. Start the high-velocity cleaner again and repeat until the submerged section has been inspected.

SEWER CLEANING REPORT

CLEANING NEEDED DATE: ____________

FOREMAN: ____________ ¼ SEC.# ______ M.H.# ________ TO M.H.# ____________

SIZE OF LINE: ____________ TYPE OF MATERIAL IN LINE: ____________ CLEANING NEEDED WITHIN ________ DAYS

HYDRANT LOCATION: ____________

COMMENTS:

SEWER CLEANING COMPLETED

FOREMAN: ____________ DATE: ____________

¼ SEC.# ____________ M.H.# ____________ TO M.H.# ________ AMOUNT OF MATERIAL: ________

FEET JETTED: ____________ FEET RODDED: ____________ FEET BUCKETED: ________

COMMENTS:

MANHOLE ADJUSTMENT REPORT

FOREMAN: ____________ ¼ SEC.# ______ M.H.# ____________ DATE: ________

MH TO BE RAISED: ______ INCHES MH TO BE LOWERED: ______ INCHES

REFORM CHANNEL: ____________ STEPS TO BE REPLACED: ______ TYPE PAVEMENT: ______

REMARKS:

MANHOLE ADJUSTMENT COMPLETED

FOREMAN: ____________ DATE: ____________

MH RAISED: ____________ INCHES MH LOWERED: ________ INCHES

CHANNELS REFORMED: ________ STEPS REPLACED: ________ PAVEMENT REPLACED: ________

Fig. 5.29 Sewer inspection follow-up work request forms

DROP JOINT:
Moving television camera from right to left past this joint will cause it to 'drop.'

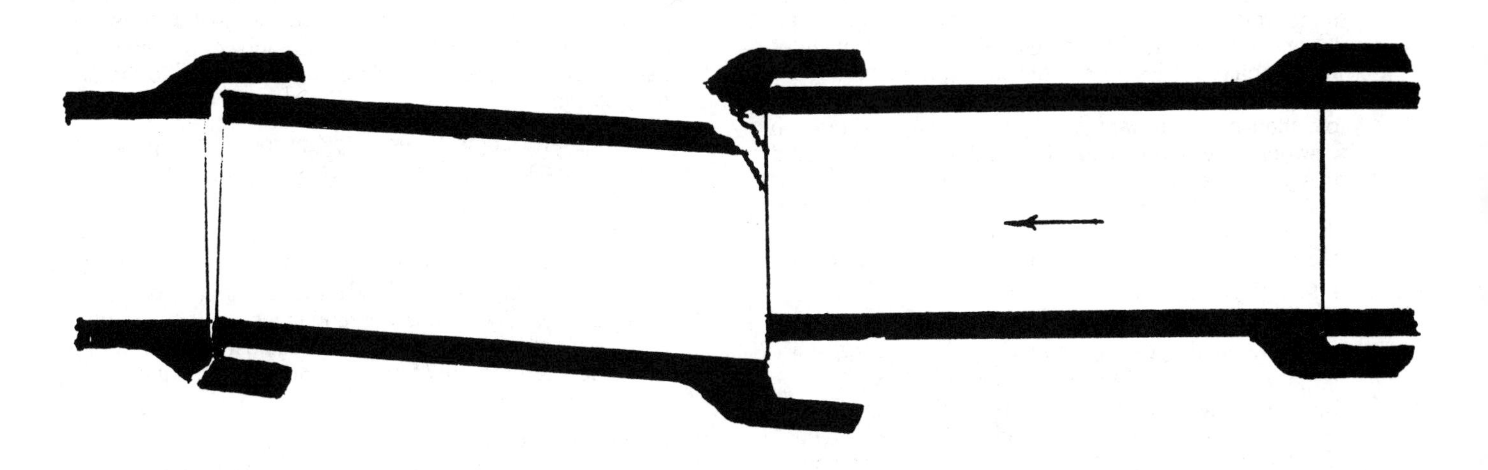

JUMP JOINT:
Television camera must be moved from right to left to pass this type of situation.

Moving the camera past situations like these can prevent the camera from being moved back out if a full obstruction is encountered later down the pipe. It is best to document conditions like these, repair them, and then retelevise the line.

Fig. 5.30 Drop joint and jump joint

3. When coming out of a situation where the camera has been under water, it is possible for the viewport to become fouled with grease or other surface debris. In most cases the decrease in visibility will be slight, but it is possible to have a foggy or indistinct picture, which can reduce the visual inspection effectiveness. There are several things that can be attempted to clear the viewport.

   a. If the distance to the downstream manhole is more than 50 or 100 feet, back the camera up until it is fully under water again. Allow the camera to stand for perhaps a minute to allow any debris on the viewport to be floated off, then pull it forward rapidly to the point where the viewport is once again out of the water. The idea behind this is to create a wave ahead of the viewport that washes surface debris aside with water brought up with the camera. This may be attempted safely several times. Do not pull the camera forward farther than its former progress.

   b. If the distance to the downstream manhole is still over 50 or 100 feet, open a fire hydrant or similar source of a large volume of water and discharge into the upstream manhole (any upstream manhole) in an attempt to wash the surface debris through the line. Pull the camera back into the water to rinse the viewport and proceed once it is clear.

   c. If the downstream manhole is not too far, move the camera down to it in the same manner as if it were under water. This means the operator moves it slowly, advising of any sudden stoppage of progress. Once at the downstream manhole, clean the viewport and move it back to the place where it came out of the water and proceed to televise the rest of the pipe.

4. Other obstructions are handled in much the same manner. Such problems as protruding service connections (protruding taps), objects sticking into a pipe, pieces of broken pipe protruding into the pipe, and heavy root developments are typical examples. Once again, the operator must determine if these can be passed without damage to the equipment or if they must be removed before televising the remainder of the line. A rule of common sense is applicable at this point.

RULE: IF A SITUATION IN A PIPE PRESENTS A RISK TO THE CAMERA AND THE TELEVISING EQUIPMENT, IT IS OBVIOUSLY A SITUATION THAT REQUIRES EXCAVATION AND REPAIR. UP TO THAT POINT, THE CAMERA HAS DONE ITS JOB AND IT CAN BE MOVED BACK OUT AND USED IN OTHER LOCATIONS WHILE THE NECESSARY REPAIRS ARE BEING MADE TO THE PIPELINE. TO RISK DAMAGE AND THE NEED TO REPAIR THE CAMERA CAN PRODUCE UNNECESSARY COSTS. THIS RISK ALSO CAN PREVENT THE USE OF THE TELEVISION EQUIPMENT WHILE CAMERA REPAIRS ARE BEING MADE.

5. Camera tumbled or off-vertical conditions should not be a problem. A properly designed camera and skid assembly should right itself in a straight pipe run. Even if a camera is inverted and the picture is upside down, moving the camera back and forth should cause it to return to the upright position.

   a. Bends and some types of offset joints in a pipe can cause a camera to get tilted. This is usually a temporary condition.

   b. If a camera does not right itself in a few feet, it is an indication that something is wrong. A cable could be tangled in the skid assembly, or a root clump has fouled the assembly. Under such circumstances, the operator should select the most rapid and safest method of getting the camera out of the line and determine the cause of the problem.

6. A foggy or indistinct picture, after getting a sharp picture in a pipe, indicates that the viewport and possibly the optical lens of the camera have become fogged. This can take place shortly after a camera has started down the pipe, or it can occur after a camera has pulled past a service with hot or very warm water dumping out of it.

   a. A cold piece of glass placed in a warm and humid atmosphere will fog up. Use of some types of anti-fog compounds may be effective, but the best solution is to warm the camera before placing it in the pipe. A glass surface that is warmer than the atmosphere it is placed in will not become foggy.

   b. When fogging occurs shortly after the camera enters the pipe, it is best to take the camera back out and put it where it can be warmed up to as much as 100°F (38°C). Avoid placing the camera in front of a heater where one side can be warmed more rapidly than the other. Possible ways of warming a camera are to put it in a warm room, or behind or under a cardboard cover in the sunlight, or in a vehicle parked in the sun with the windows closed.

   c. When fogging takes place in a pipe after passing a service with hot water flowing out, simply stop the camera and wait for a few minutes. The warmth will be picked up by the glass parts and de-fogging will take place. Also the hot water may be cooled by the discharge of water from a fire hydrant.

   d. Steam resulting from water striking the hot light assembly can be ignored. It will not last long and provides no serious problem to the picture.

   e. Use a liquid dish detergent on the lens to prevent fogging and also to cause grease to slip off the lens.

7. Do not allow grease to accumulate on the light head. If the heat from the light ignites the grease, the light head could be damaged beyond repair. Also the fumes from the burning grease that come out the manhole have a horrible odor.

8. Be aware of the possibility that the TV camera could encounter explosive gas mixtures in a sewer. The heat from the light head may be sufficient to ignite certain explosive gas mixtures. Therefore, it is very important that a municipality or contractor effectively ventilate the manholes and

sewer and use gas detection equipment prior to allowing manhole access to employees or equipment (see Chapter 4, Section 4.41, "Atmospheric Hazards"). *DO NOT OPERATE A TV CAMERA IN ANY LINE THAT TESTS POSITIVE FOR EXPLOSIVE GASES.*

9. If a sewer is flowing so full that the TV camera will be submerged, plug the sewer at the entrance to the upper manhole and televise the line as soon as the water drops. *BE CAREFUL THAT THE WASTEWATER HELD BEHIND THE PLUG DOES NOT BACK UP INTO SOMEONE'S HOME OR FLOW OUT THE TOP OF A MANHOLE.*

   To avoid plugging upstream manholes and the possibility of flooding someone's home, a pump may be used to pump the wastewater. Place a pump with a long discharge hose at the nearest upstream manhole with the discharge hose going to the downstream manhole. Run the pump to lower the depth of water backing up and reduce the chances of backing up wastewater into a home. This procedure also can be used to perform pipe maintenance, replace pipe, or make pipe joint repairs.

QUESTIONS

Write your answers in a notebook and then compare your answers with those on page 255.

5.32L List the types of conditions that the TV camera may encounter which require special precautions.

5.32M Why should you *NEVER* attempt to force a TV camera past an obstruction?

5.32N If a TV camera encounters a bad joint that it cannot pass, how can the remainder of the line be televised?

5.32O When coming out of a situation where the camera has been under water and the viewport is fouled with grease or other surface debris, what would you do?

5.32P What would you do if you were operating the TV monitor and observed a foggy or indistinct picture shortly after the TV camera entered the pipe?

5.324 Shutting Down After Completion of the Pipe Run

When the TV operator is able to see the end of the pipe or the hardware in the downstream manhole, the operator continues with caution. When the downstream manhole is reached, the camera is stopped. At this point, the inspection is completed. Add two feet (the remaining distance to the center of the downstream manhole) to the footage reading and record this information on the TV log as the total distance of the run. Shutdown procedures can now start.

The first act of the shutdown procedure is to turn off all power to the camera and light. *POWER MUST BE OFF BEFORE ANYONE ATTEMPTS TO HANDLE THE CAMERA OR PULL IT OUT OF THE PIPE.*

The standard light used on a sewer inspection television camera is a quartz-iodine type. These lamps, similar to a projection lamp, achieve extremely high temperatures (in some cases as high as 1,600°F or 870°C). The lamp and its reflector must be permitted to cool to a safe level before anyone handles the camera.

CAUTION: MOST CAMERAS AND MONITORS CONTAIN COMPONENTS THAT CAN BE DAMAGED BY FREQUENT OR SUSTAINED LOW POWER CONDITIONS. AVOID SIMPLY SHUTTING DOWN THE POWER GENERATOR AND ALLOWING THE ENGINE TO COAST TO A STOP WHILE POWER SWITCHES TO THE EQUIPMENT ARE STILL IN THE "ON" POSITION. SHUT OFF, LOCK OUT, AND TAG SWITCHES BEFORE SHUTTING DOWN THE GENERATOR.

5.325 Final Check of Television Survey and Footage Test

Before removing TV equipment from the manhole and turning in the television log sheet (Figures 5.24 and 5.27), every entry should be checked for readability, accuracy, and cleanliness. Poorly written entries or errors in the coded entries can cause costly losses. Dirty logs can make some of the notes and numbers unreadable. (Figures 5.18, 5.19, 5.20, and 5.21 show a computerized inspection report.)

1. Some types of footage meters can develop large errors during a television operation. In most instances this is due to slippage in the measuring device or frequent backing and forwarding of the camera. To determine if such an error has taken place, compare the footage meter distance with the distance between manholes shown on the wastewater collection system map. If a difference of more than one percent exists,[6] use a tape and measure the distance between the manholes. Also be aware of the fact that maps may contain distance errors. When repairs are necessary, distance errors can be very critical. If the TV operator determines that the errors are too erratic, the operator should see that repairs are made to the footage meter and the line should be retelevised. Corrections on line distances, tap locations, and other items should be routed to the mapping section so maps can be corrected.

   a. If a difference of more than one percent in the total distance measured and the footage recorded on the log is evident, add an error factor to the log. This error factor will be the amount of the difference. Give the error factor a sign that shows how the television log should be corrected.

[6] *See Appendix, "Applications of Arithmetic to Collection Systems," Section A.12, Solutions to Problems, D. Slope or Grade, for an example of how to calculate the percent error in the footage meter.*

b. Establishment of the error factor is important when it becomes necessary to excavate some point for repairs to the sewer. The correct distance permits a person to determine that some defect observed by the camera has to be a certain distance from the upstream manhole. If the error factor were not available, a person might dig at the logged footage, find that there was an error and then not know in which direction or how far exploratory excavation should be made.

c. A good practice is to repair the footage meter and re-televise the line to be sure all distances are correct.

5.326 Removal of TV Equipment From Manholes and Preparation for Next TV Run

After the inspection report has been checked and found satisfactory, the TV equipment is removed from the manholes and made ready to be moved to the next section to be televised.

1. The operator winches the camera forward until the tow bridle is observed coming into the manhole. The tasks listed below are performed in sequence.

   a. Remove the fair lead hardware from the manhole.

   b. Move the pull winch stand from over the manhole to permit access into the manhole.

   c. If possible, remove the camera without entering the manhole by using the pull cable.

   d. Observe manhole safety precautions (see Sections 4.3, "Routing Traffic Around the Job Site," and 4.4, "Classification and Description of Manhole Hazards") if the manhole must be entered.

   e. The tag line winch cable is next disconnected from the tag line bridle on some types of equipment. The operator tosses the hardware from this winch cable back into the pipe where it will not have a chance to tangle with the power cable strain relief and hookup hardware. When this task is complete, the order is given "RETRIEVE THE TAG LINE." The tag line winch is then used to rewind the winch cable.

   f. The next step is to disconnect the power cables. Install the special plug or cap to prevent debris and moisture from getting to the electrical contacts at the end of the power cables. These plugs or caps are installed by the pull winch operators before the next step is taken.

   g. When the tag line winch cable has been pulled in far enough to prevent the ends of it from getting tangled with the power cable, the cable strain relief of the power cable is unhooked from the tag line bridle. The order can then be given to "RETRIEVE THE POWER CABLE."

2. While rewinding the tag line and the power cable, the equipment is cleaned. This consists of wiping the cables free of any bits of paper or other debris as they are returned to the winch drum or cable reel (if used).

   a. Wiping is best done with a piece of foam rubber or a felt pad. Use of a cloth with threads can cause pieces of the cloth to be torn free and to wind around the cables.

   b. While wiping the returning cables, wind them evenly onto the drum or reel and do not let the cable ball up in one spot. A winch or cable reel permitted to ball up may become tangled and present problems when used on the next job.

WARNING

NEVER ALLOW A WINCH CABLE TO BE PULLED RAPIDLY BETWEEN THE BARE FINGERS OR BARE HAND. EVEN A LEATHER GLOVE IS NOT SAFE FOR THIS ACTIVITY. SMALL SINGLE STRANDS OF THE CABLE CAN BREAK AND STICK OUT WITH THE ABILITY TO RIP BARE SKIN. A SCRATCH FROM A CABLE THAT HAS BEEN IN A SEWER ALMOST **PROMISES SERIOUS INFECTION.**

3. If the camera and skid assembly are fouled with debris, they should be cleaned before storing or starting another run. Large clumps of debris are usually removed with a stick used as a scraper. A water jet from a garden hose is used to finish the job. Be careful if the water is cold because the light bulbs can be blown out. Wipe the viewport dry as a final job.

   a. Where petroleum-base residues are encountered and cause a camera and skid assembly to become coated, paint thinner or an approved solvent may be used to clean a camera. Do not use gasoline. Extreme fire hazard can result with the use of gasoline and gasoline can be harmful to seals on the camera. The camera should be thoroughly rinsed with water after cleaning. Keep cleaning chemicals away from the seals. Apply cleaning chemicals with a cloth and wipe away the chemicals immediately after cleaning.

      Whenever petroleum-base residues are observed in a sewer, their location should be recorded and forwarded to your supervisor. The supervisor can determine the source of the discharge and decide if it is illegal. Also be aware that certain petroleum products can be ignited by the heat from the head lamp.

   b. Do not use carbon tetrachloride or chlorthane chemicals for cleaning a camera or other parts. Many of the polyesters used in seals and coatings of a camera and a cable are destroyed by these chemicals. Carbon tetrachloride is also hazardous to your health.

NOTE: Crawler tractor units must be backed to the upstream manhole for removal. Larger units must be lifted out of the manhole using a mechanical crane/hoist on the truck.

QUESTIONS

Write your answers in a notebook and then compare your answers with those on page 255.

5.32Q What must be done before turning in the television log sheet?

5.32R Why should the distance measured by the footage meter be verified?

5.32S What is the first step of the shutdown procedure when the TV run has been completed?

5.32T Before the power cables are retrieved back up the sewer line, how are the electrical contacts at the end of the power cables protected from debris and moisture?

5.32U How would you clean the tag line and power cable while they are being rewound?

5.33 Purchase of Closed-Circuit TV Equipment

When your agency decides to purchase closed-circuit TV equipment, the selection of a vendor, the type of equipment, accessories needed, and total overall operational cost are serious considerations. Perhaps the best place to start is to talk to other collection system operators and consider their experiences and recommendations to help make your decisions. Keep in mind that your particular needs are of primary importance. A piece of equipment must meet your needs and not someone else's needs. Equipment manufacturers' shows at national meetings such as the annual Water Environment Federation conference provide an excellent opportunity to inspect the equipment of many manufacturers and also to discuss your needs with the manufacturers' representatives. If possible, arrange for a demonstration of the equipment you are considering purchasing so you and your crews can become familiar with the actual operation and performance of the equipment. See the Appendix at the end of this chapter (Section 5.9) for details on desirable features of equipment.

QUESTION

Write your answer in a notebook and then compare your answer with the one on page 255.

5.33A How would you select the best closed-circuit TV equipment for your agency?

5.34 Photographing the Television Monitor Picture

Permanent still photographic records of conditions in a wastewater collection system are frequently made by photographing the video picture in the television monitor. Any of several types of cameras may be used to take photographs; however, it is important to understand some details of photographic work in order to get consistently good results.

1. Only high-speed film should be used. While slower film can take photographs, the capacity to stop the action of water motion in a pipeline is lost and can result in blurred parts of the photo.
2. The television monitor is a weak source of light, especially where large parts of the video picture are dark. Because of lighting problems, the camera must be held as close to the monitor as possible. This will require the use of a close-up lens on most cameras and the use of the smallest f-stop possible (resulting in the largest possible iris opening with the fastest possible shutter speed).
3. Time exposures of a television monitor are not recommended because the results are difficult to understand.
4. The most popular camera used for taking still photographs of the inside of a pipe shown on a television monitor is the Polaroid. In addition to having a convenient and uncomplicated close-up kit available, it uses the fastest film available on the market in black and white (3,000 ASA).

   Polaroid also is able to provide the resulting photograph in a matter of 15 seconds and ensures that the scene desired is obtained.

   a. Because Polaroid cameras have an automatic control for shutter speed and iris opening that is sensitive to the light entering the window of an electric eye on the camera front, only the light from the TV monitor should be permitted to fall on the camera front. This often requires the use of a hood to eliminate all other light between the camera and the monitor.
   b. Where a hood is not used, all background light in the monitor viewing room should be eliminated when using a Polaroid. The glass front of the monitor tube is highly reflective and any background light will be reflected back into the photographic camera, resulting in both faulty camera shutter adjustment and poor picture quality.
   c. Black and white Polaroid film is so fast that it is possible for the camera to stop the sequencing action of the television system. The average sewer inspection television system erases and builds a new picture 60 times per second. Between each picture the monitor may be dark or practically dark when the exceptionally high-speed shutter of the Polaroid camera records a picture. When this happens, the operator can only try again with the camera. A dark or blurred picture does not necessarily indicate any defect of the equipment or procedure.
   d. Where videotape readout is not provided, all still photographs should be marked with the date and information regarding the line and footage location of the scene in the picture. In cases where a written television log is being kept, the photograph should be logged and the log entry number marked on the photograph. Most of these entries are made on the backs of the photographs after they are fully dry and can be handled without damage. Photographs can be inserted into individual clear plastic protectors and clipped or stapled to the written log sheet after each television run.

5.35 Use of Newer TV Equipment (Also see Appendix, Section 5.10, "Additional TV Inspection Equipment Information," page 260.)

In Sections 5.31, 5.32, 5.33, and 5.34 we discussed equipment, staffing, procedures, purchase, and photographing with closed-circuit TV equipment. The industry is continually striving to produce better and more cost-effective equipment. Another consideration is to make the jobs of collection system operators easier and safer.

The previous sections emphasized one type of TV equipment to indicate to you the knowledge and skills necessary to televise a sewer. This section will inform you of some of the newer equipment available to televise sewers. Your equipment manufacturer will provide you with instructions on how to operate and maintain your TV equipment.

Some TV cameras can be inserted into and retrieved from sewers without an operator ever having to enter a manhole. Once the TV camera is properly positioned in the sewer to be televised, the TV operator focuses the camera from the station in front of the picture monitor (TV screen). The iris is automatically focused.

Modern closed-circuit TV systems produce a TV picture in color instead of black and white. Color reduces eye strain on persons who must observe the TV picture or review videotapes all day. Infiltration problems are easier to evaluate in color due to a better perception of the rate of infiltration.

After a TV camera is properly positioned in the sewer to be televised, the TV operator is the only person needed to televise a sewer. The TV operator regulates both the power winch (pull winch) and power reel (tag line or return winch) while viewing the TV picture monitor. If the camera approaches a tight spot or starts to get stuck, the power winch in front of the TV operator will start to slow down. An alert operator can stop the camera before it gets stuck. If the camera does get stuck or if the power winch cable breaks, the camera can be pulled back by the power reel. A much stronger cable is used on the power reel in order to retrieve the camera if the camera gets stuck.

5.36 Use of 3D Photography

Another method of inspecting sewers is by the use of 3D photography. To use this technique, a camera is placed on skids and pulled through the sewer. As the camera travels through the sewer, a photograph is taken every three or four feet. A special viewer is used to produce a three-dimensional photo of the sewer. This procedure is much cheaper than using closed-circuit television. A disadvantage is that if you need more photos or details, you don't find out until a couple of days later after the film has been developed and studied.

QUESTIONS

Write your answers in a notebook and then compare your answers with those on page 255.

5.34A How can photographs be made of a problem inside a pipe?

5.34B What kind of film should be used to take photos of the video picture in the TV monitor?

5.34C What kind of camera is most commonly used to take photos of the video picture in the TV monitor?

5.34D What information should be attached to a photograph taken of the inside of a pipe?

Please answer the discussion and review questions next.

DISCUSSION AND REVIEW QUESTIONS

Chapter 5. INSPECTING AND TESTING COLLECTION SYSTEMS

(Lesson 3 of 5 Lessons)

Write the answers to these questions in your notebook before continuing. The question numbering continues from Lesson 2.

12. What information can be obtained from a closed-circuit television inspection of a sewer?
13. What must a collection system crew do *BEFORE* starting to set up the TV inspection equipment?
14. How is the TV inspection camera moved through a sewer?
15. How does a TV inspection crew communicate with each other?
16. What conditions require that special precautions must be taken when inspecting sewers with a TV camera?
17. How is a TV camera removed from the sewer after an inspection run is completed?
18. Why must the distance recorded by a footage meter be compared with the actual measured distance?

CHAPTER 5. INSPECTING AND TESTING COLLECTION SYSTEMS

(Lesson 4 of 5 Lessons)

5.4 SMOKE TESTING

5.40 Purpose

Smoke testing of wastewater collection systems (Figure 5.31) is used to determine:

1. The sources of entry to the collection system of surface waters, also called surface inflow. This includes rain or storm waters, street or surface wash waters, and in some cases, irrigation waters.
2. Positive proof that buildings or residences are connected to a wastewater collection system.
3. Location of certain types of illegal connections to a wastewater collection system. Included with these connections are such things as roof leaders or downspouts, yard drains, industrial drains, and some types of cooling tower water drains.
4. Location of broken sewers due to settling of foundations, manholes, and other structures.
5. Location of "lost" manholes and diversion points.

Use of smoke for testing sewers is best done when the groundwater is low so any cracks will leak smoke.

5.41 Equipment

All of the following pieces of equipment are used or are applicable to a smoke testing operation:

1. Smoke blower unit (Figure 5.32). This is usually a squirrel cage blower with a gasoline engine and belt drive. The average blower capacity will be between 1,700 and 3,000 cubic feet per minute (CFM). The blower will have a rubber gasket underneath the base that permits it to be set over and force a blast of air into an open manhole.
2. Pipe plugs (Figure 5.33) to isolate two sections of pipe between three adjacent manholes. These plugs may be mechanical or inflatable.

   Where there are high wastewater flows that prevent safe plugging of a line to be smoke tested, heavy canvas or curtains with weights or sandbags on the bottom may be used to confine the smoke to the section being tested (Section 5.43).

3. Smoke bombs (Figure 5.34) of the three-minute or five-minute duration type.

> WARNING: USE ONLY MAILABLE SMOKE BOMBS SPECIFIED TO BE SAFE AND TO HAVE NO TOXIC OR RESIDUAL EFFECTS. NEVER USE MILITARY TYPES OF COLORED OR SIGNAL SMOKE BOMBS FOR THIS APPLICATION DUE TO EXPLOSIVE DANGER AS WELL AS DANGER FROM CONTACT TO OPERATORS AND PROPERTY.

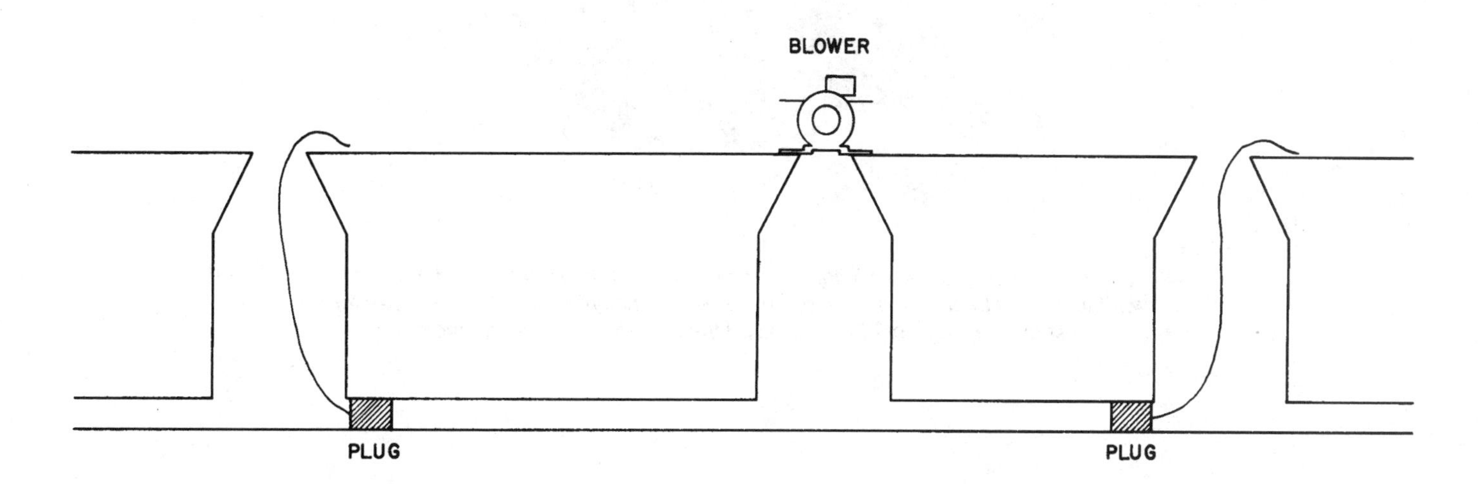

Fig. 5.31 Smoke testing

Fig. 5.32 Manhole smoke blower. Fiberglass base fits over opening of manhole. Small port in base is used for suspending a smoke bomb on a wire or chain.

(Courtesy of Superior Signal Company, Inc.)

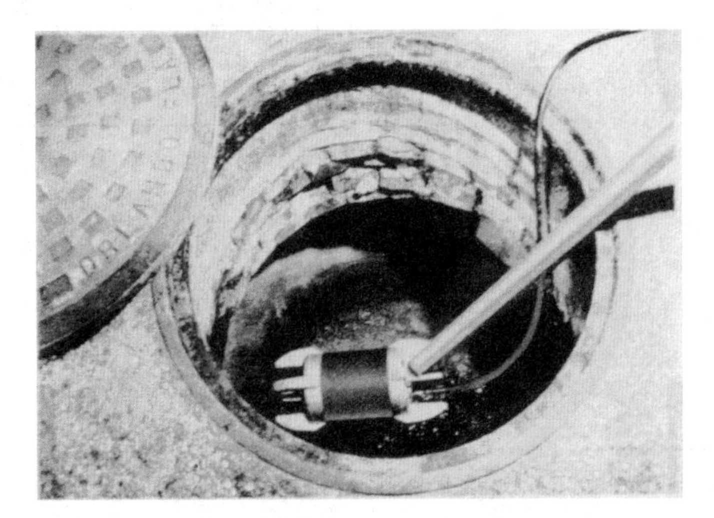

Fig. 5.33 Pipe plug. An inflatable pipe plug that can be installed and removed without having to enter a manhole. This provides maximum safety and speed for smoke testing work by eliminating the need to enter the manhole.

Blower

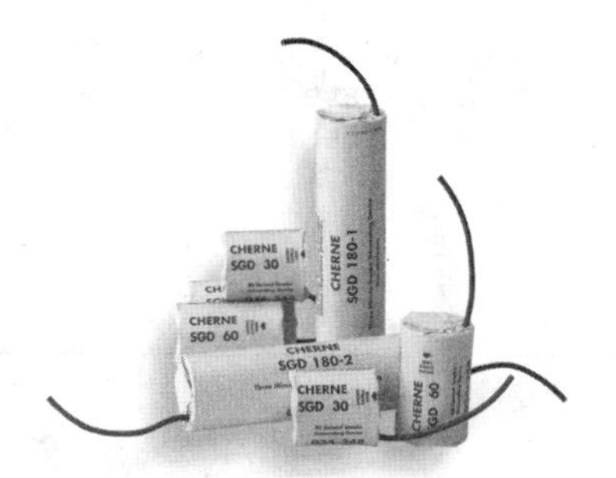

Smoke Bombs

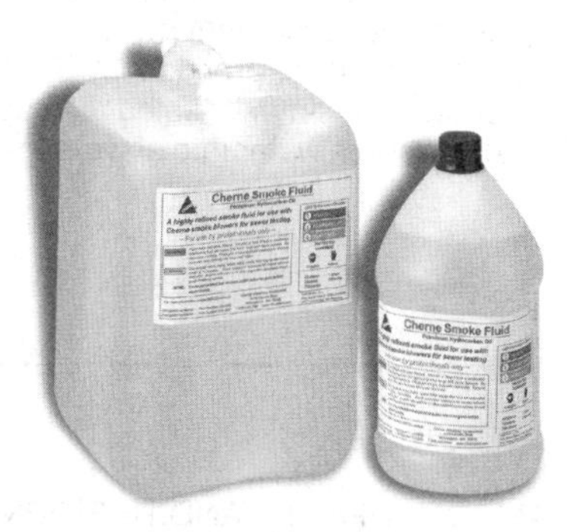

Smoke Fluid

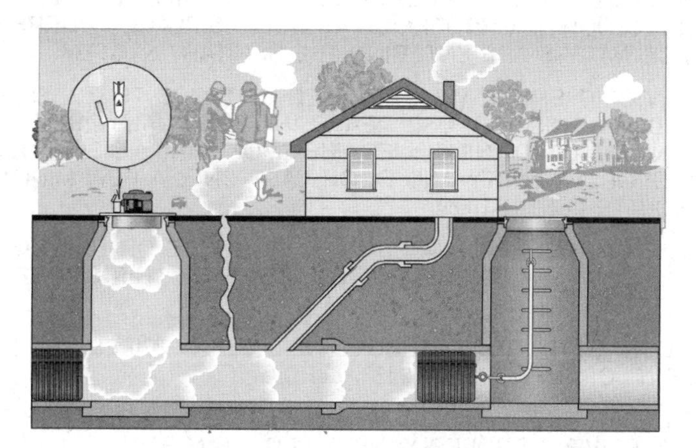

Traditional Smoke Bomb Operations

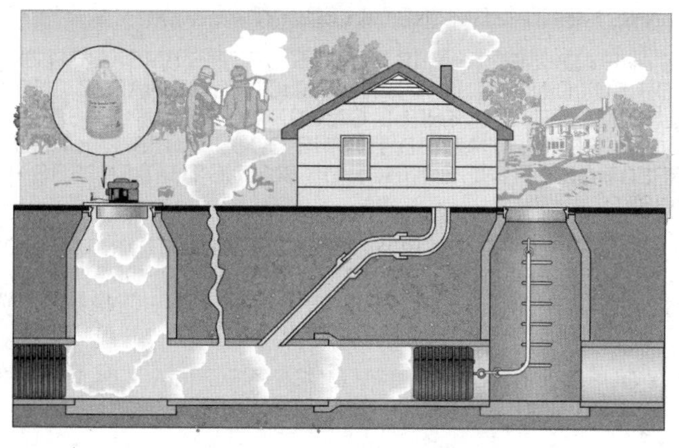

Liquid Smoke Fluid Operations

Fig. 5.34 Smoke testing operations

(Permission of Cherne Industries)

5.42 Preparations

Smoke testing of wastewater collection systems has the potential to affect the occupants of buildings connected to the line being tested. Such factors as defects in the sewer system of the building, dry traps, defective wax ring under commodes, vents terminated in an attic, or cleanout plugs missing from sewers can cause smoke to enter the building. Adequate preparation plus good advance public relations are necessary to avoid panic or severe alarm if workers or residents suddenly see smoke in a building or their home.

1. Warn the public in advance when and where smoke testing is planned. A typical public announcement is shown in Figure 5.35. This type of announcement should be published in a local paper and should be prominently displayed and distributed in the area to be smoke tested. Occupants should be requested to run some water into sewers that are not used regularly before the test starts to fill traps that may have dried out.
2. *THE LOCAL FIRE AND POLICE DEPARTMENTS MUST BE THOROUGHLY BRIEFED ON THE SMOKE TESTING OPERATION AND ITS SAFETY.* In addition, the fire department should be advised of the area to be smoke tested at the start of each day of testing. If possible, it is a good idea to have a vehicle from the fire department with one or two members of that department accompanying the smoke testing operation.
3. All operators who will participate in the smoke testing should be fully trained and briefed daily in the proper method of handling persons who discover smoke in their homes or smoke coming up all over their yard and around their house. Anticipate that such a circumstance will occur and that it can have a potential for generating considerable concern among certain types of individuals.

PUBLIC NOTICE

TO ALL OCCUPANTS AND RESIDENTS:

THE CITY WILL BE CONDUCTING LEAK TESTS IN THE SANITARY SEWER SYSTEM BY BLOWING SMOKE INTO THESE PIPES. THIS SMOKE WILL REVEAL SOURCES OF SEWER ODORS IN YOUR NEIGHBORHOOD AS WELL AS PLACES WHERE STORM AND OTHER SURFACE WATERS ARE ENTERING THE SANITARY SYSTEM.

A SPECIAL NONTOXIC SMOKE WILL BE USED IN THESE TESTS. THIS SMOKE IS MANUFACTURED FOR THIS PURPOSE, LEAVES NO RESIDUALS OR STAINS AND HAS NO EFFECTS ON PLANT AND ANIMAL LIFE. THE SMOKE HAS A DISTINCTIVE, BUT NOT UNPLEASANT, ODOR. VISIBILITY AND ODOR LAST ONLY A FEW MINUTES, WHERE THERE IS ADEQUATE VENTILATION.

BECAUSE THE PLUMBING APPLIANCES IN YOUR HOUSE OR BUILDING ARE CONNECTED TO THE SANITARY SEWER SYSTEM, SOME OF THIS SMOKE MAY ENTER YOUR HOUSE IF THE:

— VENTS CONNECTED TO YOUR BUILDING'S SEWER PIPES ARE INADEQUATE, DEFECTIVE, OR IMPROPERLY INSTALLED.

— TRAPS UNDER SINKS, TUBS, BASINS, SHOWERS AND OTHER DRAINS ARE DRY, DEFECTIVE, IMPROPERLY INSTALLED, OR MISSING.

— PIPES, CONNECTIONS AND SEALS OF THE WASTEWATER DRAIN SYSTEM IN AND UNDER YOUR BUILDING ARE DAMAGED, DEFECTIVE, HAVE PLUGS MISSING, OR ARE IMPROPERLY INSTALLED.

ALL RESIDENTS ARE ADVISED THAT IF TRACES OF THIS SMOKE OR ITS ODOR ENTER YOUR HOUSE OR BUILDING, IT IS AN INDICATION THAT GASES AND ODORS FROM THE SEWER ALSO MAY ENTER. THESE CAN BE BOTH UNPLEASANT AND DANGEROUS AS WELL AS A HEALTH HAZARD TO THE OCCUPANTS. LOCATION, IDENTIFICATION AND CORRECTION OF THE SOURCE OF SMOKE THAT ENTERS YOUR HOUSE IS URGENTLY ADVISED.

WHILE THE CITY WILL RENDER ALL POSSIBLE COOPERATION, THE CORRECTION OF ANY DEFECTS IN THE PIPES AND SEWER ON PRIVATE PROPERTY IS THE RESPONSIBILITY OF THE OWNER. THE SERVICES OF A PROFESSIONAL PLUMBER ARE ADVISED.

IF YOU HAVE ANY QUESTIONS OR DESIRE MORE INFORMATION, PLEASE TELEPHONE THE CITY OFFICES DURING REGULAR OFFICE HOURS.

CITY MANAGER

Fig. 5.35 Typical public announcement of planned smoke testing

a. Operators on the team should be drilled in the concept that any smoke entering a building or residence reveals that gases from the sewer can and do enter the building. Prompt correction of the problems by the owner is necessary for the health and safety of the building's occupants.

b. Where smoke is coming from a drain in a plumbing fixture, first inspect the trap water seal to determine if the trap is dried out or is defective. Where a trap is dried out and unlikely to be used, suggest that the traps be filled with mineral oil. Do not put mineral oil into a trap that will be used occasionally.

c. A missing wax ring under a toilet is easy to identify. Tell a householder how to solve the problem or suggest hiring a plumber to replace the wax ring.

4. Under some conditions, an observer and recorder may need to enter private property; therefore, adequate consideration of, and good public relations with, owners and occupants must be exercised.

   In a city that had a rash of Peeping Tom reports, women were employed as observers and reporters during smoke testing activities. While an irate and disturbed citizen might accuse a male city employee of compromising the privacy of her bedroom window, it is unlikely that she would make the same charge about a female. Advance publicity, identifiable clothing for operators, signs on equipment, and good public relations all help to avoid upsetting the public.

5. A city or agency attorney should be advised of the smoke testing methods and some of the possible adverse results. Although there is no basis for damage or tort (civil) claims from a building owner for smoke disturbance or alleged damage to interiors, such claims may require defense by an attorney. Discuss with the attorney what kind of information and reports are needed if any incidents or difficulties arise during smoke testing.

6. The crew that does smoke testing should inspect the area to be tested in advance in order to become familiar with the area and prepare for anticipated problems.

5.43 Staffing

The average smoke testing crew consists of five persons: one person in charge (Maintenance Operator II or Operator II) and four observers and recorders (Maintenance Operator I or Operator I).

When there is a high flow in the pipeline to be tested and a chance of surcharging and overflowing an upstream manhole above the plugged off test section, a sixth person is needed to keep watch on the manhole.

When a manhole that must be watched appears to be subject to surcharge and overflow, use of a sandbag on a rope is perhaps the best answer. The sandbag plug may be pulled up long enough to allow the manhole to drain, then replaced as required. For best results, sandbags are tied in the middle so the chance of loss is decreased as much as possible.

Another method of closing off a pipe run for smoke testing where high flows exist is the use of curtains. These are pieces of canvas with weights on the bottom or a similar material with a rigid frame member along the top edge. They are lowered on a rope until they touch the surface of the flow and then they are tied in place. These curtains must be tight around the edges to prevent smoke leaks.

5.44 Operation

At the start of the operation, the smoke blower is located over the manhole (Figure 5.36) and the ends of the pipes plugged at the next adjacent manholes. Smoke testing should not be attempted on windy days. Even a very light breeze can disperse a wisp of smoke before it is visible at a vent or the source of a leak.

1. The blower is started and allowed to force air into the pipe system for 5 to 10 minutes or longer (depending on volume of sewer being tested) before any smoke bomb is lit.

   The purpose of forcing air into the pipe system is to get an air flow established. Since air has weight and inertia, it sometimes takes even longer than 5 or 10 minutes to achieve a flow of air through a pipe system. Once a flow of air is properly established, the air will carry the telltale smoke along with it more effectively than having the blower try to push a wad of smoke into a pipe system.

2. Once the air flow is established, one or more smoke bombs are lit and lowered into the manhole at the blower using metal chains or short lengths of wire. Hooks made of coat hanger wire are often used for this purpose. Be sure to inspect the area and monitor the atmosphere to confirm safe conditions before inserting any ignition source into a manhole.

   Smoke bombs have a cardboard cylinder with the smoke-making chemicals inside and a single paper cover on the outside. The cardboard cylinder has perforations where the smoke escapes at frequent locations and these can be located with fingertip pressure. The perforated holes provide suitable places for a wire hook to be inserted in the smoke bomb.

   To avoid potential problems if it becomes necessary to retrieve a lighted smoke bomb from a manhole, the bomb can be placed in a wire basket and lowered into the manhole. It could be easily retrieved without anyone entering the manhole.

3. As soon as the bomb has been lit and put into the manhole, the observers and recorders move out according to pre-

Fig. 5.36 Blower installed over manhole for smoke test

arranged plans. One observer takes one side of the street and moves away from the manhole being smoked, observing the buildings on that side of the street. Another observer moves in the same direction on the other side of the street. The other two observers move in the opposite direction after inspecting the immediate area outside of the manhole.

a. Where proper building plumbing exists, smoke will come from the roof vents of the building drain system.

b. The observers must look at all other parts of the building (inside and outside) and its grounds (front, back, and side yards) for other indications of smoke from illegal connections.

c. Before moving on to the next house or building, the observer takes a careful look at the foundation of the building, particularly where the house service pipe would usually pass under the foundation footing.

4. The observer will cover one side of the distance between the manhole smoked and the next adjacent manhole that has been plugged.

5. In the event the smoke supply of the first bomb runs out before the area is fully observed, a second or additional bombs may be needed. One five-minute bomb is usually all that is needed (100,000 cubic feet of smoke) for a two-city-block run, or approximately 600 feet of pipe run. Where there are time-consuming conditions in the inspection process, as many as five or six bombs may be necessary.

5.45 Recording

Observers should record every building and the observed results of the smoke test. House numbers are generally sufficient for the recording purpose. Typical entries might include:

1. "265 Brown Street—smoke from vents only." This is a no-problem house as far as the smoke test was concerned.

2. "269 Brown Street—smoke from ground at building foundation, SE corner in front." This reveals what could be a source of inflow to the sewer as a result of the building sewer broken at the foundation. This is not uncommon when such pipes are laid shallow and often butted up under the house foundation. When the foundation settled, the pipe was crushed or disjointed.

3. "273 Brown Street—no smoke from vents." There may be a dip or a trap in the house sewer to the lateral. If none of the rest of the buildings beyond this house have smoke showing at vents, this is an indication that the lateral sewer has a dip in it that has created a trap. Another possibility would be a stoppage in the sewer that could result in wastewater backing up into the house.

4. "277 Brown Street—smoke from rain gutters." The roof leaders apparently are connected to the building drains. This is a source of inflow.

5. Any location where smoke is showing should be fully described and recorded. If smoke shows at the windows inside an unoccupied building, it should be recorded with a note stating that the building appeared to be vacant or may have had occupants who were absent at the moment. When this occurs, the observer should place a copy of the public announcement (Figure 5.35) where someone entering the house will find it. Letting such a person know that any odd odor found in the building could come from the sewer is a point of safety as well as concern. Also notify the health department or police to be sure corrective action is

started immediately and that people are not allowed to live in the house until the problem is corrected. Making people vacate and stay out of a building is the responsibility of the health department and police department, not the job of the collection system operator.

Typical results from smoke testing are shown in Figures 5.37 and 5.38. Whenever possible, photograph illegal situations. Photos are a valuable record.

Results of the smoke test should be recorded on a form such as the one shown in Figure 5.39.

5.46 Other Considerations When Smoke Testing

1. Smoke testing may be used to locate points of inflow or exfiltration, but does not measure the amount of inflow or exfiltration. Keep in mind the fact that the location of smoke on the ground surface does not necessarily reveal where the smoke is escaping underground, but only the point of exit at the ground surface.
2. Basement sump pumps have little chance of detection by smoke even when they have discharge connections to the building sewer. This is due to the usually required use of check valves in any sump pump tie-in to a building sewer and also the fact that the intake is submerged in water, which forms a trap.
3. Where smoke appears at the vent pipe of a house, it is positive evidence that the house is connected to a sewer that is being smoked. Service districts that have monthly billings to property owners connected to a wastewater collection system thus have definite proof of a connection.
4. Where no smoke appears at a house vent, it does not mean that such a house is not connected to the sewer. Tests have shown that there are many houses with drain grade defects in service connection pipes, or many even have a trap installed in such a pipe. Many buildings that are connected to combined sewers have traps in the service line. Some buildings that may have been on a septic tank in the past may still be on such a tank with a connection from the discharge of the tank to the sewer. Under these circumstances, a well-designed tank with baffles as required or an old tank with collapsed pipes and baffles may not permit the smoke to pass through.
5. There are some cities and counties in the United States where modern enforcement of plumbing codes fails to adequately protect a service district. The admission of smoke into a building from sewer smoke testing usually indicates a health or plumbing code violation in the building's utilities. If smoke is found in a building, look for a dry trap from a basement shower, toilet, or drain that has not been used recently *BEFORE* deciding there is a violation.
6. One way to avoid the potential problems associated with putting lighted smoke bombs in manholes is to use the type of blower/ventilator that allows the bomb to be inserted in the suction of the ventilator. Once the bomb is inserted, the ventilator distributes smoke into the collection system. The lighted bomb itself never enters the manhole or collection system. A good procedure is to test the sewer for flammable and explosive gases before smoke testing.

QUESTIONS

Write your answers in a notebook and then compare your answers with those on pages 255 and 256.

5.4A What is the purpose of smoke testing?

5.4B List the equipment needed for a smoke testing operation.

5.4C What precautions should be taken before starting a smoke testing program?

5.4D Why is the blower started and allowed to force air into the pipe system for 5 to 10 minutes before a smoke bomb is lit?

5.4E Where should observers look for smoke?

Please answer the discussion and review questions next.

DISCUSSION AND REVIEW QUESTIONS

Chapter 5. INSPECTING AND TESTING COLLECTION SYSTEMS

(Lesson 4 of 5 Lessons)

Write the answers to these questions in your notebook before continuing. The question numbering continues from Lesson 3.

19. Why are sewers smoke tested?
20. Who should be notified before smoke testing begins?
21. How would you handle a scared and irate citizen whose home was filled with smoke from a smoke testing operation?

Building rain gutter connected to sewer system.

Cleanout for building sewer that is uncovered and unsealed.

Smoke from building sewer vent indicates smoke test is effective.

Fig. 5.37 Smoke test results

Smoke reveals the foundation settled and cracked or broken sewer pipe. Storm water from adjacent roof leader created open channel into the break in the sewer pipe.

Smoke from yard drain illegally connected to wastewater collection system.

Fig. 5.38 Smoke test results

SEWAGE COLLECTION DIVISION
I/E/I/ SUPPORT UNIT
SMOKE TEST REPORT

M.H. ________ M.H. ________

DATE ____________ TIME __________ P.S. ____________ SEWER LENGTH ________ DIA. _____

ATLAS PAGE ______ TEST# __________

WEATHER ________ BY ____________

NAME ____________________________

SMOKE INTENSITY

LIGHT ☐ MEDIUM ☐ HEAVY ☐

RESULTS

☐ 01: POS. CATCH BASIN CONN.

☐ 02: REPLACED C.O. CAP

☐ 03: DEFECTIVE C.O. RISER

☐ 04: DEFECTIVE SANITARY

☐ 05: DEFECTIVE HOUSE PIPING

ADDRESS ____________________________________

____

COMMENTS ____

Fig. 5.39 Smoke test report form

(Permission of Miami-Dade Water and Sewer Department, Miami, FL)

CHAPTER 5. INSPECTING AND TESTING COLLECTION SYSTEMS

(Lesson 5 of 5 Lessons)

5.5 DYE TESTING

5.50 Purpose

Dye testing is being used to establish positively if certain facilities or fixtures are connected to a wastewater collection system. Illegal connections can be identified using dye tests. If wastewater is overflowing or reaching a creek, river, or other body of water, dye testing can be used to identify the overflowing or leaking sewer. Also, dyes are used to reveal interconnections between sanitary and storm sewers. Specially manufactured dyes are available for this type of testing. Examples of typical situations where dye tests would be appropriate include:

1. Buildings that may not show smoke at vents during smoke tests due to dips or traps in the service connection pipes can be dye tested,
2. Where a yard drain or storm drain is suspected of being tied into the building sewer or a lateral sewer, a dye test is useful, and
3. Any suspected situation of inflow or surface drainage may be checked out by using a dye.

Other uses of dye testing include estimating the velocity of flow and testing for infiltration and exfiltration. By pulling a bag of dye up a sewer and stopping at short intervals, an area of exfiltration can be located that could not be seen on a TV inspection, such as an open joint that is in the line and on grade.

5.51 Equipment

Contact the water pollution control agency to determine if there are any regulations regarding the use of dyes.

Two types of safe and harmless but effective dyes are available for dye testing. Powder in cans or containers is measured by a spoon or small dipper. Tablets of the dye are slower to dissolve than the powder form, but are less messy and are sometimes more desirable than the powder for this reason. The dye and a manhole hook are the only pieces of equipment needed. Regardless of the type of dye, dissolve it in the flow. A tablet may sink into a sump or wet well and not circulate with the usual flow.

CAUTION: Some dyes may leave a stain if spilled. These stains can be very difficult to remove.

5.52 Operation

For the purpose of maintaining good public relations, the public should be notified of your plans for dye testing programs. This is especially so when it becomes necessary to perform work within private property.

While one operator applies the dye to the suspected location, another operator maintains a watch at the next manhole downstream from the location.

1. Where a plumbing fixture is used, such as a toilet bowl or wash basin, the water is turned on and the dye powder or tablet is dropped directly into the drain.
2. Where there is no immediate supply of water, such as a roof gutter or storm drain in dry weather, pouring a bucket of water with dye powder is suggested. The amount of water and dye needed depends on the distance to the next manhole and the existing flow.

3. Based on an assumed velocity of flow, an estimate may be made of the expected flow time to the downstream manhole. Allow plenty of time because the dye often takes much longer than expected.
4. Use of powdered dye can be difficult and messy on a windy day. When the wind blows, either pre-mix the dye in water or enclose a quantity of the powdered dye in either facial tissue or toilet paper. Once wind scatters a powdered dye, the dye is impossible to collect and nearby residents of homes and businesses can become very upset when the dye lands on their property, gets wet, and causes stains on their buildings, autos, clothes, and landscaping.
5. When a number of dye tests are to be conducted on the same line or section of a sewer system, the dye testing should start at the facility farthest downstream and progressively work upstream for the other dye tests. Otherwise, if you dye the facilities upstream first, the flow is then contaminated with dye and you then must wait several hours or until the next day to conduct additional tests.
6. When the tests are completed, record whether or not the service is connected to the sewer.

QUESTIONS

Write your answers in a notebook and then compare your answers with those on page 256.

5.5A What is the purpose of dye testing?

5.5B What equipment is needed for a dye test?

5.6 PIPELINE *LAMPING*[7]

5.60 Purpose

Lamping sewers from manhole to manhole has been done for many years. The purpose of lamping is to establish that a section of pipe is straight and open or that it is not. Lamping also permits an inspector to visually examine conditions in a pipe within viewing distance of the manhole, usually only a few feet from the manhole. Larger sewers are easier to lamp than small sewers.

5.61 Equipment

A bright source of light and a mirror are all that are needed for simple lamping. In some cases, two or more mirrors are used to direct sunlight from the surface of the ground into the manhole to another mirror and then into the line.

5.62 Staffing

Four persons are usually needed for pipeline lamping, provided the two workers topside can easily contact another worker if help is needed (one supervisor and three sewer service workers or operators).

5.63 Operation

Be sure the manhole is safe to enter by following the instructions in Sections 4.3, "Routing Traffic Around the Job Site," and 4.4, "Classification and Description of Manhole Hazards."

1. One operator enters a manhole and either looks up the pipeline directly or looks up the sewer using a mirror. For smaller sewer lines, a mirror may be required. The operator in the manhole can use the mirror both to reflect sunlight and at the same time to look into the sewer.
2. A second operator enters the next manhole up or down the line and shines a bright light directly down the pipe toward the first operator. The light or mirror is moved around to be sure the light is directed down the pipe. If the first operator can see the light, the line is considered open and straight. If the light is not seen, the line is considered either "off-line" or "obstructed."
3. When the first operator is able to see the light, an operator topside is notified who signals the crew directing the light. Signals or voice communications must be established between crews when the light is directed down the sewer and when it is seen at the other end of the sewer.
4. Best lamping results are obtained by rotating the light source around the inside of the pipe. This method will produce more information than a stationary light.
5. Another device used for lamping is an inverted (upside-down) periscope with a changeable focal length from zero to 500 feet. A light source is introduced in the same manhole where the periscope is inserted. The results are quite graphic and the optical system can be calibrated to produce suitable readings within acceptable tolerances.

5.64 Comments

Lamping is an economical and fast method of determining if a sewer line is straight and clear of obstructions. However, in small-diameter sewers it has limited use and alternative methods must be used to establish their condition. More popular methods include low-pressure air testing and CCTV inspection. In large sewers where the flow can be diverted or held, the sewer can be walked and inspected using a good light or a hand-held camera.

Lamping should never be used as the sole means of acceptance testing of new sewer lines. The existence of cracks, infiltration, or other pipe problems that do not affect pipe alignment are difficult, if not impossible, to detect by this method.

QUESTIONS

Write your answers in a notebook and then compare your answers with those on page 256.

5.6A What is the purpose of pipeline lamping?

5.6B Should lamping be used to locate sources of infiltration?

5.7 AIR AND WATER TESTING

If you have a need to test sewers for leaks using air or water testing methods, refer to Chapter 7, Section 7.45, "Air Testing," and Section 7.46, "Water Testing."

Please answer the discussion and review questions next.

END OF LESSON 5 OF 5 LESSONS
ON
TESTING AND INSPECTING COLLECTION SYSTEMS

[7] *Lamping.* *Using reflected sunlight or a powerful light beam to inspect a sewer between two adjacent manholes. The light is directed down the pipe from one manhole. If it can be seen from the next manhole, it indicates that the line is open and straight.*

DISCUSSION AND REVIEW QUESTIONS

Chapter 5. INSPECTING AND TESTING COLLECTION SYSTEMS

(Lesson 5 of 5 Lessons)

Write the answers to these questions in your notebook. The question numbering continues from Lesson 4.

22. What is the main reason for conducting a dye test instead of a smoke test?
23. Under what circumstances would you lamp a sewer?

SUGGESTED ANSWERS

Chapter 5. INSPECTING AND TESTING COLLECTION SYSTEMS

ANSWERS TO QUESTIONS IN LESSON 1

Answers to questions on page 188.

5.0A Wastewater collection systems should be inspected and tested to:

1. Identify existing and potential problem areas,
2. Evaluate the seriousness of detected problems,
3. Locate the position of problems, and
4. Provide clear, concise, and meaningful reports to supervisors regarding problems.

5.0B Leaks are a problem because they allow:

1. Infiltration that uses valuable pipe and plant capacity,
2. Exfiltration that could pollute soil, groundwater, and surface water, and
3. Root intrusion into sewers.

Answers to questions on page 189.

5.1A The purpose of inspecting wastewater collection systems is to look for the causes of problems or the start of potential problems. Once a problem has been identified, its exact location must be determined. The size and kind of problem must be identified so the problem can be solved.

5.1B Major sources of wastewater collection system problems include:

1. Design-related deficiencies,
2. Improper installation,
3. Inadequate sewer-use ordinances,
4. Improper inspection and enforcement,
5. Changing patterns of activities and population shifts,
6. Disaster or contingency situations,
7. Problems of recurrent nature,
8. Problems characteristic of the region,
9. Poor coordination between local agencies, and
10. Problems related to an old and frequently neglected wastewater collection system.

Answers to questions on page 191.

5.1C Inflow and infiltration must be controlled to prevent hydraulic overloads of the collection system or the wastewater treatment plant.

5.1D The sources of inflow include (1) deliberate connections, and (2) surface drainage.

5.1E Sources of infiltration can be identified by the use of closed-circuit television, flow metering devices, smoke testing, dye testing, and visual inspection.

5.1F Infiltration problems can be corrected or eliminated by (1) sealing using pressure grouting, (2) digging up and replacing pipe, and (3) inserting a plastic or fiber liner.

5.1G Exfiltration must be controlled because it can cause pollution of groundwaters.

ANSWERS TO QUESTIONS IN LESSON 2

Answers to questions on page 194.

5.2A Manholes should be inspected to be sure the lid is at the proper elevation, the manhole is structurally sound, and the manhole is performing its intended job.

5.2B One person can inspect a manhole, but assistance may be needed to route traffic or to enter the manhole.

5.2C When inspecting a manhole, examine the following items:

1. Elevation of lid,
2. Insects under lid or in manhole,
3. Surfaces and joints inside manhole
   a. Cracks or breaks
   b. Infiltration
   c. Joints
   d. Offsets
   e. Roots
   f. Grease
   g. Gravel or debris
   h. Turbulence
   i. Grout bed of frame
   j. Steps
   k. Shelf
   l. Flow
   m. Corrosion
4. Noises that indicate infiltration or turbulence from cracked or broken pipes,
5. Manhole ring, and
6. Fit of lid.

5.2D Manholes should be inspected once every one to five years with those in areas of heavy traffic being inspected more often.

ANSWERS TO QUESTIONS IN LESSON 3

Answers to questions on page 218.

5.3A Testing and inspecting collection systems must be done thoroughly and on a regular basis to reduce the number of stoppages and failures.

5.30A Closed-circuit television inspection provides the opportunity to actually see the problems in a sewer, locate the problems, and evaluate the effectiveness of corrective action.

5.30B The following kinds of information can be obtained from a television inspection of a collection system:

1. Inspect conditions and determine location of problem areas,
2. Look for damage to sewers,
3. Search for unrecorded connections,
4. Locate inflow and infiltration,
5. Inspect installed and/or repaired taps or pipes,
6. Evaluate effectiveness of solutions and equipment, and
7. Locate buried or lost manholes.

5.31A A collection system televising crew usually consists of three people, plus additional help when working in heavy traffic and congested areas.

Answers to questions on page 223.

5.32A The TV camera should *ALWAYS* be pulled from the upstream manhole to the downstream manhole for the sake of consistency. Also there is a lesser chance of the camera becoming stuck in older lines. Pulling the camera toward the downstream manhole also helps to prevent a buildup of rags and debris in front of the lens.

5.32B When a TV inspection crew arrives at a sewer to be inspected with a TV camera, *SAFETY SHOULD BE THEIR FIRST CONSIDERATION.* Safety cones and flags should be properly placed to protect the crew and the equipment. The crew should wear hard hats and *REFLECTIVE SAFETY VESTS,* the manholes should be inspected for hazardous atmosphere and conditions, and the manholes should be ventilated.

5.32C The TV pull winch is located at the downstream manhole so the winch cable can pull the TV camera from the upstream to the downstream manhole.

5.32D The proper skids are important because the camera lens must be as close as possible to the center of the pipe when the camera moves down the pipe.

5.32E The footage meter should be set to read zero at the center of the manhole.

5.32F An essential rule that must never be violated while the TV camera is moving through the line is:

> The operator must *NEVER* stop looking at the monitor picture. Any obstruction that could wedge or damage the camera or the light must be seen in time to stop the camera's forward movement.

5.32G If a camera and skid assembly become wedged or locked in a pipe and the pull cables break, the sewer must be dug up to recover the camera.

Answers to questions on page 230.

5.32I The TV camera must be stopped while the operator records data because the operator must stop looking at the TV monitor (screen) while writing the information.

5.32J A Polaroid camera is used to photograph infiltration points, bad offsets, damaged laterals, root growths, and broken pipes. The photo provides an excellent record for supervisors to schedule repair work and for repair crews to visualize the problem to be corrected.

5.32K Some of the advantages of using a videotape over a Polaroid camera include:

1. The operator's voice can describe the problem while people watch the picture,
2. The videotape records motion, such as infiltration flows, and
3. If a pipe is off grade, the videotape can show the extent and magnitude of the defective area.

Answers to questions on page 237.

5.32L Conditions the TV camera may encounter that require special precautions include:

1. Offset joints that restrict the pipe diameter can cause a camera to wedge tightly in place,
2. When the TV camera goes under water it becomes impossible to see obstructions,
3. When the TV camera emerges from being under water, the picture can be foggy or indistinct, which can reduce the effectiveness of the visual inspection,
4. Obstructions such as protruding service connections, objects sticking into the pipe, pieces of broken pipe, and roots require special consideration and care so the camera will not become wedged in the pipe,
5. If the camera is out of position, a properly designed camera and skid should right itself,
6. A foggy or indistinct picture, after getting a sharp picture in the pipe, indicates that the viewport and possibly the optical lens have become fogged,
7. An accumulation of grease on the light head,
8. Sewers that flow more than half full require a plug at the upper manhole, and
9. Presence of explosive gas mixtures in sewer line.

5.32M A TV camera must *NEVER* be forced past an obstruction because if it becomes wedged tight, the sewer must be dug up to recover the camera.

5.32N When a TV camera encounters a bad joint that it cannot pass, the joint must be repaired before the remainder of the line can be televised. Another approach would be to pull the camera from the opposite direction toward the bad joint. Be sure your footage measurements are correct and that everyone understands what you did.

5.32O If the viewport is very dirty with grease or other surface debris after being under water, there are three possible procedures for clearing the viewport:

1. Back the camera up until it is under water and allow it to sit for a minute or so to allow debris to be floated off,
2. Open a fire hydrant and discharge water into the upstream manhole in an attempt to wash away the debris, and
3. Move the camera to the downstream manhole, clean the viewport, and move the camera back to where you wish to continue televising the line.

5.32P If a foggy or indistinct picture was observed shortly after the TV camera entered the pipe, try to determine what caused the problem and correct the situation. In this case, the glass in the lens was probably cold. Remove the camera from the pipe and try to warm the entire camera by placing it in a warm room, or behind or under a cardboard in the sunlight, or in a vehicle parked in the sun with the windows closed.

Answers to questions on page 238.

5.32Q Before turning in the television log sheet, every entry should be checked for readability, accuracy, and cleanliness.

5.32R The distance measured by the footage meter must be verified so that all important distances recorded will be correct or can be corrected in order to help repair crews and future inspection crews.

5.32S The first step of the shutdown procedure is to turn off all power to the camera and light. Power must be off before anyone attempts to handle the camera or pull it out of the pipe. Shut off all switches *BEFORE* shutting down the generator to avoid low-power damage to certain components of the camera and monitor.

5.32T The electrical contacts at the end of the power cables are protected from debris and moisture by placing a special cap or plug over the end of the contacts before the cable is retrieved back up the line.

5.32U To clean the tag line and power cable while they are being rewound, use a piece of foam rubber or a felt pad. Don't use a cloth with threads because pieces can be torn loose and wrap around the cables. Protect your hands with heavy gloves. Even a leather glove is not safe because small single strands of cable can break and stick out, scratch your hand, and cause a serious infection.

Answer to question on page 239.

5.33A To select closed-circuit TV equipment, talk to other collection system operators and consider their experiences and recommendations to help make a decision. If possible, arrange for a demonstration of the equipment being considered for purchase.

Answers to questions on page 240.

5.34A A photograph can be made of a problem inside a pipe by photographing the video picture in the television monitor or by using 3D photography.

5.34B Only high-speed black and white film should be used to take photos of the video picture in the TV monitor.

5.34C The most popular camera used to take photos of the video picture is the Polaroid.

5.34D Information attached to a photograph should include:

1. Date,
2. Identification of line,
3. Footage location of the scene in the picture, and
4. Log entry number (depends on recordkeeping system).

ANSWERS TO QUESTIONS IN LESSON 4

Answers to questions on page 247.

5.4A The purpose of smoke testing is to determine:

1. Sources of entry to the collection system of surface waters,
2. Proof that buildings or residences are connected to a wastewater collection system,
3. Location of certain types of illegal connections,
4. Location of broken sewers, and
5. Location of lost manholes and diversion points.

5.4B Equipment needed for a smoke testing operation includes:

1. Smoke blower unit,
2. Pipe plugs (mechanical or inflatable), and
3. Smoke bombs.

5.4C Before starting a smoke testing program, take the following precautions:

1. Warn the public in advance when and where smoke testing is planned,
2. Notify local fire and police departments,
3. Train all operators how to handle persons who discover smoke in their homes,
4. Prepare operators to respect the property and privacy of owners and occupants,
5. Discuss with city or agency attorney testing methods and possible adverse results, and
6. Inspect area to be tested.

5.4D Air is forced into the pipe system by a blower 5 or 10 minutes in advance to achieve a flow of air through the pipe system.

5.4E Observers should look for smoke from:

1. Roof vents,
2. Building foundations, especially where the house sewer passes under the foundation,
3. Yard drains,
4. Rain gutters, and
5. Inside buildings.

ANSWERS TO QUESTIONS IN LESSON 5

Answers to questions on page 251.

5.5A The purpose of dye testing is to determine if certain facilities or fixtures are connected to the wastewater collection system, such as buildings that don't show smoke from vents during smoke tests and yard drains or storm drains. Internal or surface inflow can also be detected by dye testing.

5.5B The only equipment needed for the dye test is the dye, which is available in either powder or tablet form, and a manhole hook.

Answers to questions on page 252.

5.6A The purpose of pipeline lamping is to determine whether or not a section of pipe is straight and open.

5.6B No, lamping should not be used to locate sources of infiltration or other pipe problems that do not affect pipe alignment.

APPENDIX

Chapter 5. INSPECTING AND TESTING COLLECTION SYSTEMS

5.8 CLOSED-CIRCUIT TELEVISION EQUIPMENT DETAILS

To review and evaluate an existing closed-circuit TV inspection program or to consider starting a new program, the following steps can help you decide whether to purchase or rent equipment and select a manufacturer or contractor.

1. Make a list of the uses of the TV inspection program and information you wish to gain from the program.
2. List the important components of the TV inspection equipment and compare the equipment of potential manufacturers and contractors.
3. Compare economics, staffing requirements, experience of others, maintenance requirements, and actual demonstrations and then select the best type of TV inspection program for your agency.

5.80 Uses of Closed-Circuit TV

5.800 Inspection

1. Examine work completed by contractors.
2. Examine taps installed by a plumber or by maintenance crews.
3. Find and pinpoint inflow/infiltration and estimate amounts.
4. Locate unrecorded and/or illegal taps such as industrial, storm water, or surface drainage.

5.801 Maintenance

1. Examine sewer problem areas such as pipe or joint rupture, condition of service connections, leaks, obstructions, corrosion, and pipe alignment.
2. Look at a sewer line to determine what type of cleaning equipment would best solve a problem such as roots, grease, or grit deposits.
3. Determine if equipment selected did the desired maintenance or repair job and produced cost-effective results.
4. Review jobs done by repair crews.

5.802 Rehabilitation

1. CCTV is an effective tool when grouting to control inflow/infiltration (see Chapter 7, Section 7.5, "Sealing Grout").
2. Sliplining. TV sewer prior to inserting a liner to look for protruding taps and severe misalignment conditions (see Volume II, Chapter 10, "Sewer Renewal (Rehabilitation)").

5.81 Experience of Others and Observations From Demonstrations

One of the best ways to determine if a manufacturer can provide you with the TV equipment you need or if a contractor can provide you with the services you need is to contact their previous clients. Talk to other collection system operators and consider their experiences and recommendations.

Always have the equipment manufacturers or contractors demonstrate their equipment. During the demonstration have the crews that will actually use the equipment try to use the equipment. Their comparisons of different equipment will be very valuable. If a contractor is demonstrating equipment, have the people who will use the results observe the demonstration and be sure they understand what is observed and what is recorded. Their comparison of the different contractors' ability to describe actual observations will be very helpful. If you are considering the purchase of TV equipment, try to rent the equipment first so you can evaluate the performance of the equipment and the results.

5.82 The Decision: Use a Contractor or Purchase Equipment

First the decision must be made whether to hire a contractor or to purchase closed-circuit TV equipment. Important factors to evaluate include what needs to be televised, how long will it take, how much money is available, and whether or not qualified operators are available to do the job.

The selection of a contractor or equipment manufacturer is usually based on the lowest qualified bidder. For this reason specifications must be carefully prepared and only qualified contractors or manufacturers should be allowed to submit bids. Recommendations and experiences of other agencies and the comments of persons participating in demonstrations should be considered when determining the qualifications of bidders.

5.83 Important Components of TV Inspection Equipment

Once the uses of a TV inspection program have been identified, the ability of the equipment of potential manufacturers and contractors to do the job can be compared. Important features of closed-circuit television equipment that should be compared are listed in Sections 5.830 through 5.838.

5.830 Camera

1. Lighting requirements, inside pipe
2. Pressure rating in psi
3. Type of connectors
4. Horizontal resolution
5. Type of circuitry
6. Iris, focus, lens adjustments
7. Depth of field at various focal settings
8. Physical specifications (shock, vibration)
9. Actual picture clarity

Television cameras for sewer line televising should meet at least the following minimum specifications:

1. Waterproof electrical and case assemblies to withstand pressure tests (external) up to 100 psi.
2. 600-line horizontal resolution.
3. Synchronous crystal control.
4. 50 degree or greater wide-angle lens, maximum distortion not to exceed 2 percent.
5. Camera lens adjustments to be made through external means.
6. Camera viewport to be of optical grade quartz, not less than $^3/_8$-inch thick.
7. Solid state electronic assembly, vidicon tube excepted, but with capacity to withstand three successive impacts of 25-Gs without damage.
8. Manufacturer to warrant camera (including vidicon tube) against defects in materials, parts, and workmanship for at least one year.
9. Manufacturer must provide adequate training for crew in field.

5.831 TV Cable

1. Tensile strength
2. Connectors
3. Length
4. Conductors, what kind, what size, how many

5.832 Video Monitor

1. Size of screen
2. Suited for mobile operation
3. Horizontal resolution
4. Picture quality

5.833 TV Cable Reel

1. Cantilever or axial mounted
2. Size
3. Type of rewind

5.834 Measure Meter

1. Accuracy
2. Placement of readout
3. Remote readout

5.835 System Control Unit

1. Protection circuits
2. Fuses
3. Circuitry lights and camera
4. Input voltage, output voltage
5. Safety of unit

5.836 Light Head Assembly

1. Type bulbs required
2. Maintenance expected on unit
3. Lamp voltage

5.837 Skids and Tow Cables

1. Durability
2. Expected wear and tear
3. Serviceability
4. Pipe size range

5.838 Miscellaneous Items

1. Video input information devices
2. Manhole jacks
3. Surface rollers
4. Power plants
5. Videotape equipment
6. Crew size

5.84 Economics

1. Initial cost and estimated life of TV unit
2. Expected maintenance costs
3. Crew requirements and labor costs
4. Expected production (feet (meters) of sewer televised per day and also feet (meters) televised per year)
5. Cost per foot (meter) (overall) of inspection
   a. Equipment Costs,

   $$\text{\$/year} = \frac{\text{Initial Cost, \$}}{\text{Estimated Life of Unit, yr}}$$

   b. Maintenance Costs, $/year

   c. Labor Costs, $/year

   d. Total Cost to Operate TV Unit, $/yr = (a) + (b) + (c)

   e. $$\text{Costs, \$/ft} = \frac{\text{(d) Total Cost, \$/yr}}{\text{(4) Expected Production, ft/yr}}$$

5.9 CODES FOR TV INSPECTION, DEFECTS, AND ACTIONS/RESPONSES

Excellent publications containing photographs illustrating standard sewer defects and terminology have been developed by the Water Research Centre, Frankland Road, Blagrove, Swindon, Wiltshire, 5H5 8YR, ENGLAND.

The codes in Table 5.1 were developed by Rich Cunningham, City of San Francisco, as an assignment of the Water Pollution Control Federation's Wastewater Collection System Committee. "Un" refers to unit of measurement (If, lineal feet of brick de-mortared, but still intact). The Water Environment Federation's MOP 7 contains a more detailed "Defects Classification Guide."

TABLE 5.1 TV INSPECTION CODE SYSTEM

Code	Description	Un	Is this Okay? Y	N
INSPECTION INFORMATION				
BDM	Brick de-mortared, but still intact	lf	☐	☐
BMS	Brick missing, backfill showing	sf	☐	☐
CAB	Camera blocked and unable to proceed further	ea	☐	☐
CAS	Camera submerged	ea	☐	☐
CAV	Cavity outside the line	ea	☐	☐
CKL	Cracking, lateral, i.e. parallel to the direction of the flow	lf	☐	☐
CKT	Cracking, transv., i.e., across the direction of the flow	lf	☐	☐
COR	Corrosion, sulfide	lf	☐	☐
CRS	Crushed	lf	☐	☐
DAM	Damage, other causes, such as chemical attack	lf	☐	☐
DEB	Debris accumulated in invert	lf	☐	☐
DIS	Distorted shape, loss of concentricity	lf	☐	☐
DWO	Dry weather overflow	ea	☐	☐
FDP	Flow depth	in	☐	☐
FIL	Infiltration flow rate	gm	☐	☐
FLO	Inflow rate	gm	☐	☐
GRS	Grease accumulation	lf	☐	☐
INV	Invert damage, such as abrasion	sf	☐	☐
JOI	Joint offset	in	☐	☐
JOP	Joint separated longitudinally	in	☐	☐
LEA	Leaking observed at deformed or missing seal	ea	☐	☐
MDJ	Mineral deposit at joint	ea	☐	☐
MHS	MH overflow or surcharging	ea	☐	☐
MIN	Mineral deposits or scaling on pipe wall	lf	☐	☐
MLS	Main line surcharging	ea	☐	☐
NFT	Inflow of tidewater	gm	☐	☐
ODR	Odor complaint	ea	☐	☐
OTH	Other	ea	☐	☐
RAT	Rat infestation	ea	☐	☐
ROA	Roach infestation	ea	☐	☐
ROO	Root intrusion	lf	☐	☐
RUO	Mechanism rusted in open position	ea	☐	☐
RUS	Mechanism rusted in shut position	ea	☐	☐
SAG	Sagged line	lf	☐	☐
SDM	Structural damage to manhole, etc.	sf	☐	☐
SER	Serviceable or okay	ea	☐	☐
SPM	Part(s) missing from structure, e.g. rungs, gaskets	ea	☐	☐
SLS	Service lateral surcharging	ea	☐	☐
SPL	Spalling	lf	☐	☐
TA1	Abandoned tap 12-3 o'clock position when seen from TV	ea	☐	☐
TA2	Abandoned tap 3-6 o'clock position when seen from TV	ea	☐	☐
TA3	Abandoned tap 6-9 o'clock position when seen from TV	ea	☐	☐
TA4	Abandoned tap 9-12 o'clock position when seen from TV	ea	☐	☐
TI1	Tap intruding 12-3 o'clock position when seen from TV	ea	☐	☐
TI2	Tap intruding 3-6 o'clock position when seen from TV	ea	☐	☐
TI3	Tap intruding 6-9 o'clock position when seen from TV	ea	☐	☐
TI4	Tap intruding 9-12 o'clock position when seen from TV	ea	☐	☐
TP1	Tap: 12-3 o'clock position when seen from TV	ea	☐	☐
TP2	Tap: 3-6 o'clock position when seen from TV	ea	☐	☐
TP3	Tap: 6-9 o'clock position when seen from TV	ea	☐	☐
TP4	Tap: 9-12 o'clock position when seen from TV	ea	☐	☐
TR1	Tap w/roots 12-3 o'clock position when seen from TV	ea	☐	☐
TR2	Tap w/roots 3-6 o'clock position when seen from TV	ea	☐	☐
TR3	Tap w/roots 6-9 o'clock position when seen from TV	ea	☐	☐
TR4	Tap w/roots 9-12 o'clock position when seen from TV	ea	☐	☐
VRM	Vermin, general	ea	☐	☐
WWO	Wet weather overflow	ea	☐	☐
FIELD INFORMATION		Un		
BMB	Basement or cellar drain backed up	ea	☐	☐
CBO	Catchbasin/Storm drain grate off, grate is missing	ea	☐	☐
CCB	Catchbasin/Storm drain culvert pipe is blocked	ea	☐	☐
CFL	Catchbasin/Storm drain full of debris	ea	☐	☐
CGB	Catchbasin/Storm drain grate covered w/debris; can't drain	ea	☐	☐
CNA	Crew needs assistance from another crew	ea	☐	☐
CNI	Contract work needs inspection at the job site	if	☐	☐
CON	Construction job in progress at site of complaint	ea	☐	☐
DEP	Depression or void in road/street	sf	☐	☐
DEW	Dewatering of job site is required	gl	☐	☐

TABLE 5.1 TV INSPECTION CODE SYSTEM (continued)

Code	Description	Un	Is this Okay? Y	N
FIELD INFORMATION (continued)				
INS	Inspection by service crew is needed	ea	☐	☐
LOS	Lost article is reported in manhole, cb, or st. drain	ea	☐	☐
MAG	Manhole above grade	ea	☐	☐
MBG	Manhole below grade	ea	☐	☐
MBU	Manhole buried or not visible	ea	☐	☐
MHO	Manhole is open/cover gone	ea	☐	☐
MHP	Manhole requires paving around rim	ea	☐	☐
MLB	Main line blocked	ea	☐	☐
MLS	Main line surcharging	ea	☐	☐
MNC	Manhole noise complaint	ea	☐	☐
ODR	Odor complaint	ea	☐	☐
ORP	Owner of property is responsible for problem	ea	☐	☐
OTR	Other type problem	ea	☐	☐
PSP	Pump/P.S. problem	ea	☐	☐
SLB	Service lateral blocked	ea	☐	☐
SLS	Service lateral surcharged	ea	☐	☐
SSR	Scheduled service is required to be performed	ea	☐	☐
SUB	Sump blocked, leading to interceptor	ea	☐	☐
TRP	Trench requires paving	sf	☐	☐
UNK	Unknown situation	ea	☐	☐
WTR	Water main broken	ea	☐	☐
ACTION/RESPONSE				
AGI	Above-ground inspection performed	ea	☐	☐
AST	Assist a crew requiring further help	ea	☐	☐
BAL	Balling style cleaning	lf	☐	☐
BGI	Below-ground inspection	ea	☐	☐
BKF	Backfill, tamp down or jet down for compaction	cy	☐	☐
BUC	Bucket machine cleaning	lf	☐	☐
BYP	Bypass: plug line upstream and pump around to downstream	lf	☐	☐
CLG	Clean grate(s) of pump station or comminutor, etc.	ea	☐	☐
CLM	Clam bucket removal of debris	cy	☐	☐
CON	Contract work is required to resolve the problem	ea	☐	☐
DCR	Dust for roaches	sf	☐	☐
DRG	Dredge/dragline cleaning	cy	☐	☐
DYE	Dye test	ea	☐	☐
EXC	Excavation	lf	☐	☐
FLU	Flush line with non-pressurized water	ea	☐	☐
GRT	Grout seal	sf	☐	☐

Code	Description	Un	Is this Okay? Y	N
ACTION/RESPONSE (continued)				
GSK	Gasket seal/reseal joints in manhole or line	ea	☐	☐
IMB	Install manhole base	ea	☐	☐
INC	Inspection done	ea	☐	☐
IST	Insituform installation performed	lf	☐	☐
JET	Hydrojet clean	lf	☐	☐
KIT	Kiting style cleaning	lf	☐	☐
LIN	Repair/patch line	lf	☐	☐
LNI	Line, new installation	lf	☐	☐
LDO	Locate dry weather overflow	ea	☐	☐
LWO	Locate wet weather overflow	ea	☐	☐
MHC	Manhole channelization	ea	☐	☐
MHI	Manhole installation	ea	☐	☐
MLA	Main line air pressure test	ea	☐	☐
MOS	Mosquito control	ea	☐	☐
MRK	Mark street surface for dig-up	ea	☐	☐
OTA	Other response/action	ea	☐	☐
OWN	Owner or other responsible party was notified	ea	☐	☐
PAD	Quiet cover/gate that causes noise	ea	☐	☐
PAV	Pave/repave street surface	sf	☐	☐
PLN	Plunge to unblock with mechanical plunger	ea	☐	☐
PLS	Plaster/Re-plaster brick or mortar	sf	☐	☐
RAT	Bait for rats	ea	☐	☐
REP	Replace cover/lid on manhole or drain	ea	☐	☐
RFC	Replace manhole frame and cover	ea	☐	☐
ROD	Mechanical rodding	lf	☐	☐
RTC	Root cutting with mechanical root shears	lf	☐	☐
RTF	Root foaming with chemical herbicide	lf	☐	☐
RTS	Root soaking with chemical herbicide	lf	☐	☐
SCO	Scooter cleaning	lf	☐	☐
SLA	Service lateral air pressure test	ea	☐	☐
SLC	Service lateral to main line connection (tap-in)	ea	☐	☐
SLP	Slip line installation	lf	☐	☐
SMO	Smoke test	ea	☐	☐
SNK	Snake to unblock stoppage in line	lf	☐	☐
STI	Structure, new installation	ea	☐	☐
STR	Structure, repair	ea	☐	☐
STV	String TV line	lf	☐	☐
TAP	Install tap/saddle	ea	☐	☐
TVI	Television inspection	lf	☐	☐
VAC	Suction debris out	ea	☐	☐

5.10 ADDITIONAL TV INSPECTION EQUIPMENT INFORMATION

Figure 5.40: Electric-powered, remote-controlled winch

This type of winch can be used at the far manhole to replace the hand winch or as the "retrieve" or "tag winch" at the truck (as shown in Figure 5.40).

Figure 5.41: Lateral inspection/sealing system

This lateral sealing device works from a tool very similar to a joint packer. The system is capable of inspecting and sealing a service lateral against infiltration for a distance of three to eight feet from the main sewer line.

Figures 5.41 and 5.42: Lateral inspection camera

This inspection camera is capable of traveling up small-diameter service laterals for approximately 80 feet. These cameras can be launched from a packer-like device from the main sewer line.

Figure 5.43: Camera with flexible head

Camera with flexible head allows operator to adjust head for closer and more detailed inspection of areas of concern.

Figure 5.44: Camera transporters (tractors)

The camera transporter, also called a tractor, is designed to carry a small camera through 8-inch to 24-inch diameter sewer lines. The electric motor-driven tractor will transport a closed-circuit television camera into a sewer line without the need of a pull winch. This device is extremely useful when televising to a "dead end" pipe or into a highly congested intersection where the stationing of equipment or operators over a sewer manhole would present a safety hazard.

Figures 5.45 and 5.46: Portable TV system for small-diameter pipe

A totally portable closed-circuit TV system that can be used in small-diameter pipe (4-inch to 6-inch diameters) and will negotiate multiple bends up to 90 degrees. The camera can be inserted into the lines through cleanouts or other openings.

Figure 5.47: Modular TV inspection system

A very compact portable TV inspection system that can easily fit into a pickup truck, van or trailer. This type of unit would be sufficient for an agency that did not have continual need for a TV system.

Figure 5.48: Miniature probe-type camera

This tiny camera, which is less than 3 inches long, can be used for inspecting building laterals. It can be inserted through a cleanout and can negotiate 90 degree bends in pipes as small as 2 inches. This camera uses a push-type conductor which is available with 150 feet of cable.

Figure 5.49: Typical camera/transporter configurations

TV inspection cameras are commonly mounted on either a skid or a tractor. Some manufacturers offer various types of systems which use interchangeable parts. This gives the collection system operator the flexibility to use different combinations of components, depending on the work that needs to be done and the types and sizes of pipes being inspected.

Figure 5.50: Combination TV/grouting equipment

Combination TV/grouting equipment allows operator to move grouting equipment to exact location where needed, observe grouting process and inspect completed grouting job.

Equipment used by operator when using TV to observe grouting operation and record observations.

Figure 5.51: TV operator's control panel

In a van, trailer or step-van unit, all the necessary controls are grouped together in a separate air conditioned room. The TV operator has full control of the camera's movement once it is placed in the sewer line.

Figure 5.52: Electronic data recording system

An alpha-numerical keyboard that allows all job data to be recorded on the videotape.

Figure 5.53: Television data log unit

This is a specially designed keyboard that can enter pipe condition information quickly onto a data cartridge. Log sheets can be printed out, work orders can be generated and rehabilitation cost comparisons can be evaluated.

Figure 5.54: Ultrasonic caliper

Ultrasonic technology can be used to measure interior pipe corrosion, debris buildup in the pipe invert, misalignment, deflection, missing manholes, and other pipeline conditions. This type of equipment uses ultrasonic transducers that are mounted on a float system to "map" the interior of the pipe by taking a series of measurements.

NOTE: For additional information on uses of TV equipment, see Chapter 10, "Sewer Renewal (Rehabilitation)," in *OPERATION AND MAINTENANCE OF WASTEWATER COLLECTION SYSTEMS,* Volume II.

Fig. 5.40 Electric-powered, remote-controlled winch

(Permission of Aries Industries, Inc.)

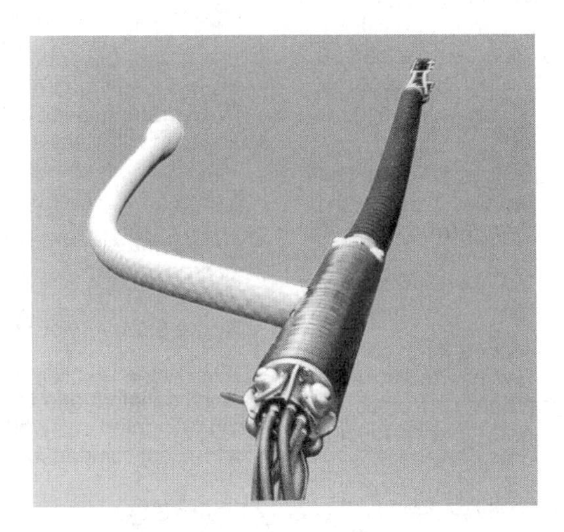

Lateral Sealing Unit
(Permission of Cues, Inc.)

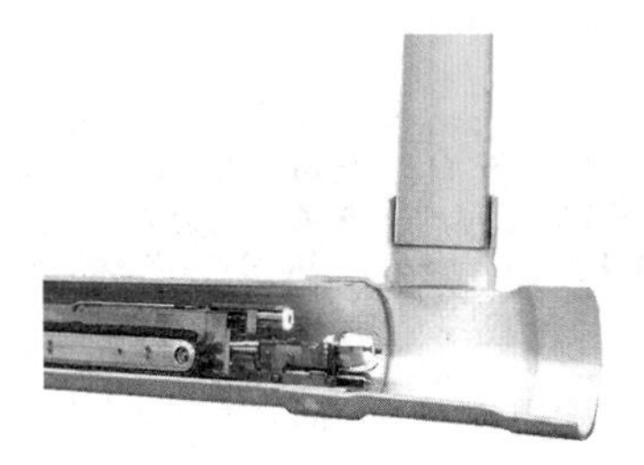
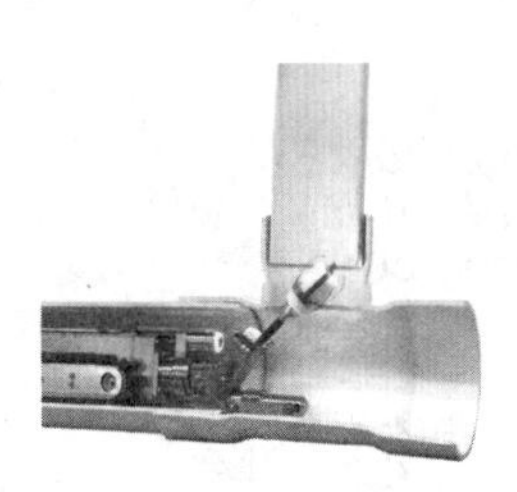
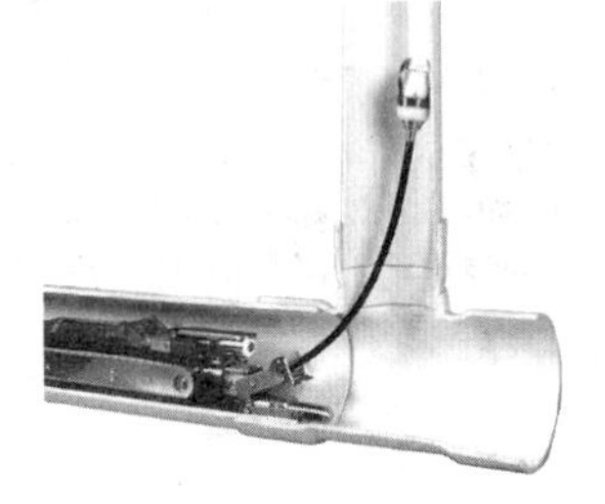

Lateral Inspection Camera

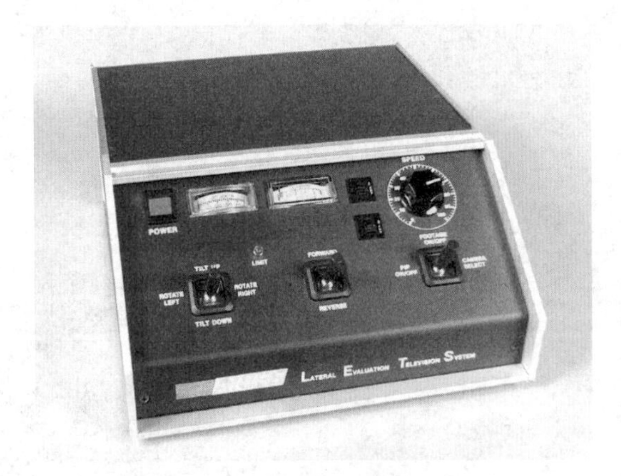

Lateral Control Unit

Fig. 5.41 Lateral sealing system
(Permission of Aries Industries, Inc.)

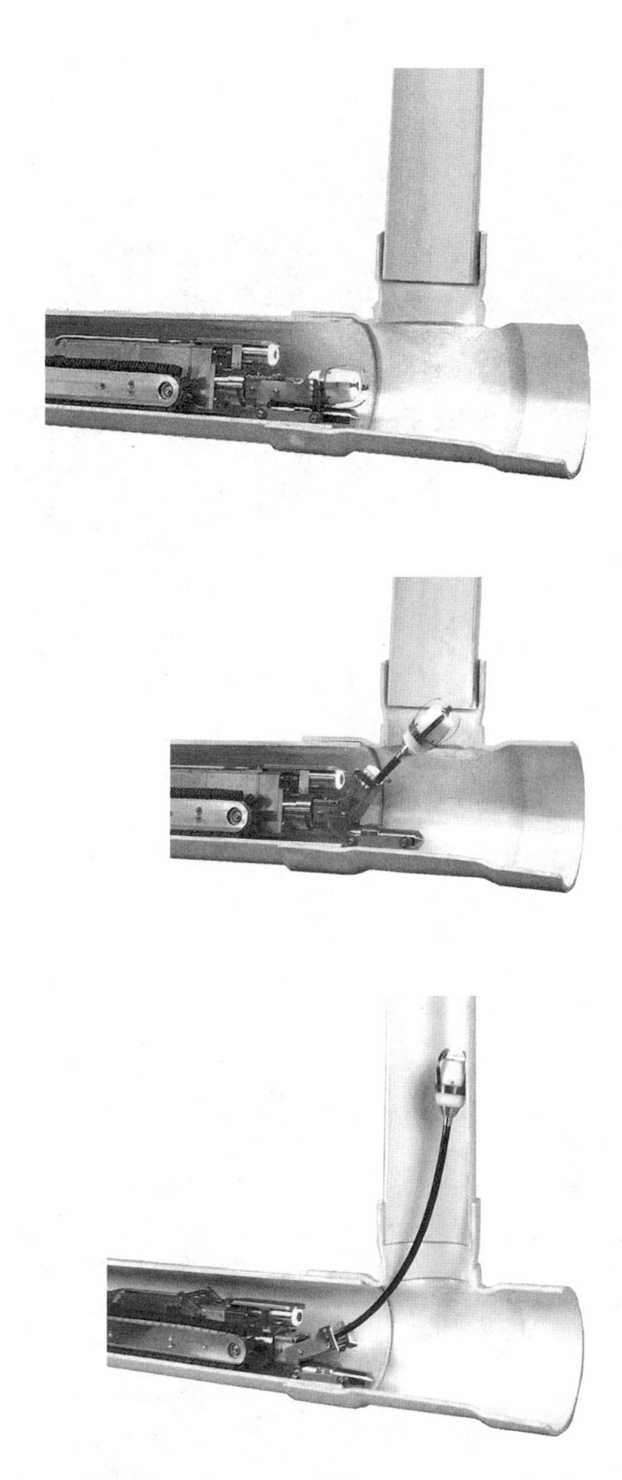

This inspection camera is capable of traveling up small-diameter service laterals for approximately 80 feet.

Fig. 5.42 Lateral inspection TV camera

(Permission of Aries Industries, Inc.)

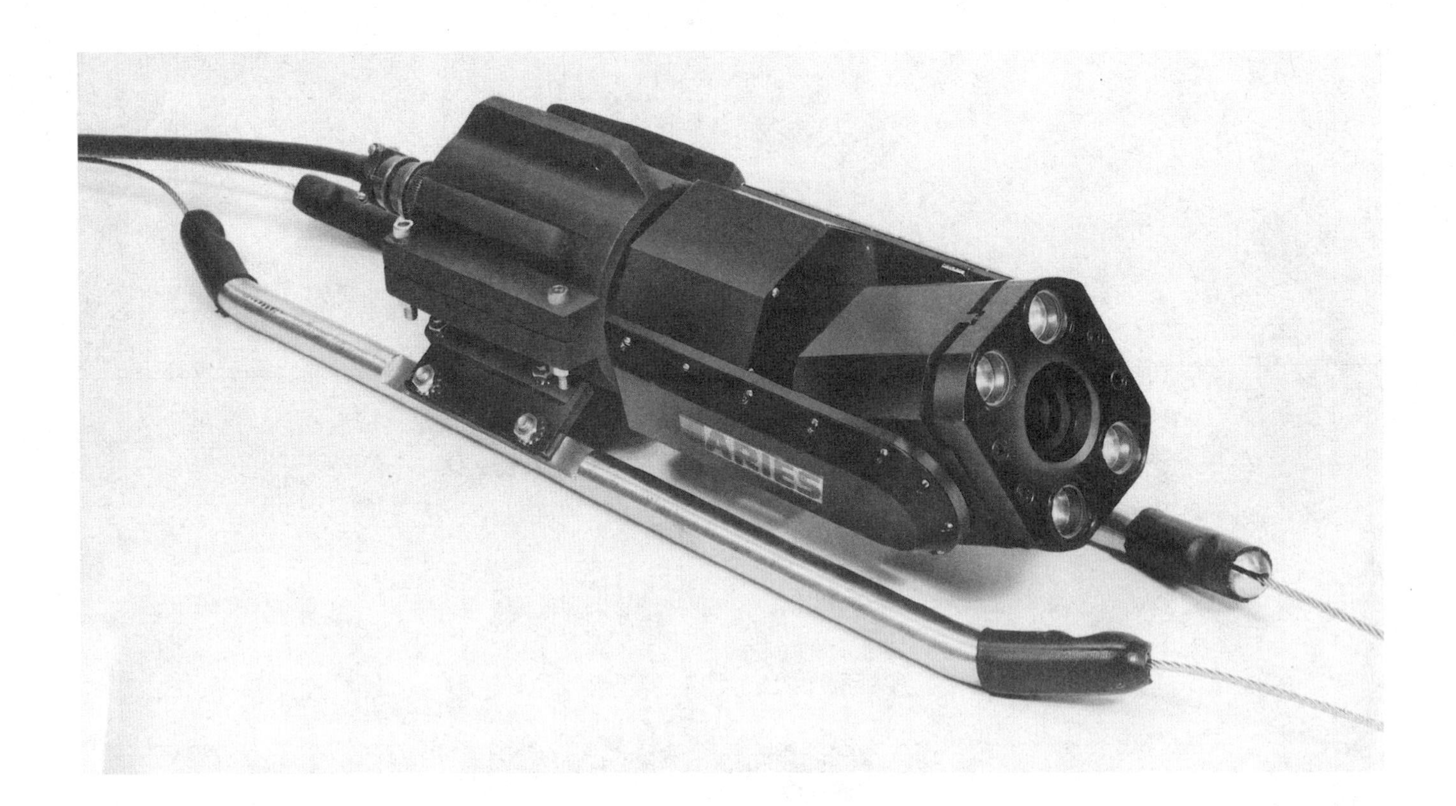

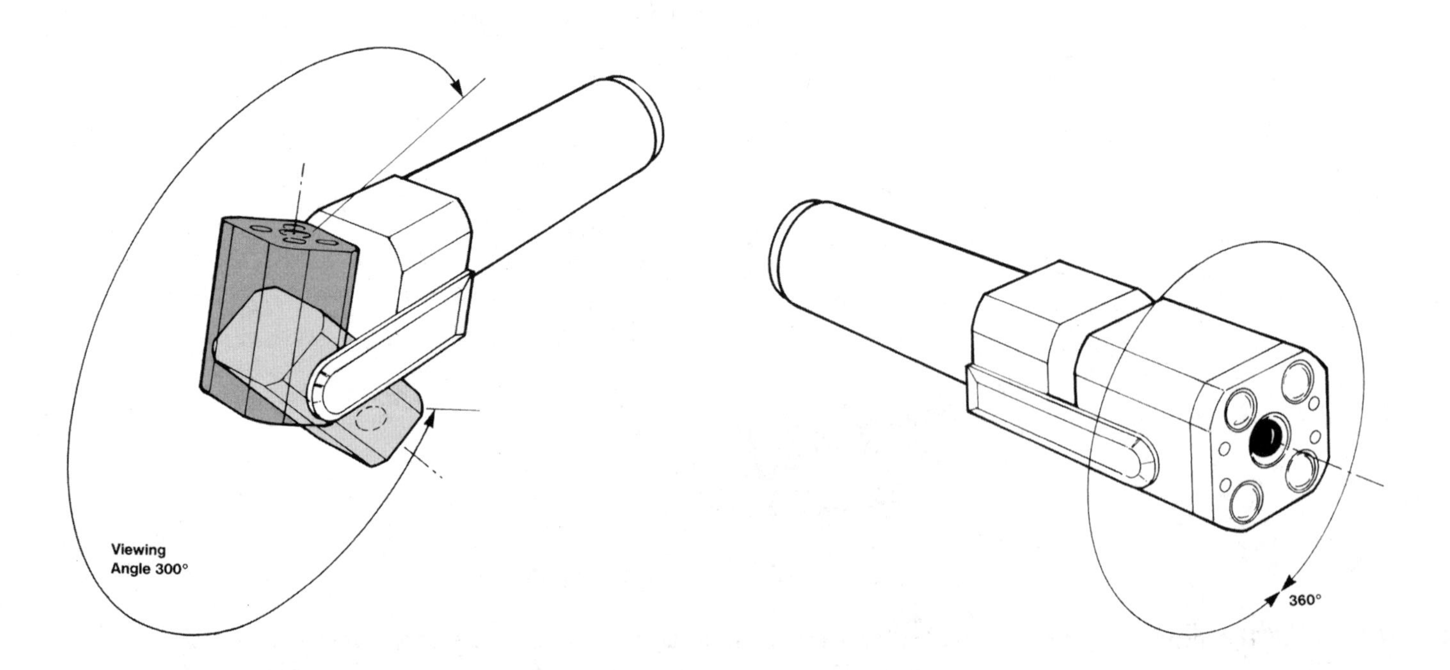

Fig. 5.43 Camera with flexible head (pan-and-tilt features)

(Permission of Aries Industries, Inc.)

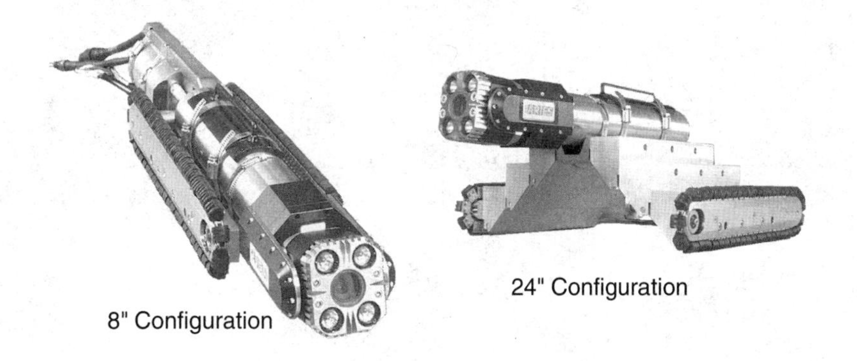

Fig. 5.44 Camera transporters (tractors)
(Permission of Aries Industries, Inc.)

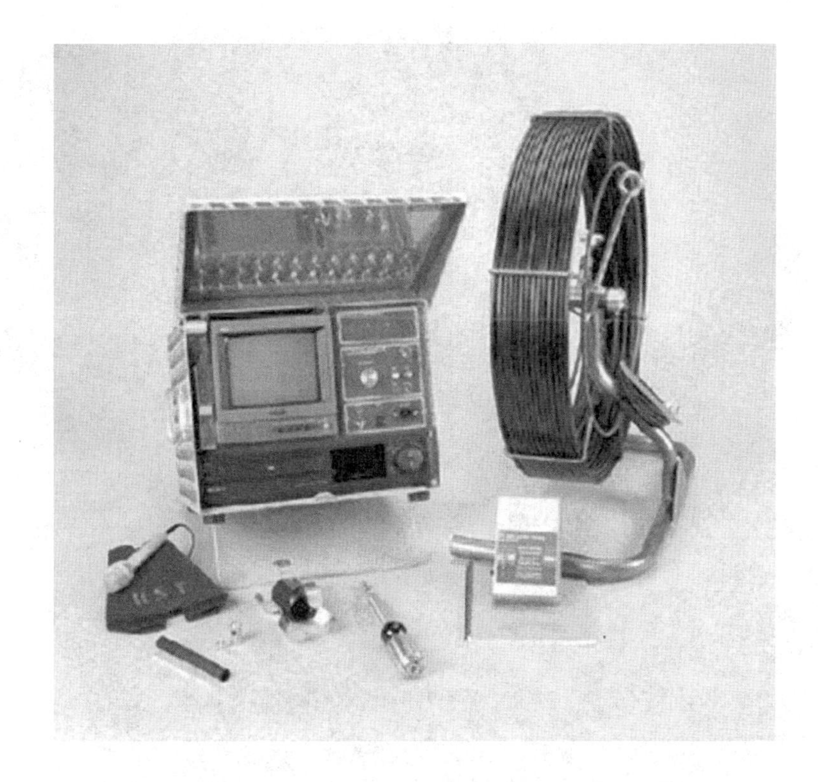

Fig. 5.45 Portable TV system for small-diameter pipe
(Permission of RST Services, Inc.)

Portable TV system for hard access locations.

Portable TV camera entering cleanout for lateral inspection.

Fig. 5.46 Applications of portable TV system in difficult to access locations

(Permission of Cues, Inc.)

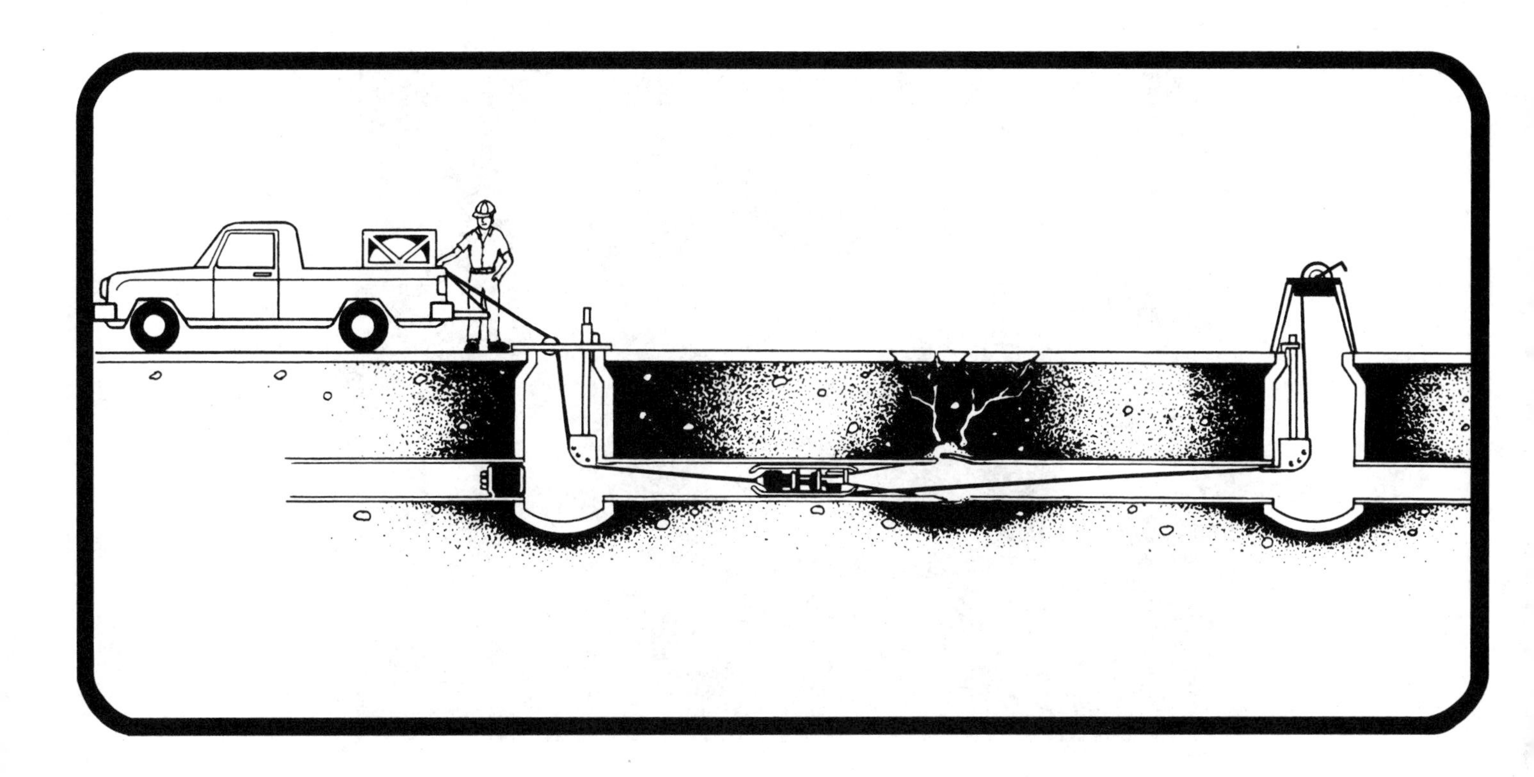

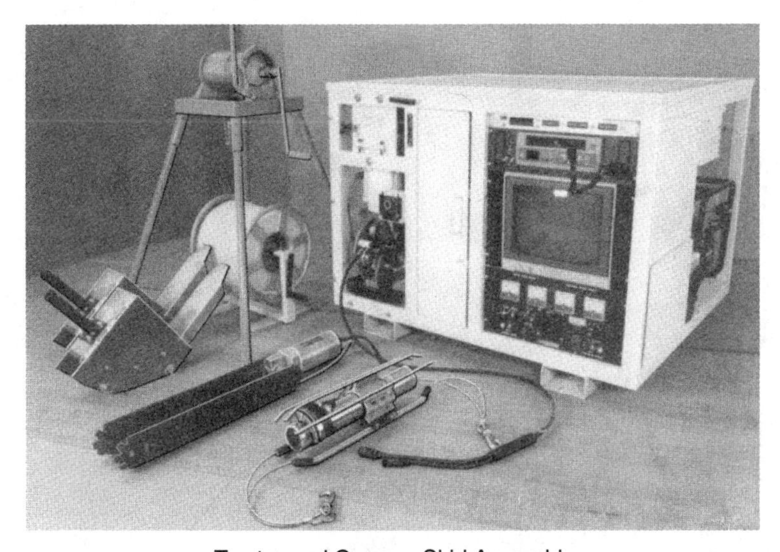

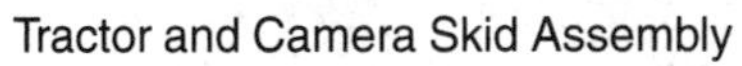

Tractor and Camera Skid Assembly

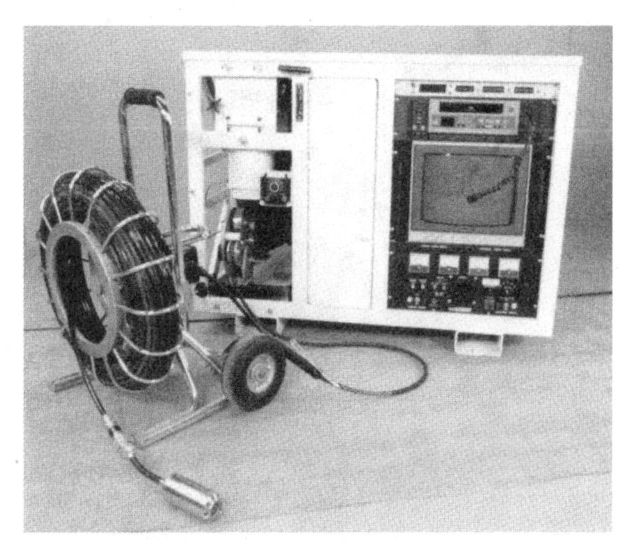

200-Foot Push Cable and Camera

Fig. 5.47 Modular TV inspection system

(Permission of Aries Industries, Inc.)

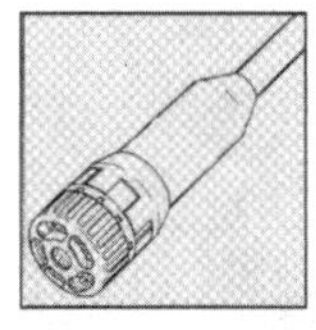

Flexiprobe

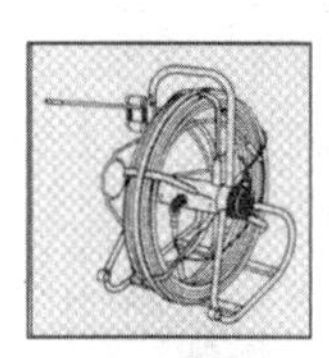

Coiler

Camera Control Unit

Light Heads

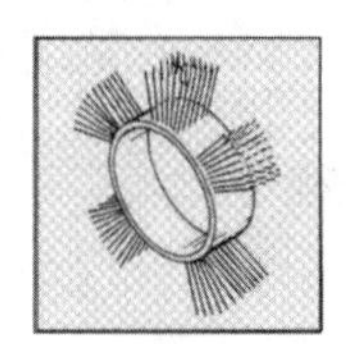

Brushes

Fig. 5.48 Miniature probe-type camera and TV inspection system components

(Courtesy of Pearpoint Inc.)

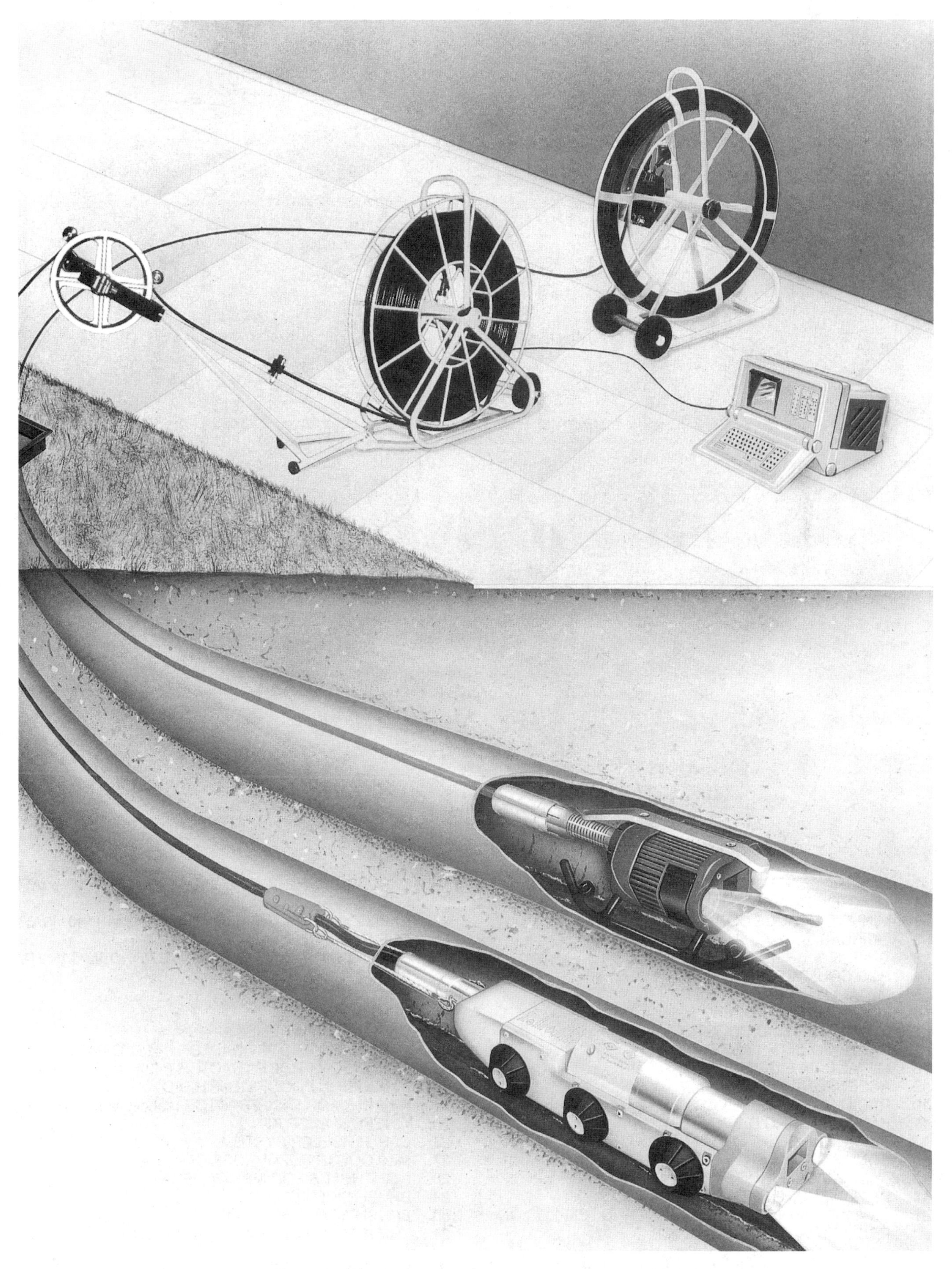

Fig. 5.49 Examples of camera/transporter configurations
(Courtesy of Pearpoint Inc.)

1. CAMERA CONTROL UNIT
1A. TRACTOR DRIVE MODULE
1B. MINI FLEXIPROBE CONTROL MODULE
2. LINK CABLE
3. T-PIECE
4. ROD COUNTER
5. METAL COILER
6. MINI COILER
6A. MINI COILER ROD COUNTER
7. CABLE COUNTER ASSEMBLY
8. METAL CABLE DRUM
9. ACCESSORY CASE
10. RADIO SONDE RECEIVER
11. P420 LOWERING DEVICE
12. THICK FLEXIBLE ROD AND EXPLOSION PROOF SONDE
13. THIN FLEXIBLE ROD
14. INTERMEDIATE FLEXIBLE ROD
15. KEVLAR REINFORCED CABLE AND EXPLOSION PROOF TRACTOR CONNECTOR
16. EXPLOSION PROOF LIGHTHEAD, P417 10 WATT FOR P415
17. EXPLOSION PROOF LIGHTHEAD, P418 40 WATT FOR P415
18. EXPLOSION PROOF COLOR FLEXIPROBE CAMERA P415
19. SKID ASSEMBLIES
    BRUSH SET
20. CABLE STRAIN RELIEF
21. EXPLOSION PROOF ADAPTOR P415 TO P420
22. EXPLOSION PROOF TRACTOR P420
23. EXPLOSION PROOF LIGHTHEAD
24. WHEELS FOR P420 TRACTOR (set of 6 wheels)
25. WHEEL SPACER KIT
26. CABLE ROLLER SYSTEM
27. MINI COLOR FLEXIPROBE CAMERA
28. LIGHTHEADS FOR MINI CAMERA
29. BRUSH SKID FOR MINI
30. BRUSH ADAPTOR 24MM LIGHTHEAD
31. THICK ROD/MINI CAMERA ADAPTOR

Fig. 5.49 Examples of camera/transporter configurations (continued)

(Courtesy of Pearpoint Inc.)

Truck-Mounted TV/Grouting Control Unit

Large Packer

Low-Void Packer

TV/Grouting Controls

Fig. 5.50 Combination TV/grouting controls and equipment

(Permission of Cues, Inc.)

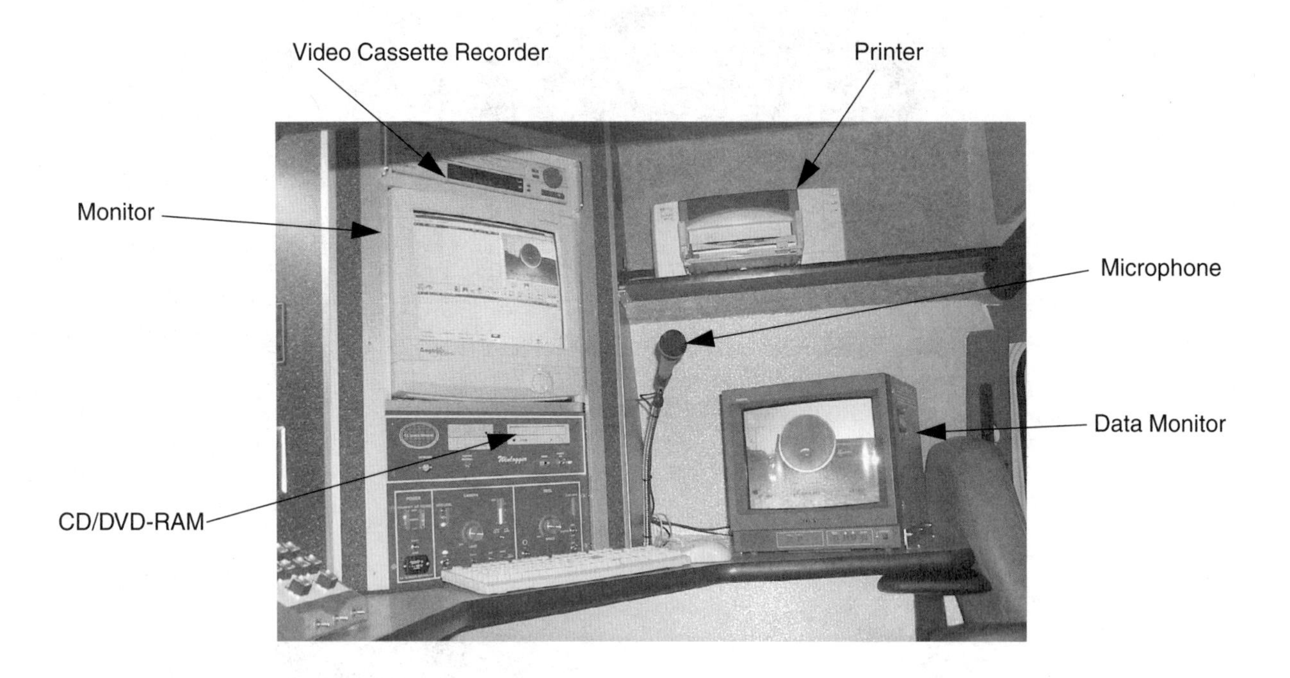

Fig. 5.51 TV operator's control panel

Keyboard for entering inspection data on TV monitor.
Information is recorded on videotape or computer-recordable disk.

Fig. 5.52 Electronic data recording system
(Permission of RST Services, Inc.)

Television Data Log Unit With Keyboard

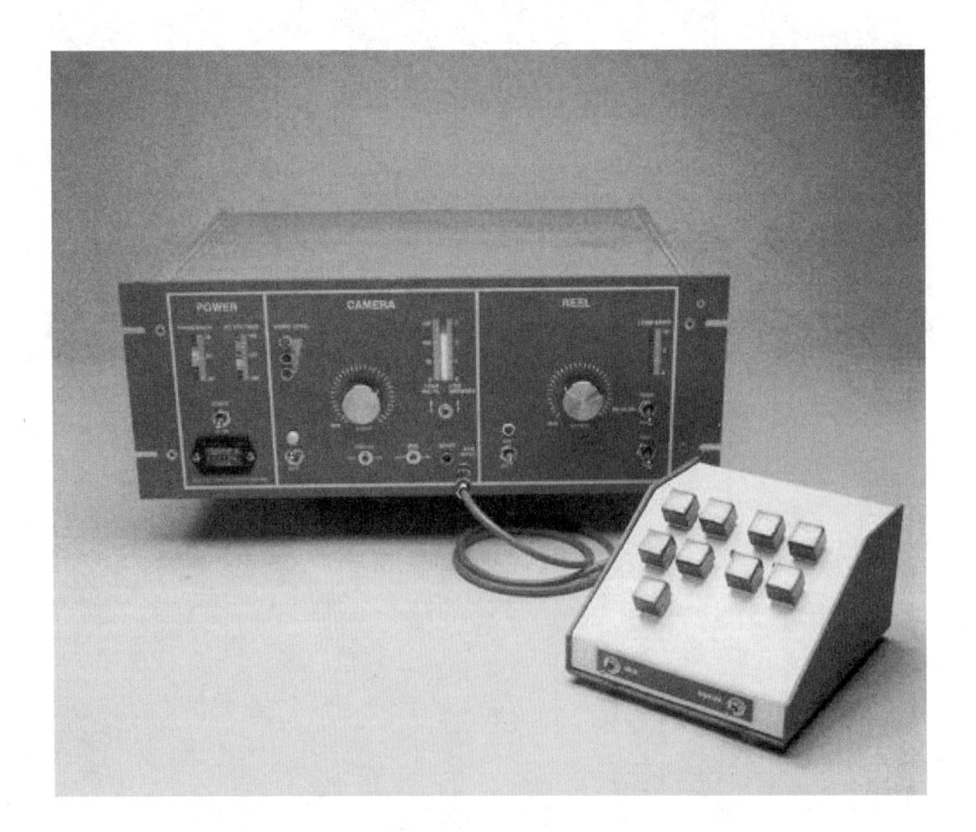

Camera Control Unit With Automatic Hand Controller

Fig. 5.53 Television data log unit

(Permission of RST Services, Inc.)

Sonic caliper. Note the location of the five crown transducers; other transducers are located on the bottom of the caliper.

Sonic caliper being used in a 42-inch sewer.

Fig. 5.54 Ultrasonic caliper

(Courtesy of Sonex)

Preparing the sonic caliper for insertion into a concrete sewer pipe. This picture shows a typical setup in an alley or a lightly traveled road.

Sonic caliper being removed from a manhole after an inspection run. This version of the caliper tool will fit through a 20-inch manhole opening.

Fig. 5.54 Ultrasonic caliper (continued)

(Courtesy of Sonex)

CHAPTER 6

PIPELINE CLEANING AND MAINTENANCE METHODS

by

Dallas Hughes

Revised by

Bob Hiler

Rick Arbour

Stephen B. Tilson

Gary Batis

TABLE OF CONTENTS

Chapter 6. PIPELINE CLEANING AND MAINTENANCE METHODS

OBJECTIVES

Chapter 6. PIPELINE CLEANING AND MAINTENANCE METHODS

Following completion of Chapter 6, you should be able to:

1. Identify types and causes of sewer stoppages;
2. Select proper methods to clear stoppages and clean sewers;
3. Determine equipment and staffing requirements for various sewer clearing and cleaning methods;
4. Set up sewer clearing and cleaning equipment safely and properly;
5. Operate and maintain sewer clearing and cleaning equipment safely and effectively;
6. Clean sewers without flooding homes and basements;
7. Record essential data regarding clearing and cleaning operations;
8. Direct sewer clearing and cleaning operations using the following equipment and techniques:
   a. Balling,
   b. High-velocity cleaners,
   c. Flushing,
   d. Sewer scooters,
   e. Bucket machines,
   f. Power rodders, and
   g. Hand rods;
9. Establish a preventive maintenance program for sewer clearing and cleaning equipment; and
10. Develop a program to control odors and corrosion from hydrogen sulfide.

WORDS

Chapter 6. PIPELINE CLEANING AND MAINTENANCE METHODS

AEROBIC (AIR-O-bick) AEROBIC

A condition in which atmospheric or dissolved molecular oxygen is present in the aquatic (water) environment.

AIR BINDING AIR BINDING

The clogging of a filter, pipe or pump due to the presence of air released from water. Air entering the filter media is harmful to both the filtration and backwash processes. Air can prevent the passage of water during the filtration process and can cause the loss of filter media during the backwash process.

AIR GAP AIR GAP

An open vertical drop, or vertical empty space, between a drinking (potable) water supply and the point of use. This gap prevents backsiphonage because there is no way wastewater can reach the drinking water. Air gap devices are used to provide adequate space above the top of a manhole and the end of the hose from the fire hydrant. This gap ensures that no wastewater will flow out the top of a manhole, reach the end of the hose from a fire hydrant, and be sucked or drawn back up through the hose to the water supply.

AIR PADDING AIR PADDING

Pumping dry air (dew point –40°F) into a container to assist with the withdrawal of a liquid or to force a liquified gas such as chlorine out of a container.

ANAEROBIC (AN-air-O-bick) ANAEROBIC

A condition in which atmospheric or dissolved molecular oxygen is *NOT* present in the aquatic (water) environment.

BOD (pronounce as separate letters) BOD

Biochemical **O**xygen **D**emand. The rate at which organisms use the oxygen in water or wastewater while stabilizing decomposable organic matter under aerobic conditions. In decomposition, organic matter serves as food for the bacteria and energy results from its oxidation. BOD measurements are used as a measure of the organic strength of wastes in water.

BALLING BALLING

A method of hydraulically cleaning a sewer or storm drain by using the pressure of a water head to create a high cleansing velocity of water around the ball. In normal operation, the ball is restrained by a cable while water washes past the ball at high velocity. Special sewer cleaning balls have an outside tread that causes them to spin or rotate, resulting in a "scrubbing" action of the flowing water along the pipe wall.

BARREL BARREL

(1) The cylindrical part of a pipe that may have a bell on one end.

(2) The cylindrical part of a manhole between the cone at the top and the shelf at the bottom.

BENCH SCALE TESTS BENCH SCALE TESTS

A method of studying different ways or chemical doses for treating water on a small scale in a laboratory.

BLOCKAGE BLOCKAGE

(1) Partial or complete interruption of flow as a result of some obstruction in a sewer.

(2) When a collection system becomes plugged and the flow backs up, it is said to have a "blockage." Commonly called a STOPPAGE.

BUCKET BUCKET

(1) A special device designed to be pulled along a sewer for the removal of debris from the sewer. The bucket has one end open with the opposite end having a set of jaws. When pulled from the jaw end, the jaws are automatically opened. When pulled from the other end, the jaws close. In operation, the bucket is pulled into the debris from the jaw end and to a point where some of the debris has been forced into the bucket. The bucket is then pulled out of the sewer from the other end, causing the jaws to close and retain the debris. Once removed from the manhole, the bucket is emptied and the process repeated.

(2) A conventional pail or bucket used in BUCKETING OUT and also for lowering and raising tools and materials from manholes and excavations.

BUCKET BAIL BUCKET BAIL

The pulling handle on a bucket machine.

BUCKET MACHINE BUCKET MACHINE

A powered winch machine designed for operation over a manhole. The machine controls the travel of buckets used to clean sewers.

BUCKETING OUT BUCKETING OUT

An expression used to describe removal of debris from a manhole with a pail on a rope. In balling or high-velocity cleaning of sewers, debris is washed into the downstream manhole. Removal of this debris by scooping it into pails and hauling debris out is called "bucketing out."

CAVITATION (CAV-uh-TAY-shun) CAVITATION

The formation and collapse of a gas pocket or bubble on the blade of an impeller or the gate of a valve. The collapse of this gas pocket or bubble drives water into the impeller or gate with a terrific force that can cause pitting on the impeller or gate surface. Cavitation is accompanied by loud noises that sound like someone is pounding on the impeller or gate with a hammer.

DEGRADATION (deh-gruh-DAY-shun) DEGRADATION

The conversion or breakdown of a substance to simpler compounds. For example, the degradation of organic matter to carbon dioxide and water.

FLOW LINE FLOW LINE

(1) The top of the wetted line, the water surface or the hydraulic grade line of water flowing in an open channel or partially full conduit.

(2) The lowest point of the channel inside a pipe or manhole. See INVERT. *NOTE:* (2) is an improper definition, although used by some contractors.

FLUSHING FLUSHING

The removal of deposits of material which have lodged in sewers because of inadequate velocity of flows. Water is discharged into the sewers at such rates that the larger flow and higher velocities are sufficient to remove the material.

HAND ROD HAND ROD

A sewer rod that can be inserted manually (by hand) into a sewer to clear a stoppage or to prevent a stoppage from developing.

HEAD HEAD

The vertical distance, height or energy of water above a point. A head of water may be measured in either height (feet) or pressure (pounds per square inch (psi)).

HERBICIDE (HERB-uh-SIDE) HERBICIDE

A compound, usually a manmade organic chemical, used to kill or control plant growth.

HIGH-VELOCITY CLEANER HIGH-VELOCITY CLEANER

A machine designed to remove grease and debris from the smaller diameter sewer pipes with high-velocity jets of water. Also called a "jet cleaner," "jet rodder," "hydraulic cleaner," "high-pressure cleaner," or "hydro jet."

HYDRAULIC CLEANING HYDRAULIC CLEANING

Cleaning pipe with water under enough pressure to produce high water velocities.

(1) Using a high-velocity cleaner.

(2) Using a ball, kite or similar sewer cleaning device.

(3) Using a scooter.

(4) Flushing.

INSECTICIDE INSECTICIDE

Any substance or chemical formulated to kill or control insects.

INSERTION PULLER INSERTION PULLER

A device used to pull long segments of flexible pipe material into a sewer line when sliplining to rehabilitate a deteriorated sewer.

INVERT (IN-vert) INVERT

The lowest point of the channel inside a pipe or manhole. See FLOW LINE.

KITE KITE

A device for hydraulically cleaning sewer lines. Resembling an airport wind sock and constructed of canvas-type material, the kite increases the velocity of a flow at its outlet to wash debris ahead of it. Also called a "parachute."

MANHOLE JACK MANHOLE JACK

A device used to guide the tag line into the sewer without causing unnecessary wear and provide support as the tag line is pulled back and forth.

MATERIAL SAFETY DATA SHEET (MSDS) MATERIAL SAFETY DATA SHEET (MSDS)

A document which provides pertinent information and a profile of a particular hazardous substance or mixture. An MSDS is normally developed by the manufacturer or formulator of the hazardous substance or mixture. The MSDS is required to be made available to employees and operators whenever there is the likelihood of the hazardous substance or mixture being introduced into the workplace. Some manufacturers are preparing MSDSs for products that are not considered to be hazardous to show that the product or substance is *NOT* hazardous.

MECHANICAL CLEANING MECHANICAL CLEANING

Clearing pipe by using equipment that scrapes, cuts, pulls or pushes the material out of the pipe. Mechanical cleaning devices or machines include bucket machines, power rodders and hand rods.

ORIFICE (OR-uh-fiss) ORIFICE

An opening (hole) in a plate, wall, or partition. An orifice flange or plate placed in a pipe consists of a slot or a calibrated circular hole smaller than the pipe diameter. The difference in pressure in the pipe above and at the orifice may be used to determine the flow in the pipe.

OXIDATION-REDUCTION POTENTIAL (ORP) OXIDATION-REDUCTION POTENTIAL (ORP)

The electrical potential required to transfer electrons from one compound or element (the oxidant) to another compound or element (the reductant); used as a qualitative measure of the state of oxidation in wastewater treatment systems. ORP is measured in millivolts, with negative values indicating a tendency to reduce compounds or elements and positive values indicating a tendency to oxidize compounds or elements.

PESTICIDE PESTICIDE

Any substance or chemical designed or formulated to kill or control animal pests. Also see INSECTICIDE and RODENTICIDE.

PIPE RODDING PIPE RODDING

A method of opening a plugged or blocked pipe by pushing a steel rod or snake, or pulling same, through the pipe with a tool attached to the end of the rod or snake. Rotating the rod or snake with a tool attached increases effectiveness.

PORCUPINE PORCUPINE

A sewer cleaning tool the same diameter as the pipe being cleaned. The tool is a steel cylinder having solid ends with eyes cast in them to which a cable can be attached and pulled by a winch. Many short pieces of cable or bristles protrude from the cylinder to form a round brush.

POWER RODDER POWER RODDER

A machine designed to remove roots, grease, and other materials from pipes. Also referred to as rodding machines.

PRECIPITATE (pre-SIP-uh-TATE) PRECIPITATE

(1) An insoluble, finely divided substance which is a product of a chemical reaction within a liquid.

(2) The separation from solution of an insoluble substance.

PREDICTIVE MAINTENANCE PREDICTIVE MAINTENANCE

The ability to identify problem areas before breakdowns or blockage of flow occurs. Predictive maintenance is the end product of effective preventive maintenance.

PREVENTIVE MAINTENANCE PREVENTIVE MAINTENANCE

Regularly scheduled servicing of machinery or other equipment using appropriate tools, tests and lubricants. This type of maintenance can prolong the useful life of equipment and machinery and increase its efficiency by detecting and correcting problems before they cause a breakdown of the equipment.

ROD (SEWER) ROD (SEWER)

A light metal rod, three to five feet long with a coupling at each end. Rods are joined and pushed into a sewer to dislodge obstructions.

RODDING RODDING

(See PIPE RODDING)

RODDING MACHINE RODDING MACHINE

(See POWER RODDER and PIPE RODDING)

RODDING TOOLS RODDING TOOLS

Special tools attached to the end of a rod or snake to accomplish various results in pipe rodding.

RODENTICIDE (row-DENT-uh-SIDE) RODENTICIDE

Any substance or chemical used to kill or control rodents.

ROTAMETER (RODE-uh-ME-ter) ROTAMETER

A device used to measure the flow rate of gases and liquids. The gas or liquid being measured flows vertically up a tapered, calibrated tube. Inside the tube is a small ball or bullet-shaped float (it may rotate) that rises or falls depending on the flow rate. The flow rate may be read on a scale behind or on the tube by looking at the middle of the ball or at the widest part or top of the float.

SCOOTER SCOOTER

A sewer cleaning tool whose cleansing action depends on the development of high water velocity around the outside edge of a circular shield. The metal shield is rimmed with a rubber coating and is attached to a framework on wheels (like a child's scooter). The angle of the shield is controlled by a chain-spring system which regulates the head of water behind the scooter and thus the cleansing velocity of the water flowing around the shield.

SEPTIC (SEP-tick) SEPTIC

A condition produced by anaerobic bacteria. If severe, the wastewater produces hydrogen sulfide, turns black, gives off foul odors, contains little or no dissolved oxygen, and the wastewater has a high oxygen demand.

SEWER BALL SEWER BALL

A spirally grooved, inflatable, semi-hard rubber ball designed for hydraulic cleaning of sewer pipes. See BALLING.

SLIPLINING SLIPLINING

A sewer rehabilitation technique accomplished by inserting flexible polyethylene pipe into an existing deteriorated sewer.

STOPPAGE STOPPAGE

(1) Partial or complete interruption of flow as a result of some obstruction in a sewer.

(2) When a sewer system becomes plugged and the flow backs up, it is said to have a "stoppage." Also see BLOCKAGE.

SWAB SWAB

A circular sewer cleaning tool almost the same diameter as the pipe being cleaned. As a final cleaning procedure after a sewer line has been cleaned with a porcupine, a swab is pulled through the sewer and the flushing action of water flowing around the tool cleans the line.

VISCOSITY (vis-KOSS-uh-tee) VISCOSITY

A property of water, or any other fluid, which resists efforts to change its shape or flow. Syrup is more viscous (has a higher viscosity) than water. The viscosity of water increases significantly as temperatures decrease. Motor oil is rated by how thick (viscous) it is; 20 weight oil is considered relatively thin while 50 weight oil is relatively thick or viscous.

WATER HAMMER WATER HAMMER

The sound like someone hammering on a pipe that occurs when a valve is opened or closed very rapidly. When a valve position is changed quickly, the water pressure in a pipe will increase and decrease back and forth very quickly. This rise and fall in pressures can cause serious damage to the system.

CHAPTER 6. PIPELINE CLEANING AND MAINTENANCE METHODS

(Lesson 1 of 5 Lessons)

6.0 HOW TO IDENTIFY PROBLEMS AND SELECT SOLUTIONS

6.00 Objectives of a Sewer Cleaning and Maintenance Program

In Chapter 5 we discussed the need for a physical inspection program for collection systems in order to develop a system inventory including the size, type, age, and condition of the system. Inspection also helps us to locate gravity sewers and their alignment. This information enables us to prepare accurate maps of the system and to identify defects in the system so that short-term maintenance requirements can be identified and long-term repair and rehabilitation programs can be planned. Once we know what we have in the ground, where it is located, and what the condition of the system is, we can develop a maintenance and cleaning program that will keep the system operating efficiently.

To operate and maintain a wastewater collection system so it will function as intended, try to strive toward the following objectives:

1. Minimize the number of stoppages per mile of sewer,
2. Minimize the number of odor complaints, and
3. Minimize the number of lift station failures.

A good maintenance program is the key to achieving these objectives. The most effective maintenance program is one that relies primarily on planning and scheduling maintenance activities so as to prevent problems from occurring. Sewer cleaning using hydraulic or mechanical cleaning methods needs to be done on a scheduled basis to remove accumulated debris in the pipe such as sand, silt, grease, roots, and rocks. If debris is allowed to accumulate, it reduces the capacity of the pipe and a blockage can eventually occur resulting in overflows from the system onto streets, yards, and into surface waters. Roots and corrosion also can cause physical damage to sewers.

Sewer cleaning should be scheduled on a regular basis; some agencies, for example, clean 100 percent of their system on a three- or five-year schedule. This approach is not necessarily the most effective way to schedule preventive maintenance because some sections will not require cleaning that often and other sections may require more frequent cleaning (such as monthly) if they are susceptible to blockages. It is important to make this schedule an active plan that is adjusted as problems are solved or problems are found.

Another important part of the maintenance program is long-term planning to identify and, if necessary, rehabilitate sewer lines that repeatedly experience problems such as blockages. Identifying problems before breakdown, failure, or blockages can occur is predictive maintenance, the result of a good preventive maintenance program. Information from the preventive maintenance program can be used over time to fairly accurately predict when lines need to be cleaned, pumps rebuilt, or other necessary work performed in advance of problems. Fixing the most serious problem areas will greatly reduce the ongoing maintenance requirements of the collection system. While we can never eliminate corrective and emergency maintenance, we can reduce the frequency, cost, impact, and consequences of these emergencies by developing and implementing an effective predictive maintenance program through preventive maintenance.

This chapter discusses how to identify the causes of stoppages and odors, how to select solutions to these problems, and how to actually solve these problems. Lift station problems and prevention of failures are presented in Chapter 8, "Lift Stations," and Chapter 9, "Equipment Maintenance," in Volume II.

6.01 Identification of Problem

Potential problems are often found during routine cleaning and maintenance. Our job is to try to prevent stoppages that will result in wastewater backing up and overflowing onto the street or into homes or businesses. Webster's Dictionary defines a stoppage as the act of stopping or arresting motion, progress or action; also, the state of being stopped or halted, such as an obstruction or a block in a water pipe. Therefore, we can consider a sewer stoppage to occur when a sewer system becomes plugged and the flow backs up. Sometimes a stoppage can occur but the flow is not completely stopped; some flow will leak or seep by the stoppage.

6.010 Types of Stoppages

For the sake of recording stoppages for follow-up maintenance work, stoppages must be identified according to the cause or type of problem. Stoppages are caused by obstructions such as roots, grease, debris, broken pipe, or a joint failure. These obstructions may require immediate removal, repair, or replacement to correct the problem.

Many stoppages are caused by people not realizing the consequences of their activities. For example, vandals have placed debris into lines and structure openings which were designed to vent the collection system. Also, they have placed materials or debris into structures designed for the purpose of performing maintenance work in the sewers. Manhole covers in isolated sewer easements have been removed by vandals. These openings offer an ideal place for vandals to cause a serious backup of wastewater in a very critical area.

Some stoppage problems created in sewer easement areas also may be caused unintentionally. This is especially true when a large piece of construction equipment begins rough grading work and this equipment knocks off the top of a manhole, spilling dirt, rocks, and other debris into the opening and causing a stoppage in the sewer. In most cases, the equipment operator is not aware of the location of the manholes because they are buried or covered with overgrowth.

Other physical stoppages are caused by obstructions found in the *BARREL*[1] of the sewer. These obstructions are sometimes found to be created by plumbers, equipment, poor installation, or by some force of nature. A plumber-made obstruction can be caused by the placement of a building sewer tap connection that protrudes into the main sewer, a poorly repaired pipe section, backfill damage to pipe and the misuse of trench compaction equipment, pipe damage due to the improper use of sewer cleaning equipment, and other such physical activities.

The forces-of-nature type of obstructions include the penetration of roots into the pipe connection or openings, which causes pipe to break and/or restricts water flow. Ground movement shifting the pipe (such as caused by earthquakes), weather changes, soil conditions underground, and the deterioration of other utility pipes are effects caused by natural forces that could produce obstructions.

Some of the most common types of debris found when removing a stoppage are a buildup of solidified grease, detergents, sticks, rags, plastic bags, broken pipe, brick, rocks, sand, eggshells, and silt, to name a few. Larger items removed from lines and manhole openings which have caused a major problem for removal are broken manhole and flushing inlet castings; concrete and asphalt rubble; steel rebars; large metal and plastic buckets; broken and lost plumber rods, snakes, and plugs; wooden posts and timber materials; barbed wire; tree limbs, stumps, and many other items.

In some instances where the *INVERT*[2] of a small-diameter sewer is connected to the invert of a large-diameter sewer that is flowing at maximum flows, hydraulic conditions can cause serious stoppages to develop. Not only do stoppages develop in the sewer, but solids can build up in a manhole and produce harmful and obnoxious gases. Most instances of stoppages caused by hydraulic conditions are found in sewers that have been extended to new developments from existing mains which were not designed for the future potential growth of a community.

A small "belly" (low spot) in a sewer line can cause debris to settle out and accumulate. Many people mistakenly believe that the wastewater will simply flow over the settled debris. However, even a small mound of particles can snag a piece of toilet paper or other solid material. As more debris accumulates at the low spot, the water in the bottom of the "belly" becomes *SEPTIC*[3] or *ANAEROBIC*[4] and hydrogen sulfide gas is produced by the decomposing debris. Wastewater flowing over the spot then causes the hydrogen sulfide gas to come out of solution, thereby producing an odor problem. If left uncorrected, accumulation of debris in these small bellies can cause a backup of wastewater in the line. A good inspection program at installation time, one that includes a closed-circuit TV inspection, prior to acceptance is the best way to prevent this problem.

Stoppages due to hydraulic conditions can be minimized by proper design. Wherever flows from one sewer meet and join the flow in another sewer, the flows should always be going downstream in the same direction as much as possible. A more frequent cleaning schedule is the best possible solution to hydraulic stoppages in many cases.

6.011 Identifying Causes of Stoppages and Problems

First, try to correct the problem as quickly as possible. Once the cause of the problem has been identified, the methods for solving the problem can be analyzed.

A mental checklist to identify the cause of a problem would include:

1. Does this line have a history of previous stoppages (a root problem line or a grease problem line)?
2. Are trees growing near the line?
3. Have new building sewers or new lateral or branch sewers been installed recently?
4. Have repairs been made recently to the sewer, to other utilities, or to the street?
5. Are there any ground or surface indications, such as a settlement or a sink hole?

Receipt of numerous odor complaints often indicates that a sewer needs to be cleaned. To determine if a sewer needs cleaning, sniff for odors, visually inspect manholes for signs of grease or debris, analyze records, and inspect the sewer using closed-circuit television. It may be necessary to clean the sewer by balling or high-velocity cleaning before televising to allow for better inspection of the sewer.

---

[1] *Barrel. The cylindrical part of a pipe that may have a bell on one end.*

[2] *Invert (IN-vert). The lowest point of the channel inside a pipe or manhole.*

[3] *Septic (SEP-tick). A condition produced by anaerobic bacteria. If severe, the wastewater produces hydrogen sulfide, turns black, gives off foul odors, contains little or no dissolved oxygen, and the wastewater has a high oxygen demand.*

[4] *Anaerobic (AN-air-O-bick). A condition in which atmospheric or dissolved molecular oxygen is NOT present in the aquatic (water) environment.*

QUESTIONS

Write your answers in a notebook and then compare your answers with those on page 383.

6.0A To operate and maintain a wastewater collection system properly, what objectives should be considered?

6.0B What happens when a stoppage or blockage occurs in a sewer?

6.0C What are the major causes of stoppages?

6.0D How would you attempt to identify the cause of a stoppage?

6.02 Methods for Cleaning and Maintaining Sewers

Sewer cleaning and maintenance methods depend on the characteristics of the wastewater being conveyed to the treatment plant, fluctuations in wastewater flows, alignment or grade of the sewer, pipe material, and the physical condition of the sewer.

Hydraulic or mechanical cleaning equipment and methods can be used to clear emergency stoppages and for performing preventive maintenance work. Hydraulic cleaning methods consist of cleaning a sewer with water under pressure that produces high water velocities. These velocities are usually high enough to wash most grit, grease, and debris found in sewers down the sewer and leave the pipe clean. Mechanical methods of clearing stoppages in sewers consist of using equipment that scrapes, cuts, pulls, or pushes the material out of the pipe.

Hydraulic cleaning equipment used today includes high-velocity cleaners, balls, kites, bags, parachutes, tires, and scooters. High-velocity cleaners rely on water under pressure to force water through a nozzle at the end of the hose to scour the pipe walls. The water pressure also serves to move the hose and nozzle through the pipe. This high-velocity water cleans the invert and the walls of the pipe and moves the debris downstream where it can be removed at a manhole. The high-velocity water cleaners also are called "jet cleaner," "jet rodder," "hydraulic cleaner," "hydro jet," or "high-pressure cleaner." Balls, kites, bags, parachutes, tires, and scooters all rely on the same principle. These devices fit into a sewer and partially block the flow. Water builds up behind the device and creates a pressure that forces water at a high velocity around the outside edge of the cleaning device. This high water velocity cleans the walls of the sewer and pushes the material and debris downstream where it can be removed at a manhole. Flushing is a hydraulic cleaning method that relies on additional water being discharged into a sewer in sufficient quantities to create high velocities and clean the sewer.

Mechanical cleaning equipment consists of bucket machines, power rodders, hand rods, and winches for pulling scrapers, porcupines, swabs, and various cutting tools through the pipe. The tools remove hardened grease, invert debris, and roots or physical obstructions such as bricks, rocks, or other large objects that find their way into the sewer system. Bucket machine operation consists of a set of specialized winches that pull a device through a pipe to collect debris. Materials are captured using a special bucket and are removed from the pipe. Rodding (either using a power rodder or a hand rodder) consists of pushing or pulling a steel rod or snake through a sewer with a special tool attached to the end. These tools are designed to cut or scrape materials from the pipe walls and are most effective on hardened grease and roots.

Both hydraulic and mechanical cleaning methods are used to maintain sewers in working condition. Chemicals and bacteria cultures also are used in sewer maintenance programs to keep lines clear.

Odor control (especially hydrogen sulfide control) requires that sewers be maintained as clean as possible. Hydraulic and mechanical cleaning methods as well as chemicals are used in odor control programs to maintain wastewater collection systems.

6.03 Selection of Solution to Problem

Some methods for cleaning sewers can also be used for removing stoppages while others cannot. For specific jobs or routine preventive maintenance, different methods are available depending on whether the problem is caused by roots, grease, sand, or some other source.

Table 6.1 outlines:

1. How to identify problems and their source,
2. Methods of correcting these problems,
3. Section in manual where solution procedures are described in detail, and
4. Comments helpful for selecting the best method.

When you have a stoppage or flow problem in a sewer, try to identify the problem. Refer to Table 6.1 to see if the problem is listed. Review the methods of correcting the problem suggested in the table, select a method, correct the problem, and evaluate the effectiveness of the solution.

This table cannot cover all problems or all equipment, but can be useful as a guide to effective and efficient collection system maintenance. The paragraphs below outline the equipment required for removing stoppages and cleaning lines, while the other sections in this chapter describe solution procedures in detail.

6.04 Equipment Required for Removing Stoppages and Cleaning Sewers

What type of equipment is best suited for removing stoppages and cleaning sewers? The answer to this question can only be determined after an analysis of the cause of the problem and the solution selected. Following is a list of the basic types of equipment used to identify and remove stoppages and clean sewers.

TABLE 6.1 CLEANING AND MAINTENANCE PROBLEMS AND SOLUTIONS

Identification of Problem	Source or Cause	Solution Method	Section	Comments
1. Stoppages — Emergency	1.a Grease	High-Velocity Cleaner	6.11	A high-velocity cleaner will open most grease stoppages.
a. Manhole overflowing b. Flooding of residences or businesses		Hand Rods, Power Rodder	6.22 6.21	Rod from downstream manhole with a 4-inch auger into stoppage. When clear, then run 6-inch or larger auger through restricted area. Initiate work order to clean with high-velocity cleaner or ball line as soon as possible. Hand rods and power rodders will usually unplug most grease stoppages.
	1.b Roots	High-Velocity Cleaner	6.11	High-velocity cleaner will usually open stoppage and restore service. Schedule TV inspection and chemical treatment. If unable to clear roots with high-velocity cleaner, power rodder may be necessary.
		Hand Rods	6.22	Rod from downstream manhole with 4-inch auger or saw. Be cautious in opening stoppages if there is a high head on the upstream manhole. Remove as much of the root mass as possible. Request TV inspection for root concentration; schedule power rodding to open line and chemically treat to control future root growth. Hand rods are effective 90 percent of the time. If cannot clear roots with hand rods, power rodder may be necessary.
		Power Rodder	6.21	Power rod line to clear stoppage. Schedule TV check, cleaning, and chemical treatment.
	1.c Debris stoppage such as rocks, lumber			Caused by broken lines, open manholes, vandalism.
		High-Velocity Cleaner	6.11	Clean line with high-velocity cleaner.
		Power Rodder	6.21	Use caution. May jam tool in line requiring a dig up to clear line and remove broken rod and tool.
2. Grease	a. Restaurant on blocked segment of sewer	a. High-Velocity Cleaner	6.11	High-velocity cleaner is an effective tool in removing grease buildups in line sizes up to 15 inches. High-velocity cleaner becomes ineffective in larger diameter pipes.
a. Stoppage causes grease buildup	b. Low velocity allowing grease buildup from home disposal unit. Problems often develop where high velocities are suddenly slowed down.	b. Balling (or tire)	6.10	Balling will remove grease deposits from pipe walls, but will not clean as effectively as properly used high-velocity cleaner.
b. TV report on routine inspection		c. Scooter (or kite)	6.13	More effective than high-velocity cleaner in lines above 18-inch diameter.
c. Observe buildup on side walls of sewer		d. Power Rodder	6.21	High-velocity cleaners are effectively used to remove grease; however, they are most extensively used and highly effective in grease removal following the use of steel tools. High-velocity cleaners should be used after complete material removal.
d. Past records		e. Chemicals	6.4	Be sure to insist on a performance contract. Do not pay until the chemical or material performs as claimed.
		f. Bacteria Cultures	6.43	Specific cultures are required for collection system maintenance.
e. Grease trap		a. Clean Trap Regularly	6.43	A regular maintenance program must be established and continued.
		b. Bacteria Cultures	6.43	Specific cultures are required for collection system maintenance.
3. Roots	a. Trees and shrubs	a. Chemicals	6.4	For long-term control, chemical treatment provides the best solution with up to three years between applications. Personnel applying root control chemicals may need to be certified and must take all recommended safety precautions when using chemicals.
a. TV report b. Poor joints or damaged pipe allow root entry		b. Power Rodder	6.21	Best suited for opening pipe with root obstruction. If rodding is the only method used, special tools are required to cut roots to the pipe walls and periodic cleaning will be required.

TABLE 6.1 CLEANING AND MAINTENANCE PROBLEMS AND SOLUTIONS (continued)

Identification of Problem	Source or Cause	Solution Method	Section	Comments
3. Roots (continued)		c. High-Velocity Cleaner	6.11	Special root cutters are available.
c. Past records	d. Repairs		Chap. 7	If TV report shows only one section of broken line or a few bad joints, dig up and repair. If a great number of defects are observed, consider pressure sealing or relining the pipe by insertion of a liner.
4. Sand, Grit, Debris a. TV report b. Grit settles during low flows c. Grit sticks to grease or slimes d. Routine inspection e. Past records	a. Eggshells, coffee grounds, bones from residential disposal units b. Broken china, bones, and glass from restaurant disposal units c. Sand, silt from poor joints and broken lines d. Lines with low flows or velocities permitting deposition of solids	a. High-Velocity Cleaner	6.11	For light concentrations of grit in small lines; not effective cleaner in lines above 15-inch diameter.
		b. Balling (or tire)	6.10	The workhorse for cleaning. Large volumes can be removed at a reasonable cost. Requires careful control in shallow lines.
		c. Scooters and Kites	6.13	More effective in larger lines. Removes some dangers of flooding in shallow lines that balling may create if not properly controlled.
		d. Bucket Machines	6.20	Used where extreme concentrations of grit and sand have loaded the line to extent that above methods are ineffective due to cost and handling of material to be removed.
5. H_2S and Odor Control a. Odor complaints b. Manhole inspection reveals line deterioration	a. Force mains b. Low flows and velocity c. Offset joints d. Bellies in line e. Drop manholes f. Manhole where trucks dump septic tank contents	a. High-Velocity Cleaner	6.11	Fast cleaning of slimes in lines up to 15-inch diameter.
		b. Balling	6.10	Best for sewers with bellies and offset joints, but expensive operation for odor control only.
		c. Scooter	6.13	Fast for larger lines.
		d. Flushing	6.12	Small lines. Usually not effective for more than one week.
		e. Plug Lifting and Vent Holes in Manhole Cover		Roofing cement makes a satisfactory hole sealer. Specially made plugs are available from vendors.
		f. Control Program	6.5	Develop program using combination of solutions.
6. System Inspection	Detects problem areas and permits realistic scheduling of preventive maintenance program	a. Closed-Circuit TV	5.3	Permits thorough inspection of system. Pinpoints damaged areas such as broken pipe segments, offset joints, lines not to grade, collapsed pipes, protruding service taps, root intrusion, deterioration of lines, and grease deposits. Informs you of required cleaning and the effectiveness of the cleaning method. If a stoppage reoccurs or you are curious regarding cause of problem, TV the line. The most useful tool in collection system maintenance.
		b. Air and Water Testing	Chap. 7	Detection of sources of infiltration and exfiltration.
		c. Smoke Testing	5.4	Locates leaks when groundwater table is below sewer. Also used to determine if buildings and residences are properly connected to collection system and to identify illegal connections. Excellent for finding odor sources in buildings.
		d. Dye Testing	5.5	Identification of illegal connections and location of overflowing or leaking manholes and sewers.
		e. Pipeline Lamping	5.6	Inspection of sewers for condition, including alignment and obstructions.
		f. Visual	5.2	Lift manhole covers and observe conditions.

6.040 Rodding

Equipment: Truck- or trailer-mounted rodding machine or hand rods with an assortment of specific cutting tools and operational accessories.

Advantages: May be used to cut roots, remove hardened grease, scrape, and dislodge certain types of materials found in sewers. This equipment is very effective in removing collection system sewer line emergency stoppages.

Limitations: Ineffective for removing sand and grit accumulations in sewers, but can loosen materials so they can be flushed out with hydraulic cleaners. Rods have a tendency to bend or coil in large-diameter pipes and are most effective in pipes that are 15 inches in diameter or smaller. When using hand rods in a sewer with a high *HEAD*[5] condition behind an obstruction, caution must be exercised when the obstruction is broken because the rods may be forced back out of the manhole.

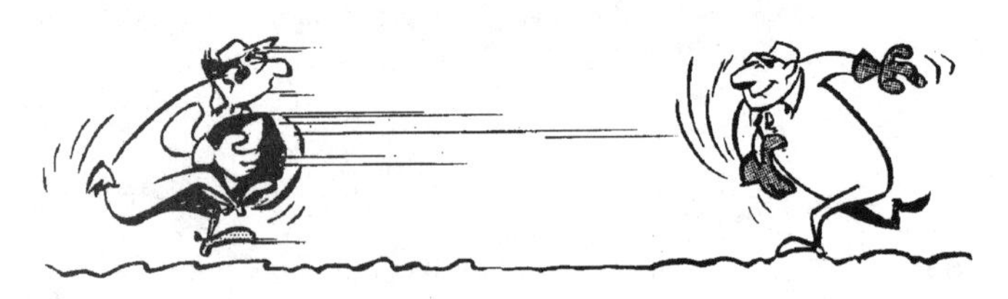

6.041 Sewer Ball or Tire

Equipment: Assortment of various sizes of sewer balls or tires to fit different diameters of sewers, tag line winch, cable, reels, water tank truck, and dump pickup. Provide water from a hydrant through a hose with an *AIR GAP.*[6]

Advantages: Hydraulic action of spinning ball and high velocity of water flowing around ball dislodge debris from pipe walls and move debris downstream. Very effective in removing heavy concentrations of sand, grit, rock, and grease from sewers.

Limitations: Dangerous to use in locations with basement fixtures and steep-grade hill areas. Strong possibility of flooding dwellings. Cannot be used effectively when sewers have bad offset joints or protruding service connections because ball can become distorted, get past these obstructions, and not do an effective cleaning job. In sewers where the ball has a tendency to get stuck, a bag full of ice can be used. When the bag becomes stuck, wait for the ice to melt sufficiently for the bag to continue downstream.

6.042 High-Velocity Cleaner

Equipment: Truck- or trailer-mounted high-velocity cleaner (hydro jet) with an assortment of nozzles, easement rollers, and operational accessories.

Advantages: Very effective preventive maintenance tool for cleaning flat, slow-flowing sewers. Efficient in removing grease, sand, gravel, grit, and debris in smaller diameter sewers. Effective in breaking up solids in manholes and washing or cleaning structures. May be used to remove emergency stoppages.

Limitations: Efficiency and effectiveness in removing debris from larger diameter lines decreases as the cross-sectional area of the pipe increases. When used by an inexperienced operator, this method may cause a backup into residences. Extra time and water consumption and an experienced operator may be necessary to remove hardened grease, hardened debris, and roots.

6.043 Bucket Machine

Equipment: Power bucket machine and power bucket machine truck loader, dump truck, and assorted operating rollers and tools.

Advantages: Best suited to remove large amounts of sand and debris from larger diameter sewers. Provides the ability to work effectively in larger pipe sizes where other cleaners cannot work efficiently.

Limitations: Setup can be time consuming and bucket operations require some advance planning. Can damage small pipes if not properly used. May not clean all debris out of large lines and other hydraulic cleaning devices should be used after bucketing a line.

6.044 Kite or Bag

Equipment: Water tank truck, dump pickup truck, and power drum winch machine.

Advantages: Very effective in moving accumulations of decayed debris and grease downstream. Capable of washing ahead of it a full pipe of deposits, including roots.

Limitations: Must use caution in locations with basement fixtures and steep-grade hill areas.

6.045 Scooter

Equipment: Water tank truck, dump pickup truck, tag line, and power winch. Provide water from a hydrant through hose with an air gap.

---

[5] *Head. The vertical distance, height or energy of water above a point. A head of water may be measured in either height (feet) or pressure (pounds per square inch (psi)).*

[6] *Air Gap. An open vertical drop, or vertical empty space, between a drinking (potable) water supply and the point of use. This gap prevents backsiphonage because there is no way wastewater can reach the drinking water. Air gap devices are used to provide adequate space above the top of a manhole and the end of the hose from the fire hydrant. This gap ensures that no wastewater will flow out the top of a manhole, reach the end of the hose from a fire hydrant, and be sucked or drawn back up through the hose to the water supply.*

Advantages: Hydraulic water action scours inside of line. Very effective in removing heavy debris and cleaning grease from line.

Limitations: When cleaning larger lines, the manholes must be designed to a larger size to receive and retrieve this equipment or the scooter must be assembled in the manhole. Must use caution in locations with basement fixtures and steep-grade hill areas.

6.046 Flushing

Equipment: Water tank truck and fire hose if fire hydrant is not convenient.

Advantages: Supplies a surge of water to move light, decaying organic matter downstream in slowly flowing sewers.

Limitations: Causes a temporary movement of debris from one point to another in the collection system. Flushing does not remedy the cause of the problem and does not move heavy debris and grit. Must use caution in locations with basement fixtures and steep-grade hill areas.

6.047 Scraper

Equipment: Wing scraper, "C" drag bucket, power bucket machine, basin machine, and dump truck or bin.

Advantages: Removes large amounts of sand, stone, rocks, and debris from larger (usually 36-inch diameter or larger) sewers.

Limitations: Sewers can be damaged. Setting up of equipment is time consuming; therefore, the operation is usually run in two or three shifts during a 24-hour period.

6.048 Chemical Root Treatment

Equipment: Water, dry chemical, or

Hydro jet (or high-velocity cleaner), foam dispersal unit, dry chemical, or

Mechanical foam-producing machine (adds compressed air to liquid), liquid chemical, water, and applicator nozzles.

Advantages: Destroys root system, decays the dead roots, and inhibits regrowth.

Chemical treatment for root control produces longer lasting results than mechanical cleaning.

Limitations: When the root tips are damaged or removed by line cleaning, chemical treatment will be less effective; therefore, no cleaning is recommended in lines prior to chemical root treatment unless extensive grease, root masses, or debris would interfere with proper application of the material.

Chemical treatment may be difficult when flow in pipe is at full capacity.

Flooding of homes or businesses may occur.

Some foaming methods require operators to hold a qualified applicator's license or certificate before applying product (refer to your state's or province's requirements).

Exposure to potentially hazardous chemicals.

6.05 Summary of Solutions

Table 6.2 shows the relative effectiveness of the possible solutions to different problems. The larger the number in the table, the more effective the solution is for a particular problem. One problem may have several effective solutions and another problem may have only one possible solution. *REMEMBER* that Table 6.2 is a summary table to provide you with an indication of the general effectiveness of solutions. *YOU* must evaluate every problem and select the best solution using available equipment.

6.06 Work Assignment

When a crew is assigned a regularly scheduled sewer cleaning job, the crew should be notified and dispatched by a written order (see Chapter 7, Section 7.04, "Work Request Form"). Crews assigned to clear emergency stoppages are notified by radio and immediately attempt to clear the stoppage. A written report is prepared after the stoppage is cleared. Work performed also must be recorded. This information provides vital documentation of the ability of the wastewater collection system to perform its intended job and also the effectiveness of equipment and crews. Volume II, Chapter 12, Section 12.72, "Recordkeeping," provides a discussion of the need for records and the value of records.

The crew chief for most cleaning operations is the Maintenance Equipment Operator.[7] The chief receives the work assignments for the day from a Maintenance Supervisor I (depends on size of agency who prepares and makes work assignments). Work assignments should include the following information for each location:

1. Location of job, including map book number, map page, and manhole number.
2. Street or easement.
3. Location of upstream and downstream manholes.
4. Line-of-sight or sound-powered telephone communications.
5. If in street, traffic conditions, or
6. If in easement,
   a. Notification of property owners.
   b. Are dogs locked up?
   c. Keep children out of work area.
7. Sewer conditions:
   a. Size,
   b. Depth, and
   c. Water supply needed.

Upon arrival at the work site, the sewers to be cleaned are located using a map.

[7] *Portions of this section and job titles are from "Manpower Requirements for Wastewater Collection Systems," by Elie Namour for EPA, Grant No. 900211, March, 1974.*

TABLE 6.2 EFFECTIVENESS OF SOLUTIONS

(Effectiveness Scale: 1 = Low, 5 = High)

Solution to Problem	TYPE OF PROBLEM				
	Emergency Stoppage	Grease	Roots	Sand, Grit, Debris	Odors
Balling [a]		4		4	3
High-Velocity Cleaning [b]	5	5	3	4	3
Flushing					2
Sewer Scooters		3		3	
Bucket Machines, Scrapers				4	
Power Rodders	5	5	5		
Hand Rods	5 [c]	3	2		
Chemicals [d]		2	4		5
Bacteria [e]		4			

[a] Kites, tires, bags, parachutes, scooters, and cones are commonly used instead of balls in large sewers (greater than 24 inches in diameter) with similar results.
[b] Effectiveness decreases as pipe diameter increases.
[c] Power rodders and high-velocity cleaners may be faster (if available) under certain conditions.
[d] Effectiveness depends on type of chemical and its intended use.
[e] Effectiveness depends on formulation of cultures.

6.07 Records

A record of all cleaning operations should be made (Figure 6.1) and filed for future reference. These records should include the date, street name or number, line size, distance, and manhole numbers or identification. Also note the kind and amount of materials removed, wastewater flow, and amount of auxiliary water used. If particular problems were encountered, these too should be noted, especially the exact location of obstructions. See Volume II, Chapter 12, Section 12.7, for details on how to prepare record forms, complete forms (Figure 6.1), analyze records, and file records.

During the types of routine cleaning operations discussed in this chapter, many manholes will be opened and used for high-velocity cleaning, balling, kiting, or flushing of sewer lines. There is no better time to take stock of these manholes and complete a Manhole Inspection Form detailing its location, condition, and any problems observed (see Chapter 5, Section 5.2, "Inspection of Manholes"). If this is done each time a manhole is opened during cleaning operations, over time the database for these structures will include up-to-date information on a high percentage of them and allow better decisions to be made in regard to routine maintenance, repair, or rehabilitation.

If pieces of broken sewer are removed, this may indicate a TV inspection is needed and that a repair may need to be made on the broken sections of pipe.

Recording traffic patterns at a site can be very helpful next time equipment must be set up at the location. Indicate where cars park (such as over manholes), traffic volume during rush hours, and whether police traffic control should be called for help before going to the site.

Computers are being used in many aspects of collection system operation, maintenance, and recordkeeping. Computer software packages are available for scheduling preventive maintenance activities, issuing work orders for repairs, keeping track of where work is done, who did the work, when, and the labor and materials required. With the proper software, any information in the computer's records can be recalled for future use. Computers are also used to keep spare parts inventories and to order spare parts when the supply runs low and before they are needed for scheduled maintenance and repairs. See Chapter 12 in Volume II for a detailed description of how computers are being used to assist in the operation and maintenance of wastewater collection systems.

When making out records, remember that you or someone else will be referring to them. The more complete the record, the easier the next operation becomes since you have a history of this sewer.

6.08 Notify Wastewater Treatment Plant

Before clearing a large septic stoppage, be sure to notify the operator on duty at the downstream wastewater treatment plant. Septic stoppages develop when the sewer has been blocked for considerable time and/or the air temperature is hot. Under these conditions the wastewater and organic solids turn black and smell like rotten eggs. If a large-diameter sewer is blocked and a large volume of wastewater backs up in the pipes, there might not be sufficient fresh water arriving at the treatment plant to dilute the septic wastewater. When a large volume of septic wastewater reaches a treatment plant, the treatment processes can become upset and fail to do their intended job. By notifying the operator in advance of the location of the stoppage and approximate volume of septic wastewater flowing toward the treatment plant, the operator can be alerted and prepare to minimize the impact on the treatment processes.

Date 12-5-02 CREW LEADER'S - DAILY REPORT Foreman BEDEGREW

NAME	HOURS	ACCOUNT NUMBER	EQUIPT. NO.	MILES	HOURS	WORK PERFORMED
WHITE	8	X5-SAD-010 754	145-401	45		BALLED SEWER LINES SHOWN BELOW
COOK						
ARANT						

	S	V	OTO	NL	DOCK	EL	REMARKS:
PARRA	8						

All work on sewer lines must be shown below

Bk.	Pg.	Line Description	SR	PM	Rod	Ball	Gr.	Rt.	
			CR	TV					Footage Totals
200	07	M/H 1-2•6-5-4-3-16				✓			
		TO PG. 06-1•15-14-13•				✓			GRIT 45 GALLONS
		11-10•9-8•7-8-10-				✓			
		12-13 TO PG. 06-1				✓			TOTAL 3823' - 6"

Fig. 6.1 Daily maintenance report

QUESTIONS

Write your answers in a notebook and then compare your answers with those on page 383.

6.0E When a grease or root stoppage causes flooding, what is the first method attempted to solve the problem?

6.0F What is the most effective method of removing grease buildups?

6.0G How can a stoppage caused by roots be cleared?

6.0H List the advantages and limitations of a high-velocity cleaner.

Please answer the discussion and review questions next.

DISCUSSION AND REVIEW QUESTIONS

Chapter 6. PIPELINE CLEANING AND MAINTENANCE METHODS

(Lesson 1 of 5 Lessons)

At the end of each lesson in this chapter you will find some discussion and review questions. The purpose of these questions is to indicate to you how well you understand the material in the lesson. Write the answers to these questions in your notebook before continuing.

1. Why should the cause of a stoppage be identified?
2. What are the causes of stoppages?
3. How would you select a method to clear a stoppage?
4. What are the advantages and limitations of using sewer balls as a method of cleaning sewers?
5. How would you determine if a sewer needed cleaning?
6. What would you do if you encountered an emergency stoppage with wastewater flowing out a manhole and down the street?

CHAPTER 6. PIPELINE CLEANING AND MAINTENANCE METHODS

(Lesson 2 of 5 Lessons)

6.1 HYDRAULIC CLEANING

Sewer cleaning methods described in this section rely on the cleansing action caused by high velocities of water in the line. High water velocities can effectively remove grease, sand, and other debris. Cleansing velocities can be obtained by allowing water pressure or head to build up in the line or by using a pump to produce the water pressure. Hydraulic cleaning methods include balling, high-velocity cleaners, flushing, sewer scooters, kites, tires, and poly pigs.

WARNING

WHEN USING THESE METHODS, CAUTION **MUST** BE TAKEN TO BE SURE BASEMENTS AND HOMES ARE **NOT** FLOODED

QUESTIONS

Write your answers in a notebook and then compare your answers with those on page 383.

6.1A List three sewer line hydraulic cleaning methods.

6.1B What is a major precaution that must be exercised when hydraulic cleaning methods are being used?

6.10 Balling

Balling is one method of removing debris from sewers. The tool is literally a "ball" of special design (Figure 6.2). The main purposes of balling or use of tires are to keep the sewer clear of debris and to maintain flow velocities of two feet per second or more to prevent solids deposition in the sewers flowing to the wastewater treatment plant.

This cleaning procedure is one of the less expensive methods and is quite popular in some cities. Usually balling is not the only cleaning method used by a sanitation agency since the ball primarily removes deposits of inorganic (grit) material lying on the bottom of the line and grease buildup inside the line. Balling can be used only in areas where the necessary water pressure behind and around the ball can be obtained without flooding basements or homes at low elevations. Flooding can occur when the elevation of the head of water on the upstream side of the ball is higher than a plumbing fixture in a home or basement. If many roots are involved, a power rodding machine probably will be needed. Power rodding and other cleaning methods and procedures will be discussed fully in other sections of this chapter.

Balls are available in sizes 6, 8, 10, and 12 inches (for metric pipe sizes, see page 583) for small lines and sizes up to 48 inches for large trunk lines. Larger sizes may be obtained by special ordering.

Balling is most commonly used in preventive maintenance programs. Under these conditions balling is very effective in reducing the possibility of stoppages developing. An effective balling program can reduce the production of hydrogen sulfide (see Section 6.5, "Hydrogen Sulfide Control") in collection systems, thus reducing damage from hydrogen sulfide-caused corrosion and the release of odors smelling like rotten eggs.

Some sewers require cleaning by balling more frequently than others. Required frequency of cleaning may vary from six months in some sluggish lines to three years in other lines. Realize that these are typical frequencies and that some sewers may require monthly cleaning while other sewers may never need cleaning. Many communities try to ball their entire collection system every year. To determine the frequency of balling for your collection system, study your agency's records. Base the frequency of cleaning the various sections of the collection system on:

1. Reducing the number and types of stoppages and complaints by analyzing records and inspecting manholes for the presence and amount of deposits;
2. Size of area the collection system serves;
3. Types of waste carried by the collection system:
   a. Residential,
   b. Commercial,
   c. Industrial, and
   d. Storm water.

6.100 Personnel and Equipment

At least three operators, one Maintenance Equipment Operator[8] and two Maintenance Operator I, are needed for cleaning lines up to 18 inches. When cleaning those lines located in easements or off the street, additional help may be required depending on the difficulty of the job.

Although some of the equipment listed in this section is specialized equipment used for balling, the majority of collection system maintenance departments will have most of the regular tools needed. Basic essentials include:

1. Water truck holding 1,000 or 2,000 gallons with hoses and air gap for hydrant use or an approved backflow preventer to protect the drinking water supply,

[8] *Job titles in this manual are from "Manpower Requirements for Wastewater Collection Systems," by Elie Namour for EPA, Grant Award No. 900211. See Special Notice on page vii at the beginning of this manual for additional information regarding job titles.*

Balling Equipment
(Courtesy of Sacramento County, Water Quality Division)

Sewer Ball
(Courtesy of the SIDU Company)

Fig. 6.2 Sewer ball and balling equipment

2. Tag line and a suitable reel to store it on (Figure 6.2) or power winch and cable (Figure 6.3),
3. Manhole jack or roller,
4. Plugs,
5. Balls (Figures 6.2 and 6.3),
6. Equipment truck,
7. Debris and grit trailer,
8. Ladder,
9. Manhole bars or lifters (manhole hooks such as the one shown between boots and rope in Figure 6.2 can cause back injuries if not used properly),
10. Buckets, rope, and manhole shovel,
11. Hip boots (Figure 6.2) and gloves,
12. Sound-powered telephones,
13. Collection system map,
14. Waterless hand cleaner and hand towels, and
15. Safety equipment. (See Chapter 4, Section 4.3, "Routing Traffic Around the Job Site," and Section 4.4, "Classification and Description of Manhole Hazards.")

Some communities use domestic water from fire hydrants instead of a water truck to obtain water to provide the necessary head for a balling operation. If fire hydrants are used, the following additional equipment will be needed.

1. Five to eight 50-foot lengths of 2½-inch fire hose.
2. Hydrant wrench.
3. Water meter.
4. Control valve for regulating flow at manhole.
5. Traffic ramp to protect hose (Figure 6.4).
6. Hydrant gate valve.[9]
7. Air gap device (Figure 6.5).

Air gap devices are required to prevent any backflow of wastewater from a manhole into a drinking water supply. The purpose of an air gap device is to provide adequate space above the top of a manhole and the end of the hose from the fire hydrant. This gap ensures that no wastewater will flow out the top of a manhole, reach the end of the hose from a fire hydrant, and be drawn back through the hose into the water supply. Suction conditions could develop if the water supply pipe near the fire hydrant should rupture and cause a washout or if any other rare condition should develop that could cause a backflow in the hose.

The tag line material may vary from plain manila rope to synthetic rope, steel cable, or plastic-impregnated steel cable. Some synthetic ropes may be adversely affected by constituents in wastewater and break in many locations with no previous evidence of deterioration. Cables are considered more reliable.

The reel on which you store this tag line should hold at least 1,000 feet. The reason will be explained later. The reel should be portable and the rope easily uncoiled and recoiled. Power winches may hold up to 2,000 feet of cable.

When cleaning lines 14 inches or larger, a power-driven reel is needed and steel cable is used for the tag line. One-quarter inch diameter cable has been used with power reels for cleaning 6-inch through 18-inch diameter sewer lines. Half-inch diameter cable is used for larger lines. Power bucket machines (Section 6.20, "Power Bucket Machines") can easily serve this dual purpose.

Because sewer balls use the principle of hydraulic head to develop high-velocity flow around the ball, precautions must be taken when determining what size winch and tag line to connect to the ball. Manufacturers, for example, recommend a maximum allowable back pressure of 5 feet of water when using balls from 6-inch diameter up through 18-inch diameter. This hydraulic head of water on the sewer ball exerts a force on the winch and cabling.

The amount of force that is exerted on the winch and cabling can be calculated using the following formula: the force in pounds is equal to the cross-sectional area in square inches times the hydraulic head in pounds per square inch. For a 6-inch diameter ball, the cross-sectional area is equal to $0.785\ D^2$ where D is the pipe diameter in inches. Therefore, the area is equal to $(0.785)(6\ in)^2 = 28.26$ square inches. A five-foot head of water is equal to 2.16 pounds per square inch (remember the formula: feet of water divided by 2.31 feet per psi). For a 6-inch diameter ball, therefore, 28.26 square inches x 2.16 pounds per square inch is equal to approximately 61 pounds of force on the winch and the tag line.

Now let's look at an 18-inch diameter ball using the same formulas: $(0.785)(18\ in)^2 = 254$ square inches x 2.16 pounds per square inch = 549 pounds of force exerted on the winch and tag line.

The above calculations are based on the full diameter of a 6-inch pipe and an 18-inch pipe. In reality, the force would be slightly less than the calculated amounts, but these illustrations show the approximate amount of force that can be generated when using sewer balls for cleaning with 5 feet of hydraulic head behind the ball.

Power-driven reels are either truck mounted or trailer mounted with stabilizer legs to hold the machines firmly in place over the manhole while in operation. Because of the terrific strain and pressure generated as water builds up behind the ball, a strong steel cable is needed. Truck-mounted winches are preferred by some agencies because this method allows the truck to tow other trailer-mounted equipment, such as exhaust blowers.

[9] *WARNING: Use a gate valve on the fire hose from the hydrant to throttle flows from a fire hydrant. Never attempt to regulate flows from a fire hydrant using the hydrant's valve because water may flow out through the weep hole and undermine the hydrant.*

Multi-sized "ball" will fit through 24-inch manhole frame. Inflate and deflate to fit size of line. Will clean larger lines up to 48 inches.

Multi-sized cleaning "ball" for lines 15 through 24 inches.

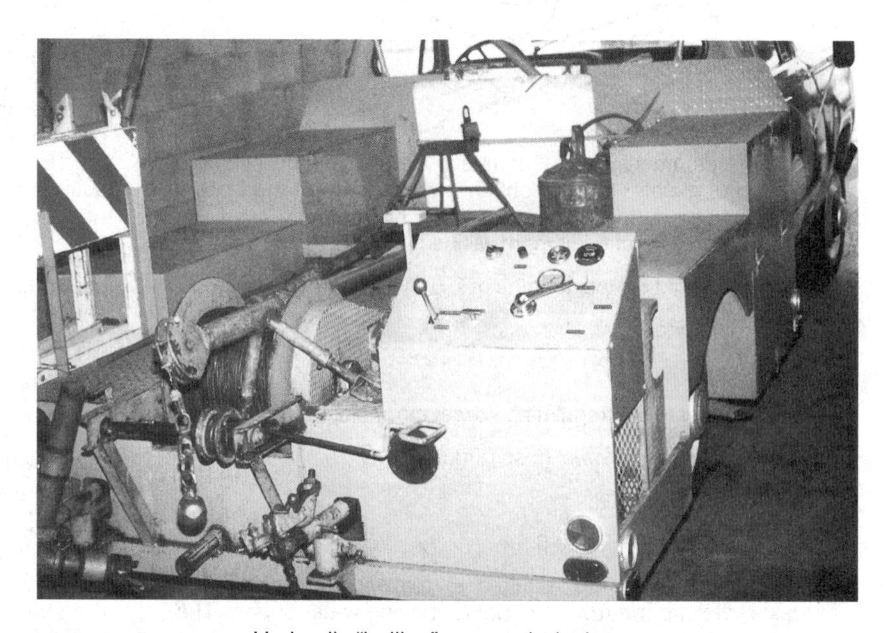

Hydraulic "balling" powered winch on truck. Used for balls and tires. Winch drum holds 2,000 feet of cable.

Fig. 6.3 Sewer balls and truck with power winch

(Courtesy of Union Sanitary District)

Fig. 6.4 Traffic ramp protecting hose from traffic
(Courtesy of Sacramento County, Water Quality Division)

Fig. 6.5 Air gap devices
(Courtesy of Union Sanitary District)

Small sewer balls are designed with diagonal ridges and grooves on the outer surface. As water builds up in the sewer in back of the ball, water will flow through these grooves and the ball will rotate. The outside design of the large balls differs somewhat, usually having a heavy tread surface to encourage a scouring action on the interior of the sewer (Figure 6.6).

The ball is resistant to punctures and is inflatable by means of a valve core. Into each side of the ball a strong metal eye lug is molded to which a clevis and swivel can be attached. The tag line is then attached to control the ball and allow it to spin. This clevis and ball bearing swivel are very essential items since they keep the line from twisting. The clevis is quickly and easily fastened to the tag line and ball.

The reel, although used primarily to hold the rope, also is used to apply a brake or drag to control the travel of the ball down the sewer.

The manhole guide jack or roller is essential to guide the tag line into the sewer without causing unnecessary wear and to give proper rolling action as the tag line is pulled back and forth or in restraining the ball.

Plugs, either mechanical or inflatable, are needed to control water coming into the working manhole from upstream. If there is insufficient flow in a sewer for balling, this method of "plugging off" the upstream water provides a method of getting an adequate supply of water to begin the balling procedure. Use a two-foot head of water on the ball to clean a line. Plugs are not needed if the flow in the sewer is sufficient to develop the necessary head behind a ball.

If the material removed from the line is to be hauled away by the balling crew, a truck is needed. This truck can easily be the same vehicle used to carry the balling equipment to and from the job site.

QUESTIONS

Write your answers in a notebook and then compare your answers with those on pages 383 and 384.

6.10A What kinds of deposits in sewers can be removed by balling?

6.10B Why is balling commonly used in preventive maintenance programs?

6.10C How frequently should a sewer be balled in a preventive maintenance program?

6.10D What is the minimum number of operators needed in a balling crew for cleaning lines up to 10 or 12 inches?

6.10E Where is the additional water obtained to provide the necessary head for a balling operation when flows in the sewer are low?

6.10F What is the purpose of the manhole guide jack or roller?

6.10G What is the purpose of an "air gap"?

6.101 Equipment Setup

Procedures for equipment setup before cleaning sewers with a ball are outlined and discussed in the next two sections. Each operator on the crew must perform the specified duties for the crew to be as efficient and productive as possible.

Before setting up the balling operation, certain safety measures need to be taken, both for handling street traffic and personal safety at and in the manholes (see Sections 4.3, "Routing Traffic Around the Job Site," and 4.4, "Classification and Description of Manhole Hazards"). Now is a good time for a tailgate safety session (Volume II, Chapter 11, Section 11.52).

TRAFFIC ROUTING

MANHOLE HAZARDS

Position the reel or power equipment so the tag line is centered in the upstream or working manhole (Figure 6.7). Attach the line to the ball using the clevis and swivel. Place the ball in the manhole by hand, unless the water is too high. If the water is high, push the ball into place with a forked rod. Insert the ball into the sewer just far enough to determine if it is inflated properly. The ball needs to be snug enough to cause good cleaning action (Figure 6.6), but not so tight the water cannot flow around the ball and carry it downstream. On the other hand, if it is too loose the ball will only float downstream with little cleaning action. When inflating the ball, be careful that it does not become too hard. The ball must be soft enough so it can squeeze around obstructions in the sewer.

Most sewer balls have two nipples on opposite sides that are used to attach the line to the ball. If one nipple breaks, use the other nipple. *DO NOT ATTACH A LINE TO ONE NIPPLE*

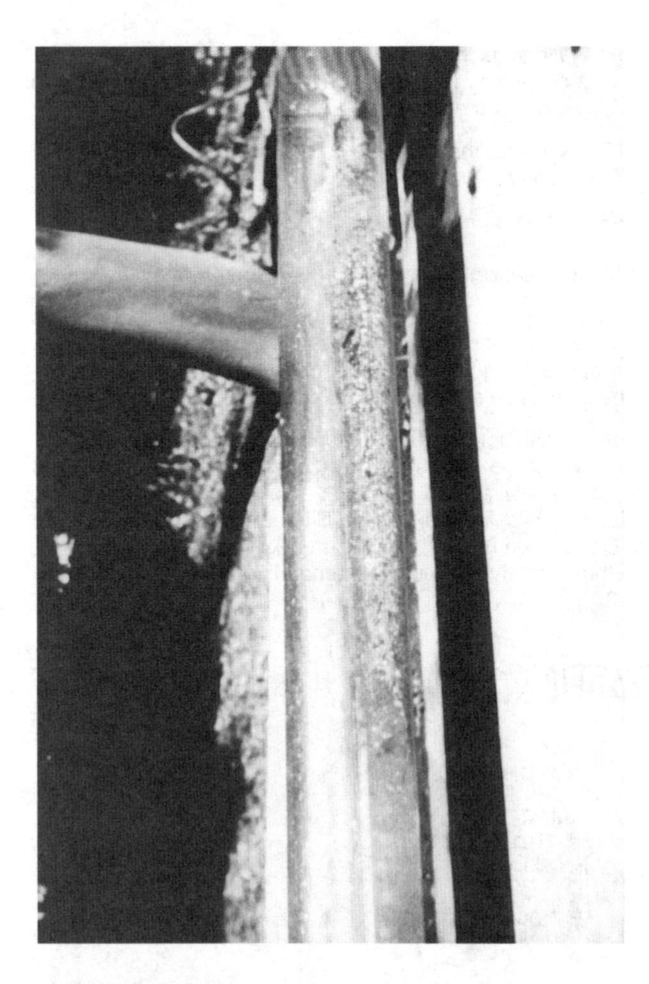

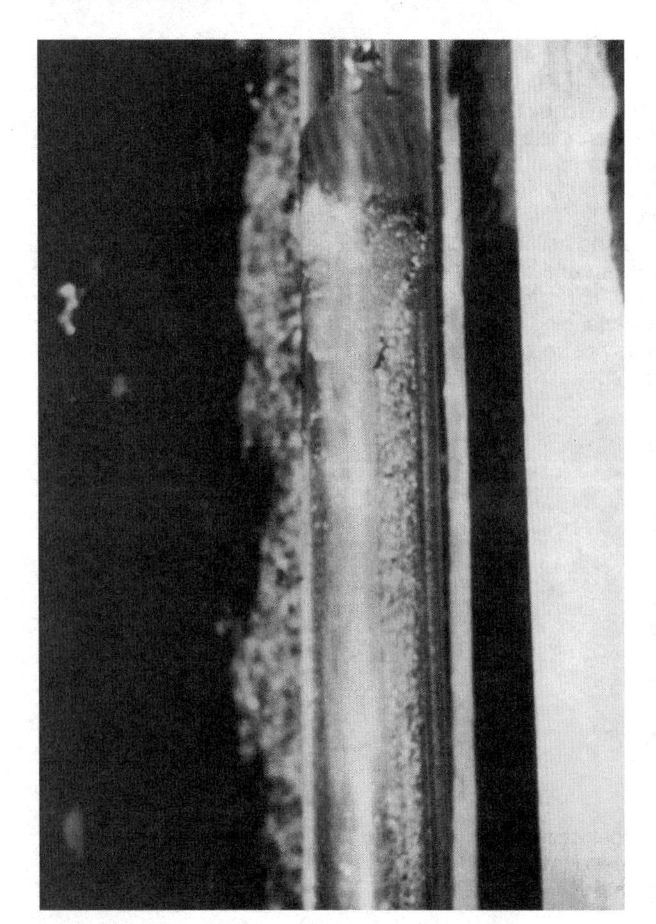

Fig. 6.6 Scouring action created by sewer ball
(Courtesy of Sacramento County, Water Quality Division)

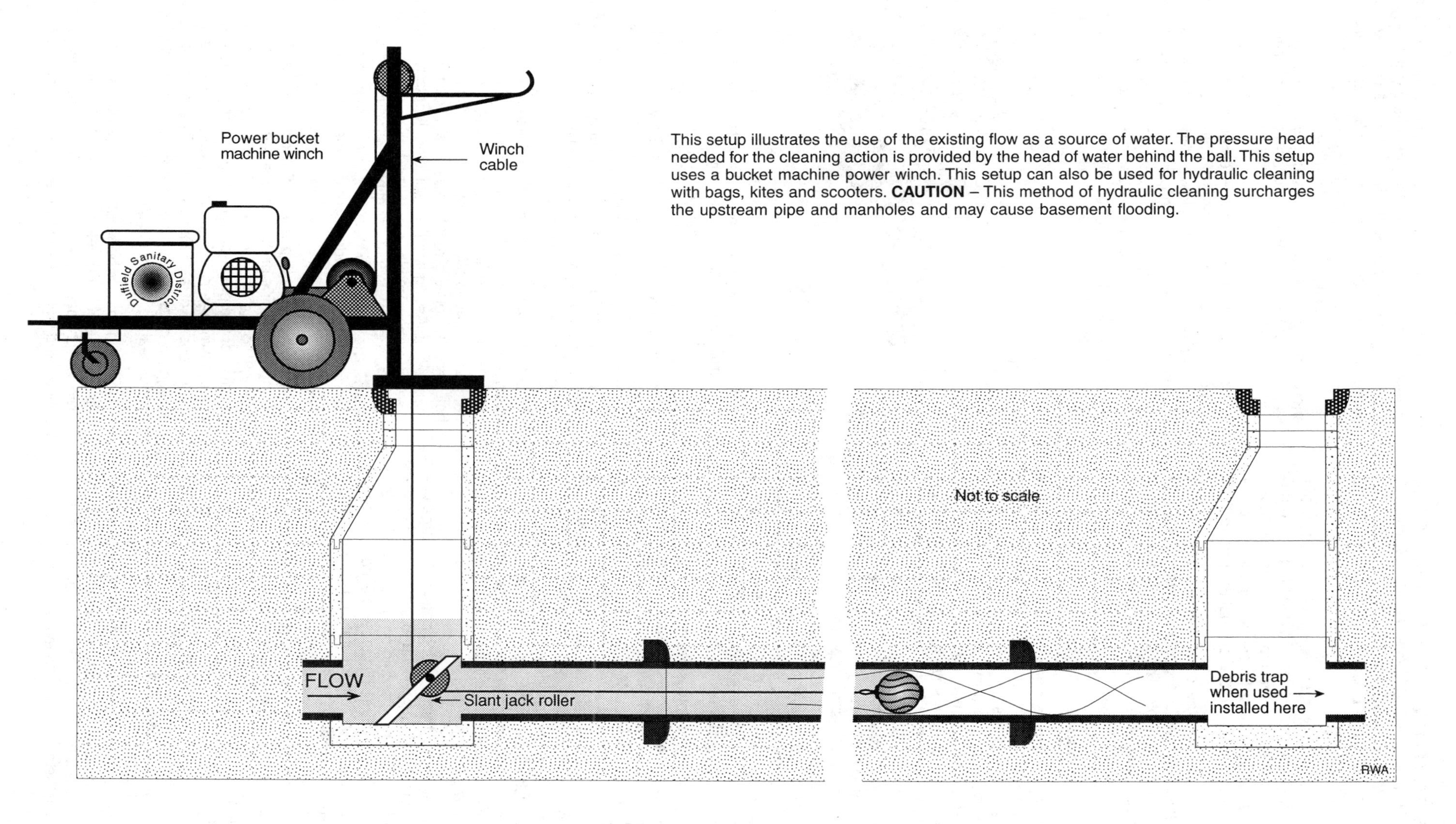

Fig. 6.7 Typical setup for hydraulic cleaning using a sewer ball
(Permission of SRECO-FLEXIBLE)

FROM THE UPSTREAM MANHOLE AND A LINE TO THE OTHER NIPPLE FROM THE DOWNSTREAM MANHOLE BECAUSE THIS WILL CAUSE A REDUCTION IN THE SWIRLING ACTION OF THE BALL AND THUS REDUCE THE CLEANING ACTION.

The manhole guide jack or roller is now installed in the bottom of the manhole and secured to allow the tag line to run freely and away from areas of the manhole that will cause abrasion. This roller or jack must be secure enough to allow strain on the tag line without lifting the jack out of position.

An operator is stationed at the downstream manhole with adequate communication with the operator running the ball and tag line at the upper manhole. The downstream operator's job is to keep the operator guiding the ball informed of conditions in the downstream manhole during the balling operation.

6.102 Cleaning Operation

The balling operation is now ready to begin. If the water supply coming into the working manhole is not enough to give proper head pressure, the plugs can be used. Insert plugs into lines coming into the manhole so the section of pipe being cleaned is isolated. Discharge water into the upstream manhole with the water truck or from a fire hydrant with an air gap over the manhole (Figure 6.5). Usually a balling operation will function very well with 2.0 feet of head in the upstream manhole to create a scouring velocity around the ball. There should be a maximum elevation set by the crew leader that the water head may reach. If plugs are used to store water in the upstream line for release to operate the ball, be sure to attach a safety line securely to them. When the plug is released, the water pressure behind it will have a tendency to "blow" it out and there is the danger of losing the plug into the downstream line and causing a major problem. Either method will create enough pressure to operate the ball properly.

As the ball begins to travel downstream, it will also be turning with water traveling around it and flushing debris ahead of it. *HOLD THE TAG LINE TIGHT AND ALLOW THE BALL TO TRAVEL SLOWLY DOWNSTREAM AT A STEADY RATE.* This procedure produces the best cleaning action. Ineffective cleaning results if the ball travels too fast. If you are short of head pressure, holding or stopping the ball travel and then releasing it at intervals will increase the head pressure and spinning action. This also causes the jet stream around and in front of the ball to break up debris deposits. The ridges of the grooves in the ball's surface also will give a scouring action to the interior of the sewer.

If the ball suddenly stops, this is an indication of a large deposit, that sufficient material has built up to stop the ball travel, the ball has hit a bad tap, or a crushed or offset pipe. By pulling back on the tag line a short distance, the ball will "egg shape" causing first a vacuum in front of it and then a sudden flushing action. This "bouncing" of the ball sometimes has to be repeated several times to remove large or stubborn deposits. If a deposit can't be removed or reoccurs frequently, complete a form requesting that the line be televised.

As the operation continues, the debris to be removed will start appearing when the ball gets within 20 to 25 feet of the downstream manhole. When the debris does begin to appear, an operator should be in the downstream manhole with a short-handled shovel, shaped to fit the channel, so that the operator can remove the debris as it appears. The accumulation of the debris may be slowed down by holding the tag line and stopping the forward progress of the ball. By releasing the ball a little at a time, the operator in the downstream manhole is able to remove all of the debris as it appears.

When balling a sewer that contains considerable debris and slimes that are producing hydrogen sulfide (rotten egg smell), a sand trap should be used. This is much safer than having an operator in the manhole shoveling debris and being exposed to the hydrogen sulfide that is being pushed down the sewer ahead of the balling operation. When using a sand trap (as shown in Figure 6.7), the pressure head behind the ball will have to be raised to maintain the cleansing velocity around the ball.

The information and procedures given for balling also can be used for cleaning with tires (Figure 6.8). When a flushing action is needed in a sewer, give slack to the tag line and the tire will tip forward from the top, thus releasing water. Pull back on the tag line and the tire will return to an upright cleaning position. Tires are not as flexible as balls and cannot be inflated or deflated like a ball for proper fitting in a pipe. Tires used for cleaning sewers can be made up in an agency's shop.

Constant communication between the reel operator (Maintenance Equipment Operator) and the operator removing material from the bottom of the downstream manhole (Maintenance Operator I) relayed through the operator above the downstream manhole is essential for the safety of the operator in the downstream manhole. If auxiliary water is being used, shut this water off at the control valve once debris starts arriving in the downstream manhole. The downstream operators should shovel this material into a bucket and lift the bucket from the manhole with a rope as quickly as possible so the reel operator can continue.

Continue balling and watch and listen for the ball at the downstream manhole. Keep it from entering the manhole until all remaining material has been removed. A strong shovel or bar can be placed over or across the incoming line to hold the ball back. When the manhole is clean, the ball can enter the manhole. If the tag line is of sufficient length to reach the next downstream manhole, insert the ball in the downstream line and the operation can continue without another complete setup. The one crew needs only to move to the next lower manhole and get ready to remove debris when it starts arriving.

When the operation is completed, disconnect the ball and rewind the tag line on the storage reel. If using a cable, attach a weight to the end of the cable to prevent twisting. With the removal of the plugs and lower guide roller, you are ready to move downstream. The debris should be loaded on the truck and disposed of at a disposal site. The manhole cover rings should be cleaned and the covers replaced and inspected for proper seating. Clean the street.

Communications between crew members can be improved with the use of hand-held radios or walkie-talkies, sound-powered telephones, or effective hand signals. Page 308 illustrates hand signals used by crews cleaning sewers.

Typical tire used for cleaning lines from 8 through 48 inches. Photo shows side of tire where head develops. Chain bridle is used for controlling tire.

Tire on left shows front of tire with pickup chain for removing tire from downstream manhole. Tire on right is used to clean lines over 24 inches and is hinged to allow tire to pass through 24-inch manhole frame.

Fig. 6.8 Tires

(Courtesy of Union Sanitary District)

HAND SIGNALS IN SEWER CLEANING

QUESTIONS

Write your answers in a notebook and then compare your answers with those on page 384.

6.10H Why must a sewer ball be properly inflated?

6.10I How is the material cleaned from the sewer by the ball removed from the collection system?

6.10J How fast should the ball move downstream?

6.10K What could cause a ball to stop moving downstream?

6.10L How could you determine what caused a ball to stop moving downstream?

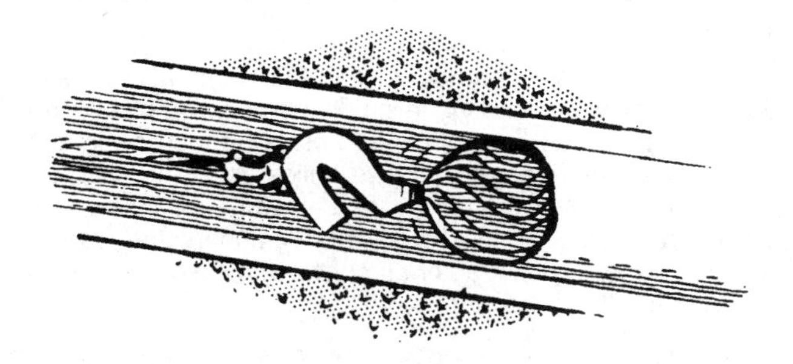

6.103 Precautions and Safety

Certain precautions have to be taken before and during all cleaning operations.

Using water pressure in the cleaning of gravity flow sewers always requires care and judgment with respect to basement fixtures and low-elevation homes. If there is any possibility of the water level behind the ball getting higher than basement fixtures (Figure 6.9), inspect these residences or business establishments for potential problems or use another method of cleaning, such as a high-velocity cleaner.

If a stoppage occurs below the ball, it must be cleared as soon as possible to avoid flooding homes or streets. First try to break the stoppage by using the ball. If this doesn't work, use a hand rod, power rodder, or high-velocity cleaner at the downstream manhole to clear the stoppage.

As previously outlined, all street work requires safety measures sufficient for the particular conditions involved. Be sure to set up barricades or other warning devices a good distance ahead of the working areas. Always use flaggers where there is a need to control traffic (see Section 4.3, "Routing Traffic Around the Job Site," and Section 4.4, "Classification and Description of Manhole Hazards"). A great deal of emphasis also must be placed on safe manhole entry using confined space procedures.

If a sewer ball becomes stuck, try to work it loose without breaking the cable. A high-velocity cleaner (Section 6.11) inserted in the downstream manhole may be able to go up the line and knock the ball loose. Another approach is to use a power rodder and try to puncture the ball.

6.104 Records

See Section 6.07, "Records."

6.105 Summary

The balling method of cleaning is a relatively easy procedure that is very effective and uses a minimum amount of equipment. Common sense operating procedures are involved and techniques can be developed and used to overcome problems common to this method of cleaning.

QUESTIONS

Write your answers in a notebook and then compare your answers with those on page 384.

6.10M What would you do if a stoppage occurred below the ball?

6.10N What precautions and safety measures should be taken during a balling operation?

6.11 High-Velocity Cleaning Machines

The use of water pressure to clean sewers dates back to the early 1900s; however, this method has been greatly improved by the development of high-velocity cleaning machines.

The fundamental idea of using a self-propelling nozzle on the end of a hose is still used. Instead of using a fire hose hooked to a hydrant, we have self-contained portable equip-

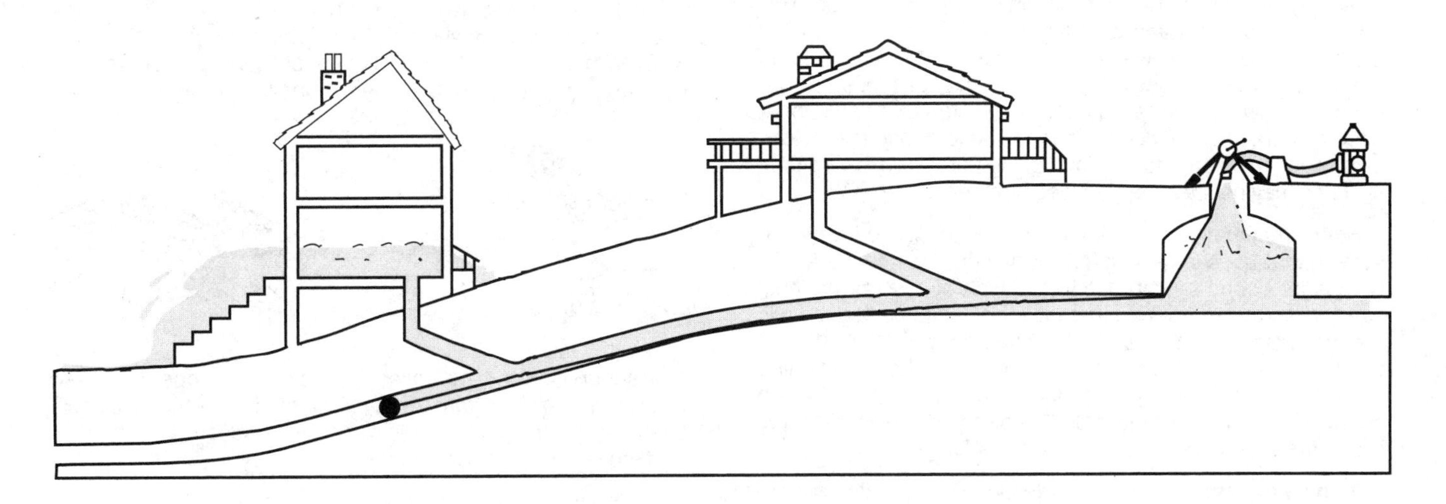

Fig. 6.9 How to flood a basement

ment, hoses, and nozzles that not only do a better job, but do it faster and cheaper.

High-velocity cleaning machines are used to:

1. Open stoppages,
2. Remove grease,
3. Clean lines of silt, sand, sludge, and other light debris, and
4. Wash manholes and wet wells.

Although the design of the high-velocity cleaning machine will vary with different manufacturers, they are essentially all the same.

6.110 Personnel and Equipment

High-velocity cleaners were developed primarily as a preventive maintenance tool. They are most effective at removing silt, sand, sludge, light grease, and other loose debris found in sewer pipes. They are not as effective as power rodding machines for removing heavy roots, hardened grease, and other hardened deposits, and are not as effective as power bucket machines in removing packed debris in large-diameter pipes. High-velocity cleaners are commonly used in daily maintenance, sometimes in combination with mechanical cleaners when an experienced operator locates specific areas where the high-velocity cleaner is not entirely effective in cleaning the sewer.

High-velocity cleaners come in a wide variety of specialized sizes for different work requirements and they can be either trailer mounted or truck mounted (Figures 6.10, 6.11, 6.12, and 6.13). The main components of a high-velocity water cleaner consist of a water supply tank, a high-pressure water pump, pump power source, and a hose reel with 500 or more feet of hose. Specialized units are available with water capacities ranging from 300 gallons to large truck-mounted units with 3,000-gallon capacity. Pump power sources can include auxiliary engines, power take off systems, or hydrostatic drives. Hose lengths vary from 500 feet to 1,000 feet.

Trailer-mounted systems usually use a 3/4-inch diameter sewer cleaning hose, deliver water flows from 25 GPM through 40 GPM, and operate at pressure ratings from 1,500 psi to 2,000 psi. Truck-mounted systems usually use a 1-inch diameter sewer hose that increases the water flow to the nozzle. Such systems can deliver 55 GPM to 80 GPM and operate at pressure ratings from 1,500 psi to 2,500 psi.

Additional specialized equipment is also used in some collection system maintenance programs for specific duties. This equipment can be rated at a higher volume of 100 GPM at 2,000 psi; the higher water flow is designed to move large amounts of debris from the pipe. Also, higher pressure systems with low volume, such as 18 GPM at 4,000 psi, can be used to work in small-diameter lines and for stoppages. However, since the water flow is limited, high-pressure low-volume systems may not clean larger lines properly.

Some high-velocity cleaners have an air conveyance system that uses large fans or positive displacement vacuum pumps for debris removal. With this type of system, material can be vacuumed from the manhole into a debris tank as it is brought back with the jet and the debris tank can later be taken to a disposal area. These systems can be either trailer or truck mounted and are generally known as combination machines or vactors. Figure 6.12 shows this type of machine with a front-mounted articulated hose reel.

To help in easement and remote cleaning operations, specially designed easement machines (Figure 6.14) have been developed for use in difficult terrain. These machines have a hose reel with hundreds of feet of operating sewer hose and they are self-powered, which permits them to get into areas where road vehicles have difficulty or are unable to pass. Remote operating controls allow operators to clean lines in these areas without damage to property or vehicles and without injury to personnel.

Various cleaning nozzles are attached to the end of the hose for different conditions found in the sewer. These nozzles force the water in the hose to exit the nozzle orifices (openings) in a specific pattern for the work needed to be done. Other operating accessories that are used along with the various nozzles include a lower manhole roller or slipper guide to protect the hose at the mouth of the pipe in the manhole, a finned nozzle extension to keep the nozzle off the pipe invert and prevent the nozzle from turning up laterals, and easement rollers for use at off-road locations or manholes with difficult access.

To prevent a large amount of debris from accumulating downstream when cleaning large-diameter sewer pipes, sand traps are placed in the manhole to block off the downstream pipe. Water entering the manhole builds up at the sand trap and drains out over the top of the trap while the solids settle out at the trap. Debris shovels and grabbers are used to remove this solid material, or air conveyance vacuums can be used to pull it from the manhole.

High-velocity cleaners operate effectively in pipes ranging from 4-inch diameter to 20-inch diameter, depending on the machine's size and capabilities. However, in pipes larger than 20 inches in diameter, the cleaning effectiveness of high-velocity cleaners decreases as the cross section of the pipe gets larger. These cleaners are most commonly used as a preventive maintenance machine to keep sewer pipes clear and pinpoint the need for mechanical cleaners in specific areas. Primarily used to clear and flush debris, deposits of materials, grease, and detergents, they do not work well in removing roots, hardened grease, and hardened deposits on the pipe invert unless special tools are used. High-velocity cleaners are frequently used for a variety of tasks that are not specifically sewer operations, for example, street washing, backup fire control, chemical spraying, and landscape watering.

To illustrate how to clean a sewer with a high-velocity cleaner, let us examine how a truck-mounted high-velocity cleaning machine would be used to clean sewers with 6-inch to 15-inch diameters. The machine is operated by one operator (Maintenance Equipment Operator). This operator will need help in order to be notified when the nozzle has reached the upstream manhole and when dirt and grit are removed from the working manhole. Follow all confined space procedures for manhole entry! (Review Section 4.4 on page 115.)

A supervisor can be called to assist a two-operator crew whenever someone must enter a manhole or other confined space. Additional assistance may be needed for traffic control. Staffing requirements include:

1. Maintenance Equipment Operator, and
2. Maintenance Operator I (one or two operators).

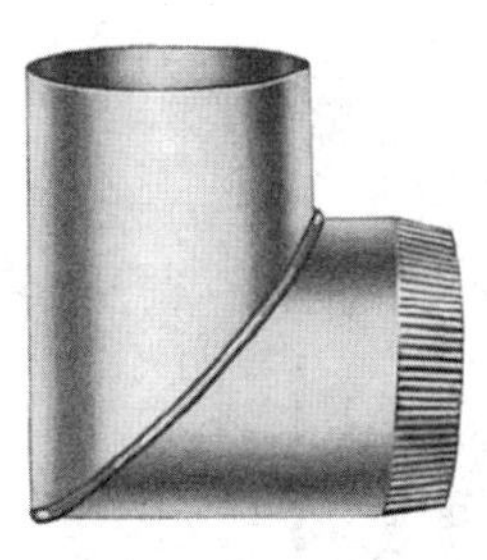

Sand trap elbow

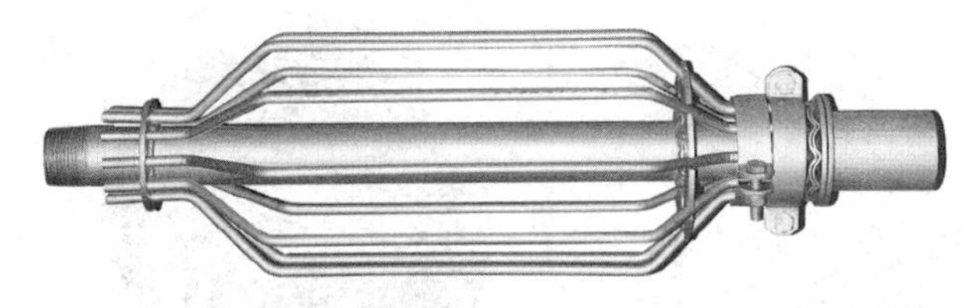

Specialty use proofing tool
(Used to be sure line does not have grease or root deposits remaining in pipe after cleaning.)

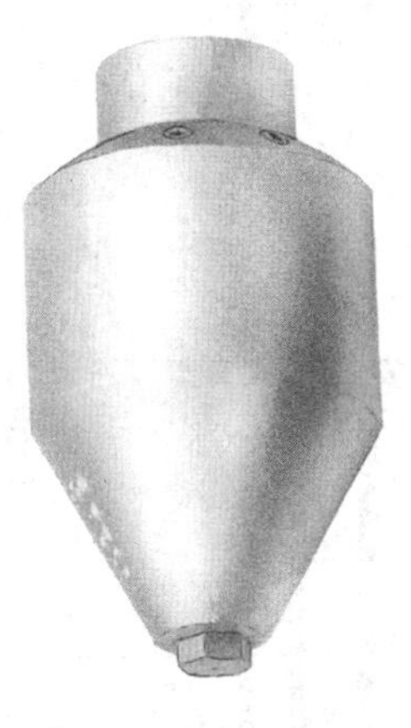

Crown obstruction
grease nozzle

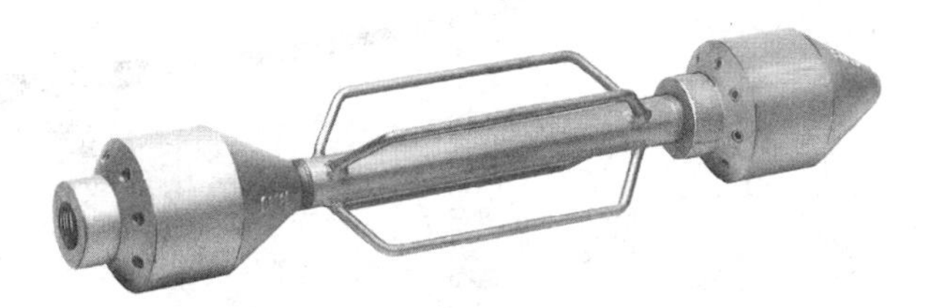

Crown obstruction tandem grease nozzle

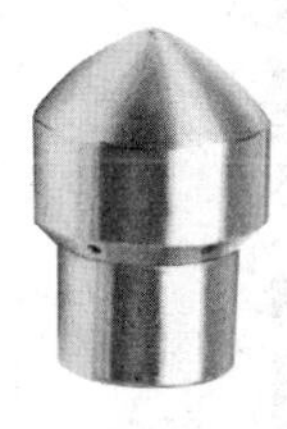

Preventive maintenance
radial nozzle

Stoppage relief
penetrator nozzle

Specialty use
teardrop nozzle

Skid-type invert obstruction nozzle

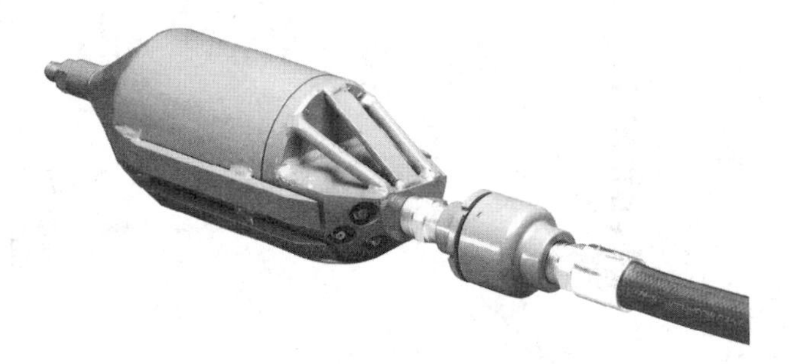

Torpedo-type invert obstruction nozzle

Fig. 6.10 High-velocity cleaning equipment

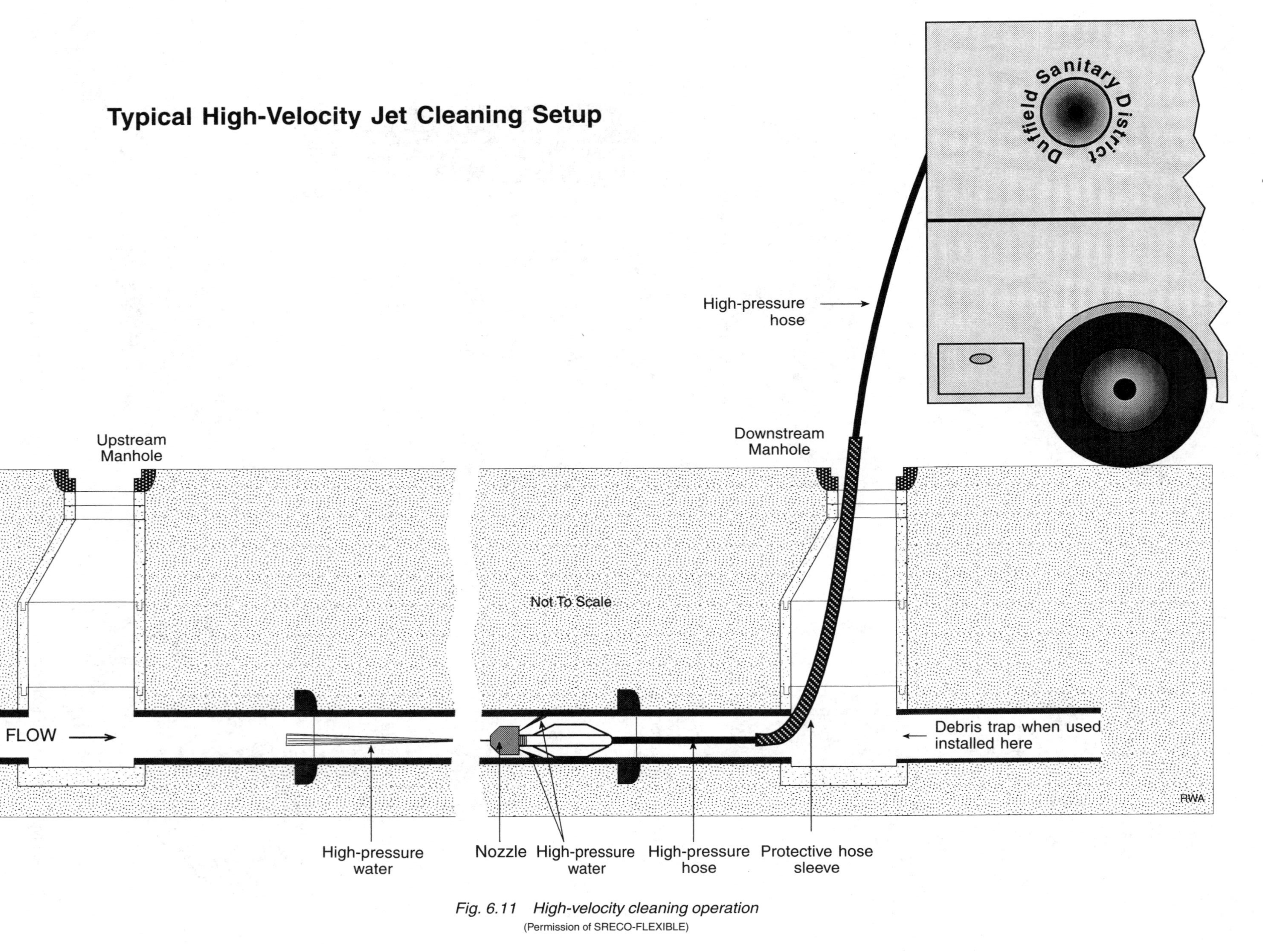

Fig. 6.11 High-velocity cleaning operation
(Permission of SRECO-FLEXIBLE)

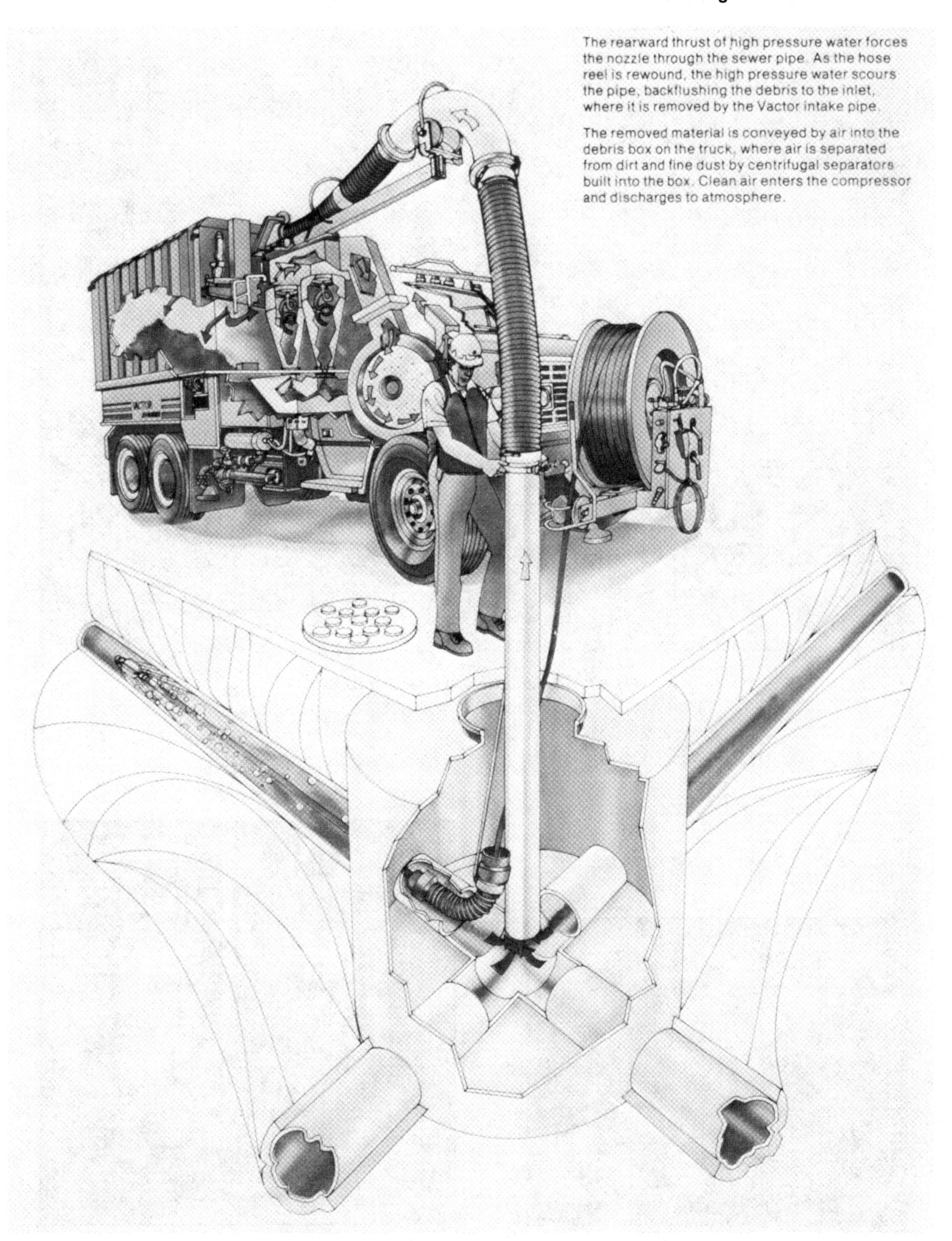

Fig. 6.12 High-velocity cleaner with vacuum device for removing grit and debris

Fig. 6.13 Combination sewer cleaning machine
(Permission of Vac-Con)

Fig. 6.14 Self-propelled easement machine
(Permission of SRECO-FLEXIBLE)

QUESTIONS

Write your answers in a notebook and then compare your answers with those on page 384.

6.11A What are the uses of high-velocity cleaning machines?

6.11B How do high-velocity cleaners clear sewers?

6.11C How many operators are required to operate a high-velocity cleaning machine?

6.111 Equipment Setup

Because the high-velocity cleaner is a mobile, self-contained unit and this method of cleaning smaller diameter sewers is relatively fast, several setups will usually be made in the course of a day's work. Basic safety measures always apply since, for the most part, this equipment will be set up in the street and traffic conditions must be considered. See Section 4.3, "Routing Traffic Around the Job Site," and Section 4.4, "Classification and Description of Manhole Hazards," for safe procedures you can use when working in or near traffic.

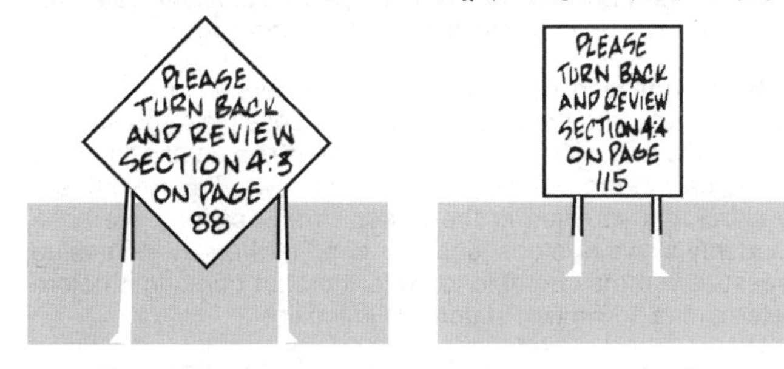

Operation of the high-velocity cleaner requires some expertise. The diagnosis of system problems, nozzle selection, proper setup, and field maintenance procedures for the cleaner are some of the areas that operators need to be aware of in order to use this equipment safely and effectively.

Proper procedures for field maintenance of the equipment can be found in the instruction books and O & M manuals that were originally provided with the equipment, or from servicing vendors. All types of high-velocity cleaners require the operator in the field to check fluid levels (such as engine and pump oil) daily, grease bearings on the hose reel and swivel, check the strainer, and inspect all gages and controls on a regular basis. In addition, operators need to be knowledgeable regarding safety, startup, and operating procedures that are specific for the machine they are using.

The high-velocity cleaner's water tank usually is filled from a fire hydrant close to the area where you will do the cleaning. When using a hydrant, a special hydrant wrench should be used to operate the main hydrant valve. *NEVER* use a pipe wrench on the hydrant valve because the valve nut is brass and can be easily damaged. A gate valve that fits on the hydrant should be installed to regulate water flow and eliminate the need to open and close the hydrant's main valve repeatedly. Fully open the hydrant's main valve and then open the gate valve to allow the water to run for a few minutes. This procedure clears the rust from the hydrant water. If rust is present in the water, it will have a brownish-red tinge to it. There should be a screen filter in the pump suction line to filter out large particles that can damage the pump and water system. This filter should be checked daily for any materials trapped in it.

When the water runs clear, close the gate valve slowly to prevent water hammer in the water main, attach the fill hose to the cleaner, and open the gate valve slowly to prevent a sudden rush of water that may bring rust into the water tank. Always use the gate valve to regulate the flow of water from the hydrant. *NEVER* throttle or control water flow from a fire hydrant with the hydrant valve because this could damage the hydrant valve. Also, throttling the flow by partially opening the hydrant valve may cause a discharge of water through the hydrant's weep hole; such a discharge could undermine the hydrant. You must have an air gap to prevent a cross connection between the fresh water flowing into the tank from the fill line and the water in the cleaner. If you are required to record the amount of water drawn from the hydrant, you will also need an approved meter with which to measure the flow.

Driving a vehicle with a fully loaded water tank can be extremely dangerous. A 1,000-gallon tank full of water weighs 8,340 pounds, plus the weight of the vehicle and cleaner. Even with baffles in the water tank, sharp turns or quick stops and starts can create driving hazards. Large trucks over 26,001 pounds now require a Commercial Driver's License (CDL), and proper licensing is mandatory. Many states may also require a license endorsement to drive "tank" vehicles.

If a given area is to be cleaned completely, start at the top or highest point in the collection system. Whether routine cleaning or spot cleaning, position the high-velocity cleaner hose reel over the center of the manhole. In areas where manholes are offset from the road, or in easements, easement rollers will need to be used to allow the hose to change direction smoothly.

Once the cleaner has been positioned, select a cleaning nozzle. Nozzles come in an assortment of shapes and sizes and have various angles of discharge. The angle of discharge is given in degrees and the most common nozzles are those with discharge angles of 15, 35, and 45 degrees. This is the degree of spray angle that the nozzle uses to clean effectively. A 15 degree nozzle will have a lower spray pattern in the pipe than a 45 degree nozzle, but will provide better thrust to pull the nozzle up the line. A 35 degree nozzle will do a better job cleaning a 10-inch diameter sewer pipe than will a 15 degree nozzle due to the steeper angle of the spray, but the 35 degree nozzle will have less thrust. Some general guidelines are described in the following paragraphs to help you select the appropriate nozzle.

There are many different kinds of nozzles (Figure 6.10) that can be used with high-velocity cleaners; however, each nozzle is designed to accomplish a specific type of task. The five basic categories of cleaning activities are preventive maintenance, stoppage relief, crown obstruction, invert obstruction, and specialty usage.

Preventive maintenance nozzles are used when there are no known problems in the pipe and a quick cleaning is scheduled. The standard radial nozzle is used for preventive maintenance work, with a 35 degree spray angle preferred by most operators. This nozzle will clean most sizes of pipe effectively and provide a good thrust to allow the nozzle to reach the next manhole.

Stoppage nozzles are used for emergency work to remove a blockage in the sewer. Generally, these nozzles have a spray angle of 15 or 35 degrees to provide the high thrust needed to pull the hose into a dry pipe as the blockage is approached from the downstream or dry end of the pipe. The head of the nozzle has a large front orifice that is designed to open a hole in the stoppage so that the nozzle head can enter the stoppage. As the head moves into and through the debris, it breaks out and blows material backward until the stoppage is relieved.

Crown obstruction nozzles are designed to remove grease, waste oils, and detergents from the top and sides of pipe walls. These nozzles are generally set at 35 or 45 degree angles to provide a sharp cutting spray to dislodge built up materials.

Invert obstruction nozzles are designed to remove debris such as sand, gravel, and sediment from the floor of the pipe. These nozzles usually have a low spray angle which directs water to the bottom of the pipe in order to loosen and move the debris that has accumulated there. Water volume plays a large role in moving materials once the nozzle dislodges them, and these nozzles will use water faster than most other nozzles. This allows the materials to be flushed from the pipe and brought to a manhole or sump for removal.

Specialty nozzles are nozzles made for specific tasks. Some examples of specialty nozzles are tandem-type nozzles for silt or sludge, teardrop nozzles at 11 degrees to climb steep grades, cable towing nozzles for camera or bucket machine cable towing, or nozzles designed to make an attaching tool operate, such as an hydraulic root cutter. These nozzles come in a variety of different angles and configurations; however, they are seldom useful for any task other than the very specific task they were designed to accomplish.

Experience will be the best teacher in learning which nozzle is the best to use in a given situation. Regardless of which nozzle you select, a finned nozzle extension should be used. The extension attaches to the sewer hose and the nozzle attaches to the opposite end of the extension. This finned extension does two things: it helps to keep the nozzle from running on the rough surface of the pipe and wearing, and it also helps prevent the nozzle from turning up a lateral or service connection or turning around in the pipe.

The water pressure at the nozzle determines the velocity of the water leaving the nozzle. The higher the pressure, the higher the velocity and thus the greater the scouring power of the water to loosen and remove deposits of grease and debris. The greater the flow of water out the nozzle, the more water is available to convey the loosened grease and debris down the sewer. The nozzle should match the volume and pressure of the cleaner on which it is being used.

Test the manhole atmosphere for sewer gas (see Section 4.4, "Classification and Description of Manhole Hazards") and install the proper size sand or debris trap in the downstream manhole. Follow confined space procedures if manhole entry is required.

Once the nozzle and extension have been attached to the sewer hose, attach the footage meter or pass the hose through the meter. Start the auxiliary engine or engage the system power take off and lower the hose into the pipe. To protect the sewer hose, use either a lower manhole roller or a slipper guide to shield the hose from the rough edges of the pipe opening in the manhole. This will not only protect the hose, but will allow smoother movement of the hose into and out of the pipe.

The first ten feet of sewer hose just behind the nozzle will be the first to wear with the impact of materials that are being moved, especially in lines where sand and gravel are being removed. To provide maximum protection for this part of the sewer hose, a braided, steel-reinforced leader hose should be used on the nozzle end.

The cleaning nozzle should always be pointed upstream of the operating manhole and laid on the invert of the manhole with a little sag in the hose. The roller or slipper guide is usually positioned after the nozzle starts up the pipe because the hose needs to move freely as the nozzle climbs the pipe under the force of the pressurized water. The cleaning operation is now ready to begin.

QUESTIONS

Write your answers in a notebook and then compare your answers with those on page 384.

6.11D When a high-velocity cleaning unit is set up in a street, what measures should be taken to warn traffic?

6.11E Why should the water tank be filled close to the working area?

6.11F Where should you start to clean the sewers in a given area or subdivision?

6.11G Which direction does the nozzle travel in the sewer?

6.112 Cleaning Operation

Turn the water valve on and start the high-pressure pump. Water now is being pumped through the hose and cleaning nozzle. With an increase in pump speed, sufficient water velocity will build up to propel the nozzle up the sewer.

An experienced operator is the best person to teach you how far to go into the sewer. Often it is possible to go several hundred feet, but it is not always safe to do so unless a history of this part of the collection system is known. First try 50 feet to check the situation in the sewer. In any case, if the hose suddenly slows or stops, don't try to go farther. By increasing pressure you may be able to go farther, but don't do it before determining how much material is in the line.

At this point, with the hose roller in place, turn the reel directional control to "in" to return the nozzle. Increase engine speed to obtain the proper psi rating for your particular equipment.

With the reel speed control, slowly start the hose reeling in until the nozzle is just ready to come back into the manhole. The amount of material you have brought back will determine how far and how fast subsequent runs can be made. Repeat this operation until the upper manhole has been reached and no more material is brought back.

During this operation several items may have to be considered and done. If this is a normal operation of routine cleaning, perhaps the entire line from manhole to manhole can be cleaned without enough material building up in the working manhole to make removal necessary until the job is finished. On the other hand, it may be necessary to clean out the channel after each "pull" if there is a lot of debris in the line. When loosened materials build up to the point of overflowing the debris trap, pull the hose and nozzle out of the manhole. Stop the pump until after the debris is cleared.

Retest the manhole for hazardous atmospheric conditions to be sure the ventilation procedures are effective. If hazardous atmospheric conditions are present, a self-contained or supplied-air breathing apparatus may be necessary in addition to ventilation. Have two operators standing by topside and follow all confined space procedures. Allow an operator properly equipped with a safety harness to enter the manhole and shovel the debris into a bucket, which in turn is pulled to the surface with a hand line. Some high-velocity cleaners have a vacuum device that removes grit and debris from manholes without anyone having to enter a manhole (Figure 6.12). After the removed grit and debris are drained, the grit may be disposed of in a landfill (if not hazardous) and the liquid may be returned to the collection system for treatment at the plant.

During the cleaning operation, care has to be taken to keep the nozzle moving and the pump operating at all times. This is especially true in small lines since you do not want to cause a stoppage. A stoppage can occur if you try to go into the line too far at a time and bring back so much material that the nozzle cannot move it downstream fast enough. The smaller lines then quickly fill with water and this in turn goes up the building service lines. If this happens on shallow lines, home damage may result from flooding. Always watch the flow coming into the working manhole. If this flow should stop, a stoppage has developed in the sewer. Pull the hose and nozzle out at once and start over. If the nozzle hits an obstruction and stops, pull the nozzle back to clear the line and then run the nozzle into the obstruction again.

The exact distance the cleaning nozzle has moved up in the sewer must be known at all times. Most machines are equipped with a footage meter; if yours is not, have some identification (a hose coupling or so many wraps of hose) so as to give an estimate of where you are in the sewer. If you know the distance between manholes, the footage counter will allow you to make the final run without someone being at the upper manhole.

The high-velocity cleaning machine also may be used for stoppages in sewers. Setup is the same as for routine cleaning. A nozzle with a forward jet as well as rear jets may be used. This forward jet provides a cutting action ahead of the nozzle. Also, the debris trap could be replaced with a special fork or pole to allow the free flow of wastewater when the stoppage is broken. Large pieces of solids will be caught on the fork or pole without causing the working manhole to back up.

Upon contact with the stoppage, if the nozzle will not penetrate the stoppage, pull the hose back a few feet by hand and let go. The nozzle will shoot forward quickly and give a pounding effect on the stoppage. When it breaks loose, retrieve the hose and nozzle quickly. After the line returns to normal, reclean the line completely. Inspect the upstream manhole and flush with the washdown hose and wash the street free of any overflowed wastewater and solids, disinfect, and complete whatever cleanup is needed. Disinfection can be accomplished by the use of a hypochlorite solution.

The high-velocity cleaning machine is quite effective in cleaning inverted siphons and working in easement areas. Special rollers are available for use where the truck cannot be positioned over the manhole. Remote control equipment is also available.

QUESTIONS

Write your answers in a notebook and then compare your answers with those on page 384.

6.11H How often should the nozzle go up the sewer and be brought back to the downstream manhole?

6.11I What is done when the loosened materials build up and overflow the debris trap?

6.11J Why must the nozzle be kept moving at all times?

6.11K What should be done if flow stops in the working manhole?

6.11L How are high-velocity cleaning machines used to clear stoppages?

6.113 Precautions and Safety

Although some of the precautions of operation have been given, there are others which soon become obvious. When setting up a vacuum device (Figure 6.12), be sure that moving overhead equipment will not come in contact with power lines or cables. Maintain at least 10 feet of clearance from overhead lines energized at 50,000 volts (50 Kv) or less; additional clearance is required for higher voltages. Also be very careful when cleaning from a manhole to a cleanout. Most often the cleaning nozzle will stop at the bottom of the cleanout. On short runs, however, it is possible to have enough speed and thrust that the nozzle will go up to the street surface and even knock small cleanout covers out of place. When cleaning up to a cleanout, there is also the chance of getting the nozzle stuck.

If this should occur, shut the machine off, pull the cleanout cover, and dislodge the nozzle with a long rod or pole.

If fresh soil or parts of the sewer pipe are washed down into the lower manhole when pulling the nozzle back, stop the operation. Continued attempts to clean the sewer could result in a large hole developing under the street and the eventual collapse of the street into the sewer. Notify your supervisor of this potentially serious problem.

While cleaning, so much material can build up below the nozzle that you cannot pull the nozzle through under pressure. Stop the pressure and pull the nozzle back through the stoppage "dead" (without water flowing out the nozzle). If this fails, try to get the nozzle to the upstream manhole and take it off, thus allowing only the hose to be pulled back through the stoppage to the downstream manhole. Replace the nozzle and go back up the line after the stoppage.

Often a bend in the line will prevent you from pulling the nozzle back under full pressure. The more pressure at the nozzle, the stiffer the hose becomes. Reduce the pressure or even turn it off for a short distance until you are past the bend.

In cold weather, particular care must be taken to keep the equipment from freezing. Special attention should be given to the high-pressure pump to make sure that it is properly drained. If the machine is equipped with a canopy over the engine and pump area, the areas can be separated by a curtain and a small electric heater can be left inside the pump area overnight or when not in use during freezing weather. Another approach to prevent freezing during cold weather is to install a piping system that allows water to circulate through the pump *AND HOSE*. Be sure to include the hose because some water always remains in the hose, no matter how much time and care are taken to drain the hose.

Do not allow an inexperienced operator to operate the machine alone. Personal safety is not so much a problem because of the design of the machine; however, costly problems can develop if care is not given to existing conditions in the collection system and those that the cleaning operation can create. Caution and experience are required to prevent "blowing toilets" with a high-velocity cleaner.

Purged plumbing, or "blown toilets," are the result of a combination of pressure, nozzle angle, and system conditions. A nozzle moving up a pipe will pull air from upstream. If the pipe is clogged with roots or grease, sufficient air cannot be pulled easily through the pipe. Instead, the air will be pulled from house vents, which in turn can pull down water in all of the plumbing P traps and the toilets. When the nozzle returns downstream, the air will be pushed upstream toward the stoppage and can exit through these same vents with sometimes dramatic results of pushing water out ahead of the air.

The best way to avoid purged plumbing situations is to use a more steeply angled nozzle at a minimum pressure when moving the nozzle up the pipe. The longer the cone of spray from the nozzle is in the pipe, the more air it will pull. A 15 degree nozzle will pull more air from upstream than will a 35 degree nozzle. In areas where there are known purged plumbing problems, use low operating pressure of around 1,000 to 1,200 psi to help reduce the air movement in the pipe. Cleaning will not be affected much by the reduced pressure, and this approach will help relieve some of the problem. If a problem continues with a higher degree nozzle and lower pressure, a closed-circuit television inspection camera should be brought in to see if there is any root or grease buildup that can be removed to help increase the diameter of the pipe and thereby allow increased air flow. A blown toilet is an indication of a problem with the vents in the building system or a belly in the sewer service to the building. Smoke testing prior to cleaning will allow you to identify potential blowback problems.

Also, do not sacrifice clean lines for footage. This equipment is capable of doing a thorough job of cleaning. There is a tendency to consider the line to be clean when you start running low on water. Take the time to get another load and make sure the line is properly cleaned.

Proofing tools are used to be sure the line does not have grease or root deposits remaining in the pipe after cleaning. A proofing tool (see Figure 6.10) is used on the end of the cleaning hose much like a finned nozzle extension. When the tool is placed on the hose and the nozzle is then connected on the end, the nozzle pulls the tool up the line. The tool is designed to pass easily through a given diameter pipe but it will stop if obstructions such as roots are present. This allows the operator to note the footage and line section for later work with a rodder or hydraulic root cutter.

Hydraulic root cutters can be used effectively on high-velocity cleaning machines to correct minor root and grease problems. These tools attach to the sewer hose. The water pressure and volume propel the root cutter up the pipe while turning a cutting blade or saw. These cutting tools are useful for light preventive maintenance duty but are not a substitute for a rodding machine in heavy root and grease masses. Caution is required when operating hydraulic root cutters and they should never be used by inexperienced operators. Dropped or misaligned joints, turns or dips in the pipe, and heavy debris can cause the tool to become stuck, possibly resulting in having to dig up the pipe to get the tool out.

Occasionally high-velocity cleaning machines are used to perform other tasks such as stringing cable through pipes for closed-circuit TV inspection cameras or power bucket machines. When towing a cable with a sewer hose, be careful that the hose is not empty when you wind it back onto the reel. If the hose flattens out as it is wrapped, then the next time the cleaner is used the hose will expand and possibly damage the hose reel. Always keep about 500 pounds of pressure in the hose when pulling a load to ensure that the hose does not flatten.

Some agencies use a TV camera to televise the sewer while using a high-velocity cleaner. This approach has been very successful to remove small patches (two to three feet long) of heavy grease. Without the TV camera, the nozzle would not have been stopped at the grease patch long enough to remove the grease. This same approach can be used to remove roots using a high-velocity cleaner rather than simply having large footage production per day and clean roots.

When relieving stoppages, avoid using the full, or surcharged, manhole to operate the high-velocity cleaning machine. It is difficult, if not impossible, to know the nozzle is secure in the invert and will not exit the operating manhole under high pressure. Nozzles under pressure and surrounded by water, as they are in a full manhole, can develop tremendous speed and power. If they exit the manhole under power they can cause serious injury to operators and damage to the machine. Always use the dry, downstream manhole to remove blockages.

Be aware of what the nozzle is doing in the pipe, and be on guard for the nozzle to turn around in larger diameter pipes and come back to the operating manhole. Debris or pieces of pipe can cause a nozzle to flip over backward and move back at the operator. Always keep an eye on the operating manhole to be sure the nozzle has not reversed direction. Never operate the nozzle in the manhole unless the nozzle is lying in the pipe invert, just outside of the pipe. When operating at pressure, the nozzle can whip and exit the pipe under tremendous force.

6.114 Records

See Section 6.07, "Records."

QUESTIONS

Write your answers in a notebook and then compare your answers with those on pages 384 and 385.

6.11M What precautions should be taken when cleaning from a manhole to a cleanout?

6.11N What would you do if so much material built up that the nozzle could not be pulled back to the downstream manhole?

6.11O What precautions must be taken to protect equipment during cold weather?

6.11P Explain how toilets are "blown" by high-velocity cleaning equipment.

6.12 Flushing

Flushing (Figure 6.15) is another method of hydraulic cleaning; it is occasionally used at the beginning of the collection system where low or sluggish flows permit the deposition of solids. This procedure may be effective in removing floatables, but not grit and other heavy solids. Flushing also is used with mechanical clearing operations such as power rodders and bucket machines. High-velocity cleaners and balling do a much better job of cleaning sewers than flushing.

During the flushing operation, observe the flow characteristics in the sewer—is the flow through the manhole slow or sluggish? A partial stoppage may restrict the flow of the flushing water being discharged into the sewer. *IF YOU OBSERVE LOWER THAN EXPECTED FLOWS AT A DOWNSTREAM MANHOLE, THIS INDICATES THAT THE UPSTREAM SEGMENT OF THE LINE NEEDS ADDITIONAL CLEANING.* As the water flows, a physical and visual check of the material in the line may reveal large amounts of sand, grease, and debris. *THIS OBSERVATION ALSO MAY MEAN THAT MORE EXTENSIVE CLEANING IS NEEDED.*

Stoppages are commonly located during the flushing operation. Material often will have built up to the point of causing a stoppage and just a sudden charge of water will break it loose, thus clearing the line. Passing this large amount of water through the line all at once also helps control rodents and insects by washing them down the line or drowning them.

Cleaning the manholes with water pressure eliminates manhole shelf buildup of street dirt and debris. In addition, cleaning the manhole with water pressure, combined with the flushing action within the sewer, will minimize the buildup of sulfide and slime in the line. This is one of the primary causes of line decay and replacement. Flushing may be used in steep hills where the use of balling or high-velocity cleaners creates a high risk of flooding homes when not properly operated.

In smaller collection systems, flushing usually must be repeated frequently to prevent many routine problems. Also, this cleaning method requires large volumes of water; therefore, before starting a flushing operation, notify the water department whose hydrants will be used as a source of water. The opening and closing of hydrants must be done with great care because improper procedures could create water quality problems in the water distribution system. In addition, notify the operator of the downstream wastewater treatment plant whenever larger than normal flows can be expected at the plant.

6.120 Personnel and Equipment

Because of the small amount of equipment and number of operators involved in flushing, the actual cost appears attractive. However, the procedure may not be cost effective because of the limited cleansing action from flushing.

A tank truck with a capacity of 1,500 to 2,000 gallons is needed. A commercial driver's license (CDL) is required to operate a vehicle of this size and weight. The truck should be equipped with a reel for holding approximately 100 feet of 1-inch ID (inside diameter) hose and a pump to give sufficient pressure to do an effective job of manhole cleaning. The nozzle type may vary from an adjustable model to a plain piece of pipe reduced to give a good cutting action. A 1/4-inch pipe nipple brazed onto a 1-inch female hose fitting provides a good stream (if the pump provides 25 to 30 psi in the line) for hosing down the manhole barrels and shelf prior to flushing the line.

A 2 1/2-inch or larger gravity discharge line from the bottom of the tank on the truck should be provided to permit a large flow of water. Location of the line must be convenient for the truck driver to position it over the manhole. If possible, the ideal location is just ahead of the left front wheel. This line should be equipped with a quick opening, full flowing valve. A short piece of old inner tube fastened just above street level will help direct the flow of water into the manhole if positioning is difficult.

The tank truck equipment should include a good filler hose (usually a short length of fire hose to reach from a fire hydrant to the top of the tank's filling hatch), hydrant wrench, water meter, air gap device, and manhole lifter. Some communities require a 2 1/2-inch water meter so that they can pay for the water used for flushing.

A debris bucket, rope, and manhole shovel are other pieces of essential equipment, plus the usual safety equipment (Chapter 4, Section 4.3, "Routing Traffic Around the Job Site," and Section 4.4, "Classification and Description of Manhole Hazards"). The equipment and tools for this operation should be arranged on the tank truck in such a manner as to be handy and within the limits of space available.

Two operators are needed and they should be clothed with hard hats, high-visibility vests, and gloves. Typical street traffic should be expected, although you probably will not be long at one manhole. Crew requirements are:

1 Maintenance Operator I, and

1 Maintenance Operator II.

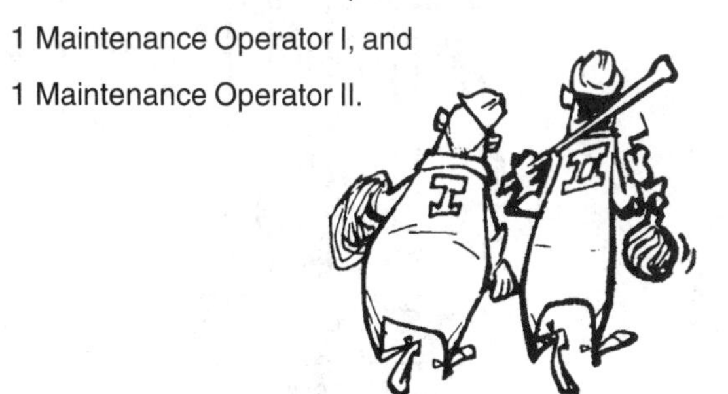

If the flushing crew consists of only two operators, neither one should ever be allowed to enter a manhole. Flushing crews usually have to enter every third or fourth manhole to remove debris. Whenever an operator enters a manhole for any reason, two operators must be topside at all times and all confined space entry procedures must be followed. The third operator could be another operator who could be readily available when needed.

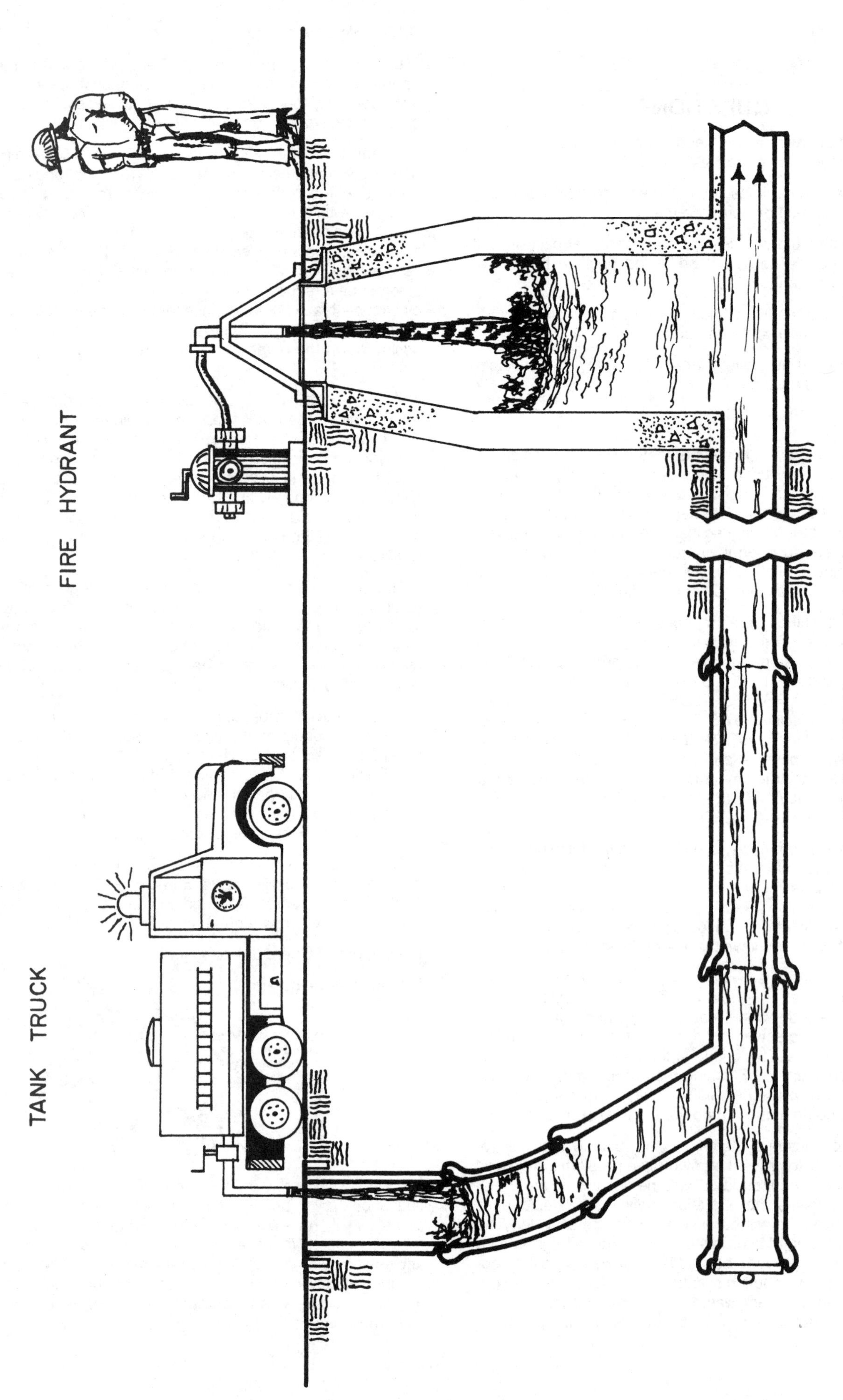

Fig. 6.15 Flushing operation

QUESTIONS

Write your answers in a notebook and then compare your answers with those on page 385.

6.12A Where in the collection system is flushing commonly used to clean lines?

6.12B What do lower than expected flows at a downstream manhole indicate during a flushing operation?

6.12C How does flushing help to control rodents and insects in sewers?

6.12D What are the staffing requirements for a flushing crew?

6.121 Equipment Setup and Flushing

Determine whether to use a tanker or fire hydrants or both as a source of water for the flushing operation. Start the flushing operation at the extreme upstream end of the collection system.

Place the water line over the manhole and protect the equipment and working area with sufficient safety cones, flags, and warning signs. Follow the directions in Section 4.3, "Routing Traffic Around the Job Site," and Section 4.4, "Classification and Description of Manhole Hazards."

A good visual inspection should be made of the manhole condition. If the water has been backed up, there will be evidence of debris in the shelf area and high water marks indicated by a grease ring. The general condition of the manhole, by number, should be part of the report and recorded at this time.

Clean the manhole first, since you will want to get more water and move on as quickly as possible after flushing the line. Use a pressure hose to clean the manhole and wash the manhole cover casting, cone, barrel, and the shelf of the manhole.

The amount of water you will dump into the line will be related to the amount of flow at the time of flushing. The entire water load may not have to be dumped in one manhole. After you have dumped some of the water, observe if there is any evidence of the water backing up.

If the flushing water backs up into the manhole, stop the addition of flushing water or reduce the application rate to prevent flooding of connected residential and commercial services. Observe the water flowing in the downstream manholes and look for signs of grease, roots, or other debris that are creating the restriction. If the restriction is caused by a heavy deposit of wastewater solids, flushing *MAY* move the cause of the restriction down the line to a manhole where it can be removed or broken up by spraying with the tank washdown hose and flushed on downstream. If the deposit is caused by roots or grease, prepare and submit a note or work order to have that segment of the line cleaned by using the proper equipment to remove the restriction.

Close the manhole and go to the next downstream manhole. Inspect this manhole as outlined in Chapter 5, Section 5.2, "Inspection of Manholes," and clean it completely with the washdown hose.

This operation will be repeated from manhole to manhole. As long as heavy flows are observed in the manholes, the operation is progressing satisfactorily. If the flows decrease, look for a stoppage by looking for flows backing up in the upstream manhole. If there is no evidence of a stoppage and flows are declining, more water will have to be introduced to maintain cleansing flows for the flushing operation.

The cleaning at each manhole with a hose takes only a few minutes, but this is the time and place where you should be alert for those items referred to in the beginning of this chapter: foreign material, grease, sand, and roots that can cause problems or indicate problems are occurring.

How many feet of sewers can be flushed by a crew in one day is difficult to estimate because of the many problems of manhole location, traffic conditions, and other obstacles. As much as 20,000 feet or more of sewer and many manholes can be flushed in a day if you do not encounter too many problems. Daily production depends on the number of manholes that have to be entered and cleaned, plus their condition and location. Whether manholes are located in rear easements and whether flushing can be accomplished by laying a fire hose from a hydrant or from the tank truck to the manhole in a backyard have considerable influence on the time required to flush a line.

6.122 Precautions and Safety

Safety measures must be strictly followed at all times, from properly protecting the equipment and working area with safety cones to checking manholes for atmospheric hazards found in sewers (toxic and explosive gases and oxygen level).

When flushing, the turbulence may liberate entrained gases from dislodged debris or slimes downstream, so care must be taken at all times.

The flushing method of cleaning presents problems very similar to those of balling operations. Be careful not to put more water into the collection system than it can handle, be careful not to cause a stoppage which in turn may flood a residence or business, and do not allow the material and debris being flushed through to accumulate. Caution is needed in all of these areas, especially when first starting a flushing program.

Care is required when opening and especially closing dry barrel fire hydrants involved in any hydraulic cleaning operation. Drinking water distribution systems in areas that typically have more sewer problems are usually just as old as the sewer. Improper fire hydrant operation can cause water hammer (if the hydrant is opened or closed too fast), which can cause major water leaks and discoloration (turbidity) in the drinking water. Using the valve in the fire hydrant to control flow is incorrect because the hydrant is designed to run either open or closed. A valve on the outlet nozzle, the end of the hose, or, if necessary, the isolation valve on the fire hydrant leg should be used to control the rate of water flow. Using the operating valve in the hydrant itself causes water to leak out of the weep holes at the bottom of the hydrant, which looks like a water leak, allows the potential for backflow into the water distribution system, and causes voids in the ground around the base of the hydrant.

6.123 Records

Since many manholes are going to be opened and cleaned during the flushing operation, there is no better time to make records of defects, leaks, roots, and other problems (see Chapter 5, Section 5.2, "Inspection of Manholes"). In addition, the amount of grease or other material visible also can be recorded. Also see Section 6.07, "Records."

6.124 Summary

As mentioned at the outset, flushing can be part of a sewer cleaning operation in conjunction with other mechanical clearing equipment, especially the power rodding machine. The power rodder can cut roots and grease from the walls and joints of pipe, but the removal of that material is dependent upon normal flows unless a flushing unit is used. Flushing is most effective during the power rodder's operation, but can follow at a later time. Flushing is not as effective as balling or high-velocity cleaning because sufficient velocities are not developed to move grease, grit, and heavy debris. A flow velocity of 2.5 feet per second will *PREVENT* the deposition of solids in a sewer, but once solids accumulate, the flow velocity from flushing is not high enough to *CLEAN* the sewer.

QUESTIONS

Write your answers in a notebook and then compare your answers with those on page 385.

6.12E Where do you start a flushing operation?

6.12F How is a manhole cleaned?

6.12G What factors influence the length of line that can be flushed in one day?

6.12H What safety precautions must be taken during a flushing operation?

6.13 Scooters

The scooter (Figure 6.16) is another cleaning tool that is very effective for cleaning wastewater collection systems by using water pressure. The scooter is designed to be self-propelled in gravity collection systems and operated similar to a balling operation. The operation is quite simple and the cost is considerably less than the cost of some other cleaning methods.

Basically the scooter is a framework on small wheels with a circular metal shield at one end; the shield is rimmed with a rubber flange. The top half of the shield is hinged and is controlled by a chain-spring system. The lower half of the shield is fastened with bolts to the front of the undercarriage. The flow of water behind and around the scooter is controlled by hinging the upper half of the shield to the scooter. This allows for hydraulic cleaning action and control of the flow of water in the line at the same time. Smaller scooters may be hinged only where the shield is attached to the scooter.

The scooter is capable of removing large objects and such other material as brick, sand, gravel, and rocks. Scooters are considered more effective in larger lines (over 18 inches) and have been used to remove grease in these lines. Even inverted siphons with angles of less than 22 ½ degrees can be cleaned with the scooter.

Since the scooter is dependent upon water pressure behind it, caution has to be used where sewers are shallow or there is a danger of flooding homes or business property because of the buildup of water in the sewer.

Scooters are available in sizes from 6 to 96 inches. The same undercarriage is used for some of these sizes. For instance, one carriage frame will adapt to shield sizes to clean 8- to 15-inch diameter pipes, and another frame will be used for 15- to 36-inch diameter pipes.

On larger sizes, stabilizing weights are used on the tail section of the carriage frame to counteract the head pressure during the cleaning operation.

The head shield on the larger sizes is designed to hinge into two or three parts. This compensates for the high head pressure when folding the head shield to allow water to pass. A spring-loaded mechanism allows the upper section to fold first, thus releasing the initial pressure. As the head pressure decreases, the middle section can be folded by continued pull on the shield control cable.

Large units are not only heavy to handle, but difficult to get through small manhole openings. By being able to fold the head shield, the scooter can be installed in most instances without dismantling. In some situations the unit may have to be assembled in the manhole.

QUESTIONS

Write your answers in a notebook and then compare your answers with those on page 385.

6.13A What causes the cleaning action of a scooter?

6.13B Scooters can be used to remove what kinds of material from sewers?

6.13C What precaution must be exercised when using a scooter?

6.130 Personnel and Equipment

Three operators are needed for safe, efficient operation of the small sizes of scooters (one Maintenance Equipment Operator and two Maintenance Operator I). Generally, sewers up to 12 inches in diameter can be cleaned by this method using a hand winch. Fifteen-inch and larger pipe diameters require power drum winch cables. With additional equipment, the operation may require an increase in personnel to handle both ends of the operation.

In addition to the scooter, the following equipment will be needed:

- A tank truck (1,000-2,000 gallons),
- Pickup truck,
- Debris traps, control lines, and storage reel,
- Upper and lower manhole jacks,
- Roller cable guides,
- Power drum winch, and
- Safety equipment (see Chapter 4, Section 4.3, "Routing Traffic Around the Job Site," and Section 4.4, "Classification and Description of Manhole Hazards").

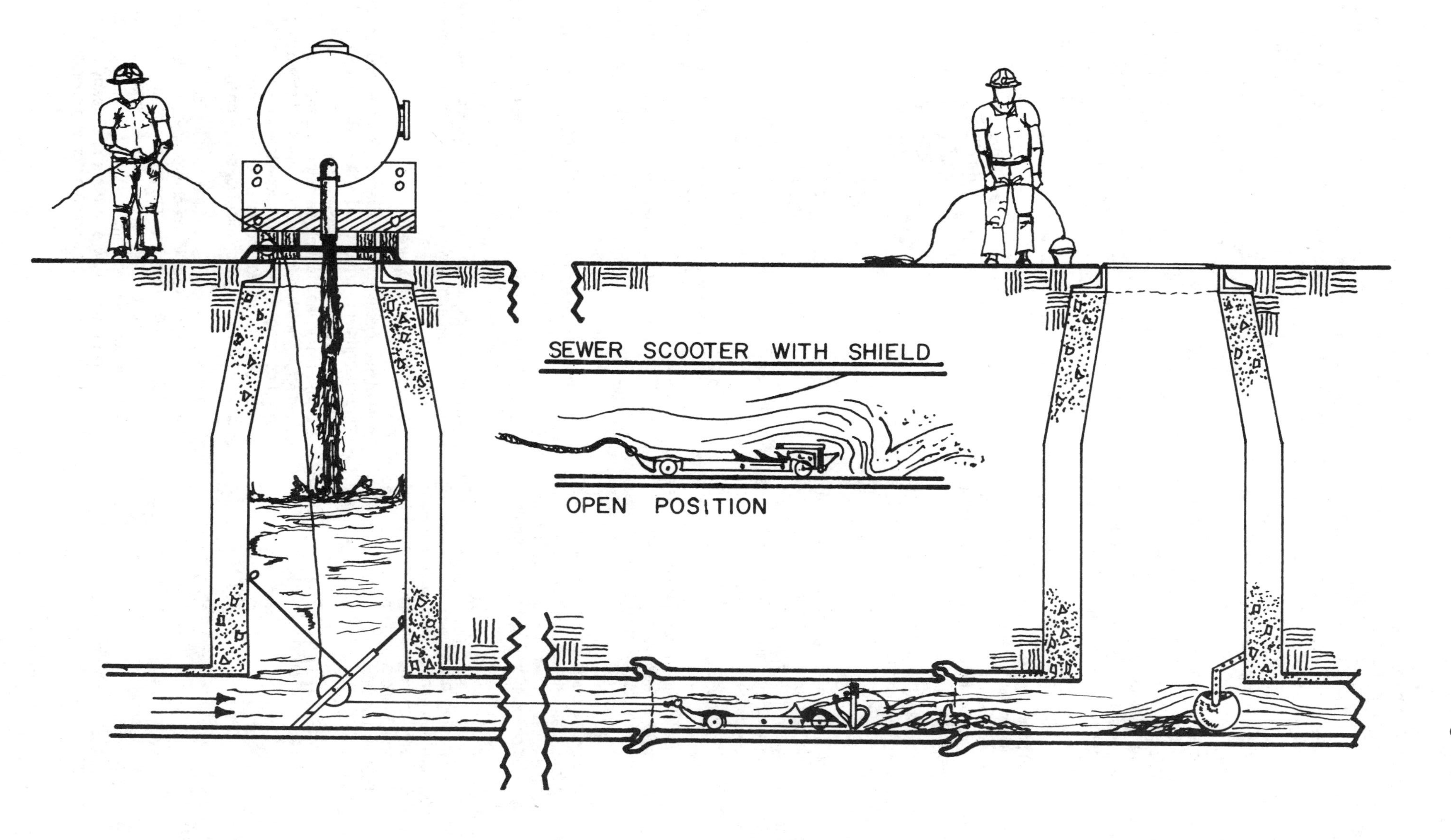

Fig. 6.16 Scooter operation

The control line may be plain manila rope or synthetic rope for small lines and steel cable on the power winch for larger lines. The synthetic rope is lighter and stronger than plain manila and will float. Synthetic rope does not mildew and will not have to be dried after use, but it has been known to deteriorate rapidly without warning due to some of the chemicals in wastewater.

The reel for storing the control line should hold 1,000 feet and be portable and easy to coil and uncoil.

When a powered drum is needed for the larger size scooters, a standard bucket machine or balling machine cable drum and winch may be used to control the scooter movement down the line. A powered drum is required because of the heavy force of wastewater built up on the shield. A double drum machine is recommended when using a second cable for tipping the head shield.

These winches are trailer or truck mounted with stabilizer legs and can be installed firmly over the manhole while in operation. Because of the size of the larger scooter and the tremendous pressure of water behind the head shield, strong steel cable is needed in place of rope for both the tag line to the carriage and the shield tipping line.

The manhole jacks or rollers are essential to guide the control lines into the sewer without causing unnecessary wear as the control lines are pulled back and forth in operating the scooter.

The debris trap is installed in the manhole downstream from the working manhole to trap any material worked loose during the cleaning operation. The trap is designed to allow water to pass and to hold back the debris. If the trap hinders the cleansing action around the shield when the scooter nears the trap, remove the trap and shovel the debris out of the sewer.

Other tools and equipment include a manhole shovel, debris bucket and hand line, boots, gloves, ladder, 100 feet of fire hose, hydrant wrench, water meter, air gap device, and manhole lid lifter.

If the debris removed from the line is to be hauled away by the cleaning crew, perhaps an additional 3/4-ton truck will be needed. Often the same truck is used for hauling the cleaning equipment, thereby reducing the overall cost of the operation.

6.131 Equipment Setup

Before starting the cleaning operation, safety measures need to be taken to warn and control traffic when working in the street and to ensure safe working conditions in manholes. Read Section 4.3, "Routing Traffic Around the Job Site," and Section 4.4, "Classification and Description of Manhole Hazards," for safe procedures.

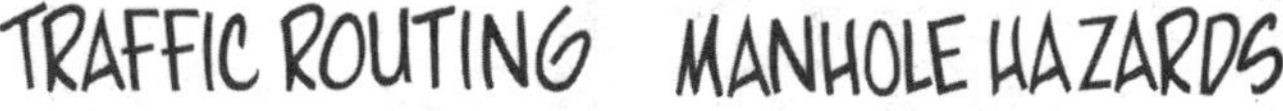

A suitable debris trap should be installed at the lower manhole so there is no delay in the operation once the scooter is installed and ready for use. Fasten the control rope or cable to the control chain of the scooter and lower the scooter into the manhole. After the head shield end of the scooter is inserted into the downstream line, install the lower manhole jack or control cable guide and secure firmly. This will allow the control cable to take the strain without rubbing on the sharp edge of the pipe during the cleaning operation. If an upper manhole jack or roller is needed, it should be installed next.

During this time the head shield of the scooter will act as a plug and a head of water will start building in the manhole. This can be controlled by pulling firmly on the control line, which in turn will fold the upper portion of the head shield and allow the excess water to flow on down the sewer. This spring folding mechanism is designed for a scooter up to 12 inches so that a single control line can be used.

QUESTIONS

Write your answers in a notebook and then compare your answers with those on page 385.

6.13D How many operators are needed on a scooter crew?

6.13E How does the debris trap installed in the downstream manhole work?

6.13F What must be done *BEFORE* equipment is set up in the street?

6.13G What is the purpose of the manhole jack or control cable guide?

6.13H During setup, how can the level of water in the manhole and sewer above the scooter be controlled?

6.132 Cleaning Operation

Let the scooter go downstream a few feet and test the folding mechanism of the head shield. Give the control line a short, smooth pull back two or three feet and hold. If the buildup of water in the manhole starts to go down, you will know the folding mechanism is working properly. If the folding mechanism does not work properly, pull the scooter back out of the manhole and look for the cause of the problem by inspecting the shield release mechanism for proper operation.

When the scooter is working properly, continue the operation by allowing slack in the control line. The slack will permit the shield to return to the upright position and the water pressure will increase behind the scooter. Allow the scooter to work its way downstream, keeping a tension on the control line. A continuous flow of water and a steady tension on the control line will ensure a smooth cleaning operation.

Attention must be given to the amount of head or water in the manhole above the invert; the best cleaning action results when the depth of water in the manhole is three times the pipe diameter. For an 8-inch line, on a flat grade, 24 inches of water above the invert of the manhole will provide sufficient head for cleaning and not flood most homes without basements. Higher wastewater levels in the working manhole increase the scooter cleaning efficiency but cause a greater risk of flooding.

The cleaning efficiency of the scooter increases as pipe size increases. This is due to the increase of head pressure and the water scouring jet action between the head shield rim and the pipe walls. This high turbulence will force the debris to move downstream ahead of the scooter. If the flow in the collection system is not sufficient for effective cleaning, you will have to add more water to the upstream manhole by using water from the tank truck or from a fire hydrant.

If the scooter slows or stops during the operation, it is probably due to an accumulation of material ahead of the scooter. This problem also may be indicated by a sudden rise of the water in the working manhole. When material builds up ahead of a scooter, give the control line a short, smooth pull back four or five feet. This will allow the upper half of the head shield to fold back and a surge of water to flush the accumulated material on downstream. Continue repeating this operation until the level of water in the working manhole stays fairly constant, at the preselected level, indicating normal operation.

Continue this operation until the scooter has reached the downstream manhole.

Someone must watch the flow of material into the lower manhole to prevent a buildup of material that could flow over the debris trap. If this occurs, loosened material is lost farther down the line and this manhole itself could become plugged. Remove the debris before the cleaning operation is allowed to continue. Debris removal requires additional help because flow control has to be maintained at the upper manhole. Notify or signal the operator of the scooter to shut off any extra water being put into the manhole and hold up the scooter's progress.

Follow confined space procedures for manhole entry and equip the operator with a safety harness, life line, hard hat, and gloves. The material will be scooped into a suitable debris bucket and brought to the surface with a hand line. Help *MUST* be topside during this operation. Vacuum devices (Figure 6.12, page 313) should be used to remove material from manholes whenever possible.

Setup for scooters sizes 15 inches and up differs from smaller scooter sizes because a second control line is used to trip the head shield and powered drum equipment is used instead of the hand line method.

If a bucket machine or balling machine winch is used to guide the scooter down the line, it will be equipped with steel cables. A 3/8-inch cable will be large enough to control the scooter and 1/4-inch cable for the control of the head shield. Half-inch cables are used for lines larger than 18 inches. The cable guides would differ in that two cables instead of one will be working in the line and they each should run over a separate roller to prevent damage to the cable.

The powered equipment usually is set up over the working manhole with the control cables running straight down into the manhole.

The bucket machine or balling machine winch, usually trailer mounted, has stabilizer legs which give a firm installation.

Trapping of material at the lower manhole also will differ on larger lines. The standard elbow-style debris trap will do an excellent job on lines up to 15 inches. In larger lines the amount of flow has to be considered. Usually all but large objects will break up and be carried with the flow. To trap these larger objects, a grill-type device or bar screen can be used which will allow the flowing water to pass through the manhole.

When the cleaning operation has been completed, remove the scooter at the lower manhole and wind the control cables on their respective drums. Clean the lower manhole of debris and remove the screening/trapping equipment. Load up the debris for disposal and clean the street area as needed.

6.133 Precautions and Safety

As with all cleaning methods, certain precautions need to be taken before and during the operation. One of the most important precautions for this method of cleaning is to determine if a buildup of water (head of water in the working manhole and in the sewer being cleaned) will flood a residence or business. The scooter requires a good head pressure to function effectively.

Before setting up all the equipment to clean large lines, examine the manhole structure to be sure the sewer scooter can be installed. Also keep in mind that more time is needed to set up and remove the larger units. Allow for this time when preparing work schedules.

Always examine the equipment cables before you start a cleaning operation and as you finish cleaning each section to avoid having a cable break when the scooter is in the middle of a sewer.

6.134 Records

See Section 6.07, "Records."

QUESTIONS

Write your answers in a notebook and then compare your answers with those on pages 385 and 386.

6.13I How can you determine if the folding mechanism is working properly when the scooter starts downstream?

6.13J What is the recommended depth of water in a manhole during a scooter cleaning operation?

6.13K What must be done if flow in the collection system is not sufficient for effective cleaning?

6.13L What could cause a scooter to slow or stop during the cleaning operation?

6.13M When material builds up in the line and slows or stops the scooter, what can be done to correct the problem?

6.13N Why should someone watch the flow in the downstream manhole during the cleaning operation?

6.13O What must be done *BEFORE* debris can be removed from a manhole?

6.13P Why should the cables be examined at the start and finish of each section?

6.14 Kites, Bags, and Poly Pigs

Kites (Figure 6.17), bags (Figure 6.17), and poly pigs (Figure 6.18) are devices commonly used to clean larger diameter sewers such as force mains. (Kites are also called parachutes.) These devices clean force mains more effectively than a sewer ball. With a small head (2 feet) on the sewer, the velocity of water flowing around the outside of the bag or through the outlet of the kite will create a cleansing velocity similar to the velocity around a sewer ball.

Force main piping is not generally susceptible to blockages like a gravity sewer. However, the capacity of the force main can be reduced when slime or other material builds up on the walls of the pipe and thereby increases friction losses in the pipe. Capacity also can be reduced by the accumulation of solids in the pipe invert.

One common method of determining the condition of the force main is by routine pump station calibration. If this is done on an annual basis, any changes in capacity and discharge head in the pump station can be identified. These changes, however, could also be the result of a worn impeller in the pump or wearing rings needing replacement. Therefore, you must first verify that the pumps are in good operating condition in order to determine that the force main is a problem. (For a detailed description of pump station inspection and calibration procedures, please see Volume II, Chapter 8, Section 8.25, "Operational Inspection.") In addition to performing the pump station calibration annually, pressure readings can be taken (if air relief valves are located along the force main) to develop a pressure profile along the pipeline. Force main pressures are an indicator of potential capacity reduction within the pipe.

Before beginning a force main cleaning operation, several factors need to be considered. For example, force mains that are under pressure from the pumps in the pump station do not have cleanouts located along the length of the pipe. With the exception of air relief or vacuum release valves, a pressurized force main is usually one continuous section of pipe from the pump station to the discharge end. Therefore, unless the force main is very short, none of the pipeline cleaning methods described in the previous sections can be used because of the lack of access.

The most common method of cleaning force mains is by using a type of polyurethane swab which is better known as a poly pig (Figure 6.18). Poly pigs are available in various densities and surface coatings. To use this method, insert the poly pig into the pipeline and then pressurize the force main behind the pig. As the device travels through the force main, it scours the inside of the pipe.

Normally the use of poly pigs requires that the pump station be shut down. Provisions must be made for handling incoming wastewater either through bypass pumping or by providing adequate short-term storage.

A launching point must be available for insertion of the pig and access at the discharge end of the force main must be available for removing the pig. Since pump stations are not commonly designed with launching points inside the station, some piping modifications may be needed to launch the poly pig. Because this is a somewhat specialized cleaning procedure, contractors who specialize in this type of cleaning method are often used.

Poly pigs are somewhat flexible and usually can negotiate around corners and bends and pass through pipe fittings such as increasers and reducers in the pipeline. However, there may be obstructions in the force main, such as a butterfly valve instead of a gate valve, which could cause the pig to become hung up. Careful review of as-built drawings (record drawings) should indicate any of these potential obstructions. In addition, some poly pigs can be equipped with a radio beacon device which allows the operator to track the location of the poly pig as it travels through the force main.

When using poly pigs for routine maintenance, time the pig's travel through the pipe and observe water flow at the receiving end. If the pig gets stuck, the water flow reduces dramatically.

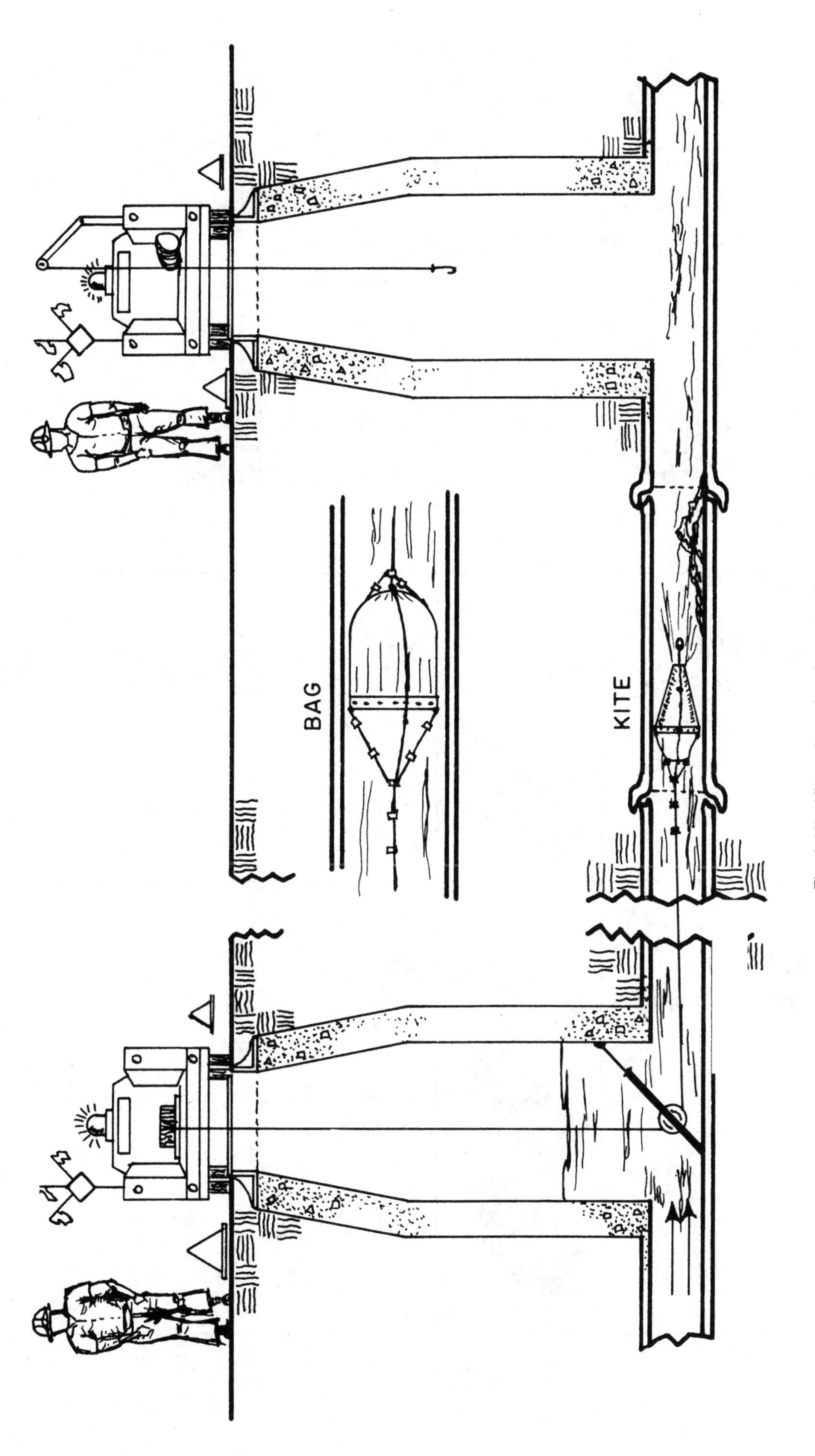

Fig. 6.17 Kite or bag cleaning operation

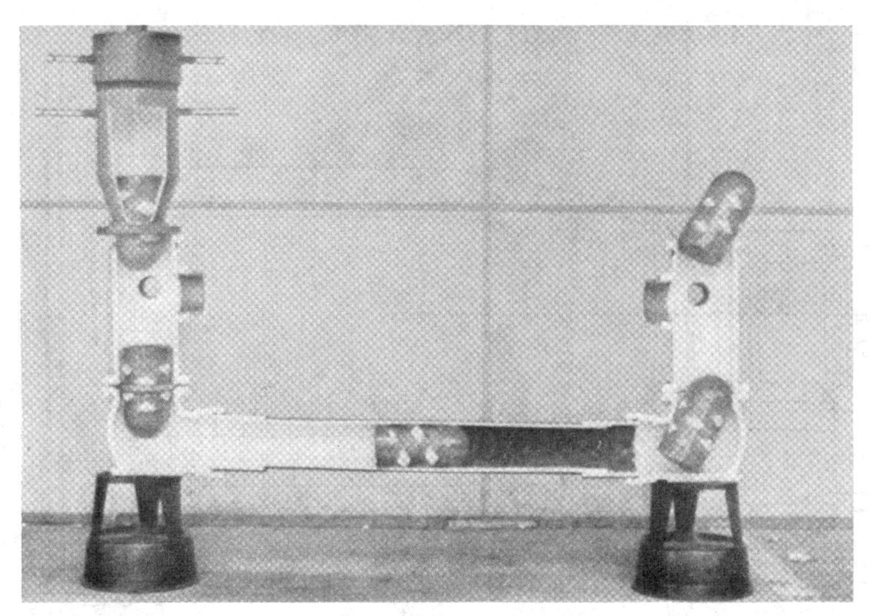

Use of poly pig in force main

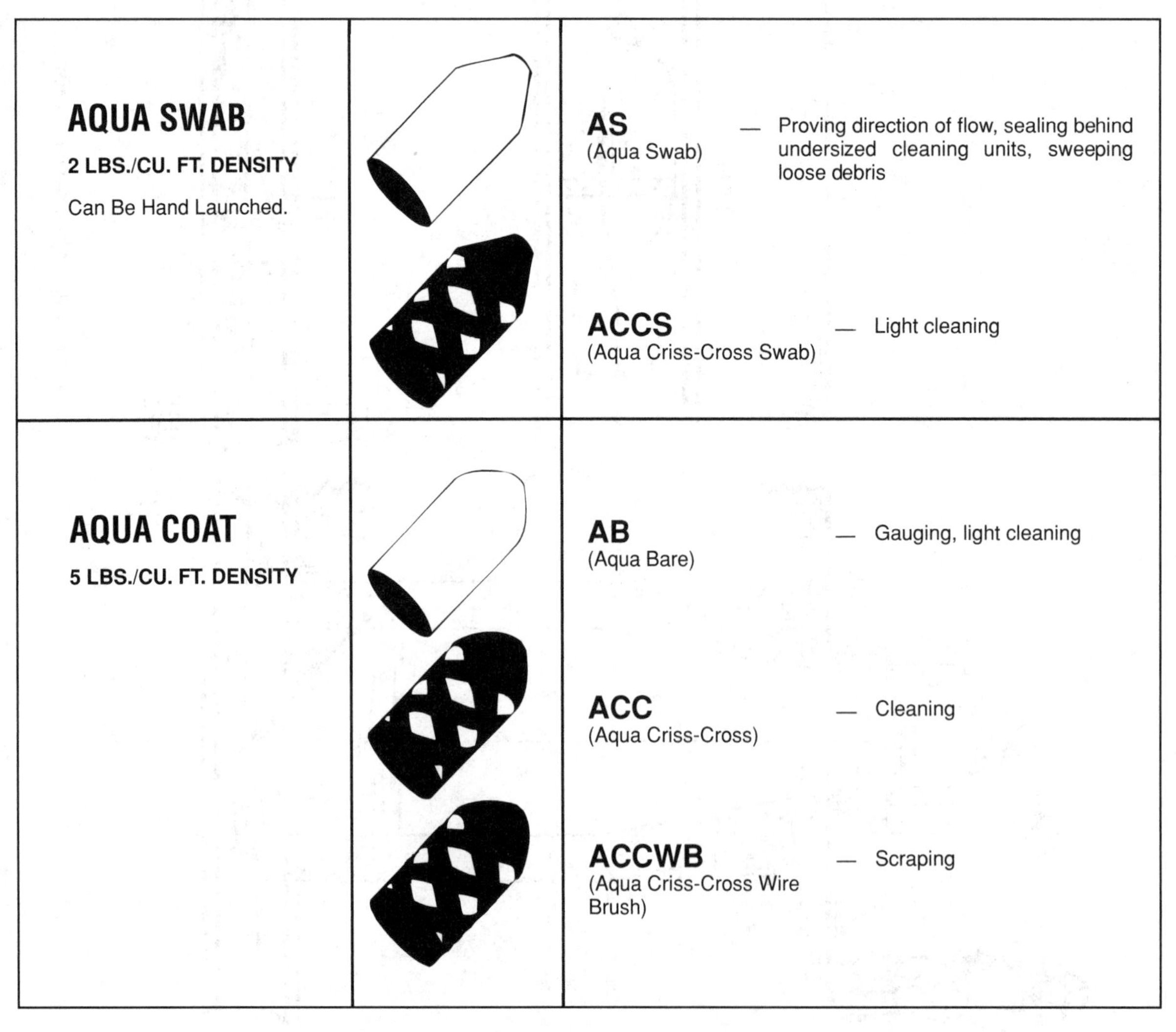

Fig. 6.18 Types of swabs and pigs

(Courtesy of Girard Industries)

Knowing the time when this happens, you can make a reasonable estimate of the pig's location. Use radio transmitters on the first pig through a long line in order to be sure there are no obstructions in the line.

In summary, the following factors need to be considered when using kites, bags, or poly pigs to clean force mains:

1. Provisions must be made for bypassing the pump station or providing alternative wastewater storage while the force main is being cleaned,
2. A launching station must be provided, either in the pump station or at the beginning of the force main itself,
3. External pumps and a water supply are needed to propel the pig through the force main,
4. The force main must be drained any time it is worked on, for example to install the launching station or to launch the pig,
5. Provisions must be made to track the pig through the force main in case it gets hung up and cannot be removed except by digging up the pipeline,
6. Some means must be provided to retrieve the pig at the end of the force main, and
7. The debris removed by the cleaning operation must be collected and taken to an appropriate disposal site.

QUESTIONS

Write your answers in a notebook and then compare your answers with those on page 386.

6.14A Instead of using a sewer ball, how are larger diameter sewers cleaned?

6.14B How are force mains cleaned?

Please answer the discussion and review questions next.

DISCUSSION AND REVIEW QUESTIONS

Chapter 6. PIPELINE CLEANING AND MAINTENANCE METHODS

(Lesson 2 of 5 Lessons)

Write the answers to these questions in your notebook before continuing. The question numbering continues from Lesson 1.

7. How do hydraulic cleaning methods clean sewers?
8. What precautions have to be taken during all hydraulic cleaning operations?
9. What would you do if a sewer ball became stuck in the middle of a sewer during a balling operation?
10. Why do sewer cleaning crews often use hand signals?
11. How fast and how far into the sewer should the nozzle of a high-velocity cleaner be allowed to go?
12. How effective is flushing?
13. What types of conditions are best suited for the use of a sewer scooter?
14. How can you determine whether a force main needs to be cleaned?

CHAPTER 6. PIPELINE CLEANING AND MAINTENANCE METHODS

(Lesson 3 of 5 Lessons)

6.2 MECHANICAL CLEANING

Wastewater collection line cleaning techniques described in this section rely on a mechanical action to clear the sewer. This clearing action results from the material in the sewer being removed by a scraping, cutting, pulling, or pushing action caused by a mechanical device or machine. Clearing techniques outlined in this section include bucket machines, power rodders, and hand rods.

6.20 Power Bucket Machines

6.200 Use of Power Bucket Machines

Solid debris is constantly entering the collection system by various means such as loose or misaligned joints and connections, manhole lids, service connections, and dischargers to the system. Sewer debris may be lumped together or spread randomly along the line. Types of deposits include silt, sand, gravel, or some types of industrial solid waste. The entry of silt, sand, gravel, and rocks into the collection system is often traced to some damage either to the sewer line or manholes, or a section where repairs were made to the system. Pipes or manholes may be broken by various construction projects, grade changes of streets, earthquakes, or excessive loads from vehicles and heavy equipment. When a pipe breaks, large quantities of debris can enter the collection system.

Sewers are purposely designed to develop sufficient velocity to provide a self-cleaning action and to convey solids through the system to the wastewater treatment plant. Suspension of the material depends on the velocity of the flow and the size and weight of the material. A velocity of 2 feet per second may not be sufficient to keep heavier solids in suspension and solids deposition will take place along the invert of the pipe.

When deposition of solids occurs in a sewer, first consideration should be given to hydraulic cleaning techniques to remove the solids from the sewer. Hydraulic cleaning techniques include balling the line (Section 6.10) and using a high-velocity water jet (Section 6.11). These techniques are used due to their simplicity, effectiveness, and the costs of removing the deposited solids when compared with other sewer cleaning techniques. Hard grease and soap may be removed by using a porcupine followed by a cleaning tool called a swab (Figure 6.19).

There are instances when the hydraulic techniques are not the best method of removing roots or deposits of solids because of the volume, size, weight, or types of material that may be found in the sewer. Keep in mind the fact that hydraulic cleaning methods become less and less effective as the line size increases. Large trunk and interceptor lines can reach 60 inches in diameter or more. Hydraulic cleaning methods are not effective in lines of this size because high-velocity cleaners, for example, can put more material into suspension than can be removed, creating a larger problem downstream.

When there is an indication that a sewer is failing and that sand or mud is entering the sewer, do not use a high-velocity cleaner. Nozzle action develops a negative pressure that can hasten the collapse of a bad section of pipe. A bucket machine can be used to remove the sand or mud when proper care is exercised. If possible, televise the sewer as soon as possible to determine the cause of the problem and the condition of the pipe.

Power bucket machines are an excellent maintenance tool for removing large amounts of debris from the sewer, especially in larger lines. These machines should not be used in small lines of 6 to 12 inches on a routine basis unless a porcupine or squeegee is being used, or unless materials are being carefully removed from a pipe collapse. Bucket machines provide a mechanical method to remove a large amount of material without allowing much of the material to escape downstream.

To clean a line after a pipe breaks, the power bucket machine can be used effectively both before and after repair to remove the heavy deposited material. Next clean the sewer hydraulically to restore it to full capacity. A damaged line must be repaired to prevent the entry of more material.

Figures 6.20 through 6.25 show power bucket machines, accessories, and operation.

QUESTIONS

Write your answers in a notebook and then compare your answers with those on page 386.

6.20A Identify two sewer line mechanical clearing techniques.

6.20B What cleaning techniques should be given first consideration when solids deposit in a sewer?

6.20C Under what conditions would hydraulic cleaning techniques not be feasible?

6.20D When should power bucket machines be used?

6.201 Personnel and Equipment

Buckets range in size from 6 inches to 36 inches in diameter. Volumetric capacity of the buckets ranges from 0.13

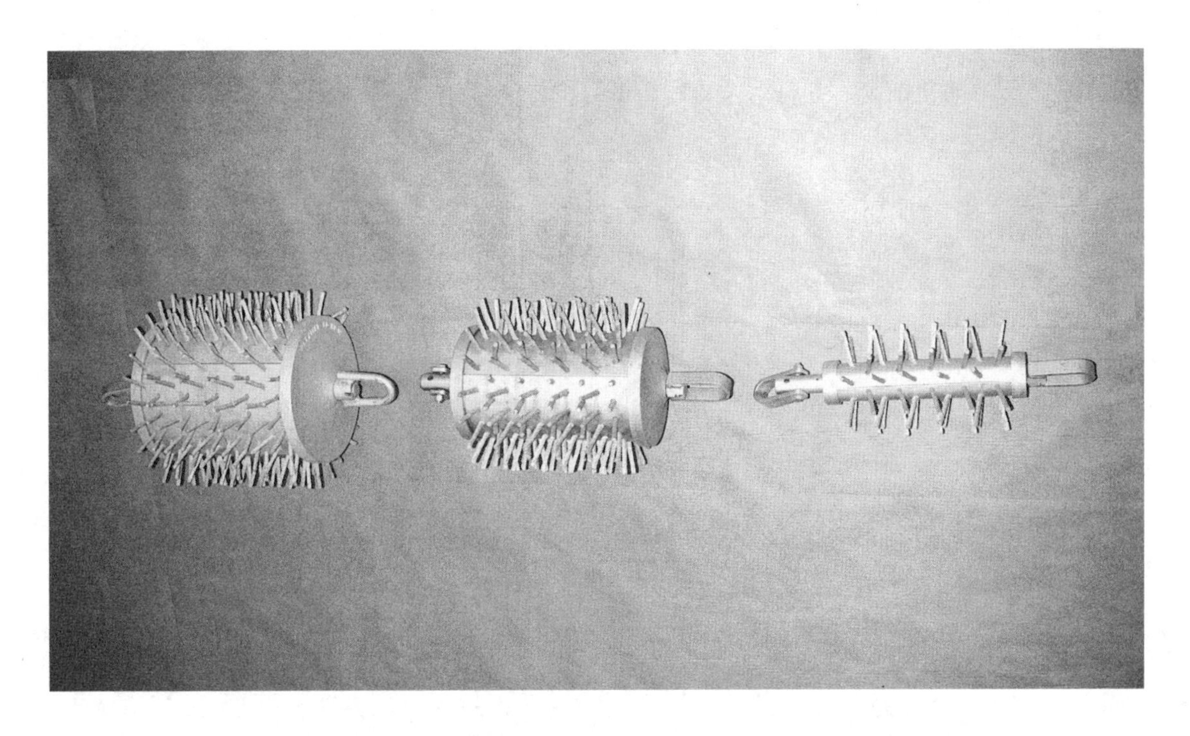

Porcupines are used to scour pipe walls of hardened debris, roots, grease, and industrial deposits.

The triple squeegee is a pull-type swab that is used after the line is cleaned with a porcupine.

Fig. 6.19 Porcupine and swab

(Permission of SRECO-FLEXIBLE)

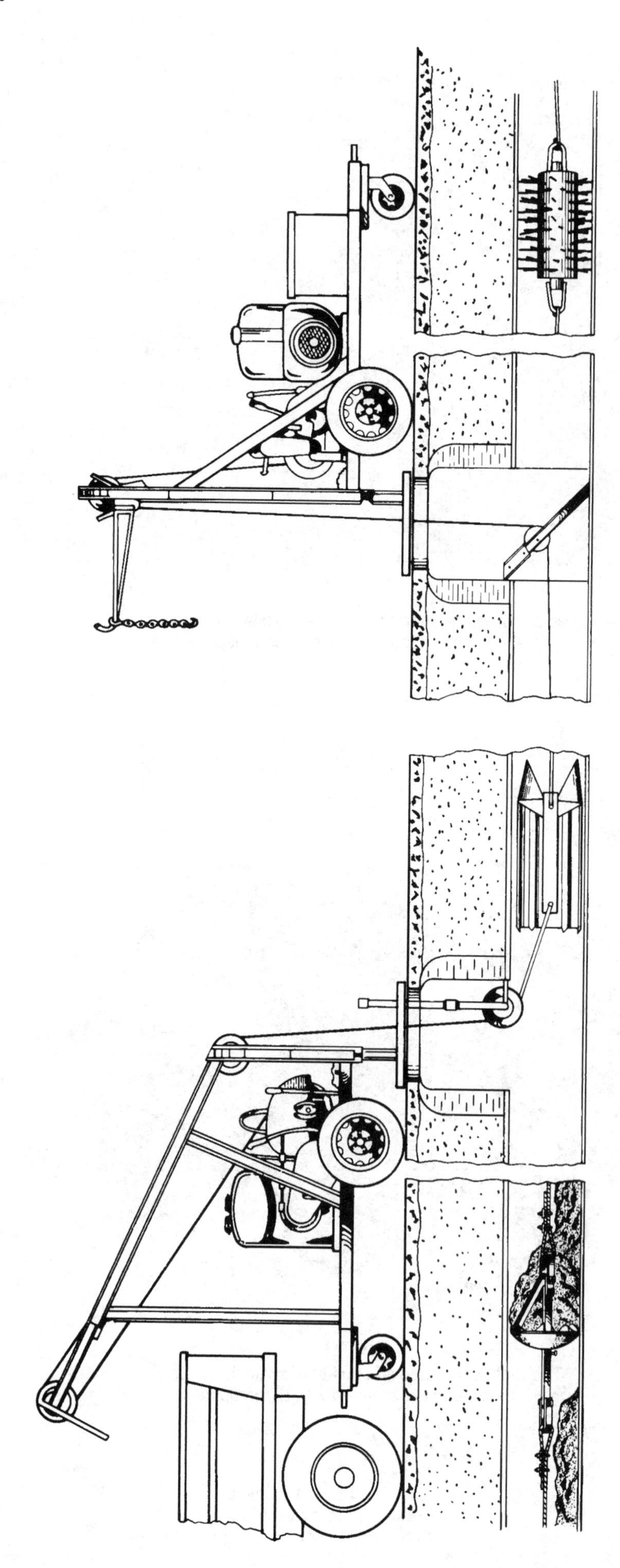

Fig. 6.20 Power bucket machine setup
(Permission of SRECO-FLEXIBLE)

Heavy-duty power truck loader bucket machine

Truck loader in operation

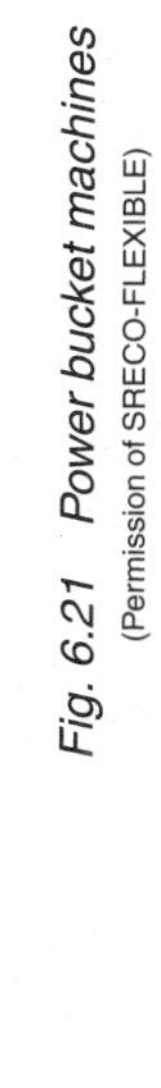

Fig. 6.21 Power bucket machines
(Permission of SRECO-FLEXIBLE)

Clam-type bucket

Roller and yoke assembly
for buckets up to 18 inches

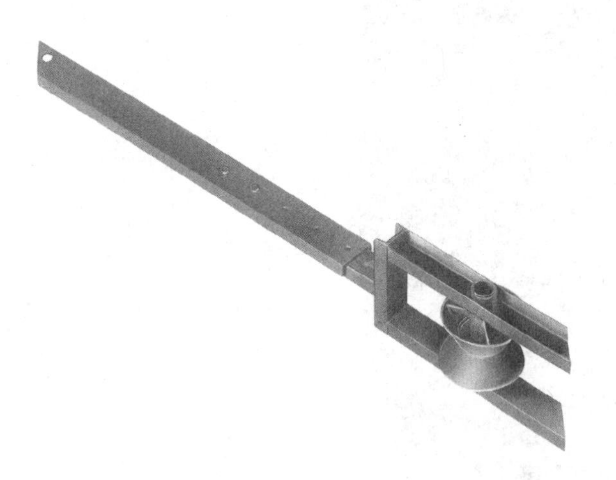

Slant jack

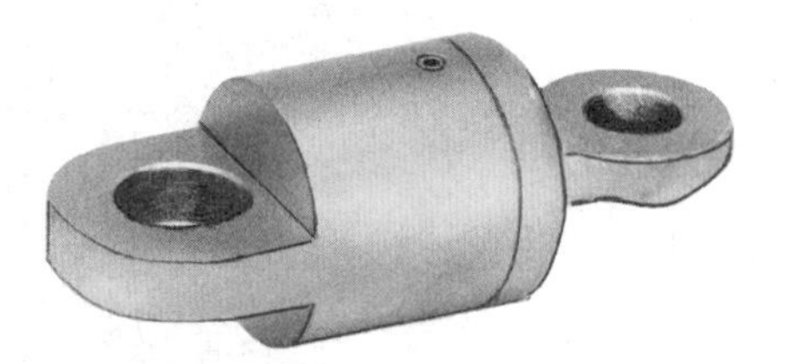

Ball bearing swivel

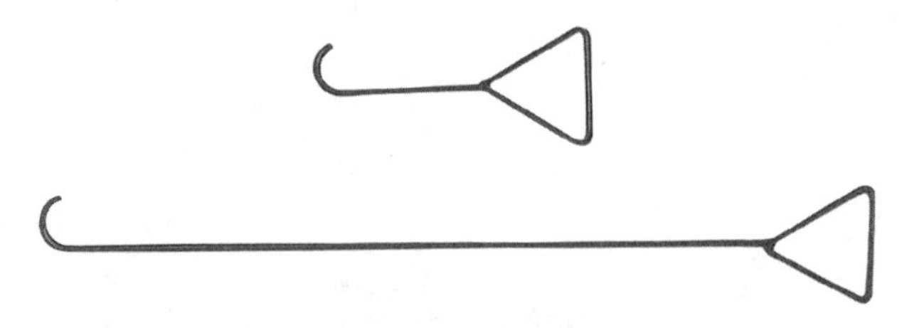

Bucket hooks

3-point jack roller

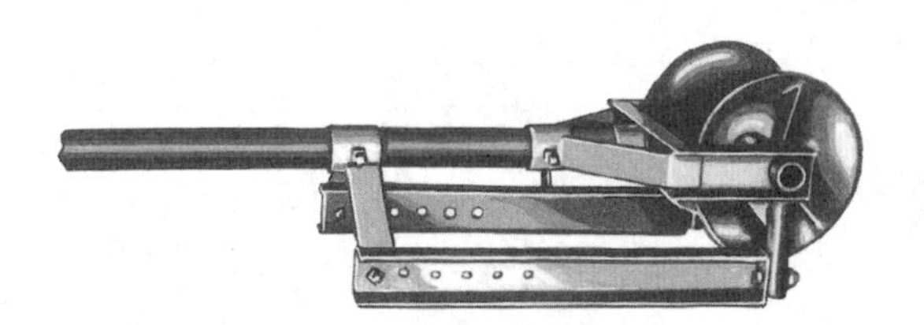

Drop yoke roller assembly

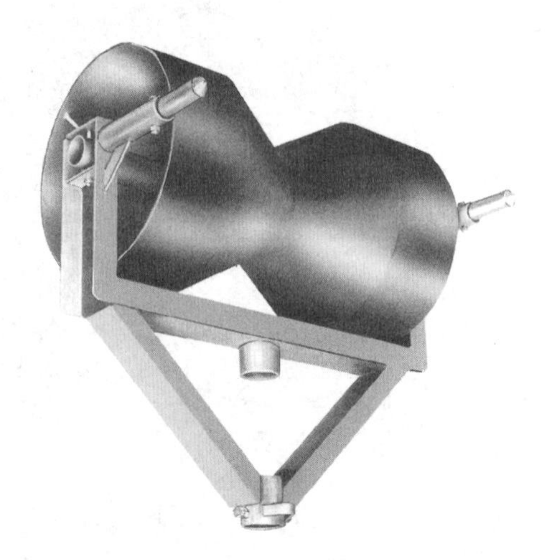

Wide flange roller and yoke assembly
for buckets 20 inches or larger

Fig. 6.22 Power bucket machine accessories

(Permission of SRECO-FLEXIBLE)

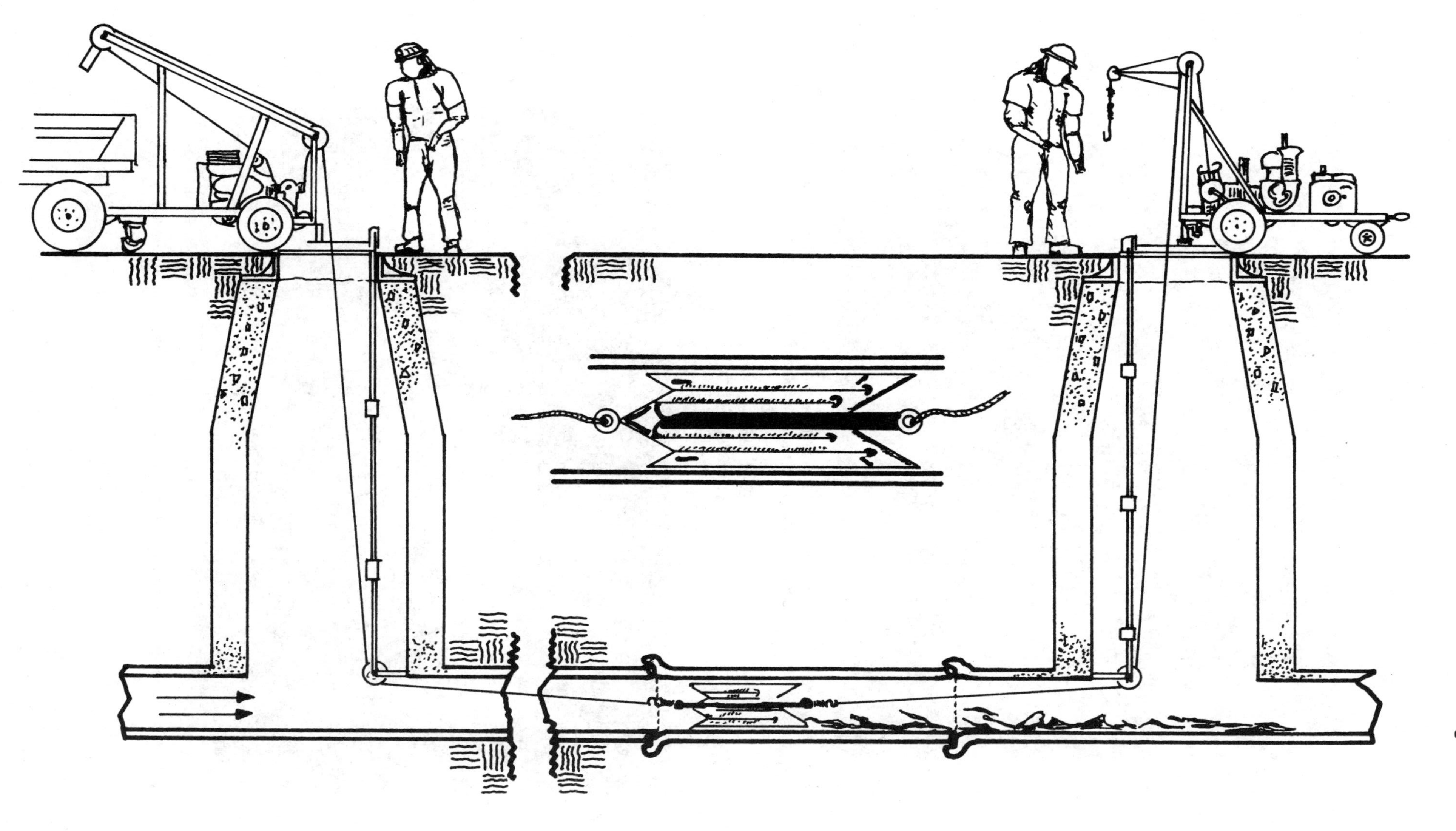

Fig. 6.23 Power bucket operation

Pull-type swab with holes in discs for flushing action. Used for cleaning lines with grease and as a follow-up cleaning after a porcupine (Figure 6.19).

Fig. 6.24 Swab with holes in discs
(Courtesy of Union Sanitary District)

Heavy-duty hydraulic winch truck used when cleaning larger lines.

Fig. 6.25 Winch truck
(Courtesy of Union Sanitary District)

cubic foot to 8.5 cubic feet. An 18-inch diameter bucket will pass through a standard 24-inch diameter manhole opening and has a volume of approximately 1.6 cubic feet.

The bucket machine is usually trailer mounted (Figure 6.21), but can be truck mounted. Trailer-mounted bucket machine units must have a pintle safety hook attached to the truck that tows the trailer to the job site. Also a wire cable and shackles should be placed around the truck frame and to the frame on the bucket machine for added safety in case the pintle hook fails.

The bucket machine units have a steel framework on which is mounted a gas engine and a drum winch. The drum is coupled to the engine through a controllable drive train, usually a chain and belt drive combination. The drum is capable of holding 1,000 feet of 1/2-inch steel cable and is mounted on the framework in such a position that it can be

centered over a manhole. The framework includes a vertical "A" frame of sufficient height to allow the cleaning bucket to be lifted above ground level.

Two machines are needed for this operation and both are basically the same design. At least one machine also will be equipped with an additional smaller drum capable of holding 1,000 feet of 1/4-inch steel cable. This drum can be operated separately from the 1/2-inch cable drum. The purpose of the small cable and drum is twofold. One is for the purpose of threading the sewer line from manhole to manhole, and the other is to have a suitable cable which can be left in the line overnight. The drum is, therefore, often referred to as an overnight drum.

One machine has a chute with rollers and shaker bar, which allow the material removed to be dumped directly into a dump truck. The machine that pulls the bucket out of the manhole will be referred to as the working machine.

With each bucket machine, equipment will be needed in the manhole to guide the cable into the sewer line without rubbing or cutting into the pipe. At the working machine, use a V-shaped roller (Figure 6.22) of sufficient size to accommodate the size bucket being used. This roller will be held in place by means of a steel pad that is firmly secured flush with the street surface by two adjustable stabilizer legs on the machine. The companion machine may use the same method or perhaps use a slant jack (Figure 6.22) in the bottom of the manhole for the cable to travel over. This would dispense with the use of the pad and roller.

A variety of tools are available for use with these machines and more are being developed, making these machines usable for purposes other than bucketing of debris. For example, root cutters and insertion pullers can be used with bucket machine power winches.

For this discussion, we will consider the use of the clamshell bucket (Figure 6.20) and porcupine tool. These are available in sizes from 6 to 36 inches in diameter, but sizes under 12 inches are rarely used.

The bucket is designed so that when it is pulled into the sewer line, the jaws are in an open position. The jaws dig into the deposits of material.

When the working machine pulls the bucket out, the jaws are forced closed by means of a slide action. Any material in or in front of the bucket is scooped up.

The porcupine tool (Figure 6.19) is a steel cylinder having solid ends with eyes cast in them to which a cable can be attached. Many short pieces of cable protrude from this cylinder like a round brush. This tool is quite effective for final cleanup since the bristles produce a scrubbing action.

Equipment needed other than that furnished with the bucket machines includes sand traps, manhole shovel, debris bucket and hand line, manhole lifter, pipe wrenches, and hand tools such as crescent wrenches and pliers.

Safety equipment includes flags, cones, barricades, traffic signs, atmospheric monitoring equipment, safety harness, lifeline, tripod, SCBA, and other safety devices the job setup may require (see Section 4.3, "Routing Traffic Around the Job Site," and Section 4.4, "Classification and Description of Manhole Hazards").

Experience has shown that many unexpected repairs occur during the clearing operation. Extra cable swivels, clevises, cable clamps, and other essential accessories should be readily available if not carried with the machines.

The number of operators needed to effectively conduct the clearing operation will vary, but at least three operators are needed and four operators are recommended. If traffic control is necessary, flaggers will also be needed. Personnel requirements are:

1. One Maintenance Equipment Operator,
2. Two Maintenance Operator I, and
3. One Maintenance Operator II.

QUESTIONS

Write your answers in a notebook and then compare your answers with those on page 386.

6.20E What is the function of the working machine in a power bucket operation?

6.20F What is the purpose of the porcupine tool?

6.20G How many operators are needed on a power bucket crew?

6.202 Equipment Setup

Whether the working machine is set up at the upstream or downstream manhole depends on field conditions. In smaller sewers with lower flows, the working machine is usually located at the downstream manhole. Under these circumstances there is little tendency for the material being removed to be washed out of the bucket. Also, working from the downstream manhole reduces the possibility of a stoppage developing in the sewer below the working area.

Often the working machine is located at the upstream manhole (Figure 6.23) when clearing debris from larger sewers. Under these conditions the debris in the full bucket tends to remain in the bucket because the bucket is being pulled against the flowing water when the full bucket is returned to the working manhole. If flows are low, the possibility exists of a stoppage developing downstream from the working area.

Once this equipment is set up, it usually remains in one position for several hours. Special emphasis is placed on safety measures taken for control of traffic and ensuring a safe environment in a manhole (see Section 4.3, "Routing Traffic Around the Job Site," and Section 4.4, "Classification and Description of Manhole Hazards").

Bucket machine clearing requires a certain amount of skill in operation of the equipment and this can be obtained only by actually working with the equipment. This method is one of the more expensive operations due to several factors, but mainly the hours involved in setting up the equipment. In the process of bucket machine clearing it is not unusual to find you cannot complete the job before the end of the day. Because of the time and effort needed to thread the line, the overnight line is left in the sewer. This will be discussed further in Section 6.203, "Operation."

During the setup operation, communication has to be established between the operators at the upstream and downstream manholes. Communication may be either by sound-powered telephones, radio, or signals as shown in Section 6.102, page 306. A foolproof system must be devised and used between winch operators because someone can easily be injured if the proper signal is not sent, received, or is misunderstood.

Position the two machines over the respective manholes (Figure 6.23); usually the working machine is at the downstream manhole and the backpull machine is at the upstream manhole. Place the pads under the stabilizer feet of the machine and jack them down so that the weight of the machine holds the pads in place. The legs also function as a method of leveling the bucket machine. Insert a sand trap or debris trap in the downstream line to stop material pushed ahead of the bucket when clearing.

The lower manhole roller is lowered into the manhole by attaching various lengths of threaded pipe, usually 2-inch pipe in 4-foot sections. With this roller properly situated (the cable and bucket must clear the top of the sewer), secure the pipe to the pad by means of a chain clamp device. With this firmly in place, the roller will not lift up when strain is applied to the cable while pulling. Give consideration to this *NOW* since the same pressure is exerted to lift up on the roller as is exerted in pulling with the cable. If there is a setup problem, roller security is one of *THE FIRST* problems encountered when the clearing operation starts.

As described previously, the upstream machine may possibly be set up with a slant jack-type roller (Figure 6.22). If so, the pad and roller, with pipe extensions, will not be needed. Again, make sure the slant jack is secure since this machine also will be exerting considerable lifting-pulling action.

To thread the sewer, a synthetic rope can be used. The upstream machine can be equipped with a small drum exactly the same as the "overnight" drum on the downstream working machine. One thousand feet of the synthetic rope can be carried on this drum. Some synthetic rope has been known to deteriorate without warning due to chemicals in wastewater.

To this rope attach a nylon parachute designed for such use and allow this parachute to float downstream to the working manhole. A high-velocity cleaner may be used to string the line also. Attach the parachute rope to the $^{1}/_{4}$-inch steel cable of that machine. A swivel must be used to keep the rope and cable from twisting. The upstream operator engages the small rope drum and pulls the $^{1}/_{4}$-inch cable upstream to the machine. Remove the rope drum from the machine so it will not interfere with the operation. Connect the $^{1}/_{4}$-inch cable and swivel to the $^{1}/_{2}$-inch cable. When ready, signal the downstream operator to engage the small drum of the machine and pull in the $^{1}/_{4}$-inch cable, thus threading the sewer line with the $^{1}/_{2}$-inch cable. The operator of the working machine will likewise remove the $^{1}/_{4}$-inch cable drum.

QUESTIONS

Write your answers in a notebook and then compare your answers with those on page 386.

6.20H Why might an overnight line be left in the sewer?

6.20I Why must a foolproof communication system be devised and used?

6.20J What is the purpose of the swivel between the parachute rope and the $^{1}/_{4}$-inch steel cable?

6.203 Operation

The clearing bucket is now ready to be attached to the cable. Experience and history of the sewer line dictate how much smaller a bucket must be than the actual diameter of the sewer. A cable swivel of sufficient size is attached to each *BAIL OF THE BUCKET*[10] by means of a clevis. The $^{1}/_{2}$-inch cables are then attached to the swivels using a clevis. Position the bucket so the jaws will be opening upstream.

The use of a bucket flipper installed on the open or jaws end of the bucket is recommended. This flipper is installed on the bail using two connecting clevises (shackles) so the flipper is in the same plane as the bails. This device will prevent the bucket from trying to go around the various rollers at right angles to the bails and jamming the bucket.

[10] *Bucket Bail. The pulling handle on a bucket machine.*

Take up the slack in the cable of the working machine until the bucket is hanging free over the manhole. Signal the upstream operator to take in the cable. Hold the brake securely on the working machine to keep the bucket from dropping into the manhole. Experience is gained quickly at this point because special care has to be taken in getting the bucket around the lower roller and into the sewer.

At the start, always make a few short pulls to check out not only the amount of material in the sewers, but also the equipment and the setup. One approach is to try an undersized bucket first to determine if it can be pulled through the sewer. If this works, a larger bucket can be tried because it will remove more material before it is full. As the bucket is pulled upstream into the material deposited in the sewer, a definite resistance can be noticed when the bucket is full.

When the bucket is full, pull the loaded bucket back out of the sewer. To start pulling the bucket back, signal the working machine operator after disengaging the machine. Hold a slight amount of brake to keep the cable from spooling too freely as the bucket is pulled downstream. If the working machine is equipped with a footage counter, watch closely as the full bucket nears the manhole. Reduce the speed of pull when the bucket enters the manhole and makes the sudden turn around the lower roller and up to the street surface. If a footage counter is not used, the operator can quickly note the cable wraps and their position on the drum. This technique also can apply when the bucket is being pulled upstream. By knowing the approximate diameter of the drum and the number of wraps or layers pulled off, the operators can judge fairly closely the distance the bucket is in the sewer. Bucket location is important to this operation.

When the bucket is pulled back out of the manhole, slowly bring the bucket up and allow it to move up the chute or truck loader (Figures 6.21 and 6.23). When the bucket reaches the top of the chute or truck loader, the bucket dumps the debris into the truck.[11]

When the working machine is equipped with a truck loading chute, the operation of dumping the bucket is done by the machine operator. An extra operator is important at this end of the operation even with the chute method of dumping. Often the operator needs help when the controls of the machine cannot be left. When extra help is not available, the danger of accidents is greatly increased. This method of clearing is time consuming and when the operation is hurried, the quality of clearing, equipment care, and safety are often unintentionally sacrificed.

The operation is now ready to be repeated. After several pulls, check the buildup of material against the sand trap in the bottom of the manhole. This material should be removed before it is lost downstream.

When the bucket of the correct size has been pulled freely from the upstream manhole to the lower manhole, it may be advisable to finish cleaning the sewer with the porcupine cleaning tool. Remove the bucket, fasten the two cables together and pull the cable upstream until the swivels are above street level. Attach the porcupine with a swivel at each end and insert it carefully into the sewer. If a slant jack is used, the porcupine may have to be put on in the manhole due to the design of most slant jacks. If a pad and roller are being used, allow the working operator to slowly pull the porcupine into the sewer and then on downstream where it can be removed.

WARNING

DO NOT PULL THE PORCUPINE UPSTREAM SINCE IT HAS A TENDENCY TO SEAL OFF THE LINE AND CAUSE THE SEWER TO BACK UP.

Pulling the porcupine upstream may cause flooding problems if there are building laterals connected into the sewer being cleaned.

Since each machine is capable of handling 1,000 feet of cable, often it is possible to pull through or skip a manhole, thus saving another time-consuming setup. Two sections can be cleared at once, provided the sections are in a straight line. A disadvantage of this procedure is that you may greatly increase pipe damage by making more trips through the same section of sewer. A short damaged section may become a long damaged section.

If the job is not complete at the end of the day, remove the bucket at the downstream working manhole. Install the 1/4-inch overnight drum and connect it with a swivel to the 1/2-inch cable of the upstream machine. Pull the 1/4-inch cable through the sewer and secure it in the upstream manhole in such manner that it cannot come loose. Then replace the manhole cover.

Remove the 1/4-inch cable and drum from the downstream working machine, lower them into the manhole and secure them.

By following this procedure, the pull line will be ready for the next day's setup and considerable setup time will be saved.

When work is complete, use the reverse of the setup procedure to remove the equipment from the sewer and the manholes. Hose off buckets and cables. Load the equipment into the trucks and prepare for the next setup or the next day's use.

Final cleanup is performed with special care given to washing down and completely cleaning the entire area where any debris or wastewater may have been deposited on the street surface during the operation.

In these instructions, the setup of this equipment has been to have the working machine at the downstream manhole and to do the actual clearing by going upstream into the material to be removed. This does not mean the reverse procedure cannot be used and it often is. However, conditions in the sewer must be kept in mind and particularly the results of whichever procedure is selected. If large amounts of debris and especially roots are encountered when pulling the clearing tool downstream, a stoppage can be caused below the working area. By working upstream as outlined, this danger is practically eliminated. However, use of improper tools and techniques also can allow debris to cause a stoppage downstream. Again, the continuous use of this equipment and the experience gained will govern the choice of the setup procedure.

---

[11] *Some regulatory agencies do not allow material from a bucket clearing operation to be dumped on a street. It must be emptied into a container or a truck.*

QUESTIONS

Write your answers in a notebook and then compare your answers with those on page 386.

6.20K What size clearing bucket should be selected?

6.20L How can you tell when the bucket is full?

6.20M Where is the debris in the bucket first dumped after the bucket is removed from the manhole?

6.20N Why is a sand trap inserted in the downstream manhole?

6.20O How can the sewer line be cleaned after a bucket has been pulled through the line?

6.20P Should the porcupine be pulled upstream or downstream?

6.20Q Why should the bucket be pulled upstream?

6.204 Precautions and Safety

Because a rather large tool is being used in the sewer, any problem encountered that would keep the bucket, porcupine, or other such clearing device from moving can cause water to build up behind it. Prevention of stoppages during clearing should be foremost in the mind of the operator. If there is any question of the clearing tool getting through, try a smaller size and determine if the sewer would allow a larger tool to pass through the sewer.

As mentioned previously, it is important to know the location of the clearing tool at all times. Footage counters are recommended, but if they are not available, keep track of the wrap of cable. In the event the clearing tool has to be dug up, it helps to know where it is stuck.

Regularly inspect cable clamps, clevises and swivels, bucket condition, cable condition, and the condition of accessories. If a cable should come loose from the bucket on the wrong end for any reason, you may not be able to pull the bucket through the debris to the manhole at the other end.

If a bucket becomes lodged, don't try to force your way through. If pulling the bucket upstream, have the downstream machine pull it back a few feet and try again. Pulling back will open the jaws and usually will let the bucket come into the problem area at a different angle. If repeated attempts fail, have the bucket pulled back a few feet, enough to open the jaws, and then hold the brake tightly on the upstream machine while the working machine again pulls. This tends to hold the jaws open and also gives a lifting effect to the bucket.

Be careful when lowering the bucket into the working manhole. Keep the slack out of the cable. Do not allow the cable to tangle or come off the lower roller by keeping the slack out of the cable. Prevent the bucket bail from folding back on the bucket when it is pulled into the sewer because there is no way that it can flip over when you start to pull the bucket.

Personal safety must be stressed when using cable-operated equipment. Be especially careful when guiding cable onto the drums. Keep your hands clear of the drum. Use a manhole hook to guide the cable. In an instant, your hand or fingers can become entangled and the operator cannot stop the machine fast enough to avoid injury. New or inexperienced operators should not operate this equipment without proper training and supervision. Some of the safe practices used are techniques acquired through repeated operations and observations of problems encountered during rewinding.

6.205 Records

Since bucket machine clearing is not repeated in the same line any more often than necessary, good records must be kept. Often at a given spot in the sewer line the bucket may "hang up." This could be caused by several possible factors, but if it is a severe problem, a repair at that point may be indicated. Cocked joints can turn into broken joints if the bucket is banged against them continuously. Misaligned joints may also cause the bucket to become lodged tightly. Let these events become part of the permanent record. If repairs subsequently are made, this too can be noted and the next time this sewer is cleaned, the operators will have this information. Also see Section 6.07, "Records."

QUESTIONS

Write your answers in a notebook and then compare your answers with those on page 386.

6.20R Why must the location of the bucket in the sewer be known at all times?

6.20S Why should the condition of the equipment be inspected periodically?

6.20T What precautions should be taken when working around cable-operated equipment?

6.20U Why should particular problems encountered when clearing a sewer be recorded?

6.21 Power Rodders

6.210 Uses of Power Rodders

Power rodding machines use a steel rod to push or pull various clearing tools through sewers (Figures 6.26 and 6.27). These machines are of various designs, but are equipped to store either continuous or sectional rod in a reel-type cage in lengths up to approximately 1,000 feet. (Note: Please refer to pages 583 and 584 for metric pipe sizes and other metric conversion factors.) This reel can be rotated to give turning action as the rod is pushed in or removed.

Power rodders can be used for the following purposes:

1. Routine preventive maintenance.
2. Scheduled clearing of:
   a. Grease deposits,
   b. Roots, and
   c. Debris accumulations in flat lines.
3. Threading cable for:
   a. Balling equipment (15-inch diameter and larger lines),
   b. TV inspection, and
   c. Bucket machines.
4. Emergency use for clearing stoppages.

Power rodders are one of the most widely used methods for clearing a wastewater collection system. Since the machines were introduced for commercial use in the early 1950s, many improvements have been made in design, operation, and clearing tools.

The power rodding machine can perform tasks that hydraulic cleaning methods have difficulty in doing, such as heavy root removal, grease and grit deposit removal, and stoppage relief. While they are effective in these areas, power rodding machines do not move silt, sand, and sludge as efficiently as hydraulic cleaning methods. Many agencies use both power rodders and high-velocity cleaners in collection system maintenance due to the respective limitations and advantages of each method of cleaning.

Although other clearing methods play an important role in maintaining the collection system, the power rodder can handle stubborn stoppages of roots, grease, and debris. The power rodder is called for when emergency crews cannot clear a stoppage with hand rods. This could be due to the type or size of stoppages or a distance too great for hand rods to be effective. The power rodder is a handy tool as support equipment to thread cables in lines for other cleaning and maintenance methods. After a sewer has been cleared with a power rodder, it should be cleaned hydraulically to restore the line to full capacity.

6.211 Classification

Power rodders can be classified according to several different characteristics, but most are either sectional or continuous type machines. There are, however, machines that use coiled rod and are called coil rodders. These are powered by either an electric motor or gas engine. The electric unit is usually equipped with $^{1}/_{2}$-inch coiled rod and is capable of clearing pipe sizes up to 8 inches in diameter and distances of as much as 250 feet. The gas-powered coil rodder is capable of clearing line sizes up to 10 inches and distances of 500 feet. A one-inch coiled rod is used on these units.

The coil rod for both gas and electric coil rodders is available in 25- and 50-foot sections with a thread and nut coupling. Both machines are usually trailer mounted and the trailer is equipped with storage boxes for tools and other equipment. The clearing tools used on the coil rodders, though basically designed the same as those used on the large rod machines, have a different type of connection for attaching to the coil rod.

While sectional and continuous power rodders can be used to accomplish similar kinds of work, they are two different types of machines. The sectional power rodder is designed to use sections of rod that are mechanically attached to each other. Rod size can be either $^{5}/_{16}$- or $^{3}/_{8}$-inch diameter, and rods are available in 36-, 39-, and 48-inch lengths. They are assembled by threading a nut into a coupling after the special hooked end of the rod is set into the coupling. Sectional rodding machines are equipped with a chain drive head that will be set up for one of the standard rod lengths and *ONLY* that length of rod can be used. However, the diameter of the rod can be changed, if desired. The rods are connected and stored in a reel inside the machine. Between 800 and 1,000 feet of rod can be stored on most of these machines.

A sectional rodder is capable of clearing lines effectively in diameters up to 12 inches and distances up to approximately 750 feet. In smaller diameter lines it may be possible to extend this to 1,000 feet, if the machine's rated capacity will allow. These distances are too long for effective clearing if the sewer contains sand or grit.

The continuous power rodder is designed to use a single continuous rod. The rod size can be one of the following: $^{5}/_{16}$ inch (0.3125 in), $^{3}/_{8}$ inch (0.375 in), or $^{1}/_{2}$ inch (0.50 in). Continuous rodders are equipped with a multi-roller drive head that moves the rod in and out of the machine. The drive head will be set up for one of these diameters, and *ONLY* that diameter can be used, regardless of length. Rod storage in these machines is usually between 1,000 and 1,500 feet, with some special application machines holding 2,000 feet of rod.

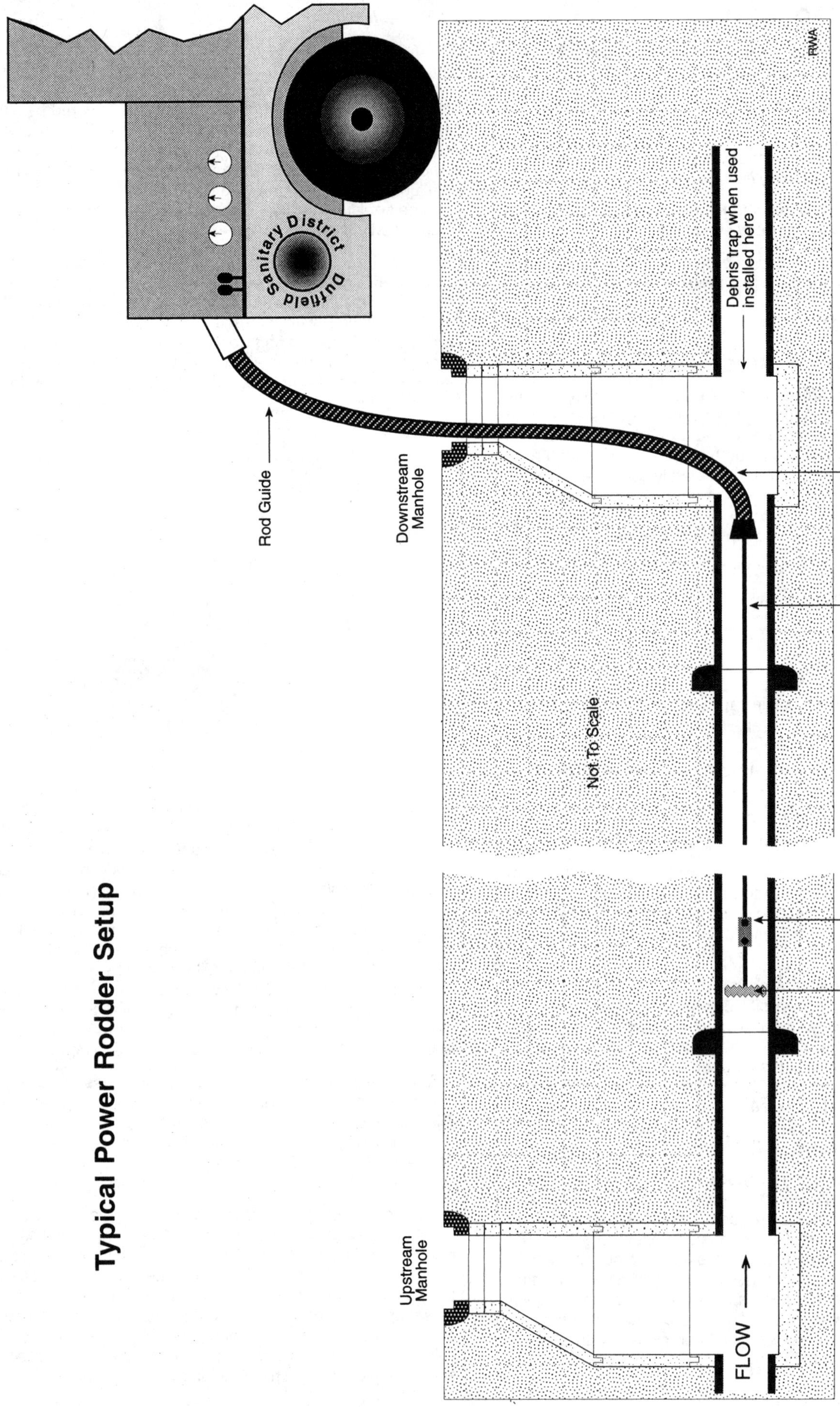

Fig. 6.26 Power rodding operation
(Permission of SRECO-FLEXIBLE)

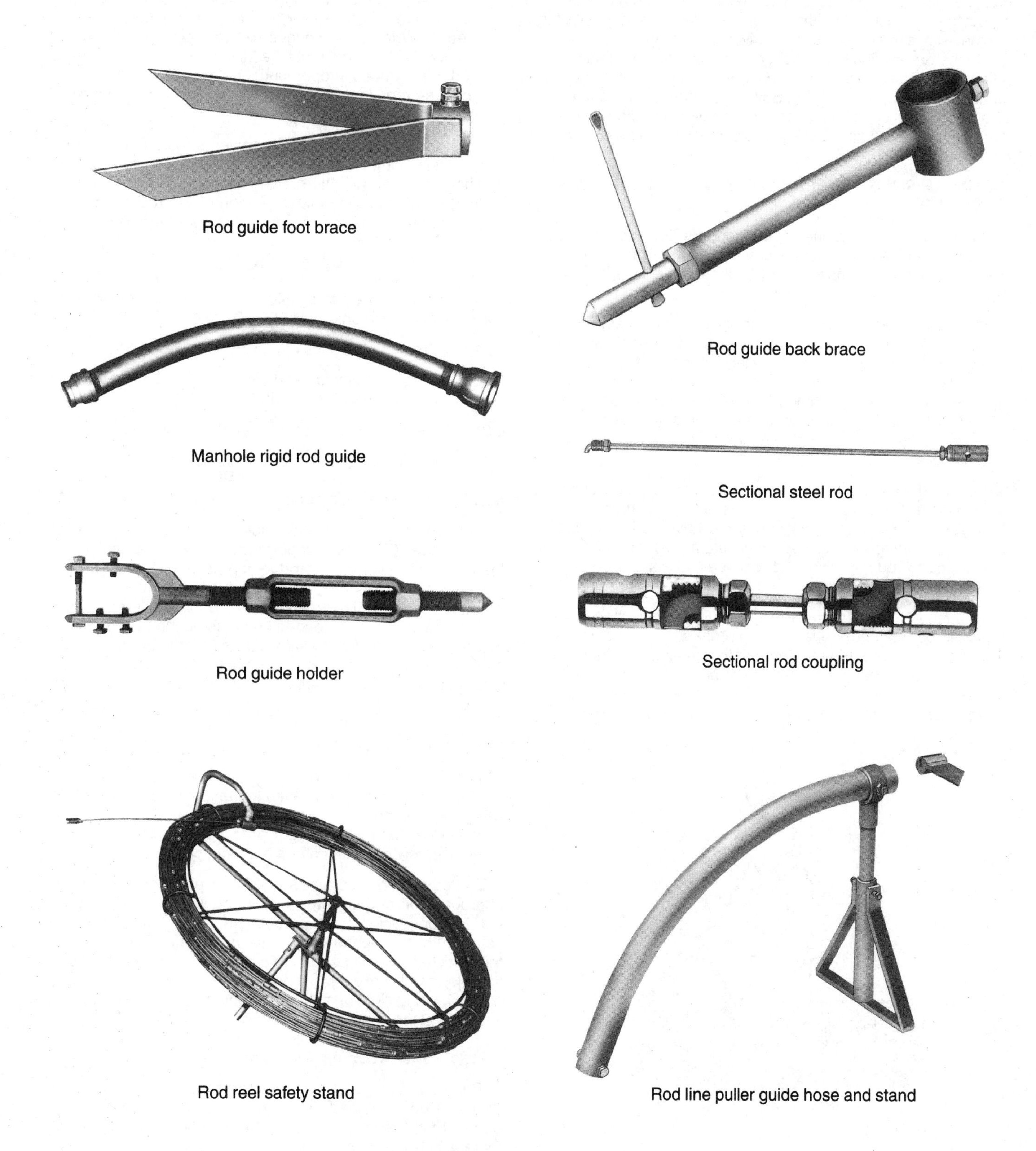

Fig. 6.27 Power drive and hand rodding accessories
(Permission of SRECO-FLEXIBLE)

Both continuous and sectional rodders use either a gas or diesel engine (between 16 and 25 horsepower) and a hydraulic system for power. (Older machines may use a smaller engine with a mechanical, belt-driven power system.) Both machines are available in a trailer- or truck-mounted configuration, with power take off (PTO) availability on the truck-mounted unit for power supply. Both types of machines can be effective in lines up to 15 inches in diameter, depending on the diameter of the rod that is used, and can operate effectively at distances of 750 to 1,000 feet.

COMPARISONS BETWEEN SECTIONAL AND CONTINUOUS RODS.

Continuous and sectional power rodders perform roughly the same work and the operational setup is similar for both, but it is important to understand why these machines are different.

1. Continuous rodding machines are quickly and easily loaded; however, when loading continuous rod, a special transfer rack is necessary to do it safely. When continuous rod needs to be changed, the entire reel must be loaded. With sectional rodding machines, coupling all the sections together, tightening, and loading a full reel may take several hours. However, it is seldom necessary to replace the entire reel all at once.
2. Sectional rods are pushed through the drive head by a positive drive. Continuous rods are pushed by rollers that can slip and cause wear. Continuous rod drive heads require daily tightening on the rollers to ensure a proper grip on the rod and to reduce roller and rod wear. Couplings on sectional rods cause more wear on the guide hose liner.
3. Continuous rodders can rod greater distances than sectional rodders and are less apt to break at great distances due to the torque transfer through the continuous length of the rod. Sectional rod couplings and nuts can work loose or increase drag friction of the rod in the pipe. Occasional inspections are necessary to keep the couplings securely attached.

4. Continuous rod must be clean as it passes through the drive rollers to prevent damage and wear. Sectional rod drive heads are more forgiving; however, sectional rod couplings can attract debris. Continuous rod is easier to clean than sectional rod but both types should be cleaned as they are being retrieved to prevent grit from going up the roller system.
5. When a sectional rod is broken, repairs can be made quickly in the field by replacing the broken section or recoupling the rod and adding a new rod only when more footage is needed. When a continuous rod breaks, repair is difficult and the broken rod must be discarded. The continuous rodder is reduced in operating length to the length remaining in the reel. In some cases a new reel must be installed after only a few breaks in the rod.
6. A broken continuous rod in a sewer is more difficult to retrieve since it has no couplings for the pick-up tool to grab. The broken end of the continuous rod cannot be reused, and possibly the remaining portion of the rod on the machine also is lost if the remaining footage is too short.

QUESTIONS

Write your answers in a notebook and then compare your answers with those on pages 386 and 387.

6.21A What are the uses of power rodders?

6.21B What size sewers and distances can be cleared with a sectional power rodder?

6.21C List the advantages of a sectional rod.

6.21D List the advantages of a continuous rod.

6.212 Personnel and Equipment

Two operators are recommended to properly operate the power rodder. One operator performs the actual operation and the assistant monitors the rod, guide hose, or far manhole to signal when the tool has reached the location where it will be removed or where it will be exchanged for a different tool. A third operator is recommended on an as-needed basis if there is a high or dangerous traffic volume, if entry into a manhole may be likely, or if heavy physical obstructions are likely to be encountered. Minimum personnel requirements are:

1. Maintenance Equipment Operator, and
2. Maintenance Operator I.

If a two-operator crew has to enter a manhole, a supervisor or someone else must be called to the job site so that two operators will be topside when the third operator enters the manhole. All confined space entry procedures must be followed whenever anyone must enter a manhole.

The equipment operator should be well trained in the use and operation of the power rodder and will be responsible as well for supervision of the proper setup, safety measures, and selection of the clearing tool to be used. The effective use of the selected tool will depend to a great extent upon the operator—how well the operator determines the problem at hand and how conscientious the operator may be in getting the sewer as clear as possible.

6.213 Sewer Rodding Tools and Uses

The same basic tools are used for either the continuous or sectional power rodder. Many have been developed on the basis of field reports and the industry continues to produce more effective new tools and techniques.

Pages 345 through 348 illustrate some of the basic tools used with rodding machines to solve different pipeline clearing problems. Tools used with the rodding machine are designed to be rotated clockwise in the pipe. Only corkscrew and boring tools can be rotated counterclockwise to back them out of accumulated material in the sewer or to get past misaligned joints.

SEWER RODDING TOOLS AND USES

(Photographs provided by SRECO-FLEXIBLE)

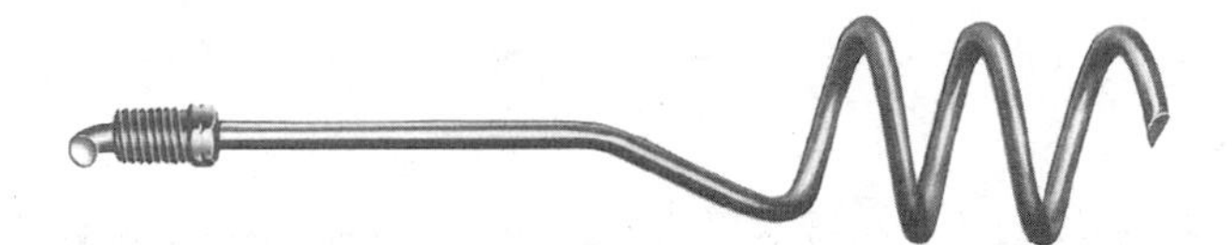

ROUND WIRE CORKSCREW

Generally used with hand rodding applications in small-diameter pipes to relieve stoppages or to thread into and break up solid deposits.

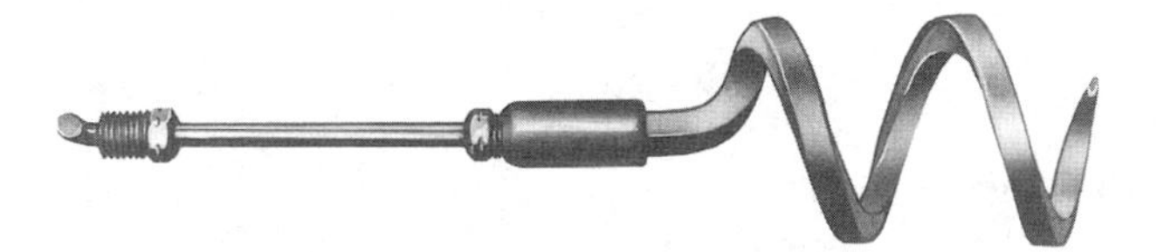

SQUARE BAR CORKSCREW

Primarily used to relieve stoppages in pipes over 6 inches in diameter. An effective stoppage tool due to the open structure of the blade that allows materials to pass through the tool.

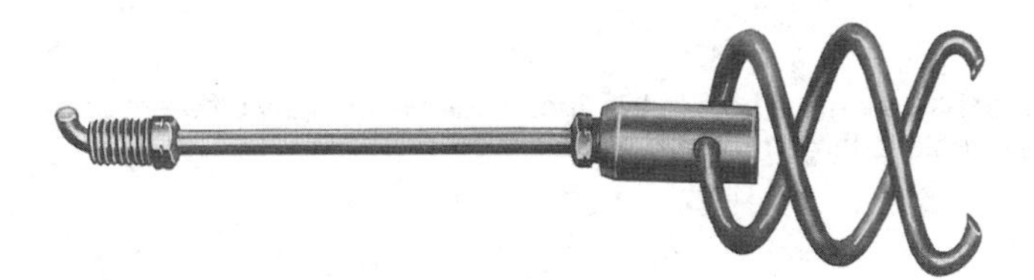

DOUBLE POINT CORKSCREW

Used to engage and retrieve root masses, cans, plastic bottles, and fabrics from a pipe. Double point allows the tool to bite into material and retrieve it.

SAND CORKSCREW

Used in pipes where sand has plugged the line. The forward screw portion pilots the tool into the sediments; the following screw portion further enlarges the access point and allows water to enter the material to loosen it.

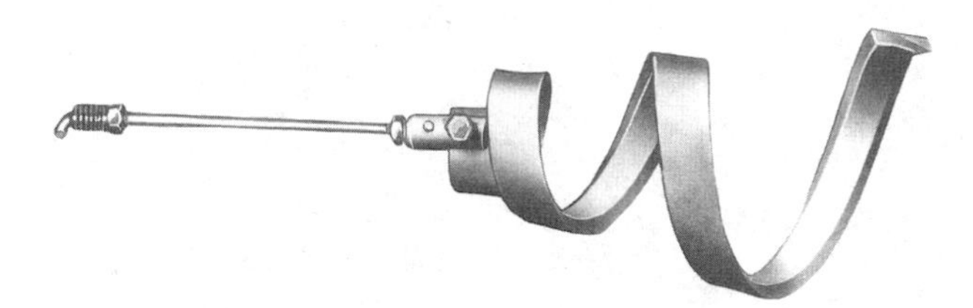

AUGER

Used to pilot a hole through roots, grease, and other solids in the pipe. Primarily used in conjunction with other tools to open a path in the line so that another tool, such as a root saw, can be used more easily. Effective in cleaning misaligned pipes because the direction of rotation can be reversed to climb over bad joints.

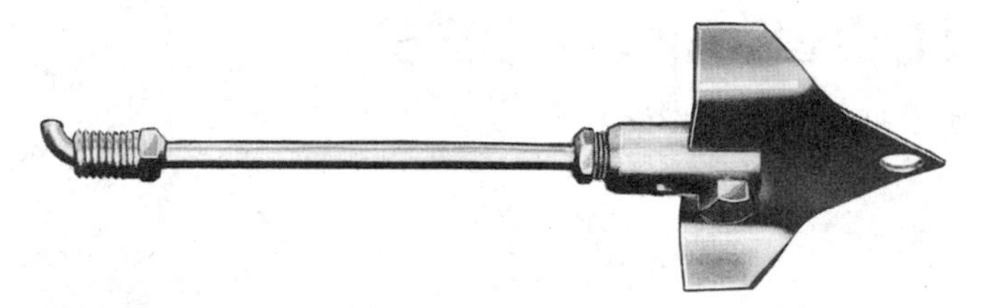

SAND LEADER

Used to rotate and remain above sand and other built-up deposits to thread a line with cable or attach a different tool from the next manhole or access point.

ROOT SAW

Used exclusively to cut through root masses in the pipe after the auger has piloted a hole in the line. Available in many different configurations for different kinds of root cutting from small curtain or "veil" roots to large root intrusions.

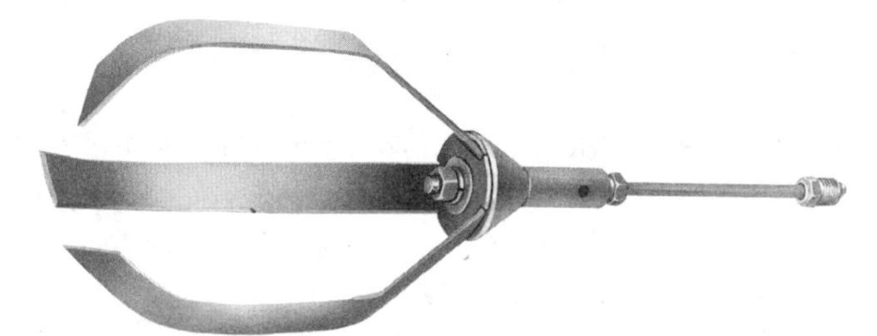

SPRING BLADE CUTTER

Used as a finishing or pull back tool after a root saw or auger has removed the bulk of roots or grease from the line. The tool is installed at the manhole away from the rodding machine, and is slowly pulled backward through the pipe at a high rotation speed. The blades are designed to scour the pipe walls to remove the balance of materials left behind. This tool should be used ONLY while being pulled.

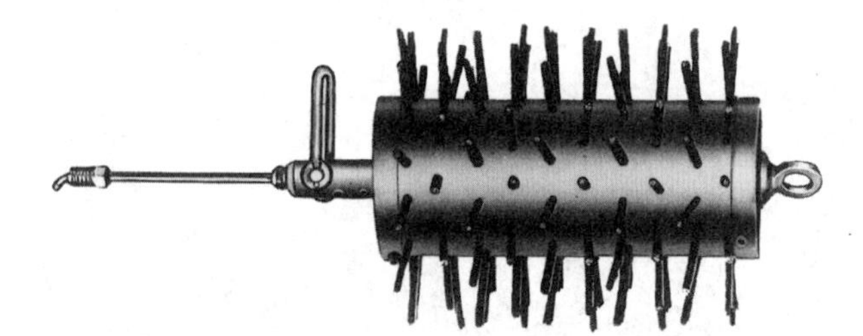

PORCUPINE

Also used as a finishing tool to scour the pipe after cleaning operations. This tool, as with the spring blade cutter, should be used ONLY while being pulled.

SPEARHEAD

Spearheads, or boring tools, are used to break up stubborn stoppages, break glass, pierce cans, and break up packed silt, sand, or industrial debris.

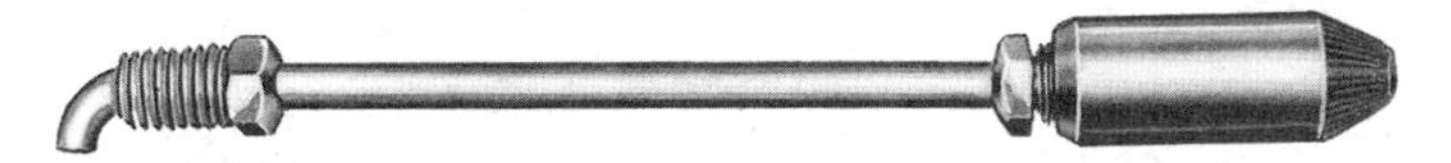

PILOT BULLET

Used primarily on the end of the string of rod to allow the rod to guide easily through the rod guide hose. Installed after the line has been cleaned and the rod is pulled back into the machine. This tool also can be used for stoppages and threading a cable in a pipe.

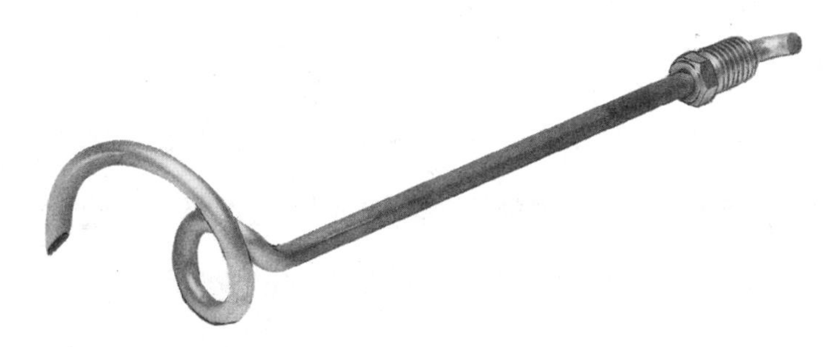

PICK-UP TOOL

Used to retrieve sectional rods that have broken in a sewer line. Tool slowly turns and locks on the rod coupling, allowing the string of rod to be removed.

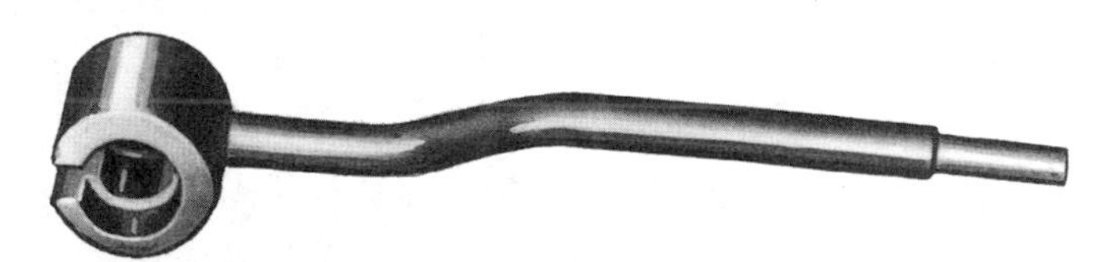

ASSEMBLY WRENCH

Used on the coupling nut to install and change tools, or change sectional rods.

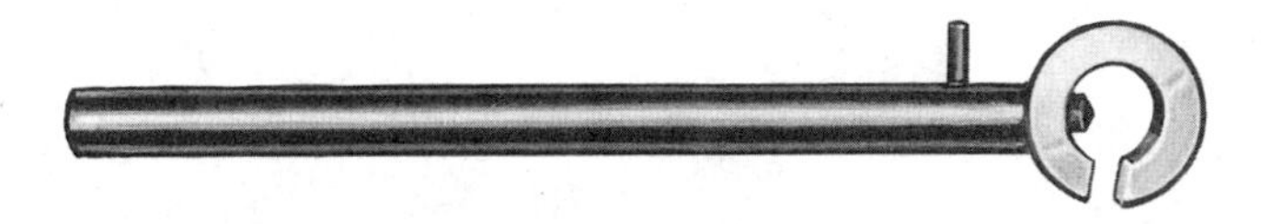

ASSEMBLY TURNING HANDLE

Used to provide a firm, safe hold on the rod coupling so that the assembly wrench can turn the coupling nut.

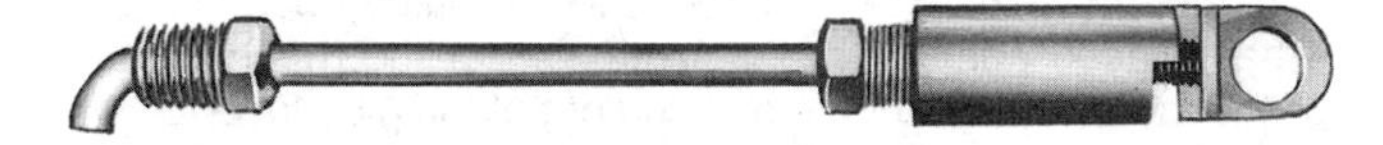

SWIVEL

Used when towing a cable through a pipe. Swivel allows the cable or rod to twist independently and does not allow knotting of the cable.

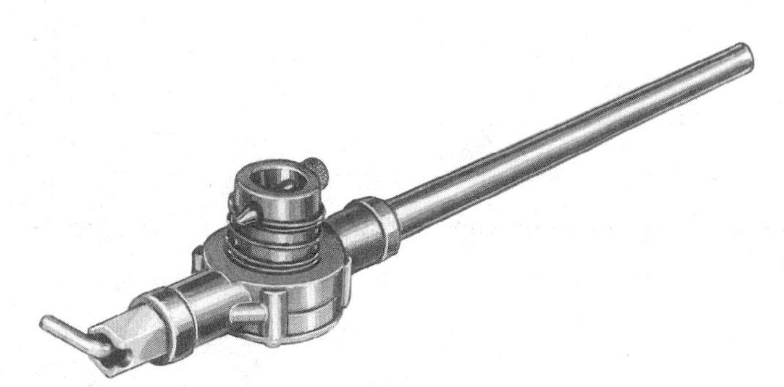

RATCHET TURNING HANDLE

Used in hand rodding applications, it allows one-way turning of the rod by a ratchet device. Ratchet can be set up to turn clockwise or counterclockwise.

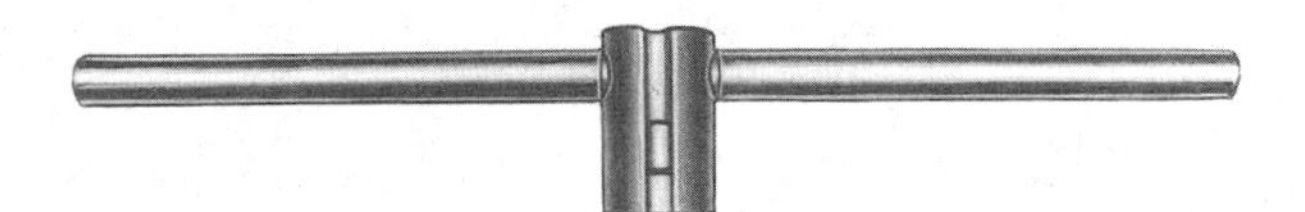

PULLOUT TOOL

Used in hand rodding applications, it allows the rod to be pushed or pulled in a pipe while the rod is being turned either by hand or with a rod turning machine.

BAR TURNING HANDLE

Used in hand rodding applications, the bar will lock onto a rod coupling allowing the operator to turn the rod safely by hand.

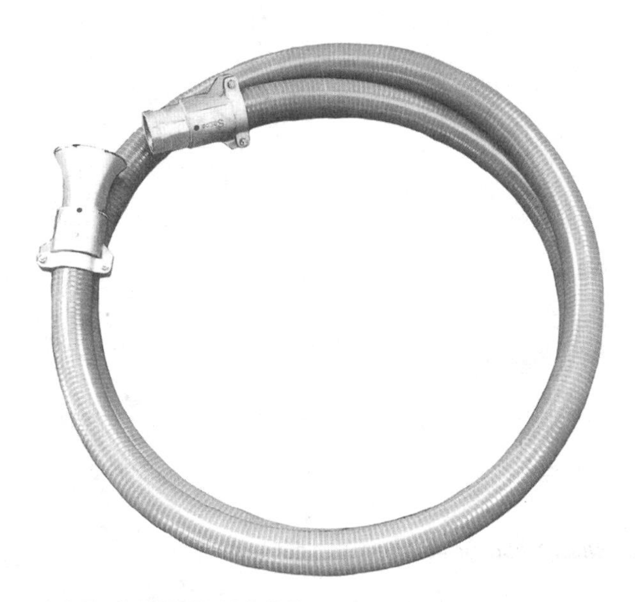

ROD GUIDE HOSE

Used on all power rodders to protect and support the rod from the machine to the bottom of the manhole.

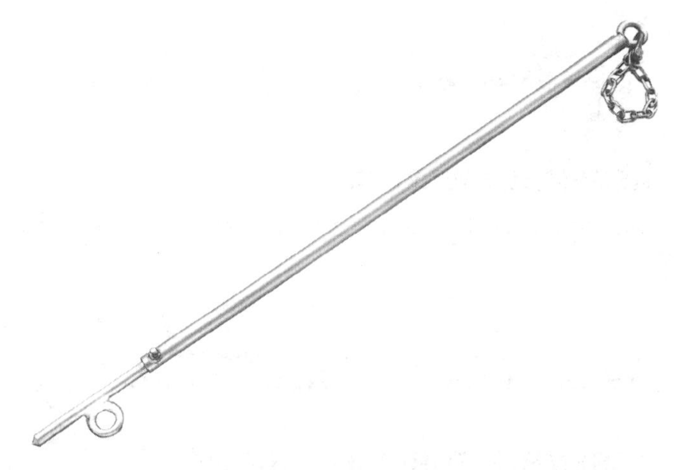

LOWER MANHOLE BRACE

Used with the rod guide hose to prevent the bell end of the hose from pushing back out of the pipe when an obstruction is encountered. This brace attaches to the guide hose bell and, when placed against the opposite wall of the manhole, prevents the guide hose from pushing out of the pipe invert.

The turning speed for each tool varies with the line size, conditions in the pipe, and the operator's level of experience. As a general guideline, corkscrew and boring tools should be rotated slowly, about 25 RPM (revolutions per minute), to avoid the potential to exit the pipe either through a service connection or an offset joint. Root saws need to turn faster to effectively cut materials, usually at a speed of about 35 to 50 RPM depending on line conditions. Finishing or pull back tools need to rotate fast, at about 55 to 60 RPM, as they are slowly pulled back.

Most tools are designed to fit well within the pipe for which they are made, for example, a 6-inch root saw will usually measure 5⅝ inches in diameter. This will leave room for the tool to make any turns or dips in the pipe without getting stuck. Finishing tools are the only tools designed to fit snugly into the pipe; however, these are only used to clean the pipe to the side wall after the bulk of the material has been removed.

QUESTIONS

Write your answers in a notebook and then compare your answers with those on page 387.

6.21E How many operators are needed for a power rodder crew?

6.21F What must a two-operator crew do when someone has to enter a manhole?

6.21G Is the power rodder more effective in large or small lines?

6.214 Equipment Setup

If a particular area or subdivision is to have roots or grease cleared with a power rodder, start at the top of the system. In so doing, any debris you may lose downstream from your work will be going into an area that you will be clearing next and not into a cleared line. Be careful that any debris lost downstream does not cause a stoppage in a downstream sewer already in poor condition.

Since most manholes are located in street areas, certain safety measures must be taken for handling traffic and ensuring safe conditions in manholes. See Section 4.3, "Routing Traffic Around the Job Site," and Section 4.4, "Classification and Description of Manhole Hazards," for details.

ALWAYS test the manhole atmosphere for explosive gases *BEFORE* rodding or high-velocity cleaning a sewer because sparks from the rodding or cleaning operation could cause an explosion.

Whether rodding upstream or downstream, the setup of the power rodder will be the same except when rodding from inlets, siphons, and other similar structures. First position the machine at the manhole so the rod guide hose is pointed in the direction you will be rodding. Most collection system operators rod only from the upstream manhole when there is no stoppage. When rodding from the downstream manhole and using small tools (spearhead blade or bullet nose), these tools will occasionally move up into house service connections instead of going up the sewer.

Install a leader tool or smooth coupling on the end of the rod before installing the rod guide hose (Figure 6.27, page 343). This leader tool will prevent the sharp end of the rod from damaging the interior of the rod guide hose. After installing the rod guide hose, run the rod out far enough from the end of the rod guide hose to remove the guide tool and install the selected clearing tool. Pull the clearing tool back near the end of the rod guide hose. A rope will be fastened to the end of the rod guide hose; use this rope to insert the clearing tool into the sewer.

Pull the power rodder away from the manhole just enough to take the slack out of the guide hose. A guide hose anchor brace or free-standing hose brace is used to support the hose, especially when working in shallow manholes where several feet of guide hose could be left unsupported between the machine and the manhole. A lower manhole rod guide brace should be used to prevent the bell end of the guide hose from pushing back out of the pipe when the clearing tool comes in contact with roots or other materials in the system. Proper positioning of this brace is important since slack rod in the manhole can cause breakage and is extremely dangerous.

Install a sand trap, root basket, or other suitable debris trap. If rodding upstream, install this trap in the downstream side of the working manhole. If rodding downstream, put it in the downstream side of the lower manhole. This will keep loosened roots, grease chunks, or other material from going farther and possibly causing a stoppage.

Different models of power rodders may add to or change this setup procedure slightly, but generally this procedure will cover the basic setup.

QUESTIONS

Write your answers in a notebook and then compare your answers with those on page 387.

6.21H Should a sewer be rodded upstream or downstream from a manhole?

6.21I What is the purpose of the leader tool that is installed on the end of the rod?

6.21J Why should a debris trap be installed on the lower side of the downstream manhole when using a power rodder?

6.215 Tool Selection and Operation

Selecting the proper tool for the job that is being done takes some experience and a knowledge of what the different tools are designed to do. Many of the tools that are used with rodding machines have a specific function which they perform best. Just as you would not use a screwdriver when a chisel is needed, you should not use a root saw to remove grease instead of using augers that are designed for grease removal. It is important to understand the function of each rodding tool and how each tool can be used with other tools for a specific job.

As previously mentioned, the operator (Maintenance Equipment Operator) will be responsible for selecting the proper clearing tool. A history of the sewer you are clearing will be of great help. If this information is not available, look for other indications of the condition of the sewer, such as grease in the manhole or trees growing close to the sewer pipe.

If the rodding machine is being set up to clear a stoppage, use a square bar corkscrew from the dry manhole downstream from the stoppage. After the stoppage has been broken loose, you should have some indication of what caused it. The next tool to use depends on whether the stoppage was caused by roots, grease, or grit. After the stoppage has been cleared, remove the loosened material from the pipe using the same procedures you would use if this were a scheduled maintenance task.

There are many occasions when it will be better to change tools or tool size from the far manhole instead of removing the rod guide hose from the manhole at the machine to do it. This procedure will save a considerable amount of time and effort when clearing many sections in a given area. When changing tools at the far manhole, stop the rotation of the tool and slowly push the rod up the line. Lift the tool to the street with a gaff (a metal rod with a hook on one end and a handle on the other end). *NEVER* touch the tool on the end of the rod until the rodding machine has been shut down. Many accidents have occurred because operators did not take the proper precautions when using rodding machines. With the rodding tool at street level, make the necessary changes (size or type of tool) and slowly pull the tool back into the manhole. Be sure to wait until the new tool reaches the invert before turning the equipment back on and continuing with the clearing operation.

To remove roots from a pipe, begin with an auger, followed by a root saw, followed by a finishing tool to clear the roots to the pipe walls. Select an auger that is approximately one size smaller than the pipe itself, for example, use a 6-inch auger for an 8-inch pipe. Run the auger up the line to the next manhole to open a hole for the next tool. At the far manhole, gaff the auger up to the street and change to a root saw. The root saw can be the same size as the pipe because the auger has opened a hole through the roots.

Lower the root saw into the manhole and begin rotation as the tool is pulled back to the rodding machine. Pulling the root saw as it cuts forces the blade of the saw to cut in the same plane and allows the rod to absorb the torque through its entire length in the pipe. The blade of the saw should rotate relatively fast (high RPM), but the tool should be pulled slowly toward the rodding machine. Once the saw tool reaches the rodding machine, many operators will return the saw to the far manhole and change to a finishing tool, such as a spring blade cutter. The saw can be moved forward to the far manhole relatively quickly because most of the material has already been removed from the line. Once the root saw gets to the far manhole, gaff it up to the street, install the spring blade cutter, and return the tool to the invert. Rotate the spring blade cutter at a fast rate and slowly bring the tool back downstream so that it can remove any stubborn materials still left on the pipe walls.

Hardened grease can be a problem, especially in areas that have a high concentration of waste grease and oil dischargers such as restaurants. This grease can build up to substantial amounts and harden to the pipe walls, making removal difficult unless rodding tools are used. Augers are generally used to remove hardened grease. As with root clearing operations, use an auger smaller than the pipe size and work the tool from the rodder toward the far manhole. When the auger reaches the far manhole, switch to an auger the same size as the pipe and pull the tool back to the machine. Most operators prefer to return the auger once again to the far manhole to attach a spring blade cutter to scour the pipe walls of the remaining material. It is a good idea to flush the line with water after the hardened grease has been removed.

In some cases grease will be built up to a point that you will need to use several augers of different sizes before you can use one the same size as the pipe. It is important to try to break out this material in small pieces so that it can be removed more easily downstream and to prevent large pieces from causing another stoppage farther downstream.

Once the appropriate cleaning tool has been selected for the job, the operation can begin. Setup of the rodding machine is the most important part of effective rodding. Improper setup can cause tool and rod damage and may even damage pipes and the rodding machine itself. It takes less time to set up the equipment properly the first time than it will take to tear everything down and set it up again.

Set up the rodder as described in Section 6.214. Thread the rod guide hose with the rod, attaching the proper tool using the assembly wrench and assembly turning handle. After lowering the guide hose into the manhole and getting the tool into the pipe, be sure the lower manhole guide brace is properly set up to prevent the end of the hose from pushing back from the pipe invert. Also be sure that all bends in the guide hose are removed by pulling the rodding machine forward. It is important to remove all of the bends in the guide hose so that the rod is not forced to rotate in a severe bend. When the rod is forced to stay in one place in a bend as it is being rotated, the rod will heat up as it is alternately stretched and compressed. If the rod becomes too hot and enters the cool wastewater, the temper of the rod metal can be damaged and the rod will be more likely to break at that point.

Once the guide hose has been set up properly, the forward speed and strength need to be set. Hydraulically powered machines manufactured since 1975 have a pressure gage with an adjustment valve that can be used to set forward pressure. The forward pressure should be limited to about 350 pounds. If there is too much pressure on the rod, the rod may bend in the pipe and break, or reverse itself causing a tangle.

Most rodding machines are equipped with a footage meter to locate the tool in the line. Be sure this meter is reset each time the tool has been placed in the line so that the location of the tool in the pipe is always known. If you encounter an obstruction that will not allow a rodding tool to pass, note the footage so that a closed-circuit television inspection camera can be used to find the trouble or mark the street so that another crew will know where to excavate to correct this problem.

Using the rodder controls, push the tool a few feet into the line and begin rotation at the proper speed for that tool. The forward speed should be fairly slow unless the pipe being cleared is well known. A slow speed of rotation is always best with any pipe clearing equipment. A cardinal rule of rodding is that the rod can always be pushed and pulled when it is not rotating. When rotating, however, it must be pushed or pulled slowly. This will allow the rod to move through any bends in the setup without heating up and breaking.

When the tool engages material, the tool will most likely stop rotating but the rodding machine will continue to turn the rod. If you do not back the tool away from the material, the rod will twist into a loop near the clearing tool. Once the loop has formed, continued rotation of the rod will cause it to coil inside the pipe like a giant spring. This condition is known as overtorque and it can damage rods to the point of breakage.

Overtorque can be a serious operational problem, especially for inexperienced operators. Rod damage due to overtorque usually is not visible until the rod breaks in the reel or gets caught in the drive head. Either of two different methods can be used to control overtorque. When working in a line with roots and grease, move the reel spin direction lever to neutral while the machine is operating to see if the reel stops quickly or begins to spin in reverse. If this happens, it means that the tool is stuck and the rod is relieving the torque pressure. Simply pull the rod back a few feet, begin rotation again, and move the tool forward. Repeat this sequence of steps until the material is penetrated.

More experienced operators often use the second approach, which is to watch the rod and pressure gage carefully as the tool is working in the pipe. The pressure gage will show an increase in pressure as the tool encounters material in the pipe. As the torque builds up and the rod begins to coil itself, an experienced operator will see evidence of this between the machine and the guide hose anchor. (When the overtorque condition becomes severe, the entire guide hose may begin to twist with the rod.) Noting these changes in pressure and the appearance of the guide hose, the operator withdraws the rod, while keeping the rod turning, until the tool clears the material and spins off the excess torque in the line. The operator then carefully moves the tool back up the pipe and repeats the same procedure until the material has been penetrated.

While avoiding overtorque sounds difficult, it is a relatively easy operation and the procedures can be mastered with some experience. Listen to the sound of the engine as it works the rodding machine during operation. If the engine pulls down, it is an indication that the tool is working in material and that you should pay attention to the rod tension and torque buildup. By looking at and feeling the controls as you operate the machine, you will quickly master the procedures for using power rodders effectively.

Some basic rules of operation that should be observed when running rodding machines include the following:

1. Do not jam the rod into an obstruction. This is the fastest way to break a rod or tool in the pipe. Smooth, steady operation is required;
2. Do not rotate the rod in one position for an extended period. Heat builds up quickly where the rod bends in the manhole. Keep the rod moving in or out of the machine while it is rotating;
3. When torque builds up, retract the tool from the material to relieve the torque and then return to the material;
4. If the rod should break, or when changing tools at the far manhole, be sure the torque is out of the rod before you handle it. Prodding the rod with a pole or shovel will cause any built up torque to spin off and allow safe handling;
5. Be sure that forward pressure is adjusted to low levels to avoid bending or breaking the rod; and
6. Do not allow any bends or "S" curves to remain in the guide hose. Rods caught in a bend for extended periods will heat up and be damaged. Pull any bends out by pulling the rodding machine forward.

When the opposite manhole can be reached without changing the tool, a faster rotation can usually be used on the way back to catch places missed on the way in.

Again, do not sacrifice clean lines for footage. One run may not be enough to get the sewer clean, especially with roots. The effectiveness of the clearing operation can be increased in problem areas by dumping or running water into an upstream manhole during the clearing operation. A flow in the sewer is essential for cooling the rotating tool and for flushing the sewer.

A slight variation of this procedure can be used when rodding from the upstream manhole if there are no stoppages. Send a small tool to the downstream manhole and then attach a full-size tool (a root saw or spring blade root cutter if roots are a problem). This technique keeps the rod straight in the sewer by pulling on the clearing tool instead of pushing, which causes the rod to coil in the sewer. Advantages of this procedure include the application of greater pressure on partial obstructions in the sewer and increased rod life.

After completing the first section of line, remove the lower manhole brace and pull the clearing tool and hose from the manhole. If debris has built up at the trap, have an operator who is properly harnessed descend into the manhole. Follow all confined space entry procedures. With the trowel or manhole scoop, clean the dislodged material out of the channel. With the bucket, remove the material from the manhole and clean up around the working area.

If you are going to the next manhole to continue clearing, usually the rod guide hose can be left attached with the clearing tool. The power rodder often is equipped with hose hangers for this purpose, so the rod guide hose and tool are ready for the next setup. This timesaving decision is up to the machine operator and depends on the specific equipment setup because a move of some distance might tend to damage the hose. Some machines are equipped with a swivel connection for the rod guide hose so a bend is eliminated in this operation. By simply removing the tool and pulling the rod back into the machine, the flexible hose can be laid around and alongside the machine. This is only one of a number of timesaving shortcuts which in the long run save considerable time and wear and tear on the equipment.

QUESTIONS

Write your answers in a notebook and then compare your answers with those on page 387.

6.21K How do you determine which clearing tool to use?

6.21L Which tool would you use to clear a stoppage?

6.21M After a stoppage has been cleared, which tool would you use?

6.21N What precaution should be taken *BEFORE* changing to another clearing tool at a manhole when the rod is still in the sewer?

6.21O Why must you always know where the clearing tool is in the sewer?

6.21P How would you determine the speed of the rod in or out and the rotation speed?

6.21Q Why is water sometimes added to the opposite manhole during power rodding?

6.216 Recovery of Broken Rods

When a rod breaks, measure the amount of rod pulled back to determine the approximate location of the broken rod. If you are using a rodder with sectional rods, it's only a matter of removing the broken stub of a rod at the coupling and installing the pick-up tool. With a continuous rod it is more complicated because you have also lost your tool coupling with the rod. Depending on the style of sectional rod coupling used, you will either bend a new top on the rod with a special bending tool or perhaps the style of coupling will require a flat spot or dimple in the rod for a set screw. Needless to say, extra couplings have to be handy.

Then a fishing expedition begins to try to pick up the broken piece. The pick-up tool is placed in the sewer the same as a clearing tool. Send the rod into the sewer slowly and with slow and steady rotation. "Feel" your way and most often you can tell when resistance is encountered. At this point, stop rotation and without letting the rod "unwind," pull steadily out of the sewer. The pick-up tool is designed to screw itself around the lost rod and become snug. Perhaps several attempts will have to be made, but the rod must be removed, even if you must dig it up. If you have a clearing tool lodged, counter-rotation of the rod reel is often all that's needed to get it loose. Go in again, only more cautiously. Use of a TV camera will help speed the recovery process.

Rods may break in the reel section of the rodder, especially with a sectional rod. *WHEN MAKING THIS REPAIR BE VERY CAREFUL*. The rod is under some tension and can be dangerous if it gets away from you.

Use gloves and keep the end of the rod away from your face. Most accidents using power rodders involve the rod or clearing tools. Extreme care and proper instruction in attaching tools and rod repair must be followed.

There will be many different problems and a standard answer cannot be given for any specific problem. What works in one situation may not in another. Experience and observation will overcome most problems and hazards. Broken clearing tools often can be picked up with another clearing tool or perhaps flushed out, particularly if a high-velocity cleaner is available.

QUESTIONS

Write your answers in a notebook and then compare your answers with those on page 387.

6.21R What tool is used to recover a broken rod?

6.21S What is done if a broken rod can't be recovered with a pick-up tool?

6.217 Precautions and Safety

> **WARNING**
>
> NEVER HANDLE THE ROD UNDER TENSION. TREAT TWISTED RODS VERY CAREFULLY. THEY ARE **VERY** DANGEROUS.

Once again, emphasis needs to be placed on speed of rod travel and rotation. These two forces combined can break or twist most rods. Most machines will be capable of holding close to 1,000 feet of rod, which is quite heavy. With the weight of the rod on a spinning reel, considerable twisting power is developed, but it also takes a few moments to get it stopped. An operator can get through some difficult stoppages or root problems with skill and patience.

Do not attempt to rod distances greater than from one manhole to the next unless absolutely necessary. Trying to avoid extra setup time can be much more dangerous than normal operation.

When working on a rodding crew, required procedures of operation must be established and followed for each individual. You must know your own job and the jobs of everyone else on the crew so everyone can be familiar with the overall operation.

With constant use, rods do suffer from fatigue; and when a rod breaks, it's likely to be in the sewer. Avoid turning the rod without rod travel in the sewer. Heat builds up very quickly at any bend when the rod is turning. Therefore keep it moving at all times if possible.

Porcupine cleaning tools are available for attachment to the power rodder. Their use is discouraged because of potential safety hazards. Other tools can do a similar job in a safer manner. If a porcupine must be used, a winch and tag line must be used for safety.

6.218 Maintenance of Rods and Equipment

The rod couplings on sectional rods should be kept tight at all times. These will work loose while rodding, and a visual inspection by the operator even while working the machine will often pick up a coupling that is loose.

Make a practice of running the rod out on top of the ground on occasion and inspect the rod for bent pieces and loose couplings. When couplings are allowed to stay loose, the rod will wear and become weak at that point. Do not throw used couplings away—they can be reused.

Continuous rods are driven by rollers pressing on the rod. A cleaning tool in the form of a clamp is furnished with the machine and has a very important function. A coarse piece of cloth or similar material is kept in the clamp too. When you are ready to pull the rod back into the machine, clamp this tool around the rod at the head of the machine and it will wipe the rod clean as it is returned to the reel. If grease and grit are allowed to pass into the pressure rollers, they will build up and cause wear on the rod.

Even with a clean rod, the pressure of its being rolled back and forth causes some wear. Therefore the torque on these drive rollers has to be checked and reset. Specifications for the particular machine are furnished by the manufacturer.

Always avoid sharp bends in the guide hose regardless of the type of machine. The metal liner can become bent and distorted. Inspect the guide hose frequently and carefully for wear, mounting, and mechanical damage such as crimping.

Be completely familiar with the clearing machine. Know what operating and maintenance procedures the unit requires. Inadequate maintenance before rodding a sewer can result in considerable time lost attempting to recover a broken rod. The necessary maintenance items can easily be done on the job:

SECTIONAL RODDER

1. Inspect for loose rod couplings,
2. Keep drive head chain tight,
3. Inspect for worn drive dogs and improperly adjusted belts, and
4. Look for "stacking" of the rod on the reel when bringing the rod back into the reel cage.

CONTINUOUS RODDER

1. Maintain proper torque on the driver rollers,
2. Examine wear on the rod guide bushings,
3. Inspect for sharp teeth on drive gears. This indicates wear or needed adjustment,
4. Look for leaky hoses and fittings on hydraulic units. Maintain the proper oil level in the hydraulic oil tank. Keep the hydraulic oil filter clean, and
5. Fluctuating oil pressure may be due to hot hydraulic oil. This may or may not be just from continuous use. Be suspicious and check for other problems.

Always keep the rodder, tools, and accessories as clean as possible. The engine should receive the maintenance and care recommended by your shop manager or the manufacturer's directions. For more information, see Section 6.32, "Rodding Machine Maintenance."

With proper maintenance and care the rodding machine will do a good job of clearing a sewer. Like any other machine it has its limitations and as experience is gained in the operation, you will know what it can or cannot do.

6.219 Records

See Section 6.07, "Records."

QUESTIONS

Write your answers in a notebook and then compare your answers with those on page 387.

6.21T Why must extreme care be used when handling a rod or clearing tool?

6.21U Why should a rod occasionally be run out on top of the ground?

6.21V Why should the rod be cleaned when it is returned to the reel?

6.21W Why must everyone on a rodding crew know their own job and the jobs of everyone else on the crew?

6.21X What maintenance should be done on the engine of the power rodder?

6.22 Hand Rods

With all of the modern equipment available, it is still necessary to have hand rods available (Figures 6.28 and 6.29). One of the first reasons is that hand rods can be used when and where the modern methods cannot be used effectively. Hand rods are a quick and simple answer for many stoppages. Hand rods are often used on "service request" trucks and for "off-hours" stoppage calls. They are used extensively for emergencies at night or during weekends. After the stoppage is cleared, thorough cleaning can be done at a later date.

Often maintenance crews will carry a reel with a hand rod on the same work truck used in conjunction with a power rodder. If when clearing a particular area the power rodder cannot be put to use, such as in an easement, the hand rod can be used. This avoids having to leave a section of sewer uncleared and likely forgotten. Also a hand rod can be used from the opposite manhole to free a stuck or broken power rod.

Other maintenance crews may do only hand rodding. This is not uncommon in sanitary districts with hilly areas or in very small districts where the total length of sewer in the system does not warrant routine clearing with other types of equipment. Other factors may be involved, but the main point is that hand rod methods are here to stay, at least for the foreseeable future.

6.220 Personnel and Equipment

The particular job often will dictate how many operators will be needed. However, if a crew is essentially doing only hand rodding, at least two operators are recommended:

1. Maintenance Operator I, and
2. Maintenance Operator II.

A three-operator crew is required if one operator needs to enter a manhole. Usually rod crews or "service request" crews consist of two operators and they never enter a manhole. If the need arises for someone to enter a manhole, all confined space entry procedures must be followed, including the need to have a third operator available *BEFORE* anyone enters a manhole.

The hand rods are regular sectional rods as used on power rodders. Usually the rod length is 36 inches. This length is easy to control and convenient for keeping track of footage or number of sections used. The rod is stored in a shallow reel approximately four feet in diameter and so constructed as to keep the rod contained on the reel. The reel is equipped with a removable tripod, which allows it to be set up quickly near the work site at an angle convenient for operation. When the reel is properly set up on the tripod, the rod can be pulled off or onto the reel allowing the reel to turn freely. Many work trucks are equipped with this reel fastened to the truck. When rodding is needed where the truck cannot be taken, such as in an easement, sections of the rod are simply removed from the reel and pulled to the working area by hand.

Hand rod trailers are available as well; they are compact and the reel is mounted at an angle convenient for use. These trailers also have a box for tools and a rack for a three-wheel, gas-powered unit to turn the rod. This power drive is essential, especially when other rodding equipment is not available or usable. The power drive has a lightweight engine with a gear reducer and control that allows the end of the rod to be attached for turning. With a hand control, the rod can be rotated to cut roots or penetrate into stoppages. Operation of a hand rodder is very similar to the operation of large power rodders, except the distance of penetration is not as great.

Other than these basic items, reel, rod, power drive, and perhaps a trailer unit, tools would include root saws, augers, cutter blades, pick-up tool, assembly wrench, turning handle, rod guide tube and extension pipe, guide tube jack, sand traps, manhole shovel, debris bucket and hand line, and a manhole pick, hook, or lid lifter.

6.221 Equipment Setup

The location of the work site will govern the measures needed to create safe working conditions. Refer to Section 4.3, "Routing Traffic Around the Job Site," and Section 4.4, "Classification and Description of Manhole Hazards," for procedures to ensure safe working conditions with regard to traffic routing and manhole work.

TRAFFIC ROUTING

PLEASE TURN BACK AND REVIEW SECTION 4.3 ON PAGE 88

MANHOLE HAZARDS

PLEASE TURN BACK AND REVIEW SECTION 4.4 ON PAGE 115

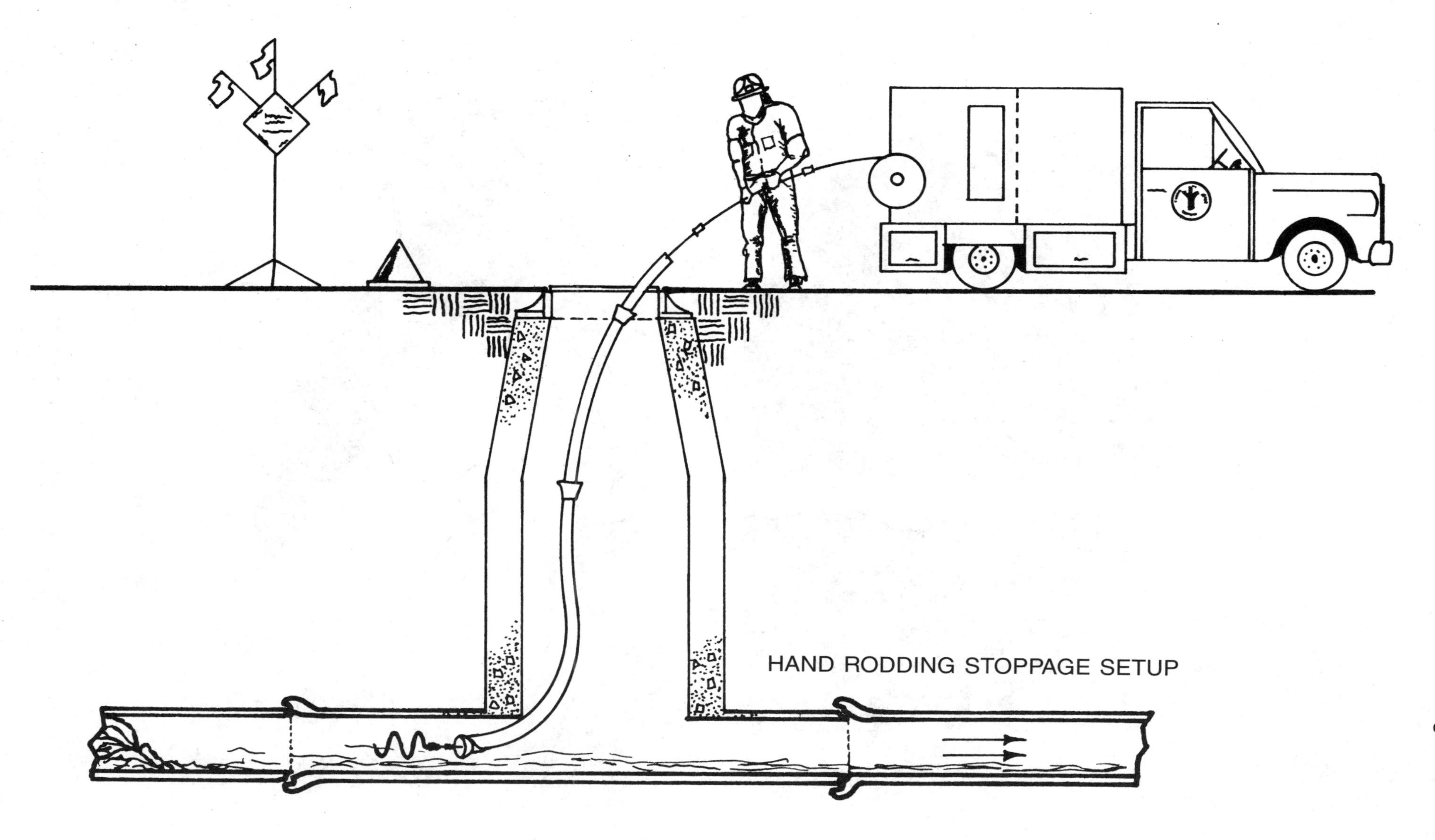

Fig. 6.28 Hand rodding operation

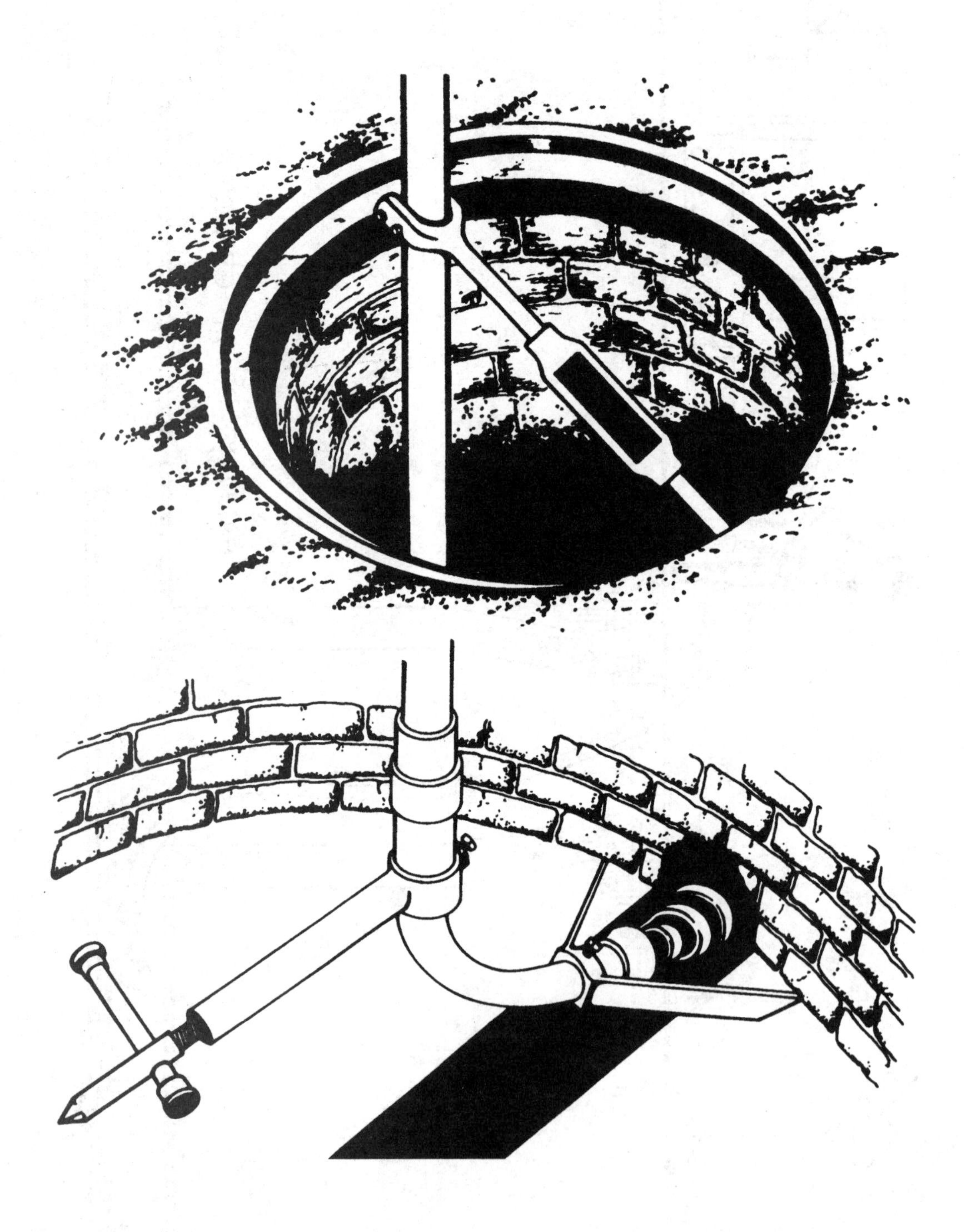

(Note the use of 2-inch pipe sections.)

Fig. 6.29 Hand rodding manhole setup
(Permission of SRECO-FLEXIBLE)

When using a hand rod it is not uncommon to have a number of feet of bare rod exposed on top of the ground or street. Extra caution must be exercised using this setup. Loose ends in a twisting operation can catch and twist and then let go with disastrous effects. Even if you are in an easement area, care must be taken not to damage flowers, shrubbery, lawns, fences, and gardens with the loose rod.

Ventilate the manhole and test for explosive gases, toxic gases, and oxygen level. Determine the depth of the manhole. Assemble sufficient pipe on the rod guide tube to allow the curved channel portion to firmly rest in the sewer and leave enough protruding above the street level to allow for working conditions, usually about 18 to 24 inches. To the exposed end, attach the flared fitting through which the rod will be fed.

Station the reel, whether truck, trailer, or tripod mounted, several feet from the working manhole. Thread the rod through the assembled rod guide tube, install the clearing tool, then insert it into the sewer, upstream or downstream, whichever method is chosen. Firmly anchor the rod guide tube to the bottom of the manhole. If going upstream, install a sand trap or debris trap in the downstream manhole. If going downstream, the trap would be installed in a like manner at the lower manhole. The guide tube brace or yoke is installed to hold the tube in place at the top of the manhole. The decision to rod upstream or downstream is determined by the rodding crew in the field. Three factors are considered when trying to clear a stoppage:

1. Usually it is easier and cleaner to rod into a dry downstream manhole than an overflowing upstream manhole.
2. Try to rod the shorter distance to the stoppage. Distances are estimated on the basis of past history, good records, and on-site observations.
3. If rodding upstream, what will happen when a stoppage breaks loose under a high head? If the operation is being observed properly, you can hear the flow coming and get out of the way. If rodding upstream into a stoppage with a high head behind it creates a safety hazard, rod from the upstream manhole. Always try to keep control of the rod at all times. See Section 6.223, "Precautions and Safety."

Setup procedures for hand rodding vary with the different problems involved, but this procedure will fit most needs.

QUESTIONS

Write your answers in a notebook and then compare your answers with those on page 387.

6.22A When are hand rods used?

6.22B How many operators are needed on a hand rodding crew?

6.22C Which has the greater penetration into a sewer, a hand rod or a power rod?

6.22D What precautions must be taken when a bare rod is exposed on top of the ground or street?

6.222 Operation

For routine clearing, the tool to be used at the end of the rod would depend somewhat on the sewer's condition. As with power rodding, if some history or prior knowledge of the situation is available, it takes away the guesswork. Both the square bar auger and the root saw often are used for routine clearing, so one of these could be used to start clearing the sewer. If roots are suspected, perhaps it would be advisable to rotate the rod from the beginning so as not to miss any. If this method is decided upon, pull off approximately 25 feet of rod from the reel and disconnect with assembly tools. Insert the end of this rod through the power drive unit, attach the tool, and start rotation—one operator running the power drive and one operator feeding the rod into the guide tube. When the rod is worked into the sewer, attach another 25 feet of rod and repeat the operation.

Admittedly, this is a slow operation. As indicated previously, in some "hard to get at" areas or where other equipment is not available, this method is still needed.

When the clearing or rodding tool has reached the opposite manhole, the rod often can be pulled out by hand, especially if you are working upstream. If not, rotate the rod with the power drive or turning handle and pull on the rod at the same time. If sufficient space is available, long lengths may be removed at a time, but care again must be given to traffic or other hazardous conditions. If the rod can be pulled by hand, simply attach it to the rod remaining on the reel and roll it up as it is pulled from the sewer. Disassemble the guide tube brace, clearing tool, and guide tube. Clean the material from the manhole and remove the debris trap. If it is necessary for an operator to enter the manhole, the operator must be properly equipped with safety harness, lifeline, and hard hat. Be sure to test for atmospheric hazards before entering the manhole and keep the manhole ventilated. Two operators must remain topside whenever anyone enters a manhole. All confined space entry procedures must be followed.

Many techniques for using hand rods can be considered in addition to the method outlined above. One of these would be "to prove" the sewer in routine clearing. This would be to simply push the tool, auger, or saw up to the next manhole and back again without turning; if it goes by hand, fine. Otherwise, push it as far as possible by hand and then turn it only enough to get the tool through. This, of course, gets footage faster, but you may also pass up roots or other material. If this method is chosen, you can work downstream much easier than upstream.

Turning handles are often used for small trouble spots rather than using the power drive unit. If roots or other obstructions are encountered where the turning handle cannot be used, then the power drive has to be used.

To clear stoppages with a hand rod, follow a procedure similar to the one described in this operation section. Whenever possible, start at the manhole below the stoppage. After installing a suitable debris trap, work an auger of a size smaller than the line size up to the stoppage. If possible, hand turn the auger into the stoppage with a turning handle until the rod is tight; then, with torque still on the rod, pull back. This usually will dislodge a stoppage. Try to remove the dislodged material so it won't cause another stoppage downstream.

A sudden charge of wastewater will be noticeable in the working manhole when a stoppage is cleared. After this water has subsided, run the tool back and forth in the stoppage area and then pull out the rods and tool. If hand turning the rod doesn't clear the stoppage, put the gasoline-driven power drive unit on the rod and use this unit. If this procedure doesn't work, try the power rodder (Section 6.21). After the stoppage has been cleared, inspect the upstream manhole to be sure the flow is normal and clean up any overflowed area.

In larger sewers the stoppage often occurs in the manhole rather than in the sewer. Clear the stoppage from the downstream working manhole and then clean the upstream manhole after the water drops.

If the stoppage takes place during normal working hours, a thorough cleaning job should be done on the sewer at that time.

Procedures for clearing and setup are essentially the same for all problems where hand rods are involved. Some shortcuts are used, of course, and many are not recommended, mostly because safety measures are sacrificed or equipment can be damaged. Some precautions are discussed in the next section.

6.223 Precautions and Safety

One particular problem occurs all too often when hand rodding; that is, having too much loose rod unattended or used in an unsafe method. Because it is time consuming to be continually fastening short sections of rod together, the tendency is to use a long section. If proper care and attention are given to this loose rod, a certain amount of looseness is acceptable. However, the "problem" at hand all too often takes precedence. Someone can be seriously injured and considerable property damage can be done should a car run over and catch hold of a loose piece of rod.

Another problem becomes evident when using the power drive unit. If too much distance is allowed between the power drive and the manhole rod guide tube, a loop can easily and suddenly be thrown in the rod and someone will be injured.

NEVER USE RUBBER GLOVES WHILE HANDLING OR GUIDING A ROD THAT IS TURNING. EVEN IF THE ROD IS WET, IT WILL SEIZE THE PALM OF THE RUBBER GLOVE AND, IF NOT STOPPED IN TIME, EITHER TWIST THE GLOVE FROM THE OPERATOR'S HAND OR CAUSE INJURY. Use a manhole hook or other device to guide a turning rod.

When working a rod upstream to a stoppage, be cautious when the stoppage is broken loose, especially if the grade of the sewer is steep. The sudden surge of stored wastewater rushing downstream can shoot up through the rod guide tube and get someone wet.

On steep grades where a root mass has been broken loose, it is not uncommon for the water pressure to literally force the rod back out the guide tube with such force it cannot be held. In that case all you can do is control the loose rod to the best of your ability.

6.224 Records

In districts where hand rodding is performed primarily in easements, it probably isn't done too often, especially if those sewers require only routine maintenance. Work in these areas must be properly recorded, the same as work in the more troublesome areas.

Operators must be continually encouraged to keep systematic and complete records. All too often we keep track of the problems and let the rest go, until they too become a problem. All clearing must be properly recorded in accordance with Section 6.07, "Records."

QUESTIONS

Write your answers in a notebook and then compare your answers with those on page 387.

6.22E What kind of tool is used at the end of a hand rod?

6.22F What would you do if hand turning a rod did not clear a stoppage?

6.22G What precautions must be considered when working a rod upstream to clear a stoppage?

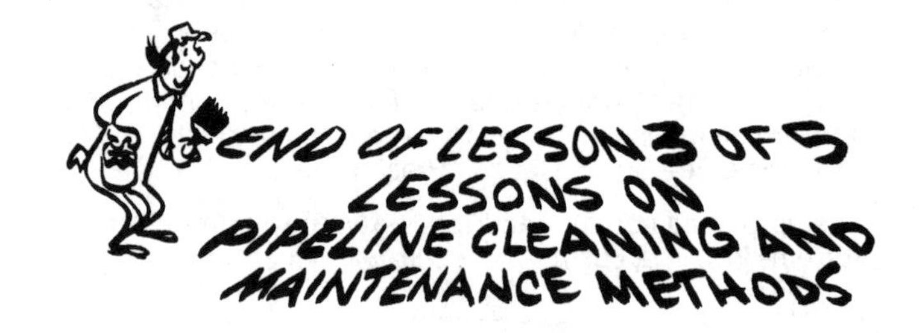

Please answer the discussion and review questions next.

DISCUSSION AND REVIEW QUESTIONS

Chapter 6. PIPELINE CLEANING AND MAINTENANCE METHODS

(Lesson 3 of 5 Lessons)

Write the answers to these questions in your notebook before continuing. The question numbering continues from Lesson 2.

15. Under what conditions would you use mechanical clearing techniques?
16. Why are power bucket machines usually not used for routine cleaning?
17. How can you determine when a bucket is full?
18. How would you select a tool for a power rodder?
19. How is a power rodder operated in a sewer?
20. When would you use a hand rod?

CHAPTER 6. PIPELINE CLEANING AND MAINTENANCE METHODS

(Lesson 4 of 5 Lessons)

6.3 CLEANING EQUIPMENT MAINTENANCE

6.30 Need for Cleaning Equipment Maintenance

Water and dirt are the two worst enemies of mechanical equipment; therefore, special attention must be given to sewer cleaning and inspection equipment. To do our jobs quickly and effectively, our equipment must be properly maintained and repaired at all times.

Equipment manufacturers provide the purchasers of their equipment with manuals that explain how to operate and maintain their equipment. Your job is to learn how to properly operate and maintain the equipment and be sure everyone follows these procedures. This section outlines the procedures for maintaining bucket machines, rodding machines (power rodders and hand rodders), and high-velocity cleaners. The ideas presented can be applied to developing preventive maintenance procedures for balling equipment, sewer scooters, and TV inspection equipment.

6.300 Objective of Cleaning Equipment Maintenance

The objectives of cleaning equipment maintenance are to *KEEP EQUIPMENT IN GOOD REPAIR TO HELP PREVENT EQUIPMENT FAILURE ON THE JOB, TO PROLONG THE LIFE OF THE EQUIPMENT, AND TO INCREASE THE EFFICIENCY AND SAFETY OF MAINTENANCE OPERATIONS.* To accomplish these objectives we need tools, time, and talent.

TOOLS: Without the proper tools, even the best mechanic cannot do a good job even with ample time.

TIME: We can have a toolbox full of the proper tools and have the "know-how" to fix things, but we never seem to have time for preventive maintenance until it is too late and the equipment breaks down. Equipment failures are costly in terms of repair costs, hazards to operators, and working time lost by crews.

TALENT: Most crews can find the "time" for preventive equipment maintenance. Try to use the information that comes from the manufacturer and the knowledge and talent that come from the experience of operating the equipment to develop and implement an effective equipment maintenance program.

6.301 Who Is Responsible?

The crew leader or the equipment operator is responsible for the proper maintenance of cleaning equipment. Report worn, broken or lost parts of equipment immediately so that replacements can be ordered. Regardless of your job, try to have the equipment used by the crew you work with properly maintained at all times. Your job will be easier and more enjoyable if you are not always fighting with equipment that doesn't operate properly or is always broken.

QUESTIONS

Write your answers in a notebook and then compare your answers with those on page 388.

6.3A Why must cleaning equipment be properly maintained at all times?

6.3B How can an operator learn how to maintain cleaning equipment?

6.3C What are the objectives of cleaning equipment maintenance?

6.3D What are the three essential factors for an effective equipment preventive maintenance program?

6.3E Who is responsible for the equipment maintenance program?

6.31 Bucket Machine Maintenance

Bucket machine clearing procedures require special care of the equipment. Breakdowns and failure of equipment often result in costly repairs.

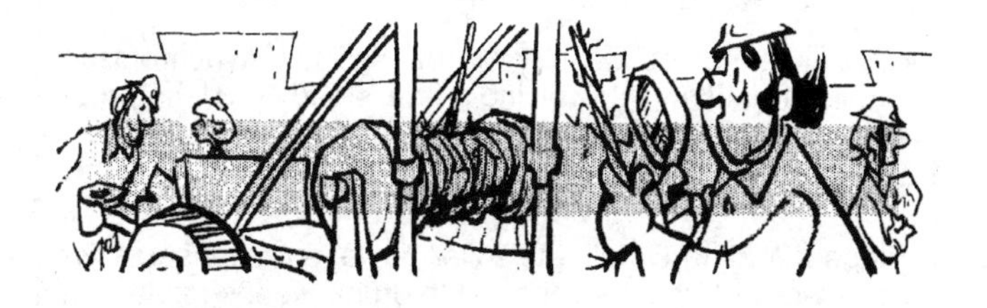

6.310 Maintenance Tools

A lot of time and effort can be saved if enough tools are kept on the job for crews at both the upstream and downstream manholes. Often these crews are working a considerable distance apart, especially when working in easements because some manholes may not be accessible in backyards.

Necessary tools include pipe wrenches, crescent wrenches, pliers, screwdrivers, Allen wrenches, hack saw, chisel, hammer, cable cutter, extra oil squirt can, grease gun, wire brush, extra cable clamps, clevis, pins, and wiping rags.

6.311 Maintenance Procedures

The following tasks must be done regularly or when indicated for effective bucket machine maintenance.

1. Keep cable scroll and level wind system (Figure 6.30) well oiled. *THE NUMBER ONE PROBLEM IS LACK OF PROPER LUBRICATION.*
2. Keep oil or grease cups full and chains lubed.
3. Inspect belt tension, chain tension, and gear clearance. *NOTE:* Record any *WEAR ON SPUR GEARS* (Figure 6.31).
4. Lube spindles of lower rollers often, especially if they are in contact with water. Make sure they have clearance and turn properly so the cable doesn't slide and cut grooves.
5. Watch for loose cable clamps and frayed cable. Tighten all cable clamps at the start of each day.
6. Look for loose rivets on buckets, badly worn bails and clevis eyes. Return these to the shop for hard-facing and repair.
7. Keep pipe threads clean and oiled on extension pipes. Keep extra couplings on exposed pipe thread.
8. Keep stabilizer legs in good operating condition. *IF A BEND STARTS, GET IT STRAIGHTENED NOW!*
9. Keep legs locked "up" when traveling.

QUESTIONS

Write your answers in a notebook and then compare your answers with those on page 388.

6.3F Why should bucket machine crews have enough tools for crews at both the upstream and downstream manholes?

6.3G What is the number one maintenance problem with the cable scroll and level wind system?

6.3H When should cable clamps be tightened?

6.3I What should be done when a bend is observed in a stabilizer leg?

6.32 Rodding Machine Maintenance

Preventive maintenance for all rodding machines is similar, but some aspects depend on the type of machine. Maintenance procedures are outlined in this section for sectional and continuous rodders and hydraulic units. Be sure to read and understand the equipment manufacturer's instructions and follow them.

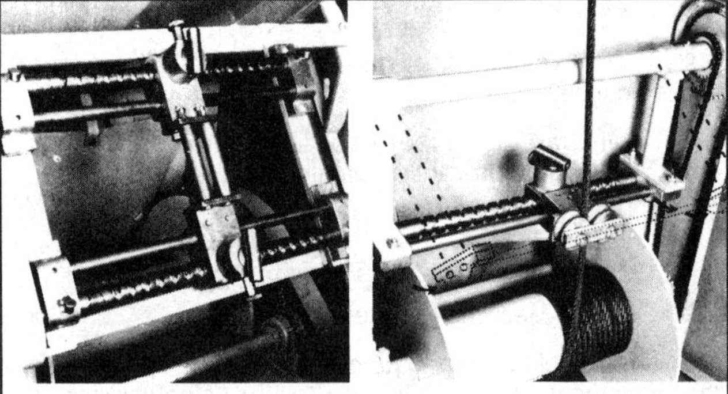

DUAL SCROLL FLOATING SCROLL

DUAL SCROLL (Truck-Loaders)

Heavily reinforced dual scroll features two synchronized scrolls and follow dogs giving a firm, steady guide to properly wrap the cable on main cable drum.

FLOATING SCROLL (Pull-in Machines)

Hinged floating scroll with roller guides compensates for each wrap of cable on the cable drum. Keeps cable in a direct line from upper cable sheave through level wind to cable drum.

Fig. 6.30 Cable scroll and level wind system

(Courtesy of Municipal & Utility Division, Rockwell International)

SAFETY BOOSTER CLUTCH SURE-LOCK CLUTCH

SAFETY BOOSTER CLUTCH

The safety operating lever (#1) controls bucket travel. At point (#2), tension is held on the operating lever at a predetermined speed of bucket travel. The tension is adjustable, and under overload conditions the belts will slip, thus protecting pipe and machine.

SURE-LOCK CLUTCH

Positive engagement of the bull gear drive to the cable drum is accomplished by means of two heavy hardened pins which lock into two sleeved, hardened holes in the bull gear. A slight pull and twist on the knob (#3) disengages the gear; a reverse twist re-engages.

Fig. 6.31 Bucket machine spur gears

(Courtesy of Municipal & Utility Division, Rockwell International)

6.320 *Maintenance Tools*

Some of the tools used to maintain rodding machines include assembly wrenches, crescent wrenches, Allen wrenches, screwdrivers, pliers, hack saw, file, oil squirt can, grease gun, and wiping rags.

6.321 *Sectional Rodder*

1. Daily
   a. Inspect tension of drive head chain. Tension must be tight. Why? If loose, drive dogs tend to ride up and over rod couplings.
      (1) Make sure chain track clearance is not too great.
      (2) *DO NOT* turn adjusting turnbuckle without loosening bearing plate.
   b. Inspect wear of drive dogs. Why? If badly cupped, rod coupling will tend to hang up.
2. Inspect
   a. Belt tension and wear.
   b. Vari-speed control.
   c. Drive chains to sprockets.
3. Forward-reverse transmission (Snow-Nabst)
   a. Oil level (use only machine oil as specified by manufacturer, *NOT MOTOR OIL*).
   b. Firm IN-OUT action, without drive head creeping. When adjusting, remember that if equipment has cast-iron bands inside, they will break easily. Some equipment has a cast-iron guide ring or support instead of a band.
4. Tight rod couplings
   a. *VISUAL INSPECTION* while in operation. Stop operation and tighten as necessary.
   b. Periodically inspect all rod couplings. Get rid of those that are severely worn or damaged and save the rest for reuse.
5. Stacking of rod on reel

   VISUAL INSPECTION by operator when returning rod to the reel.
   a. Proper adjustment of rod guide tube and rollers prevents jamming or couplings catching on travel out.
   b. Wear on funnel of rotating guide tube.
   c. Wear on tube rollers.

6. Brake band or disc brake blocks
   a. Inspect for wear.
   b. Replace before rivets cause drum damage.
7. Rod hose guides
   a. Do not shove rod out without at least a coupling on the end, preferably a bullet tool.
   b. Keep slack out of hose guide when working shallow manholes.
   c. Lubricate periodically with light, penetrating oil or transmission fluid to keep flexible for use.

6.322 *Continuous Rodder*

1. Proper torque on the drive rollers frequently depends on rod wear. Inspect for slippage while in use. Adjust torque when needed or wear will be excessive on rod and rollers. Also, performance of rodder will be poor.
2. In combination with gears for drive head, check for:
   a. Tooth clearance of gears.
   b. Sharp or pointed teeth.
   c. Worn gear bushings.
   d. Broken gears.
   e. Slack on reverse.
   f. Gear chatter.
3. Wear of rod guide bushings. Worn bushings tend to shave the rod.
4. Keep rod clearing tool in use, replace pads often. Grit combined with grease tends to shorten rod life.
5. Rod hose guides
   a. Do not shove rod out without at least a coupling on the end, preferably a bullet too.
   b. Keep slack out of hose guide when working shallow manholes.
   c. Lubricate periodically with light, penetrating oil or transmission fluid to keep flexible for use.

6.323 *Hydraulic Unit on Rodding Machine*

1. Look for and repair leaky fittings and hoses.
2. Keep oil filter clean (some filters have a warning-type gage).
3. Maintain proper hydraulic oil level (this is particularly important in hot weather).
4. Replace cracked, chafed, or worn hoses.

For more information, see Section 6.218, "Maintenance of Rods and Equipment."

QUESTIONS

Write your answers in a notebook and then compare your answers with those on page 388.

6.3J Why must the tension on the drive head chain of a sectional rodder be tight?

6.3K What does the torque on the drive rollers of a continuous rodder depend on?

6.3L Hydraulic units require what maintenance?

6.33 High-Velocity Cleaner Maintenance

To properly operate and maintain a high-velocity cleaner, the operator must be completely familiar with all aspects of the unit. Read and understand the manufacturer's instructions for proper maintenance procedures.

6.330 Maintenance Tools

Tools should include pipe wrench, large crescent wrench, grease gun (special grease recommended for pressure swivel), screwdriver, a good grade of plastic tape, plus other hand tools found in a well-equipped toolbox.

Special tools needed to maintain the pressure pumps are provided by the manufacturer.

6.331 Continuous Considerations

Whenever you are operating a high-velocity cleaner, always be alert for:

1. Odd or suspicious sounds from:
   a. Intake valve seats,
   b. Relief valve leaking (squeal), and
   c. Belt slippage.

2. Higher or lower than usual RPM to maintain pressure. May be caused by:
   a. Intake valve partially closed,
   b. Air leaking into system,
   c. Plugged *ORIFICE*[12] in cleaning nozzle,
   d. Belts slipping, or
   e. Faulty bypass pressure valve.
3. Sluggish reel return. May be caused by:
   a. Low hydraulic oil supply,
   b. Filter needs replacing,
   c. External or internal leak in a hydraulic control, or
   d. Foreign particles in the directional valve. (Filter should catch particles.)

6.332 Daily

1. Keep all equipment and accessories clean.
   a. Tool compartments
   b. Engine compartment
      (1) Wipe up oil and grease
      (2) Paint rust spots when they appear
2. Hold Tank
   a. Drain to prevent:
      (1) Rust, and
      (2) Sand or dirt deposits. Always flush hydrants briefly before filling tank.
   b. Clean tank strainers
3. Oil Levels
   a. Engine
   b. Pressure pump
   c. Hydraulic oil tank
4. Inspect belt tension and packing glands. (See Chapter 9, "Equipment Maintenance," Section 9.3, "Pumps," in Volume II.)
5. Lube drive chains and 90-degree or in-line high-pressure swivels to the hose reel.
6. Repair splits in hose, or replace as necessary.
7. Look for worn or plugged orifices in nozzle.
8. Inspect finned nozzle extension. The finned nozzle extension helps keep the nozzle off the bottom of the pipe, helps keep it centered in the pipe, and helps to prevent the nozzle from turning up service connections, other pipes, or manholes where the nozzle could get tangled in the manhole rungs.

6.333 Cold Weather Maintenance

1. Drain pump daily. Use air pressure to drain pump if necessary.
2. Cover rear canopy openings. A small electric heater will help keep equipment from freezing.

6.34 Equipment Engines

Equipment engines require special attention so they will operate when needed. Develop a regular maintenance program and follow the manufacturer's recommendations. The following items are important considerations for proper care of engines used to operate cleaning equipment.

1. Always be sure oil and water levels are in the proper range *BEFORE* starting the unit.
2. Use a fresh supply of the proper grade of clean gasoline.
3. Change oil and air filters according to the manufacturer's recommendations.
4. Exercise (operate) equipment weekly if not used.
5. Use the proper type of oil in engines, transmission, and for lubrication.

---

[12] *Orifice (OR-uh-fiss). An opening (hole) in a plate, wall, or partition. An orifice flange or plate placed in a pipe consists of a slot or a calibrated circular hole smaller than the pipe diameter. The difference in pressure in the pipe above and at the orifice may be used to determine the flow in the pipe.*

6. Keep battery terminals clean and battery charged for engines with electric starters, especially during winter months.

QUESTIONS

Write your answers in a notebook and then compare your answers with those on page 388.

6.3M Why must the operator of a high-velocity cleaner be completely familiar with the unit?

6.3N When operating a high-velocity cleaner, you should always be alert for what kinds of problems?

6.3O What oil levels should be checked daily?

6.3P What items must be considered during cold weather maintenance of high-velocity cleaners?

Please answer the discussion and review questions next.

DISCUSSION AND REVIEW QUESTIONS

Chapter 6. PIPELINE CLEANING AND MAINTENANCE METHODS

(Lesson 4 of 5 Lessons)

Write the answers to these questions in your notebook before continuing. The question numbering continues from Lesson 3.

21. Why should pipeline cleaning equipment be maintained?
22. How would you maintain pipeline cleaning equipment?
23. What maintenance items should be considered continuously when you are operating a high-velocity cleaner?
24. How would you maintain equipment motors used to operate cleaning equipment?

CHAPTER 6. PIPELINE CLEANING AND MAINTENANCE METHODS

(Lesson 5 of 5 Lessons)

6.4 CHEMICALS
by Walt Driggs and John Brady

6.40 Uses of Chemicals

Chemicals can be very helpful aids for cleaning and maintaining wastewater collection systems (Figure 6.32). Proper application of effective chemicals can be used to control roots, grease, odors, concrete corrosion, rodents, and insects. The purpose of this section is to indicate how to determine if chemicals might be effective and how to select, apply, and evaluate the effectiveness of chemicals. The keys to successful use of chemicals are proper preparation and proper application. Where specific chemicals are mentioned, trade names are provided solely for illustrative purposes.

6.41 How To Select Chemicals

6.410 Words of Caution

Many people are reluctant to try chemicals because of past unfortunate experiences involving exaggerated claims. Did you ever hear the story about the salesman selling a compound of caustic and aluminum shavings? He claimed his chemical would unplug stoppages. The question was asked, "What will this chemical do to roots?" He replied, "When you pour this chemical into a manhole, it starts to boil and churn. The aluminum shavings will move so fast they will cut the roots off at the edge of the pipe." This is an example of a false claim.

Look out for exaggerated claims by promoters. No chemical will stop odors, remove roots, or repair breaks in sewers. Some chemicals may cause more damages than benefits. For example, you may clear a kitchen drain trap with half a cup of

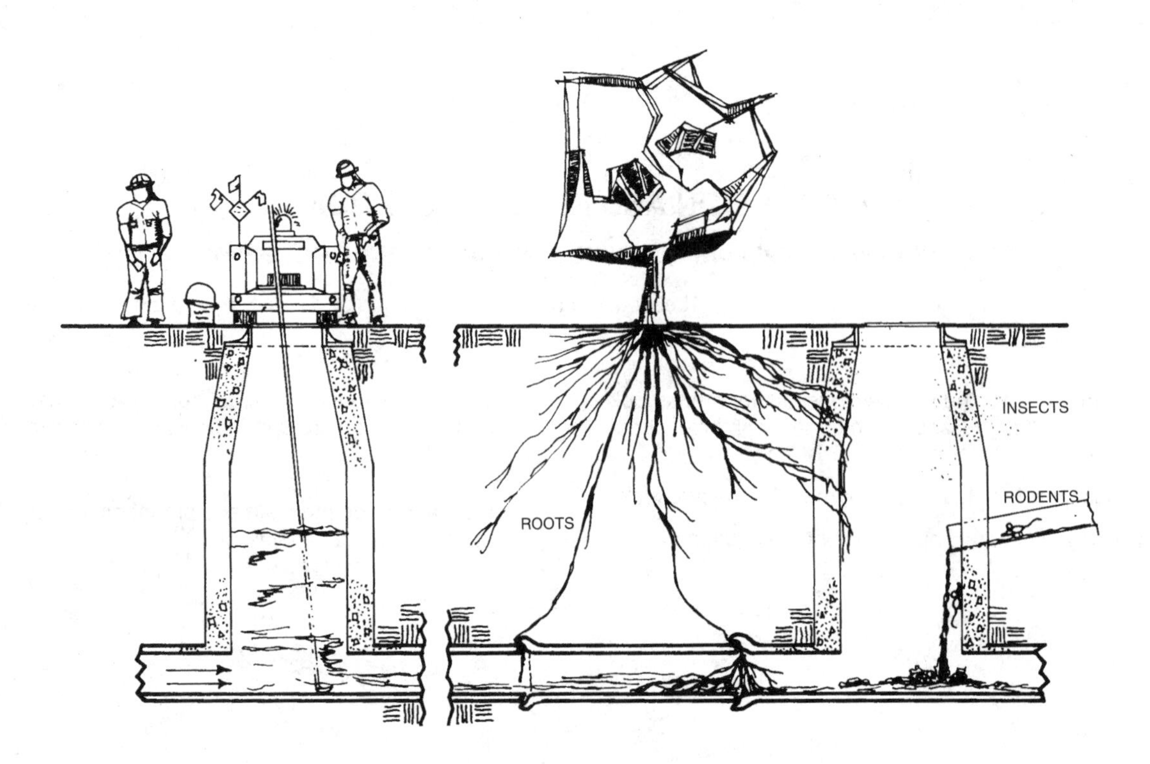

Fig. 6.32 Chemical operation

chemical. To clear a similar stoppage in a 36-inch sewer would require tons of chemicals. This procedure would be very expensive and could cause extreme hazards downstream.

Two very important factors must be remembered when evaluating and using chemicals:

1. Chemicals will not clear stoppages or blockages in sewer mains. Be wary of any chemical that makes this claim. How does the chemical move 100, 200, or 300 feet down a plugged sewer to the stoppage at the proper concentration?
2. Laboratory or "desk top" demonstrations work because chemical concentrations are adequate for the amount of water involved. Apply these concentrations in the field and the volume of water in a sewer could require a large amount of costly chemicals to achieve the required concentration at the proper location. Ask your chemical supplier to calculate the exact amounts of chemicals needed to provide the desired chemical concentrations. For example, how much chemical is necessary to produce effective heat ranges in the sewer to dissolve grease or burn roots?

Don't get trapped into thinking you will at least be doing something if you buy a chemical. Very often you are merely contributing to the promoter's income but not helping your situation. Contact water pollution control regulatory agency officials before using a chemical. Some chemicals are banned, while others are approved for appropriate use.

6.411 A Selection Procedure

Ideally, chemicals you wish to use have been tested by a respectable laboratory and the results are favorable. Many universities, states, federal agencies, and private testing laboratories conduct impartial tests of new chemicals before they are sold. For assistance and guidance regarding the application of chemicals to your particular problem, contact the U.S. Environmental Protection Agency, state water pollution control agencies, and professional associations for recommendations. If a chemical has not been tested to your satisfaction, don't buy it.

All chemicals of the herbicide, insecticide, fungicide, or rodenticide types require registration under the U.S. Environmental Protection Agency, Pesticides Regulation Division. These chemicals must have an approval number issued by EPA before use to prevent harmful contamination of a watercourse.

If a chemical does not require an EPA approval number and is untested, attempt to have a recognized laboratory perform the necessary tests. Otherwise, you must design your own testing program or risk throwing your agency's funds down the sewer.

WARNING

Never put any chemicals containing metal chips or filings into a sewer. This material will most likely settle out in the sewer without reacting or, if it does react, may produce hydrogen gas. Hydrogen gas increases the possibility of an explosive atmosphere developing in the sewer.

If you must develop your own chemical testing program, here are some steps to follow that are fair to your agency and the chemical company.

1. Ask the vendor to provide documentation of claims in writing. Realize that only favorable results will be provided. Also ask for the names and telephone numbers of a few customers who have used the products. A telephone call to operators who have used the chemicals will reveal if the chemicals are still being used successfully and if the costs are reasonable. Phone calls provide the opportunity for you to ask questions and also to obtain information people might be reluctant to put in writing.
2. Calculate the cost per foot of sewer to be treated before buying and applying the chemical. Costs may be too high even if the chemical works as claimed.
3. Work with the vendor to select a segment of sewer to serve as a test section (at least between two or more manholes). Develop a procedure for evaluating the effectiveness of the chemicals in the test section over a period of time recommended by the vendor. If the vendor refuses to demonstrate the product without charge and with an authoritative evaluation other than their own—*BE VERY CAUTIOUS*.
4. Take photographs or videotape the existing problems in a test section of sewer *BEFORE* the chemicals are applied.
5. Agree in writing with the vendor on the results that should be produced by the chemicals and the method of payment for the chemicals. Determine payment on the basis of effectiveness of chemicals.
6. Apply chemicals as directed under the supervision of the vendor. Recalculate and verify dosage rates applied.
7. Retelevise and photograph or videotape the treated test section of sewer in accordance with your agreement with the vendor.
8. Be sure to keep good records before, during, and after testing.
9. If the chemicals worked as claimed or agreed upon, pay the bill. If the chemicals did not work and there were no agreements regarding partial payments for partial effectiveness, then don't pay. No work, no pay.
10. Be wary of companies that offer a "money back guarantee." How do you get your money back if you are not satisfied? What happens if the company goes bankrupt? A reputable company with a good credit rating and bank references may be worth the risk.
11. Sometimes the selection of a "control" sewer may be helpful. No chemicals are applied to problems such as roots in this sewer. The purpose of this line is to provide a means of comparing problem development (such as root growth) in lines with and without chemicals.

If your agency is too small or without the resources to conduct an extensive testing and evaluation program as outlined in this section, try to use as many of the concepts discussed as possible to protect your agency.

Any company not willing to stand behind their products and subject them to an agreed upon evaluation does not deserve to use your valuable time trying to sell you questionable products. When your agency has a firm policy regarding the testing of chemicals before purchase, the effectiveness of your chemical program will be greatly improved.

QUESTIONS

Write your answers in a notebook and then compare your answers with those on page 388.

6.4A Chemicals can be used to control what types of problems in wastewater collection systems?

6.4B Why are some people reluctant to use chemicals to solve sewer problems?

6.4C Why might doing nothing be better than trying an untested chemical?

6.4D Why should metal chips or filings not be put into a sewer?

6.4E What precautions should be taken before developing a testing program?

6.42 Roots

6.420 Causes of Root Problems

Let's try to understand how to approach the cure for roots in your wastewater collection system by learning the causes of root problems.

One city in the United States had a serious problem. The residents were complaining about the high rate of stoppages in the wastewater collection system, the flooding of their homes, and the high costs they were having to pay to replace their home service lines.

The city council at one of their meetings discussed the complaints, heard the city manager tell of the high cost of keeping the flow within the pipe, and decided there was one simple answer to this problem—let a contract for the removal of all the trees in the rear easements (estimated cost of removal—$300,000) and pass an ordinance that would prohibit any replanting.

The next city council meeting was attended by almost the entire town. The public told the city council that they were not going to give up the shade in their side and rear yards. "There must be a way to keep our shade trees that have increased the value of our property and made our rear yards a beautiful place to enjoy and also to keep wastewater flowing in the collection system," commented one city resident. Somehow shade trees and wastewater collection system sewers must be able to occupy the same easement.[13]

The remedy or cure for roots was not as simple as the city council had thought, or did they just try the most obvious and not try to seek the right cure for the problem of roots?

We are going to learn more about root control, but first we should try to understand the causes of root problems.

Let's think of the farmer of 1900 and some of the ways he combatted weeds in his field. He hoed and disced and some of the more progressive farmers were starting to treat the soil with chemicals that would control the weeds and allow the crops to flourish. Effective use of the proper chemicals on roots can help avoid a lot of hard work today.

Roots growing near sewers are drawn to the sewer lines by vapor trails that escape into the soil through very small cracks in the pipe or at section joints. The tiny hair-like roots enter the sewer line one cell at a time and quickly grow in thickness as the root mass develops. Roots thrive in the sewer atmosphere, which is rich in the essential nutrients, nitrogen, phosphorus, and potassium (N-P-K).

Cutting or pruning sewer roots causes thicker, faster, and stronger regrowth. The rate of root regrowth increases with each new "pruning." This is true of both main and lateral roots. If you have ever pruned a tree, you know that after each pruning the tree develops new growth just below each cut. Roots grow in a similar way—each time a root is cut, it will add new growth laterally and it will increase in diameter.

The entry point of roots in a sewer main is almost always above the flow line at a joint or pipe defect. Roots that grow from the top of the pipe (Figure 6.33) are called "veils." This kind of growth catches passing grease and other debris and soon forms a significant barrier to the flow in the line. Gradually, as the size of the root mass increases, it can cause a stoppage and may eventually block the pipe completely. Once established in the sewer pipe, the entry root becomes the main stem or trunk, separating the pipe as it grows in diameter. This main stem can eventually break the pipe or open up the joint and cause an inflow or infiltration problem as the pipe is forced apart (Figure 6.34).

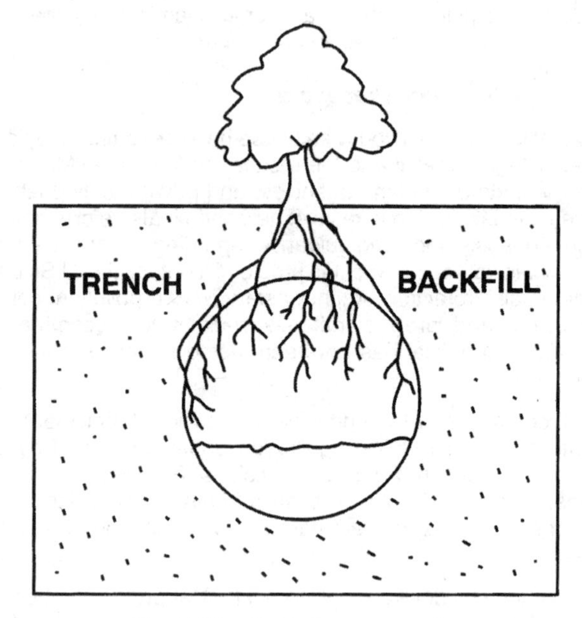

Fig. 6.33 Roots in a sewer pipe

In lateral connections, where the incline is often steep from the house to the main line elevation in the street, the steady flow of nutrients in the household wastewater quickly drains to the main. This encourages the growth of very long, thick roots that grow the entire length of the lateral into the main. These root clumps grow into the main from the lateral in a form commonly called "tails." When the line is completely backed up and manholes are surcharged or overflowing, there is little choice but to cut the roots out to prevent property damage and environmental fines from overflows.

---

[13] *Good planning might have placed the wastewater collection system in the street depending on distances, costs, and service requirements. This would not have prevented problems caused by roots, but would have placed the sewer lines a greater distance from trees planted in front yards than from trees in backyards if the sewer was in the backyard. Usually fewer trees are planted in front yards than backyards.*

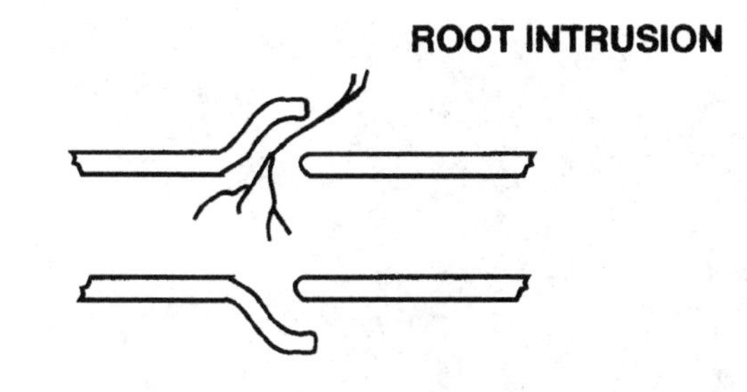

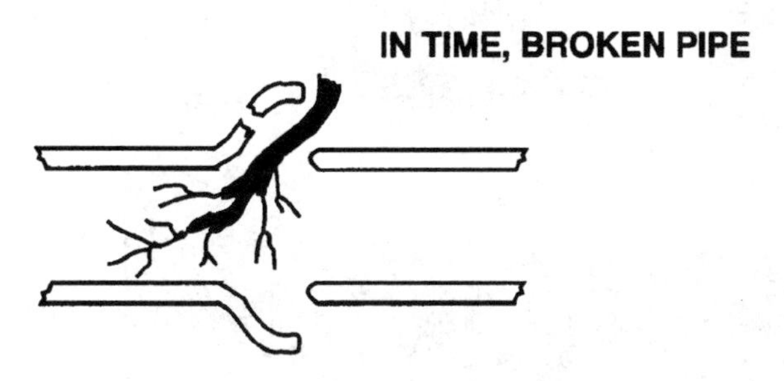

Fig. 6.34 Root breaking pipe

6.421 Information Needed From Vendor or Manufacturer

When you are starting a chemical program for root control, the following information must be obtained from the manufacturer:

1. Recommended herbicide and EPA approval number. Verify information with EPA and other regulatory agencies because rulings sometimes change when new experience is gained.
2. Recommended method of application such as flooding, spraying or foam. Sometimes the application of a combination of two herbicides is more effective than only one herbicide.
3. Concentration of chemical and amount required to treat a selected sewer line. Also learn the correct way of determining the specified concentration for other sizes and lengths of sewers for future reference.
4. Time of contact of chemical with roots, or time required for the sewer to be flooded or blocked with chemical to achieve root kill.
5. Preferred meteorological conditions at time of application. Rates of herbicide uptake through roots are largely determined by rates of plant transpiration or water loss from the leaves of the plant.
6. Cost for chemicals and application, including labor and equipment.
7. Hazards to wastewater collection system operators working downstream from section where chemicals will be applied.
8. Effects upon the downstream wastewater treatment plant by the use of the chemical.
9. Effects on receiving waters of wastewater treatment plant effluent.
10. Material Safety Data Sheet (MSDS).

6.422 Application (Flooding and Foaming)

This is the area where many chemicals fail to perform to the manufacturer's claims and make many maintenance people leery of new chemical products. If specific claims are made as to results to be expected, then the chemical manufacturer should be willing and able to provide *SPECIFIC* instructions as to method of application, concentration of chemical, and required contact time. Directions for the application of some chemicals suggest that you "dump one can of chemicals in the sewer and if that doesn't work, dump in another can." Haphazard instructions of this type will usually produce haphazard results.

To apply a chemical to control roots, first televise the section to be treated to determine existing conditions. No high-velocity or mechanical cleaning is recommended in lines prior to chemical root treatment unless excessive grease, root masses, or debris in the lines would interfere with proper application of the chemicals. When the root tips are damaged or removed by sewer line cleaning, chemical treatment will be less effective. Also, infiltration of water into the sewer line may be expected to wash the chemicals off the roots and decrease the effectiveness of root kills. Treatment of roots during a period when infiltration rates are low usually is more effective.

FLOODING (Figure 6.35)

Flooding sewers and building service laterals to control roots can only be done in low-flow sewers. This method is most commonly used in sewer lines ranging four inches to six inches in diameter. In larger diameter sewer lines (six to twelve inches in diameter), application of a chemically active foam is often used to control roots.

When ready to apply the chemicals, isolate the system to be treated by plugging the upper and lower ends of the sewer line to be treated (usually a segment of line between manholes). Mix the chemicals according to the manufacturer's directions and safety precautions, and slowly meter the chemicals into the test section of the line. If the line is to be flooded, allow the mixture of water and chemical to fill the sewer and manhole to a level three to six inches above the top of the sewer pipe in the upstream manhole. Leave the line plugged and the chemicals in the line for the specified contact time. Maintain constant observation of the system above the plug to prevent flooding of a home or business.

One method of reducing chemical costs is to reuse the chemicals in downstream sections of the sewer. Plug the lower sections of the line and allow the chemicals to flow downstream to the next section of line requiring treatment. A small addition of chemicals may be necessary to maintain the desired concentration. Some dilution of the original chemical batch usually occurs when the chemical slug flows down to the next test section.

Another method of recycling used chemicals is to pump the chemicals from the test section into a tank and reapply them in another section. Some additional chemicals may have to be added to maintain the desired dosage in the next test section. Be sure to follow the manufacturer's recommendations regarding safe handling of chemicals at all times.

FOAMING (Figure 6.36)

Foaming root control products can be used to destroy existing roots in sewers and prevent regrowth for periods of three to five years. These foam products contain two active ingredients. The first chemical is a fumigant that penetrates the cell walls of existing roots in the line and immediately causes the start of a slow process of decay; cell by cell the root mass collapses and decays. The second type of chemical is a growth

Dry chemical is added through a manhole, cleanout, or toilet.

Fig. 6.35 Flooding method
(Permission of ROOTX)

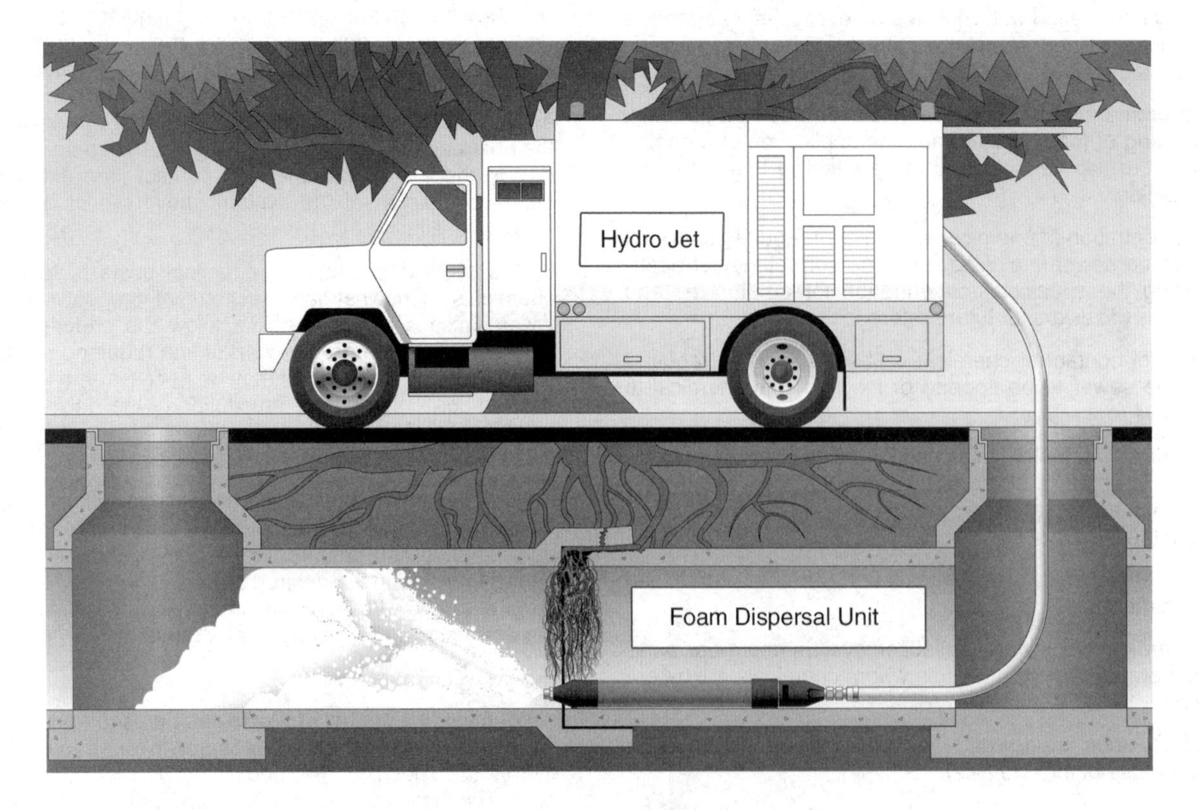

Dry chemical is added to the foam dispersal unit.

Fig. 6.36 Foaming method
(Permission of ROOTX)

inhibitor that attaches itself chemically to the surfaces of the pipe and to all organic material (the root mass) in the pipe. Note that both ingredients are needed for good root control, the first to kill existing roots and cause them to decay, and the second to prevent or slow future root growth.

Foam can be applied to the roots inside the sewer lines in one of two ways: by spraying a layer of foam on the roots (Figure 6.37, top two photos) or by entirely filling a section of the line with foam (Figure 6.37, bottom two photos). For effective treatment, the foam must come into contact with as many root surfaces as possible, including those inside root masses.

Foam spraying involves mixing the dry chemical with water and pouring the mixture into a foam dispersal unit (Figure 6.36). The foam dispersal unit is attached to a high-velocity cleaner or hydro jet. A two-stage nozzle on the dispersal unit first travels through the root mass to wash off greases while opening up the mass for better contact of foam with the root strands. Immediately following this quick process, the two-stage nozzle opens up to deliver foam through the same hose as it is retrieved evenly along the entire line from manhole to manhole. For best spraying results, the orifice on the dispersal unit nozzle must be adjusted to deliver the required gallons per minute of spray at high-velocity cleaner engine idle speed.

Another technique for applying foam to control roots in sewers consists of blocking off a section of sewer between two manholes and pumping enough foam into the section to completely fill the line and any connecting laterals, if desired. (In most cases, foaming of the laterals eliminates any protruding roots that may have developed in the laterals as a result of prior use of cutting tools.) Bypass pumping and cleaning prior to treatment generally are not required.

In this method, a specialized foam-making machine (Figures 6.38 and 6.40) is required to mechanically produce the needed quantity of foam and to deliver it into the sewer line with the required amount of pressure by means of a foam control nozzle (Figure 6.39). Water and the liquid root control chemicals stored on the foam-making machine are mixed together during the foaming operations and compressed air is injected into the mixture to assist in creating foam. The foam is then pumped through a discharge hose, completely filling the main line from manhole to manhole. Foam is compressed against all pipe surfaces, into cracks and joints, and is forced up connecting sewers for maximum contact of all roots. Upon contact with the foam, the roots in the sewer line stop growing and begin to die. Trees are not harmed.

It should be noted that the average main line cost for foaming root control per foot for 6- and 8-inch (15- and 20-cm) lines is lower than the cost of rodding equipment over a three-year period. Cost factors include access to the line in easements, seasonal hazards, scope of the project, availability of maps and seasonal local workers, and special criteria set by the agency.

Only properly trained operators should be permitted to apply foam products for root control in sanitary sewers. In some states and provinces, formal certification is required. The root control program also must meet local, state, and federal regulations and reporting requirements for use of the specific product(s), and the foam manufacturer's instructions for use must be closely observed.

6.423 Evaluation

Root control by chemicals is not as fast as removing roots by cutting them off with a power rodder, but chemical control is more permanent. Effective chemicals can control roots in a sewer for as long as two to five years.

To evaluate the effectiveness of a chemical root control program, televise both the test and control sections one week, one month, and three months after treatment to evaluate the effectiveness of the chemicals. With proper chemicals and application, root control by the use of effective chemicals is a very desirable preventive maintenance program and is cost effective. Evaluating costs can easily be done over the test period. Compare the cost of a two-person rodding crew cleaning the lines once a year with the costs of the chemical treatment program. If desired, calculate the same comparison over a 3-, 4-, or 5-year period of time.

6.424 Safety Precautions

Before using any chemicals, make sure the field crew has thoroughly read the Material Safety Data Sheet (MSDS) and the label on the chemical container and understands the type of chemical, reactions that will occur, safety measures for handling and applying, equipment used for mixing and application, action to be taken if the chemical is spilled or splashed on operators or materials, and cleanup. Research may be needed to discover if any special application licenses or permits are needed or whether your insurer has any special requirements to be met. Action in case of an accident includes first aid treatment to be given and how to inform a doctor of the exact chemical compounds involved if a person needs medical aid so the doctor will know how to neutralize the chemical and properly treat the victim.

When handling or mixing any chemical, always wear protective equipment as specified on the MSDS (rubber gloves, an apron, goggles or a face mask). Do not enter manholes without proper precautions (Chapter 4, Section 4.3, "Routing Traffic Around the Job Site," and Section 4.4, "Classification and Description of Manhole Hazards"). Spills of root control chemicals on lawns or gardens will kill plants and sterilize the soil for long periods of time.

WARNING

1. Do not mix unknown chemicals.
2. Do not add a strong acid or alkali (base) to a chemical unless instructed to mix the chemicals.
3. Do not work with chemicals until you have reviewed the MSDS.

6.425 Other Methods of Root Control

This section has emphasized the control of roots in sewers by the use of chemicals. There are other methods of root control.

First, the best way to control roots is to install sewers that don't leak. Modern pipe materials and joints can be installed without leaks so roots can't enter the sewer. In older sewers where there is the potential of root intrusion, methods of root control include:

1. Eliminating deep-rooted trees.
2. Inserting a liner in the sewer (Volume II, Chapter 10, "Sewer Renewal (Rehabilitation)").
3. Not allowing trees to be planted over sewers.
4. Removing roots and sealing sewer (Chapter 7, Section 7.5, "Sealing Grout").
5. Frequent clearing of roots from sewer.

Another possible approach to killing roots in sewers is by the use of scalding hot water. Roots can be thermally killed in

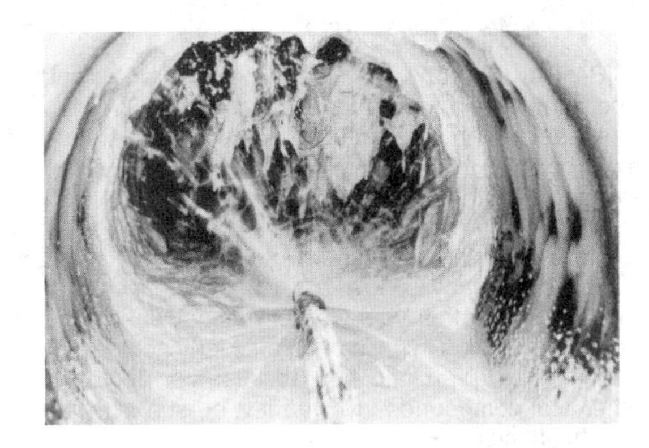

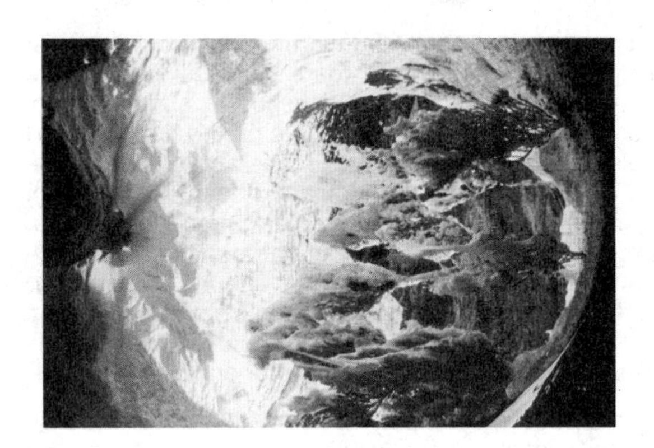

Foam Spraying Application
(large-diameter pipes, 12-inch diameter or larger)

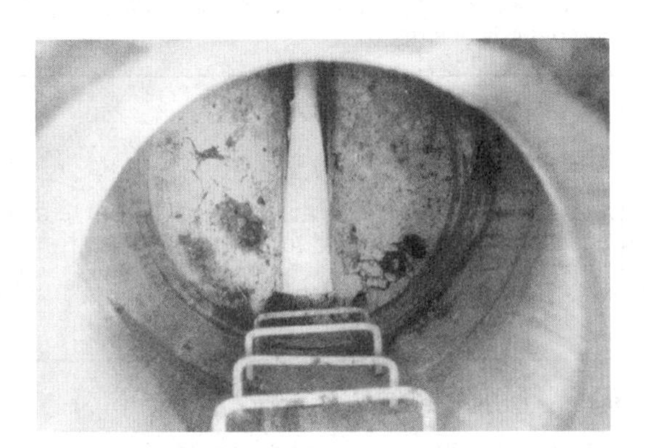

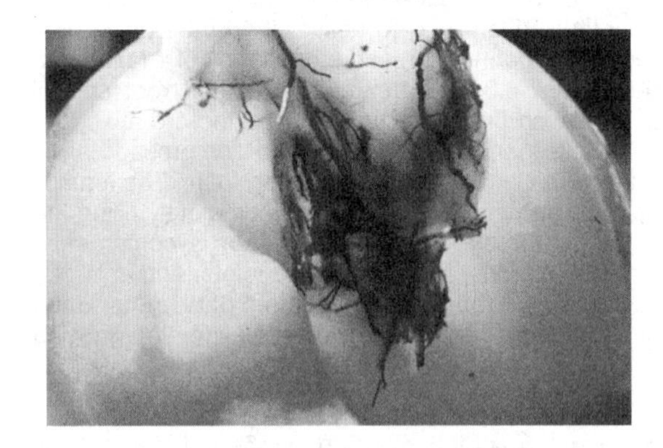

Foam Filling Application
(small-diameter pipes, 12-inch diameter or smaller)

Fig. 6.37 Foam application techniques for root control in sewers
(Permission of Douglas Products)

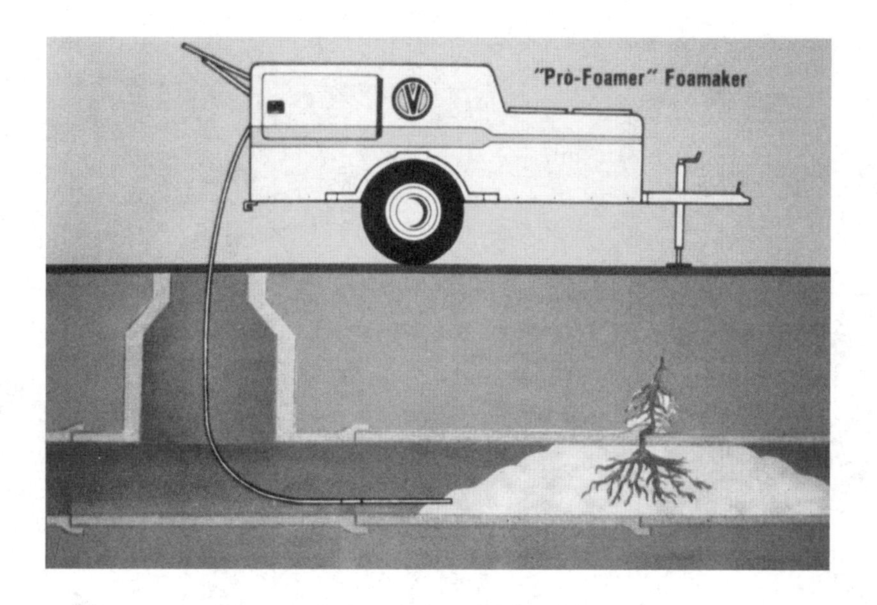

Fig. 6.38 Mechanical foam making equipment
(Permission of Douglas Products)

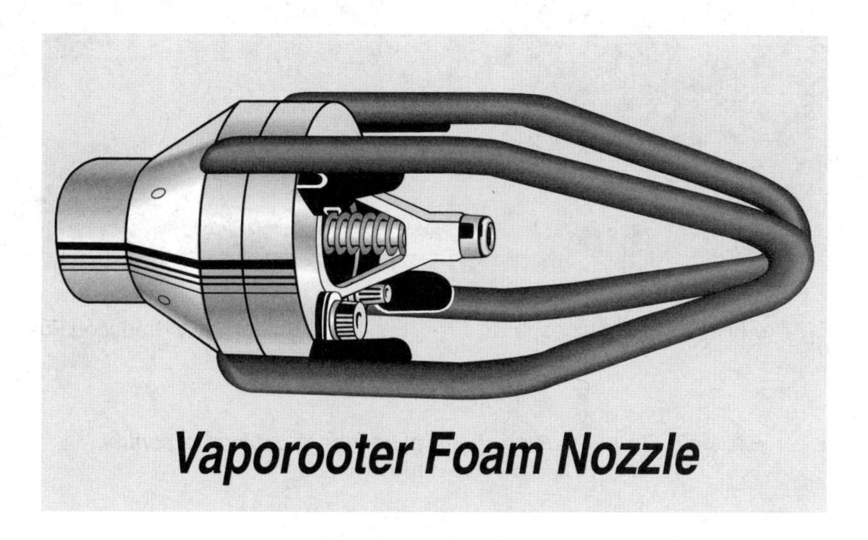

Fig. 6.39 Jet foam nozzle
(Permission of Douglas Products)

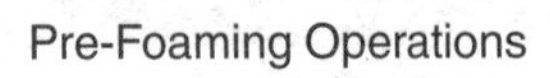
Pre-Foaming Operations

Mechanically Produced Foam

Fig. 6.40 Truck-mounted chemical treatment unit for root control

(Permission of Duke's Root Control, Inc.)

sewers by flooding one manhole section at a time with scalding hot (82° to 98°C or 180° to 210°F) water for 30 minutes. A truck-mounted packaged steam generator is required. The advantages of this system include a good root kill at any time of

the year, the killing of rodents and insects, and the liquefying of grease. Limitations of this approach include high energy costs and the problem of handling or disposing of the liquefied grease.

6.426 Additional Reading

An excellent paper on the evaluation of chemicals for root control is "Chemical Control of Tree Roots in Sewer Lines," by John F. Ahrens, Oliver A. Leonard, and Neal R. Townley in Journal Water Pollution Control Federation, Volume 42, Number 9, pp. 1643-1655 (September 1970). Also see "Thermal Kill of Roots in Sanitary Sewers," by James R. Conklin, Deeds & Data, Water Pollution Control Federation, February 1977, p. 4.

QUESTIONS

Write your answers in a notebook and then compare your answers with those on pages 388 and 389.

6.4F What happens when roots in a sewer are cut?

6.4G What information must be obtained from the manufacturer before starting a chemical program for root control?

6.4H How can chemicals be applied for root control?

6.4I How can homes or businesses become flooded during a chemical operation?

6.4J How can chemicals used for root control be recovered and recycled?

6.4K What safety precautions should be used when chemically treating roots?

6.43 Grease

Grease and soap problems used to be confined to sewer lines that served restaurants or industrial dischargers. These materials were controlled to some degree by grease traps or grease interceptors. With the widespread use of home garbage disposal units, the problem has now spread throughout the collection system.

Problems commonly develop when greases, oils, and soaps cool and solidify and form a coating or deposit on the walls of a sewer. The rate of buildup of the grease deposit depends on the amount of grease carried by the wastewater, the flow and velocity of the wastewater, and the size of the sewer. If the sewer changes slopes (steep to flat), intersects with another sewer, or has dips or other alignment irregularities, then these factors can encourage the development of grease deposits. Sewers larger than 18 inches in diameter tend to have fewer grease problems than smaller sewers because they usually flow more than half full with velocities high enough to prevent the buildup of grease.

Most uniform building codes or sewer-use ordinances require commercial and industrial dischargers to install grease traps or grease interceptors if grease is in their wastewater discharges. These units are similar to a septic tank and are designed to allow floatables (greases, oils, and soaps) to accumulate on the surface and heavy solids to settle to the bottom. The floatables and solids must be removed on a regular basis.

Responsibility for cleaning and maintaining grease traps and grease interceptors depends on community policy, but usually it is the responsibility of the discharger. Unfortunately, many dischargers do not know how to or do not bother to properly clean and maintain grease traps and grease interceptors. Some industries do an excellent job because contents of the traps can reveal that problems have developed in their industrial processes that are causing a loss of profits and also causing an increased sewer-use billing.

Grease traps and interceptors must be thoroughly cleaned. Merely pumping the liquid out of the trap or interceptor does not remove the grease or solids. Frequency of cleaning may vary from twice a month to once every six months, depending on the amount of grease in the wastewater and flow conditions.

Field evaluations of grease traps and grease interceptors have revealed that usually they are undesirable and ineffective for the following reasons:

1. Dischargers (such as restaurants) do not properly clean and maintain the facilities.
2. In spite of the installation and maintenance of the facilities, grease problems still develop in smaller sewers. Problems develop because the facility is too small, is improperly maintained, or the wastewater temperature was so high that the grease did not solidify until it reached the sewer.
3. Undesirable side effects are produced by grease traps and interceptors in the collection system and the wastewater treatment plant.
   a. Solids deposited on the bottom of the trap or interceptor start to decompose. This decomposition produces odors in the collection system, exerts a greater load on the wastewater treatment plant, and increases the sewer-use charges.
   b. Decomposing material from the trap or interceptor can flow into the wastewater collection system and increase the production of hydrogen sulfide (see Section 6.5, "Hydrogen Sulfide Control").
   c. When the trap or interceptor is cleaned, a large mass of undesirable material must be disposed of in a sanitary landfill.

In summary, we may be better off without grease traps and grease interceptors because without them grease gets into the sewer where collection system operators have the knowledge and equipment to handle grease problems better than anyone else. The ultimate solution to grease problems is to educate the sewer users that grease, oil, fat, and soap should not be discharged into the sewer. These materials should be separated, salvaged, and recycled, if possible, like aluminum cans and waste papers.

Effective grease and soap control with methods available to the collection system maintenance operator include the high-velocity cleaner, hydraulic balling, power rodding, and bacteria cultures.

Many chemicals have been sold in the past and are being sold today to control grease and soap in sewers. These chemicals are sold as bio-acids, digesters, enzymes, bacteria cultures, catalysts, caustics, hydroxides, and neutralizers. Many of the compounds are claimed to control odors, remove grease, eliminate roots, and accelerate activity in biological treatment processes in treatment plants. By scanning any trade journal, you can easily develop an extensive list of these products. Many have their proper use in specific locations under specified conditions.

CAREFUL EVALUATION OF THE RESULTS PRODUCED BY THESE CHEMICALS AND BACTERIA CULTURES SHOULD BE MADE IN COMPARISON WITH MECHANICAL AND HYDRAULIC CLEANING METHODS USED FOR SEWER LINE MAINTENANCE. The ideal control of many troublesome materials is at the source by educating the public that grease, oils, coffee grounds, and eggshells should be disposed of in the garbage can along with paper, tin cans, and plastic containers. This practice would eliminate some of the homeowners' plumbing problems and make the job of the collection system and treatment plant operators much easier. The only recommendation we offer regarding these types of chemicals is to develop a stringent evaluation program to test these products as outlined in Section 6.411.

6.44 Odors

Odors have long been associated with collection systems and wastewater treatment plants and are primarily caused by the production of hydrogen sulfide (Section 6.5, "Hydrogen Sulfide Control"). Some industrial discharges produce offensive odors that are not caused by hydrogen sulfide. Since the control of hydrogen sulfide is presented in Section 6.5, this section will emphasize overall odor control.

Most odors can be controlled in a properly designed, cleaned, and maintained collection system. Odor problems usually are due to low-velocity flows, long transmission lines in the collection system, high temperatures, and poorly maintained collection systems. If these problems can be corrected, odor problems would be eliminated. Odor control is usually obtained by controlling the production of hydrogen sulfide (Section 6.5). Masking agents are used to mask or overpower an undesirable odor. They do not control the production of undesirable odors. In some cases a combination of both odor control and masking agents is used to combat odor problems.

One of the most common causes of odors results when the air flow in the sewer above the flowing wastewater is stopped at a siphon or by a small sewer joining a large sewer so low that an air seal is produced. These problems can generally be cured with air lines or air jumpers, which are pipes that take air off the top of the inlet structure and return it at the end of the siphon.[14] Wet wells are seldom the site of substantial sulfide generation, even though they are often the place where odors are released from wastewater. An air line or air bypass around the lift station can be used to control odors released from a wet well if the force main is not too long.[15]

6.440 Masking Agents

Numerous masking agents are available on the market; they generally consist of a water base and an oil carrier of perfume scents to overpower or mask the objectionable odor. Instead of the sewer smelling like rotten eggs, you can now have your choice of mint, pine, banana, Chanel #32, or an assortment of other aromas to mask sewer odors.

Obviously a masking agent does not correct the cause of the odor problem. If the original problem is caused by hydrogen sulfide, you still have the other problems created by high concentrations of hydrogen sulfide that include a poisonous or toxic gas that is flammable and explosive. Also, hydrogen sulfide can be converted to sulfuric acid, which is very corrosive. Therefore you haven't accomplished anything except to hide a potentially hazardous situation. Probably every agency in the country has used masking agents for the control of odors to satisfy a complaining citizen, but they have not corrected the source of the problem.

[14] *See GRAVITY SANITARY SEWER DESIGN AND CONSTRUCTION (MOP FD-5). Obtain from Water Environment Federation (WEF), Publications Order Department, 601 Wythe Street, Alexandria, VA 22314-1994. Order No. MFD5. Price to members, $25.00; nonmembers, $35.00; plus shipping and handling.*

[15] *See PROCESS DESIGN MANUAL FOR SULFIDE CONTROL IN SANITARY SEWERAGE SYSTEMS, pages 6-8 to 6-10, U.S. Environmental Protection Agency. Obtain from National Technical Information Service (NTIS), 5285 Port Royal Road, Springfield, VA 22161. Order No. PB-260479. EPA No. 625-1-74-005. Price, $41.00, plus $5.00 shipping and handling per order.*

6.441 *Information Needed From Manufacturer*

Use of a masking agent requires the following information:

1. Recommended deodorant or masking agent.
2. Method of application—drip feed, spray, atomizers, or blowers.
3. Feed rates and amount of masking agent required to meet feed rates. Time of application, such as late afternoon and early evening, or periods of maximum odor production.
4. Cost of masking agent.
5. Effect on wastewater treatment plants with biological treatment processes.
6. Effect on waters receiving effluent from treatment plant.
7. Safety precautions for operators.
8. Any restrictions on use, especially in residential areas.

6.442 *Application of Masking Agents*

The manufacturer's procedures must be followed exactly to achieve the desired effectiveness. Excessive use of any chemical is discouraged because this practice is the same as throwing money down the sewer. Application may be simply removing the masking agent from a one- to 55-gallon container using a hose and needle valve. Allow the masking agent to drip into a manhole or lift station wet well at a rate of so many drops per minute. More elaborate equipment and procedures include the use of timers, blowers, atomizers, and specially designed nozzles.

6.443 *Evaluation*

Very careful consideration must be given to these applications because the masking agent is quite expensive. Use may be justified for short time periods until the cause of the problem can be corrected. Remember, masking agents do not eliminate or correct the source of the problem.

6.444 *Measurement of Odors*

To evaluate the effectiveness of an odor control program, gas detection equipment is available. The U.S. Bureau of Mines approves equipment capable of measuring the concentrations of both ammonia and hydrogen sulfide in parts per million.

6.45 Corrosion

Control of corrosion is discussed extensively in Section 6.5, "Hydrogen Sulfide Control." Use of chemicals to control hydrogen sulfide generation is contained in Section 6.54, "Chemical Control of Hydrogen Sulfide."

6.46 Rodents and Insects

Some collection systems are infested with rodents and insects. Rats are a serious problem in some areas because of the threat to public health and the destruction of food. Effective control of this problem requires assistance from the local health department or animal control agency for the development and implementation of a cooperative eradication program. These agencies can establish a vector (a bug, insect, fly, or rodent that can transmit infection) control program and can recommend types of poisons and when and where they should be distributed. Tight sewers with proper traps and connections and good manholes can limit rodent problems.

Insect control can be achieved by an effective maintenance program flushing slow or stagnant lines periodically during the warm summer months or whenever insects from sewers are a problem. A serious insect problem can be caused by small flies that breed and live in trickling filters and sewers. Cooperation with the local health department can expedite an effective control program.

Control of insects in manholes is best achieved by an effective maintenance program that keeps manholes clean. Use of water to wash and flush manholes can keep them clean. Spraying manholes with an insecticide for dangerous insects and washing down the manhole are good practices if insects are a problem. A white latex-base coating has been used successfully in manholes to control roaches and other insects.

QUESTIONS

Write your answers in a notebook and then compare your answers with those on page 389.

6.4L What are the causes or sources of grease problems?

6.4M How can odors be controlled from a wastewater collection system?

6.4N What are two serious shortcomings of the use of masking agents to overcome odors?

6.4O Why should insects and rodents in a wastewater collection system be controlled?

6.5 HYDROGEN SULFIDE CONTROL
by William F. Garber

6.50 Why Control Hydrogen Sulfide?

Hydrogen sulfide (H_2S) is one of the most serious problems confronting wastewater collection system operators. The hotter the weather, the flatter the sewers, and the longer the flow time to the wastewater treatment plant, the worse the problem becomes. Some of the problems created by H_2S gas include:

1. Paralysis of the respiratory center and death of collection system operators,
2. Rotten egg odors,
3. Corrosion and possible collapse of sewers, structures, and equipment,
4. Loss of capacity of the sewer, and
5. A flammable and explosive gas under certain circumstances.

6.51 How Is Hydrogen Sulfide Produced?

The major portion of the sulfide content in wastewater arises from the conversion of sulfate (SO_4^{2-}) to sulfide (S^{2-}) by bacteria in the slime in sewers. Materials containing proteins in the wastes are an equally important food for bacteria that produce hydrogen sulfide. If any sulfur-containing compound is available as a food for slimes that reduce oxygen in the wastewater, the bacteria can produce odorous, reduced sulfur compounds, including hydrogen sulfide.

The relationship between hydrogen sulfide (H_2S) and the sulfide ion (S^{2-}) in wastewater is tremendously sensitive to pH. The lower the pH, the more hydrogen sulfide is present as compared to sulfide. Therefore in terms of odor and corrosion control, low pH values in wastewater are most undesirable.

Hydrogen sulfide is not only odorous but, upon escape to the sewer walls, it can be converted to sulfuric acid by other bacteria. In some cases, solutions of sulfuric acid in excess of 40 percent have been found on the walls of the sewer. Where such heavy acid concentrations develop, washing of such areas once a day by such action as rising wastewater flows does not appear to be sufficient to prevent corrosion.

Another important effect of low pH values in a wastewater actively generating sulfide (such as occurs with slug discharges of acids) is the potential for sudden development of a toxic atmosphere. Deaths have occurred as a result of such conditions. Remember that H_2S is about equivalent in toxicity to hydrogen cyanide. Levels of hydrogen sulfide toxicity are as follows[16]:

1. 20 ppm for 8 hours, not dangerous but will produce headache and irritation of eyes and nose.
2. 50 ppm for 8 hours, severe symptoms of poisoning such as damage to eye tissues.
3. 700 ppm, rapid loss of consciousness.
4. 1,000 ppm, quickly fatal.

The ppm concentration of hydrogen sulfide in the sewer atmosphere is quite different from that in the wastewater. A concentration of 1 mg/*L* (ppm) in turbulent wastewater can quickly produce a concentration of 300 ppm in an unventilated atmospheric space. (Also see Section 4.41, "Atmospheric Hazards.")

In a sewer or waste channel flowing partially full, there will almost always be some oxygen present in the air space above the flowing wastewater. However, sulfide forms and is present in the flowing wastewater. Anaerobic reactions produce sulfide in the slimes while sulfide oxidation is occurring in the surface of the flow. The net sulfide present represents the difference between the amount produced and the amount oxidized.

a. In general, the faster and more turbulent the flow, the more oxygen is dissolved, and thus the more sulfide is oxidized to sulfate (SO_4^{2-}). However, manhole turbulence and ventilation often cause corrosion damage in areas immediately downstream from a manhole.

b. Conversely, the more sewer slimes arising from more wetted walls, from heavy sand and debris in the bottom, and from slower flows, the higher the residual sulfide concentration.

6.52 Control of Hydrogen Sulfide

Proper design of a collection system, using the current understanding of sulfide generation, can often eliminate serious odor and corrosion problems by controlling the release of sulfide. Designers must realize that the higher the wastewater temperature, the greater the production of hydrogen sulfide. Thus designs acceptable in cooler climates may be unacceptable in warmer climates.

Another factor of importance in terms of maintenance is the presence of slimes in sewers. Sulfate is split by bacteria living in the sewer slimes. The more area or habitat available for such slimes, the more sulfide produced and the greater the odor, corrosion, and maintenance problems. Cleaner sewers harbor fewer slime bacteria and greatly reduce the problems caused by sulfide.

Generally speaking, there are only two positive methods to control the production and destructive effects of hydrogen sulfide. These are:

1. Design according to the relationship outlined by Dr. Pomeroy[17] so that sulfide release will be minimized, and
2. Use construction materials that the sulfuric acid will not attack. Odor is still a problem and positive methods of odor control are required.

If design cannot be close to the conditions required by the Pomeroy relationships, careful and routine sewer cleaning must be carried out. Excess sand and grit in the sewers not only use up sewer capacity, but slow velocities and provide more slime concentrations for the generation of sulfide. Wastes with strengths (*BOD*[18]) or temperatures greater than design also cause corrosion and odors in spite of the use of this design method, as will factors such as turbulence at joints and wash-off of corrosion products. In cases where industrial wastes with high BOD and temperatures are encountered, a sewer-use ordinance should be enforced to control these discharges. However, its use should be considered in every installation since it would help minimize both corrosion and odor production.

---

[16] *Gas concentrations are given in parts per million parts (ppm) by volume.*

[17] *PROCESS DESIGN MANUAL FOR SULFIDE CONTROL IN SANITARY SEWERAGE SYSTEMS, U.S. Environmental Protection Agency. Obtain from National Technical Information Service (NTIS), 5285 Port Royal Road, Springfield, VA 22161. Order No. PB-260479. EPA No. 625-1-74-005. Price, $41.00, plus $5.00 shipping and handling per order.*

[18] *BOD (pronounce as separate letters). **B**iochemical **O**xygen **D**emand. The rate at which organisms use the oxygen in water or wastewater while stabilizing decomposable organic matter under aerobic conditions. In decomposition, organic matter serves as food for the bacteria and energy results from its oxidation. BOD measurements are used as a measure of the organic strength of wastes in water.*

The use of materials that will be unaffected by the sulfuric acid resulting from sulfide release is an obvious answer to the corrosion problem. Clay pipe, plastic pipe, the lining of reinforced concrete pipe and structures with materials such as PVC, furane, coal tar epoxy and stainless steel, epoxy-lined asbestos-cement pipe, and corrugated steel pipe with asphalt and asbestos linings are materials that have been proposed and/or are offered for this purpose. Clay pipe and reinforced concrete pipe lined with PVC sheets keyed to the concrete are the only materials with a use record long enough at this time to give reliable data for design use. The PVC lining is subject to damage from hard materials floating in the sewer. Where corrosion will occur over some predictable period, sacrificial lining of pipes with high-calcium carbonate mortars is used. The entire pipe also can be made using limestone sand and aggregate and high-calcium carbonate mortars. Comparative tests have shown that pipe made entirely with limestone aggregate will last three to five times longer than pipe made with ordinary concrete.

Provision of corrosion resistance still leaves the problem of odor control. Where design cannot provide for the natural oxidation of sulfide and corrosion-resistant materials of construction are necessary, carefully designed odor control facilities must be installed. Odor control at treatment plant wet wells may be achieved by withdrawing odorous air and passing it up through trickling filters or bubbling it up through aeration tanks in activated sludge plants. This odorous air is moist and often causes corrosion problems in the transmission ducts.

In addition, methods such as activated carbon towers and packed absorption towers have been successfully used for such odorous air. Masking has been used as a short-term solution for sudden, severe odor problems. Treatment of the exhausted odors is complicated by the fact that the gas is made up of stench materials and solvents. Removal of both major components is necessary.

Compressed air has been successfully injected into force mains by some operators to control problems caused by hydrogen sulfide. This approach is simple and effective in many, but not all, situations. Turbulent conditions are critical to dissolve oxygen in the air into the wastewater to successfully prevent sulfide buildup and to oxidize sulfide already present in the wastewater.[19] Aeration of wastewater in lift station wet wells is helpful to keep septic conditions and production of hydrogen sulfide to a minimum. The procedure is especially effective during low-flow periods. Be very careful with aeration in a wet well that you do not cause pump *CAVITATION*[20] and cause pumps to become *AIR BOUND.*[21]

Chemical treatment (Section 6.54) of the wastewater is used to control hydrogen sulfide. Chemicals are expensive but can be effective over a short time period. Zinc and iron salts *PRECIPITATE*[22] the sulfide. Lime dosage at about 8,000 mg/*L* over an hour will kill the slimes over periods of one day to about two weeks, depending upon conditions. Use of lime can produce large quantities of lime sludge, which can cause disposal problems. Other materials providing a pH greater than about 11 or 12 over a half-hour period can be used instead of lime. Chlorination with chlorine or hypochlorite is effective but expensive. Neutralization of the sulfuric acid on the sewer walls by lime dusting or ammonia release has been used. The process is expensive and results have been variable.

AS A GENERAL SUMMARY, HOWEVER, CHEMICAL TREATMENT CANNOT BE CONSIDERED TO BE A PERMANENT ANSWER FOR MOST LOCATIONS. IT PROVIDES SHORT-TERM RELIEF BUT IS GENERALLY TOO EXPENSIVE FOR CONTINUOUS USE.

The collection system operator must urge engineers to consider the available theory in designing new collection systems. Then, operators themselves must regularly strive to keep the facilities as close to the design conditions as possible by implementing a cleaning schedule and controlling the waste materials discharged to the collection system. Where this is not possible, resistant materials of construction, odor treatment, and chemical dosage in severe situations may be necessary. The maintenance operator's options are limited, but knowledge of the possible causes of trouble and of the available solutions enable the operator to make the most logical use of available forces and procedures.

Lawrence[23] forcefully pointed out that sewer maintenance problems vary with each location's wastewater temperature, strength, and age as well as design factors. His paper summarizes in considerable detail the factors of importance and which portions of the country are particularly susceptible to maintenance problems.

6.53 Collection System Maintenance and the Treatment Plant

Sewer maintenance and sewer design have a definite and important effect on wastewater treatment. This is particularly true of the south and southwest and where separate sewers

---

[19] *For details on the design of systems to inject compressed air into force mains, see PROCESS DESIGN MANUAL FOR SULFIDE CONTROL IN SANITARY SEWERAGE SYSTEMS, pages 5-1 to 5-9, U.S. Environmental Protection Agency. Obtain from National Technical Information Service (NTIS), 5285 Port Royal Road, Springfield, VA 22161. Order No. PB-260479. EPA No. 625-1-74-005. Price, $41.00, plus $5.00 shipping and handling per order.*

[20] *Cavitation (CAV-uh-TAY-shun). The formation and collapse of a gas pocket or bubble on the blade of an impeller or the gate of a valve. The collapse of this gas pocket or bubble drives water into the impeller or gate with a terrific force that can cause pitting on the impeller or gate surface. Cavitation is accompanied by loud noises that sound like someone is pounding on the impeller or gate with a hammer.*

[21] *Air Binding. The clogging of a filter, pipe or pump due to the presence of air released from water. Air entering the filter media is harmful to both the filtration and backwash processes. Air can prevent the passage of water during the filtration process and can cause the loss of filter media during the backwash process.*

[22] *Precipitate (pre-SIP-uh-TATE). (1) An insoluble, finely divided substance which is a product of a chemical reaction within a liquid. (2) The separation from solution of an insoluble substance.*

[23] *Lawrence, C.H., "Sewer Corrosion Potential," Journal Water Pollution Control Federation, Volume 37, 1067-1091, August 1965.*

with long flow lines are constructed. However, with secondary treatment, these factors should be examined by everyone. By keeping deposits and slimes to a minimum and flow velocities high enough, sewer maintenance can affect the *DEGRADATION*[24] of wastewater. Sewer design is, of course, very important (Chapter 3), but collection system operators must try to make what is available produce the best possible results. Degradation of the wastewaters in the collection system produces several undesirable effects at a wastewater treatment plant:

1. Odor and corrosion problems in plant facilities, and
2. Initial or immediate oxygen demand on biological treatment processes.

Table 6.3 is an example of degradation in the wastewater that occurred in a wastewater collection system and had a substantial effect on treatment plant operations. It also shows that such degradation can be measured by either the *OXIDATION-REDUCTION POTENTIAL (ORP)*[25] or the dissolved sulfide content.

TABLE 6.3 DISTANCE TO TREATMENT PLANT vs. DISSOLVED SULFIDE (mg/*L*) (EFFECTIVE BOD 304 mg/*L*)

Distance Upstream of Treatment Plant, Miles	ORP, E_H* in MV	Dissolved Sulfide, mg/liter
9	+258	0.2
6	+122	0.6
5	+ 70	1.1
3	+ 50	1.2
0	+ 48	1.8

* E_H is the potential required to transfer electrons from the oxidant to the reductant in millivolts, MV.

A low ORP and a high dissolved sulfide content will produce odors and corrosive conditions.

The initial oxygen demand at the start of a biological treatment process may arise in most part from the septic condition of wastewater arriving at the treatment plant. At Los Angeles the initial oxygen demand is in the order of 50 mg/*L*-hour. This means that a substantial amount of oxygen must be available at the moment the primary effluent reaches the aeration basins in order to reestablish *AEROBIC*[26] conditions and to reduce problems from the release of odors.

QUESTIONS

Write your answers in a notebook and then compare your answers with those on page 389.

6.5A What problems are created by the presence of hydrogen sulfide?

6.5B What are the sources of hydrogen sulfide in sewers?

6.5C List the factors that you consider important in the production of hydrogen sulfide.

6.5D How does collection system maintenance affect the wastewater treatment plant?

6.54 Chemical Control of Hydrogen Sulfide

In the preceding pages you have learned that hydrogen sulfide (H_2S) is toxic, can cause corrosion, and in high concentrations is flammable and explosive under certain conditions. Hydrogen sulfide is produced by sulfur-reducing anaerobic organisms that can thrive in a wastewater collection system. These sulfur-reducing organisms require the following environment to thrive:

1. The absence of dissolved oxygen (DO) in the wastewater;
2. The presence of sulfate or sulfur compounds. These compounds are present in most wastewaters or are produced by the degradation of materials containing proteins;
3. The greater the strength (higher BOD) of the wastewater, the greater the rate of sulfide generation;
4. The warmer the temperature of the wastewater, the more favorable the environment; and
5. The slower the flow velocity and the dirtier (more debris in) the sewer, the better the conditions for hydrogen sulfide generation.

Chemical control of hydrogen sulfide generation may consist of use of one of the eight chemicals listed below in order of popularity of application:

1. Chlorine (Cl_2),
2. Hydrogen peroxide (H_2O_2),
3. Pure oxygen (O_2),
4. Air,
5. Lime ($Ca(OH)_2$),
6. Sodium hydroxide (NaOH),

[24] *Degradation (deh-gruh-DAY-shun). The conversion or breakdown of a substance to simpler compounds. For example, the degradation of organic matter to carbon dioxide and water.*

[25] *Oxidation-Reduction Potential (ORP). The electrical potential required to transfer electrons from one compound or element (the oxidant) to another compound or element (the reductant); used as a qualitative measure of the state of oxidation in wastewater treatment systems. ORP is measured in millivolts, with negative values indicating a tendency to reduce compounds or elements and positive values indicating a tendency to oxidize compounds or elements.*

[26] *Aerobic (AIR-O-bick). A condition in which atmospheric or dissolved molecular oxygen is present in the aquatic (water) environment. If no dissolved oxygen is present, anaerobic conditions exist.*

7. Iron salts (ferrous sulfate ($FeSO_4$), ferrous chloride ($FeCl_2$) and ferric chloride ($FeCl_3$)), and

8. Calcium nitrate ($Ca(NO_3)_2$ or agricultural fertilizer.

6.540 Chlorine (Cl_2)

The most widely used chemical for hydrogen sulfide control has been chlorine injected into the wastewater flow at various points along the collection system. Sewer lift stations often serve as injection sites due to the availability of the site and the required utilities, such as water and power, to operate the chlorination equipment.

Chlorine doses of 10 to 20 mg/*L* have been effective in controlling the production of hydrogen sulfide in many collection systems. Doses of chlorine at these levels are toxic to most organisms in the wastewater. When these aerobic organisms are inactivated or killed, they no longer require oxygen or exert a BOD. This permits the wastewater to remain aerobic longer and delays the start of anaerobic decomposition, which produces sulfide. Chlorine also is toxic to the organisms that convert sulfate to sulfide.

Chlorination may be accomplished by the use of chlorine gas and chlorinators or by the addition of a hypochlorite solution. Hypochlorite is available as a dry powder, in tablet form, or as a liquid (usually as sodium hypochlorite).

When considering the chlorination of a wastewater collection system, beware of the following problems:

1. Chlorine is a highly toxic, oxidizing chemical. Leaving this chemical unattended in or adjacent to residential areas with the constant possibility of leaks may result in serious injury to the public and to private property.

2. Chlorine requires application equipment including evaporators (if use rates are high), chlorinators, injectors, solution diffusers in the sewer line, and possibly an air gap water system with separate injector water pumps. All of this equipment must be properly maintained.

3. For effective results, the chlorine solution must be added at a point of high turbulence where it mixes instantly with the wastewater flow.

4. Under extremely low flow conditions or if the solution diffuser is improperly designed, chlorine could be released from the diffuser and cause serious damage to the chlorination diffusion equipment as well as create a hazard to collection system and treatment plant personnel.

5. If chlorine dosage is to be properly controlled to meet collection system sulfide generation rates or daily variation of flows, then flow metering and chlorine feed control equipment will be required.

6. Chlorine costs are increasing rapidly. Large amounts of energy are required to produce chlorine. If a chlorine shortage occurs, chlorine may not be available for hydrogen sulfide control in collection systems.

7. *NEVER* apply an overdose of chlorine near a wastewater treatment plant. Chlorine is toxic to organisms and can kill aerobic organisms in secondary biological treatment processes.

CHLORINE SAFETY

Any person using chlorine should be thoroughly trained in safe procedures for storing and handling chlorine. Operators must receive instruction in the use of self-contained breathing apparatus and the use of chlorine container repair kits.

The safe handling of chlorine is discussed in *OPERATION OF WASTEWATER TREATMENT PLANTS*, Volume II, Chapter 14, in this series of manuals. For additional information on chlorine safety, read:

1. *CHLORINE MANUAL*, Sixth Edition. Obtain from The Chlorine Institute, Inc., 2001 L Street, NW, Suite 506, Washington, DC 20036. Pamphlet 1. Price to members, $15.00; nonmembers, $30.00; plus 10 percent of order total for shipping and handling.

2. *CHLORINE SAFE HANDLING BOOKLET.* Obtain from PPG Industries, Inc., Chemicals Group, One PPG Place, Pittsburgh, PA 15272. No charge.

6.541 Hydrogen Peroxide (H_2O_2)

Hydrogen peroxide was first used in the wastewater field for providing additional oxygen (O_2) to the activated sludge process in wastewater treatment plants. Hydrogen peroxide has been found beneficial in collection systems for hydrogen sulfide and odor control. Wastewater can be kept aerobic by the application of large amounts of hydrogen peroxide.

The hydrogen peroxide you buy in the drugstore for medical purposes has a strength of 3 percent, while the hydrogen peroxide used in a collection system may have a strength from 35 to 50 percent.

APPLICATION OF HYDROGEN PEROXIDE

Your hydrogen peroxide supplier should be very helpful in establishing your application program. Control of hydrogen sulfide using hydrogen peroxide relies on two major factors:

1. Conditioning Program. To inactivate and control slimes, dose with sodium hydroxide (Na(OH)) or calcium hydroxide ($Ca(OH)_2$). Lime ($Ca(OH)_2$) dosage at about 8,000 mg/*L* over an hour will kill the slimes over periods of one day to about two weeks, depending on conditions. However, disposal of lime sludge must be considered. Slimes are not affected by reasonable doses of hydrogen peroxide. Slimes should be controlled before the beginning of the application of hydrogen peroxide. Periodically the slimes may have to be controlled during the regular hydrogen peroxide application program.

2. Application Program. Effective control of hydrogen sulfide by hydrogen peroxide requires the feeding of sufficient hydrogen peroxide during the daily variation of flows to be sure sufficient amounts of hydrogen peroxide are fed to the wastewater to neutralize the hydrogen sulfide and maintain aerobic conditions. Unlike chlorine, excessive hydrogen peroxide is not harmful to the wastewater treatment plant. However, excessive dosage of any chemical is throwing money down the sewer.

Recommended doses:

Hydrogen Peroxide Concentration, %	Dose Rate, mg/*L* H_2O_2 per mg/*L* of total H_2S
35	13 to 15 mg/*L*
50	11 to 13 mg/*L*

Occasionally repeat the preconditioning dosage of sodium hydroxide or calcium hydroxide (lime) to control slime growth.

The actual method of applying hydrogen peroxide will be recommended by the supplier. Usually either a drip feed or pumping system is used and neither system is as expensive to install or as complex as a chlorination system.

Hydrogen peroxide, like chlorine, has the following limitations:

1. Must be isolated and protected from access by the public, especially children,
2. Is very corrosive and must not be contaminated by foreign material, and
3. Containers must be protected from the sun to prevent heat buildup.

HYDROGEN PEROXIDE SAFETY

1. Follow manufacturer's directions carefully.
2. Wear protective clothing, boots, goggles, and rubber gloves. Hydrogen peroxide must not come in contact with skin or eyes.
3. Avoid getting hydrogen peroxide on your clothing or shoes and especially combustible materials because drying may cause a fire.
4. In case of a spill or a fire, use *WATER ONLY* to dilute the hydrogen peroxide or put out the fire.
5. Keep flames, open lights, and sparks away from hydrogen peroxide containers and the application system.

6.542 Pure Oxygen (O_2)

With the great thrust to clean up the environment and the desire to control water pollution, many new treatment plants are using pure oxygen activated sludge systems. Pure oxygen systems are widely used by various industries and became known to the public through the space program.

If a wastewater treatment plant is producing oxygen for the activated sludge process, additional oxygen can be produced for introduction into the main intercepting sewers to control the production of hydrogen sulfide. Injection of gaseous oxygen into wastewater by using compressors and diffusers helps maintain aerobic conditions in the sewer.

Limitations of using pure oxygen include:

1. Must be kept free of petroleum products,
2. Uses specially designed compressors, dissolution diffusers, and shutdown and purge control systems to prevent ignition hazard that could touch off the explosive condition of the oxygen, and
3. Must be isolated from public access.

Pure oxygen systems will not be a common alternative to control sulfide in gravity collection systems. These systems have been used effectively to inject oxygen into wastewater in force mains. When evaluated for use in collection systems, an extensive engineering study and design will be required.

6.543 Air

Air is obviously the cheapest chemical available because it is free. There are no hazards to operators involved in handling air. Another advantage of applying air at lift station wet wells is that the resulting turbulence keeps deposits of solids from accumulating during low flows.

A limitation of applying air is the noise from air compressors. The most serious problem using air is the difficulty of transferring sufficient oxygen from the air to the wastewater to produce and maintain aerobic conditions.

6.544 Lime ($Ca(OH)_2$)

Lime can be used to control hydrogen sulfide. Addition of lime can increase the pH of the wastewater to a level where hydrogen sulfide will not exist. A disadvantage of this approach is the problem of disposing of lime sludges.

6.545 Sodium Hydroxide (NaOH)
by Blake Anderson

Sodium hydroxide (caustic soda) has been used extensively by the Los Angeles County Sanitation Districts (LACSD) for many years to control sulfide production in the wastewater collection system for flows under 5 CFS. Periodic treatments of a sewer in which sulfide is produced will usually maintain control. Treatment consists of feeding enough sodium hydroxide into a sewer to raise the pH of the wastewater to 12.5 or greater for a half-hour period. The high pH inactivates the slime layer in the sewer and sulfide production stops. In the average sewer with a sulfide problem, this treatment is necessary every two weeks during the summer. In a few cases more frequent treatment is necessary.

When testing indicates that a sewer is beginning to produce sulfide at an unacceptable level, then treatment is scheduled. "Unacceptable level" is very subjective and difficult to define. In some sewers after treatment the sulfide concentration falls way off and then suddenly jumps back to pretreatment levels after a couple of weeks. This level is clearly unacceptable and treatment should be scheduled. In other sewers the recovery of sulfide production to pretreatment levels may begin slowly at first after a week or so. These slightly increasing levels can indicate that higher production will shortly return and treatment should be scheduled before problems develop. Only experience will tell you what to expect from a particular sewer.

The sodium hydroxide is hauled to the treatment site in a 2,500-gallon vacuum truck[27] purchased by the LACSD for this particular purpose. Parked over an open manhole, the truck feeds the sodium hydroxide through a *ROTAMETER*[28] and into the manhole for 30 minutes. One hundred to 125 gallons of 25 percent sodium hydroxide are required for each cubic foot per second of wastewater flow. The high-pH wastewater travels down the sewer and will inactivate the slime layer as long as the pH remains high. Dilution from subsequent sidestreams and neutralization by acidic substances will eventually bring the pH to near neutral levels.

---

[27] *At one time a 500-gallon tank placed in an old dump truck was used to transport the sodium hydroxide. This equipment is fairly inexpensive and is ideally suited for the needs of a small collection system.*

[28] *Rotameter (RODE-uh-ME-ter). A device used to measure the flow rate of gases and liquids. The gas or liquid being measured flows vertically up a tapered, calibrated tube. Inside the tube is a small ball or bullet-shaped float (it may rotate) that rises or falls depending on the flow rate. The flow rate may be read on a scale behind or on the tube by looking at the middle of the ball or at the widest part or top of the float.*

Sodium hydroxide is purchased by the LACSD in large bulk quantities at 50 percent concentration. The sodium hydroxide is diluted to 25 percent concentration in a 15,000-gallon storage tank and held there for future use. Store the sodium hydroxide in the dilute form to reduce its *VISCOSITY*[29] and to keep it from crystallizing at lower temperatures.

Sodium hydroxide also is obtained free by the LACSD as "spent caustic" from industries. This sodium hydroxide is usually at a 3 to 6 percent concentration and is usually combined with other materials. The spent caustic is tested occasionally to determine if it contains the materials of the particular industry in high enough concentrations to be harmful to operators, the environment, or to the wastewater collection system.

The major disadvantage of using spent caustic is the large quantity required, usually 2,000 gallons per treatment. Hauling back and forth between the industry and various treatment sites takes a great deal of driver and truck time and usually only one treatment per load is possible. Purchased sodium hydroxide is much more concentrated and several treatments per truck load can be made.

A major consideration when using sodium hydroxide is the side effect that high pH can have on wastewater treatment plants using biological processes. If the sodium hydroxide enters a downstream treatment plant at a high enough pH, an upset of a biological process can occur. If a pH of near 9 or more is anticipated, then *BENCH SCALE TESTING*[30] should be done to determine the effects of a high pH slug entering the treatment plant. Consult your staff engineer or a consulting engineer for advice in this area.

The LACSD has used several methods to control the effects of high pH on its activated sludge secondary treatment plants. First, there is simple dilution. At some treatment plants the flow is so great that the high-pH wastewater coming from the treated sewer line is diluted to near neutral pH levels. Second, there is diversion around the plant. At some treatment plants that have been built to relieve sewer capacity downstream, it is possible to pull stop logs and divert the high-pH wastewater around the plant and on to a downstream plant for a short period. Third, there is interception. At one treatment plant the early morning flows are very low. The sodium hydroxide is fed at the treatment site. The 2,500-gallon treatment vacuum truck along with an additional 4,000-gallon vacuum truck intercept the high-pH wastewater three hours later, just before it enters the treatment plant. This is accomplished by dropping a stop log in the line to dam the sewer and then pumping the high-pH wastewater out of the sewer and into the trucks when it arrives.

Safety is another important consideration when using sodium hydroxide. This is a caustic material that will cause burns. Handling it should be done wearing a face shield and gloves. When treating a sewer line, it is important to coordinate with all your maintenance crews and with construction contractors to avoid harming them.

The chief advantage of sodium hydroxide is its lower price compared to other chemicals that require continuous application. Even with the cost of operators and equipment included, the LACSDs have found it much cheaper than the other principal chemical methods. The chief disadvantage is the amount of attention that has to be given to hauling and applying sodium hydroxide. Some of the other chemical methods probably require less attention and time. The economic benefits of using sodium hydroxide compared to other chemicals should be determined for each particular collection system. Labor, materials, and equipment should all be included in such an analysis.

LACSD is now using magnesium hydroxide instead of sodium hydroxide to control corrosion in sewers. Magnesium hydroxide is less hazardous to handle.

6.546 Iron Salts

Many metal salts will react with dissolved sulfide to form metal sulfide precipitates, thus preventing the release of hydrogen sulfide gas to the atmosphere. For environmental and economic reasons, iron salts in the form of ferrous sulfate ($FeSO_4$), ferrous chloride ($FeCl_2$), or ferric chloride ($FeCl_3$) are the iron solutions most often used to control hydrogen sulfide in wastewater collection systems. This method of treatment is effective because the metal sulfide precipitates formed are highly insoluble, which means they won't dissolve again in the flowing wastewater.

The dosage required of any one or a combination of iron salts is heavily dependent upon the controlled sulfide level desired. The actual dosage rate needed to maintain the desired level must be determined by field testing of a particular system. Complete control is difficult to achieve, but feed rates that produce iron concentrations of 2.5 to 20 mg/*L* in the wastewater in the sewer commonly achieve 90 percent reduction of sulfide. Application of a single form of iron salt or a combination of solutions may require concentrations or dose rates of greater than 4.5 total iron to 1.0 total dissolved sulfide.

The storage and handling of these chemical solutions require special equipment and the careful following of appropriate safety precautions. All three forms of iron chemicals are corrosive acid solutions. A face shield, rubber gloves, an apron, and boots should be worn when transferring or working with ferrous or ferric solution feeding equipment. If you are splashed with these solutions, immediately flush the affected areas with water. An emergency eyewash and deluge shower should be installed at all storage sites. When these chemicals are spilled, they will leave a dark brown stain after being flushed with water.

Ferrous solution storage tanks are usually constructed of fiberglass, concrete, or steel and must have a rubber lining. PVC or rubber-lined piping is used to transfer or convey solutions to the point of application. Pumps must also be plastic or rubber lined to prevent corrosion damage. The transfer of solutions from tankers to storage facilities is usually accomplished by *AIR PADDING*[31] of the tanker.

Solution feed or metering equipment consists of small diaphragm or piston pumps. The length of the pump stroke may be adjusted to discharge a known volume of solution on each stroke. Use of a variable-speed motor on the pump enables the operator to increase or decrease the number of pump strokes per minute to achieve the correct dosage rate.

---

[29] *Viscosity (vis-KOSS-uh-tee). A property of water, or any other fluid, which resists efforts to change its shape or flow. Syrup is more viscous (has a higher viscosity) than water. The viscosity of water increases significantly as temperatures decrease. Motor oil is rated by how thick (viscous) it is; 20 weight oil is considered relatively thin while 50 weight oil is relatively thick or viscous.*

[30] *Bench Scale Tests. A method of studying different ways or chemical doses for treating water on a small scale in a laboratory.*

[31] *Air Padding. Pumping dry air (dew point –40°F) into a container to assist with the withdrawal of a liquid or to force a liquified gas such as chlorine out of a container.*

The metered dosage of solution flows in PVC pipe to the point of application and is discharged directly into the wastewater in the sewer. The normal turbulence of the wastewater flow provides adequate mixing of the iron solution with the wastewater.

For additional reading on the use of iron salts to control hydrogen sulfide, see *SULFIDE IN WASTEWATER COLLECTION AND TREATMENT SYSTEMS*. Obtain from American Society of Civil Engineers (ASCE), Book Orders, PO Box 79404, Baltimore, MD 21279-0404. Order No. 681. ISBN 0-87262-681-4. Price, $53.00.

6.547 Calcium Nitrate ($Ca(NO_3)_2$)

Calcium nitrate, an agricultural fertilizer, has been used successfully in sewers to prevent the formation of hydrogen sulfide. The presence of nitrate in wastewater represents a supplemental oxygen source for bacteria. Therefore, it will not be consumed while aerobic conditions exist. However, when the wastewater flowing in a sewer becomes anaerobic or anoxic (the condition when nitrate is present), the bacteria use the oxygen in the nitrate as a source of oxygen rather than using the oxygen in the sulfate as a source of oxygen. Therefore, if sufficient nitrate is present, hydrogen sulfide will not be produced.

6.548 Other Chemicals

Potassium permanganate has been tried and found not to be cost effective in most areas. Ozone generators have been used on a small scale and also found not to be cost effective. Power requirements to produce ozone from the atmosphere or even with pure oxygen available to supply the generator have been very costly to date.

6.55 Summary

Many factors are important for effective control of hydrogen sulfide. You must become aware of the fact that your job involves more than the obvious cleaning of sewers by removing sand and obstructions. You have an active biological system in the sewers. Control of this system will not only affect you directly in regulating corrosion and odor problems, but also will have a direct bearing on the downstream wastewater treatment plant. You may aid treatment plant digestion by accomplishing some of the breakdown in the sewers, but you also may substantially increase the immediate oxygen demand on biological treatment processes. Collection system operators need to reach beyond the complexities and difficulties of their mechanical problems and recognize the need to consider the biological ecosystem in the wastewater collection system too.

The best method of hydrogen sulfide control is proper collection system design and maintenance. If chemical control is required, chlorine and hydrogen peroxide appear the most feasible. Aeration may be helpful when properly applied at selected locations. Selection of a chemical should be based on an analysis of the extent of the problem and a comparison of the costs and effectiveness of the alternatives.

Also see Section 8.33, "Troubleshooting Force Main Odors," in Volume II.

QUESTIONS

Write your answers in a notebook and then compare your answers with those on page 389.

6.5E What are three potential methods of chemical control of sulfide?

6.5F Why is chlorine effective in controlling sulfide?

6.5G How does hydrogen peroxide control sulfide generation?

6.5H Applying pure oxygen may be a feasible way to control sulfide under what conditions?

6.5I How do metal salts applied to wastewater in a collection system control hydrogen sulfide gas?

6.5J Why does the storage and handling of metal salt chemical solutions require special equipment and safety precautions?

Please answer the discussion and review questions next.

DISCUSSION AND REVIEW QUESTIONS

Chapter 6. PIPELINE CLEANING AND MAINTENANCE METHODS

(Lesson 5 of 5 Lessons)

Write the answers to these questions in your notebook. The question numbering continues from Lesson 4.

25. Why is the method of chemical application important and why must it be done correctly?
26. How would you decide whether or not to select an untested chemical? Describe a testing and evaluation procedure.
27. Why must insects and rodents be controlled in a wastewater collection system?
28. What factors contribute to the production of hydrogen sulfide in a collection system?
29. How would you attempt to control odors from a wastewater collection system?

30. What safety precautions must be exercised when using chlorine to control sulfide?
31. Explain the disadvantages of using chemical methods to control sulfide.
32. Why is the proper operation and maintenance of a wastewater collection system important to the operator of a wastewater treatment plant?

SUGGESTED ANSWERS

Chapter 6. PIPELINE CLEANING AND MAINTENANCE METHODS

ANSWERS TO QUESTIONS IN LESSON 1

Answers to questions on page 289.

6.0A To operate and maintain a wastewater collection system properly, try to achieve the following objectives:

1. Minimize the number of stoppages per mile of sewer,
2. Minimize the number of odor complaints, and
3. Minimize the number of lift station failures.

6.0B When a stoppage or blockage occurs in a sewer, wastewater will back up and overflow onto the street or into homes and businesses.

6.0C Major causes of stoppages include obstructions such as roots, grease, debris, broken pipe, or joint failures. Vandals, construction work, the forces of nature, and intersecting flows also can cause stoppages.

6.0D To identify the cause of a stoppage, try to identify the source of the problem by determining if:

1. The line has a history of previous stoppages (a root or grease problem line),
2. Trees are growing near the line,
3. New building sewers or new lateral or branch sewers have been installed recently,
4. Repairs have been made recently to the sewer or to the street, such as street resurfacing, or
5. Changes in the ground surface indicate a problem has developed.

Answers to questions on page 296.

6.0E When a grease or root stoppage causes flooding, high-velocity cleaning is the first method suggested to attempt to clear the stoppage.

6.0F High-velocity cleaners are an effective tool for removing grease buildups in sewers up to 15 inches in diameter. A parachute or bag may be a better method for larger diameter sewers.

6.0G A stoppage caused by roots can be cleared by either a power rodder or a hand rodder. A high-velocity cleaner also is effective with a special head and proper operation.

6.0H Advantages:

1. Very effective in cleaning flat, slow-flowing sewers,
2. Efficient in removing grease, sand, gravel, grit, and debris in smaller diameter sewers,
3. Effective in breaking up solids in manholes and washing structures, and
4. May be used to eliminate emergency stoppages in sewers.

Limitations:

1. Efficiency and effectiveness in removing debris from larger diameter sewers decreases as the cross-sectional area of the pipe increases,
2. When used by an inexperienced operator, may cause a backup into residences, and
3. Extra time and water and an experienced operator may be necessary to remove hardened grease, hardened debris, and roots.

ANSWERS TO QUESTIONS IN LESSON 2

Answers to questions on page 297.

6.1A Sewer line hydraulic cleaning methods include:

1. Balling,
2. High-velocity cleaner,
3. Flushing,
4. Sewer scooter,
5. Kites,
6. Tires, and
7. Poly pigs.

6.1B When hydraulic cleaning methods are being used, care must be taken to be sure basements and homes are not flooded.

Answers to questions on page 303.

6.10A Balling can remove deposits of inorganics (grit) and grease from sewers.

6.10B Balling is commonly used in preventive maintenance programs because it is very effective in reducing stoppages and economical in comparison with other cleaning methods.

6.10C The frequency of balling a sewer in a preventive maintenance program may vary from six months in some sluggish lines to three years in other lines. To determine the frequency of balling, study your district records and base the frequency of cleaning on:

1. Number of stoppages and complaints,
2. Size of area the collection system serves, and
3. Type of wastes carried by the collection system.

6.10D Three operators are needed for a balling crew:

1. Maintenance Equipment Operator,
2. Maintenance Operator I, and
3. Maintenance Operator I.

6.10E Additional water can be obtained from fire hydrants or a water truck to provide the necessary head for a balling operation when flows in the sewer are low. Mechanical or inflatable plugs are sometimes installed in the upstream sewer as a method of getting an adequate supply of water to begin the balling procedure.

6.10F The manhole jack or roller is essential to guide the tag line into the sewer without causing unnecessary wear and to give proper rolling action as the tag line is pulled back and forth or in restraining the ball.

6.10G The purpose of an "air gap" device is to prevent the backflow of wastewater into a drinking water supply.

Answers to questions on page 309.

6.10H A sewer ball must be properly inflated to cause good cleaning action, but not so tight the water cannot flow around the ball and carry it downstream. If the ball is too loose, it will only float downstream with little cleaning action.

6.10I Material cleaned from the sewer is caught in the downstream manhole by using a shovel with the spade shaped like the bottom of the sewer. The material is shoveled into a bucket and then removed from the manhole.

6.10J The ball should move downstream slowly at a steady rate.

6.10K A ball could stop moving downstream if it encounters:

1. A large deposit,
2. A buildup of material in the line below the ball,
3. A bad tap, or
4. Crushed or offset pipe.

6.10L To determine what caused the ball to stop moving downstream:

1. Try to "bounce" the ball back and forth in the line to remove large or stubborn deposits, or
2. Consult TV log of line. If line has not been televised or not recently, request that the line be televised.

Answers to questions on page 309.

6.10M To remove a stoppage below a ball, use a power rodding machine, hand rod, or a high-velocity cleaner at the downstream manhole.

6.10N The following precautions and safety measures should be taken during a balling operation:

1. Avoid flooding basements, homes, or businesses,
2. Safely route traffic around the job site, and
3. Practice manhole safety from start to finish.

Answers to questions on page 315.

6.11A High-velocity cleaning machines are used to:

1. Open stoppages,
2. Remove grease,
3. Clean lines of debris, and
4. Wash manholes and wet wells.

High-velocity cleaners are frequently used for a variety of tasks that are not specifically sewer operations, for example, street washing, backup fire control, chemical spraying, and landscape watering.

6.11B Nozzles attached to the end of the hose create high water velocities that clean the sewers.

6.11C One person can operate the high-velocity cleaning machine. The crew consists of two or more persons, depending on field conditions.

Answers to questions on page 316.

6.11D Traffic can be warned of high-velocity cleaning units set up in a street by use of:

1. Rotating beacon on top of cab of truck,
2. Flag stands with "Crew Working" signs,
3. Safety cones, and
4. Flaggers.

6.11E The water tank should be filled close to the working area because it is difficult and dangerous to drive a truck with a tank full of water.

6.11F To clean the sewers in a given area or subdivision, start at the top or highest point in the collection system.

6.11G Always direct the nozzle upstream in the sewer.

Answers to questions on page 317.

6.11H Send the nozzle up the line and bring it back to the downstream manhole enough times to reach the upstream manhole and return without bringing back debris.

6.11I When loosened materials build up and overflow the debris trap, pull the hose and nozzle out of the manhole. Retest the manhole atmosphere. Allow an operator to enter the manhole and shovel debris into a bucket for removal from the manhole. Debris also can be removed by the use of a vacuum device if one is available.

6.11J The nozzle must be kept moving at all times or it could become stuck in the line and cause flooding in homes.

6.11K If flow stops in the working manhole, pull the hose and nozzle out at once and start over again to remove the built-up material that is causing the stoppage.

6.11L High-velocity cleaning machines clear stoppages by using the forward jet to provide cutting action ahead of the nozzle.

Answers to questions on page 319.

6.11M When cleaning from a manhole to a cleanout, be careful not to knock the cleanout cover out of place or get the nozzle stuck.

6.11N When the nozzle becomes stuck, try:

1. Pulling the nozzle back with the water pressure off, and
2. Moving nozzle to the upstream manhole, removing nozzle, and pulling the hose back.

6.11O During cold weather, be sure to keep equipment from freezing. The high-pressure pump must be properly drained, the equipment can be heated with a heater, or water can be circulated through the pump and hose.

6.11P "Blown toilets" are the result of a combination of pressure, nozzle angle, and system conditions. A nozzle moving up a pipe will pull air from upstream. If the pipe is clogged and not enough air can be pulled easily through the pipe, the air will be pulled from house vents, which in turn can pull down water in all of the plumbing P traps and the toilets. When the nozzle returns downstream, the air will be pushed upstream toward the stoppage and may exit through these same vents with sometimes dramatic results of pushing water out ahead of the air.

Answers to questions on page 321.

6.12A Flushing is commonly used at the beginning of the collection system where low or sluggish flows permit the deposition of solids.

6.12B Lower than expected flows at a downstream manhole during a flushing operation indicate that there is a restriction in the upstream segment of the line and that additional cleaning is required.

6.12C Flushing helps control rodents and insects in sewers by sending a sudden surge of water down the line that can wash them out or drown them.

6.12D A flushing crew can consist of two operators *PROVIDED* neither of them enters a manhole. If a manhole must be entered for any reason, such as to remove debris, a third operator must be available so two operators remain topside to retrieve the operator in the manhole if an accident occurs. Confined space entry procedures must be followed whenever anyone must enter a manhole.

Answers to questions on page 322.

6.12E The flushing operation starts at the extreme upstream end of the collection system.

6.12F Manholes are cleaned with a pressure hose by washing the manhole barrels and shelf, usually prior to flushing the line.

6.12G Factors that influence the length of line that can be flushed in one day include:

1. The number of manholes that have to be entered and cleaned,
2. Location of manholes—street or backyard easement, and
3. Whether water for flushing is provided by the tank truck or a fire hydrant and its location with respect to the location of the manhole.

6.12H Safety precautions that must be taken during a flushing operation include:

1. Traffic considerations,
2. Atmosphere in manholes (toxic and explosive gases and oxygen level), and
3. Avoid flooding of residences and businesses.

Answers to questions on page 322.

6.13A The cleaning action of a sewer scooter results from the high velocity of water flowing around the shield of the scooter.

6.13B Scooters are capable of removing large objects and such other material as brick, sand, gravel, and rocks. They are considered more effective in larger lines (over 18 inches) and have been used to remove grease in these lines.

6.13C When using a scooter, care must be taken where sewers are shallow to avoid flooding of homes or businesses because of the buildup of water in the sewer.

Answers to questions on page 324.

6.13D Three operators are needed for the safe and efficient operation of a scooter crew. When cleaning lines 15 inches in diameter or larger or in heavy traffic, additional crew members may be needed.

6.13E The debris trap installed in the downstream manhole is designed to allow water to pass and hold back the debris.

6.13F Before setting up equipment in the street, safety measures need to be taken to warn and control traffic and to ensure safe working conditions in manholes.

6.13G The purpose of the manhole jack or control cable guide is to allow the control cable to take the strain without rubbing on the sharp edge of the pipe during the cleaning operation.

6.13H During setup, the level of water in the manhole and sewer above the scooter can be controlled by pulling firmly on the control line to fold the upper portion of the head shield and allow the excess water to flow down the sewer.

Answers to questions on page 326.

6.13I To determine if the folding mechanism is working properly when the scooter starts downstream, give the control line a short, smooth pull back two or three feet and hold. If the buildup of water in the manhole starts to go down, you will know the folding mechanism is working properly.

6.13J The recommended depth of water in a manhole during a scooter cleaning operation is three times the diameter of the pipe. The deeper the water the more effective the cleaning action, but care must be exercised to avoid flooding homes and businesses.

6.13K If the flow in the collection system is not sufficient for effective cleaning, add more water to the upstream manhole by using water from the tank truck or from a fire hydrant.

6.13L If the scooter slows or stops during the cleaning operation, it is probably due to an accumulation of material ahead of it.

6.13M When material builds up ahead of a scooter, give the control line a short, smooth pull back four or five feet. This will allow the upper half of the head shield to fold back and a surge of water to flush the accumulated material on downstream.

6.13N Someone must watch the flow of material into the lower manhole to prevent a buildup of material that could flow over the debris trap.

6.13O Before debris can be removed from a manhole,

1. Shut off any extra water being put into the upper manhole,
2. Hold up the progress of the scooter,
3. Follow confined space procedures for manhole entry, and
4. Operator entering manhole must have proper safety equipment.

6.13P Cables should be examined at the start and finish of each section to avoid having a cable break when the scooter is in the middle of a sewer.

Answers to questions on page 329.

6.14A Instead of using a sewer ball, larger diameter sewers are cleaned using kites, bags, tires, or a scooter.

6.14B Poly pigs are frequently used to clean force mains.

ANSWERS TO QUESTIONS IN LESSON 3

Answers to questions on page 330.

6.20A Sewer line mechanical clearing techniques include:

1. Bucket machines,
2. Power rodders, and
3. Hand rods.

6.20B Hydraulic cleaning techniques should be given first consideration for removing solids deposited in sewers because of simplicity and costs.

6.20C Hydraulic cleaning techniques may not be the best solution for removing large amounts of debris, especially in larger size pipes. Nozzle action from a high-velocity cleaner can develop a negative pressure that could hasten the collapse of a bad section of pipe. Hydraulic cleaning methods also can put more material into suspension than can be removed downstream.

6.20D Power bucket machines should be used mainly to remove debris from a break or other debris that cannot be removed by hydraulic cleaning techniques. They should not be used as a routine cleaning tool on a regular basis.

Answers to questions on page 337.

6.20E The working machine pulls the bucket out of the manhole.

6.20F The porcupine tool is quite effective for final cleanup since the bristles produce a scrubbing action.

6.20G A power bucket crew should consist of at least three operators and four operators are recommended.

Answers to questions on page 338.

6.20H An overnight line might be left in a sewer line overnight if the job is not completed before the end of the day. The overnight line is used to speed start-up the next day and avoid the time and effort to thread the sewer again.

6.20I A foolproof communication system must be devised and used because someone can easily be injured if the proper signal is not sent, received, or is misunderstood.

6.20J The purpose of the swivel is to keep the rope and cable from twisting.

Answers to questions on page 340.

6.20K Usually the size of clearing bucket is determined by the size of the sewer being cleared. Experience and history of the sewer may dictate the use of a bucket smaller than the actual sewer size. Also, the size of the manhole opening limits the maximum size of the clearing bucket.

6.20L As the bucket is pulled upstream into the material deposited in the sewer, a definite resistance can be noticed when the bucket is full. Always make a few short pulls to check out not only the amount of material, but also the equipment and the setup.

6.20M When permitted by the governing regulatory agency, the debris in the bucket is first dumped on the street and then loaded and hauled away or dumped into a loading chute that moves the debris directly into a dump truck.

6.20N A sand trap is inserted in the downstream manhole to catch debris loosened by the bucket but not retained in the bucket.

6.20O The line may be cleaned with a porcupine cleaning tool after a bucket has been pulled through the line.

6.20P The porcupine should be pulled downstream because it can seal off or plug the line and cause the wastewater to back up in the line.

6.20Q Usually the bucket should be pulled upstream to minimize stoppages that could develop when large amounts of debris or roots are encountered. However, depending on conditions at the site and the experience of the operator, it may sometimes be advisable to pull the bucket downstream.

Answers to questions on page 341.

6.20R The location of the bucket in the sewer line must be known at all times so if it ever becomes stuck, it can be dug up.

6.20S The condition of the equipment should be inspected regularly to avoid breakdowns. If a cable should come loose from the bucket, it could be very difficult to retrieve the bucket.

6.20T When working around cable-operated equipment, keep your hands clear of the drum. Use a manhole hook to guide the cable. In an instant, your hand or fingers can become entangled and the operator cannot stop the machine fast enough to avoid injury.

6.20U Locations of problem areas must be recorded so future cleaning crews will know what to expect and hopefully avoid serious problems. Also, it may become necessary to repair a sewer and this too should be noted for future reference.

Answers to questions on page 344.

6.21A Power rodders can be used for:

1. Routine preventive maintenance,
2. Scheduled clearing of grease deposits, roots, and debris accumulations in flat lines,
3. Threading cable for other cleaning equipment, and
4. Emergency use for stoppages.

6.21B Sectional power rodders are capable of clearing lines with diameters up to 12 inches and distances up to approximately 750 feet.

6.21C Advantages of sectional rod:

1. When a rod is broken, this section can be replaced,
2. Repair can be done quickly in the field, and

3. The broken section can be replaced with a new section at once or the rod can be recoupled, adding new rod only when more footage is needed.

6.21D Advantages of continuous rod:

1. Greater distances can be obtained, and
2. Loading the machine with new rod is fast and simple.

Answers to questions on page 349.

6.21E Two operators are needed to properly operate the power rodder and three are recommended.

6.21F When a two-operator crew has to enter a manhole, they must obtain assistance before anyone enters the manhole. Two operators must be topside at all times when someone enters a manhole and all confined space entry procedures must be followed.

6.21G Power rodders are more effective in smaller lines.

Answers to questions on page 349.

6.21H A sewer may be rodded either upstream or downstream from a manhole.

6.21I The purpose of the leader tool installed on the end of the rod is to prevent the sharp end of the rod from damaging the interior of the guide hose.

6.21J A debris trap is necessary to keep loosened roots, grease chunks, or other material from going farther downstream and possibly causing a stoppage.

Answers to questions on page 352.

6.21K To determine which clearing tool to use,

1. Study the history of the line being cleared,
2. Look for grease in the manhole, and
3. Look for trees growing close to the sewer.

6.21L Use a square bar corkscrew to try to clear the stoppage from the dry manhole downstream of the stoppage.

6.21M After a stoppage has been cleared, examine the material at the downstream manhole and try to determine if the stoppage was caused by grease, grit, or roots. When the probable cause of the problem has been identified, then select the best tool to correct the problem.

6.21N When changing to another clearing tool, *NEVER* take hold of the clearing tool until it has stopped rotating and the power rodder is shut down. Also be careful the rod is not coiled in the sewer and won't unwind from this torque when the tool is moved about.

6.21O The location of the clearing tool in the sewer must be known at all times so if the tool ever gets stuck, you will know exactly where to dig to open the sewer and recover the tool.

6.21P The speed of the rod in or out and the rotation speed depend primarily on existing conditions in the sewer and also on the experience of the operator.

6.21Q Water is sometimes added to the opposite manhole during power rodding if existing flows are insufficient in order to keep the rod cool and to increase the effectiveness of the clearing operation.

Answers to questions on page 352.

6.21R Use a pick-up tool to recover a broken rod.

6.21S Broken rods must be recovered. If they can't be recovered by a pick-up tool or flushed out with a high-velocity cleaner, the line must be dug up to recover the tool.

Answers to questions on page 354.

6.21T Use extreme care when handling rods and clearing tools because they may be under tension and can be very dangerous if they get away from you.

6.21U Occasionally the rod should be run out on the top of the ground and inspected for bent pieces and loose couplings. When couplings are allowed to stay loose, the rod will wear and become weak at that point.

6.21V The rod should be cleaned when it is returned to the reel to prevent the buildup of grease and grit, which will cause wear on the rod.

6.21W Everyone on a rodding crew must know their own job and everyone else's job so everyone can be familiar with the overall operation.

6.21X The engine of the power rodder should receive the maintenance and care recommended by the shop manager or the manufacturer's directions.

Answers to questions on page 357.

6.22A Hand rods are used to clear many stoppages, especially during the night or on weekends.

6.22B A hand rodding crew usually consists of two operators. They never enter manholes. When a manhole must be entered, a third operator must be available and all confined space entry procedures must be followed.

6.22C A power rod has a greater penetration into a sewer than a hand rod.

6.22D When a bare rod is exposed on top of the ground or street, care must be taken that the rod does not injure someone or damage equipment and landscaping.

Answers to questions on page 358.

6.22E The kind of tool used at the end of a hand rod depends on the sewer's condition and the type or source of the problem we are trying to correct.

6.22F If hand turning a rod does not clear the stoppage, put the gasoline-driven power drive unit on the rod and use this unit. If this procedure doesn't work, try the power rodder.

6.22G When working a rod upstream to clear a stoppage, be careful that the surge of wastewater that is released when the stoppage is broken does not:

1. Shoot up the rod guide tube and get someone wet, and
2. Force the rod back out the guide tube.

ANSWERS TO QUESTIONS IN LESSON 4

Answers to questions on page 359.

6.3A Cleaning equipment must be properly maintained at all times so we can do our jobs quickly and effectively.

6.3B An operator can learn how to maintain cleaning equipment from the manufacturer's directions and from experience.

6.3C The objectives of cleaning equipment maintenance are to keep equipment in good repair to prevent equipment failure on the job, to prolong the life of the equipment, and to increase the efficiency and safety of maintenance operations.

6.3D The three essential factors for an effective equipment *PREVENTIVE* maintenance program are *TOOLS*, *TIME*, and *TALENT* (knowledge).

6.3E The crew leader or equipment operator is responsible for the proper maintenance of cleaning equipment.

Answers to questions on page 360.

6.3F Bucket machine crews should have enough tools for operators at both the upstream and downstream manholes because often the crews are a considerable distance apart, especially when working in easements.

6.3G The number one maintenance problem with the cable scroll and level wind system is lack of proper lubrication.

6.3H Cable clamps should be tightened at the start of each day.

6.3I When a bend is noted in a stabilizer leg, get it straightened now.

Answers to questions on page 361.

6.3J If the tension on the drive head chain is loose, the drive dogs tend to ride up and over rod couplings.

6.3K The torque on the drive rollers of a continuous rodder depends on the rod wear.

6.3L Maintenance of hydraulic units includes:

1. Looking for and repairing leaky fittings and hoses,
2. Keeping oil filter clean,
3. Maintaining proper hydraulic oil level, and
4. Replacing cracked, chafed, or worn hoses.

Answers to questions on page 363.

6.3M The operator of a high-velocity cleaner unit must be familiar with all aspects of the machine so the operator can properly operate and maintain the unit.

6.3N When operating a high-velocity cleaner, always be alert for:

1. Odd or suspicious sounds,
2. Higher or lower than usual RPM to maintain pressure, and
3. Sluggish reel return.

6.3O Each day check the oil levels of:

1. Engine,
2. Pressure pump, and
3. Hydraulic oil tank.

6.3P Cold weather maintenance includes:

1. Draining pump daily, and
2. Covering rear canopy openings.

ANSWERS TO QUESTIONS IN LESSON 5

Answers to questions on page 366.

6.4A Chemicals can be used to control roots, grease, odors, corrosion, rodents, and insects.

6.4B Some people are reluctant to use chemicals to solve sewer problems because of past unfortunate experiences.

6.4C Doing nothing might be better than trying an untested chemical because even if the chemical didn't work, you might have to pay chemical costs and equipment and labor costs for application.

6.4D Metal chips and filings should not be put into a sewer because they will probably settle out in the sewer without reacting. If they do react, they may produce hydrogen gas, which could produce an explosive mixture.

6.4E Before developing a testing program, calculate the cost per foot of sewer tested for buying and applying the chemical and check with appropriate testing agencies to determine if the chemical will be harmful to the wastewater treatment plant or a watercourse.

Answers to questions on page 373.

6.4F When roots are cut in a sewer, the roots will send out new growth just back from the cutting. Also, the root will increase in diameter and break the pipe or open up the joint even more.

6.4G Before starting a chemical program for root control, obtain the following information from the manufacturer:

1. Recommended herbicide and EPA approval number,
2. Method of application,
3. Concentration and amount of chemical needed,
4. Contact time,
5. Preferred meteorological conditions,
6. Total cost for chemicals and application,
7. Hazards to downstream operators,
8. Effects on downstream wastewater treatment plant,
9. Effects on receiving waters of plant effluent, and
10. Material Safety Data Sheet (MSDS).

6.4H Chemicals can be applied for root control by flooding and foaming techniques.

6.4I Homes and businesses can become flooded during the chemical contact period of a chemical operation because the section being chemically treated is plugged off from the rest of the system. Wastewater will build up behind the plugged upstream manhole and could cause flooding.

6.4J Chemicals used for root control can be recovered and recycled by carefully moving the chemical operation down the sewer. After one contact period is over, plug the lower end of the next downstream section to be chemically treated. Release the chemicals from the upstream section and they will be "caught" in the downstream section. Also the chemicals can be pumped from one section into a tank and transported to the next section to be chemically treated.

6.4K When chemically treating roots, always read the MSDS and the label on the chemical container, wear protective clothing, and know what to do if an accident occurs.

Answers to questions on page 375.

6.4L Sources of grease in wastewater collection systems include restaurants, industries, and homes with garbage grinders.

6.4M Odors from a wastewater collection system can be controlled by proper design, cleaning, and maintenance methods.

6.4N Use of masking agents to overcome odors does not:

1. Correct the problem or cause of the odors, but
2. Mask the odor of hydrogen sulfide so you may not be aware of the presence of this toxic, flammable, and explosive gas. Also, hydrogen sulfide can be converted to sulfuric acid, which is very corrosive.

6.4O Insects and rodents in a wastewater collection system should be controlled for the safety and protection of collection system operators and the public.

Answers to questions on page 378.

6.5A Problems created by hydrogen sulfide include:

1. Paralysis of the respiratory center and death of operators,
2. Rotten egg odors,
3. Corrosion and possible collapse of sewers, structures, and equipment, and
4. Loss of capacity of the sewer.

6.5B Hydrogen sulfide in sewers comes from the reduction of sulfate (SO_4^{2-}) to sulfide (S^{2-}) by bacteria in the slimes on sewers. Materials containing proteins in the wastes are an equally important food for bacteria that produce hydrogen sulfide. Any sulfur-containing compound that is available as a food for slimes that reduce oxygen in the wastewater can produce odorous reduced-sulfur compounds, including hydrogen sulfide.

6.5C Factors that are important in the production of hydrogen sulfide include:

1. Design,
2. Cleaning procedures,
3. Maintenance program,
4. Materials,
5. Slimes,
6. pH,
7. BOD,
8. Temperature,
9. Slope of sewer,
10. Flow (full capacity?),
11. Velocity and turbulence, and
12. Age of system.

6.5D Collection system maintenance has a definite effect on the rate of degradation of wastewater in the collection system. Undesirable effects of degradation on the wastewater treatment plant include:

1. Odor and corrosion problems in plant facilities, and
2. Initial or immediate oxygen demand on biological treatment processes.

Answers to questions on page 382.

6.5E Potential methods of chemical control of sulfide include:

1. Chlorine,
2. Hydrogen peroxide,
3. Pure oxygen,
4. Air,
5. Lime for pH control, and
6. Sodium hydroxide for pH control.

6.5F Chlorine is effective in controlling sulfide because it is toxic to aerobic bacteria that can cause anaerobic conditions and also toxic to organisms that convert sulfate to sulfide.

6.5G Hydrogen peroxide controls sulfide generation by keeping wastewater aerobic (sulfide forms only under anaerobic conditions).

6.5H Applying pure oxygen may be a feasible way to control sulfide in intercepting sewers when the wastewater treatment plant uses pure oxygen in the activated sludge process.

6.5I Metal salts applied to wastewater in a collection system control hydrogen sulfide gas by the formation of metal sulfide precipitates, thus preventing the release of hydrogen sulfide gas.

6.5J The storage and handling of metal salt chemical solutions require special equipment and safety precautions because they are corrosive acid solutions.

CHAPTER 7

UNDERGROUND REPAIR

by

William Puthuff, Jim Hicks, Mel Arts, John Dolenga,

John Cavoretto, and Gene Turney

Revised by

Bob Hiler

TABLE OF CONTENTS

Chapter 7. UNDERGROUND REPAIR

LESSON 3

OBJECTIVES

Chapter 7. UNDERGROUND REPAIR

After completion of Chapter 7, you should be able to:

1. Safely repair or construct sewer lines and manholes,
2. Determine the need for shoring and properly select and use shoring,
3. Determine and check pipeline grade,
4. Describe the need for and duties of a wastewater collection system construction inspector,
5. Inspect a sewer under construction for proper:
   a. Bedding materials and construction,
   b. Pipe laying procedures, and
   c. Backfilling and compaction,
6. Test the ability of a wastewater collection system to withstand inflow/infiltration,
7. Contact utility agencies with underground facilities near a construction or repair project *BEFORE* excavation starts,
8. Excavate, repair, and backfill service lines,
9. Excavate, repair, and backfill main lines,
10. Install sewers spanning excavations,
11. Seal leaky sewers by grouting,
12. Raise a manhole frame and cover to grade,
13. Repair and install manhole bottoms, and
14. Keep accurate records and prepare necessary reports.

WORDS

Chapter 7. UNDERGROUND REPAIR

ANGLE OF REPOSE ANGLE OF REPOSE

The angle between a horizontal line and the slope or surface of unsupported material such as gravel, sand, or loose soil. Also called the "natural slope."

ANNULAR (AN-you-ler) SPACE ANNULAR SPACE

A ring-shaped space located between two circular objects. For example, the space between the outside of a pipe liner and the inside of a pipe.

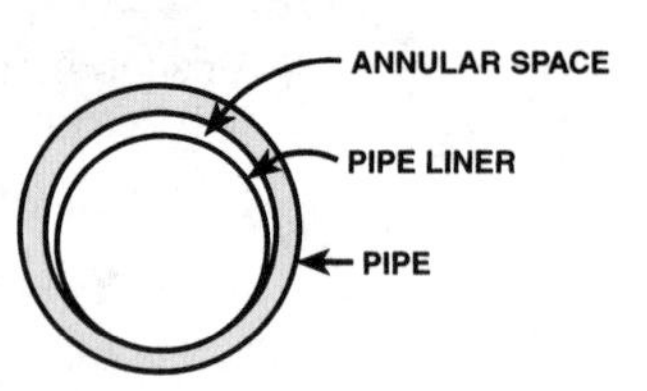

CATALYST (CAT-uh-LIST) CATALYST

A substance that changes the speed or yield of a chemical reaction without being consumed or chemically changed by the chemical reaction.

COMPETENT PERSON COMPETENT PERSON

A competent person is defined by OSHA as a person capable of identifying existing and predictable hazards in the surroundings, or working conditions which are unsanitary, hazardous or dangerous to employees, and who has authorization to take prompt corrective measures to eliminate the hazards.

CROSS BRACES CROSS BRACES

Shoring members placed across a trench to hold other horizontal and vertical shoring members in place.

FRICTION LOSS FRICTION LOSS

The head lost by water flowing in a stream or conduit as the result of the disturbances set up by the contact between the moving water and its containing conduit and by intermolecular friction.

GUNITE GUNITE

A mixture of sand and cement applied pneumatically that forms a high-density, resistant concrete.

HYDROPHILIC (HI-dro-FILL-ick) HYDROPHILIC

Having a strong affinity (liking) for water. The opposite of HYDROPHOBIC.

HYDROPHOBIC (HI-dro-FOE-bick) HYDROPHOBIC

Having a strong aversion (dislike) for water. The opposite of HYDROPHILIC.

PENTA HOSE PENTA HOSE

A hose with five chambers or tubes.

POLYMER (POLY-mer) POLYMER

A long chain molecule formed by the union of many monomers (molecules of lower molecular weight). Polymers are used with other chemical coagulants to aid in binding small suspended particles to larger chemical flocs for their removal from water.

PREVENTIVE MAINTENANCE UNITS PREVENTIVE MAINTENANCE UNITS

Crews assigned the task of cleaning sewers (for example, balling or high-velocity cleaning crews) to prevent stoppages and odor complaints. Preventive maintenance is performing the most effective cleaning procedure, in the area where it is most needed, at the proper time in order to prevent failures and emergency situations.

SAND EQUIVALENT SAND EQUIVALENT

An ASTM (American Society for Testing and Materials) test for trench backfill soils. The test uses a glass tube and the soil (backfill) is mixed with water in the tube and shaken. The tube is placed on a table and the soil allowed to settle according to particle size. The settlement of the soil is compared to the settlement of a standard type of sand grain sizes.

SHEETING SHEETING

Solid material, such as wooden 2-inch planks or $1^{1}/_{8}$-inch plywood sheets or metal plates, used to hold back soil and prevent cave-ins.

SLOPE SLOPE

The slope or inclination of a sewer trench excavation is the ratio of the vertical distance to the horizontal distance or "rise over run."

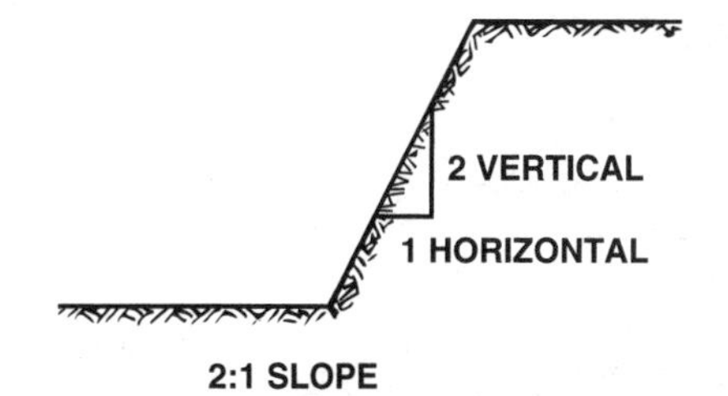

2:1 SLOPE

SPOIL SPOIL

Excavated material such as soil from the trench of a sewer.

STRINGERS STRINGERS

Horizontal shoring members, usually square, rough cut timber, that are used to hold solid sheeting, braces or vertical shoring members in place. Also called WALERS.

SURCHARGE SURCHARGE

Sewers are surcharged when the supply of water to be carried is greater than the capacity of the pipes to carry the flow. The surface of the wastewater in manholes rises above the top of the sewer pipe, and the sewer is under pressure or a head, rather than at atmospheric pressure.

UPRIGHTS UPRIGHTS

Vertical shoring members that may be solid (SHEETING) or spaced from 2 to 8 feet apart to prevent cave-ins.

VACUUM TEST VACUUM TEST

A testing procedure that places a manhole under a vacuum to test the structural integrity of the manhole.

WALERS (WAY-lers) WALERS

Horizontal shoring members, usually square, rough cut timber, that are used to hold solid sheeting, braces or vertical shoring members in place. Also called STRINGERS.

WELL POINT WELL POINT

A hollow, pointed rod with a perforated (containing many small holes) tip. A well point is driven into an excavation where water seeps into the tip and is pumped out of the area. Used to lower the water table and reduce flooding during an excavation.

CHAPTER 7. UNDERGROUND REPAIR

(Lesson 1 of 4 Lessons)

7.0 UNDERGROUND WORK

This chapter discusses repair of sewer service lines, main lines, and manholes.

7.00 Safety Considerations

Underground work can be very hazardous. Whenever excavation is necessary to repair a sewer line or manhole, precautions must be taken to protect operators from cave-ins and also from dangerous gases and fumes. Precautions during excavation include adequate shoring and ventilation. Other safety considerations include protecting

1. Operators from traffic,
2. Public from injury,
3. Operators from injury, and
4. Public and private property, including other utilities.

Also traffic must be routed safely around the job site. See Chapter 4, Section 4.3, "Routing Traffic Around the Job Site," and Section 4.4, "Classification and Description of Manhole Hazards," for additional details.

7.01 Types of Underground Work

Repair and new construction work consist of working on or installing service lines and manholes located on private property or streets and main lines and manholes located in rear easements and streets. Wastewater collection system repairs are necessary to correct damage to pipe or to remove an obstruction that is preventing the proper functioning of the collection system. Damage to pipe may occur through natural forces, but a large percent of repair jobs may be caused by poor design or poor workmanship during construction and later additions to the system.

7.02 Demand for Repair

A request to repair a line may originate from a user who is not being provided service due to collection system failure or from the agency's inspection or *PREVENTIVE MAINTENANCE UNITS*.[1] If the request for repair originates from an agency's emergency service crew, they may have restored service and in so doing discovered a damaged pipe that must be repaired or an obstruction that must be removed.

Work request forms (see Section 7.04) provide an organized approach for other crews to notify a repair crew that a line needs repairs. Whenever an emergency crew, TV inspection crew, or preventive maintenance crew discovers a situation needing repair, a work request form is completed and delivered to a supervisor. The supervisor reviews work requests and assigns them to a repair crew. Examples of various types of defects located and identified by emergency crews or preventive maintenance crews are described in the following sections. Some agencies enter work requests into a computer which reviews tasks and priorities and then assigns jobs to each crew. These computer assignments are reviewed by a supervisor before the crews leave the maintenance yard.

7.03 Sources of Damages and Stoppages

Sewer lines may have to be dug up because they are damaged or blocked. Causes of these problems are described and shown in this section.

7.030 Roots

Roots often enter pipes through cracks and cause stoppages. Figure 7.1 shows a cracked pipe that was dug up with roots growing around it. A probable cause of the pipe damage in Figure 7.1 is improper bedding (such as large stones) that caused the pipe to crack and allowed roots to enter the pipe.

[1] *Preventive Maintenance Units. Crews assigned the task of cleaning sewers (for example, balling or high-velocity cleaning crews) to prevent stoppages and odor complaints. Preventive maintenance is performing the most effective cleaning procedure, in the area where it is most needed, at the proper time in order to prevent failures and emergency situations.*

These roots may prevent inspection or cleaning equipment from passing through a sewer.

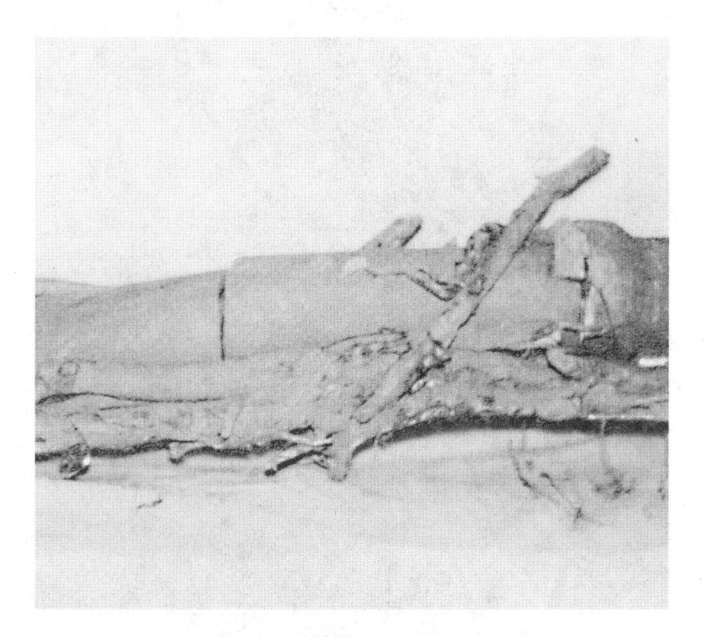

Fig. 7.1 Cracked pipe

7.031 Improper Taps

Improper taps can prevent building, branch, and lateral sewers from functioning as intended. Common types of tap defects include:

1. Protruding taps,
2. Cracks or holes that allow root entry, leaks, and concrete to enter the sewers,
3. Insufficient opening in tap, and
4. Misalignment of tap.

A tap protruding into a main line can cause material to build up and stop the flow in the main line (Figure 7.2). Also it can prevent inspection and cleaning equipment from passing by the tap. To repair a protruding tap, remove a section of the main line containing the tap and install a factory-made wye or tee and connect it to the building sewer. A protruding tap may be prevented by cutting a hole in the main line with a can-type cutter and installing a proper saddle.

An improper tap cut in a main line can allow cement to flow inside the main line during installation and cause problems in the future (Figures 7.3 and 7.4). Cracks, holes, or pipe broken at a tap allow root intrusion, infiltration, and exfiltration.

Protruding or improper taps and laterals, improper offset joints, and chemical deposits can be removed by a mechanical cutting system (Figure 7.5). These devices are used in smaller pipes and also can be used with television inspection systems.

7.032 Damage by the Public or Other Utility Agencies

Other utility agencies or the public can cause damage by underground construction, illegal taps, or vandalism. Figures 7.6 and 7.7 show drain pipes installed by a homeowner to drain the yard and house gutters. Other agencies can run lines and ducts through sewer lines. Borings for foundation tests can penetrate sewer lines. Also, telephone pole and guy wire augers can damage sewers.

7.04 Work Request Form

Upon discovery of a sewer line needing repair, a work request form should be completed (Figure 7.8). Information on the form is essential for establishing repair priorities and scheduling work. The work request form should provide the following information:

1. Name of crew chief,
2. Date,
3. Street address near requested repair,
4. Map location,
5. Manhole numbers on each side of damaged pipe,
6. Size of pipe,
7. Depth of pipe,
8. Distance from both upstream and downstream manholes to damaged pipe,

9. Location of pipe (street or easement),
10. Work requested, and
11. Sketch of site with a North arrow.

7.05 Assignment of Repair Job

When the repair or construction supervisor (maintenance supervisor) receives the work order form, the supervisor reviews the job and assigns it to a crew leader. If work loads are heavy, the crew leader may assign a priority to work orders. Those repairs that are not urgent may be given a low priority and the work is done at a later date. When a crew leader is as-

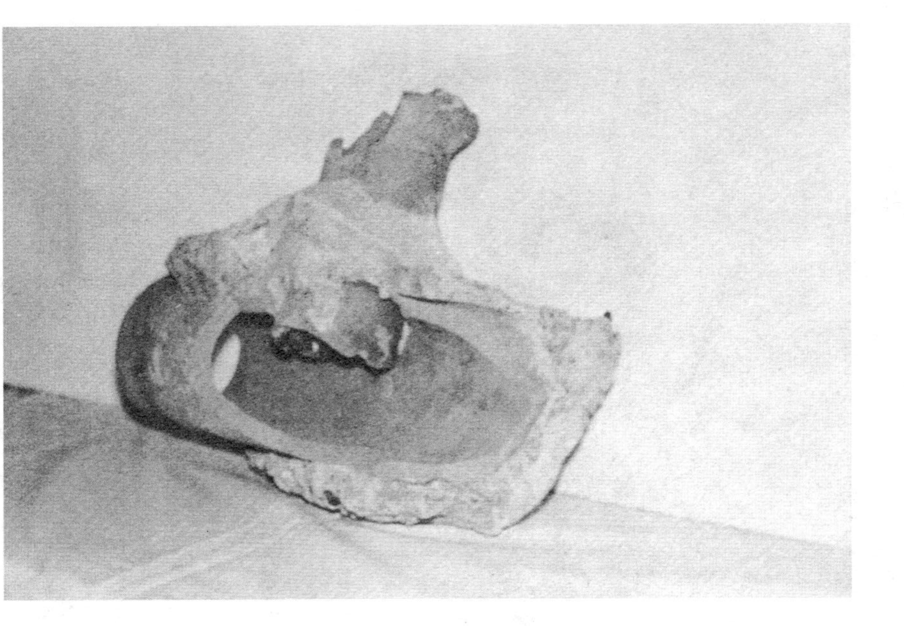

Fig. 7.2 Protruding tap

Fig. 7.3 Improper tap—note cement in main line

Fig. 7.4 Improper tap—note cement poured around tap

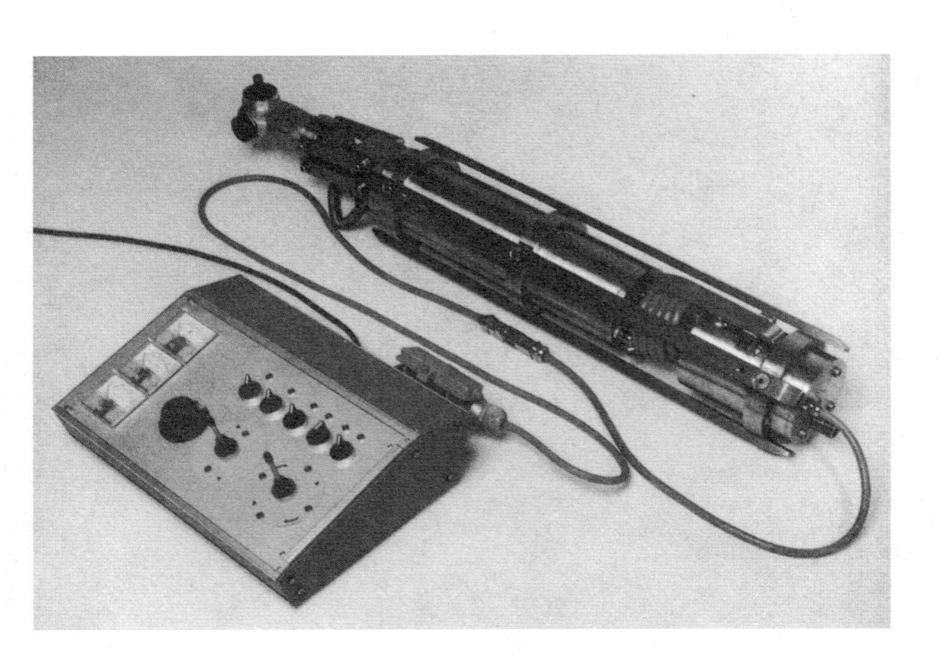

Fig. 7.5 Mechanical cutting system
(Reproduced with permission of Beleke Technologies)

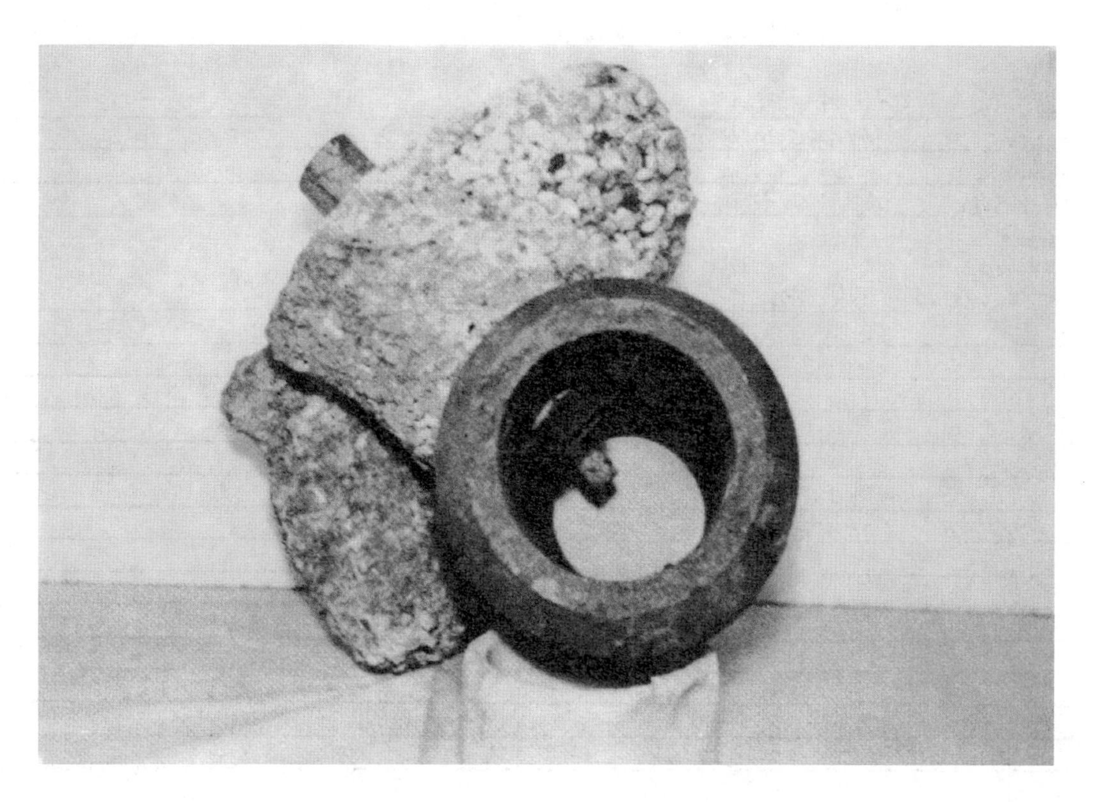

Fig. 7.6 Illegal drain tap

Fig. 7.7 Illegal drain tap

Crew Chief SWANSON Date 9-26-02

Address MARCONI AVE. Book 269 Page 05

Manhole 8 to 7

Pipe Size 6" VCP Depth 6'9" - 9'8" Distance 292

Street [ ] Easement [X]

Work Requested REPLACE TAP W/ROOTS

27' FROM M/H 8 → 7

PIC. #1

M/H 8 LOCATED AT 3609 MARCONI AVE.

Approved '

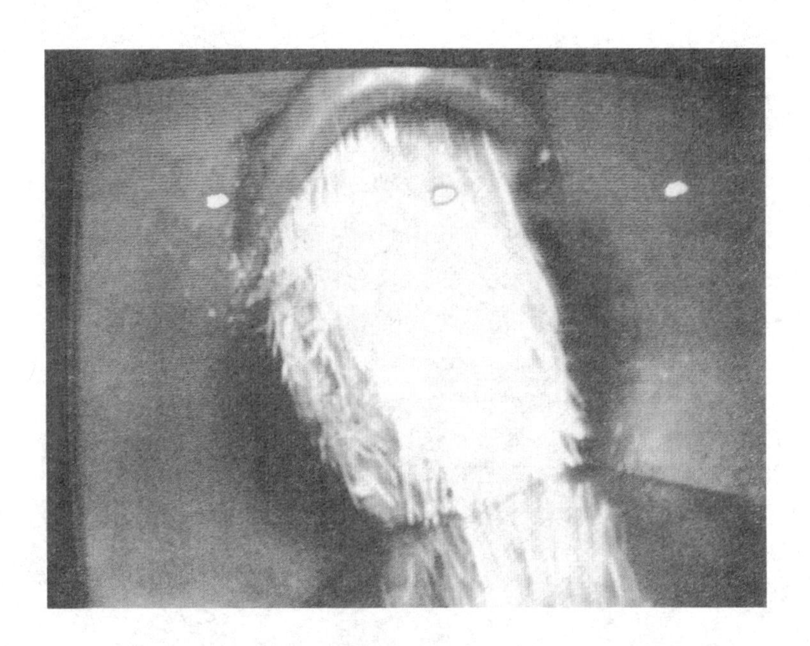

Fig. 7.8 Work request form and photo of roots in tap that must be removed and replaced

signed a work order, the job should be evaluated and the crew prepared to accomplish the repair job.

7.050 Investigate and Plan Job

Before starting a repair job, the crew leader or supervisor in charge of the work should investigate the job site and plan the work. Either the crew leader or the supervisor should make notes on conflicts with utilities, traffic patterns, parked vehicles, or private property that might have to be moved, such as trees or fences, and soil conditions if known.

7.051 Meet With Utility Companies

All utility companies in the area should be contacted. Each company should meet the crew leader at the job site and locate and mark utilities buried near the work area. Some underground utility companies are so small that their personnel cannot come out to the site and mark locations. If this occurs, inspect their maps in their office and do your own marking of the utilities. Utility companies will not accept responsibility for damage unless their personnel located their utilities. Most companies insist their personnel be on the job site until their facilities are uncovered and no conflict exists. When all underground utilities are located and marked, some thought should be given to overhead utility conflicts. The type of equipment used on the job depends on the potential problems encountered when doing the job.

7.052 Public Relations

Contact should always be made with the property owners in the area making them aware of any inconvenience that might occur. If excavation is going to block a driveway into a business, contact the owner or manager and try to schedule your work so little or no conflict will exist. Good public relations are very important and should be considered at all times.

7.053 Preconstruction Meeting

A tailgate safety and preconstruction meeting should be held with all operators on the project (Figure 7.9). Discuss what is to be accomplished and the duties of each person. What is to be accomplished means telling the crew what repair is necessary, such as a tap replacement, cleanout installation, or main line repair. The duties of each person will be the usual work assignment, unless special tools or materials are required for the specific job assignment. The preconstruction discussion focuses on what to expect and what to do if the unexpected occurs.

All safety regulations and work procedures for the particular job should be thoroughly explained. The notes taken by the crew leader are valuable and should be used at this time. Operators' safety, pedestrian safety, and underground conflicts should be thoroughly reviewed so there will be no surprises when the crew arrives on the job site and starts to work. For more information, read Chapter 11, "Safety/Survival Programs for Collection System Operators" in Volume II.

QUESTIONS

Write your answers in a notebook and then compare your answers with those on page 502.

7.0A Why are underground repair and new construction very hazardous?

7.0B Why is underground repair necessary?

7.0C How are underground damages discovered?

7.0D What are some causes of damages and stoppages that require underground repair?

7.0E What is the purpose of a work request form?

7.0F What should a crew leader do before starting a repair job?

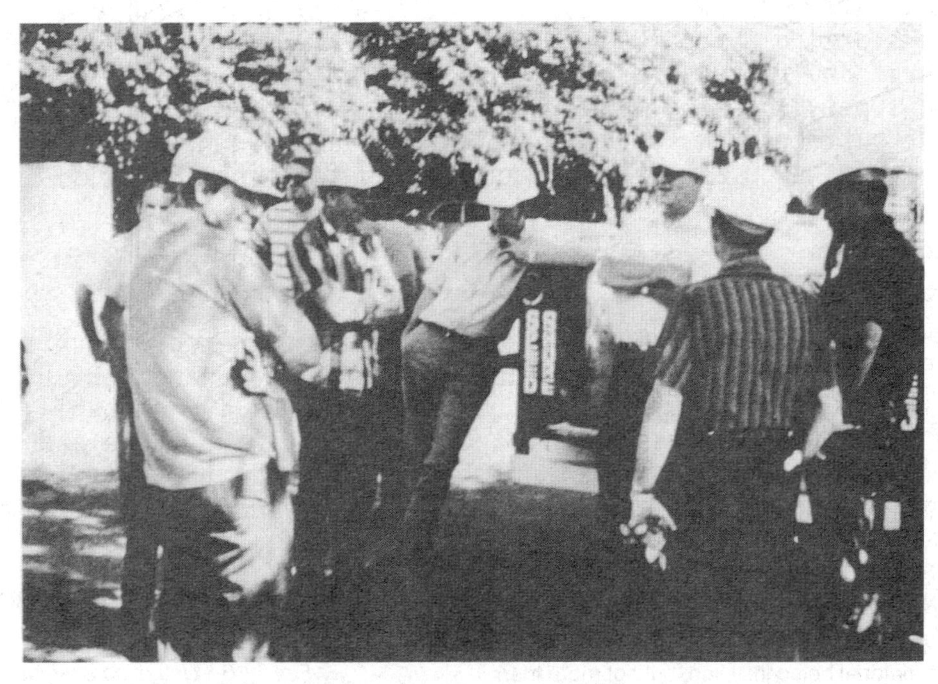

Fig. 7.9 Tailgate safety and preconstruction meeting

7.1 SHORING

Whenever excavation is necessary for repairs or new construction, shoring may be necessary to protect operators from cave-ins. The need for shoring and shoring requirements depend on many factors, such as depth of trench, width of trench, type of soil (clay, loam, sand), soil conditions (compaction, moisture), and nearby activities that could cause vibrations. Another important consideration when determining shoring requirements is the length of time the excavation is expected to be open. If an excavation is going to be open for an extended period, you would want heavier bracing to provide additional protection against vibrations from traffic and rains that could cause flooding.

Shoring requirements are dictated by laws and codes and are strictly enforced by the regulatory agencies. Lack of shoring and shoring failure are the major causes of underground construction deaths. For example, seven people died and there were 67 lost-time injuries related to construction shoring problems in California during one year. By comparison, there were no deaths and 203 lost-time injuries to operators in the wastewater collection field during this same time period. How can we prevent these accidents? First, learn about the shoring laws and codes, types of shoring, and proper use of shoring, and then apply this knowledge. Also develop a safety program and an approved OSHA (Occupational Safety and Health Act) shoring class. The Williams-Steiger Occupational Safety and Health Act (OSHA) of 1970 contains shoring regulations, and they are strictly enforced. Penalties for violating the regulations include warnings, fines, and even prison sentences for flagrant violations. Contact your local safety office for all regulations pertaining to any excavations done by you or your crews. (Also see Section 7.2, "Construction," for detailed shoring information.)

7.10 Shoring Requirements

Typical shoring requirements include:

1. Shoring systems in trenches shall consist of uprights held rigidly opposite each other against the trench walls by jacks or horizontal cross members (cross braces) (Table 7.1) and, if required, longitudinal members (stringers) as hereinafter provided. Uprights shall not exceed 15 degrees from the vertical.

2. Uprights in trenches over 10 feet deep shall be not less than 3-inch by 8-inch material, and shall be at least 2-inch by 8-inch material in trenches less than 10 feet deep. (See Table 7.2 on page 407 for additional details and for requirements applying to running ground.) Uprights shall extend from above the top of the trench to as near the bottom as permitted by the material being installed, but not more than 2 feet from the bottom.

3. Cross braces shall consist of steel screw-type trench jacks with a foot or base plate on each end of pipe or timbers placed horizontally and bearing firmly against uprights or stringers. Wood braces shall be no smaller than those listed in Table 7.1.

TABLE 7.1 SIZES OF WOOD AND PIPE BRACES IN TRENCHES[a]

Width of Trench, Feet (Incl.)	Size of Wood Braces	Size of Pipe Braces	Aluminum Hydraulic Shores
1 - 3	4" x 4"	$1^1/_2$" STD[b]	2"
3 - 6	4" x 6"	2" STD	2"
6 - 8	6" x 6"	2" STD	2"
8 - 10	6" x 8"	3" STD	2"
10 - 12	8" x 8"	3" STD	3"

Wider trenches shall have cross braces of correspondingly larger dimensions.

[a] Consult with your local safety agency for sizes of wood and pipe cross braces required in your area.

[b] For metric pipe sizes and other metric conversion factors, please refer to page 583.

NOTE: In place of the above shoring systems, the use of properly maintained screw-type metal jack trench shoring units with equivalent strength is acceptable (Figure 7.10).

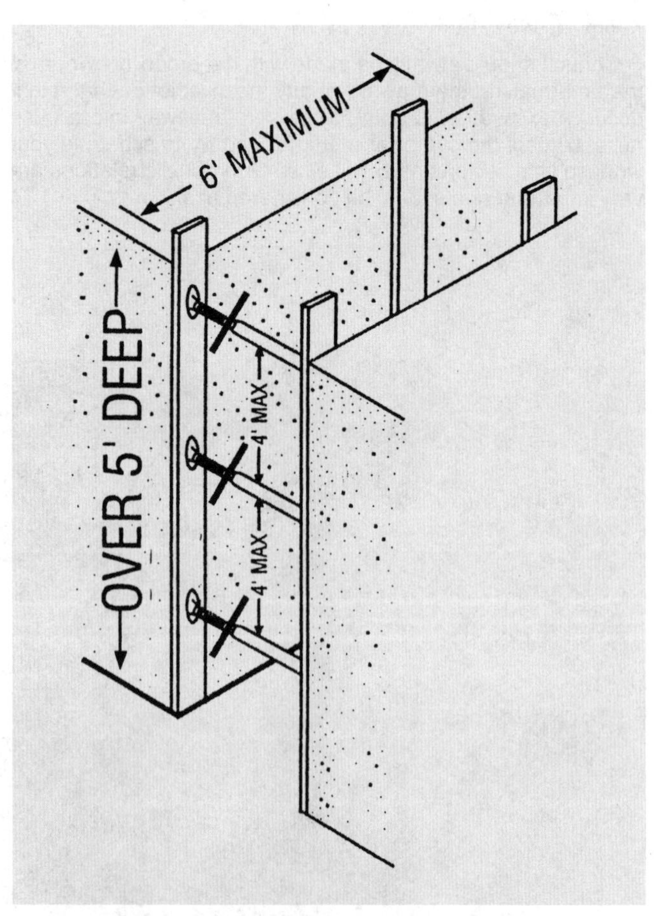

NOTE: Some safety regulations require shoring for depths over 4 feet.

Fig. 7.10 Use of screw-type metal jack trench shoring units

4. The minimum number of horizontal cross braces, either screw jacks or timbers, required for each pair of uprights shall be determined by the number of 4-foot zones or segments into which the depth of trench may be divided. One horizontal cross brace shall be required for each of these zones, but in no case shall there be less than two cross braces or jacks. Trenches, the depths of which cannot be divided equally into these standard 4-foot zones or segments, shall have an extra horizontal cross brace supplied for the short remaining zone, if such zone is greater than two feet. In no case, however, shall the vertical spacing of horizontal cross braces be spaced greater than 5 feet center to center. Minor temporary shifting of horizontal cross bracing will be permitted, when necessary, for the lowering of materials into place. Be sure no one is allowed in the trench while the cross braces are temporarily shifted or moved. If the diameter of a pipe is greater than four feet, the lower cross braces require alternate considerations.

5. The dimensions and spacing of the elements of the shoring system shall be governed by the depth of the trench, type of soil encountered, and other special conditions of the site, but in no case shall they provide less strength than the members listed in Tables 7.1 and 7.2. These dimensions and spacings are to be considered as a minimum requirement.

QUESTIONS

Write your answers in a notebook and then compare your answers with those on page 502.

7.1A Shoring requirements depend on what factors?

7.1B What are the main causes of cave-in deaths?

7.1C What do shoring systems in trenches consist of?

7.1D Uprights must extend how close to the bottom of a trench?

7.1E How is soil stability classified?

7.11 Responsibility

All excavation work shall at all times be under the immediate supervision of a competent person (such as a supervisor or crew leader) with authority to modify the shoring system or work methods as necessary to provide greater safety. This person will frequently examine the material under excavation and improve the shoring or methods beyond the minimum requirements, as necessary to ensure protection of operators from moving ground. Improved shoring or methods may be required if there is evidence of excessive shifting, displacement, or removal of cross bracing.

Someone must be on the job site and in charge at all times. This person not only accepts responsibility, but is authorized to change any predetermined shoring system when soil conditions change. Material handling, traffic pattern changes, public relations, repair work excavation, backfill, and compaction are directed by the person in charge. Some of the listed jobs will have to be delegated or they won't be done effectively.

Tailgate safety meetings and the preconstruction meetings are very important. Everyone on the job site must know exactly what to do and what is expected to be accomplished.

7.12 Placing of Spoil[2] (Excavated Material)

The spoil (excavated material) should be placed two feet or more back from the edge of the trench (Figure 7.11). Try to keep the area between the trench and the spoil area clear of loose material. If not kept clear, material such as rocks, tools, and pipe could be knocked into the excavation, or someone could trip on the equipment and fall into the trench. Keeping

TABLE 7.2 DIMENSIONS AND SPACINGS OF ELEMENTS OF SHORING SYSTEM

Soil Stability	Depth, Feet[a]	Uprights: Size, Inches	Uprights: Horizontal Spacing, Feet	Cross Braces: Size, Inches	Cross Braces: Horizontal Spacing, Feet	Stringers: Size, Inches	Stringers: Vertical Spacing, Feet
Hard compact	5 - 10	2 x 8	8	4 x 4	8	Where Indicated	
	Over 10	3 x 8	6	4 x 6	6	Where Indicated	
Unstable	5 - 10	2 x 8	4	4 x 4	4	Where Indicated	
	5 - 10	2 x 8	2	4 x 4	2	Where Indicated	
	Over 10	3 x 8	Solid	4 x 6	6	4 x 6	4
Running	4 - 8	2 x 8	Solid	4 x 4	6	4 x 6	4
	Over 8	3 x 8	Solid	6 x 6	6	6 x 6	4

Soil Stability may be classified as hard compact, running, or unstable.
HARD COMPACT soil is all earth material not classified as running or unstable.
RUNNING soil is earth material whose angle of repose[b] is approximately zero, as in the case of soil in a nearly liquid state, or dry, unpacked sand which flows freely under slight pressure.
UNSTABLE soil is earth material that has a significant angle of repose, but because of vibration, moisture or other reasons, cannot be depended upon to remain in place without extra support.

[a] *NOTE:* Some safety regulations require shoring for *ALL* depths over 4 ft.
[b] Angle of Repose. The angle between a horizontal line and the slope or surface of unsupported material such as gravel, sand, or loose soil. Also called the "natural slope."

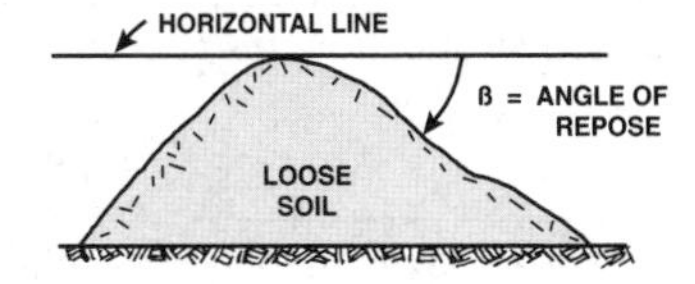

[2] *Spoil. Excavated material such as soil from the trench of a sewer.*

Fig. 7.11 Keep area near trench clear

This photo shows two unsafe conditions:

1. Shores do not extend above the ground surface, and
2. Spoil is too close to edge of trench.

the area clear is not only good for the safety of operators, but reduces the chances of damage to new pipe or other conduits and utilities, which could cause thousands of dollars worth of damages. Also, keeping spoil away from the edge of the trench tends to reduce the load on the side walls, thus reducing the loads on the sheeting, uprights, stringers, and cross bracing.

7.13 Shoring Regulations[3]

7.130 Operator Safety

All trenches five feet in depth and deeper must be effectively shored to protect operators against the hazard of moving ground. Trenches less than five feet in depth also must be shored when examination indicates hazardous ground movement may be expected.

Special provisions must be made to prevent injury to operators engaging in installation of shoring. This may be done by providing and requiring the use of special devices such as long-handled jacks that will allow upper cross braces to be placed from the ground surface before operators enter the trench. At those points in deep trenches requiring additional braces, operators should then progress downward, protected by cross braces that have already been set firmly in place. The reverse procedure should be followed when removing shoring.

7.131 Soil Conditions and Trench Location

The type of shoring installed depends on the soil conditions and location of the trench. You cannot shore an excavation alongside a railroad track the same as one in an open area because of vibrations created by passing trains (Figure 7.12). When the decision is made on the type of shoring to use, remember that soil conditions change and an alternate shoring system should be readily available so little work time will be lost if a change must be made.

7.132 Soil Classification

Although soil types are commonly described as hard compact, running, or unstable, the more formal classification system described below also is widely used. The soil classifications in this section are listed and described in decreasing order of stability.

Type A-25 Soil:

Cohesive soils with an unconfined compressive strength of 1.5 tons per square foot or greater. These are some of the examples of cohesive soils: clay, silty clay, sandy clay, clay loam, and sandy clay loam. Cemented soils such as caliche and hardpan are also considered Type A. However, no soil is Type A if:

- The soil is fissured.
- The soil is subject to vibration from heavy traffic, pile driving, or similar effects.
- The soil has been previously disturbed.

---

[3] *These regulations are from the California Code of Regulations, Construction Safety Orders, Title 8, Subchapter 4, Article 6. Other states have similar regulations. You should contact your state industrial safety officer to obtain the shoring regulations for your state. Some states require shoring in all trenches greater than four feet in depth.*

Fig. 7.12 Shoring alongside a railroad track

- The soil is part of a sloped, layered system where the layers dip into the excavation on a slope of four horizontal to one vertical (4H:1V) or greater.
- The material is subject to other factors that would require it to be classified as a less stable material.

Type B-45 Soil:

- Cohesive soil with an unconfined compressive strength greater than 0.5 ton per square foot but less than 1.5 tons per square foot.
- Granular cohesionless soils including: angular grave (similar to crushed rock), silt, silt loam, sandy loam, and, in some cases, silty clay loam and sandy clay loam.
- Previously disturbed soils except those which would otherwise be classed as Type C soil.
- Soil that meets the unconfined compressive strength or cementation requirement for Type A, but is fissured or subject to variation.
- Dry rock that is not stable.
- Material that is part of a sloped, layered system where the layers dip into the excavation on a slope less steep than four horizontal to one vertical (4H:1V) but only if the material would otherwise be classified as Type B.

Type C-60 Soil:

- Moist, cohesive soil or a moist, dense, granular soil that does not fit into Type A or Type B classification, and is not flowing or submerged.
- This material can be set with a nearly vertical outside surface and will stand supportive long enough to allow the vertical shores to be properly installed.
- The competent person must monitor the excavation for signs of deterioration of the soil as indicated by, but not limited to, freely seeping water or flowing soil entering the excavation or below the sheeting.
- An alternate designed for less-stable Type C soil where there is evidence of deterioration.

Type C Soil:

- Cohesive soil with an unconfined cohesive strength of 0.5 ton per square foot or less.
- Granular soils including gravel, sand, and loamy sand.
- Submerged soil or soil from which water is freely seeping.
- Submerged rock that is not stable.
- Material in a sloped, layered system where the layers dip into that excavation on a slope of four horizontal to one vertical (4H:1V) or steeper.

7.133 Soil Testing

A *COMPETENT PERSON*[4] shall conduct the soil testing. The competent person must be able to perform a series of tests and be capable of identifying existing and predictable hazards. The soil test is divided into two phases, visual and manual testing.

VISUAL TESTS. Visual analyses are conducted to determine qualitative information regarding the excavation site in general, the soil adjacent to the excavation, the soil forming the sides of the open excavation, and the soil taken as samples from excavated material.

- Observe samples of the excavated soil and the soil in the sides of the excavation. Estimate the range of particle sizes

[4] *Competent Person. A competent person is defined by OSHA as a person capable of identifying existing and predictable hazards in the surroundings, or working conditions which are unsanitary, hazardous or dangerous to employees, and who has authorization to take prompt corrective measures to eliminate the hazards.*

and the relative amounts of each particle size. Soil that is primarily composed of fine-grained material is cohesive material.

- Observe soil as it is excavated. Soil that remains in clumps when excavated is cohesive. Soil that breaks up easily and does not stay in clumps is granular.
- Observe the sides of the open excavation and the surface area next to the excavation. Crack-like openings, such as tension cracks, can indicate fissured material. If chunks of soil spall (chip or flake) off a vertical side, the soil could be fissured. Small spalls are evidence of moving ground and are indications of potentially hazardous situations.
- Observe the area next to the excavation and the excavation itself for evidence of existing utility and other underground structures and look for indications of previously disturbed soil.
- Observe the opened side of the excavation to determine whether layers slope toward the excavation. Estimate the degree of slope of the layers.
- Observe the area next to the excavation and the sides of the opened excavation for evidence of surface water, water seeping from the sides of the excavation, or the location of the level of the water table.
- Observe the area next to the excavation and the area within the excavation for sources of vibration that may affect the stability of the excavation face.

MANUAL TESTS. Manual analyses of soil samples are conducted to determine quantitative as well as qualitative properties of soil and to provide more information in order to classify soil properly.

- Plasticity. Mold a moist or wet sample of soil into a ball and attempt to roll it into threads as thin as 1/8-inch in diameter. Cohesive material can be successfully rolled into threads without crumbling. For example, if at least a two-inch length of 1/8-inch thread can be held on one end without tearing, the soil is cohesive.
- Thumb Penetration. The thumb penetration test can be used to estimate the unconfined compressive strength of cohesive soils. Type A soils with an unconfined compressive strength of 1.5 tons per square foot can be readily indented by the thumb; however, they can be penetrated by the thumb only with very great effort. Type C soils with an unconfined compressive strength of 0.5 ton per square foot can be easily penetrated several inches by the thumb, and can be molded by light finger pressure.
- Dry Strength. If the soil is dry and crumbles on its own or with moderate pressure into individual grains or fine powder, it is granular (any combination of gravel, sand, or silt). If the soil is dry and falls into clumps that break up into smaller clumps, but the smaller clumps can only be broken up with difficulty, it may be clay in any combination with gravel, sand, or silt. If the dry soil breaks into clumps that do not break up into small clumps and that can only be broken with difficulty, and there is no visual indication the soil is fissured, the soil may be considered unfissured.

7.134 Sloping Excavation Walls

In place of a shoring system, the sides or walls of an excavation may be sloped or benched if this method is considered more economical and the job can be done safely. Where sloping is a substitute for shoring that would otherwise be needed, excavation side walls should be on a slope of ¾ foot horizontal to one foot vertical except where the instability of material requires a slope of greater than ¾:1 (¾ vertical to 1 horizontal).[5] Read Section 7.202, "Trench Shoring," for more details.

7.135 Access to Construction Work Area

Convenient and safe means must be provided for operators to enter and leave the excavated area. Access shall consist of a standard stairway, ladder, or ramp securely fastened in place at suitably guarded or protected locations where people are working. The means of access and egress (use of ladders, for example) must not be more than 25 feet from a worker and must be provided in excavations four feet or more in depth.

7.136 Pedestrian Safety

When working near curbs and sidewalks, keep a pedestrian way open on the sidewalks. If this area becomes covered with material and equipment, the pedestrians will be in the work area or walking across someone's yard; neither situation is acceptable. If necessary, erect barriers to keep pedestrians and children in safe, permissible areas. Be especially careful if children are playing or watching near the job site.

QUESTIONS

Write your answers in a notebook and then compare your answers with those on page 502.

7.1F What should be the authority of someone supervising excavation work?

7.1G Where should spoil or excavated material be placed?

7.1H All trenches deeper than _____ feet must be shored.

7.1I How can operators be protected if shoring is not practical?

7.1J Why must a walkway be provided for pedestrians?

7.14 Types of Shores

Several types of shores have been tested and approved by OSHA; each type has its good and bad points. Selection of types of shoring depends largely on soil conditions in your area.

7.140 Hydraulic Shores

Hydraulic shores are used frequently due to their ease of installation and removal. They are usually not used on jobs when they will be required to be left installed in one segment of trench for a time period greater than five days due to the possibility of the hydraulic pressure bleeding off during a longer period of installation.

The standard aluminum hydraulic cylinder (Figure 7.13) is 2 inches in diameter. Each shore has two cylinders and two 7-foot rails equal in strength to 3" x 10" timbers. Each rail is 8 inches wide and 1½ inches deep. The complete unit weighs 70 pounds and can spot-brace trenches 22 to 48 inches wide. The two cylinders are connected by a high-pressure, flexible hose (Figure 7.14) that allows fluid to be pumped into the cylinders to expand them to the desired length. Recommended fluid to use in hydraulic shores is one quart shoring fluid (water-soluble oil) to 19 quarts water for each five-gallon mixture.

[5] *Slope of ¾:1.*

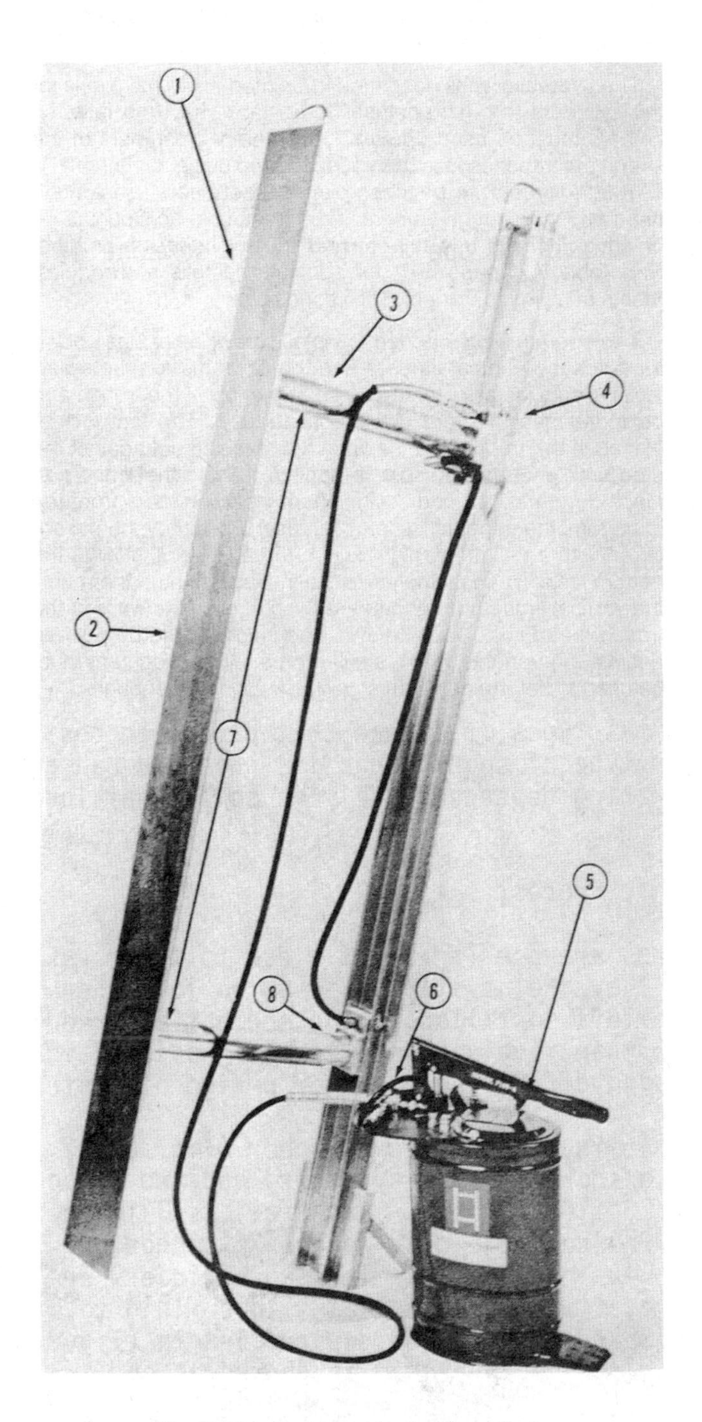

Fig. 7.13 Aluminum hydraulic cylinder
(Courtesy of SPEED-SHORE)

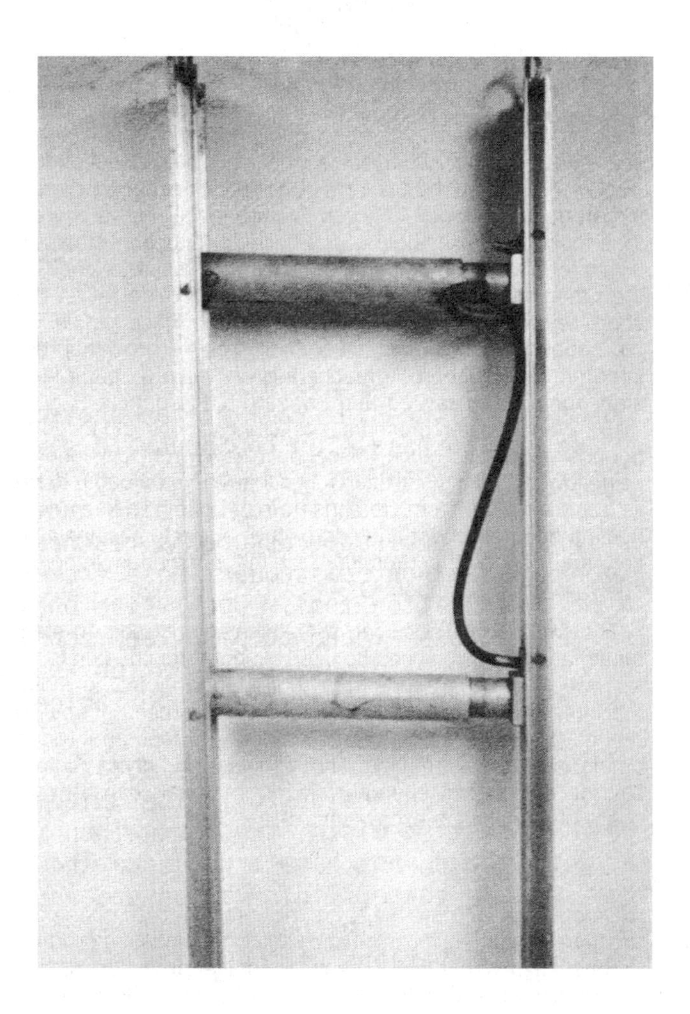

Fig. 7.14 Aluminum hydraulic cylinder and pressure hose

1. Aluminum alloy material
2. Flanged side rail
3. Hydraulic cylinder
4. Offset pins connecting side rails and hydraulic cylinders
5. Pump
6. Pressure gage
7. 4-foot cylinder spacing
8. Cylinder pad

Never use plain water without shoring fluid. The shoring fluid acts as a lubricant, protects the shores from corrosion damage and mineral deposits, and lowers the freezing point of the liquid mixture. Always follow the shoring fluid manufacturer's recommendations.

Installation of the standard hydraulic shore is usually a one-operator job. Place shore across trench with one rail on top of the other in the trench. Connect pump hose to male fitting on shore (top cylinder). Open valve on pump. The installation tool (Figure 7.15, page 413) is from 30 to 96 inches long, depending on depth of trench in which you are installing the shores. This tool is approximately a one-half inch diameter steel rod with a tee handle on one end, a hook and pressure release on the other end. This tool is hooked onto the side rail of the shore and held firmly. With the other hand, grasp handle on the opposite side rail and lower shore into trench. When desired position of shore is reached, close valve on pump and start pumping by raising and lowering pump handle (Figure 7.16). When desired pressure is reached, usually 750 to 1,000 pounds per square inch, disconnect hose from cylinder by using lower end of installation tool. Open valve on pump to relieve pressure in hose. Coil hose around pump to keep connecting end free of dirt and sand. Move pump to next installation and repeat the procedure at the prescribed intervals shown in Table 7.2, page 407.

The two-inch cylinder can be reinforced with square steel tubing and extensions can be used to shore trenches up to 14 feet wide. For excavations from 10 feet up to 24 feet in width, the three-inch aluminum hydraulic cylinder is used. If you do not have experience shoring trenches 10 feet and wider, obtain correct information and engineering advice before starting the project. Improper shoring materials can be very costly on trenches or excavations of this size.

7.141 Screw Jacks

Screw Jacks (Figure 7.17) are inexpensive to purchase but are time consuming to install and usually are not used on large production jobs. For minor repairs in smaller excavations, they are more practical than other types of shores. Screw jacks must be placed from the top down, and those who have used them know this is a slow process. Screw jacks are made of iron and come in two pieces. They are available with a 1½- or 2-inch threaded shaft and wing nut that is used for adjustment to fit trench width. One end of the shaft is plain and the other has a ball-socket for a shoe that goes against the shoring. The screw length on the 1½-inch shaft varies from 10 to 18 inches. The 2-inch shaft has a screw length of 18 inches only.

The other part of the screw jack consists of a shoe and ball joint with a male fitting to fit inside a 2-inch pipe. This fitting has a hole to secure the desired length of pipe to it. Cut the length of pipe required, usually about 8 inches shorter than the trench width. Drill a hole through one end large enough to accept a cotter pin. All jacks should be assembled before excavation is started. Extra pipe should be available if needed. The weight of each jack varies from 13 to 23 pounds, depending on length of pipe used.

The hydraulic jacks described in Section 7.140 have rails for the cylinders to push against. Screw jacks have no rails, so timbers must be used (Figure 7.18). Vertical members of the shoring in trenches less than 10 feet deep use a minimum 2" x 8" rough timbers. In trenches over 10 feet deep, use a minimum of 3" x 8" rough timbers. Timbers should be Douglas Fir or equal. Always use full cut and rough dimension timbers. See Table 7.2, page 407, for additional details and requirements applying to "running" soil conditions.

Two operators are needed to install screw jacks. One operator holds the vertical timbers and the other installs the screw jacks. Begin installing screw jacks at the top of the trench and no more than two feet down from the top. The first jack is placed in the trench and the wing nut screwed out against the pipe until enough pressure is applied to hold the timbers in place (Figures 7.18 and 7.19). When installing jacks from top to bottom, the upper jacks tend to loosen up; if they are not secured with a nail, they might fall on the operator installing the bottom jack. When all the jacks are in place, tighten them until the vertical timbers (uprights) are flush against the walls of the trench (see Figure 7.20 below). Each jack should be pushed against the uprights with a similar force. Good judgment must be used to determine when screw jacks are tight enough.

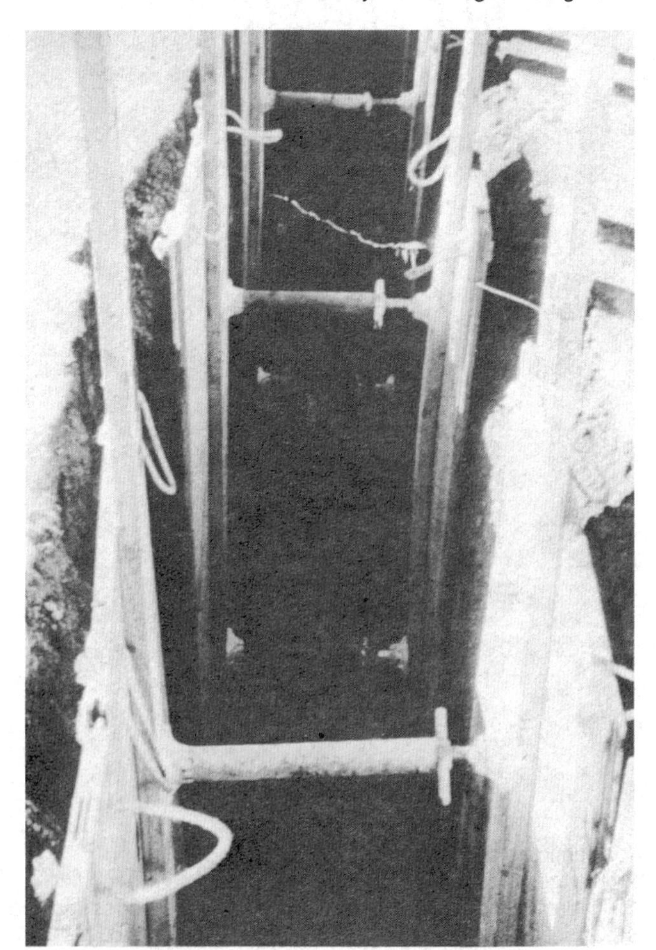

Fig. 7.20 Screw jacks holding vertical timbers or uprights against wall

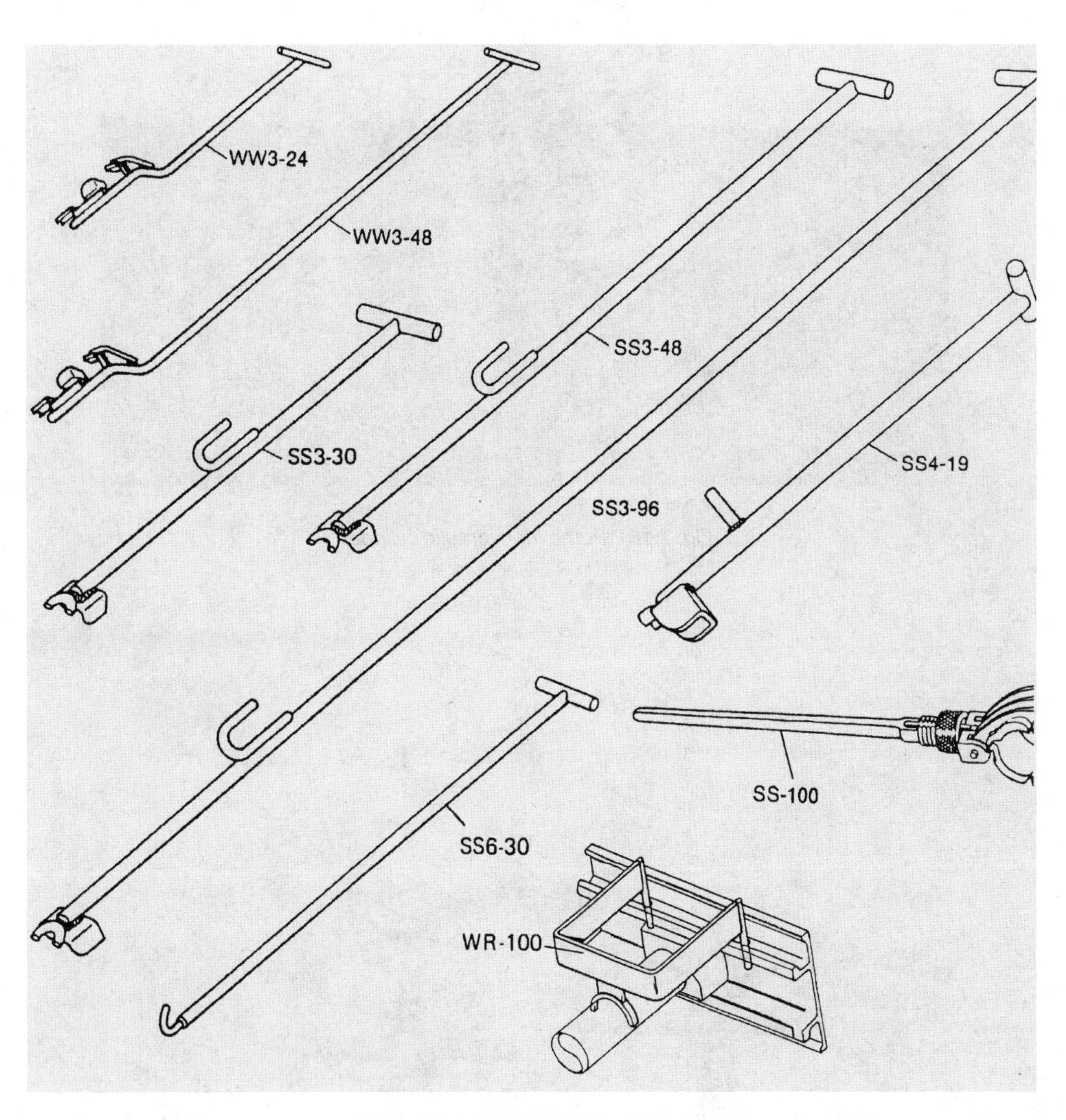

SPEED SHORE ACCESSORY PARTS LIST
(U.S. PATENT NO. 3224)

Part No.	Description	Part No.	Description
SS3-30	Speed Shore Release Tool – 30" (76.2 CM)	SS6-96	Removal Hook – 96" (243.8 CM)*
SS3-48	Speed Shore Release Tool – 48" (121.9 CM)	WW3-24	Water Release Tool – 24" (61.0 CM)
SS3-96	Speed Shore Release Tool – 96" (243.8 CM)	WW3-48	Water Release Tool – 48" (121.9 CM)
SS4-19	Single Shore Release Tool – 19" (48.2 CM)	SS-100	Speed Shore Cylinder Wrench
SS6-30	Removal Hook – 30" (76.2 CM)	WR-100	Water Rack
SS6-48	Removal Hook – 48" (121.9 CM)*	*Not pictured	

Fig. 7.15 Installation and release tools
(Courtesy of SPEED-SHORE)

Fig. 7.16 Pump for hydraulic cylinder
(See Figure 7.13, item 5, pump)

Fig. 7.17 Screw jacks

Fig. 7.18 Screw jacks with timber uprights

WARNING: Never use jacks for access ladders. Always use a ladder.

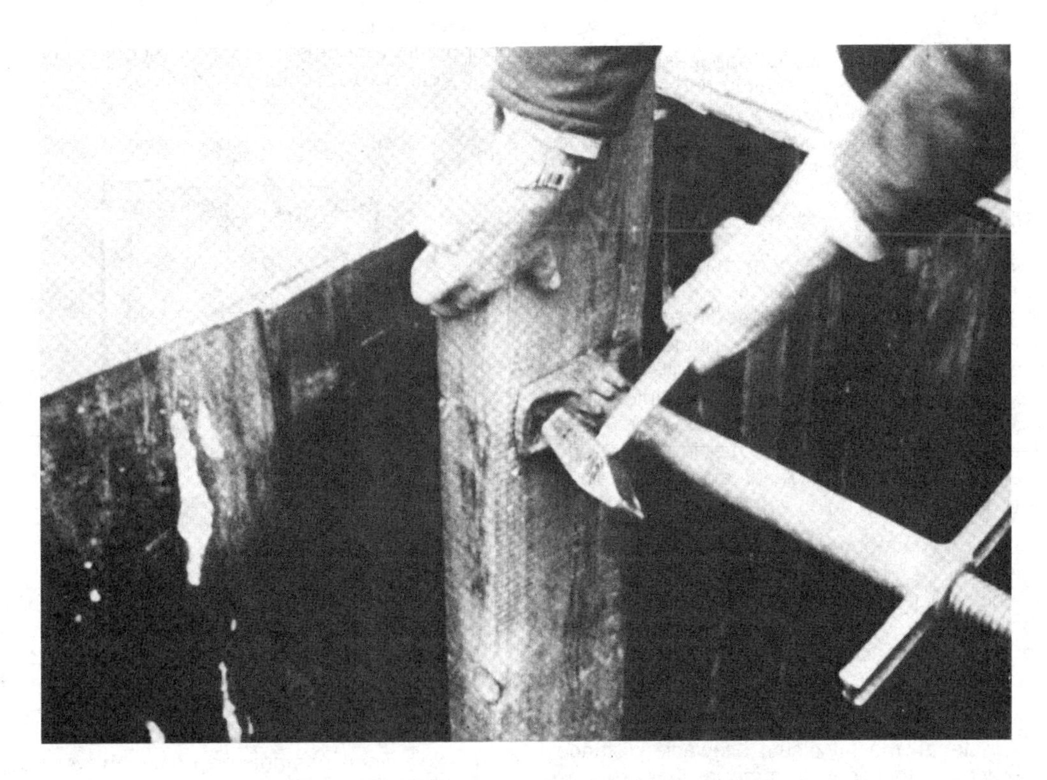

Fig. 7.19 Nailing screw jack into timber upright

7.142 *Air Shores*

Air shores (Figure 7.21) are new by comparison to the other shores discussed. They are a pneumatic shoring cross brace made for trenches where shoring is required.

Air shores are not equipped with rails like hydraulic shores. Like screw jacks, they are used only as the cross brace that can be rapidly set to hold shoring members in place.

Five sizes of air shores are designed for various trench widths and excavations. The 18- to 27-inch span air shore weighs 10 pounds. The largest air shore has a 60- to 96-inch span and weighs 30 pounds. Air shores are activated or expanded by compressed carbon dioxide (CO_2) gas.

Installing or removing air shores requires two operators. One or two additional operators are needed to hold the sheets and braces in place on each side of the trench while the air shore is being installed. The shore is held in place or positioned by one operator. The other person actuates the carbon dioxide valve and expands the air shore ram against the shoring to a pressure of 200 to 300 psi. When the shore has been expanded, the operator positioning the air shore rotates the locking collar and inserts a pin behind the locking collar and through the ram to form a mechanical lock. The carbon dioxide is then bled off from the ram and the air shore becomes a mechanical brace.

To release the air shore, carbon dioxide gas pressure is applied to the air shore until the ram expands enough to remove the locking pin. Next the carbon dioxide pressure is reduced until the shore is free and can be removed along with the sheeting or timbers used to shore the trench.

7.143 *Solid Sheeting*

Occasionally excavation is necessary in materials that will not stand vertical and, because of the location, it is not practical to slope the trench. Under these circumstances you must use a solid sheeting method (Figure 7.22). Timbers used in this method must be full-dimension rough-cut timbers. Minimum sizes are 2 x 8-inch rough timber for trenches up to 8 feet deep and 3 x 8-inch rough timber for trenches over 8 feet deep. Metal plates can be used as solid sheeting if properly engineered. See Table 7.2, page 407, for additional details and requirements applying to "running" conditions.

The horizontal shoring members are generally referred to as "stringers" or "walers." When purchasing shoring material, buy the square 6 x 6-inch rough timber (Figure 7.23), or the 8 x 8-inch rough timber because finished timber is smaller and not as strong. Either hydraulic or air shores or screw jacks can be used as cross braces.

The same rule applies on placing the cross braces as in hard compacted material: the maximum allowable distance between cross braces is four feet. In material that requires solid sheets, the sheeting should be driven into the excavation bottom one foot or more and kept below the bottom of excavation as it goes down. All timbers used must be free from splits, large knots, or loose knots. In some areas suitable timbers are extremely difficult to find. For this reason, metal sheeting may be better. Whether to rent or purchase shoring depends on expected needs and economics.

7.144 *Cylinder Shoring*

Cylinder shoring can be made from 1/4-inch or 5/16-inch thick steel plates that are four feet wide. These sheets are rolled into a four-foot high cylinder with a four-foot inside diameter (Figure 7.24). The top and bottom of each cylinder are reinforced with a welded angle iron rolled to fit the cylinders. Each cylinder section weighs either 550 (1/4-inch plate) or 675 (5/16-inch plate) pounds. This weight usually limits the use of cylinder shoring to street locations or areas where cranes or hydro lifts can be used to position the cylinders. Cylinders four feet high and four feet in diameter are easy to transport and handle, but are confining on large jobs. Larger cylinders can be made and used to rebuild or construct a new manhole.

One end of each section of a cylinder is provided with three male alignment pins approximately two inches in diameter and spaced at one-third intervals around the bottom circumference of the cylinder. The top or opposite end of the cylinder has three holes to accept the male locking pins. As the cylinder sections are stacked on top of each other, the locking pins prevent rotation or slipping of the sections. The pins thus maintain vertical alignment and structural safety at the section connections. Two hooks are installed inside the cylinders to make moving the cylinders easier.

The excavation is usually dug as deep as practical within the safety limitations. The bottom of the hole is made as level as possible and the first cylinder is set in the hole. From here on down it is all hand excavation. An operator and a ladder are put into the cylinder (Figure 7.25) and a bucket and rope are used to move the soil to the top of the ground. The cylinder is kept level as it goes down. When the proper depth is reached, another cylinder is set on the first one. This process is repeated until the excavation is completed.

Some agencies use square boxes instead of cylinders for shoring under these conditions. Another way to remove excavated soil from the inside of a cylinder or a box is to use a vacuum device like a Vac-All. Operators inside the cylinder loosen soil and guide it into the suction tube. This procedure works very well with dry soils and some wet soils. Problems develop when attempting to vacuum wet clays and any wet soil during freezing conditions because of freezing in the tube.

To reverse the procedure, backfill is dumped into the cylinders until about two feet of the bottom cylinder is filled and

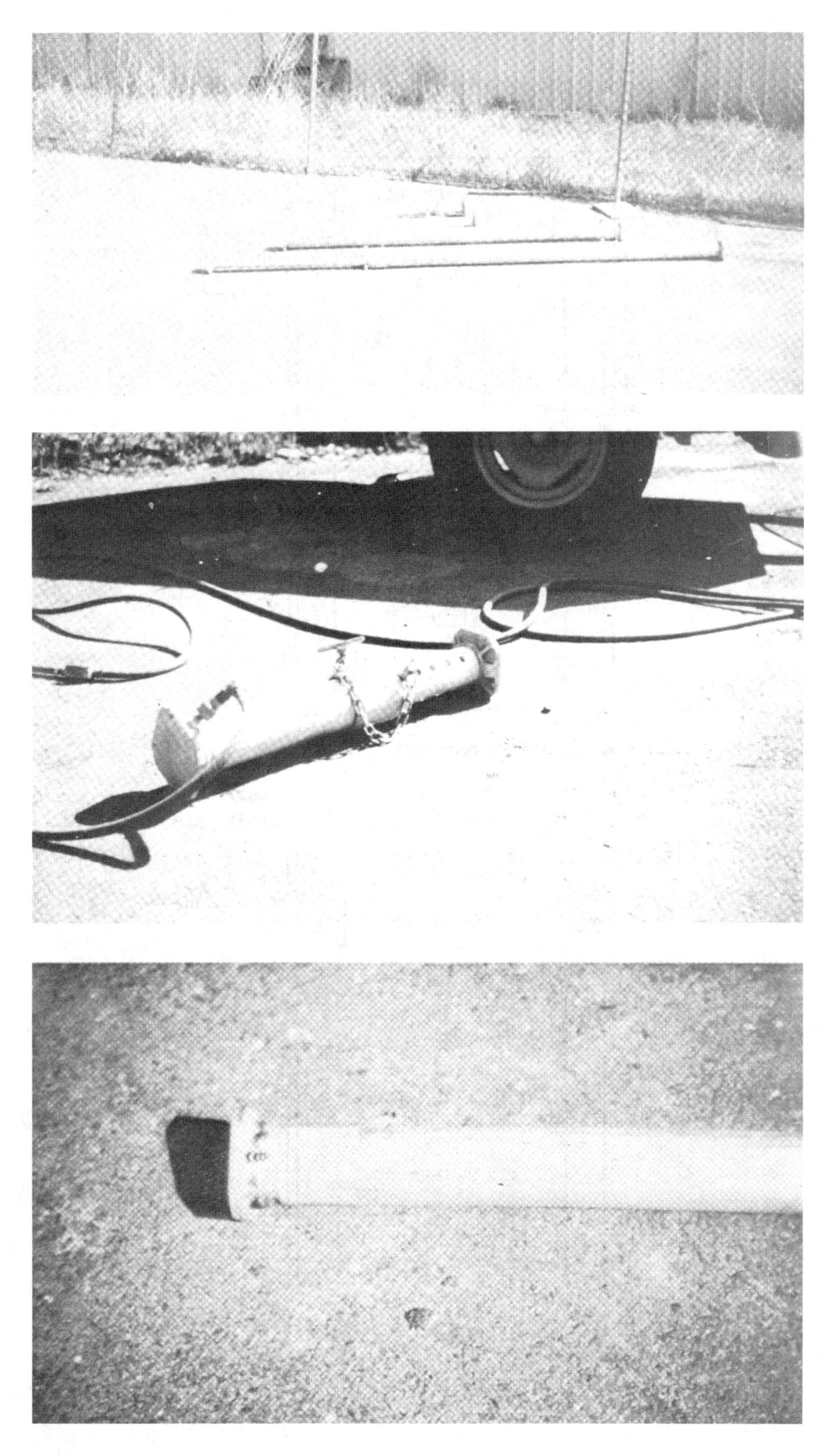

Fig. 7.21 Air shores

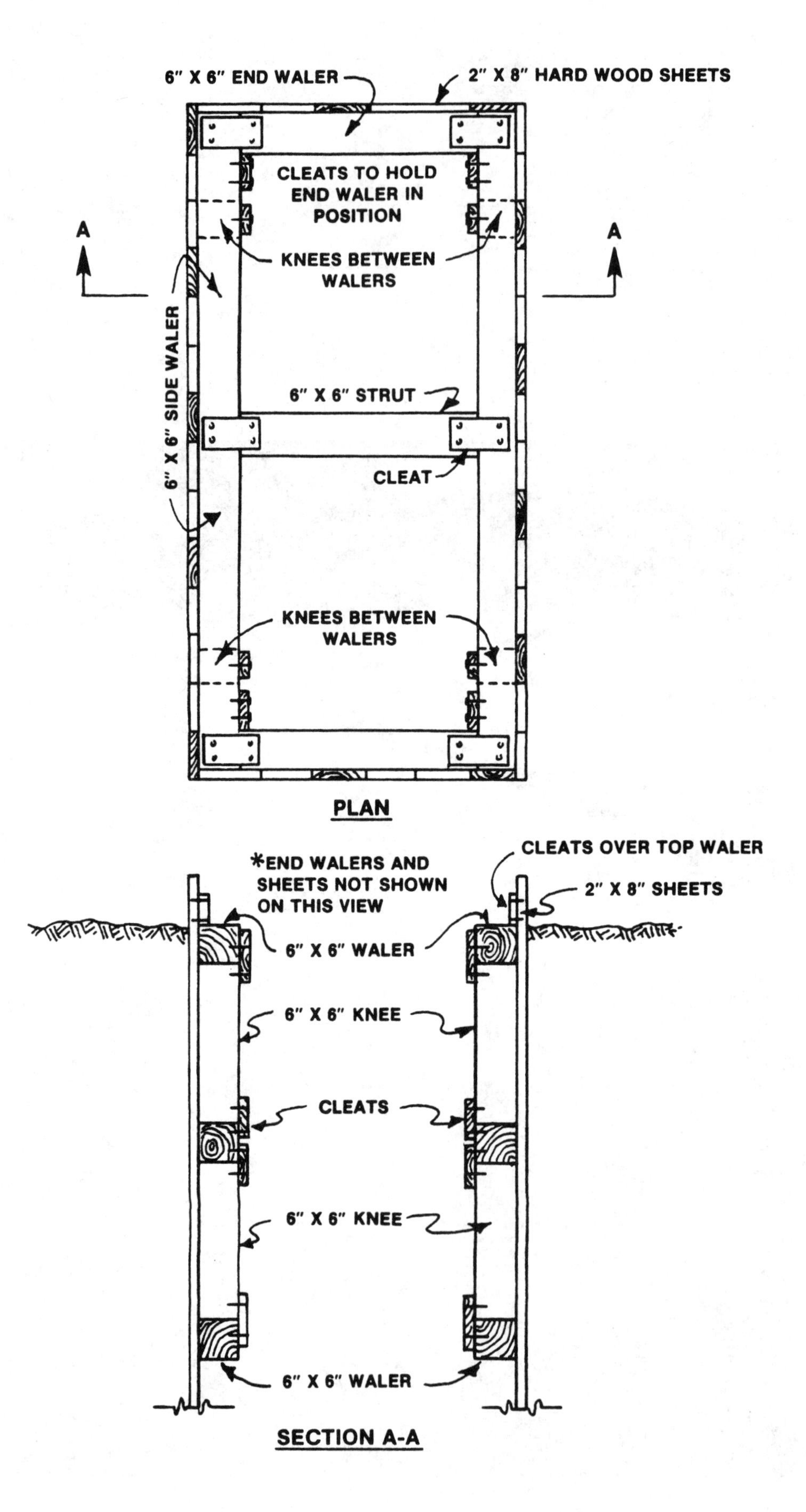

Fig. 7.22 Solid sheeting

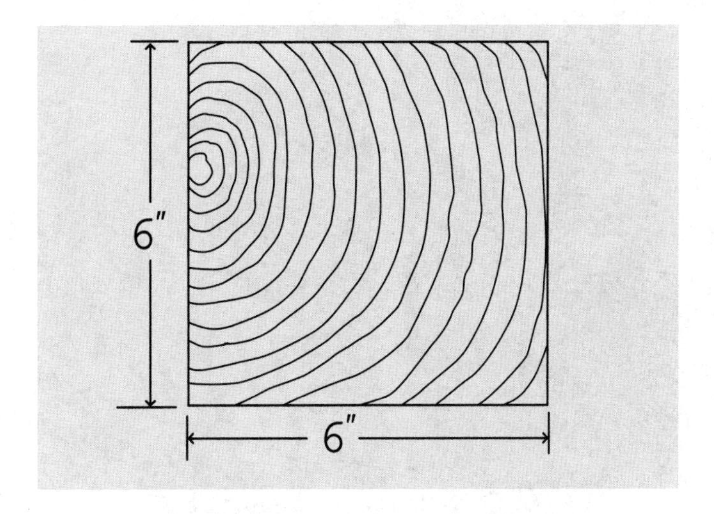

Fig. 7.23 Square 6 x 6-inch rough timber

Fig. 7.24 Cylinder shoring

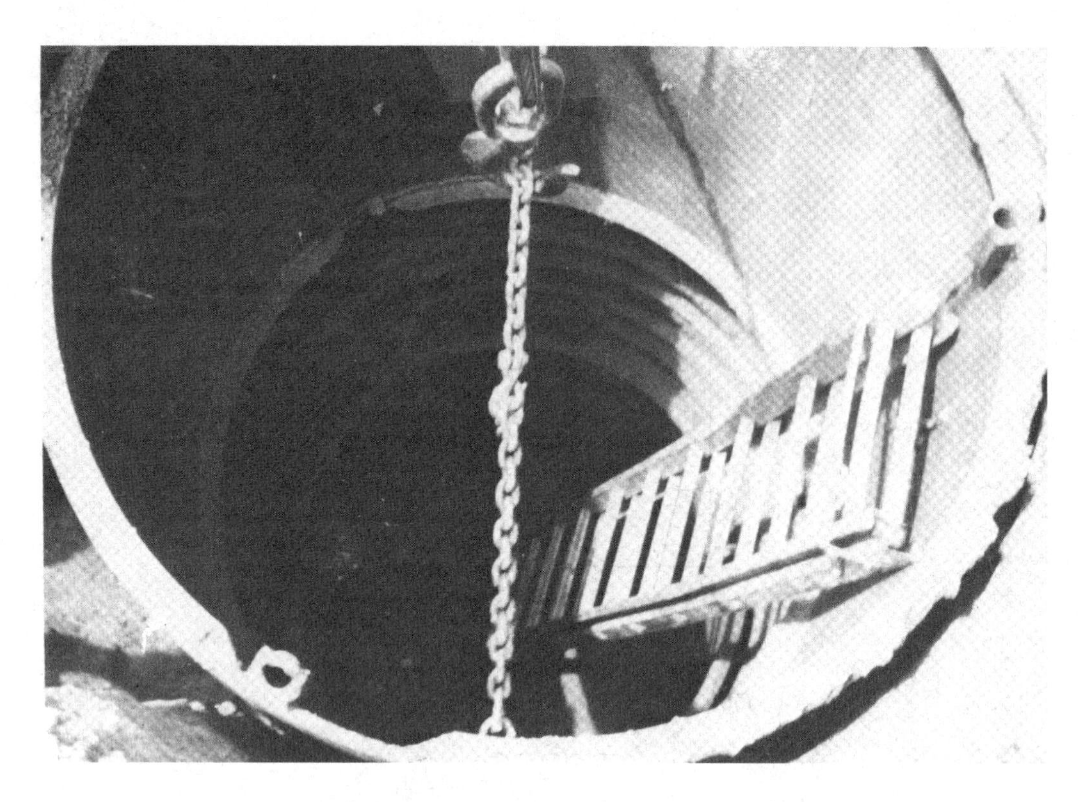

Fig. 7.25 Ladder inside cylinder

compacted. A chain is connected to the inside hooks of the bottom cylinder and all the cylinders are lifted up about two feet. Continue to add backfill in two-foot layers, compact, and raise the cylinder until the bottom edge is at the level of fill each time. Remove the top cylinder when the entire cylinder is raised to the level of the ground. Repeat this procedure of dumping fill into the remaining cylinders in two-foot layers until all cylinders are raised to the ground level and the excavation is filled to grade. (The depth of soil that can be compacted properly in each layer depends on the capability of the compacting equipment being used.)

The cylinder method is time-consuming, but requires less time than use of sheeting, uprights, stringers, and cross bracing. A limitation is that only the inside diameter of the cylinder is the working area. This approach has been used to make a new connection or repair a small portion of a main line in areas where this method was the best technique. If the working area is small and only a small portion of the main line needs to be exposed, cylinders can be very practical.

A vertical bore slightly larger than the cylinders can be made if the soil is stable enough to hold until the cylinders can be placed in the hole. This procedure saves the time required for hand excavation and is much faster. Excavations as deep as 30 feet have been completed very successfully using this method.

7.145 Shield Shoring

Shield shoring (Figure 7.26) is used during excavation and installation of various types of sewers in areas and soil conditions that present a combination of problems. Shield shoring consists of a rectangular box that must be open on the top and bottom and usually the ends are open too. The size of the shield depends on the size of sewer line being installed and the amount of space needed to install the pipe. Shields can vary in width from 30 inches to 10 feet and in length from 20 to 30 feet. Figure 7.27 illustrates the use of a shield shoring box.

Another type of shield shoring is an adjustable box that can provide protection for either trench or manhole work. Boxes of this type are certified by their manufacturer for use within specified excavation depths and widths. These boxes are usually fabricated in such a way that two boxes may be stacked to increase working depth and they are certified by the manufacturer for this use.

The city of Tacoma, Washington, has developed a shoring crib that was fabricated in their shop and can be installed in an excavation in less than 30 minutes. The four pictures in Figure 7.28 show how the crib is installed.

If the excavation is dug with a trencher, the shield is connected to the trencher and pulled along as the trench is dug (Figure 7.29). Pipe is laid as the trench is completed. If a backhoe or other bucket-type equipment is used to excavate the trench, the shield is stationary until a portion of the trench is dug; then the shield is hooked onto the excavation equipment and pulled into position.

Another type of shield has hydraulic rams that expand and make the rectangular box longer or contract to make it smaller. The purpose of the rams is to move the shield with its own power. In the expanded position, pipe is bedded and laid; then the rams are contracted and backfill is placed behind the shield. The rams are then expanded and force the shield ahead.

The initial cost for any shield is very expensive, but often may be cheaper than solid sheeting in the long run.

Fig. 7.26 Shield shoring

1. Hauling shield shoring box to job site

2. Unloading shield shoring box at job site

Fig. 7.27 Shield shoring box

3. Shield shoring box installed in trench

4. Operators working inside shield shoring box

Fig. 7.27 Shield shoring box (continued)

1. Dig trench.

2. Insert crib.

Fig. 7.28 Installation of a shoring crib

3. Install sheeting.

4. Shoring completed.

Fig. 7.28 Installation of a shoring crib (continued)

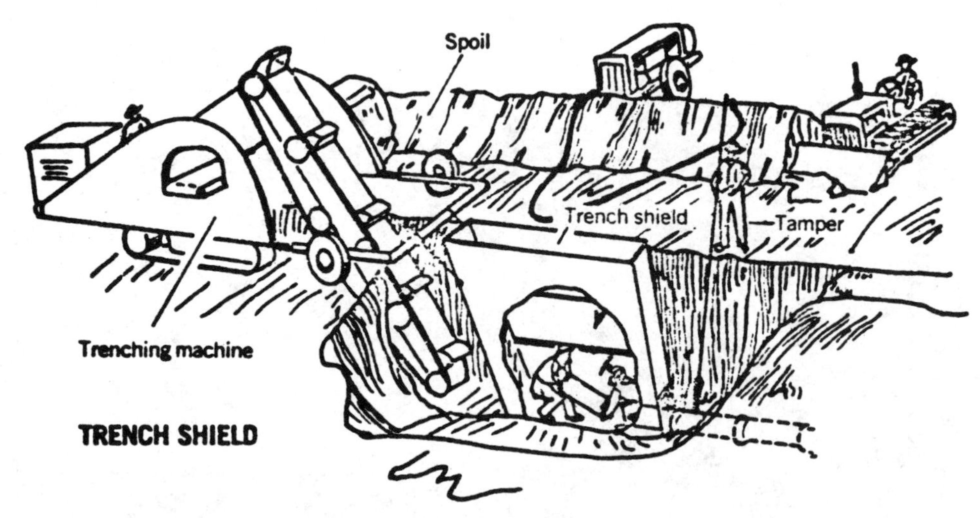

Fig. 7.29 Trench shield being pulled by trenching machine

7.15 Shoring Safety Reminder

This section has identified the common shores and their uses. Selection of shores depends on several factors and the job you are trying to accomplish.

The next portion of this chapter, Section 7.2, "Construction," contains additional detailed information about safe excavation procedures, selection and use of various types of shoring, and shoring size requirements for excavations in hard compact soil.

PROTECT YOUR OPERATORS AT ANY COST. Attend all trench safety lectures and seminars. Contact your own safety council and see what information is available and what is required. Be sure that everyone complies with all regulations. *ALL TRENCHES AND EXCAVATIONS ARE UNSAFE UNTIL PROPERLY SHORED, USED AND BACKFILLED* (Figure 7.30).

QUESTIONS

Write your answers in a notebook and then compare your answers with those on page 502.

7.1K List the different types of common shores.

7.1L What are the main advantages of hydraulic shores?

7.1M What are the main advantages and limitations of screw jacks.

7.1N When must solid sheeting be installed?

7.1O How is material excavated in cylinder shoring?

7.1P When is cylinder shoring used?

7.1Q Under what conditions is shield shoring used?

7.1R What is the major concern during a shoring operation?

END OF LESSON 1 OF 4 LESSONS
on
UNDERGROUND REPAIR

Please answer the discussion and review questions next.

DISCUSSION AND REVIEW QUESTIONS

Chapter 7. UNDERGROUND REPAIR

(Lesson 1 of 4 Lessons)

At the end of each lesson in this chapter you will find some discussion and review questions. The purpose of these questions is to indicate to you how well you understand the material in the lesson. Write the answers to these questions in your notebook before continuing.

1. What safety items must be considered when excavation is required to repair a sewer located in a street?
2. Why might a sewer line have to be dug up?
3. Why should other utility companies be notified before any excavation?
4. How can operators in trenches be protected from cave-ins?
5. How can you determine the shoring requirements in your area?
6. What are the different types of shoring and the advantages and limitations of each type?

Fig. 7.30 All trenches are UNSAFE until properly shored, used and backfilled

CHAPTER 7. UNDERGROUND REPAIR

(Lesson 2 of 4 Lessons)

7.2 CONSTRUCTION

7.20 Excavation and Shoring

Most sanitary and storm sewers are constructed within excavated trenches.[6] The following sections will outline the more important aspects of excavation for sewer construction and the shoring of these excavations to eliminate damage to adjacent property and to provide a safe work area during construction.

7.200 Excavation Equipment

The type of equipment to be used for excavating trenches for construction of sewers will be determined to a great extent by the size and depth of the sewer, type of material to be excavated, and the ground surface conditions. The more common types of excavation equipment and the conditions leading to their use are described in the following sections.

TRENCHING MACHINES. Trenching machines are a type of heavy equipment with a series of steel buckets mounted on an endless chain or a wheel. The buckets empty excavated material onto a conveyor belt that discharges it to the side of the trench. When used at moderate depths in stable types of soil where the vertical trench walls can be held with a standard type of shoring, trenching machines are quite productive.

BACKHOES. The most commonly used type of equipment for trench excavations is the backhoe. They are manufactured in many sizes with varying depth capabilities and bucket sizes. Backhoes have the ability to excavate trenches with either vertical or sloping sides and, with the addition of a cable sling, can be used to lower large-diameter sewer pipe into a trench, thus eliminating the need for a separate piece of equipment. A backhoe is particularly suitable for excavating through rocky and unstable types of soils. Small, rubber-tired backhoes are quite mobile and are often used for shallow and restricted excavations that are often encountered in easements.

CLAMSHELLS AND DRAGLINES. Clamshell and dragline buckets can be installed on most cranes. Clamshells are seldom used as the primary means of trench excavation due to their low productivity. However, this type of equipment is used when vertical lifts of excavated material in tight quarters are required, such as between shoring and at the headworks of sewer tunnels. Draglines, on the other hand, are most productive in open areas where the bucket may be cast into a wide sloping trench and the excavated material dropped to the side.

BORING AND TUNNELING MACHINES. Boring and tunneling machines are used for construction of sewer lines when they are too deep for trench excavations or the surface over the sewer cannot be disturbed.

SPECIAL EQUIPMENT. Special equipment including air compressors, hydraulic splitters, jackhammers, dynamite, water trucks, and dump trucks are sometimes needed to assist the previously discussed types of excavation equipment when large or solid rocks must be removed from the trench. Blasting of rocks is a hazardous operation requiring the services of specialists.

When high groundwater levels are encountered during trench excavation, the water can be removed from the excavation by use of sump pumps or by use of a system of *WELL POINTS*,[7] a header pipe, and a central pump. The number and spacing of well point connections to the header pipe depend on the ability of water to flow through the soil and the amount of water to be removed. Three-foot intervals are commonly used with a six-inch diameter header pipe and two-foot intervals with an eight-inch diameter header pipe. Whenever groundwater is controlled by lowering the water table, subsidence (dropping) of the ground in the nearby area may occur and cause damage to structures. After work is completed, the groundwater level is allowed to return to its original level at approximately the same rate it was lowered. This is accomplished by regulating the pumping rate used to remove the water. This type of dewatering often reduces the hazards of excavation in unstable soils.

Front-end loaders are used to place bedding and backfill material and to load excess excavated material into dump trucks for hauling away from the job site.

7.201 Location of Utilities

Most wastewater collection systems will be constructed next to or below existing underground utilities in existing streets.

---

[6] *NOTE: In some unusual circumstances, the use of excavated trenches is not appropriate or feasible and it may be necessary to construct sewers above ground. For instance, if surface conditions do not allow open excavation, if the correct grade would be too deep for trenching equipment, or if inverted siphons are impractical for spanning an extensive watercourse, above-ground sewers may be the appropriate alternative. It is also occasionally necessary to construct sewers in tunnels. These conditions are very unusual, however, and will not be described in this manual.*

[7] *Well Point. A hollow, pointed rod with a perforated (containing many small holes) tip. A well point is driven into an excavation where water seeps into the tip and is pumped out of the area. Used to lower the water table and reduce flooding during an excavation.*

Even though most construction plans will attempt to show the location of existing utilities, the information is usually based on utility companies' records and is subject to inaccuracies.

Extreme care must be taken to protect underground utilities when you are excavating near them. Most utility companies will, upon request, send a representative to show you the location and approximate depth of their underground utility. Centers for providing communication between a party planning to excavate and utility companies have been established in many localities. Many of these centers use the designation USA, an abbreviation for Underground Service Alert, to list their telephone number in the "white pages" of the local telephone directory. Use this service before you begin excavating for sewer construction and excavate carefully in the vicinity of an underground utility. These precautions will help eliminate costly delays and repairs caused by digging into or through a utility. Property damage and injuries can also be caused by the rupture of a utility conveying a hazardous material or electricity. Determining which party (contractor, sewer agency, or utility company) is responsible for payment of the cost of damages resulting from an excavation for a sewer can be difficult, time consuming, and expensive. Therefore, **TAKE THE TIME TO PREVENT ACCIDENTS.**

7.202 Trench Shoring

Hazardous trenching procedures are a prime cause of death and injury. The National Institute of Occupational Safety and Health (NIOSH) reports approximately 75 deaths and over 1,000 work-related injuries a year are caused by construction excavation cave-ins.

As a result of this high incidence of injury and death from excavation and trench cave-ins, the Federal Occupational Safety and Health Act (OSHA) established regulations for the protection of workers in excavations and trenches. This Act also requires each state to adopt similar regulations. For example, the State of California sets forth its regulations in Title 8, Sections 1503 and 1539-1547 of the Construction Safety Orders portion of the California Code of Regulations. The following sections contain excerpts from *TRENCH AND EXCAVATION SAFETY GUIDE*,[8] a document based on these regulations.

SUPERVISION. All work in an excavation must be supervised by a "competent person". A "competent person" is defined by OSHA as a person capable of identifying existing and predictable hazards in the surroundings, or working conditions that are unsanitary, hazardous, or dangerous to employees, and who has authorization to take prompt corrective measures to eliminate the hazards.

HAZARDS. Remove trees, poles, boulders, and similar objects that may be hazardous to workers. Do not allow work in or near the excavation until a qualified person has determined that no hazards to workers exist from possible moving ground. Inspect excavations after rainstorms, thaws, or other events that may influence the stability of the soil and increase hazards before workers are allowed to enter the excavation. Protect workers who enter excavations five feet deep or more with a system of shoring, sloping trench walls, benching, or equivalent alternative methods. When necessary, provide similar protection for workers in excavations less than five feet deep (four feet in some areas).

SPOIL. Dump excavated material (spoil) far enough from the trench so that it does not fall back into the trench. When trenches are five feet deep or more, locate the spoil at least two feet back from the edge and on only one side of the trench. Leave the other side open for access. Do not try to contain the spoil by any method that will disturb the soil already in place (such as driving stakes).

ACCESS. Provide a safe and convenient way for workers to enter and leave the excavation. In trenches four feet deep or more, provide a safe means of access, such as a stairway, ramp, or ladder, within 25 feet of any work area in the excavation. Don't climb up or down shoring braces for access.

CROSSINGS. Install trench crossings with standard guardrails and toeboards when the excavation is more than 6 feet deep. Don't try to jump across a trench.

UNDERMINING. Do not excavate beneath the level of the base of an adjacent foundation, retaining wall, or other structure until a qualified person has determined that the earthwork will not create a hazard to workers. Support undermined sidewalls so they will support anticipated loads. If the excavation endangers the stability of adjoining structures, shore, brace, or underpin those structures.

RETAINING WALLS. Do not use an existing wall or structure as a retaining wall until it has been determined that it will safely support the additional expected loads.

REMOTE WORK LOCATIONS. Provide barriers to prevent workers from falling into excavations. Barricade or cover all wells, pits, shafts, and caissons. Backfill temporary wells, pits, and shafts as soon as the operation is completed.

WATER ACCUMULATION. Use diversion ditches, dikes, and other methods to prevent surface water from entering the excavation and to drain surrounding areas.

VIBRATION OR SUPERIMPOSED LOADS. Use additional bracing to strengthen shoring in excavations located near streets, railroads, or other sources of vibration and external loads. Take similar precautions when excavations are made in areas that have been previously filled.

SHORING SYSTEMS. Provide devices that allow the upper cross braces to be set in place from ground level. In deep trenches where additional braces are needed, workers should proceed downward, protected by cross braces already set in place at the top. When removing shoring, use the reverse procedure by starting at the bottom and working upward to the top.

STANDARD SHORING SYSTEM. Install shoring in accordance with Tables 7.3 and 7.4 and Figures 7.31 through 7.37, or according to plans prepared by a civil engineer registered in the state where the work will be performed.

INSPECTIONS. Excavations and the adjacent areas must be inspected on a daily basis by a "competent person" for evidence of potential cave-ins, protective system failures, hazardous atmospheres, or other hazardous conditions. The inspections are only required if an operator exposure is anticipated.

[8] *TRENCH AND EXCAVATION SAFETY GUIDE, CAL/OSHA Communications, no longer in print.*

TABLE 7.3 METAL-WOOD SHORING FOR HARD COMPACT SOIL[a]

UPRIGHTS			BRACES (STRUTS) AT 8' ON CENTERS				STRINGER (WALER)
			ALUMINUM PIPE		STD. STEEL PIPE		
Depth (feet)	Horiz. Spacing (feet)	Wood Size (inches)	Min. ID (inches)	Max. Excav. Width (feet)	Min. ID (inches)	Max. Excav. Width (feet)	Wood Size (inches)
5	8	3x8	$2\frac{1}{2}$ ($3\frac{1}{2}$)	8 (10)	$1\frac{1}{2}$	3	
to	4	2x10	$2\frac{1}{2}$ ($3\frac{1}{2}$)	8 (14)	$1\frac{1}{2}$	3	4x4
7	2	2x8	$2\frac{1}{2}$ ($3\frac{1}{2}$)	8 (20)	$1\frac{1}{2}$	3	4x4
Over 7	8	4x10	$2\frac{1}{2}$ ($3\frac{1}{2}$)	6 (8)	2	6	
to	4	3x10	$2\frac{1}{2}$ ($3\frac{1}{2}$)	9 (11)	$2\frac{1}{2}$	12	6x8
10	2	3x8	$2\frac{1}{2}$ ($3\frac{1}{2}$)	12 (16)	3	15	6x8
Over 10	8	6x8	$2\frac{1}{2}$ ($3\frac{1}{2}$)	6 (7)	2 ($2\frac{1}{2}$)	8 (12)	
to	4	4x8	$2\frac{1}{2}$ ($3\frac{1}{2}$)	8 (10)	2 ($2\frac{1}{2}$)	10 (11)	8x8
12	2	3x8	$2\frac{1}{2}$ ($3\frac{1}{2}$)	10 (15)	$2\frac{1}{2}$ (3)	13 (15)	8x8
Over 12	8	6x8	$2\frac{1}{2}$ ($3\frac{1}{2}$)	5 (6)	2 ($2\frac{1}{2}$)	6 (10)	
to	4	4x10	$2\frac{1}{2}$ ($3\frac{1}{2}$)	7 (9)	2 ($2\frac{1}{2}$)	8 (12)	8x10
15	2	3x10	$2\frac{1}{2}$ ($3\frac{1}{2}$)	9 (13)	$2\frac{1}{2}$ (3)	13 (15)	8x10
Over 15	8	6x10	$2\frac{1}{2}$ ($3\frac{1}{2}$)	4 (5)	$2\frac{1}{2}$ (3)	8 (12)	
to	4	4x12	$2\frac{1}{2}$ ($3\frac{1}{2}$)	6 (8)	$2\frac{1}{2}$ (3)	10 (15)	6x12
20	2	3x12	$2\frac{1}{2}$ ($3\frac{1}{2}$)	8 (11)	$2\frac{1}{2}$ (3)	12 (15)	6x12

- Metal pipe braces must be schedule 40, standard steel pipe or equivalent.
- Timber must be "selected lumber."
- Timber members of equivalent "section modulus" may be used for uprights and stringers shown in these tables.
- See page 431 for screw jack installation.
- Numbers in parentheses indicate maximum safe span for a specified diameter pipe.
- Tables may be modified by a civil engineer.
- Metal sheeting or other material equivalent to the strength of the wood members may be used.
- Place stringers to develop maximum strength (long side horizontal).

[a] *TRENCH AND EXCAVATION SAFETY GUIDE*, CAL/OSHA Communications.

TABLE 7.4 HYDRAULIC SHORING FOR HARD COMPACT SOIL[a]

	UPRIGHTS		STRINGERS (WALERS)		BRACES (STRUTS)		
Depth (feet)	Horizontal Spacing (feet)	Size Aluminum Rail	Size Aluminum Rail	Vertical Spacing (feet)	Hydraulic Cylinders	Horiz. Spacing (feet)	Max. Excv. Width (feet)
5	8	8" Wide	6" Wide	5	2" ID-$2\frac{1}{2}$" OD	8 cc	12 20
to		Standard	Standard				
7	*	**	**				***
Over 7	8	8" Wide	6" Wide	5	2" ID-$2\frac{1}{2}$" OD	8 cc	9 20
to		Standard	Standard				
12	*	**	**				***
Over 12	6	8" Wide	6" Wide	5	2" ID-$2\frac{1}{2}$" OD	6 cc	9 20
to		Standard or	Standard or				
16	*	HD	8" Wide HD	5	2" ID-$2\frac{1}{2}$" OD		***
Over 16	6	8" Wide	6" Wide	4	2" or 3" ID	4 cc	9 20
to		Standard or	Standard or		or		
20	*	HD	8" Wide HD	4	$2\frac{1}{2}$" or $3\frac{1}{2}$" OD		***

* Plywood may be used behind uprights.
** See Hydraulic Shoring Association Manual for strength of rails.
*** Use a $3\frac{1}{2}$ x $3\frac{1}{2}$ x 3/16" steel oversleeve to Std. 2" ID. No steel oversleeve required on 3" ID.

- Tables may be modified by a civil engineer.

[a] *TRENCH AND EXCAVATION SAFETY GUIDE*, CAL/OSHA Communications.

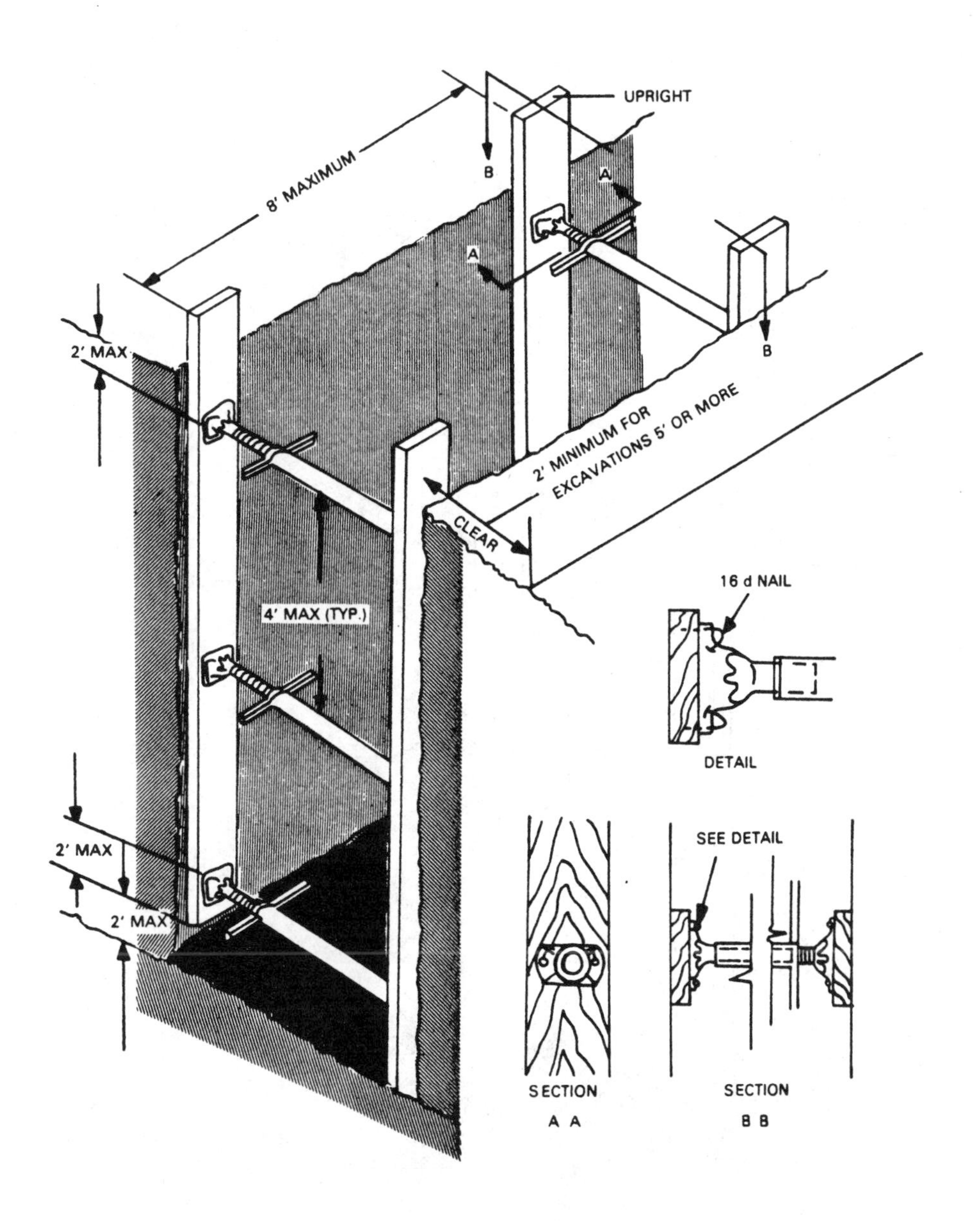

NOTE USE OF TRENCH SCREW JACKS HOLDING UPRIGHT PLANKS IN PLACE

Fig. 7.31 Minimum shoring requirement in hard compact soil
(Reproduced from *TRENCH AND EXCAVATION SAFETY GUIDE*, Cal/OSHA Communications)

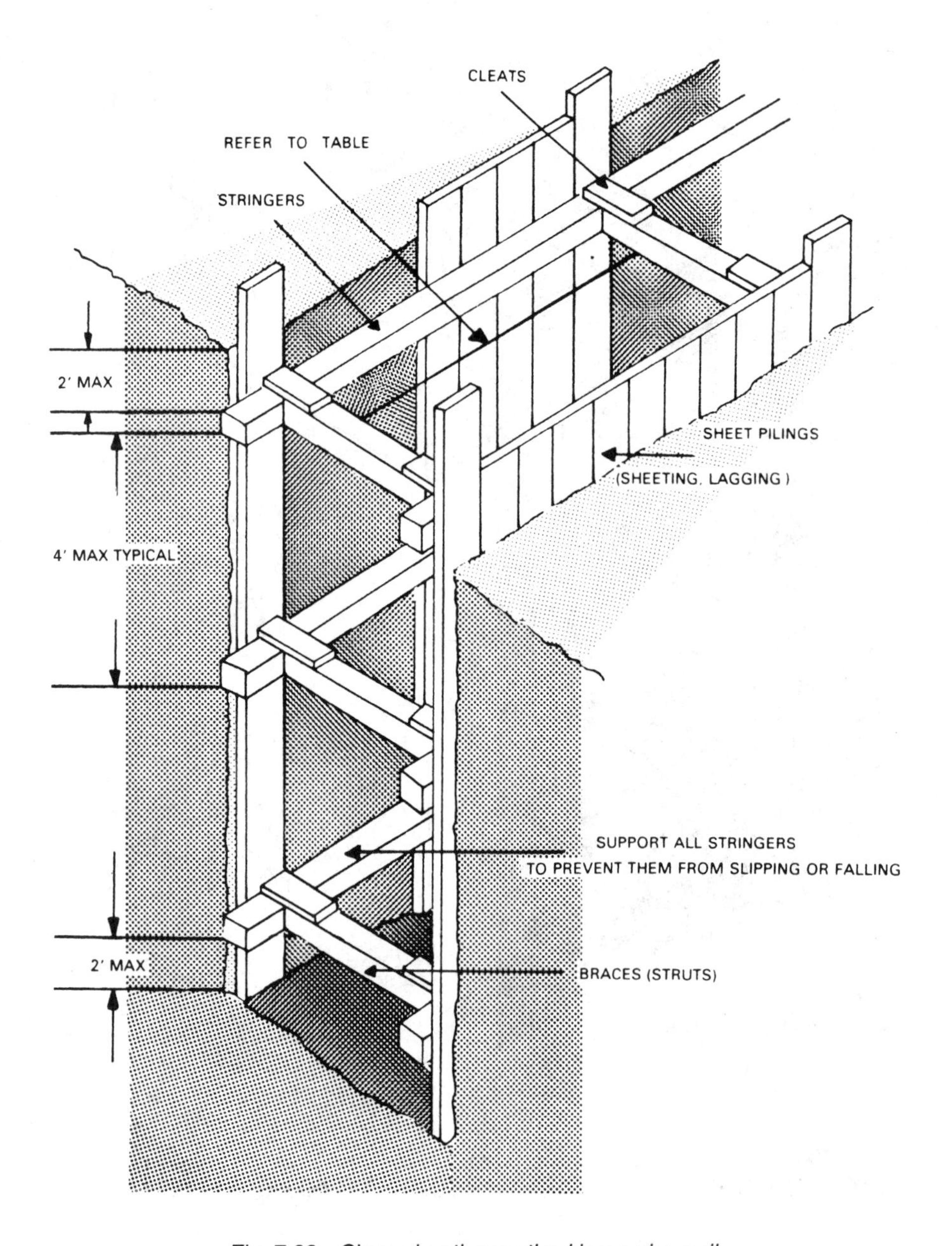

Fig. 7.32 Close sheeting method in running soil

(Reproduced from *TRENCH AND EXCAVATION SAFETY GUIDE*, Cal/OSHA Communications)

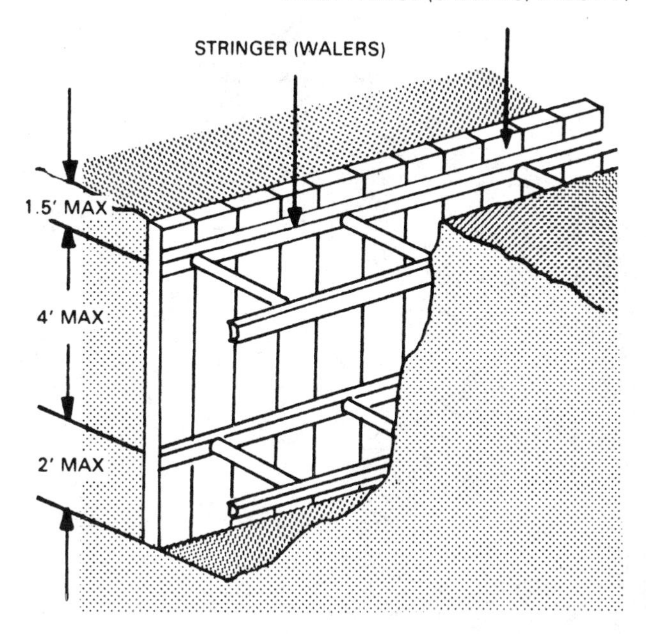

Fig. 7.33 Close sheeting method in running soil
(Reproduced from *TRENCH AND EXCAVATION SAFETY GUIDE*, Cal/OSHA Communications)

Fig. 7.34 Close sheeting shoring

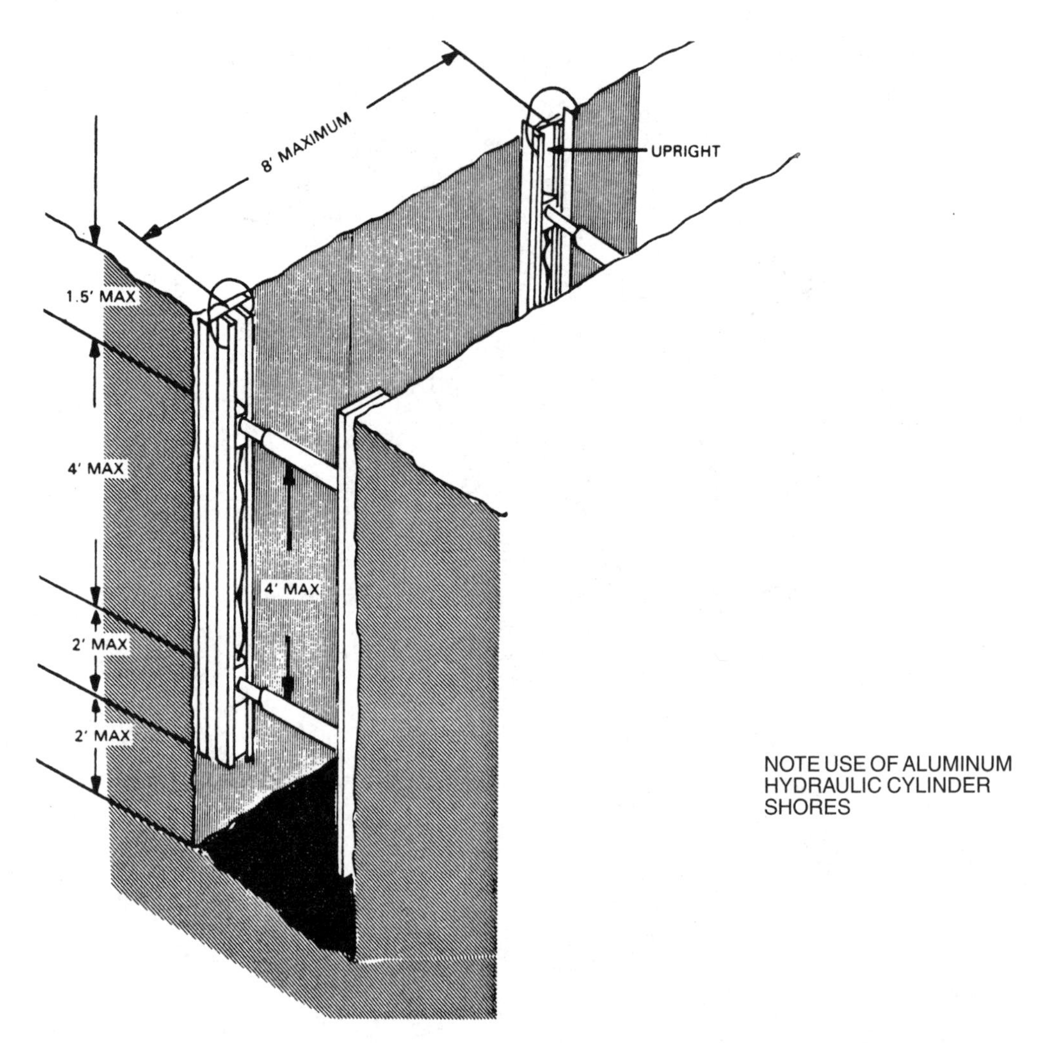

NOTE USE OF ALUMINUM HYDRAULIC CYLINDER SHORES

Fig. 7.35 Minimum shoring requirement in hard compact soil—hydraulic shore
(Reproduced from *TRENCH AND EXCAVATION SAFETY GUIDE*, Cal/OSHA Communications)

Fig. 7.36 Hydraulic cylinder shores

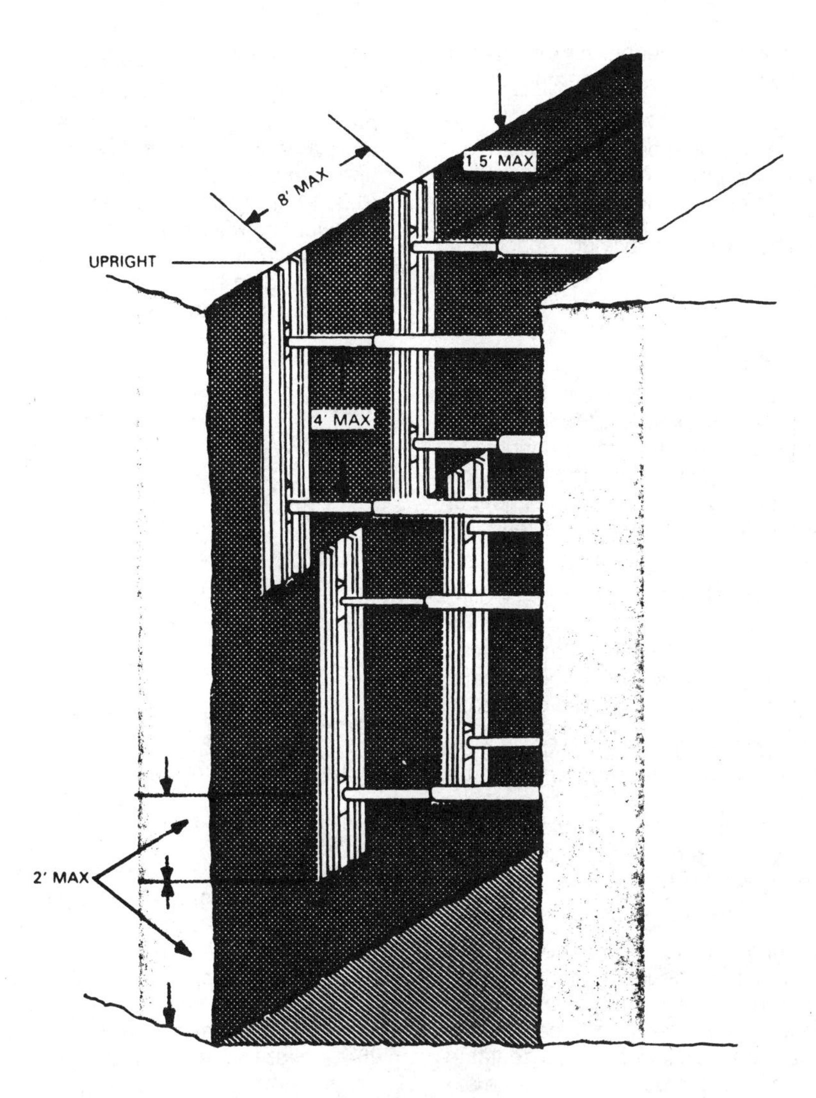

Fig. 7.37 Typical installation in hard compact soil—hydraulic shore
(Reproduced from *TRENCH AND EXCAVATION SAFETY GUIDE*, Cal/OSHA Communications)

ATMOSPHERIC MONITORING. Atmospheric monitoring prior to entering an excavation more than four feet deep is required if a hazardous atmosphere can reasonably be expected to exist. Excavations in landfill areas and in areas where hazardous substances are stored nearby, or in which connections or openings are being made to in-service sewers or manholes are examples of situations where a hazardous atmosphere could exist. Emergency rescue equipment such as breathing apparatus and safety harness/lifelines may also be required. If rescue equipment is required, it must be attended. Confirm these requirements with your local safety agency.

NOTE: Contact your local safety agency regarding shoring regulations that apply to your activities. Some safety regulations require shoring whenever trench depths are 4 feet or greater and others require shoring for depths 5 feet or greater.

Shoring must be composed of:

- Solid wood sheeting or wood sheet-piling not less than 2 inches thick.
- Plywood at least $1\,^1/_8$ inches thick.
- Wood uprights at least 2 inches by 8 inches (planks).
- Wood braces and diagonal shores at least 4 inches by 4 inches and not subjected to compressive stress in excess of values given by the following formula:

  $S = 1{,}300 - (20\ L/D)$

  Maximum Ratio (L/D) = 50

  L — length, unsupported (in inches)

  D — least side of the timber (in inches)

  S — allowable stress (in pounds per square inch of cross section)

- Wedge or cleat diagonal shores (struts) must be installed at the bulkhead end (partition at end of trench). If diagonal shores bear on the ground, they should not impose loads in excess of the test-determined soil-bearing values. (Allow for the horizontal component of force.) Do not place diagonal shores at an angle greater than 45 degrees from the horizontal. Securely anchor tie rods when they are used to restrain the top of sheeting (planks) or other restraining systems. Assume that there is full loading due to groundwater when using tight sheeting or sheet piling (unless full loading is prevented by weep holes, drains, or other methods).

- Provide additional stringers, ties, and bracing to allow temporary removal of individual supports. Required thickness of sheeting and spacing of shores are as follows:

Minimum Rough Thickness of Sheeting or Lagging	Maximum Spacing of Shores
2 inches	4 feet
3 inches	7 feet

TRENCH SHORING SYSTEMS. Do not slope a shored trench in excess of 15 degrees from the vertical. Make uprights at least 2 inches in nominal thickness. Plywood panels at least 1 1/8 inches thick may be installed behind uprights to hold loose material not likely to impose heavy loads. Extend uprights to the top of the trench and to within at least 2 feet of the bottom. If running soil (sand) is encountered, extend uprights to the bottom of the trench.

CROSS BRACES. Always use at least two cross braces. Install one horizontal brace for each 4-foot zone or partial zone measuring 2 feet or more. Use metal screw-type trench jacks with a base on each end or timbers placed horizontally against the uprights or stringers. Hydraulic braces may also be used.

PROTECTIVE SHIELDS AND WELDING HUTS (Figure 7.38). Plans for protective shields and welding huts must be prepared by a registered civil engineer. Construct protective shields and welding huts out of steel or other materials providing equivalent strength. They must provide at least as much protection as adequate shoring would provide.

SLOPING OR BENCHING SYSTEMS. When sloping is used as a substitute for shoring, the slope should be at least 3/4 horizontal to one vertical, unless the instability of the soil requires a flatter slope. Exceptions: In hard compact soil where the depth of the excavation or trench is 8 feet or less, make a vertical cut of 3.5 feet with a slope of 3/4 horizontal to one vertical. In hard, compact soil where the depth of the excavation or trench is 12 feet or less, make a vertical cut of 3.5 feet with a slope of one horizontal to one vertical. When benching in hard, compact soil, use a slope ratio of 3/4 horizontal to one vertical, or flatter, with a first vertical rise of 3.5 feet maximum and all subsequent vertical rises at a maximum of 2 feet. See Figure 7.39 for typical sloping and benching systems.

Exact slope angles and benching dimensions depend on the type of soil that is being excavated. OSHA has established three soil classifications, Type A, Type B, and Type C, in decreasing order of stability (see Section 7.132, "Soil Classification," page 408). A competent person must examine the work site and determine the soil classification to determine requirements for a specific site.

QUESTIONS

Write your answers in a notebook and then compare your answers with those on page 503.

7.2A What factors determine the type of equipment to be used for excavating trenches for construction of sewers?

7.2B How can high groundwater levels encountered during trench excavation be removed from the excavation?

7.2C What is the definition of a "competent person" who supervises trench excavation?

7.21 Controls

Controls must be used to ensure that sewer lines and appurtenances are constructed as planned. Precision is important when the slope of a typical 8-inch diameter sewer line is approximately a 3/16-inch drop for each 4-foot length of pipe (0.004 ft/ft). Even slight deviations from the designed slope of a sewer line will cause sags that interfere with the normal flow of wastewater and will cause maintenance problems in the future, including stoppages and the generation of toxic, odorous, and corrosive hydrogen sulfide gas.

7.210 Control Points

Using a sewer plan and profile, surveyors set control points, usually at 50-foot intervals, to guide accurate excavation. The control points are stakes driven into the ground or nails and washers (shiners) driven into pavement at a designated offset from the sewer center line as shown in Figure 7.40 (page 439). The precise elevation of a control point is determined by its relation to a previously determined control elevation called a "bench mark." Figure 7.41 illustrates the field notes of a survey to place and determine the elevation of the controls for the sanitary sewer set forth in the plan and profile illustrated by Figure 7.42.

Note that the control points have been set on a 5-foot right offset from the sewer center line as you face upstream with respect to the wastewater flow. The size of the sewer and method of trench excavation will determine the best side and offset distance for the controls. Also note that the distance of the control point from the starting point along the sewer is expressed in terms of 100-foot stations and usually to one-hundredths of a foot. For example, a manhole 425.26 feet from the beginning of construction would be located at station (sta)4+25.26. Construction controls placed at 50-foot intervals would be designated 0+00, 0+50, 1+00, 1+50 and so on.

Surveyors mark the offset of the control points from the sewer center line and the amount of cut in feet from the top of the control point to the invert of the sewer on a grade stake (or guard stake) or on the pavement exactly opposite the control. In Figure 7.40 the markings on the guard stake are as follows:

5 — Offset 5 feet
0+00 — Station 0+00 or distance from starting point
C — Cut
7^{99} — Cut 7.99 feet or depth of sewer invert below hub.

The cut from the top of the control points to the invert of the sewer is calculated by knowing the given elevation of the beginning point of the sewer to be constructed, the sewer slope (grade), the distance from the beginning point to the control

Fig. 7.38 Protective shields and welding huts

(1) In lieu of a shoring system, the sides or walls of an excavation or trench may be sloped, provided equivalent protection is thus afforded. Where sloping is a substitute for shoring that would otherwise be needed, the slope shall be at least 3/4 horizontal to 1 vertical unless the instability of the soil requires a slope flatter than 3/4 to 1.

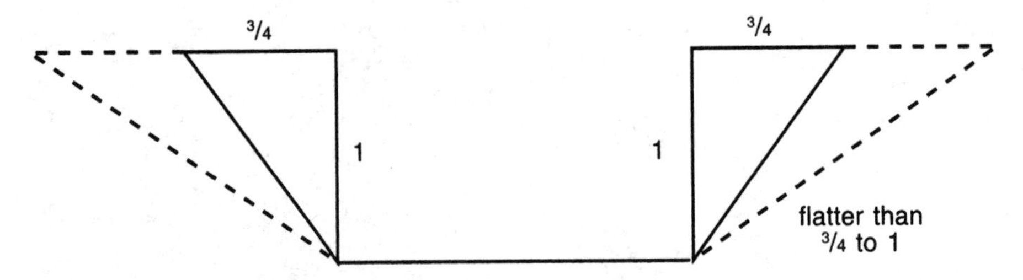

EXCEPTIONS: In hard, compact soil where the depth of the excavation or trench is 8 feet or less, a vertical cut of 3½ feet with sloping of 3/4 horizontal to 1 vertical is permitted.

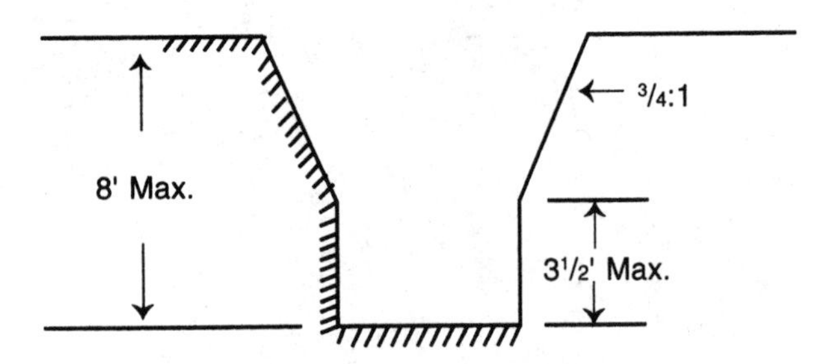

In hard, compact soil where the depth of the excavation or trench is 12 feet or less, a vertical cut of 3½ feet with sloping of 1 horizontal to 1 vertical is permitted.

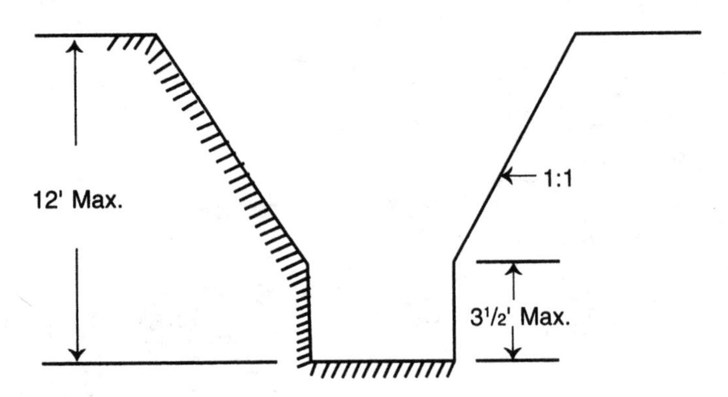

(2) Benching in hard, compact soil is permitted provided that a slope ratio of 3/4 horizontal to 1 vertical, or flatter, is used.

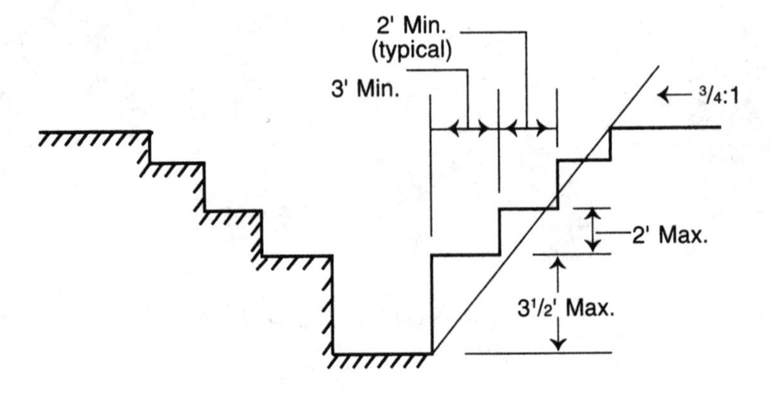

Fig. 7.39 Sloping or benching systems

(Reproduced from *CONSTRUCTION SAFETY ORDERS*, Title 8, Section 1541, California Code of Regulations)

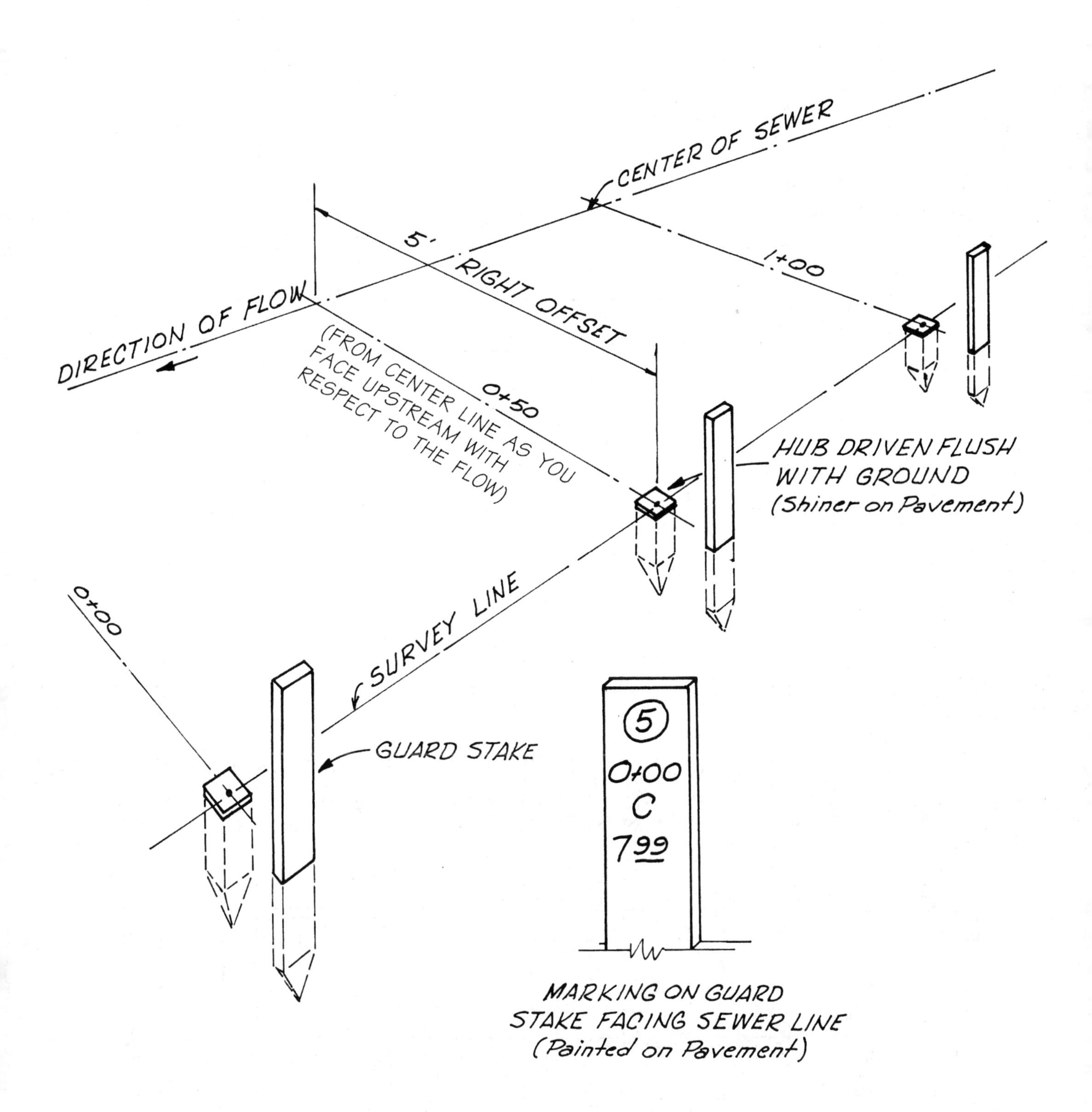

Fig. 7.40 Control points for sewer construction

6" LATERAL AND 4" BUILDING SEWER
EXTENSION TO 19105 BROOKVIEW DRIVE

OBJ.	B.S.	H.I.	F.S.	ELEV.	B.M.
INV. 8"					263.47
	13.85	277.32			
SANITARY M.H. RIM			5.65	271.67	
0+00			5.69	271.63	
+50			5.57	271.75	
1+00			5.38	271.94	
1+19			5.31	272.01	
+50			5.20	272.12	
+75			5.06	272.26	
2+75			5.34	271.98	
3+05			5.43	271.89	
M.H. RIM			5.64	271.68	
0+00				271.63	
	5.41	277.04			
STORM M.H. RIM			5.62	271.42	
(DIP)				6.46	
INV.				264.96	
PALMTAG M.H. RIM			5.43	271.61	
(DIP)				4.56	
				267.05	

NOTE: CONSTRUCTION CONTROLS SET 5' RIGHT OF SEWERLINE.

6/28/02 DRY & WARM
L.T. ∅
H.G. ⊼Ⅲ

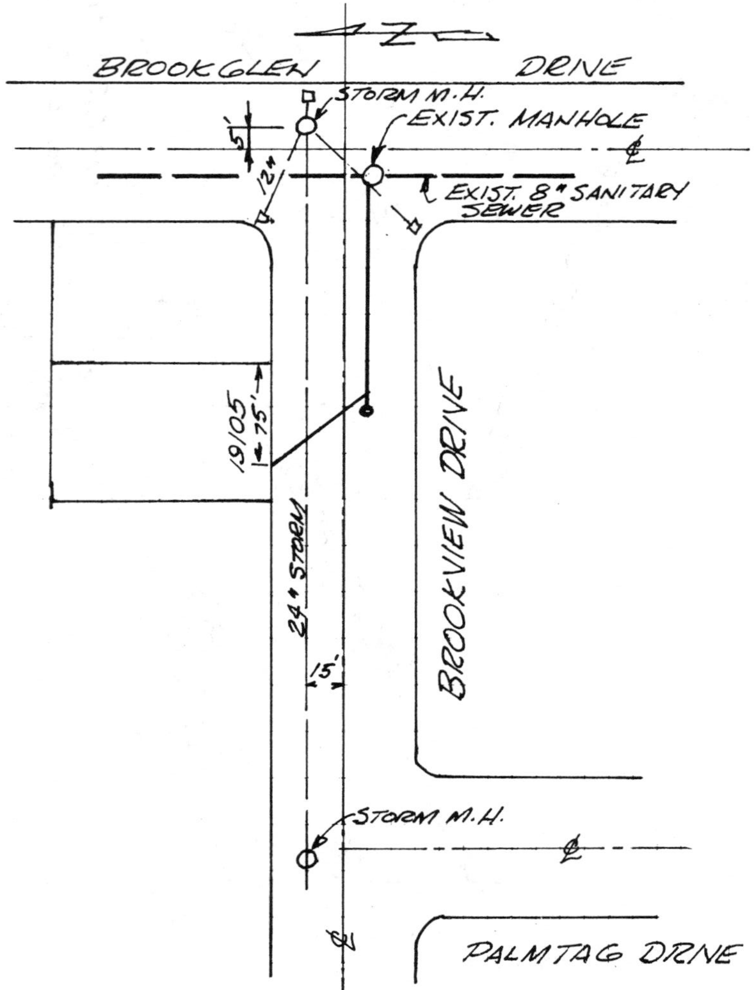

Fig. 7.41 Field notes

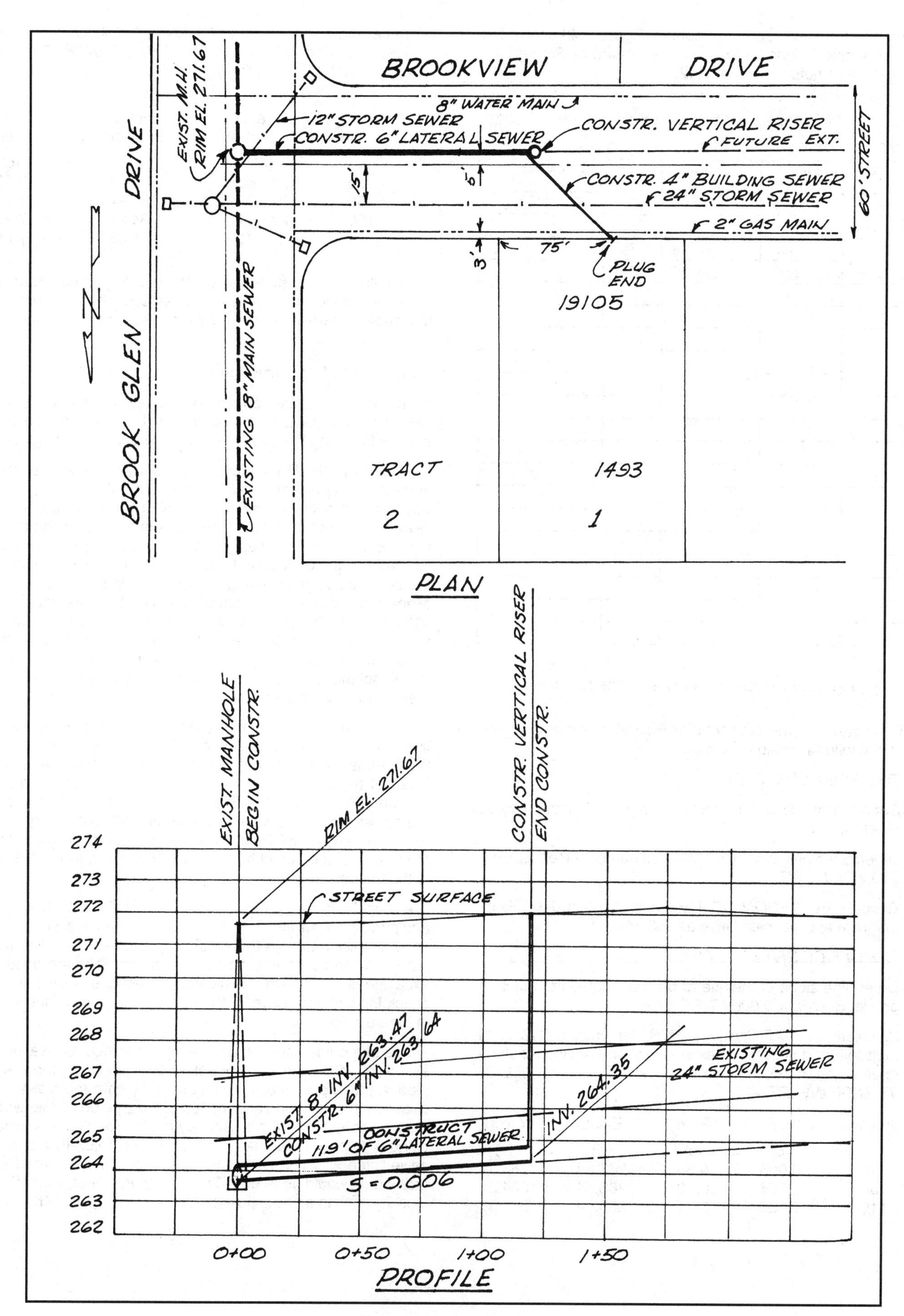

Fig. 7.42 Plan and profile

point, and the elevation of the control. Figure 7.43 shows the cut sheet calculations for the sewer illustrated by Figure 7.41 using the field notes in Figure 7.42.

PROJECT 6" LATERAL SEWER EXTENSION JOB NO. ______

LOCATION BROOKVIEW DRIVE DATE ______

Station	Elevation of Point	Elevation of Invert	Cut to Invert	Remarks
5' RIGHT 0+00	271.63	263.64	7.99	EXIST. MANHOLE AT BROOK GLEN DRIVE
0+50	271.75	SLOPE 0.006 263.94	7.81	
1+00	271.94	264.24	7.70	
1+19	272.01	264.35	7.66	CONSTR. VERTICAL RISER

Fig. 7.43 Cut sheet for sewer construction

To prepare a sewer cut sheet, as illustrated above by Figure 7.43, the following steps are taken:

IN THE FIELD, SURVEYOR:

1. Sets control hub or shiner 5 feet to right of line and records *STATION.*
2. Determines elevation of established hub or shiner, *ELEVATION OF POINT.*
3. Records *ELEVATION OF INVERT* for Station 0+00 Exist. manhole at Brookglen Drive as 263.64.

IN THE OFFICE OR IN THE FIELD:

4. Determine slope of the line to be constructed from plan or profile, *SLOPE = 0.006 FT/FT.*
5. Calculate *ELEVATION OF INVERT* for each *STATION* by multiplying Station (a) times slope (b) and adding result (c) to elevation of existing manhole invert (d) to obtain *ELEVATION OF INVERT* (e).

Station		Slope		Rise		Exist.		Invert
(a)	x	(b)	=	(c)	+	(d)	=	(e)
50		0.006		0.30		263.64		263.94
100		0.006		0.60		263.64		264.24
119		0.006		0.71		263.64		264.35

6. To calculate *CUT TO INVERT,* take *ELEVATION OF CONTROL* (a) and subtract *ELEVATION OF INVERT* (b), to obtain *CUT TO INVERT* (c).

Station	Elevation of Control		Elevation of Invert		Cut to Invert
	(a)	–	(b)	=	(c)
0+00	271.63		263.64		7.99
0+50	271.75		263.94		7.81
1+00	271.94		264.24		7.70
1+19	272.01		264.35		7.66

By following this step-by-step procedure, you can learn to calculate cuts to inverts and others can easily check your work and show you where any errors were made in calculations.

7.211 Use of Construction Controls

The line and grade controls for the sewer are used first to determine the location and depth of the sewer trench and second for laying the sewer pipe to the proper line and grade.

STRING LINE. The project supervisor will place steel rods immediately adjacent to the control point (stake or nail) and will mark on each rod an even number of feet of cut to the sewer invert. For example, if the cut from a control point to the invert is 7.99 feet, the supervisor would mark a 9.00-foot cut by measuring 1.01 feet up from the control point to a point on the steel rod. A grade string line then would be stretched between three consecutive control points on the steel rods. The string should be perfectly straight; any departure from a straight line indicates an error has been made in either the control survey or setting of the string line. Figure 7.44 illustrates setting of the grade string line and the three-point method of detecting errors.

The excavation equipment operator uses the grade string line to control the depth of cut. In most instances sewer trenches are excavated a predetermined distance below the invert of the sewer to provide for the thickness of the pipe's wall and for imported bedding under the pipe. For instance, if the sewer pipe has a wall thickness of 0.10 foot and a 0.25-foot thick bedding will be used, a 9-foot cut from the grade line to the pipe invert would become a 9.35-foot cut to the bottom of the trench.

A trenching machine, as described in Section 7.200, is equipped with an indicator that follows along the string line when the operator has the excavating bucket line on line and grade. When backhoes, clamshells, or draglines are used to excavate a trench, the equipment operator's assistant will check the line and grade by the use of a grade pole illustrated in Figure 7.45.

The string line used for controlling the trench excavation is also frequently used as a control guide when laying the sewer pipe. A grade pole is used to measure the calculated distance between the string line and the top of the pipe being laid when the pipe invert is on the prescribed grade. For example, when using a 9-foot string line to lay an 8-inch (0.67-ft) diameter sewer pipe with a 0.12-foot wall thickness, the length of the grade pole would be (9.00 – 0.67 – 0.12) 8.21 feet. Controlling pipe laying with a grade pole is illustrated in Figure 7.46.

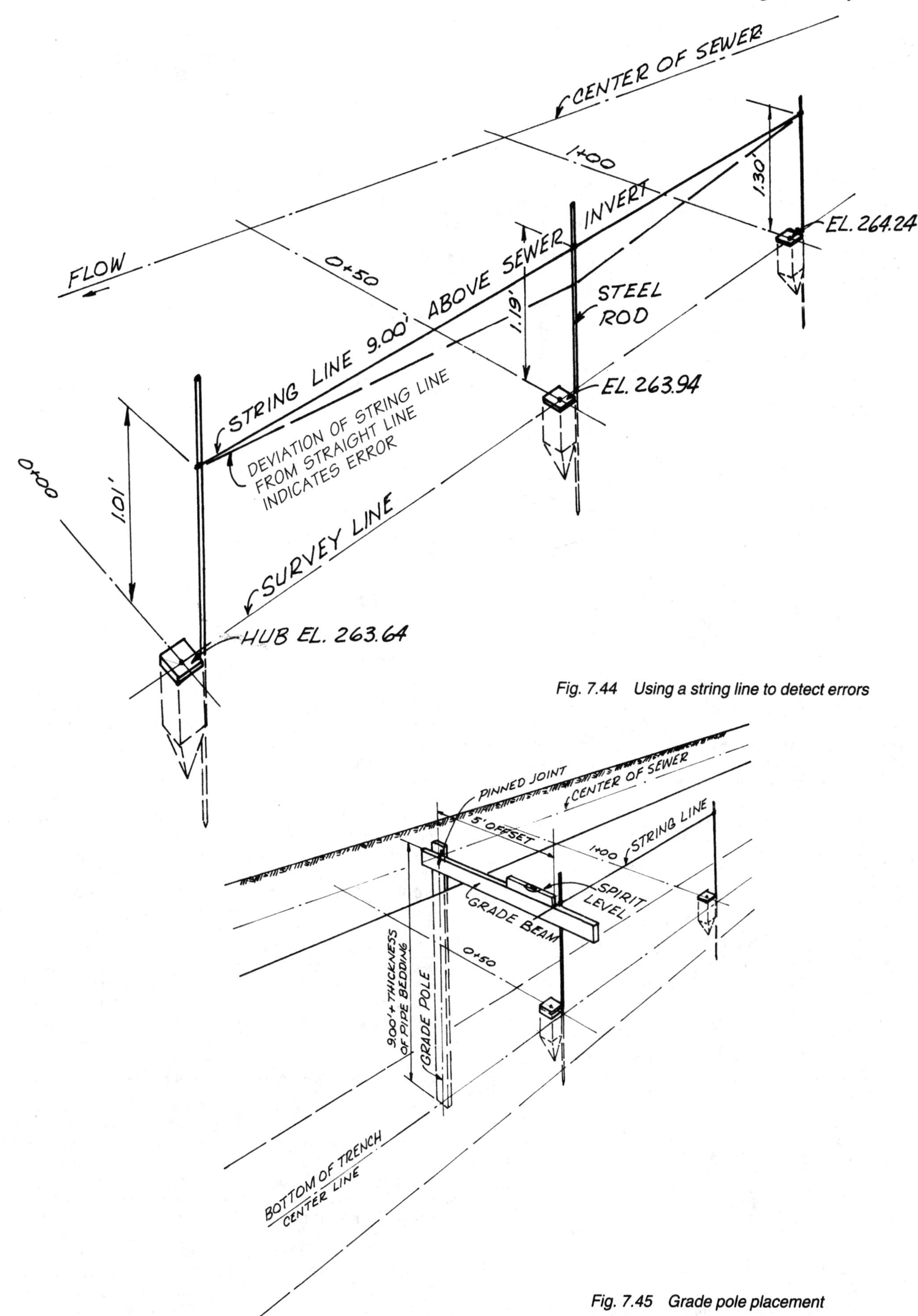

Fig. 7.44 Using a string line to detect errors

Fig. 7.45 Grade pole placement

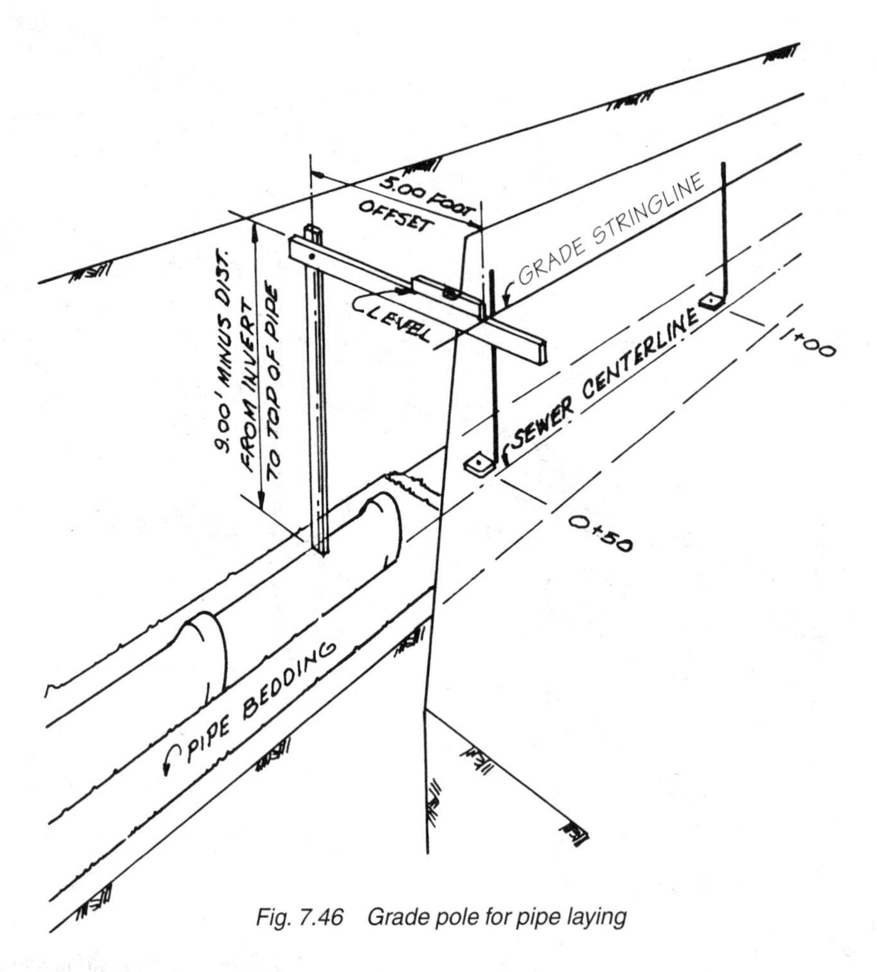

Fig. 7.46 Grade pole for pipe laying

BATTER BOARD. Batter boards can be used as illustrated in Figure 7.47 to control the laying of sewer pipe after the trench has been excavated. However, this type of control is difficult to set up, particularly for wide trenches, and the boards interfere with placement of the pipe.

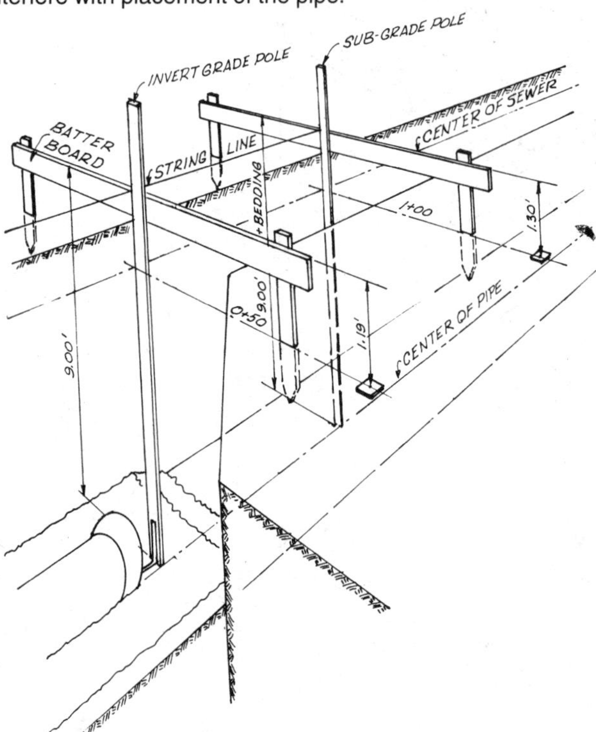

Fig. 7.47 Grade control using batter boards

FIXED-BEAM LASER. A fixed-beam laser projects a bright red, highly focused beam of light along the line of the sewer to be constructed at a determined distance above the sewer invert grade. The beam is intercepted by a target creating a small bright spot of light that serves as a reference point. Refer to the manufacturer's instructions to set the laser for the desired grade of the sewer line. Position and level the laser on the sewer line in one of the several locations illustrated in Figure 7.48. Use the control points described in Section 7.210 to position the laser on the designated line and grade. Measure the projected laser beam from another control point at least 100 feet ahead of the laser and make the appropriate adjustments to its position. Several types of targets are illustrated in Figure 7.48; use one of these to determine the proper depth of trench excavation and pipe position. When the laser is positioned in the sewer line for pipe laying control, check the beam for deflection by atmospheric conditions and take appropriate ventilation measures to eliminate any inaccuracy.

7.22 Sewer Pipe Bedding and Initial Backfill

Sewer pipe bedding is the material upon which the pipe is laid and which serves as its foundation. Initial backfill is the material surrounding the sewer pipe. The type of bedding material, initial backfill, and the care used in placement greatly affect the ability of a pipe to support the load created by backfilling the trench. The following sections briefly describe types of bedding and initial backfill materials and their ability to assist sewer pipe in supporting backfill loads. For more information on pipe bedding and initial backfill, refer to Chapter 9 of *GRAVITY SANITARY SEWER DESIGN AND CONSTRUCTION*, published by the Water Environment Federation (WEF), 601 Wythe Street, Alexandria, VA 22314-1994.

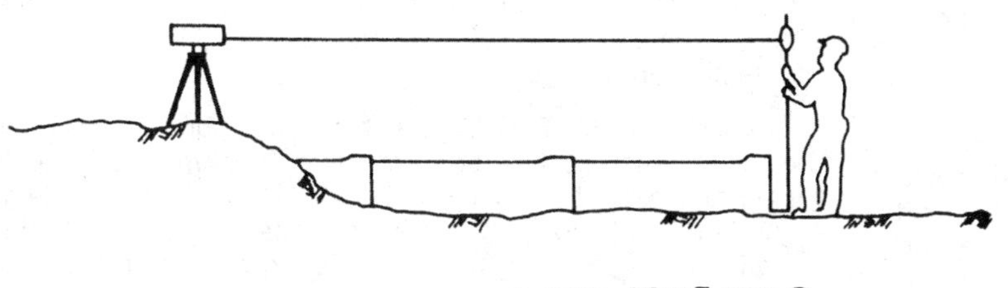

A SEWER LASER CAN BE SET UP ON A TRIPOD OR A THREE-POINT TRIVET PLATE IN THE EXCAVATION, ABOVE IT OR ON THE PIPE. THE LASER TARGET IS MOUNTED ON A POLE AND ADJUSTED TO GIVE THE DISTANCE FROM THE BEAM TO THE PIPE INVERT. A LEVEL VIAL ON THE POLE INDICATES A VERTICAL POSITION.

THE VERSATILITY AND FLEXIBILITY OF A SEWER LASER PERMITS A VARIETY OF OPEN EXCAVATION SET-UPS WITH THE BEAM PROJECTED DOWN THE CENTER LINE OF THE PIPE OR OVER-THE-TOP.

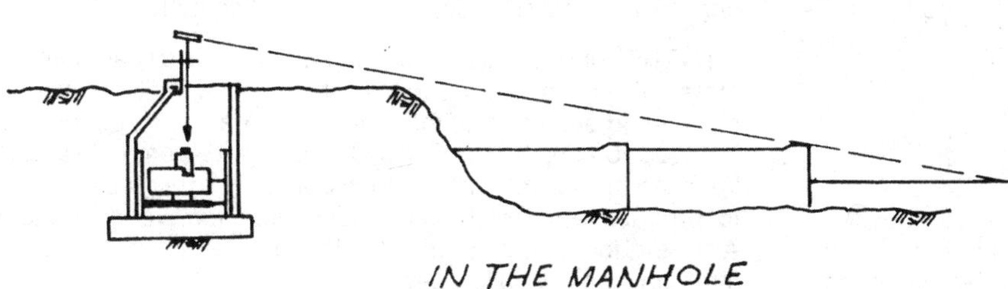

A SEWER LASER CAN BE SET UP IN A MANHOLE UTILIZING A TRANSIT TO SET THE SEWER LINE ACCURATELY. THE TRANSIT IS PLUMBED OVER THE LASER ON A MOUNT THAT CLAMPS TO THE MANHOLE EDGE. THE LASER BEAM IS PROJECTED ALONG THE PIPE CENTER LINE.

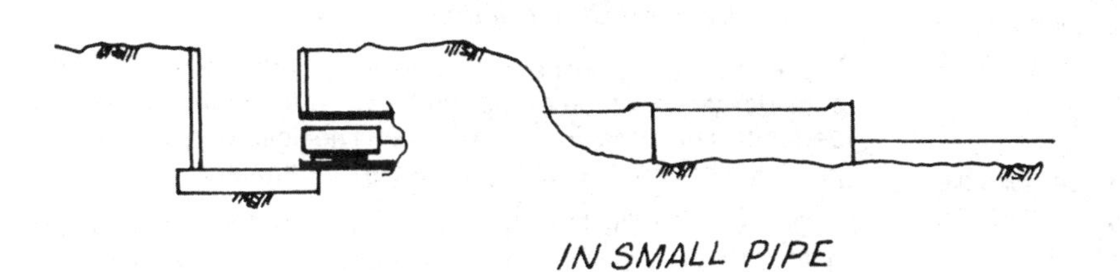

SOME SEWER LASERS CAN BE SET DIRECTLY INSIDE IN PIPES AS SMALL AS 6 INCHES IN DIAMETER. THIS ALLOWS FAST SET UPS THE SECOND DAY AS WELL AS THE VERSATILITY TO MEET SITUATIONS IN WHICH THE LASER CANNOT BE SET UP IN A MANHOLE.

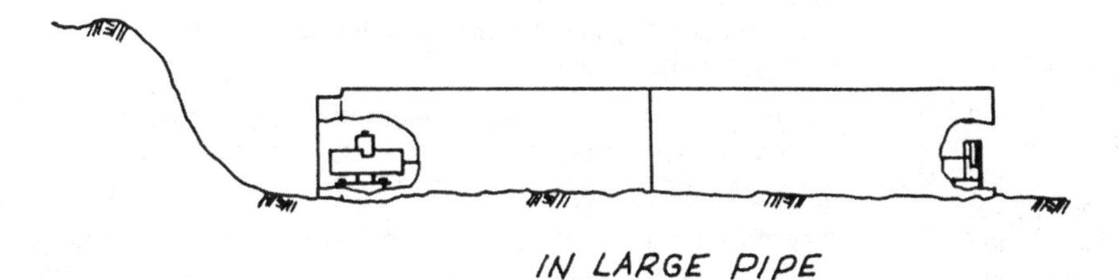

FOR LARGE PIPE, A LASER CAN BE SET UP DIRECTLY ON THE INVERT OF THE PIPE USING THE THREE-POINT TRIVET PLATE.

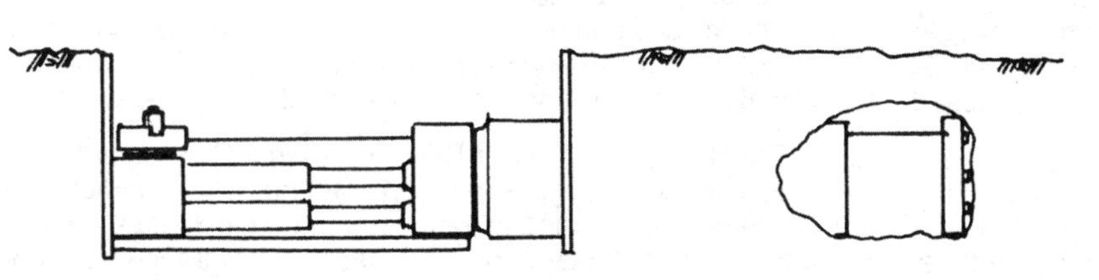

ELECTRONIC SELF-LEVELING SEWER LASERS CAN ALSO BE USED TO PROVIDE LINE AND GRADE CONTROL IN PIPE JACKING OPERATIONS. THE LASER IS SET UP IN THE JACKING PIT AND THE TARGET IS MOUNTED ON THE CUTTING SHIELD.

Fig. 7.48 Grade control using fixed-beam laser

Also, manufacturers of most sewer pipe will, upon request, supply information on the effects types of bedding and initial backfill have upon the load supporting capacity of their pipe.

7.220 Effects of Bedding and Initial Backfill on Rigid Pipe

Manufacturers of rigid sewer pipe usually will specify the load supporting strength of their pipe based on a "three-edged bearing test" conducted in a laboratory and usually expressed as pounds per lineal foot of pipe. For example, if the standard strength of 8-inch diameter vitrified clay pipe is 2,200 pounds per foot, it means the pipe will support this load without cracking.

Similarly, types of sewer pipe bedding have been given a standard class designation indicating the ability of the bedding to assist sewer pipe in supporting trench backfill loads. These bedding designations or ratings take into account the three-edge bearing test load. The rating is called a "bedding factor" and the three-edge bearing load is given a value of 1.

CLASSES OF BEDDING. The standard classes of rigid sewer pipe bedding and their load factors or "bedding factors" are shown in Figure 7.49. For example, an 8-inch vitrified clay pipe that has a three-edge bearing load supporting strength of 2,200 pounds per foot will have a supporting strength of (2,200 lbs/ft x 1.5) = 3,300 pounds per foot when laid on a Class C type of bedding.

FACILITATION OF PIPE LAYING. In addition to assisting sewer pipe in supporting trench backfill loads, a granular type bedding, such as Classes B and C, provides an easy means of establishing a true grade on an imperfect and undercut trench bottom. Also, granular material can easily be arranged to provide holes for pipe bells when laying sewer pipe and to provide for the specified surface contact between the pipe barrel and bedding material.

For these reasons, Classes B and C are the most commonly used types of sewer pipe bedding. Class D is impermissible bedding which means that this type of bedding should not be used unless specific construction conditions require it.

7.221 Effects of Bedding and Initial Backfill on Flexible Pipe

Flexible pipes depend on the soil surrounding them to help support the load of trench backfill. As the weight of backfill exerts pressure on the pipe, it flexes or deflects, as illustrated in Figure 3.23 (page 65). The sides of the pipe move outward against the soil. Unless the surrounding soil or bedding resists this flexing at some point, the load will simply deflect the pipe to the point of collapse. Therefore, the material and compaction of the initial backfill around a flexible pipe are very important to ensure that it will not be deflected beyond the allowable amount when the trench backfill is applied and compacted. As a rule of thumb, deflection of flexible pipe should not exceed five percent of the inside diameter of the pipe.

For a given load on a flexible pipe, the deflection of the pipe will decrease in proportion to the stiffness of the pipe and the compaction of the soil. Also, as the pipe stiffness diminishes, the need for a greater support from the enveloping soil increases. Data on long-term deflection of flexible pipe in different bedding and backfill conditions can usually be obtained from pipe manufacturers. Also, information on flexible pipe bedding classifications and deflection can be found in *GRAVITY SANITARY SEWER DESIGN AND CONSTRUCTION*, published by the Water Environment Federation (WEF).

7.222 Bedding Materials

Some comments on the sewer pipe bedding and initial backfill materials illustrated in Figure 7.49 will be helpful in understanding their use.

GRANULAR MATERIAL. Crushed rock aggregate, with 1/4-inch to 3/4-inch particle size, is the material most commonly used to construct the Classes B and C bedding. It has a fairly high bedding factor (1.5 to 1.9), is moderately priced, and is relatively easy to work with when laying pipe. This type and size of material is superior to sand or pea gravel for bedding and initial backfill material. In some locations water may flow through the bedding material in the trench to downstream manholes where the water may infiltrate into the collection system.

Greater care must be taken in compacting sand bedding material than when aggregate is used. Also, sand has a tendency to move if water jetting is used to compact the subsequent trench backfill or if the trench area is re-excavated at a later time to construct other facilities. Pea gravel, due to its smooth and somewhat circular shape, also has some of the undesirable characteristics of sand bedding material.

NATIVE MATERIAL. Sometimes it is not practical to import material to construct a bedding or the soil conditions are unsuitable. In such cases, the native soil in a trench bottom can be used to construct a Class C bedding. Extreme care must be taken in excavating the trench bottom to true grade in order to produce an undisturbed and compacted trench bottom. Also, the low bedding factor of 1.1 requires the use of high-strength pipe or shallow sewer depths.

7.223 Bedding Construction

Some sewer pipe bedding and initial backfill construction considerations become apparent as you study the classes of bedding illustrated in Figure 7.49. The following comments point up some of the less apparent considerations.

CLASS A BEDDING. When a concrete bedding is used, the rigid pipe must be supported (usually on concrete blocks or clay bricks) and held to true line and grade when the concrete is placed under the pipe. This is not an easy task since the pipe, particularly in smaller diameters, will tend to float out of alignment and grade.

The concrete can be poured into the trench without use of forms providing the bottom is undisturbed, clean, and compact, and the trench is not too wide.

When a concrete backfill is used over a granular bedding, the problem of the pipe floating during placement of the concrete is lessened. However, uniform placement is essential to maintaining alignment. Also, a waterproof liner may have to be placed over the top of the granular bedding to stop a rapid flow of water from the concrete into the bedding.

CLASSES B AND C BEDDING. When a Class B or C bedding is constructed, it is essential that the bottom and sides of the trench be cleaned of all loose material prior to the placement of the granular bedding material. Initial trench backfill of native material for a height of 6 inches above the pipe should be free of rocks, mechanically compacted in about 6-inch layers with the proper moisture being maintained in the backfill material. Since it is difficult and expensive to compact native backfill under and around the pipe, granular material, which requires no compaction, is quite often substituted for the native backfill material.

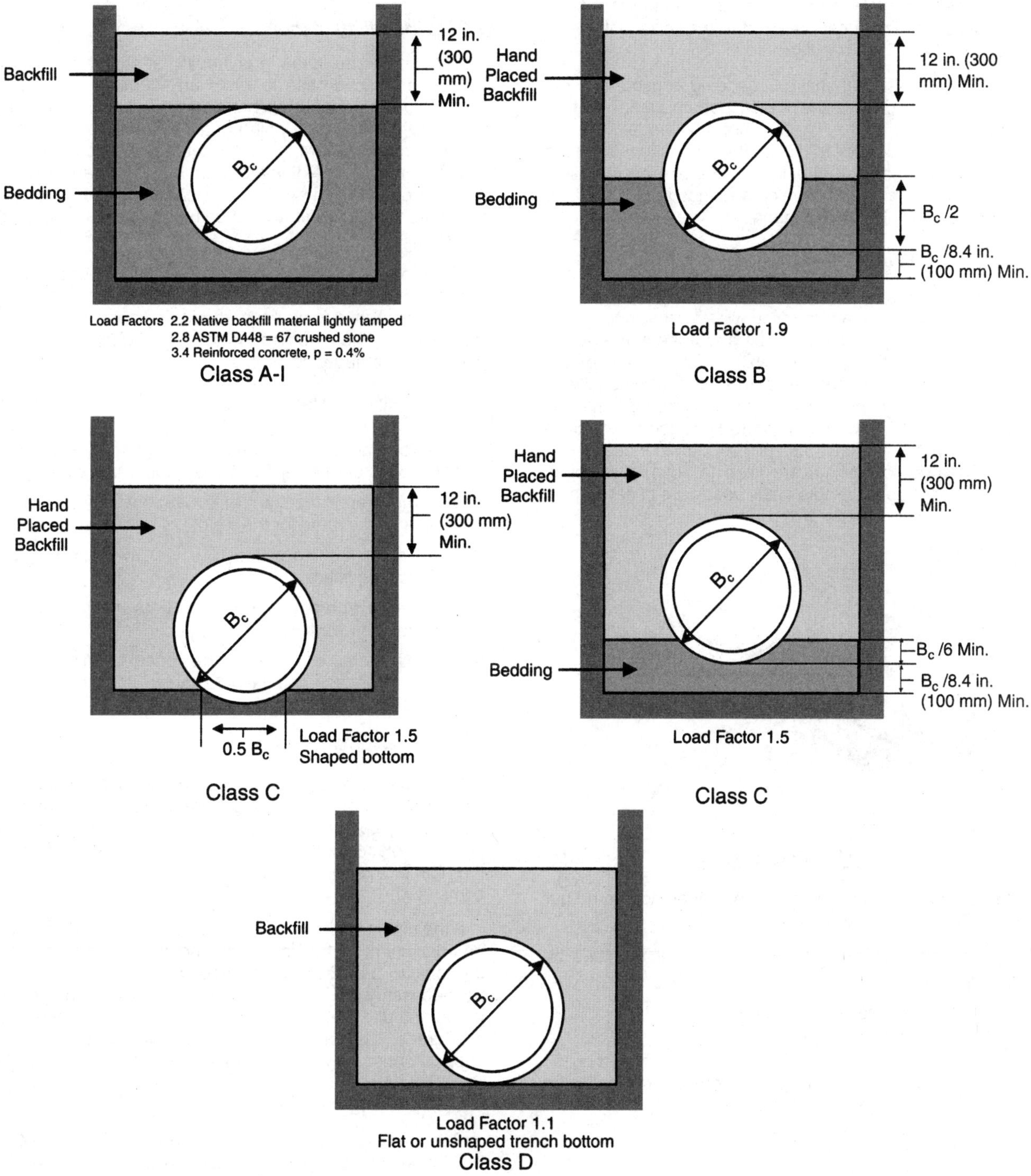

Backfill
12 in. (300 mm) Min.
Bedding
B_c
Load Factors 2.2 Native backfill material lightly tamped
2.8 ASTM D448 = 67 crushed stone
3.4 Reinforced concrete, p = 0.4%
Class A-I

Hand Placed Backfill
12 in. (300 mm) Min.
Bedding
B_c
B_c /2
B_c /8.4 in. (100 mm) Min.
Load Factor 1.9
Class B

Hand Placed Backfill
12 in. (300 mm) Min.
B_c
0.5 B_c
Load Factor 1.5
Shaped bottom
Class C

Hand Placed Backfill
12 in. (300 mm) Min.
Bedding
B_c
B_c /6 Min.
B_c /8.4 in. (100 mm) Min.
Load Factor 1.5
Class C

Backfill
B_c
Load Factor 1.1
Flat or unshaped trench bottom
Class D

Fig. 7.49 Classes of bedding for rigid sewer pipe

Care should be exercised in using granular materials for pipe bedding and initial backfill. If the material contains significant voids or spaces, silty or sandy soils tend to move into these voids. This soil migration occurs particularly where groundwater conditions exist which may cause excess loss of pipe support or line and grade change in the pipe after installation is complete. Well-graded material with sufficient fines (small, grit-like particles) can be used to minimize soil migration into the bedding and backfill material.

CLASS D BEDDING. If Class D bedding is used, it is extremely important to have the trench bottom cleaned of all loose material and at a true grade. Care must be taken in the excavation of bell holes not to disturb the foundation under the pipe barrel. In no instance should the pipe rest on the joints while attempts are made to develop a foundation for the pipe barrel using the initial backfill material.

7.224 Special Pipe Foundations

In most instances, a trench bottom will support the combined load of bedding material, pipe, wastewater, and trench backfill since their total weight will be about the same as the material excavated from the trench for construction of the sewer. However, sometimes extremely poor soil conditions (such as fill area) combined with large-diameter pipe may require a firm, constructed foundation to keep the sewer on a true grade. In these situations, pile-supported concrete or timber cradles (saddle piling) are used to support the pipe. Design and construction of this rarely used type of sewer pipe support is beyond the scope of this manual.

QUESTIONS

Write your answers in a notebook and then compare your answers with those on page 503.

7.2D How can accurate excavation of sewer trenches be guided?

7.2E What is the purpose of a grade string line?

7.2F What is the purpose of sewer pipe bedding?

7.2G Why should extreme care be taken in excavating the trench bottom to true grade?

7.23 Pipe Laying

Most manufacturers of sewer pipe furnish instructions for laying and joining their pipe and these instructions can be obtained by writing to the manufacturers. Thoroughly understand these instructions before attempting to lay pipe. Contact the local manufacturer's representative if you need more help.

The following comments provide a general overview of pipe laying and joining procedures and are generally applicable to all types of sewer pipe materials. (*NOTE:* Pipe sizes and other measurements are given in English units. Please see pages 583 and 584 for metric pipe sizes and for conversion factors to calculate metric measurements.)

7.230 Handling of Pipe

Care should be taken in unloading and storage of the pipe at the construction site to avoid any damage to the pipe or joints. No pipe should be handled by hand unless it can be done safely without dropping and damaging the pipe or hurting the operators. Proper lifting equipment should be used in removal of the large-diameter pipe from the transport vehicle to the ground. The barrel should be supported on a block of wood at the spigot end (the smooth end opposite the "bell") for its protection.

The proper equipment should be used to lower the pipe into the trench. When large-diameter pipe is being laid, the handling equipment should have adequate capacity for lifting and maneuvering the pipe into the trench. The pipe sling or hook should not damage the pipe.

The following safety rules apply to the handling of sewer pipe:

1. Use barricades, traffic cones, and high-level flags to identify the work area when unloading pipe within streets.
2. Use the proper mechanical equipment, cable, slings, and closed hooks rated for the weight of pipe being lifted.
3. Keep crane booms at least 10 feet away from overhead high voltage lines.
4. Make sure operators wear hard hats and stand clear of pipe being moved.
5. If pipe is stored in a street at the construction site, surround it with barricades and flashing lights. Block the pipe to prevent rolling.

7.231 Placement of Pipe

Small-diameter pipe (up to 10 inches in diameter) can be placed directly on the bedding material with the bell end of the pipe facing upstream. Larger diameter pipe should be supported in place with a lifting hook or sling with the bell end of the pipe facing upstream until the jointing operation has been completed.

If the pipe bedding material is properly compacted to grade, no adjustment of the pipe grade should have to be made after the jointing process has been completed. Sometimes it may be necessary to lay large-diameter pipe above the design grade and use the impact force and weight of the pipe to bring it to grade. Care must be used, however, to provide uniform support under the entire length of pipe.

7.232 Joining of Pipe

The method of joining pipe will vary with the type of joint. Most pipes for sanitary wastewater collection systems are furnished with a bell on one end, a spigot on the other end, and a resilient type of gasket to make the joining of the bell of the downstream pipe with the spigot of the upstream pipe watertight. Most resilient gaskets are in the form of an O ring sealed in a groove on the spigot or of mating rings precast in the bell and on the spigot. The basic method of joining pipe with resilient joints is:

1. Clean off the bell and spigot,
2. Lubricate the gasket and seat on the pipe as prescribed by the manufacturer,
3. Guide the spigot end of the pipe into the bell,
4. Push or pull the pipe into place,
5. Inspect the joint for proper seating of the gasket, and
6. Check the alignment and grade of pipe.

When pipe up to 10 inches in diameter is laid in a trench, the joining can usually be done by hand. When pipe larger than 10 inches in diameter is laid, a mechanical method is used to either push or pull the pipe and gasket into place. If the pipe is pushed into place, the pushing force should be applied to the face of the pipe by the use of blocking to prevent damaging the pipe. The pushing force can be supplied by a bar used as a lever with one end shoved into the trench bottom or by the bucket of a backhoe. Larger diameter pipe is usually joined by the use of a cable and winch either placed along the top of the pipes to be joined or within their interior. Care must be used to prevent pulling the previously laid pipe apart at its joint. Three methods of joining sewer pipe with resilient types of joints are illustrated in Figure 7.50.

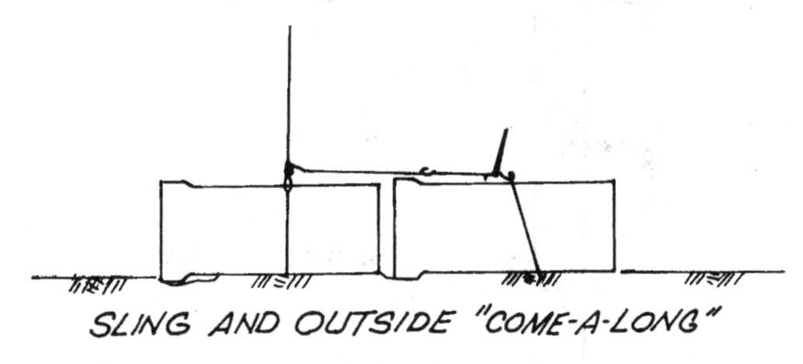

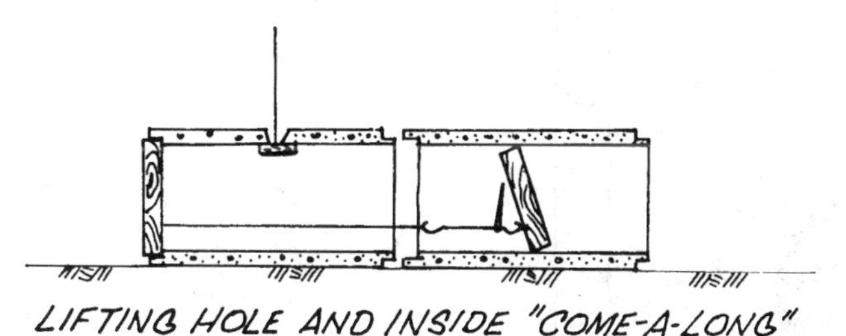

Fig. 7.50 Methods of pipe joining

(Courtesy of *CONCRETE PIPE INSTALLATION MANUAL*, American Concrete Pipe Association)

Manufacturers of pipe with other than bell and spigot joints will furnish instructions on joining their pipe.

7.233 Protection of Pipe

Once the sewer pipe has been laid to a true grade and alignment, it should be protected from displacement or breakage until the trench is backfilled. This protection is particularly important with respect to smaller diameter and rigid types of pipes such as concrete and vitrified clay pipe. Protection can be provided by immediate placement of initial backfill material.

Plug the upper end of the sewer pipe at the end of the work day or whenever pipe laying is to be stopped for several hours. This action will keep material out of the pipe that is difficult to remove once the sewer is completed and also keep out children and small animals. If the possibility exists that the trench could be flooded, do not plug the ends of the pipe completely. A plugged pipe in an open trench may float if the trench becomes flooded and would have to be relaid.

7.234 Completion of Pipe Bedding

Class A and Class B bedding require that the backfill be carefully tamped under and around the pipe to a height of 12 inches above the top of the pipe. When bedding smaller diameter pipe, do not displace the pipe by backfilling and compacting on one side and then on the other side. Also, take care not to lift the pipe from its true grade by over-compacting the backfill under the center of the pipe.

Be very careful when compacting backfill over vitrified clay pipe because this pipe is brittle and may be shattered by the use of too heavy a tamper, or by too strong a stroke from a large tamping machine, or by a heavy vibrating compactor.

7.24 Sewer Appurtenances

This section provides an overview of construction procedures for the sewer appurtenances that collection system operators may have to construct occasionally. Construction procedures for the other sewer appurtenances listed in Section 3.33 can be learned by observing construction by contractors and by talking with public works inspectors.

7.240 Construction of Manholes
(Refer to Figure 3.16, page 55, and Figure 7.51)

After a hole has been excavated for a manhole, the soil must be prepared to provide a suitable foundation. Compact the soil below the manhole base. If the existing soil does not provide a suitable base, gravel or crushed aggregate can be used.

On straight sewers, the pipe is either laid through or terminated at the manhole. Flexible joints are placed in the pipe just outside the area of the manhole base to allow for any differential settlement between the base and pipe. Water stops should be installed where pipes connect to the manhole to prevent infiltration and exfiltration.

The manhole base is either precast (Figure 7.51) or made of unreinforced portland cement concrete poured into a basin formed in the bottom of the manhole excavation. The base should be about one foot greater in diameter than the outside of the manhole barrel (refer to Figure 3.16, page 55). To shape the channel in the manhole base for smaller diameter sewers, lay in a short section of pipe and pour concrete to the crown (top) of the pipe. The bottom half of the sewer line channel is formed into the base of the manhole. The short section of the upper half of the sewer pipe is broken out after completion of the manhole. If precast manhole barrels are used, a form

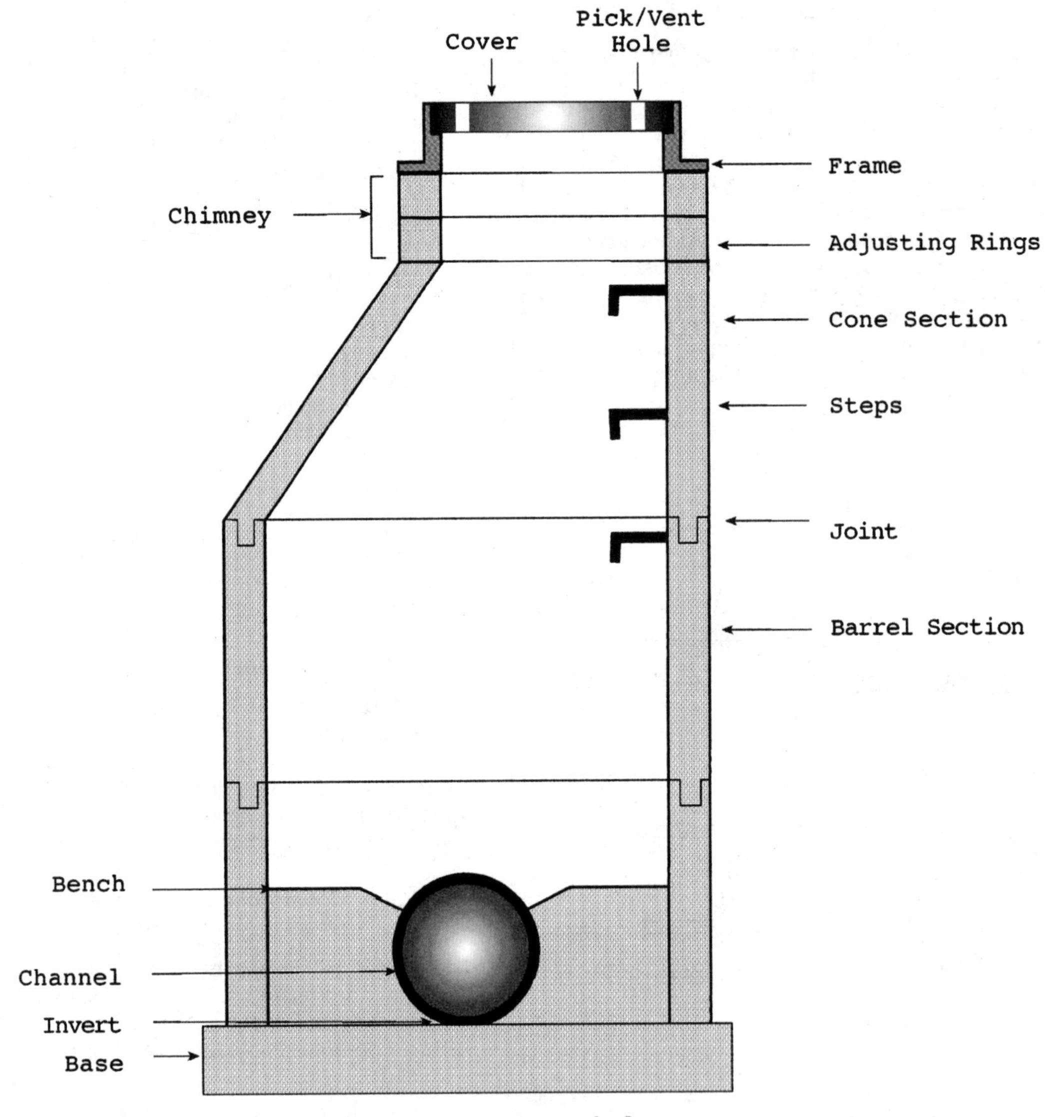

(For more details, refer to Chapter 3, Figure 3.16, page 55.)

Fig. 7.51 Manhole

matching the bottom tongue of the barrel should be pressed into the manhole base and made level before the concrete hardens. The finished channel, however formed, should be U-shaped and as deep as the invert of the pipe.

After the concrete manhole base has hardened enough to support the weight of the manhole barrel and cone (this takes about four hours and hardening can be accelerated with minor sacrifice of strength by the addition of calcium chloride, in an amount not exceeding two percent of the dry weight of the cement content, to the concrete mixture), fill the preformed groove in the manhole base with cement mortar and set the first precast concrete manhole barrel in place. Successive manhole barrels are set on top of the preceding barrel after the groove in the preceding barrel is filled with cement mortar.

Continue in this manner until the combined height of the precast manhole cone and cast-iron frame reaches the planned height of the manhole.

The precast manhole cone section is placed in the same manner as the barrels using cement mortar in the tongue and groove joints. Since a combination of manhole barrels and the top cone and frame seldom make exactly the planned manhole elevations, provisions should be made for possible adjustment with precast concrete grade rings or brick masonry to join the top of the manhole cone to the cast-iron frame at the desired elevation.

In some areas, manhole cones are constructed with brick and mortar instead of using a precast section. A disadvantage of using brick is that the area around the manhole cannot be backfilled immediately because backfilling must wait until the mortar has set.

Fiberglass-reinforced polyester manholes of 48-inch round, rib-stiffened barrels are another type of prefabricated manhole. Segments, one-third of a circle and 4 feet long, are glued together to form the barrel. Fiberglass transition cones, sewer pipe connectors, and ladders are provided. The junction with the sewer pipe may be sealed by either attaching a special pipe connector or by packing concrete into the entrance hold and around the entering pipe. Standard cast-iron frames and covers are placed on top of the transition cone. A ledge at the cone top allows room for grade adjustments to be built with iron rings or bricks.

If the manhole is being constructed in an unfinished street, the cast-iron frame and cover are usually left off the manhole and a steel plate is placed over the manhole opening until after the street has been paved. Steel plate covers are more durable than plywood covers and can be located with a metal detector when buried. Then the pavement is removed from around the manhole opening, the steel plate is removed, and cement mortar is used to attach the cast-iron frame to the manhole cone or grade rings. The void between the manhole frame and the pavement is then filled with paving material and compacted to grade.

Care should be taken to ensure tight joints between the sections of the manhole base, barrel, and cone to make them watertight, especially where the groundwater level is above the sewer during any time of the year. An asphaltic mastic joint material is manufactured for use in place of cement mortar when leakage of water through the manhole joints is anticipated.

When constructing a manhole in an easement, it is a good practice to raise the top of the manhole cover about nine inches above normal ground elevation. This will make the manholes visible and less subject to introduction of dirt and surface water through the manhole cover and into the sewer. Manhole covers may be bolted to the frame to prevent unauthorized removal and the dumping of debris into the manhole.

Since a typical precast concrete manhole barrel or cone will weigh about 2,500 pounds, adequate lifting and placing equipment are required, including a crane or front-end loader, and a "spider" to secure the manhole section to the lifting cable.

Cast-in-place concrete manhole barrels and cones are sometimes used; however, their construction details are beyond the scope of this manual.

7.241 Construction of Cleanouts

Vertical lateral sewer cleanouts at the terminals of lateral and branch sewers (see Figures 3.14 and 3.15, page 53) are constructed with a long radius 1/4 bend[9] of the same diameter and preferably of the same material and jointing as the sewer pipe. They are supported by a compacted granular bedding material (1/4-inch to 3/4-inch crushed aggregate) placed up to the top of the 1/4 bend. Full-diameter pipe risers, again preferably of the same material and jointing as the sewer pipe, are placed on top of the 1/4 bend to a height about 8 inches below the ground surface. Some agencies prefer the construction of risers using a wye and 1/8 bend[10] for ease of maintenance and the preparation of bedding and placement of backfill. A fitting that is a combination wye and 1/8 bend is available; it is called a "combo."

Hold the vertical pipe riser in place while the trench backfill is compacted. After compaction of the backfill, place a cast-iron frame around the riser pipe with the top set to the ground surface. Pour a portland cement concrete collar around the frame and pipe. Note from Figures 3.14 and 3.15 that a gasket of fibrous material is placed between the concrete collar and the riser pipe so that any settling of the cast-iron frame and concrete collar will not be transmitted to the riser pipe.

[9] *1/4 Bend. A 1/4 bend is equal to a 90° elbow (1/4 of 360°).*
[10] *1/8 Bend. A 1/8 bend is equal to a 45° elbow (1/8 of 360°).*

7.25 Sewer Trench Backfill and Compaction

Sewer trench backfill, in most instances, is placed around and over the sewer pipe and up to ground surface. The essentials of proper backfilling are: (1) protection of the pipe from movement and breakage from impact or crushing from backfill loading, (2) filling of the excavation, and (3) restoration of the ground surface.

7.250 Backfill Materials

The ideal sewer trench backfill material is one that is immediately available, free of any objects that will damage the sewer pipe, and easily compacted. Most backfill materials will not be ideal, and compensation for this fact must be made in the backfilling procedures.

Backfill materials are generally classified as native or imported, with subclassifications under each.

NATIVE BACKFILL. Native backfill material, as the name implies, is the material excavated from a sewer trench and, in many instances, is the best material for backfill. This material is immediately available since it is generally stored alongside the trench, and it is not too difficult to remove any large rocks or broken pieces of pavement that are in the material and that might damage the pipe. However, the native material may not be easily compacted and some judgment must be exercised in determining if and how the material can be suitably compacted, or if it will be better to import backfill material. This is particularly important in the initial backfilling of flexible, solid-wall plastic pipe and reinforced plastic mortar pipe where compaction is essential to good performance. Soils of clay and silt are particularly difficult to compact to required densities.

In most respects, sand is the ideal backfill material and is used as a standard against which the compactability of other types of backfill material is measured. Laboratory tests can determine the *SAND EQUIVALENCY*[11] of a representative sample of the backfill material, and a trained inspector can determine the approximate degree of equivalency of other samples in the field.

IMPORTED BACKFILL. Imported trench backfill is quite expensive since there is the cost of hauling and disposing of the native material excavated from the trench combined with the cost of purchasing, excavating, and hauling the imported material. For this reason, the use of native backfill material, even with special compacting procedures, should be carefully weighed against the cost of imported material.

Selection of the type of imported backfill material should be based on the ideal qualities listed in this section and modified to balance the cost of the imported material with the cost of obtaining the desired results.

Material being considered for importation should be examined at its source and tests should be made to determine its sand equivalency.

Suitable imported backfill materials range from crushed aggregate through sand and gravel materials to randomly excavated soil with an absolute minimum sand equivalent of 15. Any material with a lower sand equivalency is probably not any better than the native material. If the sand equivalent is greater than 30, the backfill can be compacted using jetting techniques. Jetting techniques cannot be used to compact clay or silt soils used for trench backfill.

An important consideration in selecting an imported backfill material is its reaction to surrounding conditions. For instance, one of the disadvantages of using sand backfill is its tendency to move when its side support is removed, such as in subsequent excavations in or near the sewer trench.

7.251 Placement of Backfill

The procedures used for placement of trench backfill, backfill compaction, and surface restoration should provide for protection of the sewer pipe and minimize backfill subsidence (settling) after surface restoration.

PROTECTION OF PIPE. If the class of bedding and initial backfill requires placement of native or imported backfill around the pipe, hand shovel the material from the ground surface to around both sides of the pipe at the same time so that the pressure of uneven backfill will not move the pipe out of alignment. Compact the initial backfill under the bottom of the pipe to eliminate any voids in this critical area of pipe support in accordance with the specifications for the class of bedding (see Figure 7.49). This type of backfilling is generally carried to about 6 to 12 inches above the pipe and must be very carefully placed and compacted to avoid damage to the sewer.

Another, less satisfactory, method of placing the backfill around the pipe is by drifting in the material. In this method the backfill material is pushed into the lower end of the trench by a bulldozer or front-end loader and allowed to slide down the slope that the previously placed material has assumed. The disadvantage of this method is that the initial backfill is not compacted and larger particles of backfill material will slide down the slope and surround the pipe faster than the fine material.

COMPLETING BACKFILL. After initial placement of backfill material around and over the pipe, mechanical equipment, such as bulldozers or front-end loaders, is used to move the backfill material from alongside to inside the trench. If the backfill material particle size is not too large and the trench not too deep, the material may be dropped vertically into the trench. Otherwise, the backfill material should be drifted into the trench as described in the preceding paragraph. Figure 7.52 illustrates a typical backfilling operation.

---

[11] *Sand Equivalent. An ASTM (American Society for Testing and Materials) test for trench backfill soils. The test uses a glass tube and the soil (backfill) is mixed with water in the tube and shaken. The tube is placed on a table and the soil allowed to settle according to particle size. The settlement of the soil is compared to the settlement of a standard type of sand grain sizes.*

Fig. 7.52 Placing backfill

The method of compacting the backfill material, as described in Section 7.252, will influence the method used for placing the material in the trench. For example, if the material is to be compacted in lifts, it will be necessary to drop the material vertically into narrow trenches or to drift and spread with mechanical equipment in wide trenches.

REMOVAL OF TRENCH SHORING. Removal of trench shoring in a safe manner requires coordination with the placement of the trench backfill. The shoring should not be removed until the need for operators to be in the trench no longer exists and the backfilling operation has started.

Hydraulic shoring can be decompressed and removed by operators standing topside as placement of the backfill material proceeds. For this reason hydraulic shoring is an excellent type of shoring for most trenches.

When steel screw-type trench or timber braces are used by choice or necessity, closer coordination with placement of the backfill is required. In this instance the backfill should be dropped or drifted to the level of the lowest trench brace. Then, after its removal, the trench should be backfilled to the level of each successive brace until all are removed. The vertical shores still will be in place, and it is usually necessary to pull them out with a crane or the bucket of a front-end loader. Backfilling equipment in the vicinity of the trench brace removal operation should be stopped to avoid a possible collapse of the trench walls while a brace is being removed by an operator in the trench.

When either timber or steel sheet piling is used for trench shoring, it may be left in place if its removal will damage the sewer pipe or adjacent facilities. However, the upper several feet of the sheeting should be cut off and removed to avoid interference with surface restoration or future underground construction.

7.252 Compaction of Backfill

Compaction of trench backfill above the initial bedding has no beneficial effect on the performance of a sewer. The load on a sewer pipe increases to a certain extent as the backfill material is made more dense by compaction. However, it is essential that trench backfill material be well compacted to avoid later settling that could cause damage to utilities crossing the trench or to structures adjacent to the trench. Subsidence could also cause depressions in the overlying street surface; these would require repaving.

The following sections outline the more common methods of compacting backfill material and measuring their effectiveness.

WATER COMPACTION. Jetting trench backfill material with water is one common method of compaction. When the type of backfill material is suitable, this type of compaction is simple, quick, and economical, but still inferior to mechanical compaction. Generally, backfill material with a sand equivalent of less than 30 is not suitable for compaction by water jetting.

Equipment for compaction by water jetting consists of 1 1/2- to 2-inch diameter jet pipe long enough to reach to within 2 feet of the top of the sewer pipe, hose, and a water supply that will produce a 10 to 15 pounds per square inch pressure at the jet nozzle.

Push the jet pipe into the backfill with the water running until the nozzle is within about 2 feet of the top of the pipe. Leave the nozzle in place until water reaches the trench surface. Usually the backfill material will stop subsiding about the time water reaches the surface. When water reaches the surface, move the jet pipe about 10 feet up the trench and to the opposite side of the trench; repeat the jetting operation. In a wide trench it may be necessary to shorten the distance between the jetting locations. As a rule it is better to over-jet rather than under-jet. Figure 7.53 illustrates compaction of a trench backfill by water jetting.

If a trench is deeper than 12 feet, the backfill material should be placed and compacted in two or more lifts or layers before filling and compacting the remainder of the trench.

Vibrating equipment is sometimes used along with the water jetting to develop a greater compaction of the backfill.

Another way to compact backfill is to flood the filled trench with water. This method is less satisfactory than water jetting and is not often used. The procedure is to make a shallow ditch along the trench backfill and flood it with water. At the same time, rod the backfill material from top to bottom until the required compaction is obtained. The rod is a 2- to 2 1/2-inch pole that is used to push the mud and water on the surface to the bottom of the trench and to fill the air voids in the backfill material.

The disadvantages of compaction with water jetting or flooding are: (1) these methods are not suitable for all types of soils; (2) they are subject to wide variations of results depending on procedures used; (3) the trench surface cannot be restored or pavement replaced until the trench backfill has dried out; and (4) because the amount of compaction is less than that obtained by mechanical methods, there may be later subsidence and pavement failures.

MECHANICAL COMPACTION. Properly performed mechanical compaction of trench backfill should be used when later subsidence of the backfill could result in expensive or inconvenient street surface restoration or damage to adjacent structures.

Equipment for mechanical compaction of backfill material consists of hand-held air tampers; gasoline-driven impactors; plate vibrators; vehicle-mounted impactors; and vibrating rollers.

When mechanical compacting equipment is used, the backfill material has to be placed in lifts or layers matching the compacting capabilities of the equipment used. For instance, 4-inch lifts are about maximum for hand-held compacting equipment; whereas, lifts up to 3 feet thick can be compacted with vehicular types of impactors, but these impactors can cause pipe damage if not used properly.

The moisture content of the backfill material to be compacted is very important. Material with insufficient moisture will tend to bulk and rebound under the compactive effort and material with too much moisture will be moved away (pumped)

Fig. 7.53 Compaction of trench backfill by water jetting

from the compactive effort and displace the adjacent material. An approximate method to determine the proper moisture content is to add water slowly to a dry sample in your hand until it can be formed into a compact ball.

Compaction of narrow trenches requires the use of handheld equipment and numerous shallow lifts. For this reason, an economical alternative to mechanical compaction of native material may be the use of imported backfill material that does not need to be compacted or can be compacted by jetting with water.

Mechanical compaction becomes more practical in wide trenches where vehicular compaction equipment with deeper lifts of backfill material can be used to produce the required compaction at a reasonable cost.

Water jetting and mechanical compaction are often used together to compact trenches excavated in a paved area. When this procedure is used, the backfill material is first jetted and then the trench is refilled to the paved surface with the rock material to be used for the base of the repaving. The trench is allowed to "dry" for several days. After drying, a vehicular impactor is used to compact the surface of the trench just prior to repaving.

The main disadvantages of mechanical compaction of trench backfill are: (1) length of time required; (2) maintenance of shoring during compaction in lifts; (3) excessive impact can damage pipe; and (4) high costs.

MEASUREMENT OF COMPACTION. The amount of compaction that has been obtained in trench backfill material is determined by measuring the density of the "in-place material" in pounds per cubic foot and comparing it with the maximum obtainable density of the same material under ideal conditions. This is called the relative compaction since the test relates the actual density to the maximum obtainable density.

Determining the in-place density of the backfill material after compaction and the maximum obtainable density of the backfill material is a complicated procedure requiring the services of a well-equipped soils testing laboratory using standard procedures developed by state departments of public works and the American Society for Testing and Materials.

A common requirement is for the compaction of the backfill material in the lower area of the trench to 90 percent of maximum and the upper 3 feet to 95 percent of maximum.

7.253 Surface Restoration

The final operation in constructing a sewer is restoring the ground surface disturbed by the construction process. Usually the surface is restored to its original condition.

IN EASEMENTS. Surface restoration over sewer trenches in easements usually requires removal of excess soil and rocks that may have been excavated from the trench, followed by bringing in topsoil in cultivated areas and replanting in landscaped areas.

IN PAVED AREAS. Surface restoration over trenches in paved areas requires the compaction of the trench surface by mechanical means, re-excavation of the backfill material (if compaction does not produce the required depth) to the subgrade of the pavement, and repaving with materials similar to the original pavement.

QUESTIONS

Write your answers in a notebook and then compare your answers with those on page 503.

7.2H Which direction should the bell end of the pipe be facing?

7.2I Once a sewer pipe has been laid to true grade and alignment, why should it be protected?

7.2J Where are vertical lateral sewer cleanouts located?

7.2K What are the essentials of proper sewer trench backfilling?

END OF LESSON 2 OF 4 LESSONS
on
UNDERGROUND REPAIR

Please answer the discussion and review questions next.

DISCUSSION AND REVIEW QUESTIONS

Chapter 7. UNDERGROUND REPAIR

(Lesson 2 of 4 Lessons)

Write the answers to these questions in your notebook before continuing. The question numbering continues from Lesson 1.

7. Why is shoring required for excavating trenches for construction of sewers?
8. How can existing underground utilities in existing streets be protected during excavation of trenches for sewers?
9. How can the high incidence of injury and death from excavation and trench cave-ins be reduced?
10. Why are the material and compaction of the initial backfill around a flexible pipe very important?
11. Why should manholes in easements have the top of the manhole cover raised above normal ground elevation?

CHAPTER 7. UNDERGROUND REPAIR

(Lesson 3 of 4 Lessons)

7.3 REPAIR OF SERVICE LINES

Building or house service lines must be repaired when the lines develop stoppages or leaks. Repair crews are notified of a repair job when they receive a work order form from their supervisor. Repair assignments must be made in writing in order to maintain records and provide a history of who made what repairs when, where, how, and why.

This section describes the steps to follow to respond to a work order request (Figure 7.54) to install a cleanout. A cleanout was required in this example because one was not installed when the sewer was built and the cleanout is needed to help when clearing a recurring root problem.

7.30 Preliminary Investigation

Upon receipt of a work order request, existing records are examined to determine if the problem has occurred in the past. Existing records may indicate the types of problems and soil conditions the repair crew could encounter in the field.

A supervisor makes a preliminary field investigation and notifies the property owner that a crew will be digging in the yard to repair the problem sewer. A brief explanation of the job to be done is good public relations, such as how much digging will be done and where in order to install a cleanout and/or replace a faulty section of sewer. The supervisor also looks for potential problems in order to alert the crew leader (Maintenance Operator II) to any unusual problems that may exist. Helpful information includes large shrubs that have to be removed or flowers and shrubs that have to be transplanted. Utility companies must be contacted and requested to locate their utilities in the vicinity of any excavation. Take photographs of work areas on private property *BEFORE* starting to work. These photographs can be very helpful if any legal claims develop regarding cracked sidewalks and foundations or damaged lawns and fences.

After the preliminary investigations, the work order request is given to a repair crew.

7.31 Job Assignment to Repair Crew

After receiving the work order request from the supervisor, the crew leader and supervisor discuss the job. How to meet safety requirements and what materials will be needed are reviewed.

The crew leader has the responsibility to inspect the personal safety equipment of the crew. Inspection must include examination of steel-toed boots, hard hats, gloves, and safety glasses (Figure 7.55).

7.32 Equipment, Materials, and Tools

Proper tools, equipment, and materials to do the job must be on the repair crew's truck before they drive to the job site.

Materials usually required for sewer repair and cleanout installation include sections of 4-inch pipe, 4-inch wyes with main barrel ranging in length from 16 to 22 inches, $^{1}/_{8}$ bends or 45° elbows, 4-inch couplings, 4-inch bushings, and plastic plugs (Figure 7.56).

Tools required by repair crews include round point, square point, and narrow cut down shovels (Figure 7.57). These shovels are used for removing sod and exposing pipe for repair or cleanout installation. The narrow shovel is usually made by cutting down the width of a round point shovel that has had the handle broken; however, they may be purchased new. A narrow shovel is used to remove dirt between the pipe and existing ditch line.

Other hand tools may include a miner's pick and mattock for cleaning under pipe or for hard dirt near pipes, pipe cutters to cut and remove pipe that is broken or must be removed for the installation of a new fitting, six-foot digging bar, tape measure, rope and a bucket for lowering and removing small hand tools from the excavation, two-pound drill hammer, cold chisel, wonder bar, electrician's chisel, $^{3}/_{8}$-inch drive ratchet and $^{5}/_{16}$-inch socket, torque wrench, battery-powered lantern, and a small hand mirror for looking upstream and downstream once the line is opened to examine for debris left during removal of the damaged section. A wide ground cover canvas is helpful to protect lawns. Hydraulic shores, screw jacks, and timbers to shore excavations are needed (see Figures 7.13 through 7.20). Also see Volume II, Chapter 11, "Safety/Survival Programs for Collection System Operators," for safety equipment, including a first-aid kit.

7.33 Arrival at Job Site

When traveling to and from the job site, seat belts must be worn. When arriving at the job site, verify that the address and map location are correct. As soon as the truck is parked, traffic cones should be placed on the traffic side of the truck. One cone belongs in front of the truck and the others at the rear of the truck. These cones alert traffic and make drivers aware of

Crew Chief **SWANSON** Date **9 - 26 - 02**
Address **MARCONI AVE.** Book **269** Page **05**
Manhole **8** to **7**
Pipe Size **6" VCP** Depth **6'9" - 9'8"** Distance **292**
Street ☐ Easement ☒

Work Requested **DIG UP SERVICE LINE AND INSTALL WYE FOR CLEANOUT.**

Approved ______________

Fig. 7.54 Computer-generated work order request to install a cleanout

the parked repair truck (Figure 7.58). Picking up the cones prior to leaving the job site also forces the truck driver to check for children and objects in front of and in back of the vehicle before pulling out. See Chapter 4, Section 4.3, "Routing Traffic Around the Job Site."

TRAFFIC ROUTING

PLEASE TURN BACK AND REVIEW SECTION 4.3 ON PAGE 88

A good practice is to wear a hard hat whenever out of the truck at a job site. Hard hats protect your head from crank-out windows on the sides of homes and buildings and from lines and low clothesline poles.

Notify the property owner that you must enter the property to repair the sewer and install a cleanout on the service line to eliminate the sewer problem. Inspect the area with the property owner. Identify any plants or shrubs that have to be transplanted or replaced.

QUESTIONS

Write your answers in a notebook and then compare your answers with those on page 503.

7.3A Why should repair assignments be made in writing?

7.3B What items should be reviewed during the preliminary investigation of a work order request?

7.3C What should be inspected before a crew drives to a repair job?

7.3D What should be done upon arrival at the job site?

7.34 Starting Repair (Excavation)

Locate the service line to be repaired. After inspecting the repair area, materials and tools can be brought into the yard. Place a wide ground cover canvas over lawn grass near the edge of the repair area. The canvas helps keep grass from being damaged or destroyed during repair and helps speed backfilling of the ditch when the job is completed.

Use a square point or a sod shovel to cut an area approximately two feet wide and eight feet long. This 2- x 8-foot section should be cut into smaller sections for ease of handling. Place the sod sections away from where dirt is going to be stacked. This is to make shoveling dirt into the trench easier. Use the round point shovel to start removing dirt. After about 12 inches of dirt have been removed, use a rod or probe to locate any possible water or gas lines in the area being excavated that may have been overlooked or forgotten by the utility agencies. In some areas a common practice is to install the water service lines and sewer service lines in the same ditch.

Proceed with caution to dig the ditch down to the sewer service line. If a water line is located, caution should be used not to hit or strike the water service line. If utilities are encountered in the excavation of a trench, support the utilities so they won't crack or break. A brace can be placed across the trench and cables extended from the brace to the utilities to provide support. If the sewer service line is deeper than five feet or if hazardous soil conditions are encountered, install adequate shoring (Section 7.1).

Clean the area alongside and under the section of sewer line to be removed to provide room for ratchet-type chain cutters (Figure 7.59).

Fig. 7.55 Personal safety equipment

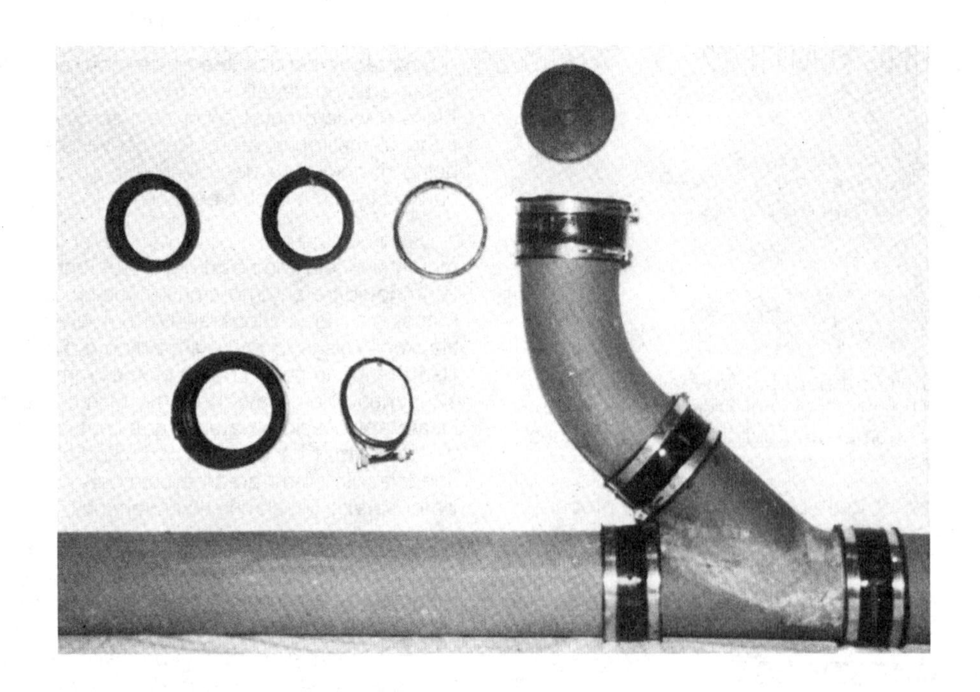

Fig. 7.56 Materials for repairs

(a)
1. Square shovel
2. Round point shovel
3. Narrow shovel
4. Probe
5. Mattock
6. Digging bar

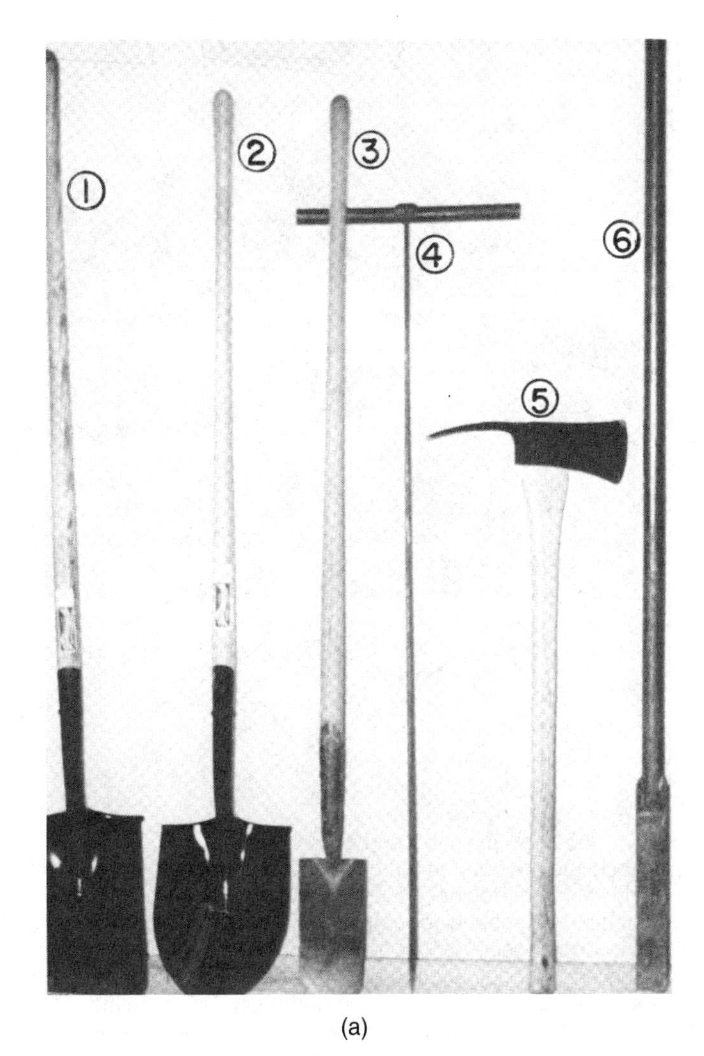

(a)

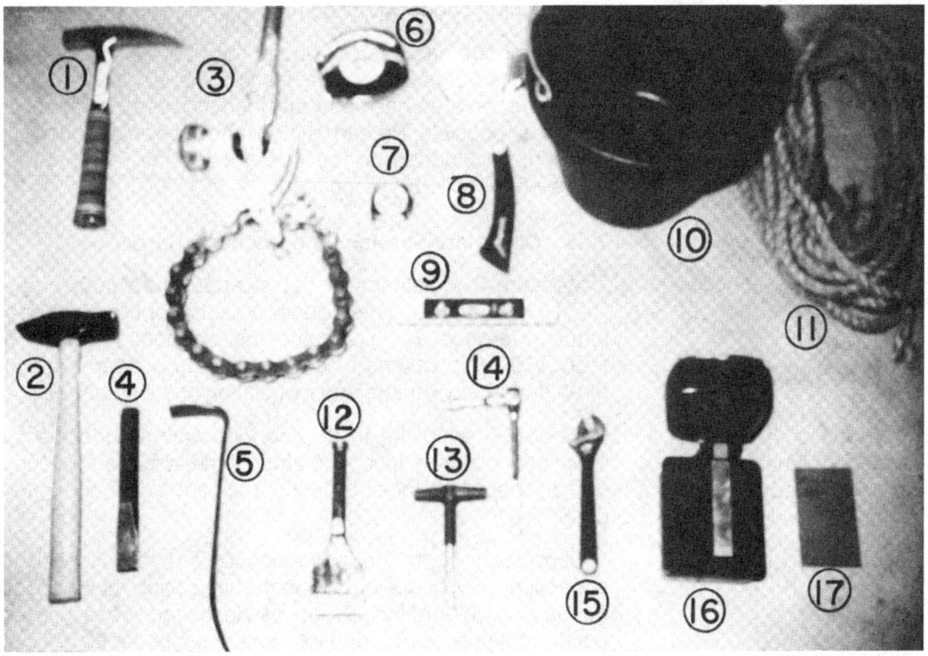

(b)
1. Miner's pick
2. 2 lb. hammer
3. Ratchet cutters
4. Cold chisel
5. Wonder bar
6. 50 ft tape
7. 10 ft tape
8. Hatchet
9. Level
10. Bucket
11. Rope
12. Electrician's chisel
13. Torque wrench
14. Ratchet and socket
15. Crescent wrench
16. Battery lantern
17. Mirror

Fig. 7.57 Tools used for repairs (b)

Fig. 7.58 Placing traffic cones around truck

7.35 Cutting Pipe

Set a ratchet cutter on the pipe, pass the end of the chain under the pipe, and attach it to the jaws on the cutter. Be sure that the chain is properly attached. Ratchet cutters are marked *CUT* and *OPEN*. To cut pipe, align the side marked *CUT* with the arrow. Apply a downward pressure on the handle until the pipe is cut. If the pipe does not cut during the first stroke, lift up on the handle until it is at about a 90-degree angle and push downward a second time. Make several cuts (usually three) so the pipe section can be removed without damaging the remainder of the pipe. After the pipe is cut, measure the distance to the next cut, such as for a wye. Install the pipe cutter at the proper location and make the next cut. After all cuts are made, remove the pipe cutter from the trench.

Remove the section of pipe that has been cut. Be careful not to cut or scratch yourself on the sharp pipe edges. Inspect the service line for any problems using a lantern and mirror. Remove any roots or other obstructions from the line and repair any breaks observed in the line.

7.36 Install Cleanout

If the service line is found to be satisfactory or after completing the repair, install a wye. A wye is installed to give access to a portion of the service line leading to the main line (Figure 7.60).

Place gravel, sand, or crushed rock in the bottom of the trench for bedding of the wye. After the wye is in place, install a 1/8 bend or 45-degree elbow in the upper leg of the wye, if necessary. Then install a section of 4-inch clay pipe into the 45-degree bend and bring it up to ground level (grade). Insert a 4-inch plug on the end of the pipe at ground level. Tighten all couplings or seal joints. Make a quick inspection of the ditch for tools before preparing the bedding and backfilling the trench.

7.37 Bedding, Backfilling, and Compaction

Proper bedding is very important. Be sure bedding is of suitable soil or material to provide a firm base. Tamp the bedding so it won't settle after the job is completed. A smooth bedding surface is essential to prevent the cleanout or pipes from shifting.

Start to backfill the ditch. A square point shovel is often used for backfilling. A water hose is required for jetting. Place the hose near the bottom of the ditch. As dirt is being backfilled into the ditch, pull the hose up to allow all backfilled soil to be soaked. This process provides compaction of the soil in the ditch and prevents extreme settling from occurring at a later date.

Care must be taken to prevent large chunks of dirt or rocks from striking the sewer service line. Pipe breakage can occur or a coupling can be knocked off, causing a leak to develop.

After all soil has been replaced into the ditch and the sod and bushes put back in place, visually inspect the repair area to be sure the area has been returned as close to original conditions as possible. Be sure all tools have been accounted for, cleaned, and loaded on the truck. Notify the resident of the home when the job is completed.

7.38 Complete Service Line Location Card

Measure the distance of the cleanout from both property lines. These measurements and a sketch of the area are for future reference to help locate the cleanout (Figure 7.61). *RECORD* observations of conditions and problems encountered during any repairs and installation of the cleanout.

Before entering the truck, carefully walk around the truck looking for children, toys, and other items, and then pick up the traffic cones. Also look inside the truck and equipment for children.

Upon returning to the maintenance yard, discuss with your supervisor any future problems that you think might develop. Potential problems include plants destroyed, trees hit, fence posts weakened, or a need for more soil after settling. This information gives the supervisor a chance to discuss these problems with the property owner and thus avoid some bad public relations at a later time.

Fig. 7.59 Ratchet-type chain cutters

QUESTIONS

Write your answers in a notebook and then compare your answers with those on page 503.

7.3E When excavating to repair a sewer line, why should the soil removed from the excavation be placed on a canvas?

7.3F What cautions should be exercised when excavating to repair a sewer line?

7.3G What is done after a section of pipe is cut and before a cleanout is installed?

7.3H What kind of bedding material is used for the wye of a cleanout?

7.3I What should be done before a trench is backfilled?

7.3J What is the purpose of the service line location card?

7.3K What should the crew leader do upon returning to the yard?

7.4 SEWER CONSTRUCTION INSPECTION AND TESTING

7.40 Need for Sewer Construction Inspection and Testing

Imagine what could happen if a new wastewater collection system was put into service and something was wrong with the system. A disastrous mess could develop and repairs would be very difficult. *PROPER TESTING* and *INSPECTION* of a new sewer system *BEFORE* placing it in service can prevent serious problems from developing and avoid considerable grief.

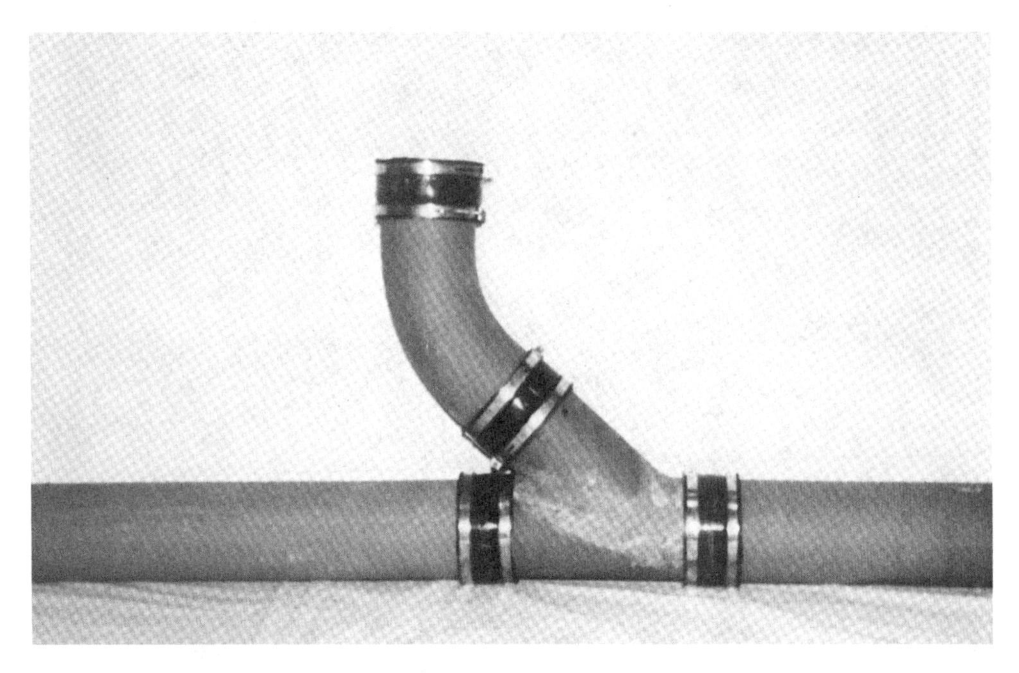

Fig. 7.60 Wye and cleanout

New wastewater collection system construction materials and procedures are inspected by a public agency-owner to:

1. Ensure compliance with the plans, specifications, and safety regulations governing the work.
2. Alert the design engineers to any modification of plans required to meet unexpected conditions found during construction.
3. Minimize the need for future maintenance of the wastewater collection system by ensuring close compliance with plans, specifications, and good construction practices.
4. Provide a record of the construction progress and "as-built" plans (record drawings) of the wastewater collection system that is buried, but cannot be forgotten. Locating portions of the system in the future can be very difficult and costly without accurate "as-builts" showing the exact location of facilities in the system.

The *PUBLIC WORKS INSPECTORS' MANUAL* written by Silas R. Birch, Jr., Inspector of Public Works, City of Los Angeles, CA, and published by BNi Building News, 1612 South Clementine Street, Anaheim, CA 92802 (price, $59.95, plus $8.00 shipping and handling), provides an excellent reference on the purposes and procedures for inspection of public works construction, including wastewater collection systems.

7.41 Duties of An Inspector

The duties of a wastewater collection system construction inspector, performed under the supervision of a registered professional engineer, are to act as an observer and reporter of materials and procedures used for constructing the system. In small towns or sanitary districts, the inspection may be shared with the design and project engineers.

Any major deviation from the plans, specifications, safety regulations, and good procedures for construction of a collection system are accurately reported by the inspector to the project engineer for remedial action. Minor deviations are usually eliminated in the field by conferences between the inspector and the contractor's superintendent of construction.

An inspector should attempt to anticipate field conditions, such as previously undetected utilities or other obstructions, that may require a change in the location of a planned sewer. A report should be submitted promptly to the project engineer in order to minimize any delays and costs caused by correcting the situation. The inspector must be sure that the installation procedures, materials, and workmanship conform with the project specifications.

Inspectors also must prepare written reports that will:

1. Provide a daily measure of construction progress, record compliance with safety regulations, and call attention to any unusual incidents;
2. Provide a monthly measure of completed work that can be used as a basis for progress payments to the contractor;
3. List the final items of work to be completed prior to acceptance of the project by the public agency-owner; and
4. Indicate the location of the buried wastewater collection system, generally submitted in the form of "as-built" plans (record drawings). Locations should be measured accurately from several permanent surface points to the critical points of the system so that later they can be located easily. Also shown on "as-builts" are the locations of all other underground utilities and facilities that are exposed during construction. Knowing the distances of other utilities from the sewer line is very helpful when making repairs in the future.

7.42 Qualifications of An Inspector

A sewer construction inspector, in order to adequately perform the duties, should have knowledge of: (1) methods and equipment used in wastewater collection systems; (2) fundamentals of construction surveying; (3) the physical properties, types, and uses of sewer construction materials; (4) procedures and equipment used in testing construction materials; and (5) laws, codes, ordinances, and standards applicable to public works construction in general and wastewater collection system construction specifically.

Address 2236 MARCONI AVE.

Parcel No. 269 - 051 - 11

Location of S/L or C/O 36' WEST OF EAST PROPERTY LINE 5' SOUTH OF NORTH PROPERTY LINE

Size 6" VCP	Information Source	Date
Easement [X]	T. V. [X]	9 - 30 - 02
Street [ ]	Parcel Book [ ]	
Commercial [ ]	Construction [ ]	
Residential [ ]	Repair [ ]	
Cleanout Yes [X] No [ ]	Ferret [ ]	
Single [ ] Dual [ ]	Other [ ]	

Use this space for diagram

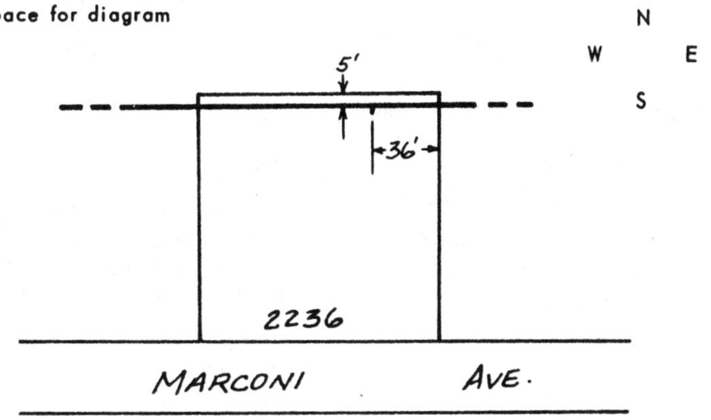

N

W E

S

5'

36'

2236

MARCONI AVE.

Date 9-3 - 02 By WELCH

Fig. 7.61 Service line location cards

A well-prepared sewer construction inspector also should have the ability to: (1) read and interpret plans, specifications, ordinances, and regulations applicable to wastewater collection system construction; (2) determine if construction materials conform to specifications; (3) keep accurate records, gather data, and prepare accurate reports; (4) make simple engineering calculations; and (5) establish and maintain cooperative working relations with those contacted in the course of work, including contractors and the general public.

7.43 Inspection and Testing of Construction Materials

Wastewater collection system construction materials should be inspected initially for compliance with specifications as close as possible to the manufacturing site. This inspection can minimize the delays and costs resulting from the replacement of rejected materials. In all situations, the materials should be inspected again for damage and defects prior to their installation. Many public agencies and manufacturers use the materials specifications published by the American Society for Testing and Materials (ASTM) as their standards for wastewater collection system materials. Copies of these specifications can be obtained from the ASTM, 100 Barr Harbor Drive, West Conshohocken, PA 19428-2959.

7.430 Sewer Pipe

Most public agencies rely on the integrity and quality control of the manufacturer of small-diameter pipe (up through approximately 24-inch diameter) to ensure production of pipe that meets specifications. However, thorough inspection of the pipe should be made after delivery to the job site to detect any damage to the pipe resulting from handling after manufacturing, including loading, transportation, and unloading.

On the other hand, most public agencies will send an observer to the plant of the manufacturer of large-diameter pipe to observe and certify to the manufacturer's testing of representative samples of pipe for compliance with specifications. As in the case of smaller diameter pipe, large-diameter pipe should be inspected after delivery to the job site to detect any damage resulting during the delivery. Some agencies hire an independent testing laboratory to test and certify that the pipe meets all specifications.

The procedures used in testing the various types of pipe used for sanitary and storm sewer construction are beyond the scope of this manual. For further information on these procedures, refer to the publications of the pipe manufacturers and the standards for testing procedures published by the American Society for Testing and Materials (ASTM), 100 Barr Harbor Drive, West Conshohocken, PA 19428-2959.

7.431 Appurtenances

As in the case of small-diameter sewer pipe, most public agencies rely upon the integrity and quality control of the manufacturer of collection system appurtenances, such as precast concrete manhole barrels, cast-iron frames and covers for manholes, and risers, wyes, and tees, to ensure compliance with specifications. On-the-job site inspection is used to detect any visible defects missed by the manufacturer or resulting from handling.

7.432 Pipe Bedding and Trench Backfill Materials

An inspector should arrange to have representative samples of sewer pipe bedding and imported trench backfill tested for compliance with the specifications for these materials. Most inspectors are not trained or equipped to perform these tests. However, by comparing samples of materials that meet their specifications with subsequent deliveries, an inspector generally can detect and reject material not meeting specifications.

Information on soil testing procedures can be obtained from the American Society for Testing and Materials and the materials and research departments of most state public works or transportation agencies.

7.433 Portland Cement Concrete

Portland cement concrete used for constructing load-bearing structures should be tested with respect to its composition, workability, strength, and density. An inspector will use the services of the agency's materials testing laboratory for conducting most of these tests.

7.44 Inspection of Construction Procedures

Proper inspection of the procedures used to construct sanitary and storm sewers using materials meeting specifications will ensure production of a wastewater collection system that will function in accordance with the engineer's design and should require a minimum of maintenance.

7.440 Location of Utilities

Be sure utility owners are asked to send a representative (locator) to mark the location of buried utilities in the vicinity of the construction you are inspecting. The importance of and procedure for locating buried utilities were described in Section 7.201.

If the actual locations of the underground utilities are different from those shown on the construction plans and require a change in the location of the proposed sewer, the inspector should immediately notify the project engineer of the location of the utilities and the need for a change of plans.

An inspector also should be familiar with the procedures to be followed to protect a utility from damage if it is exposed during construction. Many utility companies want to have their own employees inspect the utility and restore any protective coating damaged by construction activities. The inspector also should determine that an exposed utility is protected from damage during trench backfilling and compaction operations.

When a sewer is constructed under gas service lines, it is a good practice to advise the utility company when the project is completed and request their inspection of building connections to each service line to determine if any gas leaks have been caused by the construction. Be sure to mark the exact location of other utilities on the "as-built" plans (record drawings).

7.441 Trench Excavation

Before starting trench excavation in an easement, obtain pictures of nearby private property to avoid having to repair or replace private property such as gardens and sidewalks that were not damaged by the trenching activities.

The following are the major items that should be inspected for conformity with construction requirements during excavation of a sewer trench:

1. Compliance with the traffic control and other regulations established by a street encroachment permit,
2. Correct line and grade for the trench excavation (refer to Section 7.21 for sewer alignment and grade control),
3. Compliance with trench excavation safety regulations (refer to Section 7.202),
4. Cleaning of trench bottom and adequate compaction of backfill in any undercut areas to provide a solid base for sewer pipe bedding,
5. Dust control during trench excavation, and
6. Public safety. Provide safe access to residences and businesses. Secure the work site at the end of the day for the safety of children and to prevent vandalism.

7.442 Pipe Laying

During the pipe laying operation, the following items of work should be inspected:

1. Sewer pipe bedding placed to prescribed depth and grade (refer to Section 7.211).
2. Sewer pipe laid to grade with joints made in accordance with manufacturer's instructions (refer to Sections 7.211 and 7.23). Each sewer pipe in place must be closely inspected for any visible defects, such as a cracked bell or an unseated gasket. The gradient of each sewer pipe should be double checked by placing a spirit level along the top of the pipe barrel.

3. The interior of the pipe should be free of bedding materials, dirt, boards, or other debris that could create problems in the wastewater collection system when it is put into service.

7.443 Construction of Appurtenances

Construction of major wastewater collection system appurtenances should be inspected for the following:

1. Location of wyes and tees in the main and lateral sanitary sewers to provide connections for designated locations of building sewers. Each location must be recorded on the "as-built" plans (record drawings) for future recovery if building sewers are constructed after backfilling the main sewer trench.
2. Manhole bases poured on firm foundation; continuous support under sewer pipes between edge of bases and undisturbed trench bottom; construction of channels to proper width and grade; placement of manhole barrels in vertical alignment with watertight joints; and placement of cast-iron frames and covers at proper grade (refer to Section 7.240 for manhole construction details).
3. Vertical riser bends should be surrounded by compacted aggregate to support the weight of the vertical piping; the vertical pipe insulated from surface loading applied to the frame and cover; and cast-iron frame and cover set to proper grade (refer to Section 7.241 for construction details).

7.444 Trench Backfilling and Compaction

An inspector should check the following aspects of sewer trench backfilling:

1. Placement of backfill material alongside the sewer pipe in a manner that will not cause lateral or vertical displacement of the pipe.
2. Placement of backfill material in a manner that will not cause heavier material to accumulate around and adjacent to the sewer pipe.
3. Placement of backfill material in lifts and with moisture content matching required compaction method.
4. Compaction of the trench backfill in the manner and amount specified (refer to Section 7.252). The amount of compaction obtained by the compactive efforts are determined by relative compaction tests beyond the scope of training and equipment provided most inspectors. However, an inspector and the project engineer should arrange for an appropriate number of such tests to determine the adequacy of compaction. With experience, an inspector should be able to develop a feel for the amount and type of compaction obtained.
5. After backfilling is completed, visually inspect the lines for leaks, displaced pipe, and other signs of damage.
6. If time or costs do not allow TV inspection of the entire newly constructed sewer line, the design engineer, the inspector, and a maintenance operator may each identify portions of the job they believe should be given a final inspection with TV. Inspection procedures using TV equipment are outlined in Chapter 5, "Inspecting and Testing Collection Systems."

7.445 Surface Restoration and Cleanup

Surface restoration should not be started until the inspection and testing of the newly constructed sewer line has been completed. If the surface of a trench is paved or landscaped prior to inspection and testing and defects are found requiring the re-laying of portions of the sewer line, the surface restoration will have to be torn up and replaced at a significant additional cost.

The manner and procedure for surface restoration will vary with the surface conditions existing prior to construction of the wastewater collection system. The purpose of restoration is to return the surface as nearly as possible to its original condition. In many instances the surface will be under the jurisdiction of an agency apart from the public agency-owner of the system and the type of, and procedure for, surface restoration will be specified by that agency. The inspector must require the contractor to follow these specifications. During surface restoration, the inspector should have manholes or other sewer openings covered and protected from the introduction of surfacing materials and the entrance of storm water runoff, which may carry silt and debris into the newly constructed sewer.

Inspection of the final cleanup of the construction area is important since it is the last opportunity to have a contractor complete those jobs left to the end before pulling the last workers off the job. The inspector also should use a list of uncompleted work developed from a final inspection as a guide for the work to be completed before recommending acceptance of a project. Section 7.49 describes the procedures for the final inspection.

QUESTIONS

Write your answers in a notebook and then compare your answers with those on pages 503 and 504.

7.4A Why are new wastewater collection system construction materials and procedures inspected?

7.4B How do most public agencies ensure production of pipe that meets specifications?

7.4C What items of work should be inspected during a pipe laying operation?

7.4D What should an inspector do after backfilling is completed?

7.45 Air Testing

The main reason for using low-pressure air testing for testing sewers is that pressurized air exerts the same pressure in all directions on the pipe during a specific moment in time. The fact that air can leak through a smaller crack than wastewater helps find vapor leaks that may attract roots. Also, air could leak out a small crack that could become a large crack in the future. The problems with using the water test include the time, money, and water wastage involved in running these tests. Extreme care must be exercised when specifying and conducting both air and water tests. In steep terrain, air tests are better than water tests because of excessive water pressures created at the lower end of a sewer line. For this reason, air testing is replacing water pressure testing of wastewater collection systems in many regions.

Two typical air test setups for sewer lines and the necessary equipment are illustrated in Figure 7.62.

7.450 Test Pressures

Sewer pipes are usually tested at air pressure around 3 to 5 psi above any outside water pressure on the pipe. This outside pressure, which may be caused by groundwater levels that are higher than the pipe, must be known or estimated. The air pressure inside the pipe during testing must be great enough to overcome the water pressure outside the pipe so that air can leak out if there are cracks and holes in the pipe. Be careful air test pressures are not too high because they could blow out pipe joints or plugs and cause injury to personnel or damage to the system.

Where a water table elevation is known and the depth of the pipe below this water table can be measured, determine the test pressure by adding to the test pressure one-half pound per square inch for each foot of water depth above the bottom of the pipe.

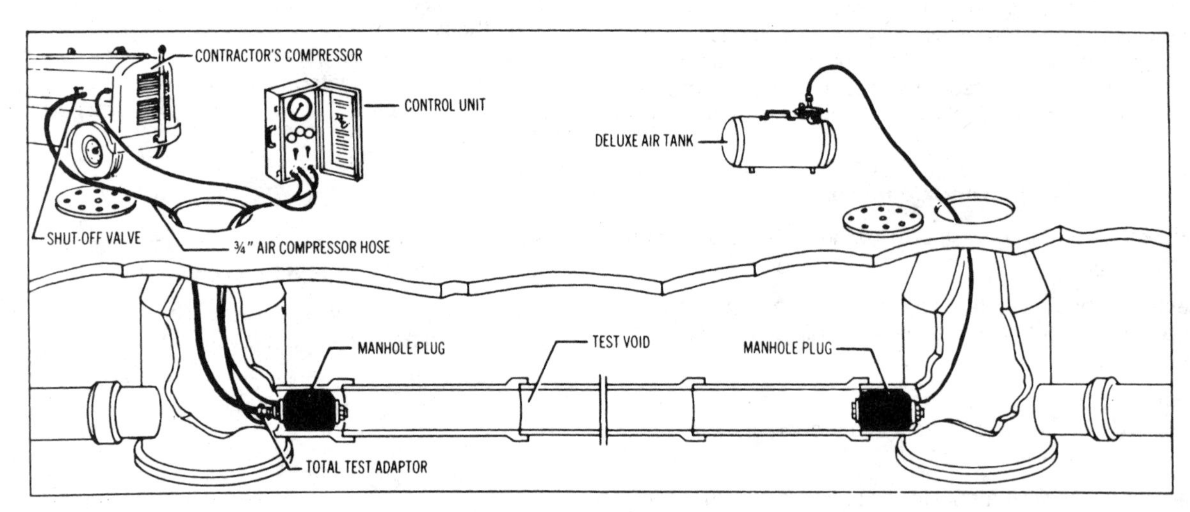

Manhole to manhole air test

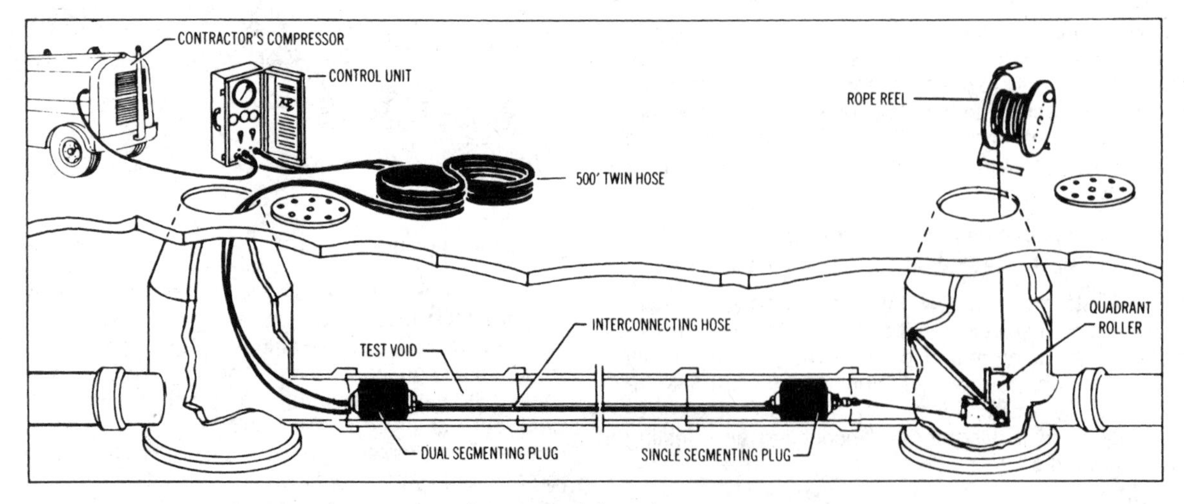

Segment of pipe air tests

Fig. 7.62 Typical air test setups

EXAMPLE: A pipe is known to lie at an average depth of 2 feet below the existing water table of the groundwater and is to be tested to a pressure of 3 psi. If we increase the test pressure one-half pound for each foot of water above it, or an additional 1 psi, the test pressure for the pipe will be 4 psi.

During the air test, a compressor pumps air into the plugged segment of the sewer and increases the air pressure inside the sewer. This higher pressure (4 psi above atmospheric pressure) is measured with a pressure gage. If leaks are present in the sewer, the air pressure in the sewer will drop back down to atmospheric pressure.

Test pressures for wastewater collection systems are seldom higher than 6 psi. A diaphragm type of pressure gage with a *LOGARITHMIC SCALE*[12] is considered the best type of gage for testing wastewater collection systems.

7.451 Volume of Air Required

A source of air pressure is required and this is usually provided by an air compressor (described in Section 7.452). An adequate capacity air compressor should be on hand before starting to pressurize a pipeline. Tables 7.5 and 7.6 can provide an estimate of the amount of air required for selected pipe sizes and required test pressures.

First, determine the pipe diameter and length of the pipe to be tested. Table 7.5 shows the volume of air in a one-foot length of pipe with the inside diameter indicated. Table 7.6 shows how much additional air in atmospheres (one atmosphere is approximately 15 psi absolute pressure) must be added to any space to achieve the indicated test pressure. The third column in Table 7.6 indicates the approximate depth of the pipe in feet below the elevation of the groundwater table. The following examples show how to determine air volumes and pressures.

TABLE 7.5 PIPE SIZES AND VOLUMES

Diameter, inches	Volume, cu ft per ft of pipe
4	0.087
6	0.196*
8	0.349
10	0.545
12	0.785
15	1.23
18	1.77
21	2.41
24	3.14
27	3.98
30	4.91
36	7.07
42	9.62
48	12.57
54	15.90
60	19.64

* Sample problem below.

TABLE 7.6 PRESSURE, ATMOSPHERES, AND DEPTHS

Test, psi*	Atmosphere Increase	Water Depth Over Pipe, ft
4	0.2666	0
5	0.3333	2
6	0.4	4

**NOTE:* For safety reasons, test pressures should not exceed 6 psi. If there is more than 4 feet of water depth over a pipe and water is not leaking into the sewer, the pipe apparently will not have problems from infiltration or exfiltration.

EXAMPLE 7

A 350-foot section of new 6-inch pipe is 4 feet below the water table and is to be given an air test.

1. From Table 7.5, 6-inch pipe has a volume of 0.196 cubic foot per foot of length. Multiply this value by the total length to be tested at one time to get the total volume of the pipe.

   Pipe Volume, cu ft = (Table 7.5 Value)(Length, ft)

   = (0.196 cu ft/ft)(350 ft)

   = 68.6 cu ft

2. For an effective 4 psi test with the pipe 4 feet below the elevation of the water table, we will need the 4 psi plus a half pound for each foot of depth, or 2 psi more. This means that our test pressure will be 6 psi (4 psi + 2 psi). For 6 psi, we have to add 0.40 atmosphere to the air that is already in the pipe, or the compressor will have to put 40 percent of the volume of the pipe into the test area.

   Volume of Test Air, cu ft = (Pipe Volume, cu ft)(Table 7.6 Value)

   = (68.6 cu ft)(0.40)

   = 27.44 cu ft

From this information, a 2 cubic feet per minute (CFM) at 75 psi compressor will take about three minutes to pressurize the length of pipe to the desired 6 psi. To determine how long a compressor will take to fill a pipe with air to the desired pressure, consult the compressor's operation and maintenance manual prepared by the manufacturer. If the difference in temperature between the air pumped into the pipe and the soil surrounding the pipe is large, consideration may have to be given to this factor when estimating the time it will take to fill the pipe with air. The restrictions or constrictions in the air line and the length of air line between the compressor and the plug often cause sufficient *FRICTION LOSSES*[13] to significantly increase the time required to pressurize the line being tested.

7.452 Air Testing Procedures

MANHOLE TO MANHOLE. Testing long lengths of pipe is usually limited to new wastewater collection systems that have intentionally plugged building sewers or no service connections.

[12] *Logarithmic (LOG-a-RITH-mick) Scale. A scale on which actual distances from the origin are proportional to the logarithms of the corresponding scale numbers rather than to the numbers themselves. A logarithmic scale has the numbers getting bigger as the distances between the numbers decrease.*

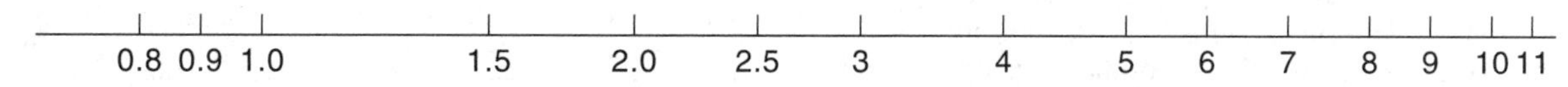

[13] *Friction Loss. The head lost by water flowing in a stream or conduit as the result of the disturbances set up by the contact between the moving water and its containing conduit and by intermolecular friction.*

1. Equipment needed:

   a. Two pipe plugs (Figure 7.63) with taps for connection of pressure gages and air hoses. For pipes up to 24 inches in diameter, these may be either mechanical plugs with rubber gaskets or inflatable plugs with rubber boots. For larger than 24-inch pipes, the plugs must either be inflatable bag stoppers or made in two or more pieces in order to be easily positioned in the pipes.

   b. An air compressor of required capacity with a hose to connect to pipe plug. There should be a shutoff valve on the hose connection to the plug. *ALL GAGES AND VALVES MUST BE LOCATED OUTSIDE THE MANHOLE TO AVOID INJURY TO ANYONE READING GAGES OR ADJUSTING VALVES IF A PLUG SHOULD BE BLOWN OUT OF THE PIPE.*

   c. A pressure gage should be connected to the line between the shutoff valve and the pipe plug to indicate pressure behind the plug. *THIS GAGE MUST BE OUT OF THE MANHOLE.*

   d. For safety reasons, the air supply line should be provided with an air safety valve or regulator. The internal air pressure in the sewer line cannot exceed 8 psi.

   e. A watch to indicate length of test in minutes and seconds is desirable where specified tests are to be made.

   f. A garden sprayer filled with soapy water and connected to a long pipe wand. With this device the operator can spray the plugs from outside the manhole.

2. The test crew should be not less than three persons. Where the air compressor does not have an automatic shutoff, one person will have to stand by to shut it down at the start of the test period.

3. Test procedure:

   a. Before testing a sewer, test your equipment. A good procedure is to carry on the equipment truck four-foot sections of PVC pipe of the sizes commonly air tested in the field. Insert a plug in each end of the PVC pipe of the diameter to be air tested and test the air test equipment. This procedure assures you that your plugs, hoses, gages, and equipment are working properly. Record these results so the accuracy of your equipment can't be questioned when performing actual field tests. Take reasonable safety precautions while testing the equipment. For example, stand clear of the pressurized pipe to prevent injury in the event a plug should be blown free.

   b. Insert plugs at each end of the pipe; expand and secure them. Any debris, such as sand or dirt, must be carefully removed from the pipe locations where the gaskets or rubber boots are to make a seal. Do not smear a heavy grease over the outer faces of the gaskets or boots to increase their capacity to seal against any uneven or rough surfaces on the inside of the pipe. The grease will reduce the capability of the plug to maintain its position. Even with proper bracing, air can bypass a greased plug.

   c. Before any air hose is connected to the plug at either end of the pipe, firmly block and wedge the plugs into place so that they cannot be pushed or blown out. To understand the need for precaution, realize that the force trying to push the plugs out is equal to the pressure (psi) times the plug surface area (sq in). Table 7.7 gives plugs' surface areas in square inches.

      Realizing the heavy forces that are exerted against the plugs by pressures they are required to withstand, adequate timbers and other bracing material must be selected with care and properly placed so that there is no possible way for the plug to work loose.

      As the plugs are inflated, the air pressure within the plug helps to seal the plug against the pipe walls to hold the plug in place and to prevent air from leaking out around the plug.

   d. After plugs are in place, connect the air hose, inflate or seal the plugs, and then pressurize the sewer to the test pressure.

   e. Supply air slowly to the section of sewer system being tested until the internal pressure is raised to 4.0 (or any other test pressure) pounds per square inch gage pressure (psig). Be careful and do not raise the air pressure much higher than 4.0 psig above the pressure outside the pipeline because it is possible to rupture pipe joints with excessive air pressure. When the test air pressure is reached, close the isolation valve and start the test.

TABLE 7.7 PLUG SURFACE AREAS

Pipe Diameter, inches	Surface Area, square inches*	If Pressure 5 psi, Force Against Plug, lbs**
4	12.56	63
6	28.27	141
8	50.26	251
10	78.54	393
12	113.10	566
15	176.70	884
18	254.47	1,272
21	346.37	1,732
24	452.39	2,262
30	706.86	3,534
36	1,017.87	5,089
42	1,385.45	6,927
48	1,809.56	9,048
54	2,290.23	11,451
60	2,827.44	14,137

* To calculate surface area, sq in = (0.785)(Diameter, in)^2

** To calculate force, lbs = (Area, sq in)(Pressure, lbs/sq in)

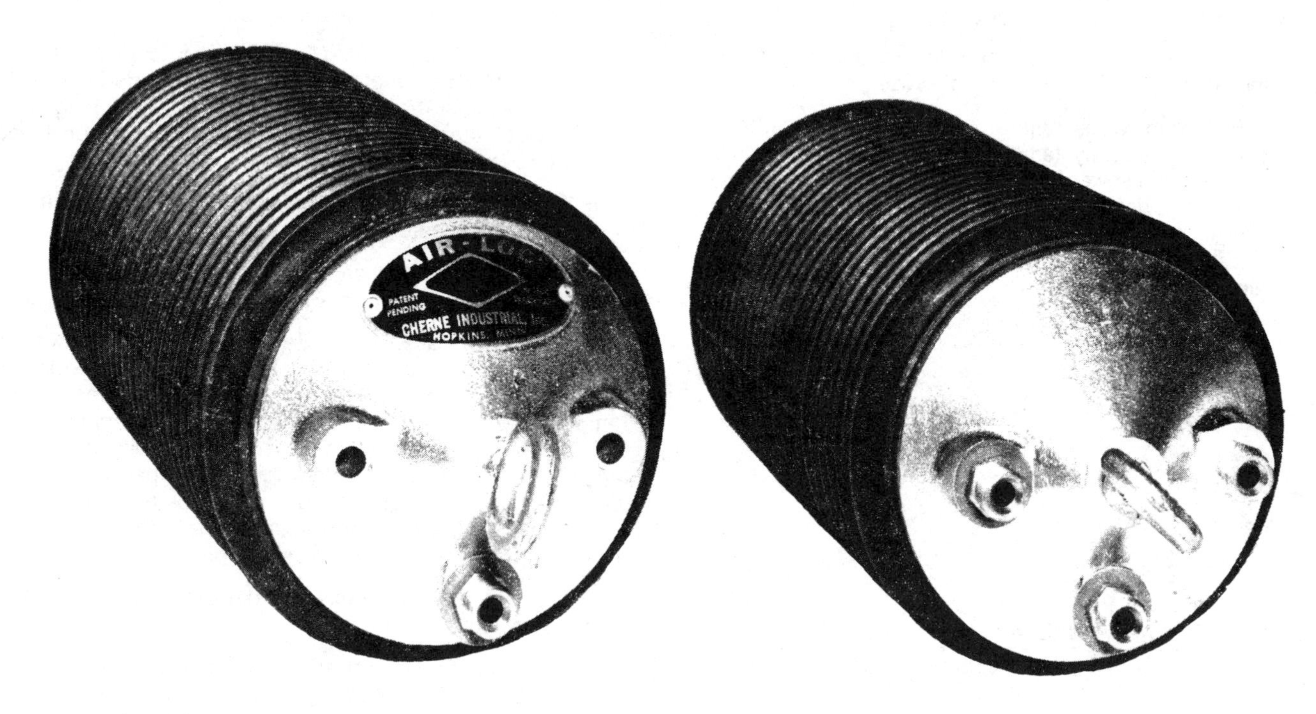

Used for (1) isolating leaks, (2) air testing between manholes, and (3) as plugs for maintenance and repair.

Fig. 7.63 Air plugs

(Courtesy of Cherne Industries, Inc.)

EXAMPLE 8

Calculate the force against each plug if a pipe is 36 inches in diameter and the test pressure is 7 psi.

Known		**Unknown**
Pipe Diameter, in	= 36 in	Force, lbs
Test Pressure, psi	= 7 psi	
Surface Area, sq in (from Table 7.7)	= 1,017.87 sq in	

Determine force against plug.

Force, lbs = (Test Pressure, psi)(Surface Area, sq in)

= (7 lbs/sq in)(1,017.87 sq in)

= 7,125 pounds

How would you like to be clobbered by a force of 7,125 pounds?

CAUTION: *NEVER ALLOW ANY PERSON TO ENTER A MANHOLE WHILE AIR IS BEING FORCED INTO A PIPE WITH PLUGS IN PLACE OR WHEN ANY PRESSURE REMAINS BEHIND PLUGS.* Remember! Entry into a manhole must be in compliance with confined space regulations. Refer to Chapter 4, "Safe Procedures," Section 4.40 of this manual for details on confined space entry requirements.

f. After an internal pressure of 4.0 psig is obtained, allow at least two minutes for air temperature to stabilize, adding only the amount of air required to maintain the 4.0 psig. Measure and record the air temperature to be sure it has stabilized.

g. After two minutes, disconnect the air supply and when the air pressure in the sewer system has decreased to 3.5 psig, start a "stopwatch" and stop it when the pressure decreases to 2.5 psig.

h. Compute the allowable time for a decrease in pressure from 3.5 to 2.5 psig using each of the two formulas below.

$$tQ = \frac{0.022\,(d^2_1L_1 + d^2_2L_2 + d^2_nL_n)}{Q}$$

$$tq = \frac{0.085}{q}\left(\frac{d^2_1L_1 + d^2_2L_2 + D^2_nL_n}{d_1L_1 + D_2L_2 + d_nL_n}\right)$$

Where: tQ = allowable time for decrease in pressure from 3.5 to 2.5 psig in seconds.

tq = same.

Q = 2.0 cubic feet per minute of air loss.

q = 0.0030 cubic foot per minute of air loss per square foot of internal pipe surface.

d = diameter in inches of sewer pipe in section of system being tested.

L = length in feet of sewer pipe being tested.

n = number of length and diameter of each pipe section. The subscript 1 represents the first pipe, 2 the second pipe, and the n could be the third pipe, then the fourth pipe and so on to the nth pipe.

i. Compare measured time for the air pressure to decrease from 3.5 to 2.5 psig and if it is greater than the lesser of the two allowable times computed from the formulas, the section of sewer system has passed the air test. Experience indicates that infiltration of groundwater into the section of a sewer system passing the low-pressure air test will be less than 200 gallons per day per inch of pipe diameter per mile of sewer.

For example, assume an air test is conducted on a section of sewer system containing 400 feet of 8-inch diameter main sewer line and 360 feet of 4-inch diameter building sewer line and the measured time for the air pressure to decrease from 3.5 to 2.5 psig is 225 seconds. Does the section of the collection system pass the air test?

$$tQ = \frac{0.022\ (d^2_1 L_1 + d^2_2 L_2)}{Q}$$

$$= \frac{0.022\ (8^2 \times 400 + 4^2 \times 360)}{2}$$

$$= 0.011\ (25{,}600 + 5{,}760)$$

$$= \underline{\underline{345 \text{ seconds}}}$$

$$tq = \frac{0.085}{q} \left(\frac{d^2_1 L_1 + d^2_2 L_2}{d_1 L_1 + d_2 L_2} \right)$$

$$= \frac{0.085}{0.0030} \left(\frac{8^2 \times 400 + 4^2 \times 360}{8 \times 400 + 4 \times 360} \right)$$

$$= 28.33 \left(\frac{25{,}000 + 5{,}760}{3{,}200 + 1{,}440} \right)$$

$$= \underline{\underline{192 \text{ seconds}}}$$

Since the measured time of 225 seconds is greater than the lesser of the allowable times (192 seconds), the section of the collection system passes the air test.

If you do considerable air testing, tables indicating allowable air losses or times for pressures to drop will be most helpful.

j. If the test fails, there is a chance that one or both of the seals made by the plugs may be defective. To determine if this is a fact, *THE PRESSURE IN THE TEST AREA IS RELEASED COMPLETELY.* A person using a garden sprayer full of soapy water with a long pipe wand can spray the plug from outside the manhole. Apply liberal amounts of the soapsuds to the seats and connections of the plugs. Be sure the bracing against the test plugs is secure. Repressurize the test section slightly. Inspect for leaks. *BE SURE TO KEEP OUT OF THE WAY OF THE PLUG IN CASE IT SHOULD BREAK LOOSE.* If the gaskets or boots are not properly seated and leak-tight against the pipe wall, depressurize the plugs and re-seat them. Continue testing in this way until you achieve an airtight seal.

k. When the air test is completed, lower the pressure in the test section before deflating either of the pipe plugs.

l. Once the test section has lost its air pressure, remove the bracing or cribbing and take out the plugs.

SEARCH FOR SPECIFIC LEAKS. Where a pipe is known to have leaks, or where a pipe in service is connected to unplugged building sewers, these small sections are tested with remote-inflation plugs. In this operation, two inflatable plugs of special design are attached to each other with a specific length of chain or steel cable (Figure 7.62, page 466). The plugs are pulled through the pipe with a rope and hand winch and stopped at each section equal to the distance between them for a test of the isolated pipe section.

1. Equipment required for this type of testing will include all of the following:

   a. An air compressor of adequate size, but usually not as large as that required for testing long lengths of pipe.

   b. An air pressure control console. This console must have two air pressure regulators, one for controlling the inflation pressures of the boots of the two plugs and one for controlling the test pressure to be injected between them. The control console must also have valving to:

      1. Inject or release the air pressure in the plug boots to isolate the section of pipe being tested,

      2. Inject air pressure into the space in the pipes between the plugs,

      3. Isolate the area between the plugs with a pressure gage showing this pressure and indicating any changes in it over a specified time period, and

      4. Release the pressure between the two plugs and then the plug pressure before moving to a new test section or removing the plugs from the sewer.

   c. Two plugs with inflation boots that are properly sized for the pipe to be tested. One of these plugs must have at least one through-plug air attachment.

   d. Three air hoses are generally used for greater speed, but two may be sufficient. One hose is used to inflate the boots of the plugs and the other hose is used to inject air to develop the test pressure between the plugs. A third hose may be used to measure the pressure in the test space between the plugs and to release the test pressure through a valve when the test at that location in the pipe is completed.

   e. Two winches to pull the plugs and hoses through the pipe to be tested. One winch is installed at each end of the sewer line to be tested so if you go too far down the line while changing plug position, you can pull the plug and hoses back to position them properly for the next test.

f. A distance marker system, usually a movable spring clamp on the pull rope and stakes or marks on the pavement. Many winches have a footage counter on the cable which records the distance.

2. Crew requirements are at least two persons plus support crew to direct traffic or provide other necessary assistance. One person operates the control console and records the test results. The second person moves the plugs with the winch each time a test is completed. A third person must be available whenever anyone must enter a manhole.

3. Prior to the testing of any pipe by this method, the sewer line should be thoroughly cleaned to remove any sand, silt, or other debris that may prevent the inflatable plugs from passing through the line and making a good seal against the pipe wall. The use of a high-velocity cleaner or balling, as described in Chapter 6, are methods used for cleaning sewer lines prior to testing. Old sewers or sewers with a history of stoppages should be inspected by closed-circuit television, as described in Chapter 5, after they have been cleaned to be certain that there are no obstructions (protruding taps or offset joints) that may prevent the movement of the air test plugs.

4. In operation, the plugs are put into the line and pulled until the rear plug is just inside the pipe at the starting manhole.

   a. Inflate the plugs to isolate the pipe area between them.

   b. Inject the desired test pressure into the area. The pressure gage measures the pressure between the plugs.

   c. The length of the test should be established on the basis of one second for each cubic foot of space between the two plugs.

   d. When a leak is detected, there is a possibility that one of the boots is not seated properly. To attempt to remove an object or bit of debris that may obstruct the boot's contact with the pipe wall, deflate boots while there is pressure between them. This may tend to blow debris out of the area of boot contact. Be extremely careful that no one will be hit by flying debris. Also be sure there is *NO SLACK* in the line between the boots or the boots could be severely damaged.

   e. Retest the same area. If a leak is still detected, record the fact and the rate of leakage as "slow," "moderate," or "rapid." Records must include the pressure drop per minute, diameter of pipe, length of pipe, location, and date tested.

   f. Deflate the plugs and move them to the next test section.

5. Keeping a good record of footages is extremely important in this type of operation. Each test location and the results must be properly and accurately recorded.

   a. The distance between the two test plugs should be established and firmly fixed before starting the test. Measure this distance from the inside sealing surfaces or the inside end of one plug to the inside end of the other.

   b. To test a section of pipe, the crew leader should review the available information on the pipe to be tested so as to select the areas to be tested and to determine the length of each test section. The inside distance between plugs is usually set at 3 feet, 6 feet, or 9 feet, plus 3 to 6 inches, depending on the distance between joints and the number of joints to be tested each time. When moving the plugs through the pipe, each move should be exactly the distance between the plugs or a little less.

   c. To select the distance between the plugs before starting the run, the supervisor should estimate whether the pipe to be tested may have many bad leaks or joints, or if only three or four problem areas are suspected. If every joint is suspected as the location of a possible leak, limit the distance to only one pipe section or less and test only one joint at a time. Where only a few, if any, joints are expected to be bad, the distance can be increased with proper coupling. Any desired distance, up to the distance between manholes, could be used. Plugs can be connected and pulled through sewers by the use of a chain or cable, which is safer than using rods.

   d. If a leak is detected by a pressure drop and more than one pipe joint is being tested, deflate the plugs and pull them back a distance equal to one pipe section. Repeat the test. If the leak is still between the plugs, deflate them and move back another pipe section length. Test again. When the leak is lost, the distance to the leak point or joint is then known because it is the next joint beyond the lead plug.

   e. The distance measurement must be maintained by the operator at the pull winch. If the plug assembly must be backed up at any time, do not move the clip on the pull cable until the test of the leak area is completed. The operator may then pull the plugs to their original position, move the clip back to relocate the plugs in the next part to be tested, and then pull them into their new position.

   If there is any question about the accuracy of the footage, use the lead plug to find the connection to the line being tested. This is done by moving the plug assembly in one-foot increments and testing after each relocation, usually done when the footage record shows the connection would be located in the next test area. When the test area cannot be pressurized, the lead plug will be within one foot of and beyond the connection.

6. The operator at the control console must always be alert to any pressure drop in the plug boots that might indicate boot perforation or leakage. When this occurs, and a boot is damaged by some object in the pipe, it must be immediately removed and repaired or replaced. Never try to test using this method with defective boots.

7. When one of the service connections to the pipe run being tested is between the plugs, the void area (space in pipe between plugs) cannot be pressurized unless the connection is blocked with another plug. The pull winch operator, who is maintaining a record of the footages tested, should be aware of the occurrence of one of the connections ahead of arrival and signal the test equipment operator.

8. Test records and footage records are usually recorded on the map with pencil marks and footage totals.

BUILDING SEWERS. There are several ways to test individual building sewers. The method you select will depend greatly on the conditions you encounter. The paragraphs that follow briefly outline the selection of a method and the proper test procedures.

1. Where a building sewer is laid in a straight line, use two inflatable plugs. Attach one plug to rigid, coupled sewer rods so that it may be pushed down the pipe with a hose for inflation at or close to the connection to the lateral sewer. The second plug will have a pressure gage and a connection for air to be passed through it from another hose.

   a. If adequate cleanouts are available, attach the first plug to the rods and push the plug into the pipe to the main or to any desired test distance. Disconnect the last section of the rod so that the rod is just far enough inside the cleanout to install a second plug. Inflate the far plug and push the hose, with closed valve, into the pipe just far enough to permit insertion of the second plug.

   b. Insert the second plug, which may be a mechanical type of plug, into the entrance of the pipe; secure it. Pressurize the space between the plugs and proceed with the air test.

   c. If the test fails, indicating a leak in the pipe section being tested, pull the far plug back a short distance and re-inflate. Repeat this procedure at short distances until no leakage is indicated. When this happens, it indicates that the far plug has been pulled far enough back and beyond the point of leakage, thus locating the approximate area of the leak.

   d. In the event a usable cleanout is not available, the pipe may be excavated and opened for test purposes where desired. A usable cleanout is one that will permit the section of pipe to be tested to be isolated from the building sewer system and its vents and traps.

2. In some testing of this type, two inflatable plugs are connected to each other with one or more rods and pushed into the pipe section. Once in position, they are used in much the same manner as the two inflatable plugs are used in testing 6-inch or larger pipes, with rods being used to relocate the plugs for each test.

3. Where available, a pipeline pressure grouting unit that has an air testing capacity can be used in conjunction with the television camera (Figure 7.64) to air test individual building sewers.

   In operation, pressurized air is pumped into the plugged line from the packer boot inflation air hose (using solenoids in the packer assembly) or through an air pass hose to the air plug in the service pipe. Pressure between the plug and the packer may be measured in the grouting system or in the pass hose of the plug. Locations of leaks in the service line may be found by moving the air plug on rods and re-inflating and testing at each different location.

7.46 Water Testing

Where conditions are appropriate, a water exfiltration test will provide an accurate test of a new sewer line's ability to convey wastewater without excessive leakage and to resist groundwater infiltration. Any building sewer connections to a sewer line being tested must be plugged. Other limitations on the use of water testing are described in Section 7.45.

7.460 Equipment Needed

1. Two watertight pipe plugs for the diameter of sewer line to be tested;

2. A source of water that will fill the length of sewer line to be tested in a reasonable period of time (a water truck or fire hydrant and hose); and

3. A tape measure and stopwatch.

7.461 Test Procedures

1. Plug the downstream end of the sewer line to be tested at its inlet into a manhole and plug the inlet sewer at the upstream manhole.

2. Fill the upstream manhole with water to a point four feet above the invert of the manhole channel, or four feet above the average depth of groundwater above the sewer line.

3. Mark the appropriate height of water in the manhole at the start of the test and record the time in hours and minutes.

4. Allow the water to drop two feet and record the elapsed time, or wait 120 minutes and mark the height of water at the end of that time.

5. Measure the diameter of the manhole at the high and low water marks and compute the volume of water that escaped (exfiltrated) from the sewer line in gallons.

6. Divide the loss of water by the time in minutes between the start and finish of the test to compute the rate of water exfiltration in gallons per minute.

7. Multiply the loss of water by 1,440 minutes per day to obtain gallons per day lost. Divide this amount by the diameter of the pipe in inches and the length in miles of the sewer line being tested to determine the rate of water exfiltration in gallons per day per inch of diameter per mile length of sewer line. The acceptable rate of water exfiltration from a sewer line is approximately 450 GPD/in/mi, or less.

8. Remove the downstream plug, exercising care with respect to the rapid flow of water that will occur in the sewer line when the test water is released. Then remove the upstream plug. If the water used in the test is to be reused for the next downstream test section, install another plug at the downstream end of the next test section before releasing water from the bottom plug of the upstream test section.

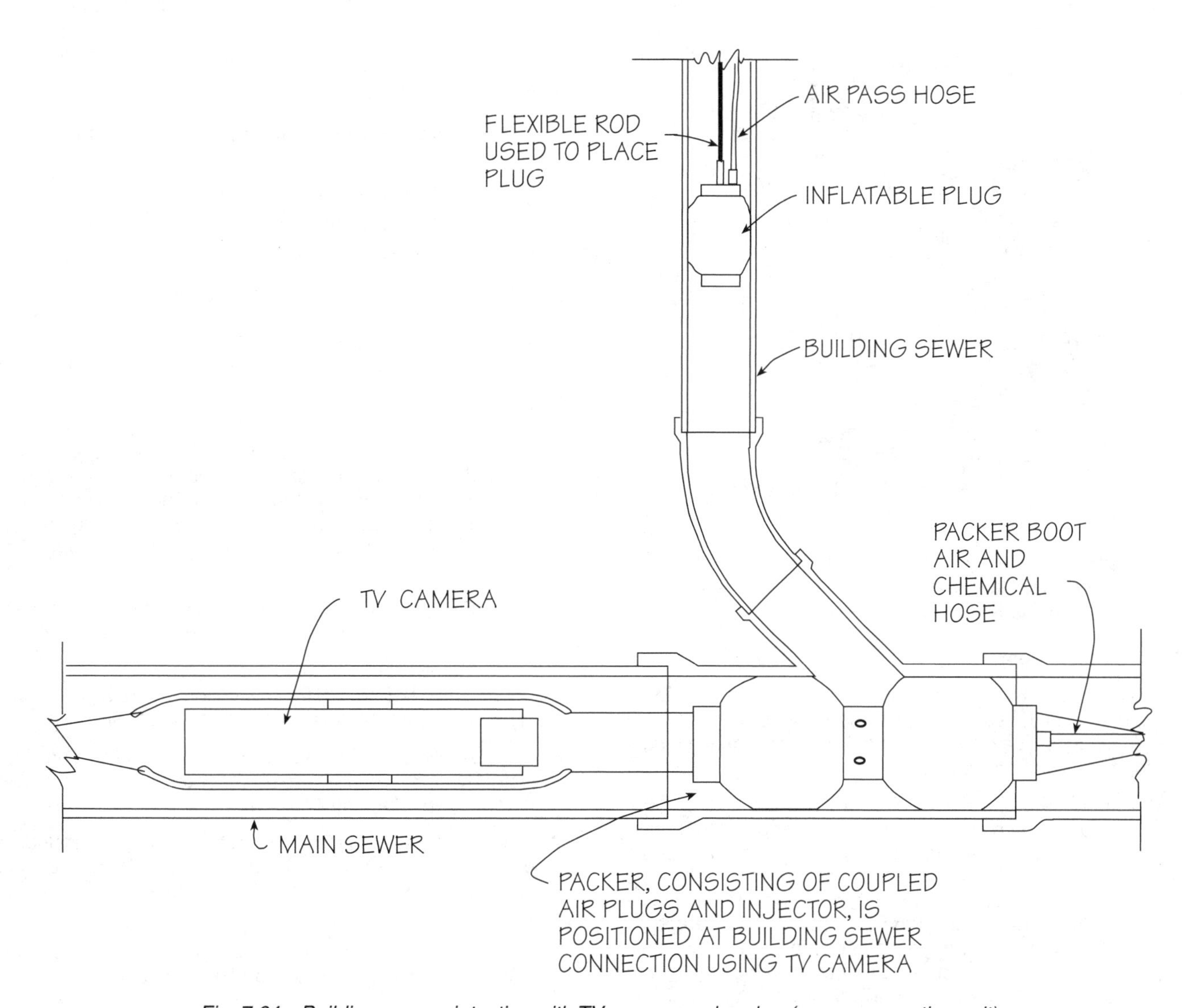

Fig. 7.64 Building sewer air testing with TV camera and packer (pressure grouting unit)

EXAMPLE 9

Compute the rate of water exfiltration from 400 feet of 8-inch diameter main sewer line and 300 feet of connecting 4-inch diameter building sewers when the water level in the upstream 4-foot diameter manhole drops 4 inches in 120 minutes.

Known		Unknown
Diameter, in	= 8 in	Exfiltration, GPD/in/mi
Length, ft	= 400 ft	
Diameter, in	= 4 in	
Length, ft	= 300 ft	
Manhole D, ft	= 4 ft	
Manhole Drop, in	= 4 in	
Time, min	= 120 min	

1. Calculate the volume of water exfiltrated (lost) during the test.

   NOTE: Surface Area, sq in = 0.785 (Diameter, in)^2

$$\text{Volume, gal} = (\text{Area, sq ft})(\text{Drop, ft})(7.48\text{ gal/cu ft})$$

$$= \frac{(0.785)(4\text{ ft})^2(4\text{ in})(7.48\text{ gal/cu ft})}{12\text{ in/ft}}$$

$$= 31.3\text{ gal}$$

2. Determine the water lost due to exfiltration in gallons per day.

$$\text{Exfiltration Rate, GPD} = \frac{(\text{Volume, gal})(60\text{ min/hr})(24\text{ hr/day})}{\text{Time, min}}$$

$$= \frac{(31.3\text{ gal})(60\text{ min/hr})(24\text{ hr/day})}{120\text{ min}}$$

$$= 376\text{ GPD}$$

3. Calculate the exfiltration in gallons per day per inch of pipe diameter per mile of pipe length.

$$\text{Exfiltration, GPD/in/mi} = \frac{\text{Exfiltration Rate, GPD}}{(\text{Diameter, in})(\text{Length, mi}) + (\text{Diameter, in})(\text{Length, mi})}$$

$$= \frac{376\text{ GPD}}{(8\text{ in})(400\text{ ft}/5{,}280\text{ ft/mi}) + (4\text{ in})(300\text{ ft}/5{,}280\text{ ft/mi})}$$

$$= \frac{376\text{ GPD}}{0.61\text{ in/mi} + 0.23\text{ in/mi}}$$

$$= 448\text{ GPD/in/mi}$$

The sewer line in this example is acceptable since its exfiltration rate is less than the acceptable rate of 450 GPD/in/mi.

If a sewer line does not pass the water exfiltration test, the search for specific leaks is usually conducted with air pressure as described in the earlier Section 7.452.

7.47 Mandrel Test

After placement and compaction of trench backfill and prior to restoration of the surface, all new sewer lines constructed of flexible types of pipe should be tested with a mandrel to measure for deflection and joint offsets.

The mandrel testing device (Figure 7.65) is a rigid cylindrical plug tapered on each end with a diameter equal to 95 percent of the average diameter of the pipe to be tested. Some agencies use 97 percent of the average diameter of the pipe for construction inspection and assume the ultimate deflection in the future will be down to 95 percent of the average diameter. The plug has pulling rings at both ends, to which pulling and tag lines are attached. Either or both the pulling and tag lines are marked so that the distance the mandrel has been pulled into a sewer can be determined.

The mandrel test is conducted by pulling the test device through a completed sewer from manhole to manhole. The mandrel device will not pass through the pipeline if a flexible type of pipe has been deflected beyond five percent of its diameter, if a rigid type of pipe has been crushed, if a joint is offset more than five percent of the pipe diameter, or if a building sewer connection is protruding into the main line. The location of the obstruction can be determined by use of the distance markings on either the pulling or tag line. The pipe must be excavated and the obstruction removed.

7.48 Acceptance Testing

This section was provided by Victor Coles, Municipal Services Company, Inc. The section describes acceptance testing procedures.

7.480 Pipe Cleaning

Prior to deflection testing and closed-circuit television (CCTV) inspection, clean the completed pipeline with a high-velocity cleaner, or other engineer-approved cleaning equipment, and provide a pipeline free of dirt, mud, rocks, or other material. Leave downstream plugs in place during cleaning and do not introduce foreign material into existing sewer lines. Conduct CCTV inspection.

7.481 Closed-Circuit Television (CCTV) Inspection

1. Acceptance criteria:
   a. No visible standing water in pipeline caused by grade defects.
   b. No pipeline structural defects observed.
   c. No pipeline installation defects observed.
   d. No infiltration observed.
2. CCTV sewer line inspection to be performed by firms that are suitably equipped, experienced, qualified, and staffed for sewer line CCTV inspection. If requested, provide a calibration tape showing various water depths for the size of pipe being videoed. Calibration tape must be on site during CCTV process.
3. Perform CCTV inspection after backfill and before surface repair. Uncover and reinstall sections of the pipe found to have defects in workmanship (including standing water caused by grade defects) as directed by owner. After repair, re-CCTV the repaired section. Do not reinstall damaged or rejected pipe.
4. Notify Engineer at least 48 hours before the CCTV inspection to allow for Engineer, at Engineer's discretion, to witness.
5. Clean lines before CCTV inspection using high-velocity cleaner, flushing balls, or other suitable means.
6. Immediately before the CCTV inspection, pour sufficient water/dye mixture in upstream manhole to produce visible flow in downstream manhole(s).
7. Provide CCTV sewer line inspection equipment including color camera system, VHS taping system, camera propulsion equipment, and van to allow witness of CCTV by inspector.
8. Use super high grade, color, VHS tapes. The tapes shall include the following video and audio information:
   a. Video:
      (1) Project number and name
      (2) Date of TV inspection
      (3) Upstream and downstream manhole numbers
      (4) Current distance along line segment
   b. Audio:
      (1) Date of TV inspection
      (2) Verbal confirmation of upstream and downstream manhole numbers and/or locations
      (3) Verbal description of pipe, size, type, and pipe joint length
      (4) Verbal description of location of each service connection
   c. Tape Identification Tag:
      (1) Manhole to manhole designation
9. If no defects are observed, submit tapes and logs to Engineer for review and acceptance.

7.482 Testing Manholes

1. Unless otherwise indicated in the contract documents, test all manholes.

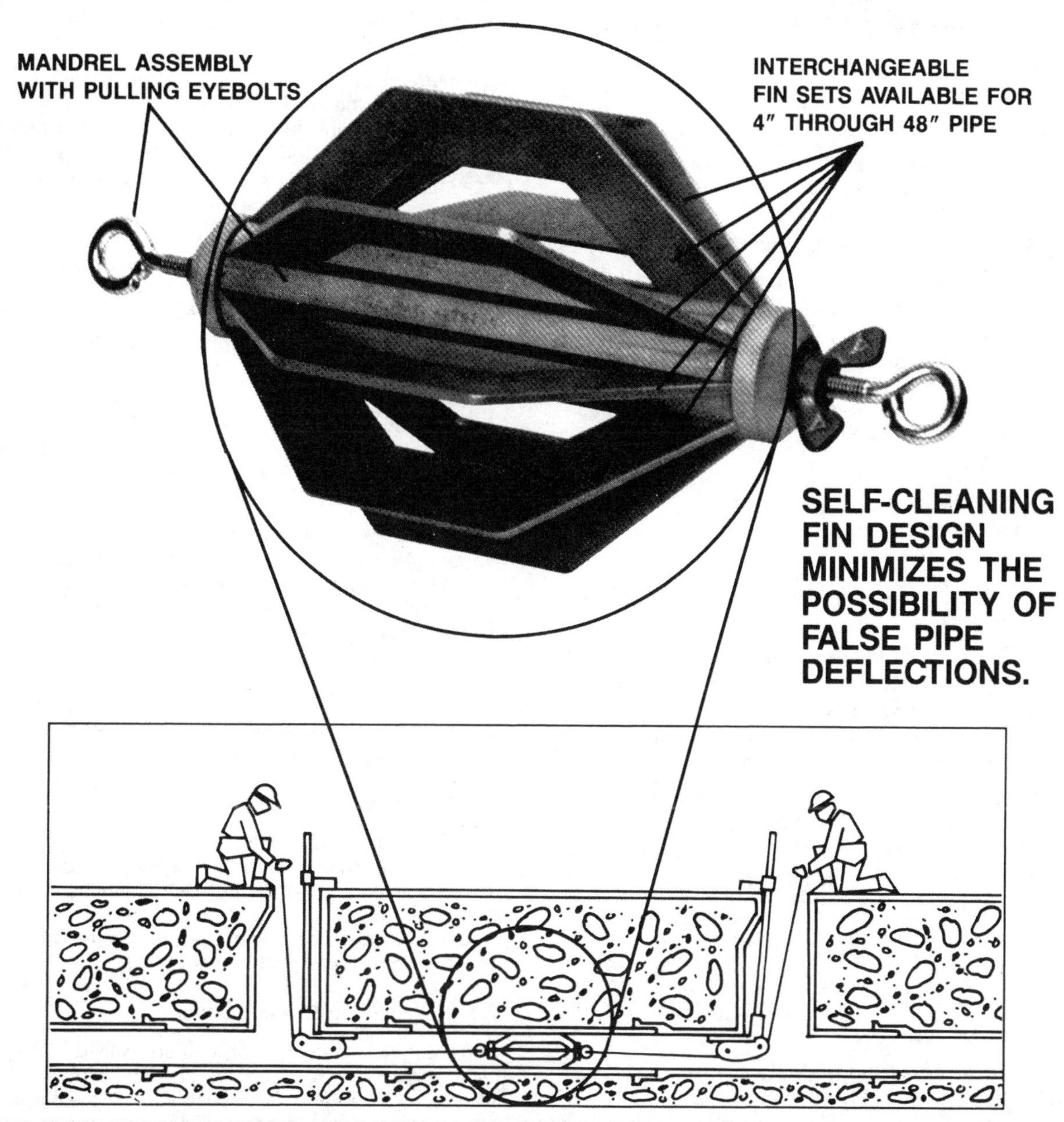

THE DEFLECTION TEST GAGE CAN BE EASILY PULLED THROUGH THE PIPELINE.

Fig. 7.65 Mandrel pipe deflection test gage

(Permission of Cherne Industries, Inc.)

2. Testing schedule.
   a. Preliminary testing: At the discretion of the Contractor, preliminary testing may be done at any time prior to installation of other utilities; however, a final test will still be required.
   b. Final testing: Perform final testing after backfilling and compaction and following installation of other utilities but prior to surface restoration. If a test fails, the manhole shall be repaired and retested at no cost to the owner.
3. Use one of the following testing methods. Test lined manholes prior to final welding of joints.
   a. Hydrostatic testing:
      (1) Pre-fill. Fill manhole with water 24 hours before the time of the test, if desired, to permit normal absorption into the walls to take place.
      (2) Plug pipes. Plug all piping, inlets, and outlets in the manhole.
      (3) Fill. Fill with water to within 1 foot (300 mm) of the ring elevation.
      (4) Requirement. An allowance of 1 gal/hr (0.40 *L*/hr) for each 1 foot (300 mm) of manhole depth is allowed. If a test fails, repair and retest at no additional cost to the contract.
   b. Vacuum testing:
      (1) Tester. Manhole *VACUUM TESTER*,[14] or approved substitution.
      (2) Procedure.
         (a) Plug lift holes with non-shrink grout and pipes with plugs.
         (b) Brace plugs to prevent them from being drawn into the manhole.
         (c) Place test head inside of the top of the cone section and inflate seal in accordance with the manufacturer's recommendations.
         (d) Draw a vacuum to 10 inches (250 mm) of mercury and shut off the vacuum pump.
         (e) Measurement. With the valves closed, measure the time for the vacuum to drop to 9 inches (225 mm).
         (f) Requirement. Time to be greater than 60 seconds for 48-inch (1.2-m) diameter manholes, 75 seconds for 60-inch (1.5-m) diameter manholes, or 90 seconds for 72-inch (1.8-m) diameter manholes.
4. Retest: If the manhole fails the initial test, make repairs and retest at no additional cost to the contract.

7.49 Final Inspection

If the inspection of each phase of the construction has been thorough, the final inspection of a wastewater collection system construction project should serve as a final check on the small items of uncompleted work remaining, particularly with respect to surface restoration and cleanup, to make the project acceptable to the public agency-owner.

Agencies should have a representative of the department that will be responsible for maintenance of the completed wastewater collection system accompany the inspector at the time of the final inspection for the purpose of pointing out any defects in the system that will cause future maintenance problems. All defects should be corrected before the job is accepted.

In large-diameter sewers, the final inspection should include a "walk through" inspection to verify that all the construction tools and debris have been removed. Final visual inspection of sewer lines may reveal standing water and other indications of grade misalignment.

A thorough and detailed list of items of work to be completed should be made at the time of the final inspection with copies being distributed to the contractor and project engineer. A copy of the "as-built" plans (record drawings) should be given to the operation and maintenance organization.

7.410 Construction Records

As previously mentioned, a sewer construction inspector is expected to submit comprehensive reports to the project engineer on the project being inspected. The following paragraphs describe the contents of these reports in more detail.

7.4100 Daily Reports

The purpose of a daily report is to provide the project engineer with an overview of the project progress and a source of information for answering questions raised later with respect to conduct of the work.

An inspector should report daily on the weather conditions, amount of work performed on the project, type of soil encountered during excavation of sewer trenches, working conditions with respect to safety regulations, number of workers, type of equipment being used, and any significant incidents that may have occurred during the day. Figure 7.66 illustrates a form used for submission of an inspector's daily report. Each agency develops forms to fit its particular reporting requirements and may include such additional items as number of yards of concrete placed and amount of sheet piling driven.

7.4101 Monthly Reports

The monthly reports of an inspector generally consist of information that will be used by the project engineer to determine the amount of work completed during the month and the amount to be paid the project contractor for completed work. Figure 7.67 illustrates a form used to submit monthly reports on the amount of completed work.

7.4102 Building Sewer Locations

In most instances wyes or tees are placed in main sewers for building sewer connections and are buried by the main sewer trench backfill before construction of the building sewer. The inspector must record the location of a wye or tee before it is buried. This is usually done by measuring the distance along the main sewer from the nearest downstream manhole and entering this data on a Building Sewer Locations form, as illustrated by Figure 7.68. The locations must be shown on the "as-built" plans (record drawings).

Also, since the part of the building sewer constructed in the street is generally buried prior to construction of the remaining part of the building sewer connecting a building to the completed sewer system, it is equally important to record the loca-

[14] *Vacuum Test. A testing procedure that places a manhole under a vacuum to test the structural integrity of the manhole.*

INSPECTOR'S DAILY REPORT

Job ______________________ No. __________ Date ______________

Contractor ______________________ Weather ______________________

	WORK	ITEM	LOCATION		QUAN
			STREET	STA. LIMITS/@STA. OR ST.	
1	Trench Exc. (1)			TO	
2	Laying 6"- - P.			TO	
3	" "- - P.			TO	
4	"4" - - P. Lots			TO	
5	B'fill, Longt. (2)			TO	
6	" , Lat. (2)			TO	
7	Manholes				
8	Compact, Longt. (3)				
9	Compact, Lat. (3)				
10	Testing				
11	Base Rock, Longt.				
12	Base Rock, Lat.				
13	Paving, Longt. (4)				
14	Paving Lat. (4)				
15	Pipe Cleaning				

(1) Type Soil Exc. ______________, (2) Type Soil B'fl ______________

(3) Type of Compaction, Longt. ______________, Lat. ______________

(4) Type Paving, Longt. ______________, Lat. ______________

Safety Check: Working Conditions ______________? Public ______________?

Comments on work conduct, personnel, equipment, materials, incidents, final inspection notes, acceptance recommendation, etc: ______________________________

Time this job ______________________ Signed ______________________

Note: Use one sheet per job. Submit original to office daily. Retain duplicate copy for inspector's diary.

Fig. 7.66 Inspector's daily report form

REPORT OF QUANTITIES INSTALLED

As of (Date): ______________________

Job: ______________________

Final Report ☐
Progress Report ☐

Installer ______________________
Engineer ______________________
Contractor ______________________

STREET	4" B.S.	6" Main	8" Main	___" Main	M.H.	F.I.	Extra Wyes	

Instructions to Inspector:
Make report in duplicate
Report progress as of last day of month, when job not complete.
Submit final report when installation has been completed.
Make notes about unusual items.

____
Inspector

Fig. 7.67 Monthly report form for completed work

BUILDING SEWER LOCATIONS

PROJECT OR TRACT NO. ____________________ UNIT NO. __________

STREET ____________________ DATE __________

FROM ____________________ TO ____________________

MAIN OFFSET FROM C/L ________ STREET WIDTH ________ SIZE OF MAIN ____________

MH No.	Station	MAIN Connections		BUILDING SEWER				Lot or Assmnt No.	* Back Flow
		Type	Size	Side of St.	Prop. Line Reference	Depth at PL	Lgth		

*NR- Not Required   OF-Overflow Device   CV-Check Valve

Fig. 7.68 Building sewer location form

tion and depth of the temporary end of the building sewer. This is usually done by measuring the distance from the end of the building sewer along the street property line to the nearest property corner and measuring the depth of the top of the end of the building sewer below the top of the curb (or surface of the ground if there is no curb). Also the end of the sewer should be located with respect to the curb line (how far inside or outside the curb and the depth from the top of the curb to the top of the sewer). These measurements also are entered onto a Building Sewer Locations record. A long piece of wood (2 x 4) placed upright at the end of a temporary sewer is very helpful in the future when attempting to locate the end of a buried sewer.

The wye, tee, and building sewer location data should be entered directly and immediately onto the appropriate forms.

7.4103 "As-Built" Plans (Record Drawings)

At the start of a sewer construction project, the inspector should obtain two sets of the plans for the project and mark one copy "Working Plans" and the other set "As-Built Plans." The working plans are used for on-the-job references and conferences.

The "as-built" plans (also called "record drawings") are for the purpose of recording daily any constructed deviations from the location or depth of the sewer system facilities or any other deviation from the plans. Changes must be recorded as soon as they are found or completed and *NEVER* put off to another day. Locations of building sewers are commonly shown on the "as-built" plans in addition to recording their location on a Building Sewer Locations form.

The actual locations of the sewer system facilities are usually shown in red pencil on the "as-built" plans using the same scale and style of drafting and notation used in drafting the plans. The "as-built" plans should not be used in place of the inspector's working plans and should be kept clean and readable.

At the conclusion of the project, the "as-built" plans should be submitted to the project engineer in completed form. These plans are either filed directly as an "as-built" record of the project or used by a draftsman to revise the tracings or used by a computer programmer to revise the computer program from which the plans are printed. Copies of these plans must be sent to field operators as soon as possible because they must work from them.

The "as-built" plans are also used for preparation of sewer location maps and should also be available to wastewater collection system operators for detailed information if needed for maintenance or repair of a sewer line.

7.411 Additional Reading

1. *GRAVITY SANITARY SEWER DESIGN AND CONSTRUCTION* (MOP FD-5). Obtain from Water Environment Federation (WEF), Publications Order Department, 601 Wythe Street, Alexandria, VA 22314-1994. Order No. MFD5. Price to members, $25.00; nonmembers, $35.00; plus shipping and handling.
2. *CLAY PIPE ENGINEERING MANUAL*. Obtain from National Clay Pipe Institute, PO Box 759, Lake Geneva, WI 53147. Price, $25.00, plus shipping and handling.
3. *CONCRETE PIPE HANDBOOK*. Obtain from American Concrete Pipe Association, 675 Grigsby Way, Cedar Hill, TX 75104. Order No. CP-01-102. Price $36.00, plus $3.50 shipping and handling.
4. *HANDBOOK OF PVC PIPE—DESIGN AND CONSTRUCTION*. Obtain from Uni-Bell PVC Pipe Association, 2655 Villa Creek Drive, Suite 155, Dallas, TX 75234. Price, $40.00.
5. *PUBLIC WORKS INSPECTORS' MANUAL*, Silas R. Birch, Jr. Obtain from BNi Building News, 1612 South Clementine Street, Anaheim, CA 92802. Price, $59.95, plus $8.00 shipping and handling.
6. *ALTERNATIVE SEWER SYSTEMS* (MOP FD-12). Obtain from Water Environment Federation (WEF), Publications Order Department, 601 Wythe Street, Alexandria, VA 22314-1994. Order No. MFD12. Price to members, $12.50; nonmembers, $17.50; plus shipping and handling.
7. "Air Testing Sanitary Sewers," Sam H. Hobbs and Lloyd G. Cherne, Journal Water Pollution Control Federation, Vol. 40, No. 4, April 1968.
8. "Testing New Sewer Pipe Installations," Roy E. Ramseier, Journal Water Pollution Control Federation, Vol. 44, No. 4, April 1972.

QUESTIONS

Write your answers in a notebook and then compare your answers with those on page 504.

7.4E What is the main reason for using low-pressure air testing for testing sewers?

7.4F What should be done before air testing a sewer?

7.4G What does a pressure drop in the boot plugs indicate?

7.4H Why should care be exercised when removing the downstream plug after water testing?

7.4I What is the purpose of a mandrel test?

7.5 MAIN LINE REPAIR

Main line repairs are required when a sewer line repeatedly shows partial or complete stoppage of flow. Partial stoppages may require only cleaning. Broken and collapsed pipes and offset joints are examples of sewers needing repairs. To describe the procedures for repairing a main line, let's replace a section of 8-inch sewer in the street.

7.50 Preliminary Field Investigation

Before starting repair of a main line, the supervisor makes a preliminary field investigation to identify potential problems, obtains necessary permits, and plans the repair job. A work order prepared by the TV inspection crew indicates that a section of 8-inch sewer needs replacement. Paint is used to mark the repair area on the street surface. All utility companies are contacted and requested to identify the location of their utilities in the area of the repair site. Representatives of the utility companies involved (water, natural gas, fuel pipelines, electrical power, and telephone) meet with the supervisor at the site and mark the location and depth of their systems. Contact the local traffic control agency if permits are needed to reroute traffic or any other problems handling traffic are anticipated. Review the "as built" drawings (record drawings) for any helpful information, such as the type of pipe and bedding. Examine any records that could reveal the elevation of groundwater. Also determine if pumping of wastewater or groundwater will be required until the repair is made.

7.51 Crew and Assignment

This main line repair job requires at least a three-operator crew. The crew chief is a collection system repair specialist (Maintenance Operator II) and a qualified backhoe operator (Construction Equipment Operator). The crew chief is helped by two assistant collection system repair specialists (Maintenance Operator I). In some states the driver of a vehicle that transports equipment to a job site may have to obtain a special driver's license.

A collection system supervisor reviews the work order and prepares a time sheet. Also the supervisor prepares a sketch of the job site showing the location of all utilities and other important information. This preliminary information, the proper procedures and safety precautions, and the necessary equipment, tools, and materials are discussed by the supervisor with the crew chief.

7.52 Materials, Equipment, and Tools

Materials required for this particular job include sections of 8-inch vitrified clay pipe, 8-inch band couplings, and sufficient crushed rock to prepare a bed for the new section of 8-inch pipe to be installed.

Equipment required for this job includes a backhoe, dump truck, tilt-bed trailer, $1\frac{1}{2}$-ton flatbed truck equipped with an air compressor to operate a pavement breaker and a pneumatic tamper, and a water truck with a 1,500-gallon capacity tank.

The dump truck pulls the trailer loaded with the backhoe. The $1\frac{1}{2}$-ton flatbed truck carries all pipe repair materials, air and hand tools, and safety equipment (shoring, personal, and traffic control). The water truck has a portable 2-inch pump mounted on the tank and carries all the hoses and pipe needed for jetting purposes during backfill. A 24-foot aluminum extension ladder and extra shoring jacks and sheeting for bigger and deeper jobs also are carried by the water truck. All portable equipment (such as equipment on trailers) pulled to a job site must use a pintle hook and have a safety cable placed around the pintle hook and portable equipment.

Before the crew leaves the yard, they inspect all equipment (including safety) and determine that all the necessary equipment, tools, and materials needed to do the job are loaded on the trucks. Hard hats should be worn by the crew at all times. See Volume II, Chapter 11, "Safety/Survival Programs for Collection System Operators," for safety equipment, including first-aid kit.

7.53 Arrival at Job Site

When the crew arrives at the job site, they locate the marked repair area and park the trucks in safe and convenient locations. The flatbed truck should be parked close to the repair area so it can be reached with air hoses and the pavement breaker. Prepare traffic control using high flags and traffic cones (see Chapter 4, Section 4.3, "Routing Traffic Around the Job Site," and Volume II, Chapter 11, Section 11.1, 10. "Traffic Mishaps") to protect equipment, the work area, and operators. Contact the homeowner and owners of parked vehicles that are in the work site area or in driveways that will be blocked during repair of the main line. Explain to these people that a major repair on the sewer line must be done now and request their cooperation. Assist them, if necessary, to minimize inconvenience while doing the job. If parked vehicles are a potential problem, the job site can be barricaded and marked the night before or very early in the morning.

QUESTIONS

Write your answers in a notebook and then compare your answers with those on page 504.

7.5A What should a supervisor do before a crew is assigned a main line repair job?

7.5B What information does the supervisor need from other utility agencies at the job site?

7.5C What items are discussed by the supervisor and crew chief before the crew starts the job?

7.5D What should the crew do before leaving the yard for the job site?

7.5E What should the crew do when arriving on the job site?

7.54 Cutting Pavement

After observing traffic flow for proper control and safety, mark off the area of the pavement to be cut with a chalk line. The cut in the pavement should be 30 inches wide and 10 feet long for this job. Before marking and cutting, verify the repair locations by measuring the distance from the designated manhole. Also measure the distance from manhole to manhole and other important distances to verify the accuracy of the sketch.

Start the air compressor and allow it to warm up. An assistant collection system repair specialist (Maintenance Operator I) puts on safety glasses, earplugs, and required safety shoes or metatarsal guards. A cutting blade or other suitable tool, depending on type of pavement, is locked into the pavement breaker. The air hose is connected and the safety chains are locked in place. The safety chain is required to prevent the air hose from whipping around if the hose connection comes uncoupled from the breaker. Some hose connections are equipped with a double-spring lock and do not require a safety chain.

Cut the pavement along the chalk lines. While the pavement is being cut, the backhoe is unloaded from the trailer and lubricated. The trailer is unhooked from the dump truck and parked. Crushed rock is dumped near the repair area for later use.

Open the upstream and downstream manholes. Observe and record the amount of flow and the depth to the top of the pipe. This depth indicates to the backhoe operator how deep to excavate. Also, this observation can reveal if the line is plugged and causing flow to back up into the manhole.

After the blacktop is cut, the pavement breaker and air hose are put away and the flatbed truck is moved to make room for the dump truck and backhoe. The hoe operator examines the work order for the location of utilities and confirms their locations in the marked area.

7.55 Excavation

Start digging with the backhoe. Load directly into the dump truck the blacktop or road surface, subbase rock, and the soil down to about two feet below the road surface. Watch for changes in soil texture or other signs that could indicate the presence of cross trenches or other underground utilities.

At this depth, determine if spoil (excavated material) is suitable for backfill. Usually the original trench backfill material will make suitable backfill material again.

Send the dump truck to the gravel plant to dispose of the excavated pavement and subbase and return with approximately three yards of 1½-inch road rock. Continue digging with the backhoe. Keep the spoil pile at least two feet from the edge of the trench; a distance of four feet is recommended. An assistant repair specialist (Maintenance Operator I) uses a square point shovel to keep the working area clean. Everyone must be alert. Watch the backhoe and the trench for signs of utilities and anything unusual. Continue excavating until approximately 1½ feet above the sewer main line that is to be repaired. Use a probe to determine the exact distance to the main and the amount of material remaining over the top of the main. The backhoe operator must be careful to avoid damaging the line. If shoring is necessary, refer to Section 7.1, "Shoring."

The backhoe continues to dig down to within six inches of the main, depending on the operator. Continue to determine the location of the main line with the probe. Do not break the main line.

If water is encountered, a sump will have to be dug. Use a trash pump to remove the water and discharge it into a manhole or catch basin. By now the dump truck has returned with the gravel and is parked out of the way.

7.56 Pipe Repair

After the shoring is installed and the trench is determined safe, place a ladder into the trench. Leave the ladder in the trench as long as an operator is in the trench.

A round point shovel is lowered on a rope to the operator in the trench to uncover the main line and locate the problem area. A narrow shovel is used to dig alongside the pipe and to dig down below the main line when necessary.

A miner's pick and wonder bar are lowered in a bucket to the operator in the trench. These tools are used to dig under the main line about a foot and a half each way from the broken bell section of the pipe that is to be replaced.

Remove all loose dirt from under the pipe before cutting the pipe for repairs. When the main line is cut, the flow drops into the trench, fills the hole, and flows out the downstream pipe. If loose dirt is not removed, this water will mix with the loose dirt and form a sloppy mess unsuitable for bedding material.

Have a sump pump and hose standing by in case water must be pumped out of the excavation. Pump to the downstream pipe or to a downstream manhole. If possible, it may be necessary to plug off flows at the upstream manhole while the pipe is being repaired. Be careful not to flood any homes, businesses, or streets whenever plugs are used. If necessary, pump wastewater from the upstream manhole to a downstream manhole.

To prepare for cutting the pipe, the operator in the trench must be wearing safety glasses, steel-toe rubber knee boots, and gloves. Carefully lower the pipe cutter into the trench. Cut the pipe approximately one foot on each side of the break in the main line. Wrap the cutter around the main and cut the pipe two or three times at both locations where the pipe is to be cut. This many cuts are necessary so the good pipe won't chip or crack when the damaged section of pipe is broken out. Use the miner's pick to break out the bad pipe. Remove the pipe and debris from the trench using a bucket and rope.

Lower a mirror and flashlight into the trench. Inspect the main line for cracks, roots, and bad taps and joints in both directions from the cut.

Measure the distance between the cut pipe ends—for example, 25 inches. Inform the operators topside of the length needed. Cut a 24½-inch long piece of pipe off a section of 8-inch diameter pipe. This leaves a ¼-inch gap at each end of the repair section to allow the centering stop of the band couplings to fill these gaps. The pipe used for repair does not have to be made of the same material as the original pipe. PVC (polyvinyl chloride) and plastic truss sewer pipes are used to repair vitrified clay pipes and are connected together using band couplings and bushings.

Lower the bedding material in buckets and place it in the void under the pipe. Leave enough room at the cut ends to slide couplings around the existing main line.

Center two 8-inch band couplings on the ends of the new pipe section and mark the location with a pencil or anything handy for a centering reference. Lubricate and slide the couplings completely over the new pipe section so the ends are even or flush.

Lower the short piece of pipe and couplings into place between the main line ends. Slide the couplings over the existing main to the reference marks on the pipe as previously discussed. The coupling's center stop should drop into the 1/4-inch gap. If the band coupling does not seat, rotate the pipe section.

Tighten the stainless steel bands with a special torque speed wrench designed for the band couplings. If the wrong wrench is used, the pipe can be damaged.

Collection system operators are continuously working with manufacturers to develop more cost-effective and permanent means of repairing sewers. Broken, crushed, or badly cracked main lines greater than 12 inches in diameter may be repaired by the insertion of a special PVC pipe. This pipe must be made for a specific segment of existing pipe needing repair. The new PVC pipe consists of four segments (long quarter sections) that are pushed or pulled to the damaged area. The segments are expanded into place with internal jacks. The four joints of the four segments will pop into place and the new pipe becomes self-supporting.

A method of repairing leaking joints is to place a neoprene seal over the entire inside joint of the pipe. The seal is placed by hand and must be wide enough to span the existing joint. An expanding stainless steel band is placed inside the neoprene seal and enlarged. The pressure provided by the expanded band holds the seal in place and prevents any leaks.

7.57 Bedding, Backfilling, and Compaction

Move bedding material into the trench and place it around the main line with a shovel. Be sure the alignment of the pipe is correct for a smooth, even flow. The bedding material should be compacted until it completely covers the pipe by 6 to 12 inches.

Clean and put away all the tools no longer needed for the job. Fill the water truck with a load of water from the nearest fire hydrant to jet the trench. Use a backhoe to ease the excavated material back into the trench.

Park the water truck close to the repair area and connect a section of 1 1/2-inch pipe to the hose from the pump. Turn on the water. Start the pump. While the water is running, push the pipe into the loose soil all the way to the bottom of the trench. As the water spreads out and works its way upward, observe the top of backfill soil settling. When water is seen breaking to the surface, pull out the pipe and move down the ditch 3 or 4 feet. Repeat this procedure until the entire ditch has been jetted. The soil should drop to about a foot below the pavement.

Back the dump truck to the trench and dump enough road rock to fill the ditch to a level about one foot above the pavement. Wheel roll the gravel with the backhoe. After rolling, the gravel should drop to a level about even with the pavement.

Start the air compressor to supply air for the pneumatic tamper to better compact the road rock to 1 or 1 1/2 inches below the pavement. Practice good housekeeping and clean up the area using a square point shovel and street broom.

Mechanical compaction methods (tampers) are often used instead of water compaction in roads and streets. The soil used for backfill should be fairly dry before it is paved over or soft spots could develop under the pavement.

7.58 Safety

Two barricades with flashers are placed over the repaired area and left there until the paving crew can complete the patch job. Some agencies will fill the hole with a temporary blacktop patch and complete the final restoration at a later date.

Complete the cleanup by washing down the street work area.

Connect the trailer to the dump truck and load and secure the backhoe. Time required and materials used are recorded (Figure 7.69). The crew proceeds to the next job after all tools are cleaned and placed in their respective vehicles and the equipment is secured and ready to go.

LABOR

Supervisor	________ Hr	$ ________
Maintenance Oper. II	________ Hr	$ ________
Maintenance Oper. I	________ Hr	$ ________
Other ________	________ Hr	$ ________

EQUIPMENT

Dump	________ Hr	$ ________
Trailer	________ Hr	$ ________
Backhoe	________ Hr	$ ________
1 1/2 Ton	________ Hr	$ ________
Compressor	________ Hr	$ ________
1 Ton	________ Hr	$ ________

MATERIALS

________	$ ________	________	$ ________
________	$ ________	________	$ ________
________	$ ________	________	$ ________
________	$ ________	________	$ ________
________	$ ________	________	$ ________
________	$ ________	________	$ ________
________	$ ________	________	$ ________
________	$ ________	________	$ ________

TOTAL COST $ ________

Fig. 7.69 Labor, equipment, and materials card

QUESTIONS

Write your answers in a notebook and then compare your answers with those on page 504.

7.5F Before cutting the street, what should the crew verify?

7.5G How can the crew tell if unidentified utilities might be in the work area?

7.5H Is spoil from an excavation suitable for backfill?

7.5I How far should the spoil pile be kept from the trench?

7.5J How is the elevation of the line being repaired determined?

7.5K How close does the backhoe dig to the main line?

7.5L How are equipment and materials lowered into the trench?

7.5M Why should all loose dirt be removed from under the main line?

7.5N How is the backfill compacted?

7.5O How is the job site protected until the paving crew arrives?

7.6 SPANNING EXCAVATIONS

Ductile iron pipe (DIP) and polyvinylchloride (PVC C-900) water pipe are installed when sewer lines cross the excavations or trenches of other utilities. Dig on each side of the excavation being crossed so that there is at least 18 inches of good, solid, undisturbed soil for the spliced-in DIP or PVC C-900 pipe to rest on with proper bedding. This procedure prevents future problems if the backfill in the excavation being spanned settles. When a pipe with joints settles, leaks can develop and eventually cause more settlement and possibly a washout.

Use crushed rock bedding material on all repairs. Place the crushed rock under the existing pipe ends and under the spliced-in section. Bring the crushed rock up to the spring line (center of the pipeline) of the pipe with attention to support of the haunch areas (lower quarter of pipe on each side). Complete the bedding by placing crushed rock all around the pipe.

Band and repair couplings are commonly manufactured for pipes up to 15 inches in diameter. Couplings and sleeves for larger diameter pipe are available as special orders from some manufacturers.

Field joints may be made for pipe when band couplings are not available or for large-diameter pipe. A field joint is made by using a cloth inserted packing (CIP) which is manufactured in rolls. CIP is a thin material which is strong and easy to cut out and handle. A neoprene packing with synthetic cloth (CIP) is better than natural rubber and cotton which was used before the development of CIP.

Cut the CIP to the length and width needed. The length should overlap each end by at least six inches. Wrap the CIP around the pipe ends and coat the overlapping packing ends with rubber adhesive.

Place one or more stainless steel straps on each side of the joint with a hand-operated banding tool. The CIP becomes the inside form so concrete can be placed around the joint. Concrete reinforcing bars, concrete netting or stucco netting may be used for reinforcement to provide additional supporting strength. Make a wooden outside form to contain the concrete. Pour the concrete and vibrate or work the concrete under and around the pipe.

After the concrete has set, finish backfilling the trench and completing the job. Once the CIP is in place and the flow is not too great, the sewer line may be placed in service.

QUESTIONS

Write your answers in a notebook and then compare your answers with those on page 504.

7.6A How can we avoid future problems caused by the settling of backfill in an excavation being spanned?

7.6B What type of bedding should be used for sewers spanning excavations?

7.6C What type of pipe material should be used when sewer lines cross the excavations or trenches of other utilities?

END OF LESSON 3 OF 4 LESSONS
ON
UNDERGROUND REPAIR

Please answer the discussion and review questions next.

DISCUSSION AND REVIEW QUESTIONS

Chapter 7. UNDERGROUND REPAIR

(Lesson 3 of 4 Lessons)

Write the answers to these questions in your notebook before continuing. The question numbering continues from Lesson 2.

12. Why should repair assignments be made using work order forms?
13. What must be done before repair work on a sewer is started?
14. When should hard hats be worn and why?
15. How would you install a cleanout in an easement?
16. Why is proper testing and inspection of a new sewer system before placing it in service very important?
17. What are the duties of a wastewater collection system construction inspector?
18. Why should pictures be obtained of nearby private property before starting trench excavation in an easement?
19. Why should all gages and valves used for air testing be located outside the manhole?
20. How would you replace a tap full of roots in a main line located in a street?
21. During excavation, how would you attempt to look for the location of unknown underground utilities?

CHAPTER 7. UNDERGROUND REPAIR

(Lesson 4 of 4 Lessons)

7.7 SEALING GROUT

Several methods or techniques are available today for internal grouting or sealing of sewer pipes. This section describes one technique—grout sealing. Other techniques are available and provide similar results. Grout sealing is used to illustrate the reasons for internal grouting or sealing and the procedures.

Grout sealing is one method of internal chemical sealing. This particular method is a process that is designed to re-form the gasket of the leaking pipe and create a seal at the site of the leak, rather than stabilize soil as in other sealing methods.

Technology is continually developing newer and more effective sealing compounds. These compounds may seal the pipe joint or crack, or seal a leak and stabilize the soil, or possibly even chemically control roots, stabilize soil and seal a leak. Before selecting a chemical compound, try to identify the problem or problems you wish to correct, determine what compounds and procedures are available, discuss their effectiveness with other operators who have used the sealant and equipment, and select what you think will do the best job for your agency. For ideas on how to select a sealing compound and application procedures, review Chapter 5, Section 5.33, "Purchase of Closed-Circuit TV Equipment," and Chapter 6, Section 6.41, "How to Select Chemicals."

7.70 Reasons for Grouting or Sealing Sewers

Of the various sewer rehabilitation techniques used today, internal sealing is one of the most widely used methods. Internal sealing is effective when the sewer line to be repaired is in an area that is unsuitable for excavation and is either infiltrating or exfiltrating through leaking joints, cracks, or small holes. Where the structural integrity of the sewer is gone due to collapsed sections or severe cracks or holes, sealing is not ideally suited to solving the problem. Under these conditions, pipe replacement repairs are necessary to ensure structural stability before sealing the remaining parts of the sewer.

Frequently, agencies attempt to stop infiltration in order to reduce flows to wastewater treatment plants, especially during heavy rainstorms. When stopping leaks into a sewer through joints that are infiltrating, you must realize that the water table around the sewer can increase in elevation. Water that once infiltrated into the sewer will now be stored in the trench bedding material. This water will build up a head on the sewer and also move up the sewer looking for more poor joints. For this reason, all joints must be tested and sealed if necessary. Also, if wastewater has been escaping from a sewer by exfiltrating through bad joints, the sealing of these joints could increase the flow to the wastewater treatment plant. These potential impacts on flows and joints must be considered when evaluating the results of a sealing program to prevent infiltration and exfiltration.

The sealing process consists of using a closed-circuit television system and a packing device. The TV camera monitors the "packer" as it is placed on leaking joints in the pipeline. When the packer is positioned correctly, a pumping system is activated to inject the sealing chemicals or grout into the leak. The use of internal chemical sealing does not require excavation, nor does it require the interruption or bypassing of wastewater except in extreme conditions. Sealing compounds have been developed which are so effective that *WELL POINTS*[15] do not have to be sunk to dewater the area and prevent washout of sealing compound.

The uses of sealing include rehabilitation of existing sewers to reduce infiltration/exfiltration and manhole sealing. Sealing processes also are used on new construction of sewer lines where the contractor is unable to pass the line acceptance test, be it either air testing, water testing, or weir measurements. Sealing in many cases avoids further delays and costly excavation.

7.71 Equipment

Sealing equipment is generally used by two groups of individuals: (1) the agency, whether it be county, municipality, or utility district, and (2) the contractor who specializes in TV inspection and sealing. Of the two groups, the contractor is the one who does a majority of the sealing projects. In either case it is a good idea to be familiar with the sealing process since most communities at one time or another will use sealing, whether through a contractor or with their own equipment.

[15] *Well Point. A hollow, pointed rod with a perforated (containing many small holes) tip. A well point is driven into an excavation where water seeps into the tip and is pumped out of the area. Used to lower the water table and reduce flooding during an excavation.*

The following list indicates the basic equipment (Figure 7.70) of a grout sealing system and briefly describes the use of each component.

AIR COMPRESSOR The system is air activated. Compressed air is used to operate the pumps and inflate the sleeve packer.

CHEMICAL PUMP The sealing chemical is pumped through hoses down into the sewer to the sleeve packer.

WATER PUMP Water (*CATALYST*[16] for the chemical) is pumped through hoses down into the sewer to the sleeve packer.

CONTROL PANEL The activation of the two pumps and the inflation and deflation of the packer are controlled from the panel. The entire sealing process is monitored and operated from this panel.

SLEEVE PACKER The actual component that does the sealing down in the sewer. This component is inflated pneumatically to isolate the leak and force the sealing chemical into the leaking area.

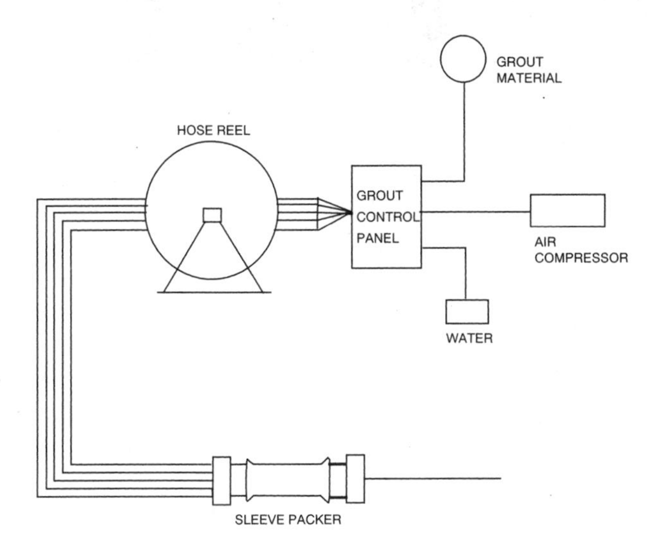

Fig. 7.70 Major components of a grout sealing system

When using the grout sealing process, it is not necessary to mix the sealing chemical. It is pumped directly out of its shipping container and the water is pumped out of its container. As long as water reaches the grout sealing chemical, the reaction will occur.

While this list covers the basics of the system, there are other items involved such as interconnect hoses, assorted valves and fittings, a hose reel, and a *PENTA HOSE*[17] to carry the air and chemicals to the packer.

7.72 Sealing Chemicals

The sealing material may be some type of *HYDROPHILIC*[18] *POLYMER*[19] or other type of chemical sealing grout that is applied to a sewer pipe joint or defect by use of the sleeve packer. The solids content is an important consideration in choosing a sealant chemical, since the sealing chemicals used may shrink to the volume of the solid material if moisture conditions are not maintained.

When the sealing material is mixed with an equal quantity of water (a special accelerator is added for winter conditions), the material expands 10 to 12 times its original volume and cures to a tough, flexible, elastic condition. The chemical will absorb the water it needs to react, which is an equal volume, and dispel the remaining water in its immediate surroundings. As is the case in actual sealing operations, there is always excess water present when sealing leaks, but this excess water is usually not a dilution factor.

Grout sealing chemicals should have the following properties:

1. Resistant to most organic solvents, mild acids, and alkali,
2. Returns to shape after repeated deformations,
3. Essentially nontoxic in cured form,
4. Does not become rigid or brittle when exposed to freeze-thaw or dry or low moisture conditions, and
5. Noncorrosive and does not contain neurotoxic ingredients in either cured or uncured form.

Figure 7.71 shows the approximate chemical quantities in fluid ounces of grout per pipe joint used to rehabilitate different types of pipes and different sized pipes. These quantities have been established by actual field experience, but may vary slightly depending on the condition of each specific joint. Factors that may affect the chemical quantities include original gasketing materials remaining in joint, open or offset condition of joint, or severity of infiltration or exfiltration.

QUESTIONS

Write your answers in a notebook and then compare your answers with those on page 504.

7.7A Under what conditions can grouting or sealing be considered an effective method to repair or rehabilitate a sewer?

7.7B What are the basic pieces of sealing equipment?

7.7C What are the desirable properties of chemical grout sealants?

7.73 Sealing Process

This next section describes the chemical grout sealing process on a typical 8-inch leaking joint. While this does not cover all aspects of sealing sewers, it does provide an adequate example of stopping the most commonly encountered leaks. Special leaks, extreme conditions, badly cracked pipes, and other situations are best handled in consultation with the equipment manufacturer.

7.730 Setup

The camera and sleeve packer are connected together with the camera turned around to look at the packer (Figure 7.72).

[16] *Catalyst (CAT-uh-LIST). A substance that changes the speed or yield of a chemical reaction without being consumed or chemically changed by the chemical reaction.*

[17] *Penta Hose. A hose with five chambers or tubes.*

[18] *Hydrophilic (HI-dro-FILL-ick). Having a strong affinity (liking) for water. The opposite of hydrophobic.*

[19] *Polymer (POLY-mer). A long chain molecule formed by the union of many monomers (molecules of lower molecular weight). Polymers are used with other chemical coagulants to aid in binding small suspended particles to larger chemical flocs for their removal from water.*

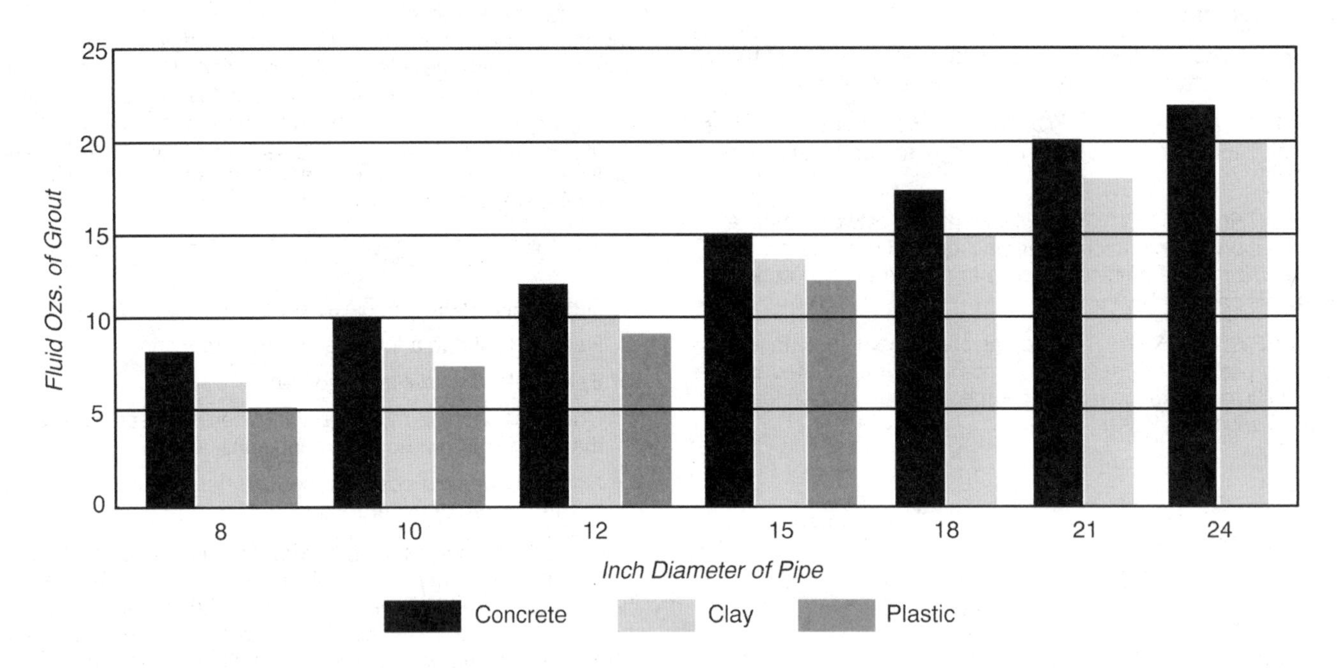

Fig. 7.71 Chemical quantities used for different pipe joints

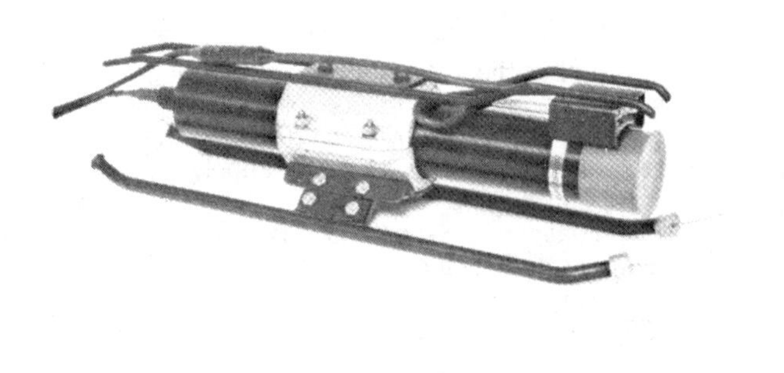

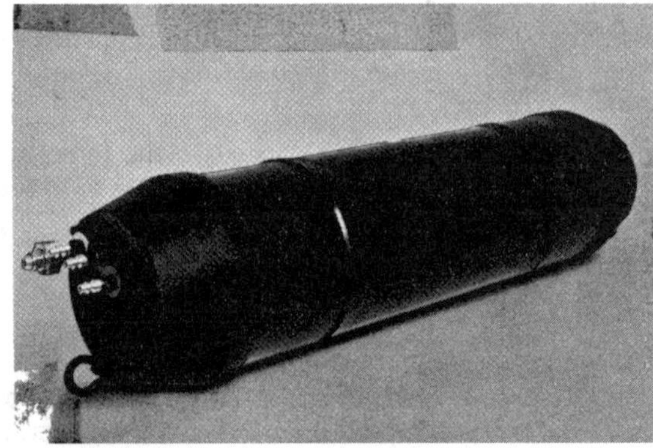

Fig. 7.72 TV camera and packer

The air and chemical hoses are then connected to the sleeve packer and each mode is tested to ensure operation before being lowered into the line.

7.731 In-Line Preparations

The camera and sleeve packer are then lowered into the manhole and guided into the sewer line as the cable winch at the opposite manhole slowly pulls the units through the line (Figure 7.73). It is better to have the camera on the downstream side of the packer because the water buildup cuts valuable time from the operation and obscures much of the visual inspection of the process. Also, the camera can reveal any grout getting around the packer. Manhole rollers and jacks are placed in the manhole to prevent gouging and scraping the hoses and cables as they move through the manhole.

7.732 Testing and Sealing

Once the sleeve packer is centered on the joint, the test may be run (Figure 7.74-1). It is necessary to test only those joints where visible leaking is not evident. All leaks must be sealed. The first step of the test is to inflate the end elements. Water is then pumped into the isolated area formed by the inflation of the end elements (Figure 7.74-2) and the pressure in the center area on the packer is monitored for an increase to test the integrity and tightness of the joint. The pressure in the center area of the packer may be set anywhere from 0 to 15 psi, depending on the depth of the line and water table condition.

Once a leak has been detected, either through visual observation or through testing, the equipment is then placed into the grout mode (Figure 7.74-2). Three (3) timers are set: one is the length of time that the chemical is injected, the second determines the reaction time from liquid to "foam," and the third timer controls the cure time. After these timers are set, the grout switch is pushed and the machine automatically pumps the chemical, allows it to react with water, and inflates the center area on the packer to squeeze the foaming chemical into the leak (Figure 7.74-3). A post-seal test may also be run to ensure the joint has been effectively sealed to desired pressure.

The entire process usually takes about 5 minutes per joint when each joint is tested, sealed, and retested. Wintertime operations will take longer due to the extra freeze-prevention precautions that are taken and due to the fact that the chemical will react more slowly as it gets colder.

7.733 Cleanup

Once the manhole to manhole section of sewer has been sealed and tested to the engineer's specifications, the camera and packer are removed from the line and cleanup of the unit takes place. The hoses and cables should be wiped free of wastewater, as should the camera and sleeve packer. It is unnecessary to disconnect the sleeve packer from the hoses since the packer is a self-sealing unit and there is no danger of the chemical sealing up in the packer if the unit is going to be used within three or four days. If the unit is to be idle for more than six days, the packer should be disconnected from its hoses and cleaned thoroughly.

It is not necessary to empty the hoses of sealing chemical unless the unit is to be idle for a long period of time, such as a wintertime layover.

7.74 Specialized Applications

7.740 Roots

The use of root-inhibiting chemicals with the sealing material is not recommended unless the roots are killed and removed prior to the sealing operation. The reason for this is that if you seal a joint with roots in the joint and then pack a chemical with a root-inhibiting or killing agent into this area, it will kill the roots. The roots will then die and leave a hole when they decay and your infiltration or exfiltration problems will be back again. Roots should be killed, removed, and then the root-inhibiting chemical can be added to the sealing chemical for use in the sealing operation.

7.741 Large-Diameter Pipe Sealing

Special equipment for sealing large-diameter pipes is designed to be lowered through a standard manhole and moved through the pipe to each joint to be sealed. The special sealing equipment does not damage or break pipes or cause settlement, assuming the line is structurally sound.

The sealing equipment consists of two separate pumping systems capable of supplying a flow of sealing materials to form a new gasket or a seal on the infiltration or exfiltration point. The sealing material passes from the pumping system through hoses to the sealing device. This device (referred to as a packer) is a cylindrical case that is smaller than the pipe size and consists of a steel ring, two inflatable end elements, and one packing element.

Special care is used to prevent damage to the pipe from excessive amounts of sealing pressure. The packer is manually positioned over the area of the infiltration or exfiltration and the two inflatable elements are expanded using precisely controlled pressures (Figure 7.75). These elements seal against the inside surfaces of the pipe to form a void (empty) area at the point of infiltration or exfiltration, which is now completely isolated from the remainder of the pipeline.

The sealing chemical is pumped into this isolated area through the hose system (Figure 7.76). The amount of sealing material will be determined by the size of the pipe and the condition of the joint. After the sealing material has been pumped into the isolated area, between the two elements, the packing element is then inflated, thus forcing the sealing material into the defective joint (Figure 7.77). The packer is held in the inflated position until the joint is cured.

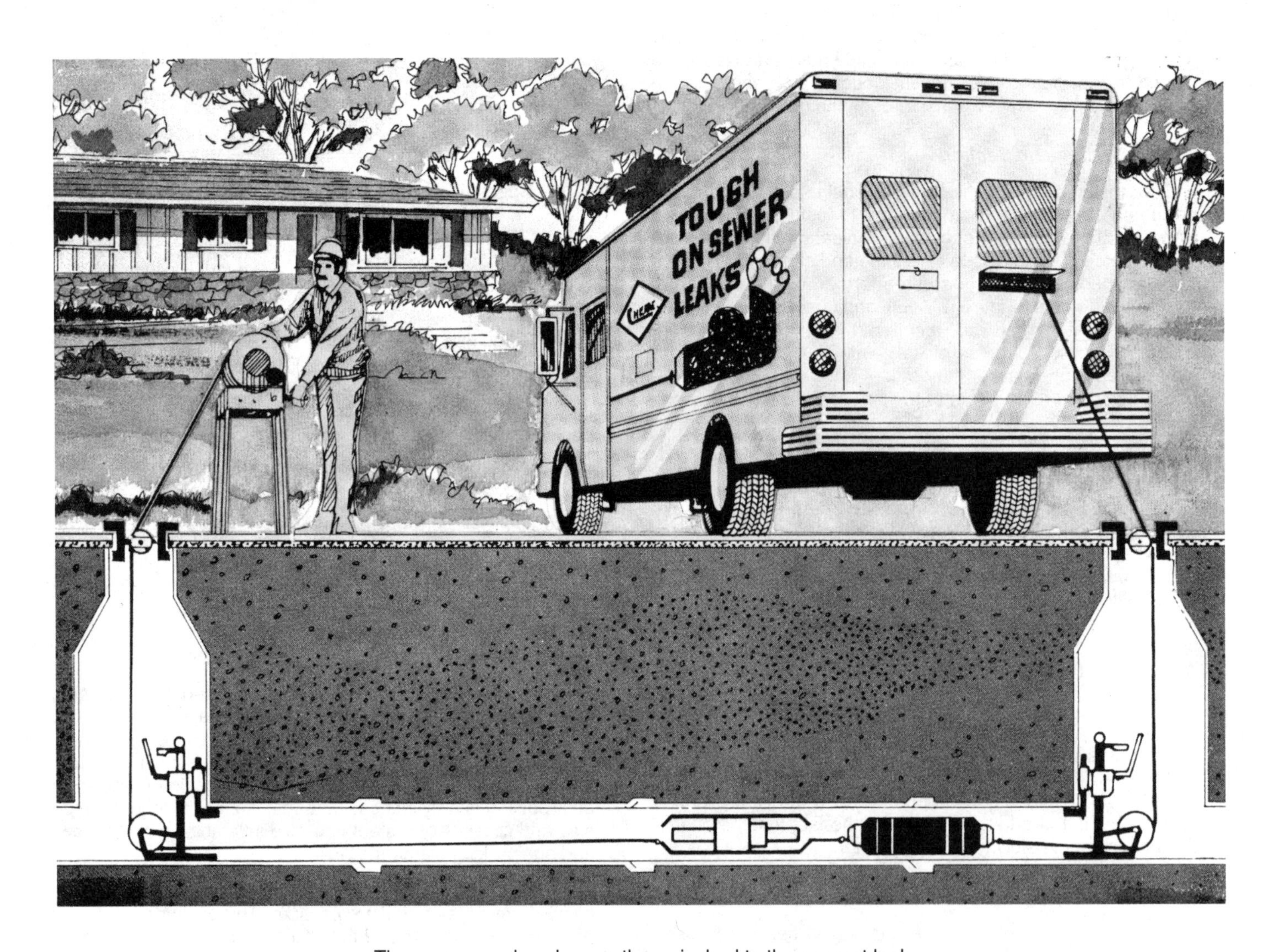

The camera and packer are then winched to the nearest leak to be tested or sealed.

Fig. 7.73 In-line operator

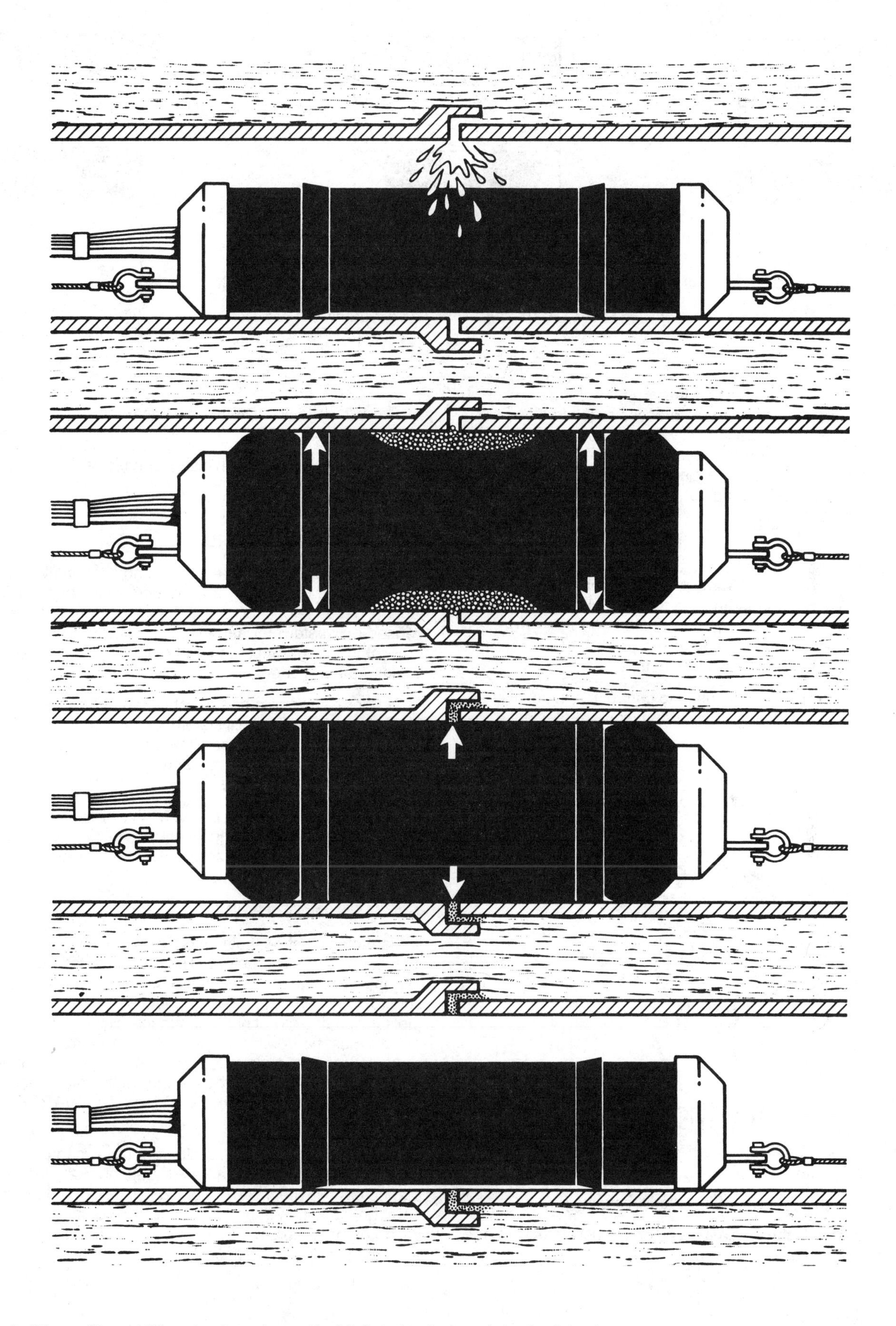

Packer ball is positioned (1) and outer edges of ball inflated to isolate defective joint. Low-pressure air introduced to isolated section confirms leak and grout material is inserted (2). Inflation of middle section of packer ball (3) compacts grout material to form effective seal. Packer ball is deflated (4) and moved to the next joint to be sealed.

Fig. 7.74 Operation of packer

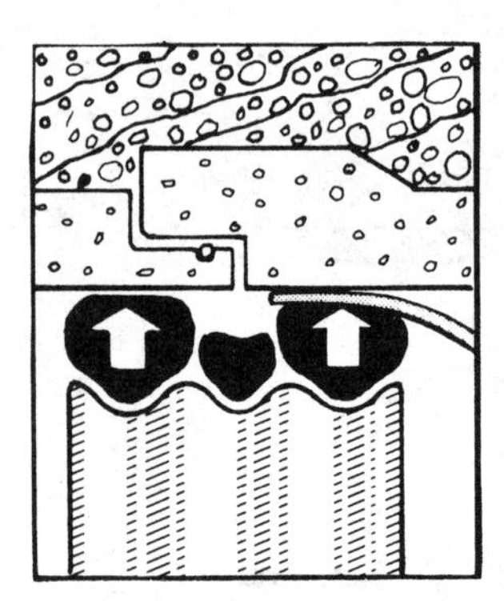

Fig. 7.75 Isolating joint

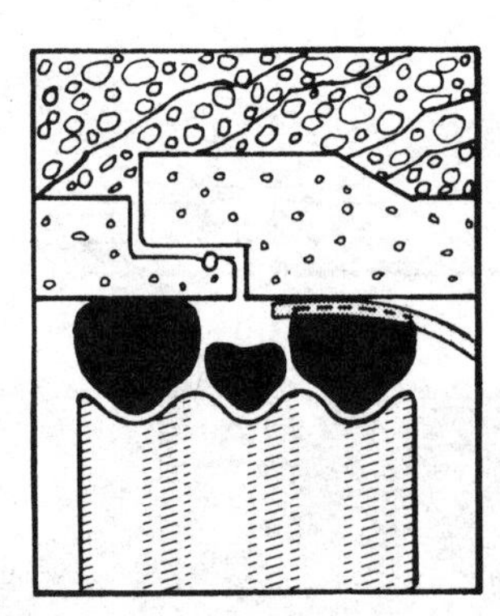

Fig. 7.76 Pumping sealing chemical

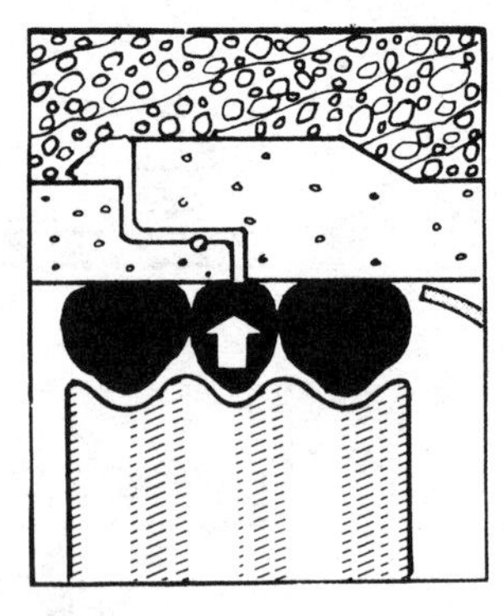

Fig. 7.77 Forcing sealing material into defective joint

The large-diameter pipe sealing process requires a two-operator crew "in the line" to effectively run the operation (Figure 7.78). Any time an operator enters a pipe, the necessary safety precautions must be exercised (see Volume II, Chapter 11, Section 11.1, 7. "Poisonous or Toxic Gases," and 8. "Oxygen-Deficient Atmospheres," and Chapter 4, Section 4.40, "Confined Spaces").

Large-diameter chemical grout sealing is effectively used in pipes ranging in diameter from 30 to 120 inches with either bell and spigot or tongue-and-groove pipe. Large-diameter pipe sealing is a very specialized process and most work is done by contractors specializing in the process. Contact the equipment manufacturer for specific details.

A similar procedure can be used to control manhole infiltration (Figure 7.79). A probe applicator injects the sealant into leaky manhole walls.

7.75 Acknowledgment

All figures in this section courtesy of Cherne Industries, Inc.

QUESTIONS

Write your answers in a notebook and then compare your answers with those on page 504.

7.7D Why is the TV camera installed facing the packer?

7.7E Why should the TV camera be on the downstream side of the packer?

7.7F Why should only those joints that are not visibly leaking be tested?

7.7G When can root-inhibiting chemicals be used with a sealant?

7.8 GUNITING[20] SEWERS

Sewers of brick, concrete, stone, and combinations of these materials can deteriorate and require repair or rehabilitation. Gunite, because of its high density, is resistant to deterioration caused by wastewater and thus is a suitable lining material to prevent further deterioration and rehabilitate sewers 30 inches in diameter and larger.

Old brick sewers in danger of collapse can be restored by properly applying gunite. Restoration by using gunite strengthens the sewer and increases flow capacity. Figure 7.80 shows an old 8-foot diameter brick sewer that was lined with two inches of reinforced gunite.

Gunite also can be shot into defective or deteriorating joints. This procedure of applying gunite to leaking joints from the inside of a sewer can effectively seal these leaks.

7.9 MANHOLE REPAIRS

Manholes must be maintained and kept in working condition just like any other structure. They must provide easy access for operators to the collection system. Frames and covers should be secure and the top should be at the same grade as the surrounding street or other surface. The manhole cone and barrels must be free from cracks and deterioration from corrosion. Bottom flow channels should be smooth and shelf areas should be located above the flowing wastewater, but sloped to prevent the trapping of debris during *SURCHARGE*[21] conditions.

Before scheduling any repair work on a manhole, carefully inspect the manhole (Chapter 5, Section 5.2, "Inspection of Manholes"). Inspect for structural failures, cracks, corrosion damage, condition of bottom, shelves, and channels, and elevation of manhole cover. When repairing a manhole, try to do all of the repair work at once.

[20] *Gunite. A mixture of sand and cement applied pneumatically that forms a high-density, resistant concrete.*

[21] *Surcharge. Sewers are surcharged when the supply of water to be carried is greater than the capacity of the pipes to carry the flow. The surface of the wastewater in manholes rises above the top of the sewer pipe, and the sewer is under pressure or a head, rather than at atmospheric pressure.*

Fig. 7.78 Sealing large-diameter pipe

Fig. 7.79 Sealing manhole walls

Fig. 7.80 Brick sewer before and after guniting
(Courtesy of Pressure Concrete Construction Company)

7.90 Raising Manhole Frame and Cover to Grade

7.900 Detection of Missing Manholes

Manholes are sometimes buried or paved over when a street is repaved. Usually this problem is discovered by a preventive maintenance crew. If the manhole is buried in the easement (back or front yard, lawn or flower bed area), the preventive maintenance crew will locate and expose the manhole lid. Try to locate missing manholes by studying the "as built" drawings (record drawings) and measuring distances from upstream and downstream manholes. Missing manholes also may be located by the use of metal detectors, ferret locators (used with hand rods or power rodders), or television. Buried railroad tracks and other metal objects can give false readings with metal detectors. The preventive maintenance crew will proceed with the work that has to be done in the manhole once it is located.

With the information on the location of the manhole, a work order is written indicating how many inches the frame and cover should be raised.

If the missing manhole is in the street or an alley and has been buried or paved over, the preventive maintenance crew will locate the manhole with a metal detector and mark the location with paint. If the buried manhole is not a pressing problem, the crew will again prepare a work order and turn the order over to the construction section. If the manhole is needed for immediate access, the preventive maintenance crew will contact a construction supervisor who will dispatch a crew with a compressor and jackhammer to expose and raise the manhole. The supervisor also will assist the preventive maintenance crew if needed to speed up the repair.

7.901 Site Preparation

Once the location of the manhole to be raised is established, the following procedure can be used to raise a manhole in an easement:

1. Take care to remove the lawn sod, shrubs, fencing, or any other obstacles that should be replaced after the manhole has been raised and the job completed.
2. Dig down and completely expose the lid.
3. Remove the lid for inspection. Be sure the casting is not offset from the grade rings or cone.
4. When the casting position has been determined, replace the lid on the casting.
5. To determine how far the manhole should be raised, use a straight rod or 2 x 4 board long enough to span the cut around the manhole. Use a level to be sure that the 2 x 4 is parallel to grade. If the distance is three inches or more, a grade ring should be used. The two common ring sizes are 3 inches and 6 inches in height and 4 inches wide (Figure 7.81).
6. Begin to remove the dirt from the outside of the casting down to the depth needed. The width of the trench around the casting should be 10 to 12 inches—wide enough for a round point shovel.
7. If the manhole is 12 inches or more below the surface of the yard, lawn, or road, then the cut of the trench around the manhole casting may have to be 18 inches or more. This is to provide a safer working area and to provide standing room alongside the casting in order to lift and remove it from the hole.

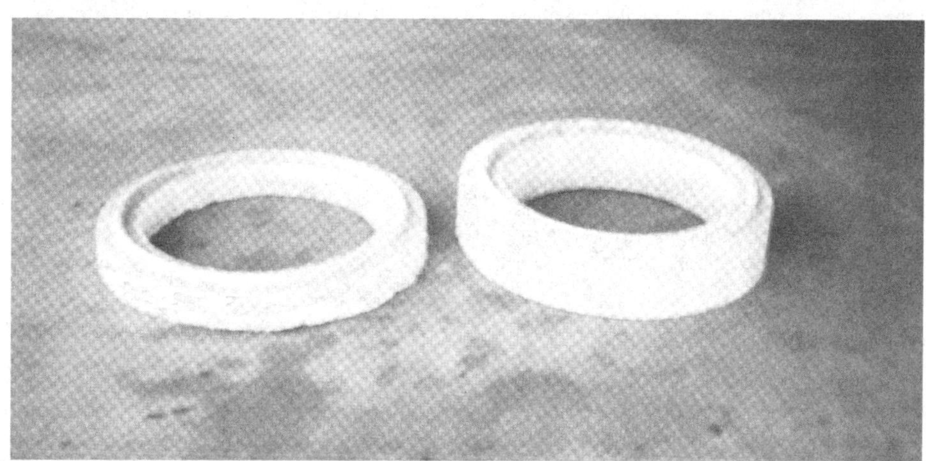

Fig. 7.81 Manhole grade rings
(Common sizes are 3 inches or 6 inches high and 4 inches wide)

Fig. 7.82 Manhole frame and cover

8. The casting is lifted from the manhole and set aside in a safe place where it will not slide back into the hole or cave in the side of the trench. *WARNING:* Manhole castings and grade rings are very heavy and care must be exercised when lifting and moving them either by hand or with lifting equipment.
9. Clean the top of the cone or grade ring and the bottom plate of the casting. To clean these surfaces, use a wire brush and scraper. Remove all dirt and any loose cement or rust.
10. Wash all surfaces including the casting plate, both sides of the grade ring, and the cone. Use water and a brush.
11. Mix the grout (suggested mix: 5 parts sand to 1 part cement type two). Blend to an even texture. Add water to the mixed sand and cement. Blend to the consistency you desire to maintain the slump you feel would be necessary for the thickness required by the manhole lid being raised.
12. If the rise is less than three inches, but more than one inch, manhole spacers should be used. Spacers may consist of small pieces of red brick, presoaked broken concrete, or small redwood blocks. These should be placed in a manner to balance the casting with the street or ground level.
13. The grade ring should be moistened on both sides just before setting the ring down on the fresh grouted cone or grade ring.
14. If the rise is more than three inches, some agencies prefer to excavate down to the manhole barrel and raise the barrel. A two-foot diameter manhole entry for any great length is very difficult to climb down or work through.

7.902 ***Raising the Frame and Cover*** *(Figure 7.82, page 495)*

1. One operator (Maintenance Operator I) using a brush or rag wets down the cement surface and is then ready for the grout.
2. The second operator (Maintenance Operator I) furnishes the first operator with the grout in a shovel or a bucket. (Note: The operator placing the grout and shaping it should always wear rubber gloves for skin protection.) Place the grout around the edge of the cone with both hands, completely covering the old surface. Use your hands to form and shape the grout to the desired height. Try to keep the grout fairly uniform. Prevent all or as much excess grout as possible from falling down into the manhole by placing cover boards over the channel and shelf in the bottom of the manhole. This procedure will be necessary when first installing grade rings to adjust the casting frame to meet existing grade.
3. After the desired height of the grout has been reached, wet down the bottom of the casting plate or frame.
4. To set the casting, two operators are needed in the hole along the outside edge of the cone. Slide the casting as you pick it up from the edge of the ditch. Center it and place it on the grouted area. Make sure the casting opening is centered with the cone opening.
5. Use a straight edge to level the casting with the surrounding surface. Using a bar or sledge hammer, tap lightly on the top edge of the casting to obtain the exact level. This procedure also will squeeze out the excess grout.
6. Feather the excess grout around the casting and cone (inside and out) using your hand or a brush. This should leave a smooth, clean finished surface. Remove all excess grout and place it on the outside edge between the casting and cone to form a seal from the outside. Cap the outside bottom of the casting plate with one inch of grout mix and overlap down to the cone or grade ring.
7. Backfill around the casting. After about four inches of soil has been placed around the casting, tamp the soil by walking on it for compaction.
8. Place the lid on the casting carefully. Continue to backfill and walk on the soil for compaction. If backfilling more than one foot of soil, use some water to improve compaction. After backfilling and compacting, clean up excess soil and haul it away to a disposal site.
9. Replace all lawn. Cut out an opening to keep the lid exposed. Replace shrubs in a manner that will not completely hide the lid or make it difficult to work in or around the manhole.
10. If fencing was removed, replace it and cut a notch out of the bottom board if it crosses the lid so the lid can be removed and replaced without any problems. Notify property owners of the notch in the fence.

Manhole lids in the street are raised following the same procedure, except we would use fast-drying cement rather than sand and cement. Compaction and backfill around the manhole with an air tamper or mechanical compactor should be to within four inches of the road level.

Use pre-packaged concrete or asphalt to fill the last four inches. Compact with an air tamper to finish grade. This procedure avoids having to leave barricades in the traffic lanes and permits traffic to resume traveling over the freshly raised manhole.

Manhole lids also may be raised by excavating around the tapered cone, removing the tapered cone, frame, and cover, installing a section of a reinforced concrete manhole barrel of desired length, replacing the cone section, and adjusting the frame and cover to the desired grade by using a minimum of grade rings or brickwork. Sometimes manhole covers have to be raised two to three feet above grade in isolated areas to keep vandals from dumping large boulders and tree stumps into manholes.

Very often manhole lids must be raised because the manhole has settled. When manholes settle, sewers connected to a manhole may crack or break and the connection joints may develop leaks. Repair these problems when raising manhole lids and conducting other manhole repairs.

Industry has developed a special manhole casting that fits into existing lids and can be adjusted vertically to correct grade. This casting allows manhole covers to be raised before streets are paved and can be tilted if the pavement slopes.

Concrete collars are frequently used around the castings of manholes located in easements and streets. The collar may be six inches or more wide and extend to a depth of a foot or more so that it bonds to the cone section of the manhole. This collar will protect the casting of manholes located in fields from being damaged by farm equipment.

Asphalt may be placed to depths greater than four inches (up to 12 to 18 inches) in streets. The asphalt is tamped in four-inch lifts and produces a dense, well-compacted asphalt against the cone section of the manhole up to the street level. This procedure usually eliminates cracking and water damage of the street surface at the casting.

If the manhole barrel and cone sections must be removed and replaced, adjust and seal the joints with a bituminous rubbery material commercially available in different widths and thicknesses. Set the material over the grooved section of the manhole barrel and then place the tongue section on top of the material to push the material into the groove. This material provides a totally impervious joint, does not have shrinkage problems, and requires much less time than the use of grout or mortar to connect and seal new sections of manhole barrels. Use mortar to set the grade ring and casting.

QUESTIONS

Write your answers in a notebook and then compare your answers with those on page 505.

7.9A Why do the lids of manholes have to be raised?

7.9B How are paved-over manholes located?

7.9C How would you determine the height a manhole lid should be raised?

7.9D When the casting is lifted from the manhole, where should it be placed?

7.9E How can the top of the cone or grade ring be cleaned?

7.9F Why should the operator placing the grout and shaping it wear rubber gloves?

7.9G What should be done with fences over manholes?

7.91 Repair of Manhole Bottoms

Most manhole repair problems are discovered by preventive maintenance crews and the repair work is assigned (using a work order) to the repair section for correction of the problem. This section describes procedures for inspecting and repairing manhole bottoms.

After all safety precautions have been taken by the three-operator crew and the manhole has been tested and purged with fresh air, the manhole is washed down from the casting down to and including the bottom. This will usually clear the walls of material such as mud, grease, or other debris. Washing also helps clean out spiders, centipedes, and other annoying pests. Run the blower at all times when there is an operator working in the manhole (see Volume II, Chapter 11, Section 11.1, 7. "Poisonous or Toxic Gases," and 8. "Oxygen-Deficient Atmospheres," and also Chapter 4, Section 4.40, "Confined Spaces").

After washing is completed, one operator (Maintenance Operator I) will enter the manhole with proper safety equipment using a secure ladder. Inspect the casting, grade rings, cone, and barrels for seepage, deterioration, and cracks. Once on the manhole bottom, inspect the condition of the channel to determine what action will have to be taken to correct any flow or leakage problems.

Problems encountered at the bottom of manholes range from *NO BOTTOM* to improper channels, such as flat bottom, basin bottom, and low inflow invert elevation to high outflow invert elevation that causes a sump or backwater problem. These problems are usually found in older sections of a wastewater collection system.

The following paragraphs describe methods that can be used to correct each type of problem.

7.910 Manholes Without Concrete Bottoms

To start repair, plug off all incoming lines. If this is not possible, bridge a channel or channels across the manhole bottom to allow flow to continue. Use half pieces of thin metal forms or plastic pipe to provide channels. Once channels are in place, probe the bottom to determine how much debris will have to be removed. Collect as much of the liquid portion of waste as possible in a bucket and remove it from the manhole for hauling off to a disposal site. Clear enough depth in the bottom of the manhole for at least two, and if possible, three feet of cobblestone, approximately four inches in diameter, for base fill. Leave at least a 10-inch space below the form or the bottom of pipes to provide a firm footing. Be careful the manhole doesn't start to settle.

Cover cobble rock with four inches of crushed rocks ($^{3}/_{4}$-inch to dust). After all this has been done and the base is determined to be solid, inspect the forms and make any necessary adjustments.

Spray the rock base with water. Mix concrete to a suitable slump. Pour the base by starting under the forms and carefully tamping under and up the side of the forms. This should prevent honeycombs and pockets. Once this is done, continue to pour until you have reached the amount needed to complete one side of the shelf and channel. The shelf slope should drop at least one inch and not more than two inches from the back wall of the manhole to the top of the edge of the channel.

Place a knee board over the finished shelf to stand on while pouring the rest of the manhole bottom. When the remainder of the bottom is finished, carefully stand on or in the pipe channel form. Remove the knee board and touch up lightly by brushing the entire bottom to remove the trowel marks and produce a brush or broom finish. The slight roughness of a broom finish reduces the chance of an operator slipping and being injured in the manhole. Forms should not be removed for at least 36 to 48 hours to allow time for the concrete to harden and cure. After the allotted time, remove the forms and inspect the channels for any honeycombs. If honeycombs are observed, use quick-setting cement to fill void areas.

To reduce the need for forms to shape the channel bottom, cut a section of plastic pipe or other type of sewer pipe in half to fit the invert and cement it into place. Advantages of this procedure include having a smooth flow channel and not having to worry about the concrete channel deteriorating and causing turbulence, which can release odors.

QUESTIONS

Write your answers in a notebook and then compare your answers with those on page 505.

7.9H Why should manhole walls be washed before inspection?

7.9I What types of problems are encountered at the bottom of manholes?

7.9J What is the first step to start repair of a manhole with no bottom?

7.9K What kind of base is provided in the bottom of a manhole without a concrete bottom?

7.9L Why should the concrete surface on the bottom of a manhole be brushed or broomed?

7.911 Manholes With Flat Bottoms

Flat-bottom manholes do not have channels through the bottom and are fairly easy to correct because the base is already in place.

1. First plug off all incoming lines, if possible. Another approach is to install a smaller pipe with the ends of the pipe sealed to the ends of the manhole inflow and outflow pipes in order to carry flow through the manhole. Clean, scrape, and brush the bottom.

2. Remove plug or plugs, form channel with half pieces of tin, plastic pipe forms, or half sections of sewer pipe, and repeat process from "no bottom manhole." Start with the procedure for pouring concrete and follow the steps through the finishing of the concrete. Remove forms 48 hours later if half sections of sewer pipe were not used.

3. Step 2 above may be replaced by use of speed-crete; once the plug or plugs are removed, the speed-crete can be used to form and shape the new channels. This process is by far the fastest and most practical procedure, with the exception of using half sections of sewer pipe.

4. Now that the channels are formed and set, start with a speed-crete and pea-gravel mixture. Mix to same consistency as concrete. Pour and trowel to desired finish. Pea-gravel is used to reduce the amount of speed-crete that would usually be needed and makes the fill-in stronger.

   NOTE: Speed-crete is not a cement. It is aluminum and plastics and has a much higher resistance to acids and sulfide, which makes it a much more practical material to use in building manhole bottoms.

   Pour the mixture to the desired height, usually 2 inches above the top of the pipes. Shape and finish one side. Now stand in the formed channel and pour and finish the shelves of the manhole.

7.912 Channel Corrections

Channel corrections are made to relieve flow problems such as the tee-shaped channel. This type channel is called "deadheading," which means the flow runs directly into the main channel flow. This causes a turbulent flow and can restrict the flow from incoming lines.

Tee-intersections can be corrected by cutting off one corner of the tee-shaped channel. Use a jackhammer and cut down below the flow line. Use speed-crete and reshape the channel into a smooth wye shape. This will channel the flow in the same direction as the main flow.

Basin channels are usually found on incomplete channel bottoms. This channel usually starts 6 to 12 inches from the wall of the manhole and then feathers out into the basin. The flow correction again can be made by forming speed-crete for the proper channel shape by hand and smoothing the surface with a wet rag or damp sponge to form a smooth finish. Fill in behind the new channel with a speed-crete and pea-gravel mixture. Form the shelves of the manhole, too.

QUESTIONS

Write your answers in a notebook and then compare your answers with those on page 505.

7.9M How are channels formed using speed-crete?

7.9N How are smooth channel surfaces obtained when using speed-crete?

7.9O Why does speed-crete have a higher resistance to acids and sulfide than concrete?

7.9P Why are channel corrections made?

7.9Q How are channel corrections made?

7.92 Inside Drops in Manholes (Figure 7.83)

A main line or house service line lateral entering the manhole at a higher elevation than the main flow line or channel causes splashing, corrosion, odors, and difficult working conditions. These lines should have inside drop pipes connected to them. The inside drop line consists of a tee attached to the incoming pipe and should fit flush against the manhole wall. A length of pipe runs from the tee to the manhole bottom and a 90-degree elbow directs the flow into the channel at the correct angle. If possible, service lines should be connected to manholes as outside drops (Figure 7.84), but inside drops are more economical after the system is completed.

7.920 Materials Used for Inside Drops

The most suitable material for pipe and fittings for all inside drop lines into a manhole up to 12 inches in diameter is plastic class 125 pipe. The class 125 plastic pipe has worked very successfully for inside drop lines. Examples of use include from 4-inch up to 12-inch inside drop lines, including discharge drops at the end of force mains.

7.921 Installation of Inside Drops

1. Use a jackhammer, drill hammer, and chisel to chip out concrete around the incoming pipe and back into the wall of the manhole at least 2 1/2 inches. Set the tee fitting into the opening at a vertical angle. The tee should line up with the pipe and fit tight against the wall.

2. Set a plumb line to the manhole bottom shelf to identify the position of a 90-degree angle. If possible, use the jackhammer and cut a channel with the proper angle to ensure correct flow with flow in the main channel.

3. Measure the distance from the tee to the 90-degree angle to determine the length of pipe needed. Glue fittings to the pipe and set in place. Recheck for alignment. Use speed-crete to cement the tee into the wall, to anchor the 90-degree angle into place, and to finish the new channel. The tee should be strapped to the wall. If the drop pipe is more than 8 feet long, this pipe also should be strapped to the wall of the manhole.

4. Force main discharge drop lines should always be double strapped to the wall around the tee fitting or a sanitary tee.

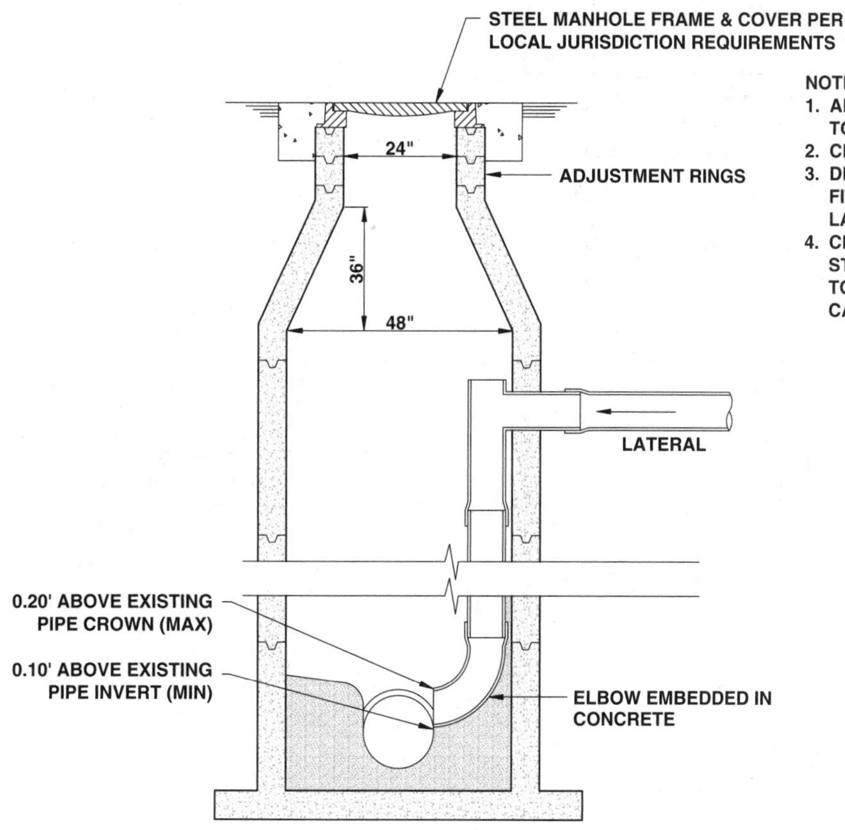

INSIDE DROP CONNECTION

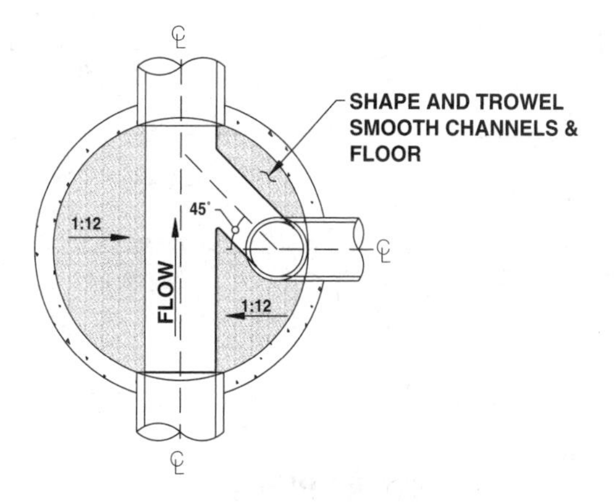

Fig. 7.83 Manhole inside drop connection

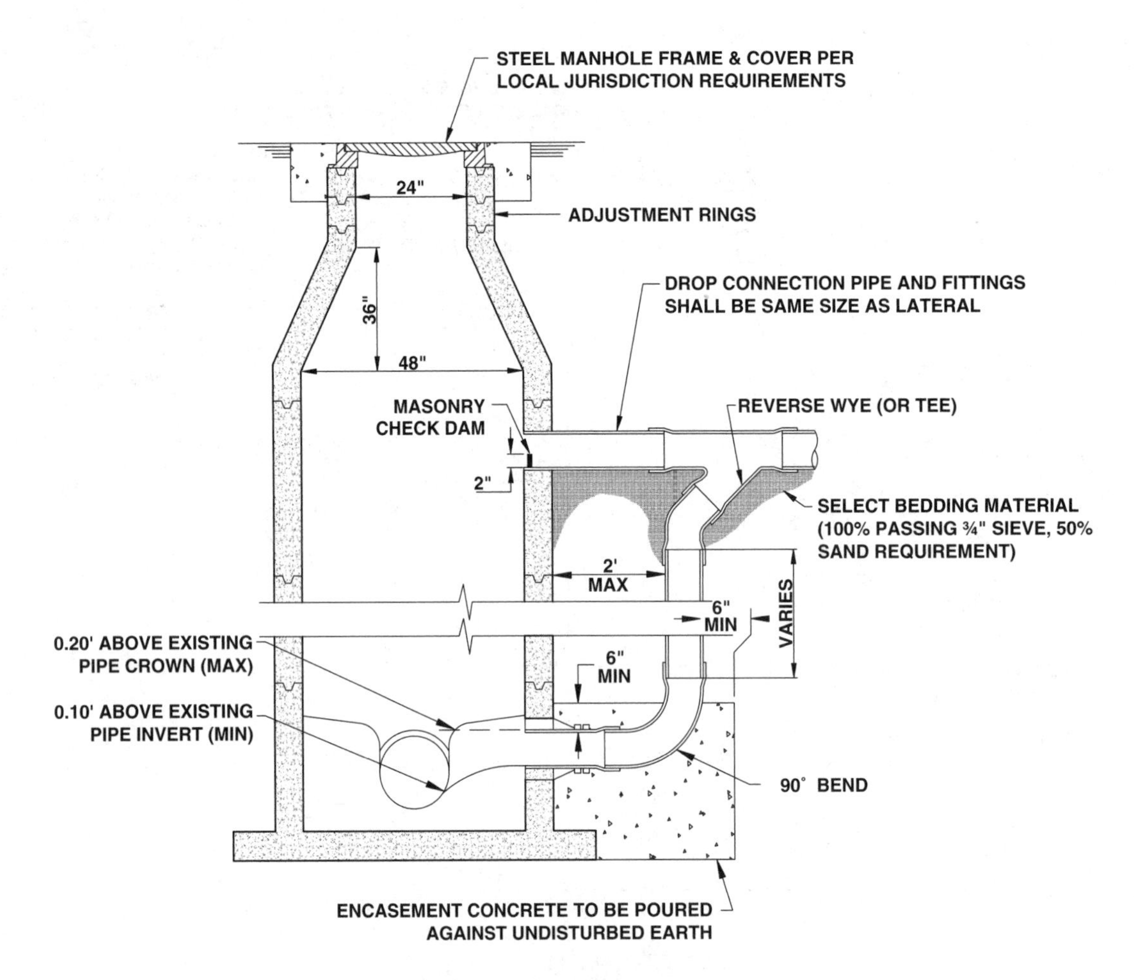

TYPICAL DROP MANHOLE

DETAIL FOR 10" OR LARGER DROP OR WHERE SPECIFICALLY CALLED OUT ON DRAWINGS

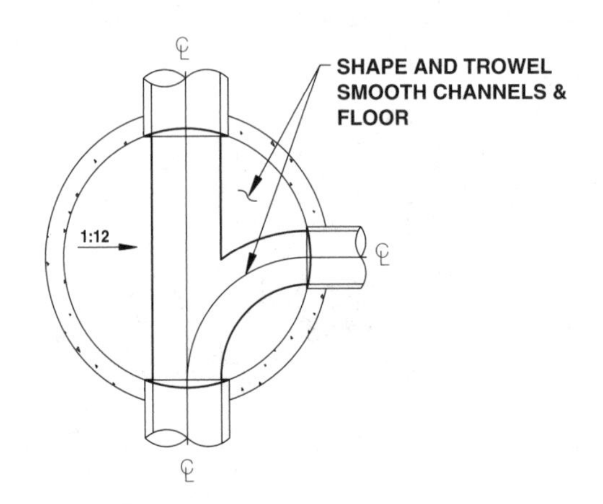

INVERT PLAN VIEW

Fig. 7.84 Manhole outside drop connection

The 90-degree angle at the bottom of the drop line should be channeled in as close to the flow line as possible to prevent any spray or splash.

7.93 Repair of Manhole Barrels

Repair of a manhole barrel depends on what is wrong with the manhole, the condition of the manhole, and the circumstances under which the repairs must be made. To repair cracks or leaks in a manhole barrel, several techniques are available: (1) patching, (2) grouting, and (3) lining.

Patching is done with a fast-drying cement and usually under adverse or wet conditions. Grouting also uses a fast-drying cement. The grout is either applied by hand or sprayed over the surface being repaired.

Lining a manhole consists of either inserting a liner or spraying a lining material over the manhole barrel. A piece of pipe can be inserted in a manhole as a liner. The pipe could be made of plastic, tile, or bricks. Grout is placed in the space between the manhole and the pipe. A liner such as epoxy or fiberglass may be sprayed over a manhole barrel. The surface of the barrel must be dry before spraying.

For additional information about methods and procedures for rehabilitating manholes, see Volume II, Chapter 10, Section 10.36, "Manholes."

7.94 Prevention of Inflow Through Manhole Covers

Inflow may enter manholes through pick holes or vent holes in manhole covers. If excessive inflow becomes a problem, one solution is to plug the vent holes. Several different styles of plugs that can be inserted in the holes to prevent inflow are available. Plugging the vent holes is not an ideal solution, however, because vent holes permit the gradual release of gases and odors.

In situations where vent holes are needed in manhole covers, inflow protector covers or inflow shields can be installed between the manhole frame and cover. These protector covers or shields can stop the inflow into the collection system. Also they are capable of controlling odors and reducing manhole rattling. The bottom sealing surface should be gasketed to prevent leakage around the outside. A special device in the bottom allows a minimum leak down rate, which prevents ponding (to avoid mosquito breeding). The leak down rates are in the range of 5 to 10 gallons per 24 hours. A gas relief valve in the center releases at a pressure of 0.5 to 1.5 psi.

The protector covers or shields may be made of stainless steel, galvanized steel, or plastic, depending on the installation. Stainless steel is recommended for areas with high levels of hydrogen sulfide. Plastic is less expensive, but may deteriorate over time. Plastic is flexible and can distort (lose its shape) when full of water and not handled properly. One operator lost a grip on a plastic cover and it fell to the bottom of the manhole where the cover got sucked into a main trunk sewer.

When installing covers or shields under manhole lids, be aware that not all manhole frames and covers are of the same configuration. Therefore, be sure the devices fit properly. A very tight fit under the manhole cover or lid is essential to hold the manhole cover in place. Inspect the covers or shields in high-traffic areas every 6 to 12 months. The constant working action of traffic on the manhole cover will wear through the plastic flange of the inserted cover or shield and then the cover or shield could fall into the manhole and possibly become stuck in the flow channel.

The cover or shield could raise the manhole cover as much as a quarter of an inch. If a snowplow is used to remove snow over the manhole cover, there could be a problem. The steel edge of the snowplow blade could catch the manhole cover and flip it out of the frame. The solution is to bevel the edges of the lid. The more expensive the cover or shield, the better the expectation of the reliability of the seal.

QUESTIONS

Write your answers in a notebook and then compare your answers with those on page 505.

7.9R What is an inside drop line?

7.9S Why are inside drop lines installed in manholes?

7.9T How are inside drop lines secured?

7.9U What material is used for inside drop lines?

7.10 NEW CONSTRUCTION

Occasionally a crew may be required to extend a line a short distance and install a new cleanout or manhole. The procedures for new installation are similar to repairs, except usually the problems created by wastewater flowing in the existing system are not present.

END OF LESSON 4 OF 4 LESSONS
on
UNDERGROUND REPAIR

Please answer the discussion and review questions next.

DISCUSSION AND REVIEW QUESTIONS

Chapter 7. UNDERGROUND REPAIR

(Lesson 4 of 4 Lessons)

Write the answers to these questions in your notebook. The question numbering continues from Lesson 3.

22. Why are sewers grouted or chemically sealed?
23. How would you seal a leaky joint in an 8-inch sewer?
24. How would you clean up after a sealing operation?
25. What types of repairs are likely to be needed in manholes?
26. How would you raise the lid or cover of a manhole?
27. How would you repair a manhole without a concrete bottom?

SUGGESTED ANSWERS

Chapter 7. UNDERGROUND REPAIR

ANSWERS TO QUESTIONS IN LESSON 1

Answers to questions on page 405.

7.0A Underground repair and new construction are very hazardous because of the threat of cave-ins and dangerous gases.

7.0B Underground repairs are necessary to replace damaged pipe or to remove difficult obstructions.

7.0C Underground damages may be discovered by homeowners when the collection system fails, by emergency crews attempting to clear a stoppage, by an inspector, or by a preventive maintenance crew.

7.0D Underground repair may be required if pipe damage or stoppages are caused by roots, cracked pipe, protruding taps, improper taps, and damage by the public, other utilities, or construction activities.

7.0E A work request form notifies a supervisor or crew leader and the crew of a repair job. The form specifies what is to be done and the job location. The crew leader and supervisor plan and schedule work on the basis of information on work request forms and other needs.

7.0F Before starting a repair job, a crew leader should visit the job site and plan the job. Potential conflicts with traffic, utilities, and other obstructions should be noted. Other utility agencies should locate their underground facilities. Property owners should be notified. Also, a preconstruction and tailgate safety meeting should be held before starting the job.

Answers to questions on page 407.

7.1A Shoring requirements depend on depth of trench, width of trench, soil conditions, nearby activities, and the laws and regulations.

7.1B Lack of shoring and shoring failure are the main causes of cave-in deaths.

7.1C Shoring systems in trenches consist of some form of sheeting or liner, uprights held rigidly opposite each other against the trench walls by jacks or horizontal cross members (braces), and, if required, longitudinal members (stringers).

7.1D Uprights shall extend as near to the bottom of the trench as permitted by the material being installed, but not more than 2 feet from the bottom.

7.1E Soil stability may be classified as hard compact, running, or unstable.

Answers to questions on page 410.

7.1F All excavation work shall at all times be under the immediate supervision of a qualified person with authority to modify the shoring system or work methods as necessary to provide adequate safety.

7.1G Spoil or excavated material must be placed one foot or more back from the edge of the trench. If the trench is over 5 feet deep, spoil must be placed more than 2 feet back from the edge of the trench.

7.1H All trenches deeper than 5 feet must be shored. In some locations trenches deeper than 4 feet must be shored.

7.1I If shoring is not practical, the sides or walls of an excavation may be sloped or benched.

7.1J Walkways must be provided for pedestrians to keep them out of the way and to protect them from being injured.

Answers to questions on page 426.

7.1K Different types of common shores include:

1. Hydraulic,
2. Screw jacks,
3. Air,
4. Solid sheeting,
5. Cylinder, and
6. Shield.

7.1L The main advantages of hydraulic shores are their ease of installation and they can be installed by one operator.

7.1M The main advantages of screw jacks are that they are inexpensive to purchase and are practical for minor repairs in smaller excavations. A limitation is that they are time-consuming to install.

7.1N Solid sheeting must be installed when excavation is in material that will not stand vertical and when it is not practical to slope the trench because of location.

7.1O Material is excavated in small cylinder shoring by hand using a shovel. In cylinders with a large enough diameter, material can be removed mechanically.

7.1P Cylinder shoring is used when the working area is small and only a small portion of the main line must be exposed.

7.1Q Shield shoring is used during excavation and installation of various types of sewers in areas and soil conditions that present a combination of problems.

7.1R The major concern during a shoring operation is the protection of the operators. Assume all trenches and excavations are unsafe.

ANSWERS TO QUESTIONS IN LESSON 2

Answers to questions on page 436.

7.2A The type of equipment to be used for excavating trenches for construction of sewers will be determined to a great extent by the size and depth of the sewer, type of material to be excavated, and the ground surface conditions.

7.2B When high groundwater levels are encountered during trench excavation, the water can be removed from the excavation by use of sump pumps or by use of a system of well points, a header pipe, and a central pump.

7.2C A "competent person" is defined by OSHA as a person capable of identifying existing and predictable hazards in the surroundings, or working conditions which are unsanitary, hazardous, or dangerous to employees, and who has authorization to take prompt corrective measures to eliminate hazards.

Answers to questions on page 448.

7.2D Using a sewer plan and profile, surveyors set control points, usually at 50-foot intervals, to guide accurate excavation.

7.2E The excavation equipment operator uses the grade string line to control the depth of cut. The string line also is used as a control guide when laying the sewer pipe.

7.2F Sewer pipe bedding is the material upon which the pipe is laid and which serves as its foundation.

7.2G Extreme care must be taken in excavating the trench bottom to true grade in order to produce an undisturbed and compacted trench bottom.

Answers to questions on page 455.

7.2H Small-diameter pipe can be placed directly on the bedding material with the bell end of the pipe facing upstream.

7.2I Sewer pipe should be protected from displacement and breakage until the trench has been backfilled.

7.2J Vertical lateral sewer cleanouts are located at the terminals of lateral and branch sewers.

7.2K The essentials of proper sewer trench backfilling are (1) protection of the pipe from movement and breakage from impact or crushing from backfill loading, (2) filling of the excavation, and (3) restoration of the ground surface.

ANSWERS TO QUESTIONS IN LESSON 3

Answers to questions on page 457.

7.3A Repair assignments must be made in writing in order to maintain records and provide a history of who made what repairs, when, where, how, and why.

7.3B Upon receipt of a work order request, examine existing records to determine if the problem has occurred in the past. During the preliminary field investigation, alert the homeowner and look for any unusual problems that may exist, such as trees or shrubs that could get in the way.

7.3C Before a crew drives to a repair job, all safety equipment should be inspected and the truck should contain the proper equipment, materials, and tools to do the job.

7.3D Upon arrival at the job site, verify that the address and map location are correct. Place traffic cones at the front and rear of the truck in accordance with local regulations. Notify the property owner and discuss care of any plants or trees that may have to be moved.

Answers to questions on page 461.

7.3E Placing soil from an excavation on a canvas protects the grass and allows for easier shoveling of soil back into the excavation.

7.3F When excavating to repair a sewer line, be careful not to damage or break other utility lines. Also, use shoring if necessary to prevent cave-ins. Ventilation equipment should be available and used when necessary.

7.3G After a pipe is cut, carefully remove the section so you don't get cut on sharp edges of the pipe, and be sure the inside of the pipe is clear before installing the cleanout.

7.3H Use gravel or sand and crushed rock in the bottom of the trench for bedding of the wye.

7.3I Before backfilling a trench, inspect the job, tighten all couplings or seal joints, and be sure there are no tools remaining in the ditch.

7.3J A service line location card must be completed to provide a record of the location of the cleanout for future reference.

7.3K The crew leader should review the job with the supervisor when it is completed. Any potential problem areas should be discussed so the supervisor can communicate with the homeowner or other agency staff.

Answers to questions on page 465.

7.4A New wastewater collection system construction materials and procedures are inspected to:

1. Ensure compliance with the plans, specifications, and safety regulations governing the work,
2. Alert design engineers of any modification of plans,
3. Minimize the need for future maintenance, and
4. Provide a record of the construction progress and "as-built" plans (record drawings).

7.4B Most public agencies rely on the integrity and quality control of the manufacturer of small-diameter pipe to ensure production of pipe that meets specifications.

7.4C During a pipe laying operation, the following items of work should be inspected:

1. Sewer pipe bedding placed to prescribed depth and grade,
2. Sewer pipe laid to grade with joints made in accordance with manufacturer's instructions, and
3. The interior of the pipe should be free of bedding materials, dirt, boards, or other debris that could create problems.

7.4D After backfilling is completed, visually inspect the lines for leaks, displaced pipe, and other signs of damage.

Answers to questions on pages 480 and 481.

7.4E The main reason for using low-pressure air testing for testing sewers is that pressurized air exerts the same pressure in all directions on the pipe during a specific moment in time.

7.4F Before air testing a sewer, test your equipment.

7.4G A pressure drop in the boot plugs may indicate a boot perforation or leakage.

7.4H Remove the downstream plug, exercising care with respect to the rapid flow of water that will occur in the sewer line when the test water is released.

7.4I The purpose of the mandrel test is to measure for deflection and offset joints.

Answers to questions on page 481.

7.5A Before a crew is assigned a main line repair job, the supervisor makes a preliminary field investigation to identify potential problems and to plan the repair job. The supervisor marks the repair area with paint and meets with representatives of utility agencies at the job site.

7.5B The supervisor needs other utility agencies to indicate locations and depths of their utilities at the job site.

7.5C Before the crew starts the job, the supervisor and crew chief review a map of the job site, the proper procedures and safety precautions, and the necessary equipment, tools, and materials.

7.5D Before leaving the yard for the job site, the crew should inspect all equipment (including safety) and determine that all the necessary equipment, tools, and materials needed to do the job are loaded on the trucks.

7.5E When arriving on the job site, the crew should carefully park the trucks, prepare traffic control, and contact homeowners and owners of parked vehicles. If parked vehicles are a potential problem, the job site can be barricaded and marked the night before or very early in the morning.

Answers to questions on page 484.

7.5F Before cutting the street, the crew should verify the location of the repair area by measuring the distances from the upstream and downstream manholes.

7.5G Unidentified utilities may be located by watching for changes in soil texture or other signs that could indicate the presence of cross trenches.

7.5H Usually spoil from an excavation is suitable for backfill because the material comes from an old backfill.

7.5I Keep the spoil pile at least two feet from the edge of the trench if the trench is over 5 feet deep and, if shallower, spoil must be at least one foot from the edge of the trench.

7.5J Use an extension probe to determine the elevation of the line being repaired.

7.5K The backhoe can dig to within six inches of the main line.

7.5L Lower equipment and materials into the trench using a bucket and rope.

7.5M All loose soil should be removed from under the main line to prevent a muddy mess and to provide room for gravel to form the bedding material for the pipe.

7.5N The backfill is usually compacted by mechanical tampers in roads and streets or by water jetting.

7.5O The job site is protected by barricades with flashers until the paving crew arrives.

Answers to questions on page 484.

7.6A Future problems that are caused by the settling of backfill in an excavation being spanned can be avoided by proper bedding. Dig on each side of the excavation being crossed so that there is at least 18 inches of good, solid, undisturbed soil for the spliced-in DIP or PVC C-900 pipe to rest on with proper bedding.

7.6B Crushed rock should be used as bedding for sewers spanning excavations.

7.6C Ductile iron pipe (DIP) and polyvinyl chloride (PVC C-900) water pipe are installed when sewer lines cross the excavations or trenches of other utilities.

ANSWERS TO QUESTIONS IN LESSON 4

Answers to questions on page 487.

7.7A Grouting or sealing can be considered as a possible means of repairing or rehabilitating sewers when the sewers are in an area that is unsuitable for excavation and is either infiltrating or exfiltrating through leaking joints, cracks, or small holes. Sealing processes also are used on new construction of sewer lines where the contractor is unable to pass the line acceptance test.

7.7B The basic pieces of sealing equipment include the air compressor, chemical pump, water pump, control panel, and sleeve packer.

7.7C Desirable properties of chemical grout sealants include:

1. Resistant to most organic solvents, mild acids, and alkali,
2. Returns to shape after repeated deformations,
3. Essentially nontoxic in cured form,
4. Does not become rigid or brittle when exposed to freeze-thaw or dry or low moisture conditions, and
5. Noncorrosive and does not contain neurotoxic ingredients in either cured or uncured form.

Answers to questions on page 492.

7.7D The TV camera faces the packer so the sealing operation can be observed.

7.7E The TV camera should be on the downstream side of the packer because the upstream water buildup cuts valuable time from the operation and obscures much of the visual inspection of the process.

7.7F Visibly leaking joints don't have to be tested because they have to be sealed anyway.

7.7G Root-inhibiting chemicals can be used with a sealant only after the roots in the sewer and joints have been killed and removed.

Answers to questions on page 497.

7.9A Manhole lids should be raised when they are below ground level or have been paved over when a street was repaved. Below-grade manholes are difficult to locate and may be subject to surface inflow.

7.9B Paved-over manholes can be located by examining "as built" drawings and measuring distances from upstream and downstream manholes by using a metal detector, or by using a TV camera inside the sewer.

7.9C To determine the height the manhole lid should be raised, place a straight rod or 2 x 4 board across the cut around the manhole and measure the distance from ground level to the manhole lid. Use a level to make sure the 2 x 4 is parallel to the ground.

7.9D Place the casting from the manhole a safe distance away where it won't slide back into the hole or cave in the side of the trench.

7.9E Clean the top of the grade ring or cone using a wire brush or scraper.

7.9F The operator placing and shaping the grout should wear rubber gloves to protect exposed skin.

7.9G If a fence was over a manhole whose lid was raised, replace the fence and cut a notch out of the bottom board so the lid can be removed. Notify property owners.

Answers to questions on page 498.

7.9H Manhole walls should be washed before inspection to clear the walls of material such as mud, grease, or other debris. Washing also helps clean out spiders, centipedes, and other annoying pests.

7.9I Problems encountered at the bottom of manholes include no bottom, flat bottom, basin bottom, low incoming invert elevation to high outgoing invert elevation that causes a sump problem, improper channels, and disjointed pipe junctions.

7.9J To start repair of a manhole with no bottom, either plug off all incoming lines and pump wastewater to a downstream manhole, or bridge a channel or channels across the manhole bottom to allow flow to continue.

7.9K The base in the manhole should consist of cobble rock and crushed rock for the new concrete bottom.

7.9L Brush the concrete surface on the bottom of a manhole to prevent a slick surface.

Answers to questions on page 498.

7.9M Channels are formed by hand when using speed-crete.

7.9N Smooth channel surfaces can be obtained using speed-crete by wiping down the walls with a wet rag or damp sponge.

7.9O Speed-crete has a higher resistance to acids and sulfide than concrete because it consists of aluminum and plastics.

7.9P Channel corrections are made to correct flow problems and produce smooth flow through the manhole.

7.9Q Channel corrections are made by removing obstructions or poor channels with jackhammers and reshaping or providing new channels using speed-crete.

Answers to questions on page 501.

7.9R An inside drop line consists of a tee, pipe, and 90-degree angle installed in a manhole to divert water from an incoming line at a higher elevation than the main flow channel at the bottom of the manhole.

7.9S Inside drop lines are installed in manholes to eliminate splashing, odors, and difficult working conditions.

7.9T Inside drop lines are strapped to manhole walls.

7.9U Plastic class 125 pipe is used for inside drop lines.

APPENDIX

OPERATION AND MAINTENANCE OF WASTEWATER COLLECTION SYSTEMS

(VOLUME I)

Final Examination and Suggested Answers

Collection System Words

Arithmetic

Subject Index

FINAL EXAMINATION

VOLUME I

This final examination was prepared **TO HELP YOU REVIEW** the material in this manual. The questions are divided into five types:

1. True-False,
2. Best Answer,
3. Multiple Choice,
4. Short Answer, and
5. Problems.

To work this examination:

1. Write the answer to each question in your notebook.
2. After you have worked a group of questions (you decide how many), check your answers with the suggested answers at the end of this exam, and
3. If you missed a question and don't understand why, reread the material in the manual.

You may wish to use this examination for review purposes when preparing for civil service and certification examinations.

Since you have already completed this course, you do not have to send your answers to California State University, Sacramento.

True-False

1. The better the job collection system operators do, the more effectively a wastewater treatment plant can do its intended job.
   1. True
   2. False
2. Local taxpayers have invested more money in underground sewers than in all the structures above ground owned by their local government.
   1. True
   2. False
3. An unpermitted discharge from a sanitary sewer system to the waters of the United States constitutes a violation of the Clean Water Act.
   1. True
   2. False
4. Inflow is the seepage of groundwater into a sewer system, including service connections.
   1. True
   2. False
5. The amount of support provided the sewer pipe by the sides of a trench decreases as the width of the trench increases.
   1. True
   2. False
6. A trained professional collection system operator demonstrates an awareness of hazards and a commitment to accomplish every task in a safe manner.
   1. True
   2. False
7. Generally, taper lengths to control traffic should be shortened to increase their effectiveness.
   1. True
   2. False
8. After a short exposure to hydrogen sulfide gas, you will lose your ability to detect hydrogen sulfide.
   1. True
   2. False
9. All collection system agencies should develop a site-specific written lockout/tagout procedure and provide training to all of the operators who may use it.
   1. True
   2. False
10. Flow metering devices are effective in areas where the sewers are surcharged.
    1. True
    2. False
11. The TV camera should always be pulled against the direction of flow to keep the front of the camera clean.
    1. True
    2. False
12. TV cameras can operate in any sewer line that tests positive for explosive gases.
    1. True
    2. False
13. Whenever a stoppage occurs, the flow is completely stopped.
    1. True
    2. False
14. The exact distance a high-velocity cleaning nozzle has moved up in the sewer must be known at all times.
    1. True
    2. False

15. Masking agents eliminate or correct the source of an odor problem.
    1. True
    2. False
16. Never use plain water without shoring fluid in a hydraulic shore.
    1. True
    2. False
17. The bell end of a pipe should be facing downstream.
    1. True
    2. False
18. Proper testing and inspection of a new sewer system before placing it in service can prevent serious problems from developing and avoid considerable grief.
    1. True
    2. False

Best Answer (Select only the closest or best answer.)

1. What is meant by operation and maintenance of wastewater collection systems?
   1. Bypassing raw wastewater from the collection system
   2. Keeping the facilities in good operating condition
   3. Keeping the trucks operating
   4. Working with other utility agencies
2. How should operators do underground sewer work?
   1. Economically
   2. Perfectly
   3. Quickly
   4. Safely
3. What is the principal cause of pipe damage that allows infiltration and exfiltration?
   1. Construction activities by other utilities
   2. Rigid pipe joints
   3. Root growth
   4. Vibration from traffic
4. Why should collection system operators have a basic knowledge of how engineers design collection systems?
   1. To design collection systems
   2. To do a better job
   3. To estimate costs of construction projects
   4. To prepare plans and specifications
5. What is a purpose of grinder pumps?
   1. To lift the pumped flows into a gravity flow system
   2. To measure the flows pumped into the collection system
   3. To pressurize the wastewater to help move it through the collection system
   4. To pump sludge into the collection system from holding tanks
6. When operators review wastewater collection system design plans and specifications, what should be the most important concern regarding manholes?
   1. Accessible location
   2. Located in creek beds
   3. Located in middle of major streets
   4. Well-ventilated location
7. What is defensive driving?
   1. A mature attitude toward driving
   2. Driving within the speed limit
   3. Following approved routes
   4. Making sure no one drives into your vehicle
8. What should be the first test performed with atmospheric analyzers?
   1. Combustible gases
   2. Inert gases
   3. Oxygen content
   4. Toxic gases
9. Why should other utility companies be contacted before opening a collection system excavation?
   1. To avoid damaging other underground utilities
   2. To coordinate repair activities
   3. To plan roadway resurfacing programs
   4. To utilize the same equipment
10. What problem is created when a leaky sewer is located under a water table?
    1. Corrosion rates of the pipe material are increased
    2. Erosive velocities develop inside the sewer pipe
    3. Force of water can cause the pipe to crack or break
    4. Infiltration can take place and occupy valuable capacity in the sewer
11. What is expected of a collection system TV operator?
    1. Ability to recognize and evaluate conditions found in pipe by TV
    2. Capability of directing the TV program
    3. Desire to keep TV camera from becoming stuck in sewer
    4. Knowledge of TV electronics
12. What information should be recorded when smoke is observed coming from the vents only of a house?
    1. This house has a defective vent
    2. This house has a leaky sewer vent
    3. This house has an empty trap
    4. This house is OK
13. Why should the operator at the downstream wastewater treatment plant be notified before clearing a large septic stoppage?
    1. To allow notification of pollution control authorities
    2. To notify agency officials of potential problems
    3. To prevent plant from being short of staff when septic slug arrives
    4. To prevent treatment processes from becoming upset and failing to do their job
14. Why should rubber gloves never be used while handling or guiding a rod that is turning?
    1. Rod can seize palm of rubber glove and cause injury
    2. Rod can wear out rubber glove too quickly
    3. Rubber gloves are too expensive
    4. Rubber gloves are waterproof and prevent lubrication
15. Why can infiltration of water into the sewer line be a problem when using chemicals to control roots?
    1. May cause sewer to leak
    2. May create adverse chemical reaction
    3. May decrease capacity of sewer
    4. May wash the chemicals off the roots

16. When must trenches less than five feet deep be shored?
    1. When changes in the weather are forecast
    2. When examination indicates hazardous ground movement may be expected
    3. When extra shoring materials are available
    4. When workers must work in trench bottom

17. Why is granular material often substituted for the native backfill material?
    1. It is lighter than native backfill
    2. It is preferred by contractors
    3. It is sturdier than native backfill
    4. It requires no compaction

18. Why should photographs be taken of work areas on private property before starting to work?
    1. In case legal claims develop regarding cracked sidewalks and foundations or damaged lawns and fences
    2. To gain experience using a camera to photograph preconstruction sites
    3. To perform public relations activities for homeowners
    4. To provide homeowners with a nice photograph of their landscaping

19. Why should all gages and valves used for air testing be located outside the manhole?
    1. For ease in reading gages and adjusting valves
    2. To avoid injury to anyone reading gages or adjusting valves
    3. To facilitate accurate reading of gages and adjusting valves
    4. To prevent water damage to gages and valves

20. Why should the soil used for backfill be fairly dry before it is paved over?
    1. Backfill soil that is fairly dry is easy to transport to the job site
    2. Fairly dry soil provides excellent compaction and compressive strength
    3. Soft spots could develop under the pavement if wet soil is used for backfill
    4. Wet soil will cause moisture to infiltrate into the sewer and use up capacity

Multiple Choice (Select all correct answers.)

1. What happens when wastewater collection systems fail due to lack of or improper operation and maintenance?
    1. Blockages occur that result in backups into homes
    2. Bypassing of raw wastewater from collection system
    3. Other underground utilities installed near the collection system can be disrupted
    4. Stoppages develop and wastewater flows out manholes into the street
    5. Streets collapse due to leaking pipes

2. Why do many wastewater collection systems now require billions of dollars of rehabilitation and upgrading?
    1. Corrosion of sewer pipes was a factor neglected during design
    2. Insufficient money was allocated and spent on preventive maintenance of wastewater collection systems
    3. Plant roots caused damage to pipe joints
    4. Problems were caused by faulty construction, poor inspection, and low-bid shortcuts
    5. Sewers were designed with inadequate flow capacities

3. What are the responsibilities of collection system operators?
    1. Controlling O & M expenses
    2. Developing and enforcing water pollution control laws
    3. Preserving the community's capital investment
    4. Protecting public health and safety
    5. Protecting the environment

4. Why is the use of pumps in collection systems avoided whenever possible?
    1. Pumps are very noisy
    2. Pumps produce odors
    3. Pumps provide an industrial appearance
    4. Pumps require extensive maintenance
    5. Pumps use costly energy to operate

5. What problems are created by hydrogen sulfide gas?
    1. Creates blockages in sewers
    2. Encourages root growth into sewers
    3. Forms corrosive sulfuric acid
    4. Is a toxic gas
    5. Smells like rotten eggs

6. Why are pipe joints one of the most critical components of the wastewater collection piping system? Because of the need to control
    1. Costs of materials
    2. Groundwater infiltration
    3. Root intrusion
    4. Trench excavation depths
    5. Wastewater exfiltration

7. Exact distances and the nature of the advance warning before a work area depend on what factors?
    1. Age of drivers
    2. Congestion
    3. Roadway conditions
    4. Tasks being performed by drivers while driving
    5. Traffic speed

8. What is a hazardous atmosphere? An atmosphere that may expose operators to the risk of
    1. Acute illness
    2. Death
    3. Electric shock
    4. Incapacitation
    5. Injury

9. Excavations and adjacent areas must be inspected on a daily basis by a competent person for evidence of what types of problems?
    1. Hazardous atmospheres
    2. Location of parked vehicles
    3. Potential cave-ins
    4. Proper completion of inspection forms
    5. Protective system failures

10. How are collection system operators usually exposed to hazardous chemicals and wastes?
    1. Driving collection system vehicles
    2. Purchase of chemicals in stores
    3. Use of chemicals every day in collection system maintenance
    4. Use of collection system as an accidental disposal method
    5. Use of collection system as an intentional disposal method

11. What are the major sources of collection system problems?
    1. Design-related deficiencies
    2. Inadequate sewer-use ordinances
    3. Improper inspection
    4. Improper installation
    5. Problems related to an old and frequently neglected wastewater collection system

12. What are the potential sources of inflow to a wastewater collection system?
    1. Basement sump pumps
    2. Infiltration
    3. Roof drains
    4. Surface drainage
    5. Yard drains

13. What are the objectives of manhole inspection?
    1. To be sure the lid is *NOT* buried
    2. To determine if location is proper
    3. To determine the proper elevations or grades around the lid
    4. To look for cracks in manhole
    5. To look for cracks where pipes are connected

14. What actions are possible from the information obtained by a closed-circuit collection system TV inspection?
    1. Establish priorities for corrective work
    2. Identify exact locations of service connections and correct as-built maps
    3. Inspect construction of house service connections to the lateral and branch sewers
    4. Provide for maximum effective and economical use of personnel and equipment
    5. Require contractor to correct defects observed in new sewers prior to acceptance

15. Why are wastewater collection systems smoke tested? To determine
    1. Locations of broken sewers due to settling of foundations
    2. Locations of certain types of illegal connections to the collection system
    3. Locations of "lost" manholes
    4. Positive proof that buildings are connected to the collection system
    5. Sources of entry to the collection system of inflow

16. What are the objectives of operating and maintaining a wastewater collection system so that it will function as intended? Minimize the number of
    1. Collection system operators needed
    2. Lift station failures
    3. Odor complaints
    4. Residences served
    5. Stoppages per mile of sewer

17. What could be the source or cause of an emergency stoppage?
    1. Debris
    2. Grease
    3. Law enforcement officers
    4. Low flows
    5. Roots

18. What kinds of provisions can be made for handling incoming wastewater when a pump station must be shut down?
    1. Adequate short-term storage
    2. Bypass pumping
    3. Reclaim and recycle wastewater
    4. Request dischargers to cease discharging
    5. Shut off water supply

19. What can the proper application of effective chemicals be used to control in wastewater collection systems?
    1. Concrete corrosion
    2. Grease
    3. Maintenance costs
    4. Odors
    5. Roots

20. What are the causes of odor problems in wastewater?
    1. Defective pipe materials
    2. High temperatures
    3. Long transmission lines
    4. Low-velocity flows
    5. Poorly maintained systems

21. What are possible causes of damages to and stoppages in sewer lines?
    1. Excessive flows
    2. Illegal taps
    3. Improper taps
    4. Roots
    5. Vandalism

22. What problems can develop if there are slight deviations from the designed slope of a sewer line?
    1. Deterioration of pipe
    2. Generation of hydrogen sulfide gas
    3. Interference with normal flows
    4. Sags
    5. Stoppages

23. Which factors may affect the chemical quantities required per pipe joint for rehabilitation?
    1. Diameter of pipe
    2. Gasketing materials remaining in joint
    3. Open or offset condition of joint
    4. Severity of infiltration or exfiltration
    5. Type of pipe material

Short Answer

1. Where else besides this training course could you look for information and help to learn more about wastewater collection systems?
2. What is the purpose of wastewater collection systems?
3. Why should collection system operators know how engineers design wastewater collection systems?
4. Why are lift stations installed?
5. Why should the slope of sewers follow the slope of the land?
6. What factors need to be considered in the selection of materials for sewer construction?
7. How would you measure the flow velocity in a collection system?

8. Which items on a vehicle and towed equipment should be inspected before leaving the yard?
9. What factors should be considered when deciding where to station flaggers?
10. What is the purpose of the advance warning area?
11. How can explosive or flammable atmospheres develop in a collection system?
12. If operators are scheduled to work in a manhole, when should the atmosphere in the manhole be tested?
13. What are the basic rules for qualified persons working with electricity?
14. How can collection system operators protect their hearing from loud noises?
15. What are some examples of improper installation of sewers?
16. How would you determine if there is proper drainage away from a manhole lid?
17. Why should gloves be worn when handling a manhole lid?
18. How does a TV inspection crew communicate with each other?
19. Under what circumstances would you lamp a sewer?
20. How would you select a method to clear a stoppage?
21. What would you do if you encountered an emergency stoppage with wastewater flowing out a manhole and down the street?
22. What precautions have to be taken during all hydraulic cleaning operations?
23. What types of conditions are best suited for the use of a sewer scooter?
24. Why are power bucket machines usually not used for routine cleaning?
25. How can you determine when a bucket is full?
26. Why should pipeline cleaning equipment be maintained?
27. What factors contribute to the production of hydrogen sulfide in a collection system?
28. How would you attempt to control odors from a wastewater collection system?
29. Why might a sewer line have to be dug up?
30. Why should other utility companies be notified before any excavation?
31. Why should repair assignments be made using work order forms?
32. How would you install a cleanout in an easement?
33. Why are sewers grouted or chemically sealed?
34. How would you raise the lid or cover of a manhole?

Problem

1. What is the flow in cubic feet per second (CFS) in a 10-inch diameter sewer flowing full? A dye test indicates that the flow velocity is 2.2 feet per second.

SUGGESTED ANSWERS FOR FINAL EXAMINATION

VOLUME I

True-False

1. True — The better the job collection system operators do, the more effectively a wastewater treatment plant can do its intended job.
2. True — Local taxpayers have invested more money in underground sewers than in all the structures above ground owned by their local government.
3. True — An unpermitted discharge from a sanitary sewer system to the waters of the United States constitutes a violation of the Clean Water Act.
4. False — Infiltration, *NOT* inflow, is the seepage of groundwater into a sewer system, including service connections.
5. True — The amount of support provided the sewer pipe by the sides of a trench decreases as the width of the trench increases.
6. True — A trained professional collection system operator demonstrates an awareness of hazards and a commitment to accomplish every task in a safe manner.
7. False — Generally, taper lengths to control traffic should be lengthened, *NOT* shortened, to increase their effectiveness.
8. True — After a short exposure to hydrogen sulfide gas, you will lose your ability to detect hydrogen sulfide.
9. True — All collection system agencies should develop a site-specific written lockout/tagout procedure and provide training to all of the operators who may use it.
10. False — Flow metering devices are *NOT* effective in areas where the sewers are surcharged.
11. False — The TV camera should always be pulled in the direction of flow to keep the front of the camera clean.
12. False — Do *NOT* operate TV cameras in sewer lines that test positive for explosive gases.
13. False — Sometimes a stoppage can occur and some flow will leak by the stoppage.
14. True — The exact distance a high-velocity cleaning nozzle has moved up in the sewer must be known at all times.

15. False — Masking agents do *NOT* eliminate or correct the source of an odor problem.

16. True — Never use plain water without shoring fluid in a hydraulic shore.

17. False — The bell end of a pipe should be facing upstream, *NOT* downstream.

18. True — Proper testing and inspection of a new sewer system before placing it in service can prevent serious problems from developing and avoid considerable grief.

Best Answer

1. 2 — Operation and maintenance of wastewater collection systems means keeping the facilities in good operating condition.

2. 4 — Operators should do underground sewer work safely.

3. 3 — Root growth is the principal cause of pipe damage that allows infiltration and exfiltration.

4. 2 — Knowing how engineers design collection systems helps collection system operators do a better job.

5. 3 — Grinder pumps pressurize wastewater to move it through the collection system.

6. 1 — When operators review wastewater collection system design plans and specifications, the most important concern regarding manholes is that they are in an accessible location.

7. 1 — Defensive driving is a mature attitude toward driving.

8. 3 — Oxygen content should be the first test performed with atmospheric analyzers.

9. 1 — Before opening a collection system excavation, contact other utilities to avoid damaging other underground utilities.

10. 4 — Leaky sewers under a water table allow infiltration to take place and occupy valuable capacity in the sewer.

11. 1 — Collection system TV operators must have the ability to recognize and evaluate conditions found in the pipe.

12. 4 — When smoke is observed coming from the vents only of a house, record that this house is OK.

13. 4 — Operators at downstream wastewater treatment plants should be notified before clearing a large septic stoppage to prevent treatment processes from becoming upset.

14. 1 — Never wear rubber gloves while handling or guiding a rod that is turning because the rod could seize the palm of the rubber glove and cause injury.

15. 4 — Infiltration of water into the sewer line could be a problem when using chemicals to control roots in that it may wash the chemicals off the roots.

16. 2 — Trenches less than five feet deep must be shored when hazardous ground movement may be expected.

17. 4 — Granular material is often substituted for native backfill material because it requires no compaction.

18. 1 — Photographs of work areas on private property should be taken before starting to work in case legal claims develop regarding damages.

19. 2 — Gages and valves used for air testing should be located outside the manhole to avoid injury.

20. 3 — Soil used for backfill should be fairly dry before it is paved over because soft spots could develop under the pavement if wet soil is used.

Multiple Choice

1. 1, 2, 3, 4, 5 — When wastewater collection systems fail due to lack of or improper operation and maintenance, blockages can occur that back up into homes, raw wastewater can be bypassed from the system, other utilities can be disrupted, wastewater can flow out manholes and into the street, and streets may collapse due to leaking pipes.

2. 1, 2, 3, 4, 5 — Many wastewater collection systems now require billions of dollars of rehabilitation and upgrading due to corrosion, insufficient money for maintenance, roots damaging pipe joints, poor construction and inspection, and inadequate flow capacities.

3. 1, 3, 4, 5 — Responsibilities of collection system operators include controlling O & M expenses, preserving the community's capital investment, protecting public health and safety, and protecting the environment.

4. 4, 5 — The use of pumps in collection systems is avoided when possible because they require extensive maintenance and because they require costly energy.

5. 3, 4, 5 — Hydrogen sulfide gas forms corrosive sulfuric acid, it is a toxic gas, and it smells like rotten eggs.

6. 2, 3, 5 — Pipe joints are one of the most critical components of the wastewater collection piping system because of the need to control groundwater infiltration, root intrusion, and wastewater exfiltration.

7. 2, 3, 5 — Exact distances and the nature of the advance warning before a work area depend on congestion, roadway conditions, and traffic speed.

8. 1, 2, 4, 5 — A hazardous atmosphere is an atmosphere that may expose operators to the risk of acute illness, death, incapacitation, or injury.

9. 1, 3, 5 — Excavations and adjacent areas must be inspected on a daily basis by a competent person for evidence of hazardous atmospheres, potential cave-ins, and protective system failures.

10. 3, 4, 5 — Collection system operators are usually exposed to hazardous chemicals and wastes when using chemicals for maintenance, when the system is used as an accidental disposal method, and when the system is used as an intentional disposal method.

11. 1, 2, 3, 4, 5 — Major sources of collection system problems include design-related deficiencies, inadequate sewer-use ordinances, and improper inspection and installation. Problems related to an old and frequently neglected system are another major source of problems.

12. 1, 3, 4, 5 — Potential sources of inflow to a wastewater collection system include basement sump pumps, roof drains, surface drainage, and yard drains.

13. 1, 3, 4, 5 — The objectives of manhole inspection are to ensure the lid is *NOT* buried, to determine proper elevations or grades around the lid, and to look for cracks in the manhole and where pipes are connected.

14. 1, 2, 3, 4, 5 — The information obtained by a closed-circuit collection system TV inspection could indicate the following types of actions: establishing priorities for corrective work, identifying exact locations of service connections, inspecting construction of house service connections, providing for effective use of personnel and equipment, and requiring the contractor to correct defects prior to acceptance.

15. 1, 2, 3, 4, 5 — Wastewater collection systems are smoke tested to determine locations of broken sewers due to settling, locations of illegal connections, and locations of "lost" manholes. Smoke testing also provides proof that the building is connected to the system and indicates any sources of inflow entry to the system.

16. 2, 3, 5 — The objectives of operating and maintaining a wastewater collection system so that it will function as intended are to minimize the number of lift station failures, odor complaints, and stoppages per mile of sewer.

17. 1, 2, 5 — Sources or causes of an emergency stoppage could be from debris, grease, or roots.

18. 1, 2 — When a pump station must be shut down, adequate short-term storage could be provided to handle incoming wastewater. Bypass pumping can also be used.

19. 1, 2, 4, 5 — In wastewater collection systems, the proper application of effective chemicals can be used to control concrete corrosion, grease, odors, and roots.

20. 2, 3, 4, 5 — Wastewater odor problems can be caused by high temperatures, long transmission lines, low-velocity flows, and poorly maintained systems.

21. 2, 3, 4, 5 — Damages to and stoppages in sewer lines can be caused by illegal taps, improper taps, roots, and vandalism.

22. 2, 3, 4, 5 — When there are slight deviations from the designed slope of a sewer line, problems can develop. These problems include stoppages that generate hydrogen sulfide gas, interference with normal flows, sags in the sewer lines, and stoppages in the sewer lines.

23. 1, 2, 3, 4, 5 — Factors that may affect the chemical quantities required per pipe joint for rehabilitation include the diameter of the pipe, gasketing materials remaining in the joint, open or offset condition of the joint, severity of infiltration or exfiltration, and the type of pipe material.

Short Answer

1. To learn more and find help regarding wastewater collection systems, contact the Water Environment Federation, state and local associations, state pollution control training agencies, colleges, and universities. Libraries can be an excellent source of useful journals and books.

2. The purpose of wastewater collection systems is to carry our wastewater away from our homes and industries to the wastewater treatment plant.

3. Collection system operators should know how engineers design collection systems so they will understand how the system works and can suggest to design engineers ways of improving the design and construction of future systems.

4. Lift stations are installed in a gravity collection system to lift (pump) wastewater to a higher elevation when the slope of the route followed by a gravity sewer would cause the sewer to be laid at an insufficient slope or at an impractical depth.

5. The slope of sewers should follow the slope of the land as closely as practical, provided the slope is adequate to produce gravity flow and the minimum scouring velocity. Deviations from ground slope could cause the sewer to become too shallow for adequate cover on the pipe or too deep for safe and economical construction.

6. Construction materials for sewers generally are selected for their resistance to deterioration by the wastewaters they convey, strength to withstand earth and surface loads, resistance to root intrusion, and their ability to minimize infiltration and exfiltration.

7. Flow velocities in sewers can be measured by the use of dyes or floats. Actual velocities should be measured in the field, rather than estimating the velocities by the use of hydraulic formulas, which give the theoretical velocity.

8. Items on a vehicle and towed equipment that should be inspected before leaving the yard include mirrors and windows, lighting system, brakes, tires, trailer hitch/safety chain, towed vehicle lighting system, and auxiliary equipment such as winches or hoists.

9. Flaggers must be located far enough in advance of the work area to allow motorists time to realize they must slow down, be alert for activity, and safely change lanes or follow a detour. Exact distances depend on traffic speed, congestion, roadway conditions, and local regulations.

10. The purpose of the advance warning area is to give drivers enough time to see what is happening ahead and adjust their driving patterns.

11. Explosive or flammable atmospheres can develop at any time in the collection system. Flammable gases or vapors may enter a sewer or manhole from a variety of legal, illegal, or accidental sources.

12. If operators are scheduled to work in a manhole, the atmosphere in the manhole should be tested before anyone enters it, preferably before the cover is even removed, and the atmospheric testing should continue for the entire time anyone is working in the manhole.

13. The basic rules for qualified persons working with electricity include:
    a. Disconnect and lock out all equipment prior to working on it,
    b. Observe common sense procedures,
    c. Use rubber mats, rubber boots, and leather gloves,
    d. Avoid wearing eyeglasses with metal frames, watches, jewelry, or metal belt buckles,
    e. Always maintain adequate clearance of body parts with live circuits,
    f. Try to work with one hand only when working around live circuits,
    g. Do not work alone,
    h. Wear tinted safety glasses, and
    i. When resetting tripped circuit breakers, always investigate the cause of the circuit breaker tripping.

14. Collection system operators can protect their hearing from loud noises by use of approved earplugs, earmuffs, and/or personal protective equipment.

15. Examples of improper installation of sewers include improper line, grade, and joint installation and shortcuts in bedding, connections, and backfilling. Other examples of improper installation include use of inferior or damaged materials.

16. To determine if there is proper drainage away from a manhole lid, lay a straightedge or board across the top of the lid. Determine if the roadway or ground surface slopes away from the lid in all directions to provide proper drainage.

17. Gloves should be worn when handling a manhole lid because the underside of a manhole lid can harbor dangerous insects such as black widow spiders, wasps, hornets, and mud daubers.

18. TV inspection crews can communicate with each other using a sound-powered telephone communication system.

19. Lamping could be used to determine whether a sewer is straight and clear, or if there are imperfections in the sewer. Lamping should not be used to look for cracks or infiltration problems.

20. To select a method to clear a stoppage, determine the cause of the stoppage and condition of the sewer, then select the most effective method on the basis of available equipment and personnel.

21. When dealing with an emergency stoppage, try to locate the stoppage, determine the cause, select the proper equipment, and clear the stoppage. Hand rods, power rodders, or high-velocity cleaners are often used to clear emergency stoppages.

22. During all types of hydraulic cleaning operations you must be very careful to avoid flooding basements and homes, properly route traffic around the work site, and observe all confined space safety precautions.

23. Sewer scooters are best suited for larger lines (over 18 inches) and are capable of removing large objects and other materials such as brick, sand, gravel, and rocks.

24. Power bucket machines usually are not used for routine cleaning of sewers because the use of power bucket machines can weaken or damage pipes.

25. You can determine when a bucket is full by noticing a definite resistance to pull at this time.

26. Pipeline cleaning equipment must be maintained in good repair to help prevent equipment failure on the job, to prolong the life of the equipment, and to increase the efficiency and safety of maintenance operations.

27. Factors that contribute to the production of hydrogen sulfide in a collection system include the presence of sulfate and other sulfur compounds, materials containing proteins, pH level, temperature of wastewater, absence of oxygen in the wastewater, amount of slimes on the pipe walls, and velocity of flow in the sewer.

28. Odors may be controlled by controlling the production of hydrogen sulfide. Most odors can be controlled in a properly designed, cleaned, and maintained collection system. Methods of odor control include proper ventilation of the sewers, regularly scheduled cleaning to control slimes, and application of proper chemicals when necessary.

29. Sewer lines may have to be dug up because they are damaged or blocked, or to retrieve equipment stuck in the line.

30. All utility companies in an area should be contacted before excavation so other underground utilities will not be damaged and possibly cause a serious injury (from electric shock or a gas explosion) to operators, the public, or property.

31. Repair assignments should be made using work order forms in order to maintain records and to provide a history of who made what repairs when, where, and why.

32. To install a cleanout in an easement, carefully dig the hole. Save any sod or lawn and try to protect vegetation. Cut pipe. Place gravel, sand, or crushed rock in bottom of trench for bedding of wye. Insert wye and cleanout. Tighten all couplings or seal joints. Prepare bedding, backfill trench, and compact backfill as trench is filled. Replace sod and vegetation and inspect repair area to be sure work area has been returned as close as possible to original conditions.

33. Sewers are grouted or chemically sealed when a sewer is either infiltrating or exfiltrating through leaking joints, cracks, or small holes. These techniques are commonly used in areas that are unsuitable for excavation.

34. Manhole lids or covers can be raised by the use of grout or fast-drying cement. Lids also may be raised by using a section of a manhole barrel or by the use of manhole castings that can be adjusted vertically.

Problem

1. What is the flow in cubic feet per second (CFS) in a 10-inch diameter sewer flowing full? A dye test indicates that the flow velocity is 2.2 feet per second.

Known		Unknown
Sewer Diameter, in	= 10 in	Flow, cu ft/sec
Flow Velocity, ft/sec	= 2.2 ft/sec	
Pipe Flowing Full Depth, in	= 10 in	

Calculate the cross-sectional area for a sewer pipe flowing full.

1. Find the value for the depth, d, divided by the diameter, D.

$$\frac{\text{d, in}}{\text{D, in}} = \frac{10 \text{ in}}{10 \text{ in}}$$

$$= 1.0$$

2. Find the correct factor for 1.0 in Table 3.4.

$$\frac{d}{D} = 1.0 \qquad \text{Factor} = 0.7854$$

3. Calculate the cross-sectional area.

$$\text{Pipe Area, sq ft} = \frac{(\text{Factor})(\text{Diameter, in})^2}{144 \text{ sq in/sq ft}}$$

$$= \frac{(0.7854)(10 \text{ in})^2}{144 \text{ sq in/sq ft}}$$

$$= 0.545 \text{ sq ft}$$

4. Calculate the rate of flow in cubic feet per second.

$$\text{Flow, cu ft/sec} = (\text{Area, sq ft})(\text{Velocity, ft/sec})$$

$$= (0.545 \text{ sq ft})(2.2 \text{ ft/sec})$$

$$= 1.20 \text{ cu ft/sec}$$

COLLECTION SYSTEM WORDS

A Summary of the Words Defined

in

OPERATION AND MAINTENANCE OF WASTEWATER COLLECTION SYSTEMS

COLLECTION SYSTEM WORDS

by

George Freeland

OPERATOR'S PROJECT PRONUNCIATION KEY

by Warren L. Prentice

The Operator's Project Pronunciation Key is designed to aid you in the pronunciation of new words. While this key is based primarily on familiar sounds, it does not attempt to follow any particular pronunciation guide. This key is designed solely to aid operators in this program.

You may find it helpful to refer to other available sources for pronunciation help. Each current standard dictionary contains a guide to its own pronunciation key. Each key will be different from each other and from this key. Examples of the differences between the key used in this program and the *WEBSTER'S NEW WORLD COLLEGE DICTIONARY*[1] "Key" are shown below:

Term	Project Key	Webster Key
sewer	SUE·er	sōō´ər
alignment	a·LINE·ment	ə·līn´mənt
infiltration	IN·fill·TRAY·shun	ĭn´fil·trā´shən

In using this key, you should accent (say louder) the syllable that appears in capital letters. The following chart is presented to give examples of how to pronounce words using the Operator's Project Pronunciation Key.

WORD	SYLLABLE			
	1st	2nd	3rd	4th
sewer	SUE	er		
alignment	a	LINE	ment	
infiltration	IN	fill	TRAY	shun

The first word, *SEWER*, has its first syllable accented. The second word, *ALIGNMENT*, has its second syllable accented. The third word, *INFILTRATION*, has its first and third syllables accented.

We hope you will find the key useful in unlocking the pronunciation of any new word.

EXPLANATION OF WORDS

The meanings of words in the glossary of this manual are based on current usage by the wastewater collection profession and definitions given in *GLOSSARY—WATER AND WASTEWATER CONTROL ENGINEERING*, prepared by the Joint Editorial Board representing APHA, ASCE, AWWA, and WEF,[2] 1981. Certain words used by wastewater collection system operators tend to have slightly different meanings in some regions of the United States. We have tried to standardize word meanings as much as possible.

---

[1] *The WEBSTER'S NEW WORLD COLLEGE DICTIONARY, Fourth Edition, 1999, was chosen rather than an unabridged dictionary because of its availability to most collection system operators. Other editions may be slightly different.*

[2] *APHA. American Public Health Association.*
ASCE. American Society of Civil Engineers.
AWWA. American Water Works Association.
WEF. Water Environment Federation.

WORDS

A

ACEOPS — ACEOPS

See **A**LLIANCE OF **CE**RTIFIED **OP**ERATOR**S**, LAB ANALYSTS, INSPECTORS, AND SPECIALISTS (ACEOPS).

ABANDONED (a-BAN-dund) — ABANDONED

No longer in use; a length, section or portion of a collection system no longer in service and left in place, underground. For example, when a house or building is razed or removed the service connection may be left open and unused.

ABATEMENT (a-BAIT-ment) — ABATEMENT

Putting an end to an undesirable or unlawful condition affecting the wastewater collection system. A property owner found to have inflow sources connected to the collection system may be issued a "NOTICE OF ABATEMENT." Such notices will usually describe the violation, suggest corrective measures and grant a period of time for compliance.

ABSORPTION (ab-SORP-shun) — ABSORPTION

The taking in or soaking up of one substance into the body of another by molecular or chemical action (as tree roots absorb dissolved nutrients in the soil).

ABSORPTION CAPACITY — ABSORPTION CAPACITY

The amount of liquid which a solid material can absorb. Sand, as an example, can hold approximately one-third of its volume in water, or three cubic feet of dry sand can contain one cubic foot of water. A denser soil, such as clay, can hold much less water and thus has a lower absorption capacity.

ABSORPTION RATE — ABSORPTION RATE

The speed at which a measured amount of solid material can absorb a measured amount of liquid. Under pressure, water can infiltrate a given volume of gravel very rapidly. The water will penetrate (or be absorbed by) sand more slowly and will take even longer to saturate the same amount of clay.

ACUTE HEALTH EFFECT — ACUTE HEALTH EFFECT

An adverse effect on a human or animal body, with symptoms developing rapidly.

ADSORPTION (add-SORP-shun) — ADSORPTION

The gathering of a gas, liquid, or dissolved substance on the surface or interface zone of another material.

AEROBIC (AIR-O-bick) — AEROBIC

A condition in which atmospheric or dissolved molecular oxygen is present in the aquatic (water) environment.

AIR BINDING — AIR BINDING

The clogging of a filter, pipe or pump due to the presence of air released from water. Air entering the filter media is harmful to both the filtration and backwash processes. Air can prevent the passage of water during the filtration process and can cause the loss of filter media during the backwash process.

AIR BLOWER — AIR BLOWER

A device used to ventilate manholes and lift stations.

AIR GAP — AIR GAP

An open vertical drop, or vertical empty space, between a drinking (potable) water supply and the point of use. This gap prevents backsiphonage because there is no way wastewater can reach the drinking water. Air gap devices are used to provide adequate space above the top of a manhole and the end of the hose from the fire hydrant. This gap ensures that no wastewater will flow out the top of a manhole, reach the end of the hose from a fire hydrant, and be sucked or drawn back up through the hose to the water supply.

AIR PADDING — AIR PADDING

Pumping dry air (dew point –40°F) into a container to assist with the withdrawal of a liquid or to force a liquified gas such as chlorine out of a container.

AIR TEST — AIR TEST

A method of inspecting a sewer pipe for leaks. Inflatable or similar plugs are placed in the line and the space between these plugs is pressurized with air. A drop in pressure indicates the line or run being tested has leaks.

AIR TEST, QUICK — AIR TEST, QUICK

(See QUICK AIR TEST)

ALIGNMENT (a-LINE-ment) — ALIGNMENT

The proper positioning of parts in a system. The alignment of a pipeline or other line refers to its location and direction.

ALLIANCE OF CERTIFIED OPERATORS, LAB ANALYSTS, INSPECTORS, AND SPECIALISTS (ACEOPS) — ALLIANCE OF CERTIFIED OPERATORS, LAB ANALYSTS, INSPECTORS, AND SPECIALISTS (ACEOPS)

A professional organization for operators, lab analysts, inspectors, and specialists dedicated to improving professionalism; expanding training, certification, and job opportunities; increasing information exchange; and advocating the importance of certified operators, lab analysts, inspectors, and specialists. For information on membership, contact ACEOPS, 1810 Bel Air Drive, Ames, IA 50010-5125, phone (515) 663-4128 or e-mail: Info@aceops.org.

ALLUVIAL (uh-LOU-vee-ul) DEPOSIT — ALLUVIAL DEPOSIT

Sediment (clay, silt, sand, gravel) deposited in place by the action of running water.

ALTERNATING CURRENT (A.C.) — ALTERNATING CURRENT (A.C.)

An electric current that reverses its direction (positive/negative values) at regular intervals.

AMBIENT (AM-bee-ent) — AMBIENT

Surrounding. Ambient or surrounding atmosphere.

AMPERAGE (AM-purr-age) — AMPERAGE

The strength of an electric current measured in amperes. The amount of electric current flow, similar to the flow of water in gallons per minute.

AMPERE (AM-peer) — AMPERE

The unit used to measure current strength. The current produced by an electromotive force of one volt acting through a resistance of one ohm.

ANAEROBIC (AN-air-O-bick) — ANAEROBIC

A condition in which atmospheric or dissolved molecular oxygen is *NOT* present in the aquatic (water) environment.

ANAEROBIC (AN-air-O-bick) DECOMPOSITION — ANAEROBIC DECOMPOSITION

The decay or breaking down of organic material in an environment containing no "free" or dissolved oxygen.

ANALOG READOUT — ANALOG READOUT

The readout of an instrument by a pointer (or other indicating means) against a dial or scale.

ANGLE OF REPOSE — ANGLE OF REPOSE

The angle between a horizontal line and the slope or surface of unsupported material such as gravel, sand, or loose soil. Also called the "natural slope."

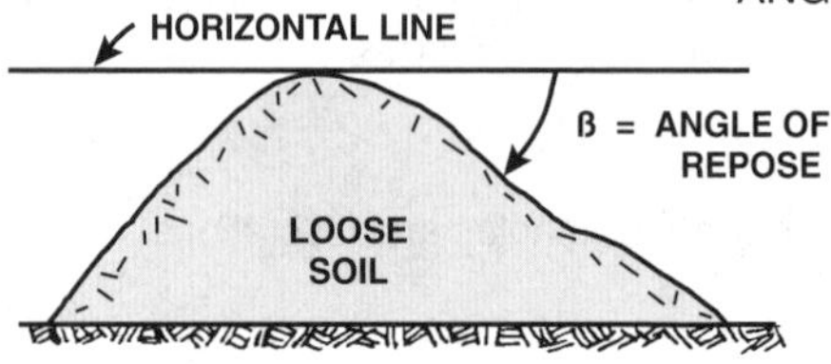

ANIMAL WASTES — ANIMAL WASTES

(1) Urine and fecal wastes of living animals.

(2) Wastes of animal tissue from meat processing (feathers included), or hospital, surgical and clinical facility wastes of animal types.

(3) Similar to (2) above, but cooked or prepared wastes of animal tissues and bones from domestic or commercial food preparation.

ANNULAR (AN-you-ler) SPACE — ANNULAR SPACE

A ring-shaped space located between two circular objects. For example, the space between the outside of a pipe liner and the inside of a pipe.

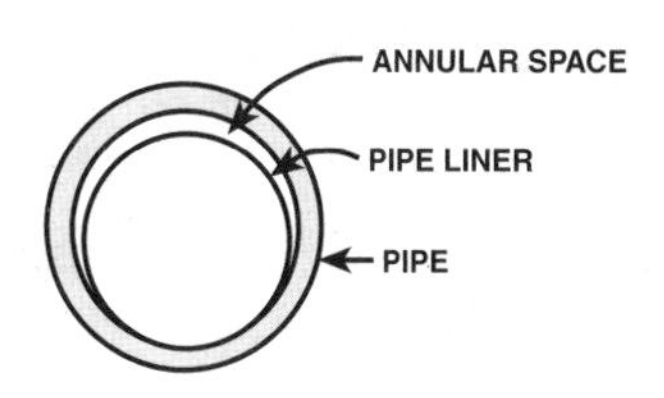

ANOXIC (an-OX-ick) — ANOXIC

Oxygen deficient or lacking sufficient oxygen.

APARTMENT COMPLEX — APARTMENT COMPLEX

One or more residential buildings at a single location. An apartment building may contain several residences with a single connection to the wastewater collection system. A complex can have several buildings with a single connection.

APPURTENANCE (uh-PURR-ten-nans) — APPURTENANCE

Machinery, appliances, structures and other parts of the main structure necessary to allow it to operate as intended, but not considered part of the main structure.

AQUIFER (ACK-wi-fer) — AQUIFER

A porous, water-bearing geologic formation. Usually refers only to materials capable of yielding a substantial amount of water.

ARCH — ARCH

(1) The curved top of a sewer pipe or conduit.

(2) A bridge or arch of hardened or caked chemical which will prevent the flow of the chemical.

ARTIFICIAL GROUNDWATER TABLE — ARTIFICIAL GROUNDWATER TABLE

A groundwater table that is changed by artificial means. Examples of activities that artificially raise the level of a groundwater table include agricultural irrigation, dams and excessive sewer line exfiltration. A groundwater table can be artificially lowered by sewer line infiltration, water wells, and similar drainage methods.

ASPHYXIATION (ass-FIX-ee-a-shun) — ASPHYXIATION

An extreme condition often resulting in death due to a lack of oxygen and excess carbon dioxide in the blood from any cause. Also called suffocation.

AUGER (AW-grr) — AUGER

A sharp tool used to go through and break up or remove various materials that become lodged in sewers.

B

BOD (pronounce as separate letters) — BOD

Biochemical **O**xygen **D**emand. The rate at which organisms use the oxygen in water or wastewater while stabilizing decomposable organic matter under aerobic conditions. In decomposition, organic matter serves as food for the bacteria and energy results from its oxidation. BOD measurements are used as a measure of the organic strength of wastes in water.

BACKFILL — BACKFILL

(1) Material used to fill in a trench or excavation.

(2) The act of filling a trench or excavation, usually after a pipe or some type of structure has been placed in the trench or excavation.

BACKFILL, BORROW — BACKFILL, BORROW

(See BORROW BACKFILL)

BACKFILL COMPACTION — BACKFILL COMPACTION

(1) Tamping, rolling or otherwise mechanically compressing material used as backfill for a trench or excavation. Backfill is compressed to increase its density so that it will support the weight of machinery or other loads after the material is in place in the excavation.

(2) Compaction of a backfill material can be expressed as a percentage of the maximum compactability, density or load capacity of the material being used.

BACKFILL, SELECT — BACKFILL, SELECT

(See SELECT BACKFILL)

BACKFLUSHING — BACKFLUSHING

A procedure used to wash settled waste matter off upstream structures to prevent odors from developing after a main line stoppage has been cleared.

BACKHOE — BACKHOE

An excavating machine whose bucket is securely attached to a hinged boom and is drawn toward the machine during excavation.

BACKWATER GATE — BACKWATER GATE

A gate installed at the end of a drain or outlet pipe to prevent the backward flow of water or wastewater. Generally used on storm sewer outlets into streams to prevent backward flow during times of flood or high tide. Also called a TIDE GATE.

BACTERIA (back-TEAR-e-ah) — BACTERIA

Bacteria are living organisms, microscopic in size, which usually consist of a single cell. Most bacteria use organic matter for their food and produce waste products as a result of their life processes.

BALLING — BALLING

A method of hydraulically cleaning a sewer or storm drain by using the pressure of a water head to create a high cleansing velocity of water around the ball. In normal operation, the ball is restrained by a cable while water washes past the ball at high velocity. Special sewer cleaning balls have an outside tread that causes them to spin or rotate, resulting in a "scrubbing" action of the flowing water along the pipe wall.

BAR RACK — BAR RACK

A screen composed of parallel bars, either vertical or inclined, placed in a sewer or other waterway to catch debris. The screenings may be raked from it.

BARREL — BARREL

(1) The cylindrical part of a pipe that may have a bell on one end.

(2) The cylindrical part of a manhole between the cone at the top and the shelf at the bottom.

BEDDING — BEDDING

The prepared base or bottom of a trench or excavation on which a pipe or other underground structure is supported.

BEDDING COMPACTION — BEDDING COMPACTION

(1) Tamping, rolling or otherwise mechanically compressing material used as bedding for a pipe or other underground structure to a density that will support expected loads.

(2) Bedding compaction can be expressed as a percentage of the maximum load capacity of the bedding material.

(3) Bedding compaction also can be expressed in load capacity of pounds per square foot.

BEDDING DESTRUCTION — BEDDING DESTRUCTION

Loss of grade, load capacity or material of a bedding.

BEDDING DISPLACEMENT — BEDDING DISPLACEMENT

Bedding which has been removed after placement and compaction. In a sewer pipe system, this can take place as a result of washouts due to infiltration, earth shifts or slides, damage from nearby excavations and/or improper backfill methods.

BEDDING FAULTS — BEDDING FAULTS

Locations where bedding was improperly applied and thus failed.

BEDDING GRADE BEDDING GRADE

(1) In a gravity-flow sewer system, pipe bedding is constructed and compacted to the design grade of the pipe. This is usually expressed in a percentage. A 0.5 percent grade would be a drop of one-half foot per hundred feet of pipe.

(2) Bedding grade for a gravity-flow sewer pipe can also be specified as elevation above mean sea level at specific points.

BEDDING, MANHOLE BEDDING, MANHOLE

(See MANHOLE BEDDING)

BEDDING MATERIAL, SELECT BEDDING MATERIAL, SELECT

(See SELECT BEDDING)

BELL BELL

(1) In pipe fitting, the enlarged female end of a pipe into which the male end fits. Also called a HUB.

(2) In plumbing, the expanded female end of a wiped joint.

BELL-AND-SPIGOT JOINT BELL-AND-SPIGOT JOINT

A form of joint used on pipes which have an enlarged diameter or bell at one end, and a spigot at the other which fits into and is laid in the bell. The joint is then made tight by lead, cement, rubber O-ring, or other jointing compounds or materials.

BELLMOUTH BELLMOUTH

An expanding, rounded entrance to a pipe or orifice.

BENCH SCALE TESTS BENCH SCALE TESTS

A method of studying different ways or chemical doses for treating water on a small scale in a laboratory.

BEND BEND

A piece of pipe bent or cast into an angular shape.

BIOCHEMICAL OXYGEN DEMAND (BOD) BIOCHEMICAL OXYGEN DEMAND (BOD)

The rate at which organisms use the oxygen in water or wastewater while stabilizing decomposable organic matter under aerobic conditions. In decomposition, organic matter serves as food for the bacteria and energy results from its oxidation. BOD measurements are used as a measure of the organic strength of wastes in water.

BIT BIT

(1) Cutting blade used in rodding (pipe clearing) operations.

(2) Cutting teeth on the auger head of a sewer boring tool.

BLOCKAGE BLOCKAGE

(1) Partial or complete interruption of flow as a result of some obstruction in a sewer.

(2) When a collection system becomes plugged and the flow backs up, it is said to have a "blockage." Commonly called a STOPPAGE.

BLOWER BLOWER

A device used to ventilate manholes and lift stations.

BLUEPRINT BLUEPRINT

A photographic print in white on a bright blue background used for copying maps, mechanical drawings, construction plans and architects' plans.

BORROW BACKFILL BORROW BACKFILL

Material used for backfilling a trench or excavation which was not the original material removed during excavation. This is a common practice where tests on the original material show it to have poor compactability or load capacity. Also called IMPORTED BACKFILL.

BRACES BRACES

(See CROSS BRACES)

BRANCH MANHOLE — BRANCH MANHOLE

A sewer or drain manhole which has more than one pipe feeding into it. A standard manhole will have one outlet and one inlet. A branch manhole will have one outlet and two or more inlets.

BRANCH SEWER — BRANCH SEWER

A sewer that receives wastewater from a relatively small area and discharges into a main sewer serving more than one branch sewer area.

BREAK — BREAK

A fracture or opening in a pipe, manhole or other structure due to structural failure and/or structural defect.

BRICKWORK — BRICKWORK

A structure made of brick, which was common in older sewers.

BRINELLING (bruh-NEL-ing) — BRINELLING

Tiny indentations (dents) high on the shoulder of the bearing race or bearing. A type of bearing failure.

BROKEN HUB — BROKEN HUB

In bell-and-spigot pipe, the bell portion is frequently called the "hub." A fracture or break in the bell portion is called a "broken hub."

BROKEN SECTION — BROKEN SECTION

A run of pipe between two joints is referred to as a "section." A fracture in a section is called a "broken section."

BUCKET — BUCKET

(1) A special device designed to be pulled along a sewer for the removal of debris from the sewer. The bucket has one end open with the opposite end having a set of jaws. When pulled from the jaw end, the jaws are automatically opened. When pulled from the other end, the jaws close. In operation, the bucket is pulled into the debris from the jaw end and to a point where some of the debris has been forced into the bucket. The bucket is then pulled out of the sewer from the other end, causing the jaws to close and retain the debris. Once removed from the manhole, the bucket is emptied and the process repeated.

(2) A conventional pail or bucket used in BUCKETING OUT and also for lowering and raising tools and materials from manholes and excavations.

BUCKET BAIL — BUCKET BAIL

The pulling handle on a bucket machine.

BUCKET MACHINE — BUCKET MACHINE

A powered winch machine designed for operation over a manhole. The machine controls the travel of buckets used to clean sewers.

BUCKETING OUT — BUCKETING OUT

An expression used to describe removal of debris from a manhole with a pail on a rope. In balling or high-velocity cleaning of sewers, debris is washed into the downstream manhole. Removal of this debris by scooping it into pails and hauling debris out is called "bucketing out."

BUILDING SERVICE — BUILDING SERVICE

A saddle or "Y" connection to a lateral or branch sewer for connection of a building lateral.

BUILDING SEWER — BUILDING SEWER

A gravity-flow pipeline connecting a building wastewater collection system to a lateral or branch sewer. The building sewer may begin at the outside of the building's foundation wall or some distance (such as 2 to 10 feet) from the wall, depending on local sewer ordinances. Also called a "house connection" or a "service connection."

BUILDING WASTEWATER COLLECTION SYSTEM — BUILDING WASTEWATER COLLECTION SYSTEM

All of the wastewater drain pipes and their hardware that connect plumbing fixtures inside or adjacent to a building to the building sewers. This includes traps, vents and cleanouts.

BYPASS — BYPASS

A pipe, valve, gate, weir, trench or other device designed to permit all or part of a wastewater flow to be diverted from usual channels or flow. Sometimes refers to a special line which carries the flow around a facility or device that needs maintenance or repair.

BYPASSING BYPASSING

The act of causing all or part of a flow to be diverted from its usual channels. In a wastewater treatment plant, overload flows should be bypassed into a holding pond for future treatment.

C

C-ZONED C-ZONED

An area set aside for commercial use.

CFR CFR

Code of **F**ederal **R**egulations. A publication of the United States Government which contains all of the proposed and finalized federal regulations, including safety and environmental regulations.

CFS CFS

Initials standing for "Cubic Feet Per Second," a measure of flow rate.

CMOM CMOM

Capacity Assurance, **M**anagement, **O**peration and **M**aintenance. A program developed by collection system agencies to ensure adequate capacity and also proper management and operation and maintenance of the collection system to prevent SSOs.

CSO CSO

Combined **S**ewer **O**verflow. Wastewater that flows out of a sewer (or lift station) as a result of flows exceeding the hydraulic capacity of the sewer. CSOs usually occur during periods of heavy precipitation or high levels of runoff from snow melt or other runoff sources.

CABLE STRAIN RELIEF CABLE STRAIN RELIEF

A mesh type of device that grips the power cable to prevent any strain on the cable from reaching the connections.

CAPILLARY (CAP-i-larry) EFFECT CAPILLARY EFFECT

Also called "wicking effect." The ability of a liquid to rise above an established level to saturate a porous solid.

CATALYST (CAT-uh-LIST) CATALYST

A substance that changes the speed or yield of a chemical reaction without being consumed or chemically changed by the chemical reaction.

CATCH BASIN CATCH BASIN

A chamber or well used with storm or combined sewers as a means of removing grit which might otherwise enter and be deposited in sewers. Also see STORM WATER INLET and CURB INLET.

CAULK (KAWK) CAULK

To stop up and make watertight the joints of a pipe by filling the joints with a waterproof compound or material.

CAULKING (KAWK-ing) CAULKING

(1) A waterproof compound or material used to fill a pipe joint.

(2) The act of using a waterproof compound or material to fill a pipe joint.

CAUTION CAUTION

This word warns against potential hazards or cautions against unsafe practices. Also see DANGER, NOTICE, and WARNING.

CAVITATION (CAV-uh-TAY-shun) CAVITATION

The formation and collapse of a gas pocket or bubble on the blade of an impeller or the gate of a valve. The collapse of this gas pocket or bubble drives water into the impeller or gate with a terrific force that can cause pitting on the impeller or gate surface. Cavitation is accompanied by loud noises that sound like someone is pounding on the impeller or gate with a hammer.

CENTRIFUGAL (sen-TRIF-uh-gull) PUMP CENTRIFUGAL PUMP

A pump consisting of an impeller fixed on a rotating shaft that is enclosed in a casing, and having an inlet and discharge connection. As the rotating impeller whirls the liquid around, centrifugal force builds up enough pressure to force the water through the discharge outlet.

CERTIFICATION EXAMINATION CERTIFICATION EXAMINATION

An examination administered by a state agency that wastewater collection system operators take to indicate a level of professional competence. In many states certification of wastewater collection system operators is voluntary. Current trends indicate that more states, provinces, and employers will require wastewater collection system operators to be "certified" in the future.

CESSPOOL CESSPOOL

A lined or partially lined excavation or pit for dumping raw household wastewater for natural decomposition and percolation into the soil.

CHECK VALVE CHECK VALVE

A special valve with a hinged disc or flap that opens in the direction of normal flow and is forced shut when flows attempt to go in the reverse or opposite direction of normal flows. Also see FLAP GATE and TIDE GATE.

CHEMICAL GROUT CHEMICAL GROUT

Two chemical solutions that form a solid when combined. Solidification time is controlled by the strength of the mixtures used and the temperature.

CHEMICAL GROUTING CHEMICAL GROUTING

Sealing leaks in a pipeline or manhole structure by injecting a chemical grout. In pipelines, the chemicals are injected through a device called a "packer." In operation, the packer is located at the leak point with the use of a television camera. Inflatable boots at either end of the packer isolate the leak point and the grouting chemicals are then forced into the leak under pressure. After allowing time for the grout to set, the packer is deflated and moved to the next location.

CHLORINATOR (KLOR-uh-NAY-ter) CHLORINATOR

A device used to regulate the transfer of chlorine from a container to flowing wastewater for such purposes as odor control and disinfection.

CHRISTY BOX CHRISTY BOX

A box placed over the connection between the pipe liner and the house sewer to hold the mortar around the cleanout wye and riser in place.

CIRCUIT CIRCUIT

The complete path of an electric current, including the generating apparatus or other source; or, a specific segment or section of the complete path.

CIRCUIT BREAKER CIRCUIT BREAKER

A safety device in an electric circuit that automatically shuts off the circuit when it becomes overloaded. The device can be manually reset.

CLEANING, PIPE CLEANING, PIPE

(See PIPE CLEANING)

CLEANOUT CLEANOUT

An opening (usually covered or capped) in a wastewater collection system used for inserting tools, rods or snakes while cleaning a pipeline or clearing a stoppage.

CLEANOUT, TWO-WAY CLEANOUT, TWO-WAY

A cleanout designed for rodding or working a snake into a pipe in either direction. Two-way cleanouts are often used in building lateral pipes at or near a property line.

CLINICAL LABORATORY CLINICAL LABORATORY

A special medical facility devoted to the identification of diseases and ailments through tests, studies and culture growths. Services of a clinical laboratory might be necessary in the event of an accidental or intentional release of contagious and/or infectious wastes into a collection system.

COAGULATE (co-AGG-you-late) COAGULATE

The use of chemicals that cause very fine particles to clump (floc) together into larger particles. This makes it easier to separate the solids from the liquids by settling, skimming, draining or filtering.

COHESIVE (co-HE-sive) COHESIVE

Tending to stick together.

COLLAPSED PIPE COLLAPSED PIPE

A pipe that has one or more points in its length which have been crushed or partially crushed by exterior pressures or impacts.

COLLECTION MAIN COLLECTION MAIN

A collection pipe to which building laterals are connected.

COLLECTION SYSTEM COLLECTION SYSTEM

A network of pipes, manholes, cleanouts, traps, siphons, lift stations and other structures used to collect all wastewater and wastewater-carried wastes of an area and transport them to a treatment plant or disposal system. The collection system includes land, wastewater lines and appurtenances, pumping stations and general property.

COLLOIDS (CALL-loids) COLLOIDS

Very small, finely divided solids (particles that do not dissolve) that remain dispersed in a liquid for a long time due to their small size and electrical charge. When most of the particles in water have a negative electrical charge, they tend to repel each other. This repulsion prevents the particles from sticking together, becoming heavier, and settling out.

COMBINED SEWER COMBINED SEWER

A sewer designed to carry both sanitary wastewaters and storm or surface water runoff.

COMBINED SYSTEM COMBINED SYSTEM

A sewer designed to carry both sanitary wastewaters and storm or surface water runoff.

COMBINED WASTEWATER COMBINED WASTEWATER

A mixture of storm or surface runoff and other wastewater such as domestic or industrial wastewater.

COMMERCIAL CONTRIBUTION COMMERCIAL CONTRIBUTION

Liquid and liquid-carried wastes dumped by commercial establishments into the wastewater collection system. Used in this context, commercial contributions are distinct from domestic and industrial sources of wastewater contributions. Examples of high-yield commercial sources are laundries, restaurants and hotels.

COMMERCIAL TELEVISION QUALITY COMMERCIAL TELEVISION QUALITY

Refers to picture quality about the level of a television picture. Commercial television pictures are considered to have resolutions between 350 and 525 lines. This is higher quality than surveillance quality and lower quality than inspection quality pictures.

COMMINUTOR (com-mih-NEW-ter) COMMINUTOR

A device used to reduce the size of the solid chunks in wastewater by shredding (comminuting). The shredding action is like many scissors cutting to shreds all the large solids in the wastewater.

COMPACTED BACKFILL COMPACTED BACKFILL

(See BACKFILL COMPACTION)

COMPACTED BEDDING COMPACTED BEDDING

(See BEDDING COMPACTION)

COMPACTION COMPACTION

Tamping or rolling of a material to achieve a surface or density that is able to support predicted loads.

COMPACTION TEST COMPACTION TEST

Any method of determining the weight a compacted material is able to support without damage or displacement. Usually stated in pounds per square foot.

COMPETENT PERSON COMPETENT PERSON

A competent person is defined by OSHA as a person capable of identifying existing and predictable hazards in the surroundings, or working conditions which are unsanitary, hazardous or dangerous to employees, and who has authorization to take prompt corrective measures to eliminate the hazards.

COMPUTED COLLECTION SYSTEM CONTRIBUTION COMPUTED COLLECTION SYSTEM CONTRIBUTION

The part of a collection system flow computed to be actual domestic and industrial wastewater. Applied to infiltration/inflow research, the computed domestic and industrial wastewater contribution is subtracted from a total flow to determine infiltration/inflow amounts.

COMPUTED COMMERCIAL CONTRIBUTION COMPUTED COMMERCIAL CONTRIBUTION

That part of a collection system flow computed to originate in the commercial establishments on the basis of expected flows from all commercial sources.

COMPUTED CONTRIBUTION COMPUTED CONTRIBUTION

A liquid or liquid-carried contribution to a collection system that is computed on the basis of expected discharges from all of the sources as opposed to actual measurement or metering. Also see ESTIMATED CONTRIBUTION.

COMPUTED DOMESTIC CONTRIBUTION COMPUTED DOMESTIC CONTRIBUTION

That part of a collection system flow computed to originate in the residential facilities based on the average flow contribution from each person.

COMPUTED FACILITY CONTRIBUTION COMPUTED FACILITY CONTRIBUTION

The computed liquid-waste discharge from a single facility based on the sources of waste flows in the facility.

COMPUTED INDUSTRIAL CONTRIBUTION COMPUTED INDUSTRIAL CONTRIBUTION

The computed liquid-waste discharge from industrial operations based on the expected discharges from all sources.

COMPUTED PER CAPITA CONTRIBUTION COMPUTED PER CAPITA CONTRIBUTION

The computed wastewater contribution from a domestic area, based on the population of the area. In the United States, the daily average wastewater contribution is considered to be 100 gallons per capita per day (100 GPCD).

COMPUTED TOTAL CONTRIBUTION COMPUTED TOTAL CONTRIBUTION

The total anticipated load on a wastewater treatment plant or the total anticipated flow in any collection system area based on the combined computed contributions of all connections to the system.

CONCENTRIC MANHOLE CONE CONCENTRIC MANHOLE CONE

Cone tapers uniformly from barrel to manhole cover.

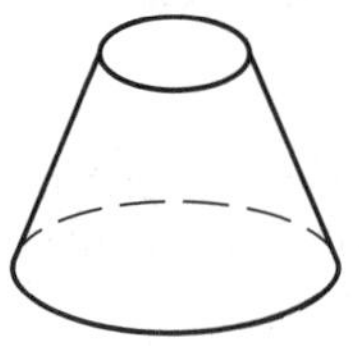

CONCRETE CRADLE CONCRETE CRADLE

A device made of concrete that is designed to support sewer pipe.

CONDUCTOR CONDUCTOR

(1) A pipe which carries a liquid load from one point to another point. In a wastewater collection system, a conductor is often a large pipe with no service connections. Also called a CONDUIT, interceptor (see INTERCEPTING SEWER) or INTERCONNECTOR.

(2) In plumbing, a line conducting water from the roof to the storm drain or other means of disposal. Also called a DOWNSPOUT.

(3) In electricity, a substance, body, device or wire that readily conducts or carries electric current.

CONDUIT CONDUIT

Any artificial or natural duct, either open or closed, for carrying fluids from one point to another. An electrical conduit carries electricity.

CONFINED SPACE CONFINED SPACE

Confined space means a space that:

A. Is large enough and so configured that an employee can bodily enter and perform assigned work; and

B. Has limited or restricted means for entry or exit (for example, tanks, vessels, silos, storage bins, hoppers, vaults, and pits are spaces that may have limited means of entry); and

C. Is not designed for continuous employee occupancy.

(Definition from the Code of Federal Regulations (CFR) Title 29 Part 1910.146.)

CONFINED SPACE, CLASS "A" CONFINED SPACE, CLASS "A"

A confined space that presents a situation that is immediately dangerous to life or health (IDLH). These include but are not limited to oxygen deficiency, explosive or flammable atmospheres, and/or concentrations of toxic substances.

(Definition from NIOSH, "Criteria for a Recommended Standard: Working in Confined Spaces.")

CONFINED SPACE, CLASS "B" CONFINED SPACE, CLASS "B"

A confined space that has the potential for causing injury and illness, if preventive measures are not used, but not immediately dangerous to life and health.

(Definition from NIOSH, "Criteria for a Recommended Standard: Working in Confined Spaces.")

CONFINED SPACE, CLASS "C" CONFINED SPACE, CLASS "C"

A confined space in which the potential hazard would not require any special modification of the work procedure.

(Definition from NIOSH, "Criteria for a Recommended Standard: Working in Confined Spaces.")

CONFINED SPACE, NON-PERMIT CONFINED SPACE, NON-PERMIT

A non-permit confined space is a confined space that does not contain or, with respect to atmospheric hazards, have the potential to contain any hazard capable of causing death or serious physical harm.

CONFINED SPACE, PERMIT-REQUIRED (PERMIT SPACE) CONFINED SPACE, PERMIT-REQUIRED (PERMIT SPACE)

A confined space that has one or more of the following characteristics:

- Contains or has a potential to contain a hazardous atmosphere,
- Contains a material that has the potential for engulfing an entrant,
- Has an internal configuration such that an entrant could be trapped or asphyxiated by inwardly converging walls or by a floor which slopes downward and tapers to a smaller cross section, or
- Contains any other recognized serious safety or health hazard.

(Definition from the Code of Federal Regulations (CFR) Title 29 Part 1910.146.)

CONTAMINATION CONTAMINATION

The introduction into water of microorganisms, chemicals, toxic substances, wastes, or wastewater in a concentration that makes the water unfit for its next intended use.

CONTRIBUTION CONTRIBUTION

Waters, wastewaters or liquid-carried wastes entering a wastewater collection system.

CORROSION CORROSION

The gradual decomposition or destruction of a material due to chemical action, often due to an electrochemical reaction. Corrosion starts at the surface of a material and moves inward, such as the chemical action upon manholes and sewer pipe materials.

COULOMB (COO-lahm) COULOMB

A measurement of the amount of electrical charge carried by an electric current of one ampere in one second. One coulomb equals about 6.25×10^{18} electrons (6,250,000,000,000,000,000 electrons).

COUPLING COUPLING

(1) A threaded sleeve used to connect two pipes.

(2) A device used to connect two adjacent parts, such as a pipe coupling, hose coupling or drive coupling.

COUPON COUPON

A steel specimen inserted into wastewater to measure the corrosiveness of the wastewater. The rate of corrosion is measured as the loss of weight of the coupon or change in its physical characteristics. Measure the weight loss (in milligrams) per surface area (in square decimeters) exposed to the wastewater per day. 10 decimeters = 1 meter = 100 centimeters.

CROSS BRACES CROSS BRACES

Shoring members placed across a trench to hold other horizontal and vertical shoring members in place.

CROSS CONNECTION — CROSS CONNECTION

(1) A connection between a storm drain system and a sanitary collection system.

(2) Less frequently used to mean a connection between two sections of a collection system to handle anticipated overloads of one system.

(3) A connection between drinking (potable) water and an unapproved water supply.

CURB INLET — CURB INLET

A chamber or well built at the curbline of a street to admit gutter flow to the storm water drainage system. Also see STORM WATER INLET and CATCH BASIN.

CURRENT — CURRENT

A movement or flow of electricity. Water flowing in a pipe is measured in gallons per second past a certain point, not by the number of water molecules going past a point. Electric current is measured by the number of coulombs per second flowing past a certain point in a conductor. A coulomb is equal to about 6.25×10^{18} electrons (6,250,000,000,000,000,000 electrons). A flow of one coulomb per second is called one ampere, the unit of the rate of flow of current.

D

DANGER — DANGER

The word *DANGER* is used where an immediate hazard presents a threat of death or serious injury to employees. Also see CAUTION, NOTICE, and WARNING.

DANGEROUS AIR CONTAMINATION — DANGEROUS AIR CONTAMINATION

An atmosphere presenting a threat of causing death, injury, acute illness, or disablement due to the presence of flammable and/or explosive, toxic or otherwise injurious or incapacitating substances.

A. Dangerous air contamination due to the flammability of a gas or vapor is defined as an atmosphere containing the gas or vapor at a concentration greater than 10 percent of its lower explosive (lower flammable) limit.

B. Dangerous air contamination due to a combustible particulate is defined as a concentration greater than 10 percent of the minimum explosive concentration of the particulate.

C. Dangerous air contamination due to the toxicity of a substance is defined as the atmospheric concentration immediately hazardous to life or health.

DATA-VIEW — DATA-VIEW

A high-speed reporting and recording system used with closed-circuit pipeline television equipment. Data-view provides digital indexing of date, job number, footages and air test pressures in the television picture itself. Where videotape recordings of television pipe inspections or pipe sealing activities are made, data-view reports are automatically recorded on the taped pictures.

DEADEND MANHOLE — DEADEND MANHOLE

A manhole located at the upstream end of a sewer and having no inlet pipe. Also called a TERMINAL MANHOLE.

DEBRIS (de-BREE) — DEBRIS

Any material in wastewater found floating, suspended, settled or moving along the bottom of a sewer. This material may cause stoppages by getting hung up on roots or settling out in a sewer. Debris includes grit, paper, plastic, rubber, silt, and all materials except liquids.

DEBRIS, INFILTRATED — DEBRIS, INFILTRATED

(See INFILTRATED DEBRIS)

DECIBEL (DES-uh-bull) — DECIBEL

A unit for expressing the relative intensity of sounds on a scale from zero for the average least perceptible sound to about 130 for the average level at which sound causes pain to humans. Abbreviated dB.

DECOMPOSED PIPE — DECOMPOSED PIPE

Pipe which has been destroyed or portions of pipe weakened by chemical actions.

DECOMPOSITION, DECAY DECOMPOSITION, DECAY

Processes that convert unstable materials into more stable forms by chemical or biological action. Waste treatment encourages decay in a controlled situation so that material may be disposed of in a stable form. When organic matter decays under anaerobic conditions (putrefaction), undesirable odors are produced. The aerobic processes in common use for wastewater treatment produce much less objectionable odors.

DEFECT DEFECT

A point where a pipe or system structure has been damaged or has a fault.

DEFECT, SURFACED DEFECT, SURFACED

(See SURFACED DEFECT)

DEFLECTED DEFLECTED

(1) Pipe which has been forced out of round by external pressures. This happens mainly to fiber and plastic pipes where backfill compaction has resulted in unequal pressures on all sides of the pipe.

(2) Pipe whose direction has been changed either to the left, right, up, or down.

DEGRADATION (deh-gruh-DAY-shun) DEGRADATION

The conversion or breakdown of a substance to simpler compounds. For example, the degradation of organic matter to carbon dioxide and water.

DESTROYED PIPE DESTROYED PIPE

Pipe which has been damaged, decomposed, deflected, crushed or collapsed to a point that it must be replaced.

DETENTION DETENTION

The delay or holding of the flow of water and water-carried wastes in a pipe system. This can be caused by a restriction in the pipe, a stoppage or a dip. Detention also means the time water is held or stored in a basin or a wet well. Sometimes called RETENTION.

DETRITUS (dee-TRY-tus) DETRITUS

The heavy, coarse mixture of grit and organic material carried by wastewater. Also called GRIT.

DEWATER DEWATER

To drain or remove water from an enclosure. A structure may be dewatered so that it can be inspected or repaired. Dewater also means draining or removing water from sludge to increase the solids concentration.

DIGITAL READOUT DIGITAL READOUT

The use of numbers to indicate the value or measurement of a variable. The readout of an instrument by a direct, numerical reading of the measured value. The signal sent to such readouts is usually an analog signal.

DIP DIP

A point in a sewer pipe where a drain grade defect results in a puddle of standing water when there is no flow. If the grade defect is severe enough to cause the standing water to fill the pipe at any point (preventing passage of air through the pipe), it is called a "trap dip," "full dip" or "filled dip."

DIRECT CURRENT (D.C.) DIRECT CURRENT (D.C.)

Electric current flowing in one direction only and essentially free from pulsation.

DISCHARGE HEAD DISCHARGE HEAD

The pressure (in pounds per square inch or psi) measured at the centerline of a pump discharge and very close to the discharge flange, converted into feet. The pressure is measured from the centerline of the pump to the hydraulic grade line of the water in the discharge pipe.

Discharge Head, ft = (Discharge Pressure, psi)(2.31 ft/psi)

DISINFECTION (dis-in-FECT-shun) DISINFECTION

The process designed to kill or inactivate most microorganisms in wastewater, including essentially all pathogenic (disease-causing) bacteria. There are several ways to disinfect, with chlorination being the most frequently used in water and wastewater treatment plants.

DISPLACED PIPE DISPLACED PIPE

A run or section of sewer pipe that has been pushed out of alignment by external forces.

DISTURBED SOIL DISTURBED SOIL

Soil which has been changed from its natural condition by excavation or other means.

DIVERSION CHAMBER DIVERSION CHAMBER

A chamber or box which contains a device for diverting or drawing off all or part of a flow or for discharging portions of the total flow to various outlets. Also called a REGULATOR.

DOMESTIC DOMESTIC

Residential living facilities. A domestic area will be predominantly residential in occupancy and is sometimes referred to as a "bedroom area" or "bedroom community."

DOMESTIC CONTRIBUTION DOMESTIC CONTRIBUTION

Wastes originating in a residential facility or dwelling. In this use, it means the type and quantity of wastes are different from commercial and industrial or agricultural wastes.

DOMESTIC SERVICE DOMESTIC SERVICE

A connection to a sewer system for hookup of a residential-type building.

DOWNSPOUT DOWNSPOUT

In plumbing, the water conductor from the roof gutters or roof catchment to the storm drain or other means of disposal. Also called a "roof leader" or "roof drain."

DOWNSTREAM DOWNSTREAM

The direction of the flow of water. In the lower part of a sewer or collection system or in that direction.

DRAGLINE DRAGLINE

A machine that drags a bucket down the intended line of a trench to dig or excavate the trench. Also used to dig holes and move soil or aggregate.

DRIFT DRIFT

The difference between the actual value and the desired value (or set point); characteristic of proportional controllers that do not incorporate reset action. Also called "offset."

DROP JOINT DROP JOINT

A sewer pipe joint where one part has dropped out of alignment. Also called a VERTICAL OFFSET.

DROP MANHOLE DROP MANHOLE

A main line or house service line lateral entering a manhole at a higher elevation than the main flow line or channel. If the higher elevation flow is routed to the main manhole channel outside of the manhole, it is called an "outside drop." If the flow is routed down through the manhole barrel, the pipe down to the manhole channel is called an "inside drop."

DRY PIT DRY PIT

(See DRY WELL)

DRY WELL DRY WELL

A dry room or compartment in a lift station, near or below the water level, where the pumps are located.

DWELLING DWELLING

A structure for residential occupancy.

DYNAMIC HEAD DYNAMIC HEAD

When a pump is operating, the vertical distance (in feet) from a point to the energy grade line. Also see TOTAL DYNAMIC HEAD, STATIC HEAD, and ENERGY GRADE LINE.

E

EARTH SHIFT — EARTH SHIFT

The movement or dislocation of underground soil or structure. Earth shift is usually caused by external forces such as surface loads, slides, stresses or nearby construction, water movements or seismic forces.

EASEMENT — EASEMENT

Legal right to use the property of others for a specific purpose. For example, a utility company may have a five-foot easement along the property line of a home. This gives the utility the legal right to install and maintain a sewer line within the easement.

ECCENTRIC MANHOLE CONE — ECCENTRIC MANHOLE CONE

Cone tapers nonuniformly from barrel to manhole cover with one side usually vertical.

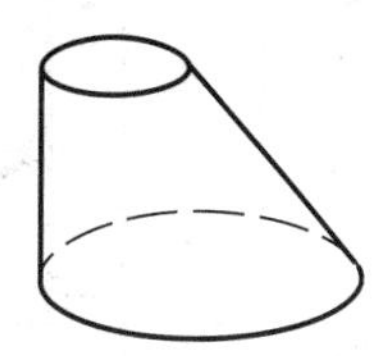

EFFLUENT — EFFLUENT

Wastewater or other liquid—raw (untreated), partially or completely treated—flowing *FROM* a reservoir, basin, treatment process, or treatment plant.

ELBOW — ELBOW

A pipe fitting that connects two pipes at an angle. The angle is usually 90 degrees unless another angle is stated. Also called an "ell."

ELECTROLYTE (ee-LECK-tro-LITE) SOLUTION — ELECTROLYTE SOLUTION

A special solution that is capable of conducting electricity.

ELECTROMOTIVE FORCE (E.M.F.) — ELECTROMOTIVE FORCE (E.M.F.)

The electrical pressure available to cause a flow of current (amperage) when an electric circuit is closed. Also called VOLTAGE.

ELECTRON — ELECTRON

(1) A very small, negatively charged particle which is practically weightless. According to the electron theory, all electrical and electronic effects are caused either by the movement of electrons from place to place or because there is an excess or lack of electrons at a particular place.

(2) The part of an atom that determines its chemical properties.

ELEVATION — ELEVATION

The height to which something is elevated, such as the height above sea level.

EMULSION (e-MULL-shun) — EMULSION

A liquid mixture of two or more liquid substances not normally dissolved in one another; one liquid is held in suspension in the other.

ENCLOSED SPACE — ENCLOSED SPACE

(See CONFINED SPACE)

ENERGY GRADE LINE (EGL) — ENERGY GRADE LINE (EGL)

A line that represents the elevation of energy head (in feet) of water flowing in a pipe, conduit or channel. The line is drawn above the hydraulic grade line (gradient) a distance equal to the velocity head ($V^2/2g$) of the water flowing at each section or point along the pipe or channel. Also see HYDRAULIC GRADE LINE.

[SEE DRAWING ON PAGE 534]

ENGULFMENT — ENGULFMENT

Engulfment means the surrounding and effective capture of a person by a liquid or finely divided (flowable) solid substance that can be aspirated to cause death by filling or plugging the respiratory system or that can exert enough force on the body to cause death by strangulation, constriction, or crushing.

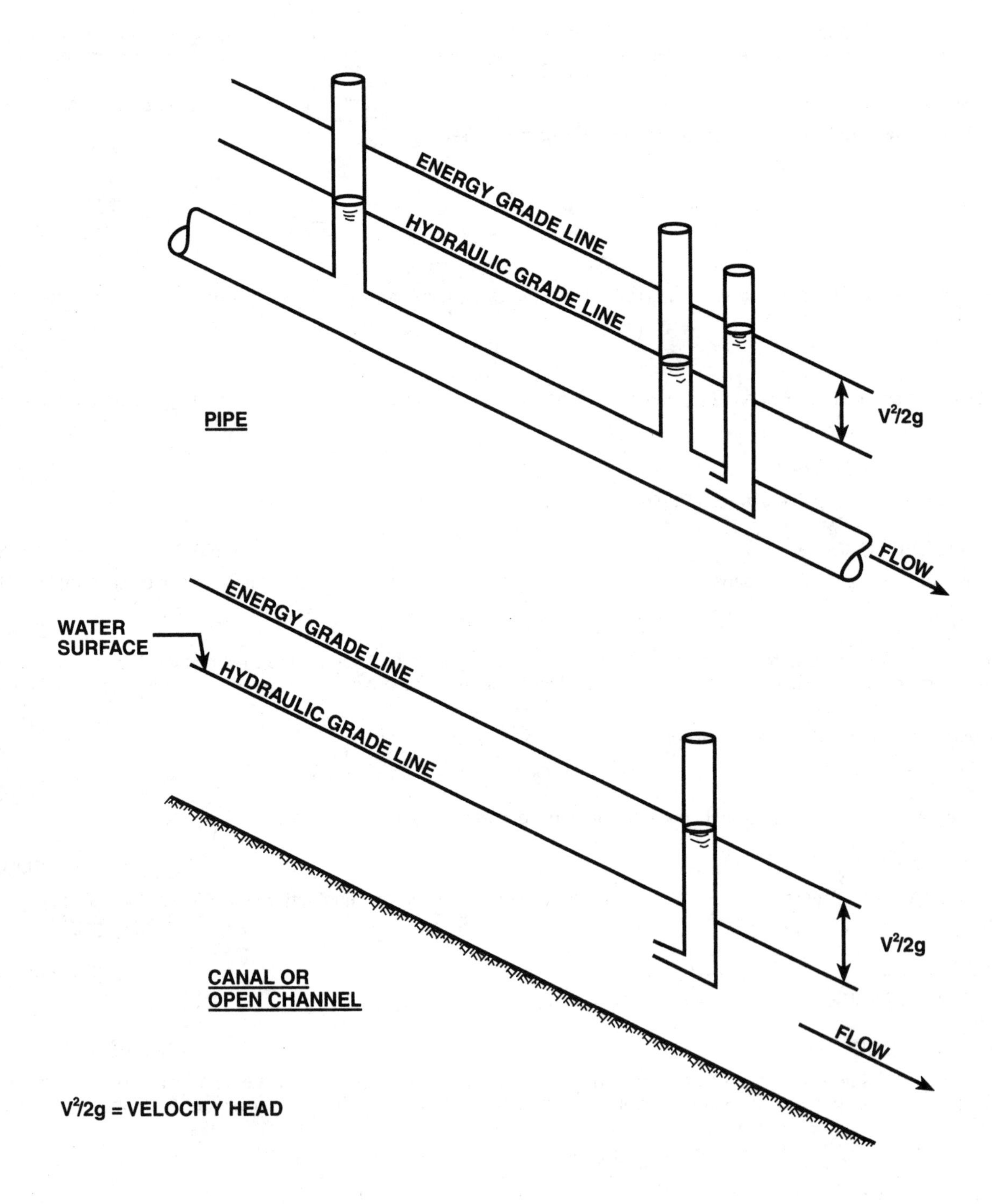

ENERGY GRADE LINE and HYDRAULIC GRADE LINE

ENTRAIN ENTRAIN

To trap bubbles in water either mechanically through turbulence or chemically through a reaction.

ESTIMATED CONTRIBUTION ESTIMATED CONTRIBUTION

A contribution to a collection system that is estimated rather than computed. The distinction between computed and estimated in such cases is difficult to specify or define. Also see COMPUTED CONTRIBUTION.

ESTIMATED FLOW ESTIMATED FLOW

A rough guess of the amount of flow in a collection system. When greater accuracy is needed, flow could be computed using average or typical flow quantities. Even greater accuracy would result from metering or otherwise measuring the actual flow.

EXFILTRATION (EX-fill-TRAY-shun) EXFILTRATION

Liquid wastes and liquid-carried wastes which unintentionally leak out of a sewer pipe system and into the environment.

EXTRADOS EXTRADOS

The upper outside curve of a sewer pipe or conduit.

EXTREMELY HAZARDOUS WASTE EXTREMELY HAZARDOUS WASTE

Any hazardous waste or mixture of hazardous wastes which, if any human exposure should occur, may likely result in death, disabling personal injury or illness during, or as a proximate result of, any disposal of such waste or mixture of wastes because of its quantity, concentration, or chemical characteristics. (Subsection 23115 of Article 2, Chapter 6.5, Division 20, of the California Health and Safety Code.) Also see HAZARDOUS WASTE.

F

FAIR LEAD PULLEY (fair LEE-d pully) FAIR LEAD PULLEY

A pulley that is placed in a manhole to guide TV camera electric cables and the pull cable into the sewer when inspecting pipelines.

FAULT FAULT

(1) A fracture in the earth's crust that leaves land on one side of the crack out of alignment with the other side. Faults are generally a result of earth shifts and earthquakes.

(2) (See DEFECT)

FILLED DIP FILLED DIP

(See DIP)

FIT TEST FIT TEST

The use of a procedure to qualitatively or quantitatively evaluate the fit of a respirator on an individual.

FLAP GATE FLAP GATE

A hinged gate that is mounted at the top of a pipe or channel to allow flow in only one direction. Flow in the wrong direction closes the gate. Also see CHECK VALVE and TIDE GATE.

FLAT FLAT

A flat is the length of one side of a nut.

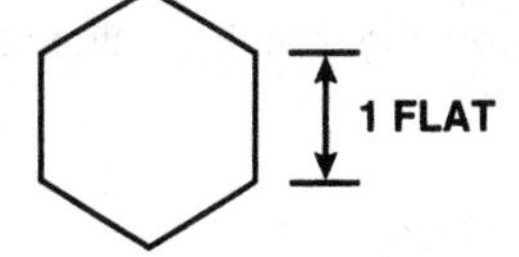

FLOAT (CONTROL) FLOAT (CONTROL)

A device used to measure the elevation of the surface of water. The float rests on the surface of the water and rises or falls with it. The elevation of the water surface is measured by a rod, chain, rope or tape attached to the float.

FLOAT LINE FLOAT LINE

A length of rope or heavy twine attached to a float, plastic jug or parachute to be carried by the flow in a sewer from one manhole to the next. This is called "stringing the line" and is used for pulling through winch cables, such as for bucket machine work or closed-circuit television work.

FLOTATION FLOTATION

(1) The stress or forces on a pipeline or manhole structure located below a water table which tend to lift or float the pipeline or manhole structure.

(2) The process of raising suspended matter to the surface of the liquid in a tank where it forms a scum layer that can be removed by skimming. The suspended matter is raised by aeration, the evolution of gas, the use of chemicals, electrolysis, heat or bacterial decomposition.

FLOTSAM (FLOAT-sam) FLOTSAM

Material floating or drifting about on the surface of a body of water.

FLOW FLOW

The continuous movement of a liquid from one place to another.

FLOW ISOLATION FLOW ISOLATION

A procedure used to measure inflow and infiltration (I/I). A section of sewer is blocked off or isolated and the flow from the section is measured.

FLOW LINE FLOW LINE

(1) The top of the wetted line, the water surface or the hydraulic grade line of water flowing in an open channel or partially full conduit.

(2) The lowest point of the channel inside a pipe or manhole. See INVERT. *NOTE:* (2) is an improper definition, although used by some contractors.

FLOW RECORDING FLOW RECORDING

A record of a flow measurement past any selected point. Usually consists of time, velocity and amount (in gallons) with maximum and minimum rates as well as the total amount over a given time period.

FLUME FLUME

An open conduit of wood, masonry, metal, or plastic constructed on a grade and sometimes elevated. Sometimes called an "aqueduct" or "channel."

FLUME, PARSHALL FLUME, PARSHALL

(See PARSHALL FLUME)

FLUSHER BRANCH FLUSHER BRANCH

A line built specifically to allow the introduction of large quantities of water to the collection system so the lines can be "flushed out" with water. Also installed to provide access for equipment to clear stoppages in a sewer.

FLUSHING FLUSHING

The removal of deposits of material which have lodged in sewers because of inadequate velocity of flows. Water is discharged into the sewers at such rates that the larger flow and higher velocities are sufficient to remove the material.

FOCAL LENGTH FOCAL LENGTH

The distance of a focus from the surface of a lens (such as a camera lens) to the focal point.

FORCE MAIN FORCE MAIN

A pipe that carries wastewater under pressure from the discharge side of a pump to a point of gravity flow downstream.

FRIABILITY (FRY-uh-BILL-uh-tee) FRIABILITY

The ability of a soil or substance to crumble under moderate or light pressure.

FRICTION LOSS FRICTION LOSS

The head lost by water flowing in a stream or conduit as the result of the disturbances set up by the contact between the moving water and its containing conduit and by intermolecular friction.

FULL DIP FULL DIP

(See DIP)

FUSE FUSE

A protective device having a strip or wire of fusible metal which, when placed in a circuit, will melt and break the electric circuit if heated too much. High temperatures will develop in the fuse when a current flows through the fuse in excess of that which the circuit will carry safely.

G

GIS GIS

Geographic **I**nformation **S**ystem. A computer program that combines mapping with detailed information about the physical locations of structures such as pipes, valves, and manholes within geographic areas. The system is used to help operators and maintenance personnel locate utility system features or structures and to assist with the scheduling and performance of maintenance activities.

GPCD GPCD

Initials standing for "Gallons Per Capita Per Day."

GPD GPD

Initials standing for "Gallons Per Day."

GPM GPM

Initials standing for "Gallons Per Minute."

GPY GPY

Initials standing for "Gallons Per Year."

GAGE GAGE

A device for checking or measuring a particular dimension of something, using specific standardized units. For example, a gage might measure the elevation of a water surface, the velocity of flowing water, the pressure of water, the amount or intensity of precipitation, and the depth of snowfall. Gages also are used to determine the location or position of equipment during installation and after operation.

GAS (SEWER) GAS (SEWER)

(See SEWER GAS)

GEL OR GELL GEL OR GELL

(1) A term sometimes applied to chemical grout. See CHEMICAL GROUT.

(2) A form of matter in a colloidal state that does not dissolve, but remains suspended in a solvent. It does not precipitate out of the solvent without the intervention of heat or of an electrolyte.

GEOGRAPHIC INFORMATION SYSTEM (GIS) GEOGRAPHIC INFORMATION SYSTEM (GIS)

A computer program that combines mapping with detailed information about the physical locations of structures such as pipes, valves, and manholes within geographic areas. The system is used to help operators and maintenance personnel locate utility system features or structures and to assist with the scheduling and performance of maintenance activities.

GEOLOGY GEOLOGY

The science that deals with the origin, history and structures of the earth, as recorded in the rocks, together with the forces and processes now operating to modify rocks.

GEOLOGY, SUBSOIL GEOLOGY, SUBSOIL

(See SUBSOIL GEOLOGY)

GEOLOGY, SUBSTRUCTURE GEOLOGY, SUBSTRUCTURE

(See SUBSOIL GEOLOGY)

GRADE GRADE

(1) The elevation of the invert (or bottom) of a pipeline, canal, culvert, sewer, or similar conduit.

(2) The inclination or slope of a pipeline, conduit, stream channel, or natural ground surface; usually expressed in terms of the ratio or percentage of number of units of vertical rise or fall per unit of horizontal distance. A 0.5 percent grade would be a drop of one-half foot per hundred feet of pipe.

GRADE RING GRADE RING

A precast concrete ring 4 to 12 inches high which is placed on top of a manhole cone to raise the manhole cover frame flush with the surface grade. Sometimes called a "spacer."

GRADIENT GRADIENT

The upward or downward slope of a pipeline.

GRANULAR GRANULAR

Any substance that appears to consist of separate granules or grains. Examples are sand and gravel.

GRAVITY GRAVITY

The attraction of the earth to any substance—solid, liquid or gas.

GRAVITY FLOW GRAVITY FLOW

Water or wastewater flowing from a higher elevation to a lower elevation due to the force of gravity. The water does not flow due to energy provided by a pump. Wherever possible, wastewater collection systems are designed to use the force of gravity to convey waste liquids and solids.

GRAVITY, SPECIFIC GRAVITY, SPECIFIC

(See SPECIFIC GRAVITY)

GREASE GREASE

In a collection system, grease is considered to be the residues of fats, detergents, waxes, free fatty acids, calcium and magnesium soaps, mineral oils, and certain other nonfatty materials which tend to separate from water and coagulate as floatables or scums.

GREASE BUILDUP GREASE BUILDUP

Any point in a collection system where coagulated and solidified greases accumulate and build up. Many varieties of grease have high adhesive characteristics and collect other solids, forming restrictions and stoppages in collection systems.

GREASE TRAP GREASE TRAP

A receptacle designed to collect and retain grease and fatty substances usually found in kitchens or from similar wastes. It is installed in the drainage system between the kitchen or other point of production of the waste and the building wastewater collection line. Commonly used to control grease from restaurants.

GRIT GRIT

The heavy mineral material present in wastewater such as sand, coffee grounds, eggshells, gravel and cinders. Grit tends to settle out at flow velocities below 2 ft/sec and accumulate in the invert or bottoms of the pipelines. Also called DETRITUS.

GRIT CATCHER GRIT CATCHER

A chamber usually placed at the upper end of a depressed collection line or at other points on combined or storm water collection lines where wear from grit is possible. The chamber is sized and shaped to reduce the velocity of flow through it and thus permit the settling out of grit. Also called a "sand catcher." See GRIT CHAMBER and SAND TRAP.

GRIT CHAMBER GRIT CHAMBER

A detention chamber or an enlargement of a collection line designed to reduce the velocity of flow of the liquid to permit the separation of mineral solids from organic solids by differential sedimentation.

GRIT CHANNEL GRIT CHANNEL

(1) An enlargement in a collection line where grit can easily settle out of the flow.

(2) The waterway of a grit chamber.

GRIT COLLECTOR GRIT COLLECTOR

A device placed in a grit chamber to convey deposited grit to a point of collection for ultimate disposal.

GRIT COMPARTMENT GRIT COMPARTMENT

The portion of the grit chamber in which grit is collected and stored before removal.

GRIT TANK GRIT TANK

A structure located at the inlet to a treatment plant for the accumulation and removal of grit.

GRIT TRAP GRIT TRAP

(1) A permanent structure built into a manhole (or other convenient location in a collection system) for the accumulation and easy removal of grit.
(2) (See SAND TRAP)

GROSS VEHICLE WEIGHT (GVW) GROSS VEHICLE WEIGHT (GVW)

The total weight of a single vehicle including its load.

GROSS VEHICLE WEIGHT RATING (GVWR) GROSS VEHICLE WEIGHT RATING (GVWR)

The maximum weight rating specified by the manufacturer for a single vehicle including its load.

GROUNDWATER GROUNDWATER

Subsurface water in the saturation zone from which wells and springs are fed. In a strict sense the term applies only to water below the water table. Also called "phreatic water" and "plerotic water."

GROUNDWATER DEPTH GROUNDWATER DEPTH

The distance of the groundwater table below the surface at any selected location.

GROUNDWATER ELEVATION GROUNDWATER ELEVATION

The elevation of the groundwater table above mean sea level at any selected location.

GROUNDWATER TABLE GROUNDWATER TABLE

The average depth or elevation of the groundwater over a selected area.

GROUNDWATER TABLE, ARTIFICIAL GROUNDWATER TABLE, ARTIFICIAL

(See ARTIFICIAL GROUNDWATER TABLE)

GROUNDWATER TABLE, SEASONAL GROUNDWATER TABLE, SEASONAL

(See SEASONAL WATER TABLE)

GROUNDWATER TABLE, TEMPORARY GROUNDWATER TABLE, TEMPORARY

(See TEMPORARY GROUNDWATER TABLE)

GROUT GROUT

A substance in a paste or liquid form which solidifies after placement or treatment. Used to fill spaces, holes or voids in other materials.

GROUT, CHEMICAL GROUT, CHEMICAL

(See CHEMICAL GROUT)

GROUTING, PRESSURE GROUTING, PRESSURE

(See CHEMICAL GROUTING)

GUNITE GUNITE

A mixture of sand and cement applied pneumatically that forms a high-density, resistant concrete.

H

HAIRLINE CRACK HAIRLINE CRACK

A stress crack in a pipe; the crack looks like a piece of hair.

HAND ROD HAND ROD

A sewer rod that can be inserted manually (by hand) into a sewer to clear a stoppage or to prevent a stoppage from developing.

HANDHOLE TRAP HANDHOLE TRAP

A device made of pipe fittings used to prevent sewer gases escaping from the branch or lateral sewer from entering a building sewer.

HAZARDOUS WASTE HAZARDOUS WASTE

Any waste material or mixture of wastes which is toxic, corrosive, flammable, an irritant, a strong sensitizer which generates pressure through decomposition, heat or other means, if such a waste or mixture of wastes may cause substantial personal injury, serious illness or harm to wildlife, during, or as a proximate result of any disposal of such wastes or mixture of wastes. The terms "toxic," "corrosive," "flammable," "irritant," and "strong sensitizer" shall be given the same meanings as given by the California Hazardous Substance Act (Chapter 13 (commencing with Section 28740) of Division 21)(Subsection 23117 of Article 2, Chapter 6.5, Division 20, of the California Health and Safety Code). Also see EXTREMELY HAZARDOUS WASTE.

HAZARDOUS WASTE HAZARDOUS WASTE

A waste, or combination of wastes, which because of its quantity, concentration, or physical, chemical, or infectious characteristics may:

1. Cause, or significantly contribute to, an increase in mortality or an increase in serious, irreversible, or incapacitating reversible, illness; or
2. Pose a substantial present or potential hazard to human health or the environment when improperly treated, stored, transported, or disposed of or otherwise managed; and
3. Normally not be discharged into a sanitary sewer; subject to regulated disposal.

(Resource Conservation and Recovery Act (RCRA) definition.)

HEAD HEAD

The vertical distance, height or energy of water above a point. A head of water may be measured in either height (feet) or pressure (pounds per square inch (psi)). Also see DISCHARGE HEAD, DYNAMIC HEAD, STATIC HEAD, SUCTION HEAD, SUCTION LIFT, and VELOCITY HEAD.

HERBICIDE (HERB-uh-SIDE) HERBICIDE

A compound, usually a manmade organic chemical, used to kill or control plant growth.

HERTZ HERTZ

The number of complete electromagnetic cycles or waves in one second of an electric or electronic circuit. Also called the frequency of the current. Abbreviated Hz.

HIGH-VELOCITY CLEANER HIGH-VELOCITY CLEANER

A machine designed to remove grease and debris from the smaller diameter sewer pipes with high-velocity jets of water. Also called a "jet cleaner," "jet rodder," "hydraulic cleaner," "high-pressure cleaner," or "hydro jet."

HOUSE CONNECTION HOUSE CONNECTION

(See BUILDING SEWER)

HOUSE SERVICE HOUSE SERVICE

(See BUILDING SERVICE)

HOUSE SEWER HOUSE SEWER

(See BUILDING SEWER)

HUB HUB

In pipe fitting, the enlarged female end of a pipe into which the male end fits. Also called a BELL.

HYDRAULIC BLOCK HYDRAULIC BLOCK

The movement of water in such a way that the flow of water from one direction blocks or hinders the flow of water from another direction.

HYDRAULIC CLEANER HYDRAULIC CLEANER

(See HIGH-VELOCITY CLEANER)

HYDRAULIC CLEANING HYDRAULIC CLEANING

Cleaning pipe with water under enough pressure to produce high water velocities.

(1) Using a high-velocity cleaner.

(2) Using a ball, kite or similar sewer cleaning device.

(3) Using a scooter.

(4) Flushing.

HYDRAULIC GRADE LINE (HGL) HYDRAULIC GRADE LINE (HGL)

The surface or profile of water flowing in an open channel or a pipe flowing partially full. If a pipe is under pressure, the hydraulic grade line is at the level water would rise to in a small tube connected to the pipe. To reduce the release of odors from sewers, the water surface or hydraulic grade line should be kept as smooth as possible. Also see ENERGY GRADE LINE.

[SEE DRAWING ON PAGE 534]

HYDRAULIC POPULATION EQUIVALENT HYDRAULIC POPULATION EQUIVALENT

A flow of 100 gallons per day is the hydraulic or flow equivalent to the contribution or flow from one person. Population equivalent = 100 GPCD or gallons per capita per day.

HYDROGEN SULFIDE GAS (H_2S) HYDROGEN SULFIDE GAS (H_2S)

Hydrogen sulfide is a gas with a rotten egg odor. This gas is produced under anaerobic conditions. Hydrogen sulfide gas is particularly dangerous because it dulls the sense of smell so that you don't notice it after you have been around it for a while. In high concentrations, hydrogen sulfide gas is only noticeable for a very short time before it dulls the sense of smell. The gas is very poisonous to the respiratory system, explosive, flammable, colorless and heavier than air.

HYDROLOGY HYDROLOGY

The applied science concerned with the waters of the earth in all their states—their occurrence, distribution, and circulation through the unending hydrologic cycle of precipitation, consequent runoff, stream flow, infiltration, and storage, eventual evaporation, and reprecipitation. Hydrology is concerned with the physical, chemical, and physiological reactions of water with the rest of the earth and its relation to the life of the earth.

HYDROPHILIC (HI-dro-FILL-ick) HYDROPHILIC

Having a strong affinity (liking) for water. The opposite of HYDROPHOBIC.

HYDROPHOBIC (HI-dro-FOE-bick) HYDROPHOBIC

Having a strong aversion (dislike) for water. The opposite of HYDROPHILIC.

I

IDLH IDLH

Immediately **D**angerous to **L**ife or **H**ealth. The atmospheric concentration of any toxic, corrosive, or asphyxiant substance that poses an immediate threat to life or would cause irreversible or delayed adverse health effects or would interfere with an individual's ability to escape from a dangerous atmosphere.

IMPELLER IMPELLER

A rotating set of vanes in a pump or compressor designed to pump or move water or air.

IMPORTED BACKFILL IMPORTED BACKFILL

Material used for backfilling a trench or excavation which was not the original material removed during excavation. This is a common practice where tests on the original material show it to have poor compactability or load capacity. Also called BORROW BACKFILL.

INDUSTRIAL TELEVISION EQUIPMENT INDUSTRIAL TELEVISION EQUIPMENT

(See INSPECTION TELEVISION EQUIPMENT)

INDUSTRIAL WASTEWATER INDUSTRIAL WASTEWATER

Liquid wastes originating from industrial processing. Because industries have peculiar liquid waste characteristics requiring special consideration, these sources are usually handled and treated separately before being discharged to a wastewater collection system.

INFILTRATED DEBRIS INFILTRATED DEBRIS

Sand, silt, gravel and rocks carried or washed into a collection system by infiltration water flows.

INFILTRATION (IN-fill-TRAY-shun) INFILTRATION

The seepage of groundwater into a sewer system, including service connections. Seepage frequently occurs through defective or cracked pipes, pipe joints, connections or manhole walls.

INFILTRATION HEAD INFILTRATION HEAD

The distance from a point of infiltration leaking into a collection system to the water table elevation. This is the pressure of the water being forced through the leak in the collection system.

INFILTRATION/INFLOW INFILTRATION/INFLOW

The total quantity of water from both infiltration and inflow without distinguishing the source. Abbreviated I & I or I/I.

INFILTRATION PRESSURE INFILTRATION PRESSURE

(See INFILTRATION HEAD)

INFLATABLE PIPE STOPPER INFLATABLE PIPE STOPPER

An inflatable ball or bag used to form a plug to stop flows in a sewer pipe.

INFLOW INFLOW

Water discharged into a sewer system and service connections from such sources as, but not limited to, roof leaders, cellars, yard and area drains, foundation drains, cooling water discharges, drains from springs and swampy areas, around manhole covers or through holes in the covers, cross connections from storm and combined sewer systems, catch basins, storm waters, surface runoff, street wash waters or drainage. Inflow differs from infiltration in that it is a direct discharge into the sewer rather than a leak in the sewer itself. See INTERNAL INFLOW.

INFLUENT INFLUENT

Wastewater or other liquid—raw (untreated) or partially treated—flowing *INTO* a reservoir, basin, treatment process, or treatment plant.

INLET INLET

(1) A surface connection to a drain pipe.

(2) A chamber for collecting storm water with no well below the outlet pipe for collecting grit. Often connected to a CATCH BASIN or a "basin manhole" ("cleanout manhole") with a grit chamber.

INORGANIC WASTE INORGANIC WASTE

Waste material such as sand, salt, iron, calcium, and other mineral materials which are only slightly affected by the action of organisms. Inorganic wastes are chemical substances of mineral origin; whereas organic wastes are chemical substances usually of animal or plant origin.

INSECTICIDE INSECTICIDE

Any substance or chemical formulated to kill or control insects.

INSERTION PULLER INSERTION PULLER

A device used to pull long segments of flexible pipe material into a sewer line when sliplining to rehabilitate a deteriorated sewer.

INSITUFORM INSITUFORM

A method of installing a new pipe within an old pipe without excavation. The process involves the use of a polyester-fiber felt tube, lined on one side with polyurethane and fully impregnated with a liquid thermal setting resin.

INSPECTION TELEVISION EQUIPMENT — INSPECTION TELEVISION EQUIPMENT

Television equipment that is superior to standard commercial quality, providing 600 to 650 lines of resolution, and designed for industrial inspection applications. Also known as INDUSTRIAL TELEVISION EQUIPMENT.

INTEGRATOR — INTEGRATOR

A device or meter that continuously measures and calculates (adds) a process rate variable in cumulative fashion; for example, total flows displayed in gallons, million gallons, cubic feet, or some other unit of volume measurement. Also called a TOTALIZER.

INTERCEPTING SEWER — INTERCEPTING SEWER

A sewer that receives flow from a number of other large sewers or outlets and conducts the waters to a point for treatment or disposal. Often called an "interceptor."

INTERCONNECTOR — INTERCONNECTOR

A sewer installed to connect two separate sewers. If one sewer becomes blocked, wastewater can back up and flow through the interconnector to the other sewer.

INTERNAL INFLOW — INTERNAL INFLOW

Nonsanitary or industrial wastewaters generated inside of a domestic, commercial or industrial facility and being discharged into the sewer system. Examples are cooling tower waters, basement sump pump discharge waters, continuous-flow drinking fountains, and defective or leaking plumbing fixtures.

INTRADOS — INTRADOS

The upper inside curve or surface of a sewer pipe or conduit.

INVERSION — INVERSION

An Insituform process in which the Insitutube or liner is turned inside out (inverted) during the installation of the liner.

INVERT (IN-vert) — INVERT

The lowest point of the channel inside a pipe or manhole. See FLOW LINE.

INVERTED SIPHON — INVERTED SIPHON

A pressure pipeline used to carry wastewater flowing in a gravity collection system under a depression such as a valley or roadway or under a structure such as a building. Also called a "depressed sewer."

J

JET CLEANER — JET CLEANER

(See HIGH-VELOCITY CLEANER)

JET RODDER — JET RODDER

(See HIGH-VELOCITY CLEANER)

JETSAM (JET-sam) — JETSAM

Debris entering a collection system which is heavier than water. Also see GRIT.

JOGGING — JOGGING

The frequent starting and stopping of an electric motor.

JOINT — JOINT

A connection between two lengths of pipe, made either with or without the use of another part.

K

KEY MANHOLE — KEY MANHOLE

In collection system evaluation, a key manhole is one from which reliable or specific data can be obtained.

KITE KITE

A device for hydraulically cleaning sewer lines. Resembling an airport wind sock and constructed of canvas-type material, the kite increases the velocity of a flow at its outlet to wash debris ahead of it. Also called a PARACHUTE.

L

LAMP HOLE LAMP HOLE

A small vertical pipe or shaft extending from the surface of the ground to a sewer. A light (or lamp) may be lowered down the pipe for the purpose of inspecting the sewer. Rarely constructed today.

LAMPING LAMPING

Using reflected sunlight or a powerful light beam to inspect a sewer between two adjacent manholes. The light is directed down the pipe from one manhole. If it can be seen from the next manhole, it indicates that the line is open and straight.

LATERAL LATERAL

(See LATERAL SEWER)

LATERAL BREAK LATERAL BREAK

A break in a lateral pipe somewhere between the sewer main and the building connection.

LATERAL CLEANOUT LATERAL CLEANOUT

A capped opening in a building lateral, usually located on the property line, through which the pipelines can be cleaned.

LATERAL CONNECTION LATERAL CONNECTION

(See BUILDING SERVICE)

LATERAL SEWER LATERAL SEWER

A sewer that discharges into a branch or other sewer and has no other common sewer tributary to it. Sometimes called a "street sewer" because it collects wastewater from individual homes.

LEAD (LEE-d) LEAD

A wire or conductor that can carry electric current.

LIFE-CYCLE COSTING LIFE-CYCLE COSTING

An economic analysis procedure that considers the total costs associated with a sewer during its economic life, including development, construction, and operation and maintenance (includes chemical and energy costs). All costs are converted to a present worth or present cost in dollars.

LIFT STATION LIFT STATION

A wastewater pumping station that lifts the wastewater to a higher elevation when continuing the sewer at reasonable slopes would involve excessive depths of trench. Also, an installation of pumps that raise wastewater from areas too low to drain into available sewers. These stations may be equipped with air-operated ejectors or centrifugal pumps. Sometimes called a PUMP STATION, but this term is usually reserved for a similar type of facility that is discharging into a long FORCE MAIN, while a lift station has a discharge line or force main only up to the downstream gravity sewer. Throughout this manual when we refer to lift stations, we intend to include pump stations.

LINER LINER

(See PIPE LINER)

LIQUID VEHICLE LIQUID VEHICLE

Water in a collection system that is used to carry waste solids. The standard toilet provides around seven gallons of water per flush as a vehicle to carry wastes through the pipe system.

LIQUOR LIQUOR

Water, wastewater, or any combination; commonly used to mean the liquid portion when other wastes are also present.

LOGARITHM (LOG-a-rith-m) LOGARITHM

The exponent that indicates the power to which a number must be raised to produce a given number. For example: if $B^2 = N$, the 2 is the logarithm of N (to the base B), or $10^2 = 100$ and $\log_{10} 100 = 2$. Also abbreviated to "log."

LOGARITHMIC (LOG-a-RITH-mick) SCALE LOGARITHMIC SCALE

A scale on which actual distances from the origin are proportional to the logarithms of the corresponding scale numbers rather than to the numbers themselves. A logarithmic scale has the numbers getting bigger as the distances between the numbers decrease.

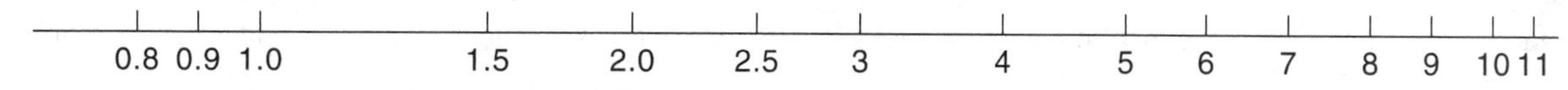

LONGITUDINAL (LAWN-ji-TWO-da-null) CRACK LONGITUDINAL CRACK

A crack in a pipe or pipe section that runs lengthwise along the pipe.

LOWER EXPLOSIVE LIMIT (LEL) LOWER EXPLOSIVE LIMIT (LEL)

The lowest concentration of gas or vapor (percent by volume in air) that explodes if an ignition source is present at ambient temperature. At temperatures above 250°F the LEL decreases because explosibility increases with higher temperature.

LUBRIFLUSHING (LOOB-rah-FLUSH-ing) LUBRIFLUSHING

A method of lubricating bearings with grease. Remove the relief plug and apply the proper lubricant to the bearing at the lubrication fitting. Run the pump to expel excess lubricant.

M

M-ZONED M-ZONED

An area set aside for manufacturing and industry.

MG MG

Initials for "Million Gallons."

MGD MGD

Initials for "Million Gallons Per Day."

mg/*L* mg/*L*

(See MILLIGRAMS PER LITER, mg/*L*)

MGY MGY

Initials for "Million Gallons Per Year."

MSDS MSDS

Material **S**afety **D**ata **S**heet. A document which provides pertinent information and a profile of a particular hazardous substance or mixture. An MSDS is normally developed by the manufacturer or formulator of the hazardous substance or mixture. The MSDS is required to be made available to employees and operators whenever there is the likelihood of the hazardous substance or mixture being introduced into the workplace. Some manufacturers are preparing MSDSs for products that are not considered to be hazardous to show that the product or substance is *NOT* hazardous.

MAIN LINE MAIN LINE

Branch or lateral sewers that collect wastewater from building sewers and service lines.

MAIN SEWER MAIN SEWER

A sewer line that receives wastewater from many tributary branches and sewer lines and serves as an outlet for a large territory or is used to feed an intercepting sewer.

MANDREL (MAN-drill) MANDREL

(1) A special tool used to push bearings in or to pull sleeves out.

(2) A gage used to measure for excessive deflection in a flexible conduit.

MANHOLE MANHOLE

An opening in a sewer provided for the purpose of permitting operators or equipment to enter or leave a sewer. Sometimes called an "access hole," or a "maintenance hole."

MANHOLE BEDDING MANHOLE BEDDING

The prepared and compacted base on which a manhole is constructed.

MANHOLE DEPTH MANHOLE DEPTH

The measurement from the top of the manhole opening to the invert or lowest point of the trough at the bottom of the manhole.

MANHOLE, DROP MANHOLE, DROP

(See DROP MANHOLE)

MANHOLE ELEVATION MANHOLE ELEVATION

The height (elevation) of the invert or lowest point in the bottom of a manhole above mean sea level.

MANHOLE FLOW MANHOLE FLOW

(1) The depth or amount of wastewater flow in a manhole as observed at any selected time.

(2) The total or the average flow through a manhole in gallons in any selected time interval.

MANHOLE FRAME MANHOLE FRAME

A metal ring or frame with a ledge to accommodate the manhole lid; located at the surface of the ground or street. Also called a "manhole ring."

MANHOLE GRADE RING MANHOLE GRADE RING

A precast concrete ring 4 to 12 inches high which is placed on top of a manhole cone to raise the manhole cover frame flush with the surface grade. Sometimes called a "spacer."

MANHOLE INFILTRATION MANHOLE INFILTRATION

Groundwaters seeping or leaking into a manhole structure.

MANHOLE INFLOW MANHOLE INFLOW

Surface waters flowing into a manhole, usually through the vent holes in the manhole lid.

MANHOLE INVERT MANHOLE INVERT

The lowest point in a trough or flow channel in the bottom of a manhole.

MANHOLE JACK MANHOLE JACK

A device used to guide the tag line into the sewer without causing unnecessary wear and provide support as the tag line is pulled back and forth.

MANHOLE, KEY MANHOLE, KEY

(See KEY MANHOLE)

MANHOLE LID MANHOLE LID

The heavy cast-iron or forged-steel cover of a manhole. The lid may or may not have vent holes.

MANHOLE LID DUST PAN MANHOLE LID DUST PAN

A sheet metal or cast-iron pan located under a manhole lid. This pan serves to catch and hold pebbles and other debris falling through vent holes, preventing them from getting into the pipe system.

MANHOLE RING MANHOLE RING

A metal frame or ring with a ledge to accommodate the lid and located at the surface of the ground or street. Also called a "manhole frame."

MANHOLE SEALING MANHOLE SEALING

The process of sealing infiltration leaks in a manhole by injecting chemical grout.

MANHOLE, SURCHARGED — MANHOLE, SURCHARGED

(See SURCHARGED MANHOLE)

MANHOLE TOOLS — MANHOLE TOOLS

(1) Special tools having conveniently short handles for working inside manholes.
(2) Special long-handled or extendable tools for removal of debris and other objects from manholes without requiring a person to enter the manhole.

MANHOLE TROUGH — MANHOLE TROUGH

The channel in the bottom of a manhole for the flow of the wastewater from manhole inlet to outlet.

MANHOLE VENTS — MANHOLE VENTS

One or a series of one-inch diameter holes through a manhole lid for purposes of venting dangerous gases found in sewers.

MANNING'S FORMULA — MANNING'S FORMULA

A mathematical formula for calculating wastewater flows in sewers.

$$Q = \frac{1.49}{n} A R^{2/3} S^{1/2}$$

Q means flow in cubic feet per second (CFS).
n means the Manning pipe or channel roughness factor.
A means the cross-sectional area of the flow in square feet (sq ft).
R means the hydraulic radius in feet (ft) where R equals A/P. P is the wetted perimeter of the channel or pipe in feet.
S means the slope of the channel or energy grade line in feet per foot (ft/ft).

MANNING'S TABLES — MANNING'S TABLES

A set of tables for finding wastewater flows in sewers.

MANOMETER (man-NAH-mut-ter) — MANOMETER

Usually a glass tube filled with a liquid and used to measure the difference in pressure across a flow measuring device such as an orifice or a Venturi meter.

MATERIAL SAFETY DATA SHEET (MSDS) — MATERIAL SAFETY DATA SHEET (MSDS)

A document which provides pertinent information and a profile of a particular hazardous substance or mixture. An MSDS is normally developed by the manufacturer or formulator of the hazardous substance or mixture. The MSDS is required to be made available to employees and operators whenever there is the likelihood of the hazardous substance or mixture being introduced into the workplace. Some manufacturers are preparing MSDSs for products that are not considered to be hazardous to show that the product or substance is *NOT* hazardous.

MEASURED FLOW — MEASURED FLOW

A flow which has been physically measured. See GAGE.

MECHANICAL CLEANING — MECHANICAL CLEANING

Clearing pipe by using equipment that scrapes, cuts, pulls or pushes the material out of the pipe. Mechanical cleaning devices or machines include bucket machines, power rodders and hand rods.

MECHANICAL PLUG — MECHANICAL PLUG

A pipe plug used in sewer systems that is mechanically expanded to create a seal.

MEGGER (from megohm) — MEGGER

An instrument used for checking the insulation resistance on motors, feeders, bus bar systems, grounds, and branch circuit wiring.

MEGOHM — MEGOHM

Meg means one million, so 5 megohms means 5 million ohms. A megger reads in millions of ohms.

MERCAPTANS (mer-CAP-tans) — MERCAPTANS

Compounds containing sulfur which have an extremely offensive skunk-like odor; also sometimes described as smelling like garlic or onions.

METERED | METERED

Measured through a meter, as a quantity of water or flow might be measured.

MICROORGANISMS (MY-crow-OR-gan-IS-zums) | MICROORGANISMS

Very small organisms that can be seen only through a microscope. Some microorganisms use the wastes in wastewater for food and thus remove or alter much of the undesirable matter.

MILLIGRAMS PER LITER, mg/*L* | MILLIGRAMS PER LITER, mg/*L*

A measure of the concentration by weight of a substance per unit volume in water or wastewater. In reporting the results of water and wastewater analysis, mg/*L* is preferred to the unit parts per million (ppm), to which it is approximately equivalent.

MILLION GALLONS | MILLION GALLONS

A unit of measurement used in wastewater treatment plant design and collection system capacities or performances. One million gallons of water is approximately equivalent to:

13,690	Cubic Feet
3.07	Acre-Feet
8,340,000	Pounds of Weight
4,170	Tons of Weight
3,785	Cubic Meters

MINERAL | MINERAL

Any substance that is neither animal nor plant. Minerals include sand, salt, iron, calcium, and nutrients.

MINERAL CONTENT | MINERAL CONTENT

The quantity of dissolved minerals in a sample of water.

MONITOR | MONITOR

(See TELEVISION MONITOR)

N

"N" FACTOR | "N" FACTOR

A coefficient value representing the pipe or channel roughness in Manning's formula for computing flows in gravity sewers. See FRICTION LOSS and MANNING'S FORMULA.

NIOSH (NYE-osh) | NIOSH

The **N**ational **I**nstitute of **O**ccupational **S**afety and **H**ealth is an organization that tests and approves safety equipment for particular applications. NIOSH is the primary federal agency engaged in research in the national effort to eliminate on-the-job hazards to the health and safety of working people. The NIOSH Publications Catalog, Sixth Edition, NIOSH Pub. No. 84-118, lists the NIOSH publications concerning industrial hygiene and occupational health. To obtain a copy of the catalog, write to National Technical Information Service (NTIS), 5285 Port Royal Road, Springfield, VA 22161. NTIS Stock No. PB86-116787, price, $103.50, plus $5.00 shipping and handling per order.

NPDES PERMIT | NPDES PERMIT

National **P**ollutant **D**ischarge **E**limination **S**ystem permit is the regulatory agency document issued by either a federal or state agency which is designed to control all discharges of potential pollutants from point sources and storm water runoff into U.S. waterways. NPDES permits regulate discharges into navigable waters from all point sources of pollution, including industries, municipal wastewater treatment plants, sanitary landfills, large agricultural feedlots and return irrigation flows.

NAMEPLATE | NAMEPLATE

A durable metal plate found on equipment which lists critical operating conditions for the equipment.

NET WASTEWATER CONTRIBUTION | NET WASTEWATER CONTRIBUTION

In a wastewater collection system, the net wastewater contribution consists of the liquid wastes and liquid-carried wastes transported by the pipelines or received by the pipelines. This value would be the only wastewater found in a collection system if all sources of infiltration, inflow and exfiltration were eliminated.

NET WASTEWATER FLOW — NET WASTEWATER FLOW

The actual wastewater flow from a collection system that reaches a wastewater treatment plant. The net wastewater flow includes the net wastewater contribution, infiltration and inflow and does not include losses through exfiltration.

NODULAR (NOD-you-lar) — NODULAR

Shaped like a rounded lump, knot or knob. An irregular enlargement.

NOMINAL DIAMETER — NOMINAL DIAMETER

An approximate measurement of the diameter of a pipe. Although the nominal diameter is used to describe the size or diameter of a pipe, it is usually not the exact inside diameter of the pipe.

NOMOGRAPH (NOME-o-graph) — NOMOGRAPH

A graphic representation or means of solving an equation or mathematical relationship. Results are obtained with the aid of a straightedge placed over important known values.

NON-PERMIT CONFINED SPACE — NON-PERMIT CONFINED SPACE

See CONFINED SPACE, NON-PERMIT.

NONSPARKING TOOLS — NONSPARKING TOOLS

These tools will not produce a spark during use. They are made of a nonferrous material, usually a copper-beryllium alloy.

NOTICE — NOTICE

This word calls attention to information that is especially significant in understanding and operating equipment or processes safely. Also see CAUTION, DANGER, and WARNING.

O

OSHA (O-shuh) — OSHA

The Williams-Steiger Occupational Safety and Health Act of 1970 (OSHA) is a federal law designed to protect the health and safety of industrial workers and collection system operators. The Act regulates the design, construction, operation and maintenance of industrial plants and wastewater collection and treatment facilities. The Act does not apply directly to municipalities, *EXCEPT* in those states that have approved plans and have asserted jurisdiction under Section 18 of the OSHA Act. *HOWEVER, CONTRACT OPERATORS AND PRIVATE FACILITIES DO HAVE TO COMPLY WITH OSHA REQUIREMENTS.* Wastewater collection systems have come under stricter regulation in all phases of activity as a result of OSHA standards. OSHA also refers to the federal and state agencies which administer the OSHA regulations.

OBSTRUCTION — OBSTRUCTION

Any solid object in or protruding into a wastewater flow in a collection line that prevents a smooth or even passage of the wastewater.

OFF LINE — OFF LINE

A run of sewer pipe between two manholes is said to be "off line" if it is not located directly under a straight line passing through the exact centers of the two manholes. Sewer alignment does not always pass through the center of a manhole, especially at junctions. Also called "misaligned."

OFFSET — OFFSET

(1) A combination of elbows or bends which brings one section of a line of pipe out of line with, but into a line parallel with, another section.

(2) A pipe fitting in the approximate form of a reverse curve, made to accomplish the same purpose.

(3) A pipe joint that has lost its bedding support and one of the pipe sections has dropped or slipped, thus creating a condition where the pipes no longer line up properly.

OFFSET INVERT — OFFSET INVERT

A trough or channel in the bottom of a manhole which is not centered in the bottom.

OFFSET JOINT — OFFSET JOINT

A pipe joint that is not exactly in line and centered. See DROP JOINT and VERTICAL OFFSET.

OFFSET MANHOLE — OFFSET MANHOLE

A manhole located to one side of a pipe run with either "Y" connections to it or the inlet and outlet pipes bent to enter and leave the manhole.

OFFSET PIPE — OFFSET PIPE

(See OFF LINE)

OFFSET TROUGH — OFFSET TROUGH

(1) When the pipe feeding into a manhole does not exactly match up with the pipe leading out of the manhole, the invert channel must be angled or curved. This is referred to as an "offset trough."

(2) (See OFFSET INVERT)

OHM — OHM

The unit of electrical resistance. The resistance of a conductor in which one volt produces a current of one ampere.

OLFACTORY (ol-FAK-tore-ee) FATIGUE — OLFACTORY FATIGUE

A condition in which a person's nose, after exposure to certain odors, is no longer able to detect the odor.

ORGANIC WASTE — ORGANIC WASTE

Waste material which comes mainly from animal or plant sources. Organic wastes generally can be consumed by bacteria and other small organisms. Inorganic wastes are chemical substances of mineral origin.

ORIFICE (OR-uh-fiss) — ORIFICE

An opening (hole) in a plate, wall, or partition. An orifice flange or plate placed in a pipe consists of a slot or a calibrated circular hole smaller than the pipe diameter. The difference in pressure in the pipe above and at the orifice may be used to determine the flow in the pipe.

OUTFALL — OUTFALL

(1) The point, location or structure where wastewater or drainage discharges from a sewer, drain, or other conduit.

(2) The conduit leading to the final disposal point or area. See OUTFALL SEWER.

OUTFALL SEWER — OUTFALL SEWER

A sewer that receives wastewater from a collection system or from a wastewater treatment plant and carries it to a point of ultimate or final discharge in the environment. See OUTFALL.

OUTLET — OUTLET

Downstream opening or discharge end of a pipe, culvert, or canal.

OVERFLOW MANHOLE — OVERFLOW MANHOLE

A manhole which fills and allows raw wastewater to flow out onto the street or ground.

OVERFLOW RELIEF LINE — OVERFLOW RELIEF LINE

Where a system has overload conditions during peak flows, an outlet may be installed above the invert and leading to a less loaded manhole or part of the system. This is usually called an "overflow relief line." Also see CROSS CONNECTION (2).

OXIDATION (ox-uh-DAY-shun) — OXIDATION

Oxidation is the addition of oxygen, removal of hydrogen, or the removal of electrons from an element or compound. In wastewater treatment, organic matter is oxidized to more stable substances. The opposite of REDUCTION.

OXIDATION-REDUCTION POTENTIAL (ORP) — OXIDATION-REDUCTION POTENTIAL (ORP)

The electrical potential required to transfer electrons from one compound or element (the oxidant) to another compound or element (the reductant); used as a qualitative measure of the state of oxidation in wastewater treatment systems. ORP is measured in millivolts, with negative values indicating a tendency to reduce compounds or elements and positive values indicating a tendency to oxidize compounds or elements.

OXYGEN DEFICIENCY — OXYGEN DEFICIENCY

An atmosphere containing oxygen at a concentration of less than 19.5 percent by volume.

OXYGEN ENRICHMENT — OXYGEN ENRICHMENT

An atmosphere containing oxygen at a concentration of more than 23.5 percent by volume.

P

POTW — POTW

Publicly **O**wned **T**reatment **W**orks. A treatment works which is owned by a state, municipality, city, town, special sewer district or other publicly owned and financed entity as opposed to a privately (industrial) owned treatment facility. This definition includes any devices and systems used in the storage, treatment, recycling and reclamation of municipal sewage (wastewater) or industrial wastes of a liquid nature. It also includes sewers, pipes and other conveyances only if they carry wastewater to a POTW treatment plant. The term also means the municipality (public entity) which has jurisdiction over the indirect discharges to and the discharges from such a treatment works.

PPM — PPM

Initials for "Parts Per Million." The number of weight or volume units of a minor constituent present with each one million units of the major constituent of a solution or mixture. Used to express the results of most water and wastewater analyses, but more recently milligrams per liter (mg/*L*) is the preferred term.

PACKER — PACKER

(See CHEMICAL GROUTING)

PACKING RING — PACKING RING

A ring made of asbestos or metal which may be lubricated with Teflon or graphite that forms a seal between the pump shaft and its casing.

PARACHUTE — PARACHUTE

A device used to catch wastewater flow to pull a float line between manholes. See FLOAT LINE.

PARSHALL FLUME — PARSHALL FLUME

A specially constructed flume or channel used to measure flows in open channels.

PATHOGENIC (PATH-o-JEN-ick) ORGANISMS — PATHOGENIC ORGANISMS

Bacteria, viruses, or cysts which can cause disease (giardiasis, cryptosporidiosis, typhoid, cholera, dysentery) in a host (such as a person). There are many types of organisms which do *NOT* cause disease and which are *NOT* called pathogenic. Many beneficial bacteria are found in wastewater treatment processes actively cleaning up organic wastes.

PEAKING FACTOR — PEAKING FACTOR

Ratio of a maximum flow to the average flow, such as maximum hourly flow or maximum daily flow to the average daily flow.

PENTA HOSE — PENTA HOSE

A hose with five chambers or tubes.

PERCOLATION (PURR-co-LAY-shun) — PERCOLATION

The movement or flow of water through soil or rocks.

PERMIT-REQUIRED CONFINED SPACE (PERMIT SPACE) — PERMIT-REQUIRED CONFINED SPACE (PERMIT SPACE)

See CONFINED SPACE, PERMIT-REQUIRED (PERMIT SPACE).

PESTICIDE — PESTICIDE

Any substance or chemical designed or formulated to kill or control animal pests. Also see INSECTICIDE and RODENTICIDE.

PHOTOGRAPHIC INSPECTIONS — PHOTOGRAPHIC INSPECTIONS

A method of obtaining photographs of a pipeline by pulling a time-lapse motion picture camera through the line. By moving the camera a specific distance at timed intervals, a sequence of photographs covering the full length of the line is obtained.

PHOTOGRAPHIC RECORDS — PHOTOGRAPHIC RECORDS

(1) The film strip from a photographic inspection.

(2) Still camera photographs of a sewer television inspection monitor.

PIEZOMETER (pie-ZOM-uh-ter) PIEZOMETER

An instrument used to measure the pressure head in a pipe, tank, or soil. It usually consists of a small pipe or tube connected or tapped into the side or wall of a pipe or tank and connected to a manometer pressure gage, water or mercury column, or other device for indicating pressure head.

PIG PIG

Refers to a poly pig which is a bullet-shaped device made of hard rubber or similar material. This device is used to clean pipes. It is inserted in one end of a pipe, moves through the pipe under pressure, and is removed from the other end of the pipe.

PIPE BEDDING PIPE BEDDING

(See BEDDING and BEDDING GRADE)

PIPE BUCKET PIPE BUCKET

(See BUCKET and BUCKET MACHINE)

PIPE CAPACITY PIPE CAPACITY

In a gravity-flow sewer system, pipe capacity is the total amount in gallons a pipe is able to pass in a specific time period.

PIPE CLEANING PIPE CLEANING

Removing grease, grit, roots and other debris from a pipe run by means of one of the hydraulic cleaning methods. See BALLING, HYDRAULIC CLEANING and KITE.

NOTE: While rodding may be used to open a pipe stoppage or to remove roots and greases, it is not known to remove grit and similar debris.

PIPE DEFLECTION PIPE DEFLECTION

(See DEFLECTED)

PIPE DIAMETER PIPE DIAMETER

The nominal or commercially designated inside diameter of a pipe, unless otherwise stated.

PIPE DIP PIPE DIP

(See DIP)

PIPE DISPLACEMENT PIPE DISPLACEMENT

(1) The cubic inches of soil or water displaced by one foot or one section of pipe.

(2) (See DISPLACED PIPE)

PIPE GRADE PIPE GRADE

The angle of a sewer or a single section of a sewer as installed. Usually expressed in a percentage figure to indicate the drop in feet or tenths of a foot per hundred feet. For example, 0.5 percent grade means a drop of one-half foot per 100 feet of length.

PIPE JACK PIPE JACK

A jack used to fasten roller guides to secure an object within a manhole.

PIPE JOINT PIPE JOINT

A place where two sections of pipe are coupled or joined together.

PIPE JOINT SEAL PIPE JOINT SEAL

(1) The tightness or lack of leakage at a pipe joint.

(2) The method of sealing a pipe coupling.

PIPE LINER PIPE LINER

A plastic liner pulled or pushed into a pipe to eliminate excessive infiltration or exfiltration. Other solutions to the problem of infiltration/exfiltration are the use of a cement grouting or replacement of damaged pipe.

PIPE PLUG PIPE PLUG

(1) A temporary plug placed in a sewer pipe to stop a flow while repair work is being accomplished or other functions are performed.

(2) In construction of a new sewer system, service saddles are sometimes installed before a building or a building lateral is in existence. Under such circumstances, a plug will be placed in the off-lead of the saddle of a "Y." In some instances, this plug may be called a "button" or a "stopper."

PIPE PLUG, INFLATABLE PIPE PLUG, INFLATABLE

(See INFLATABLE PIPE STOPPER)

PIPE PLUG, MECHANICAL PIPE PLUG, MECHANICAL

(See MECHANICAL PLUG)

PIPE RODDING PIPE RODDING

A method of opening a plugged or blocked pipe by pushing a steel rod or snake, or pulling same, through the pipe with a tool attached to the end of the rod or snake. Rotating the rod or snake with a tool attached increases effectiveness.

PIPE RUN PIPE RUN

(1) The length of sewer pipe reaching from one manhole to the next. See BARREL.

(2) Any length of pipe, generally assumed to be in a straight line.

PIPE SEAL PIPE SEAL

(See PIPE JOINT SEAL)

PIPE SEALING PIPE SEALING

(See CHEMICAL GROUTING, GUNITE or PIPE LINER)

PIPE SECTION PIPE SECTION

A single length of pipe between two joints or couplers.

PIPE SINK PIPE SINK

(See DIP)

PLAN PLAN

A drawing showing the *TOP* view of sewers, manholes and streets.

PLANT PLANT

(See WASTEWATER TREATMENT PLANT)

PLANT HYDRAULIC CAPACITY PLANT HYDRAULIC CAPACITY

The flow or load, in millions of gallons per day (or portion thereof), that a treatment plant is designed to handle.

PLANT, TREATMENT PLANT, TREATMENT

(See WASTEWATER TREATMENT PLANT)

PLUG PLUG

(See INFLATABLE PIPE STOPPER and MECHANICAL PLUG)

PNEUMATIC EJECTOR (new-MAT-tik ee-JECK-tor) PNEUMATIC EJECTOR

A device for raising wastewater, sludge or other liquid by compressed air. The liquid is alternately admitted through an inward-swinging check valve into the bottom of an airtight pot. When the pot is filled compressed air is applied to the top of the liquid. The compressed air forces the inlet valve closed and forces the liquid in the pot through an outward-swinging check valve, thus emptying the pot.

POLE SHADER POLE SHADER

A copper bar circling the laminated iron core inside the coil of a magnetic starter.

POLLUTION POLLUTION

The impairment (reduction) of water quality by agricultural, domestic or industrial wastes (including thermal and radioactive wastes) to a degree that the natural water quality is changed to hinder any beneficial use of the water or render it offensive to the senses of sight, taste, or smell or when sufficient amounts of wastes create or pose a potential threat to human health or the environment.

POLYELECTROLYTE (POLY-ee-LECK-tro-lite) POLYELECTROLYTE

A high-molecular-weight substance that is formed by either a natural or synthetic process. Natural polyelectrolytes may be of biological origin or derived from starch products, cellulose derivatives, and alignates. Synthetic polyelectrolytes consist of simple substances that have been made into complex, high-molecular-weight substances. Often called a POLYMER.

POLYMER (POLY-mer) POLYMER

A long chain molecule formed by the union of many monomers (molecules of lower molecular weight). Polymers are used with other chemical coagulants to aid in binding small suspended particles to larger chemical flocs for their removal from water.

POPULATION EQUIVALENT (HYDRAULIC) POPULATION EQUIVALENT (HYDRAULIC)

A flow of 100 gallons per day is the hydraulic or flow equivalent to the contribution or flow from one person. Population equivalent = 100 GPCD or gallons per capita per day.

PORCUPINE PORCUPINE

A sewer cleaning tool the same diameter as the pipe being cleaned. The tool is a steel cylinder having solid ends with eyes cast in them to which a cable can be attached and pulled by a winch. Many short pieces of cable or bristles protrude from the cylinder to form a round brush.

POTABLE (POE-tuh-bull) WATER POTABLE WATER

Water that does not contain objectionable pollution, contamination, minerals, or infective agents and is considered satisfactory for drinking.

POWER RODDER POWER RODDER

A machine designed to remove roots, grease, and other materials from pipes. Also referred to as rodding machines.

PRECIPITATE (pre-SIP-uh-TATE) PRECIPITATE

(1) An insoluble, finely divided substance which is a product of a chemical reaction within a liquid.

(2) The separation from solution of an insoluble substance.

PRECIPITATION PRECIPITATION

(1) The total measurable supply of water received directly from clouds as rain, snow, hail, or sleet; usually expressed as depth in a day, month, or year, and designated as daily, monthly, or annual precipitation.

(2) The process by which atmospheric moisture is discharged onto a land or water surfaces.

(3) The separation (of a substance) out in solid form from a solution, as by the use of a reagent.

PRE-CLEANING PRE-CLEANING

Sewer line cleaning, commonly done by high-velocity cleaners, that is done prior to the TV inspection of a pipeline to remove grease, slime, and grit to allow for a clearer and more accurate identification of defects and problems.

PREDICTIVE MAINTENANCE PREDICTIVE MAINTENANCE

The ability to identify problem areas before breakdowns or blockage of flow occurs. Predictive maintenance is the end product of effective preventive maintenance.

PRESENT WORTH PRESENT WORTH

The value of a long-term project expressed in today's dollars. Present worth is calculated by converting (discounting) all future benefits and costs over the life of the project to a single economic value at the start of the project. Calculating the present worth of alternative projects makes it possible to compare them and select the one with the largest positive (beneficial) present worth or minimum present cost.

PRESSURE GROUTING PRESSURE GROUTING

(See CHEMICAL GROUTING)

PRESSURE HEAD — PRESSURE HEAD

(1) The height of a water surface above a specific point of reference. Usually measured in feet and tenths of a foot.

(2) The head represented by the expression of pressure over weight (p/w), where p is pressure (lbs/sq ft) and w is weight (lbs/cu ft).

PRESSURE MAIN — PRESSURE MAIN

(See FORCE MAIN)

PREVENTIVE MAINTENANCE — PREVENTIVE MAINTENANCE

Regularly scheduled servicing of machinery or other equipment using appropriate tools, tests and lubricants. This type of maintenance can prolong the useful life of equipment and machinery and increase its efficiency by detecting and correcting problems before they cause a breakdown of the equipment.

PREVENTIVE MAINTENANCE UNITS — PREVENTIVE MAINTENANCE UNITS

Crews assigned the task of cleaning sewers (for example, balling or high-velocity cleaning crews) to prevent stoppages and odor complaints. Preventive maintenance is performing the most effective cleaning procedure, in the area where it is most needed, at the proper time in order to prevent failures and emergency situations.

PROBE — PROBE

A T-shaped tool or rod that is pushed or driven down through the soil to locate underground pipes and utility conduits. Also see SOUNDING ROD.

PROFILE — PROFILE

A drawing showing the *SIDE* view of sewers and manholes.

PROMOTED — PROMOTED

The mixture of resin and catalyst ready to cause (promote) curing in place.

PROTRUDING SERVICE — PROTRUDING SERVICE

The connection of a building lateral to a main sewer line whereby a hole is cut in the main and the end of the building lateral is allowed to extend into the main.

PROTRUDING TAP — PROTRUDING TAP

(See PROTRUDING SERVICE)

PUMP — PUMP

A mechanical device for causing flow, for raising or lifting water or other fluid, or for applying pressure to fluids.

PUMP PIT — PUMP PIT

A dry well, chamber or room below ground level in which a pump is located.

PUMP STATION — PUMP STATION

Installation of pumps to lift wastewater to a higher elevation in places where flat land would require excessively deep sewer trenches. Also used to raise wastewater from areas too low to drain into available collection lines. These stations may be equipped with air-operated ejectors or centrifugal pumps. See LIFT STATION.

PUTREFACTION (PEW-truh-FACK-shun) — PUTREFACTION

Biological decomposition of organic matter with the production of foul-smelling products associated with anaerobic conditions.

PUTRESCIBLE (pew-TRES-uh-bull) — PUTRESCIBLE

Material that will decompose under anaerobic conditions and produce nuisance odors.

Q

QUADRANT — QUADRANT

Any of the four more or less equal parts into which something can be divided by two real or imaginary lines that intersect each other at right angles. In television inspection of pipes, for example, the picture of the pipe can be divided into fourths (upper-left, lower-left, upper-right, and lower-right quadrants) to identify the location of observed objects.

QUALITATIVE FIT TEST (QLFT) — QUALITATIVE FIT TEST (QLFT)

A pass/fail fit test to assess the adequacy of respirator fit that relies on the individual's response to the test agent.

QUANTITATIVE FIT TEST (QNFT) — QUANTITATIVE FIT TEST (QNFT)

An assessment of the adequacy of respirator fit that relies on the individual's response to the test agent.

QUICK AIR TEST — QUICK AIR TEST

The same as a quick test with a packer for chemical grouting except that air pressure is used in place of liquid for a faster test and greater accuracy. Also see CHEMICAL GROUTING.

QUICK TEST — QUICK TEST

Use of a packer designed for chemical grouting to pressure test any selected small area of pipeline. Also see CHEMICAL GROUTING.

QUICKSAND — QUICKSAND

Sand that has lost its grain-to-grain contact by the buoyancy effect of water flowing upward through the voids. Such material, having some of the characteristics of a fluid, possesses no load-bearing value.

R

"R" FACTOR — "R" FACTOR

Refers to pipe or channel "Roughness Factor." See FRICTION LOSS, "N" FACTOR, and ROUGHNESS COEFFICIENT.

R-ZONED — R-ZONED

Areas established for residential occupancy.

RAIN — RAIN

Particles of liquid water that have become too large to be held by the atmosphere. Their diameter generally is greater than 0.02 inch and they usually fall to the earth at velocities greater than 10 fps in still air. See PRECIPITATION.

RECORDINGS, FLOW — RECORDINGS, FLOW

(See FLOW RECORDING)

RECORDINGS, PHOTOGRAPHIC — RECORDINGS, PHOTOGRAPHIC

(See PHOTOGRAPHIC RECORDS)

RECORDINGS, VIDEO — RECORDINGS, VIDEO

(See VIDEO LOG)

RECORDINGS, WRITTEN — RECORDINGS, WRITTEN

(See TELEVISION INSPECTION LOG)

REDUCTANT — REDUCTANT

A constituent of wastewater or surface waters that uses either free (O_2) or combined oxygen in the process of stabilization.

REDUCTION (re-DUCK-shun) — REDUCTION

Reduction is the addition of hydrogen, removal of oxygen, or the addition of electrons to an element or compound. Under anaerobic conditions (no dissolved oxygen present), sulfur compounds are reduced to odor-producing hydrogen sulfide (H_2S) and other compounds. The opposite of OXIDATION.

REGULATOR — REGULATOR

A device used in combined sewers to control or regulate the diversion of flow.

RELIEF BYPASS — RELIEF BYPASS

(See BYPASS and OVERFLOW RELIEF LINE)

RELIEF LINE — RELIEF LINE

(See OVERFLOW RELIEF LINE)

RESISTANCE — RESISTANCE

That property of a conductor or wire that opposes the passage of a current, thus causing electric energy to be transformed into heat.

RETENTION — RETENTION

(1) That part of the precipitation falling on a drainage area which does not escape as surface stream flow during a given period. It is the difference between total precipitation and total runoff during the period, and represents evaporation, transpiration, subsurface leakage, infiltration, and, when short periods are considered, temporary surface or underground storage on the area.

(2) The delay or holding of the flow of water and water-carried wastes in a pipe system. This can be due to a restriction in the pipe, a stoppage or a dip. Also, the time water is held or stored in a basin or wet well. This is also called DETENTION.

ROD GUIDE — ROD GUIDE

A bent pipe inserted in a manhole to guide hand and power rods into collection lines so the rods can dislodge obstructions.

ROD (SEWER) — ROD (SEWER)

A light metal rod, three to five feet long with a coupling at each end. Rods are joined and pushed into a sewer to dislodge obstructions.

RODDING — RODDING

(See PIPE RODDING)

RODDING MACHINE — RODDING MACHINE

(See POWER RODDER and PIPE RODDING)

RODDING TOOLS — RODDING TOOLS

Special tools attached to the end of a rod or snake to accomplish various results in pipe rodding.

RODENTICIDE (row-DENT-uh-SIDE) — RODENTICIDE

Any substance or chemical used to kill or control rodents.

ROOF LEADER — ROOF LEADER

A downspout or pipe installed to drain a roof gutter to a storm drain or other means of disposal.

ROOT, SEWER — ROOT, SEWER

Any part of a root system of a plant or tree that enters a collection system.

ROOT MOP — ROOT MOP

When roots from plant life enter a sewer system, the roots frequently branch to form a growth that resembles a string mop.

ROTAMETER (RODE-uh-ME-ter) — ROTAMETER

A device used to measure the flow rate of gases and liquids. The gas or liquid being measured flows vertically up a tapered, calibrated tube. Inside the tube is a small ball or bullet-shaped float (it may rotate) that rises or falls depending on the flow rate. The flow rate may be read on a scale behind or on the tube by looking at the middle of the ball or at the widest part or top of the float.

ROUGHNESS COEFFICIENT — ROUGHNESS COEFFICIENT

A value used in Manning's formula to determine energy losses of flowing water due to pipe or channel wall roughness. Also see FRICTION LOSS, "N" FACTOR and MANNING'S FORMULA.

ROUGHNESS FACTOR — ROUGHNESS FACTOR

(See ROUGHNESS COEFFICIENT)

RUNOFF — RUNOFF

That part of rain or other precipitation that runs off the surface of a drainage area and does not enter the soil or the sewer system as inflow.

S

SCADA (ss-KAY-dah) SYSTEM — SCADA SYSTEM

Supervisory **C**ontrol **A**nd **D**ata **A**cquisition system. A computer-monitored alarm, response, control and data acquisition system used by operators to monitor and adjust their wastewater treatment processes and facilities.

SSES — SSES

Sewer **S**ystem **E**valuation **S**urvey.

SSO — SSO

Sanitary **S**ewer **O**verflow.

SADDLE — SADDLE

A fitting mounted on a pipe for attaching a new connection. This device makes a tight seal against the main pipe by use of a clamp, adhesive, or gasket and prevents the service pipe from protruding into the main.

SADDLE CONNECTION — SADDLE CONNECTION

A building service connection made to a sewer main with a device called a saddle.

SAND EQUIVALENT — SAND EQUIVALENT

An ASTM (American Society for Testing and Materials) test for trench backfill soils. The test uses a glass tube and the soil (backfill) is mixed with water in the tube and shaken. The tube is placed on a table and the soil allowed to settle according to particle size. The settlement of the soil is compared to the settlement of a standard type of sand grain sizes.

SAND TRAP — SAND TRAP

A device which can be placed in the outlet of a manhole to cause a settling pond to develop in the manhole invert, thus trapping sand, rocks and similar debris heavier than water. Also may be installed in outlets from car wash areas. Also see GRIT CATCHER.

SANITARY COLLECTION SYSTEM — SANITARY COLLECTION SYSTEM

The pipe system for collecting and carrying liquid and liquid-carried wastes from domestic sources to a wastewater treatment plant. Also see WASTEWATER COLLECTION SYSTEM.

SANITARY SEWER — SANITARY SEWER

A pipe or conduit (sewer) intended to carry wastewater or waterborne wastes from homes, businesses, and industries to the POTW (**P**ublicly **O**wned **T**reatment **W**orks). Storm water runoff or unpolluted water should be collected and transported in a separate system of pipes or conduits (storm sewers) to natural watercourses.

SATURATED SOIL — SATURATED SOIL

Soil that cannot absorb any more liquid. The interstices or void spaces in the soil are filled with water to the point at which runoff occurs.

SCALE — SCALE

A combination of mineral salts and bacterial accumulation that sticks to the inside of a collection pipe under certain conditions. Scale, in extreme growth circumstances, creates additional friction loss to the flow of water. Scale may also accumulate on surfaces other than pipes.

SCOOTER — SCOOTER

A sewer cleaning tool whose cleansing action depends on the development of high water velocity around the outside edge of a circular shield. The metal shield is rimmed with a rubber coating and is attached to a framework on wheels (like a child's scooter). The angle of the shield is controlled by a chain-spring system which regulates the head of water behind the scooter and thus the cleansing velocity of the water flowing around the shield.

SCUM — SCUM

(1) A layer or film of foreign matter (such as grease, oil) that has risen to the surface of water or wastewater.

(2) A residue deposited on the ledge of a sewer, channel, or wet well at the water surface.

(3) A mass of solid matter that floats on the surface.

SEAL — SEAL

(See PIPE JOINT SEAL)

SEALING SEALING

(See CHEMICAL GROUTING)

SEASONAL WATER TABLE SEASONAL WATER TABLE

A groundwater table that has seasonal changes in depth or elevation.

SEDIMENT SEDIMENT

Solid material settled from suspension in a liquid.

SEDIMENTATION (SED-uh-men-TAY-shun) SEDIMENTATION

The process of settling and depositing of suspended matter carried by wastewater. Sedimentation usually occurs by gravity when the velocity of the wastewater is reduced below the point at which it can transport the suspended material.

SEISMIC (SIZE-mick) SEISMIC

Relating to an earthquake or violent earth vibration such as an explosion.

SELECT BACKFILL SELECT BACKFILL

Material used in backfilling of an excavation, selected for desirable compaction or other characteristics.

SELECT BEDDING SELECT BEDDING

Material used to provide a bedding or foundation for pipes or other underground structures. This material is of specified quality for desirable bedding or other characteristics and is often imported from a different location.

SEPTIC (SEP-tick) SEPTIC

A condition produced by anaerobic bacteria. If severe, the wastewater produces hydrogen sulfide, turns black, gives off foul odors, contains little or no dissolved oxygen, and the wastewater has a high oxygen demand.

SEPTIC TANK SEPTIC TANK

A system used where wastewater collection systems and treatment plants are not available. The system is a settling tank in which settled sludge is in intimate contact with the wastewater flowing through the tank and the organic solids are decomposed by anaerobic bacterial action. Used to treat wastewater and produce an effluent that is usually disposed of by subsurface leaching.

SERVICE SERVICE

Any individual person, group of persons, thing, or groups of things served with water through a single pipe, gate, valve, or similar means of transfer from a main distribution system. Also see BUILDING SERVICE.

SERVICE CONNECTION SERVICE CONNECTION

(See BUILDING SEWER)

SERVICE ROOT SERVICE ROOT

A root entering the sewer system in a service line and growing down the pipe and into the sewer main.

SEWAGE SEWAGE

The used household water and water-carried solids that flow in sewers to a wastewater treatment plant. The preferred term is WASTEWATER.

SEWER SEWER

A pipe or conduit that carries wastewater or drainage water. The term "collection line" is often used also.

SEWER BALL SEWER BALL

A spirally grooved, inflatable, semi-hard rubber ball designed for hydraulic cleaning of sewer pipes. See BALLING.

SEWER CLEANOUT SEWER CLEANOUT

A capped opening in a sewer main that allows access to the pipes for rodding and cleaning. Usually such cleanouts are located at terminal pipe ends or beyond terminal manholes. Also called a FLUSHER BRANCH.

SEWER GAS — SEWER GAS

(1) Gas in collection lines (sewers) that results from the decomposition of organic matter in the wastewater. When testing for gases found in sewers, test for lack of oxygen and also for explosive and toxic gases.

(2) Any gas present in the wastewater collection system, even though it is from such sources as gas mains, gasoline, and cleaning fluid.

SEWER JACK — SEWER JACK

A device placed in manholes which supports a yoke or pulley that keeps wires or cables from rubbing against the inlet or outlet of a sewer.

SEWER MAIN — SEWER MAIN

A sewer pipe to which building laterals are connected. Also called a COLLECTION MAIN.

SEWER SYSTEM — SEWER SYSTEM

(See COLLECTION SYSTEM)

SEWERAGE — SEWERAGE

System of piping with appurtenances for collecting, moving and treating wastewater from source to discharge.

SHEAVE (SHE-v) — SHEAVE

V-belt drive pulley which is commonly made of cast iron or steel.

SHEETING — SHEETING

Solid material, such as wooden 2-inch planks or 1 1/8-inch plywood sheets or metal plates, used to hold back soil and prevent cave-ins.

SHIELD — SHIELD

(1) A fabricated protective crib made of steel or aluminum plate. The shield is placed on the bottom of open trenches in unstable soil areas where conventional protective shoring is not sufficient. From inside the shield, workers can safely install or repair pipelines.

(2) A device used to protect workers from sources of electrical, mechanical, or heat energy.

SHIM — SHIM

Thin metal sheets which are inserted between two surfaces to align or space the surfaces correctly. Shims can be used anywhere a spacer is needed. Usually shims are 0.001 to 0.020 inch thick.

SHORING — SHORING

Material such as boards, planks or plates and jacks used to hold back soil around trenches and to protect workers in a trench from cave-ins. Also see SHEETING.

SILTING — SILTING

Silting takes place when the pressure of infiltrating waters is great enough to carry silt, sand and other small particles from the soil into the sewer system. Where lower velocities are present in the sewer pipes, settling of these materials results in silting of the sewer system.

SINK — SINK

(See DIP)

SIPHON — SIPHON

A pipe or conduit through which water will flow above the hydraulic grade line (HGL) under certain conditions. Water (or other liquid) is first forced to flow or is sucked or drawn through the pipe by creation of a vacuum. As long as no air enters the pipe to interrupt flow, atmospheric pressure on the liquid at the elevated (higher) end of the siphon will cause the flow to continue.

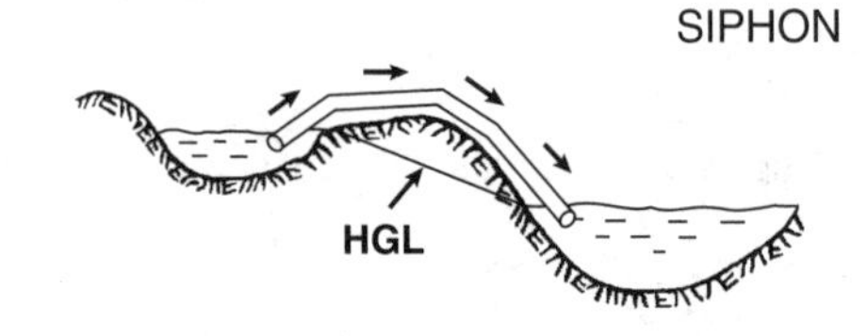

SLEEVE — SLEEVE

(1) A pipe fitting for joining two pipes of the same nominal diameter in a straight line.

(2) A tube into which a pipe is inserted.

(3) A device to protect a shaft at its bearing or wear points.

SLIPLINING SLIPLINING

A sewer rehabilitation technique accomplished by inserting flexible polyethylene pipe into an existing deteriorated sewer.

SLOPE SLOPE

The slope or inclination of a sewer trench excavation is the ratio of the vertical distance to the horizontal distance or "rise over run." See GRADE (2).

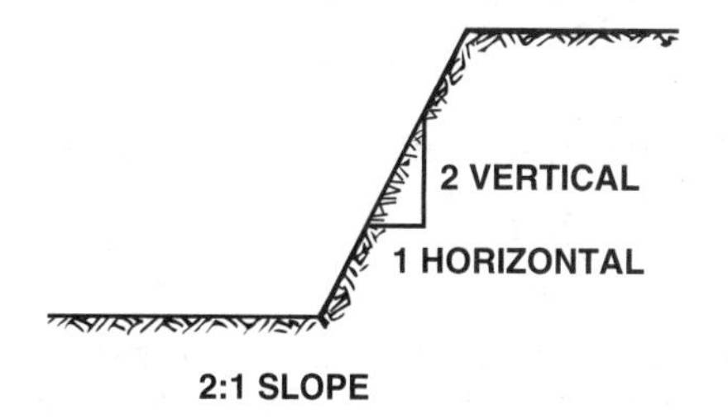

2:1 SLOPE

SLUDGE (sluj) SLUDGE

(1) The settleable solids separated from liquids during processing.

(2) The deposits of foreign materials on the bottoms and edges of wastewater collection lines and appurtenances.

SMOKE TEST SMOKE TEST

A method of blowing smoke into a closed-off section of a sewer system to locate sources of surface inflow.

SNAKE SNAKE

A stiff but flexible cable that is inserted into sewers to clear stoppages; also known as a "sewer cable."

SOAP CAKE or SOAP BUILDUP SOAP CAKE or SOAP BUILDUP

A combination of detergents and greases that accumulate in sewer systems, build up over a period of time, and may cause severe flow restrictions.

SOIL DISPLACEMENT SOIL DISPLACEMENT

Movement of soil from one place to another. Generally accompanies SILTING of a sewer system. Where infiltration is taking place and silt is carried into a sewer system, such silt or soil is removed from the ground around the sewer pipe and the result is soil displacement.

SOIL PIPE SOIL PIPE

(1) A type of wastewater or service connection pipe made of a low grade of cast iron.

(2) In plumbing, a pipe that carries the discharge of toilets or similar fixtures, with or without the discharges from other fixtures.

SOIL POLLUTION SOIL POLLUTION

The leakage (exfiltration) of raw wastewater into the soil or ground area around a sewer pipe.

SOIL STABILIZATION SOIL STABILIZATION

Injection of chemical grouts into saturated or otherwise unstable soil. The process seals out water and prevents further instability.

SOUNDING ROD SOUNDING ROD

A T-shaped tool or shaft that is pushed or driven down through the soil to locate underground pipes and utility conduits. Also see PROBE.

SPECIFIC GRAVITY SPECIFIC GRAVITY

(1) Weight of a particle, substance or chemical solution in relation to the weight of an equal volume of water. Water has a specific gravity of 1.000 at 4°C (39°F). Wastewater particles or substances usually have a specific gravity of 0.5 to 2.5.

(2) Weight of a particular gas in relation to the weight of an equal volume of air at the same temperature and pressure (air has a specific gravity of 1.0). Chlorine has a specific gravity of 2.5 as a gas.

SPHERULITIC (SFEAR-you-LIT-tick) SPHERULITIC

A rounded aggregate of particles; or, crystals in the form of a ball.

SPOIL SPOIL

Excavated material such as soil from the trench of a sewer.

SPRING LINE SPRING LINE

Theoretical center of a pipeline. Also, the guideline for laying a course of bricks.

STABILIZE STABILIZE

To convert to a form that resists change. Organic material is stabilized by bacteria which convert the material to gases and other relatively inert substances. Stabilized organic material generally will not give off obnoxious odors.

STATIC HEAD STATIC HEAD

When water is not moving, the vertical distance (in feet) from a specific point to the water surface is the static head. Also see DYNAMIC HEAD.

STATION STATION

A point of reference or location in a pipeline is sometimes called a "station." As an example, a building service located 51 feet downstream from a manhole could be reported to be at "station 51."

STATION, LIFT STATION, LIFT

(See LIFT STATION)

STILLING WELL STILLING WELL

A well or chamber which is connected to the main flow channel by a small inlet. Waves and surges in the main flow stream will not appear in the well due to the small-diameter inlet. The liquid surface in the well will be quiet, but will follow all of the steady fluctuations of the open channel. The liquid level in the well is measured to determine the flow in the main channel.

STOPPAGE STOPPAGE

(1) Partial or complete interruption of flow as a result of some obstruction in a sewer.

(2) When a sewer system becomes plugged and the flow backs up, it is said to have a "stoppage." Also see BLOCKAGE.

STORM COLLECTION SYSTEM STORM COLLECTION SYSTEM

A system of gutters, catch basins, yard drains, culverts and pipes for the purpose of conducting storm waters from an area, but intended to exclude domestic and industrial wastes.

STORM RUNOFF STORM RUNOFF

The amount of runoff that reaches the point of measurement within a relatively short period of time after the occurrence of a storm or other form of precipitation. Also called "direct runoff."

STORM SEWER STORM SEWER

A separate pipe, conduit or open channel (sewer) that carries runoff from storms, surface drainage, and street wash, but does not include domestic and industrial wastes. Storm sewers are often the recipients of hazardous or toxic substances due to the illegal dumping of hazardous wastes or spills created by accidents involving vehicles and trains transporting these substances. Also see SANITARY SEWER.

STORM WATER STORM WATER

The excess water running off from the surface of a drainage area during and immediately after a period of rain. See STORM RUNOFF.

STORM WATER INLET STORM WATER INLET

A device that admits surface waters to the storm water drainage system. Also see CURB INLET and CATCH BASIN.

STRETCH STRETCH

Length of sewer from manhole to manhole.

STRINGERS STRINGERS

Horizontal shoring members, usually square, rough cut timber, that are used to hold solid sheeting, braces or vertical shoring members in place. Also called WALERS.

STRUCTURAL DEFECT STRUCTURAL DEFECT

A flaw or imperfection of a structure or design which was built into a project, pipeline or other collection system appurtenance.

STRUCTURAL FAILURE STRUCTURAL FAILURE

A condition that exists when one or more components of a system break down or fail to perform as expected. A structural failure may result from defective parts or design or may result from other circumstances that occur after the completion of construction.

SUBSIDENCE (sub-SIDE-ence) SUBSIDENCE

The dropping or lowering of the ground surface as a result of removing excess water (overdraft or overpumping) from an aquifer. After excess water has been removed, the soil will settle, become compacted and the ground surface will drop and can cause the settling of underground utilities.

SUBSOIL GEOLOGY SUBSOIL GEOLOGY

The study of soil conditions existing below the surface of the ground at any selected site.

SUBSYSTEM SUBSYSTEM

An extensive underground sewer system connected to the main collection system, but not considered part of the main system. An example might be the underground sewer system of a mobile home park.

SUCKER RODS SUCKER RODS

Rigid, coupled sewer rods of metal or wood used for clearing stoppages. Usually available in 3-ft, 39-in, 4-ft, 5-ft and 6-ft lengths.

SUCTION HEAD SUCTION HEAD

The *POSITIVE* pressure (in feet or pounds per square inch (psi)) on the suction side of a pump. The pressure can be measured from the centerline of the pump *UP TO* the elevation of the hydraulic grade line on the suction side of the pump.

SUCTION LIFT SUCTION LIFT

The *NEGATIVE* pressure (in feet or inches of mercury vacuum) on the suction side of a pump. The pressure can be measured from the centerline of the pump *DOWN TO* (lift) the elevation of the hydraulic grade line on the suction side of the pump.

SURCHARGE SURCHARGE

Sewers are surcharged when the supply of water to be carried is greater than the capacity of the pipes to carry the flow. The surface of the wastewater in manholes rises above the top of the sewer pipe, and the sewer is under pressure or a head, rather than at atmospheric pressure.

SURCHARGED MANHOLE SURCHARGED MANHOLE

A manhole in which the rate of the water entering is greater than the capacity of the outlet under gravity flow conditions. When the water in the manhole rises above the top of the outlet pipe, the manhole is said to be "surcharged."

SURFACE RUNOFF SURFACE RUNOFF

(1) The precipitation that cannot be absorbed by the soil and flows across the surface by gravity.
(2) The water that reaches a stream by traveling over the soil surface or falls directly into the stream channels, including not only the large permanent streams but also the tiny rills and rivulets.
(3) Water that remains after infiltration, interception, and surface storage have been deducted from total precipitation.

SURFACED DEFECT SURFACED DEFECT

A break or opening in a sewer pipe where the covering soil has been washed away and the opening or break is exposed on the ground surface.

SURFACED VOID SURFACED VOID

A dip or depression in the ground that appears when silting has taken place to a degree that a void is caused in the subsoil. Through successive cave-ins, the void reaches the surface of the ground.

SURVEILLANCE TELEVISION EQUIPMENT SURVEILLANCE TELEVISION EQUIPMENT

Economical closed-circuit television equipment designed for surveillance or security work in commercial facilities. Picture resolutions generally range from 250 to 350 lines.

SUSPENDED SOLIDS SUSPENDED SOLIDS

(1) Solids that either float on the surface or are suspended in water, wastewater, or other liquids, and which are largely removable by laboratory filtering.
(2) The quantity of material removed from wastewater in a laboratory test, as prescribed in *STANDARD METHODS FOR THE EXAMINATION OF WATER AND WASTEWATER*, and referred to as Total Suspended Solids Dried at 103–105°C.

SWAB SWAB

A circular sewer cleaning tool almost the same diameter as the pipe being cleaned. As a final cleaning procedure after a sewer line has been cleaned with a porcupine, a swab is pulled through the sewer and the flushing action of water flowing around the tool cleans the line.

T

TWA TWA

See TIME WEIGHTED AVERAGE (TWA).

TAG LINE TAG LINE

A line, rope or cable that follows equipment through a sewer so that equipment can be pulled back out if it encounters an obstruction or becomes stuck. Equipment is pulled forward with a pull line.

TAP TAP

A small hole in a sewer where a wastewater service line from a building is connected (tapped) into a lateral or branch sewer.

TELEMETERING EQUIPMENT TELEMETERING EQUIPMENT

Equipment that translates physical measurements into electrical impulses that are transmitted to dials or recorders.

TELEMETRY (tel-LEM-uh-tree) TELEMETRY

The electrical link between the transmitter and the receiver. Telephone lines are commonly used to serve as the electrical line.

TELEVISION INSPECTION TELEVISION INSPECTION

An inspection of the inside of a sewer pipe made by pulling a closed-circuit television camera through the pipe.

TELEVISION INSPECTION LOG TELEVISION INSPECTION LOG

A record of a pipeline television inspection which provides date, line location, footage distances, pipe quadrant locations and descriptions of all conditions observed in the inspection. When this log is written, it is called a "written recording." When it is voice recorded on a tape, it is called a "voice tape recording." If the picture is recorded with a videotape recorder with audio remarks, it is called a "video-voice inspection record." Where data-view reporting is used, it is called a VIDEO LOG.

TELEVISION MONITOR TELEVISION MONITOR

The television set or kinescope where the picture is viewed on a closed-circuit system.

TEMPORARY GROUNDWATER TABLE TEMPORARY GROUNDWATER TABLE

(1) During and for a period following heavy rainfall or snow melt, the soil is saturated at elevations above the normal, stabilized or seasonal groundwater table, often from the surface of the soil downward. This is referred to as a temporary condition and thus is a temporary groundwater table.

(2) When a collection system serves agricultural areas in its vicinity, irrigation of these areas can cause a temporary rise in the elevation of the groundwater table.

TERMINAL CLEANOUT TERMINAL CLEANOUT

When a manhole is not provided at the upstream end of a sewer main, a cleanout is usually provided. This is called a "terminal cleanout" or a FLUSHER BRANCH.

TERMINAL MANHOLE TERMINAL MANHOLE

A manhole located at the upstream end of a sewer and having no inlet pipe. Also called a DEADEND MANHOLE.

TEST BORE TEST BORE

A hole or bore made to sample and determine the structure of underground soil conditions.

THRUST BLOCK THRUST BLOCK

A mass of concrete or similar material appropriately placed around a pipe to prevent movement when the pipe is carrying water. Usually placed at bends and valve structures.

TIDE GATE TIDE GATE

A gate with a flap suspended from a free-swinging horizontal hinge, usually placed at the end of a conduit discharging into a body of water having a fluctuating surface elevation. The gate is usually closed because of outside water pressure, but will open when the water head inside the pipe is great enough to overcome the outside pressure, the weight of the flap, and the friction of the hinge. Also called a BACKWATER GATE. Also see CHECK VALVE and FLAP GATE.

TIME WEIGHTED AVERAGE (TWA) TIME WEIGHTED AVERAGE (TWA)

The average concentration of a pollutant (or sound) based on the times and levels of concentrations of the pollutant. The time weighted average is equal to the sum of the portion of each time period (as a decimal, such as 0.25 hour) multiplied by the pollutant concentration during the time period divided by the hours in the workday (usually 8 hours).

TOPOGRAPHIC (TOP-o-GRAPH-ick) GEOLOGY TOPOGRAPHIC GEOLOGY

A study of the rock and soil formations of an area for purposes of mapping underground conditions with identifications and elevations. Often called a "geological survey."

TOPOGRAPHY (toe-PAH-gruh-fee) TOPOGRAPHY

The arrangement of hills and valleys in a geographic area.

TOTAL CONTRIBUTION TOTAL CONTRIBUTION

All water and wastewater entering a sewer system from a specific facility, subsystem or area. This includes domestic and industrial wastewaters, inflow and infiltration reaching the main collection system.

TOTAL DYNAMIC HEAD (TDH) TOTAL DYNAMIC HEAD (TDH)

When a pump is lifting or pumping water, the vertical distance (in feet) from the elevation of the energy grade line on the suction side of the pump to the elevation of the energy grade line on the discharge side of the pump.

TOTAL FLOW TOTAL FLOW

The total flow passing a selected point of measurement in the collection system during a specified period of time.

TOTALIZER TOTALIZER

A device or meter that continuously measures and calculates (adds) a process rate variable in cumulative fashion; for example, total flows displayed in gallons, million gallons, cubic feet, or some other unit of volume measurement. Also called an INTEGRATOR.

TRAP TRAP

(1) In the wastewater collection system of a building, plumbing codes require every drain connection from an appliance or fixture to have a trap. The trap in this case is a gooseneck that holds water to prevent vapors or gases in a collection system from entering the building.

(2) Various other types of special traps are used in collection systems such as a GRIT TRAP or SAND TRAP.

TRAP, DIP TRAP, DIP

(See DIP)

TRAP, HANDHOLE TRAP, HANDHOLE

(See HANDHOLE TRAP)

TRENCH JACK TRENCH JACK

Mechanical screw device used to hold shoring in place.

TRUNK SEWER TRUNK SEWER

A sewer that receives wastewater from many tributary branches or sewers and serves a large territory and contributing population. Also see MAIN SEWER.

TRUNK SYSTEM TRUNK SYSTEM

A system of major sewers serving as transporting lines and not as local or lateral sewers.

TURBID TURBID

Having a cloudy or muddy appearance.

TURBIDITY UNITS (TU) TURBIDITY UNITS (TU)

Turbidity units are a measure of the cloudiness of water. If measured by a nephelometric (deflected light) instrumental procedure, turbidity units are expressed in nephelometric turbidity units (NTU) or simply TU. Those turbidity units obtained by visual methods are expressed in Jackson Turbidity Units (JTU) which are a measure of the cloudiness of water; they are used to indicate the clarity of water. There is no real connection between NTUs and JTUs. The Jackson turbidimeter is a visual method and the nephelometer is an instrumental method based on deflected light.

TV LOG TV LOG

A written record of the internal pipe conditions observed during a sewer line TV inspection.

TWO-WAY CLEANOUT TWO-WAY CLEANOUT

An opening in pipes or sewers designed for rodding or working a snake into the pipe in either direction. Two-way cleanouts are most often found in building lateral pipes at or near a property line.

U

U-TUBE U-TUBE

(1) A pipe shaped like a U that is constructed in a force main to raise the dissolved oxygen concentration in the wastewater.

(2) U-tube manometers are used to indicate the pressure of a gas or liquid in a contained area, such as a pipeline or storage vessel.

UNDERMINED UNDERMINED

(1) A condition that occurs when the bedding support under a pipe or manhole has been removed or washed away. Conditions leading to or causing this are believed to be the presence of excess water during backfill. Other causes are horizontal boring operations, excavations adjacent to the pipe or manhole and exfiltration or infiltration at drop joints.

(2) When flow through a broken section of pipe carries away soil around the break leaving a void or empty space, the surfaces over the void are said to be "undermined."

UNDISTURBED SOIL UNDISTURBED SOIL

Soil, at any depth, which has not been excavated or disturbed by excavation or construction.

UPRIGHTS UPRIGHTS

Vertical shoring members that may be solid (SHEETING) or spaced from 2 to 8 feet apart to prevent cave-ins.

UPSTREAM UPSTREAM

The direction against the flow of water; or, toward or in the higher part of a sewer or collection system.

V

V-NOTCH WEIR V-NOTCH WEIR

A triangular weir with a "V" notch calibrated in gallons per minute readings. By holding the weir in a pipe with rubber seals forcing a flow to pass through the "V," a measure of the gallonage flowing through the pipe can be read on the basis of the depth of water flowing over the weir.

VAC-ALL VAC-ALL

Equipment that removes solids from a manhole as they enter the manhole from a hydraulic cleaning operation. Most of the wastewater removed from the manhole by the operation is separated from the solids and returned to the sewer.

VACUUM TEST VACUUM TEST

A testing procedure that places a manhole under a vacuum to test the structural integrity of the manhole.

VEGETABLE WASTES VEGETABLE WASTES

Vegetable matter entering a collection system. This term is usually used to distinguish such types of waste from animal, industrial, commercial and other types of waste solids.

VEHICLE, LIQUID VEHICLE, LIQUID

(See LIQUID VEHICLE)

VELOCITY HEAD VELOCITY HEAD

The energy in flowing water as determined by a vertical height (in feet or meters) equal to the square of the velocity of flowing water divided by twice the acceleration due to gravity ($V^2/2g$).

VENTS, MANHOLE VENTS, MANHOLE

(See MANHOLE VENTS)

VENTS, WASTELINE SYSTEM VENTS, WASTELINE SYSTEM

(See WASTELINE VENT)

VERTICAL OFFSET VERTICAL OFFSET

A pipe joint in which one section is connected to another at a different elevation, such as a DROP JOINT.

VIDEO INSPECTION VIDEO INSPECTION

A television inspection.

VIDEO LOG VIDEO LOG

A magnetic tape picture recording of a television inspection where data-view reporting has been included as part of the visual record. Also see DATA-VIEW.

VIDEOTAPE VIDEOTAPE

A magnetic tape for recording television pictures. Standard tapes also have a capacity to record a voice with the picture, or an "audio" accompaniment.

VISCOSITY (vis-KOSS-uh-tee) VISCOSITY

A property of water, or any other fluid, which resists efforts to change its shape or flow. Syrup is more viscous (has a higher viscosity) than water. The viscosity of water increases significantly as temperatures decrease. Motor oil is rated by how thick (viscous) it is; 20 weight oil is considered relatively thin while 50 weight oil is relatively thick or viscous.

VOID VOID

A pore or open space in rock, soil or other granular material, not occupied by solid matter. The pore or open space may be occupied by air, water, or other gaseous or liquid material. Also called an "interstice" or "void space."

VOLTAGE VOLTAGE

The electrical pressure available to cause a flow of current (amperage) when an electric circuit is closed. Also called ELECTROMOTIVE FORCE (E.M.F.).

VOLUTE (vol-LOOT) VOLUTE

The spiral-shaped casing which surrounds a pump, blower, or turbine impeller and collects the liquid or gas discharged by the impeller.

W

WALERS (WAY-lers) WALERS

Horizontal shoring members, usually square, rough cut timber, that are used to hold solid sheeting, braces or vertical shoring members in place. Also called STRINGERS.

WARNING WARNING

The word *WARNING* is used to indicate a hazard level between *CAUTION* and *DANGER.* Also see CAUTION, DANGER, and NOTICE.

WASTELINE CLEANOUT WASTELINE CLEANOUT

An opening or point of access in a building wastewater pipe system for rodding or snake operation.

WASTELINE SYSTEM WASTELINE SYSTEM

(See BUILDING WASTEWATER COLLECTION SYSTEM)

WASTELINE TRAP WASTELINE TRAP

(See TRAP (1))

WASTELINE VENT — WASTELINE VENT

Most plumbing codes require a vent pipe connection of adequate size and located downstream of a trap in a building wastewater system. This vent prevents the accumulation of gases or odors and is usually piped through the roof and out of doors.

WASTEWATER — WASTEWATER

A community's used water and water-carried solids that flow to a treatment plant. Storm water, surface water, and groundwater infiltration also may be included in the wastewater that enters a wastewater treatment plant. The term "sewage" usually refers to household wastes, but this word is being replaced by the term "wastewater."

WASTEWATER COLLECTION SYSTEM — WASTEWATER COLLECTION SYSTEM

The pipe system for collecting and carrying water and water-carried wastes from domestic and industrial sources to a wastewater treatment plant.

WASTEWATER FACILITIES — WASTEWATER FACILITIES

The pipes, conduits, structures, equipment, and processes required to collect, convey, and treat domestic and industrial wastes, and dispose of the effluent and sludge.

WASTEWATER TREATMENT PLANT — WASTEWATER TREATMENT PLANT

(1) An arrangement of pipes, equipment, devices, tanks and structures for treating wastewater and industrial wastes.

(2) A water pollution control plant.

WATER HAMMER — WATER HAMMER

The sound like someone hammering on a pipe that occurs when a valve is opened or closed very rapidly. When a valve position is changed quickly, the water pressure in a pipe will increase and decrease back and forth very quickly. This rise and fall in pressures can cause serious damage to the system.

WATER TABLE — WATER TABLE

The upper surface of the zone of saturation of groundwater in an unconfined aquifer.

WATER TABLE DEPTH — WATER TABLE DEPTH

(See GROUNDWATER DEPTH)

WATER TABLE ELEVATION — WATER TABLE ELEVATION

(See GROUNDWATER ELEVATION)

WATER TABLE HEAD — WATER TABLE HEAD

(See PRESSURE HEAD)

WAYNE BALL — WAYNE BALL

A spirally grooved, inflatable, semi-hard rubber ball designed for hydraulic cleaning of sewer pipes. See BALLING and SEWER BALL.

WEIR (weer) — WEIR

(1) A wall or plate placed in an open channel and used to measure the flow of water. The depth of the flow over the weir can be used to calculate the flow rate, or a chart or conversion table may be used to convert depth to flow.

(2) A wall or obstruction used to control flow (from settling tanks and clarifiers) to ensure a uniform flow rate and avoid short-circuiting.

WELL POINT — WELL POINT

A hollow, pointed rod with a perforated (containing many small holes) tip. A well point is driven into an excavation where water seeps into the tip and is pumped out of the area. Used to lower the water table and reduce flooding during an excavation.

WET PIT — WET PIT

(See WET WELL)

WET WELL — WET WELL

A compartment or tank in which wastewater is collected. The suction pipe of a pump may be connected to the wet well or a submersible pump may be located in the wet well.

WETTED PERIMETER WETTED PERIMETER

The length of the wetted portion of a pipe covered by flowing wastewater.

PIPE

A B

WATER

WETTED PERIMETER = DISTANCE FROM A to B

WICK EFFECT WICK EFFECT

(See CAPILLARY EFFECT)

X

(NO LISTINGS)

Y

"Y" CONNECTION "Y" CONNECTION

Another name for a BUILDING SERVICE.

Z

ZONE OF SATURATION ZONE OF SATURATION

(1) Where raw wastewater is exfiltrating from a sewer pipe, the area of soil that is moistened around the leak point is often called the "zone of saturation."

(2) The area of soil saturated with water.

APPLICATIONS OF ARITHMETIC TO COLLECTION SYSTEMS

by

Kenneth D. Kerri

TABLE OF CONTENTS

APPLICATIONS OF ARITHMETIC TO COLLECTION SYSTEMS

OBJECTIVES

APPLICATIONS OF ARITHMETIC TO COLLECTION SYSTEMS

Following completion of this section, you should be able to:

1. Perform simple addition, subtraction, multiplication, and division,
2. Calculate perimeters, circumferences, areas, and volumes,
3. Analyze mathematical problems and solve them, and
4. Solve collection system problems regarding:
   a. Excavations and paving,
   b. Chemical doses,
   c. Electricity,
   d. Costs,
   e. Hydraulics,
      (1) Slopes
      (2) Velocities
      (3) Flow rates
      (4) Wet wells
      (5) Pumps, and
   f. Maps and blueprints.

APPLICATIONS OF ARITHMETIC TO COLLECTION SYSTEMS

A.0 WHY ARITHMETIC?

Actually, most collection system operators use very little arithmetic when doing their day-to-day jobs. If you can add, subtract, multiply, and divide, you can solve all of the problems in this section. An electronic calculator can do the addition, subtraction, multiplication, and division for you if you can write the formula or equation to be solved.

Collection system operators need to know arithmetic in order to calculate slopes of trenches or pipes and also velocities and flows in sewers. Other examples of uses of arithmetic include the determination of proper chemical doses to control roots or odors. When estimating the cost of a sewer repair job, the costs must be calculated for budgeting purposes. To order bedding or backfill material, the appropriate quantities have to be determined so there are no unnecessary delays from insufficient materials or excess material remaining when the job is completed. Calculations are important when calibrating and comparing performance of new pumps and flowmeters with specifications. From time to time equipment performance must be reverified. As you can see, periodically certain jobs that collection system operators must do require some knowledge of and ability to do simple arithmetic. After you have worked an actual collection system arithmetic problem in the field, you should verify your calculations, examine your answer to see if it appears reasonable, and, if possible, have another operator review your work before taking any action.

In this section we have tried to show you how to solve problems actually encountered by collection system operators in the field. Hopefully, questions on certification and civil service exams will not be any more difficult.

An objective of this section is to provide you with a quick review of some basic arithmetic principles and examples of typical collection system problems. This appendix is not intended to be an arithmetic textbook. Some operators will be able to skip over the review section on addition, subtraction, multiplication, and division. Others may need more help in these and other areas. Basic arithmetic textbooks are available at every local library or bookstore and can be referred to if needed. Deserving special mention is *BASIC MATH CONCEPTS FOR WATER AND WASTEWATER PLANT OPERATORS*,[1] which provides considerable detail on how to do basic calculations.

A.1 BASIC ARITHMETIC[2]

In this section we are going to provide you with basic arithmetic problems involving addition, subtraction, multiplication, and division. You may work the problems "by hand" if you wish, but we recommend you use an electronic pocket calculator. The operating or instructional manual for your calculator should outline the step-by-step procedures to follow. All calculators use similar procedures, but most of them are slightly different from others.

We will start with very basic, simple problems. Try working the problems and then comparing your answers with the given answers. If you can work these problems, you should be able to work the more difficult problems in the text of this training manual by using the same procedures.

A.10 Addition

2	6.2	16.7	6.12	43
3	8.5	38.9	38.39	39
5	14.7	55.6	44.51	34
				38
				39
2.12	0.12	63	120	37
9.80	2.0	32	60	29
11.92	2.12	95	180	259
4	23	16.2	45.98	70
7	79	43.5	28.09	50
2	31	67.8	114.00	40
13	133	127.5	188.07	80
				240

A.11 Subtraction

7	12	25	78	83
− 5	− 3	− 5	− 30	− 69
2	9	20	48	14
61	485	4.3	3.5	123
− 37	− 296	− 0.8	− 0.7	− 109
24	189	3.5	2.8	14
8.6	11.92	27.32	3.574	75.132
− 8.22	− 3.70	− 12.96	− 0.042	− 49.876
0.38	8.22	14.36	3.532	25.256

[1] *BASIC MATH CONCEPTS FOR WATER AND WASTEWATER PLANT OPERATORS and APPLIED MATH FOR WASTEWATER PLANT OPERATORS by Joanne Kirkpatrick Price. Obtain from CRC Press LLC, Attn: Order Entry, 2000 Corporate Boulevard, NW, Boca Raton, FL 33431-9868. Prices: BASIC MATH (TX8084) $49.95; APPLIED MATH (TX69892) $67.95.*

[2] *Major portions of this appendix are taken from OPERATION OF WASTEWATER TREATMENT PLANTS, Volume I, Appendix, "How To Solve Wastewater Treatment Plant Arithmetic Problems." This manual is available from the Office of Water Programs, California State University, Sacramento, 6000 J Street, Sacramento, CA 95819-6025. Price for 10-chapter Volume I, $22.00.*

A.12 Multiplication

(3)(2)[a]	= 6	(4)(7)	= 28
(10)(5)	= 50	(10)(1.3)	= 13
(2)(22.99)	= 45.98	(6)(19.5)	= 117
(16)(17.1)	= 273.6	(50)(20,000)	= 1,000,000
(40)(2.31)	= 92.4	(80)(0.433)	= 34.64
(40)(20)(6)	= 4,800		
(4,800)(7.48)	= 35,904		
(1.6)(2.3)(8.34)	= 30.6912		
(0.001)(200)(8.34)	= 1.668		
(0.785)(7.48)(60)	= 352.308		
(12,000)(500)(60)(24)	= 8,640,000,000 or 8.64 x 10^9		
(4)(1,000)(1,000)(454)	= 1,816,000,000 or 1.816 x 10^9		

NOTE: The term, x 10^9, means that the number is multiplied by 10^9 or 1,000,000,000. Therefore 8.64 x 10^9 = 8.64 x 1,000,000,000 = 8,640,000,000.

[a] (3)(2) is the same as 3 x 2 = 6.

A.13 Division

$$\frac{6}{3} = 2 \qquad \frac{48}{12} = 4$$

$$\frac{50}{25} = 2 \qquad \frac{300}{20} = 15$$

$$\frac{20}{7.1} = 2.8 \qquad \frac{11{,}400}{188} = 60.6$$

$$\frac{1{,}000{,}000}{17.5} = 57{,}143 \qquad \frac{861{,}429}{30{,}000} = 28.7$$

$$\frac{4{,}000{,}000}{74{,}880} = 53.4 \qquad \frac{1.67}{8.34} = 0.20$$

$$\frac{80}{2.31} = 34.6 \qquad \frac{62}{454} = 0.137$$

$$\frac{250}{17.1} = 14.6 \qquad \frac{4{,}000{,}000}{14.6} = 273{,}973$$

NOTE: When we divide $\frac{1}{3}$ = 0.3333, we get a long row of 3s. Instead of the row of 3s, we "round-off" our answer so $\frac{1}{3}$ = 0.33. For a discussion of rounding off numbers, see Section A.95, "Significant Figures."

A.14 Rules for Solving Equations

Most of the arithmetic problems we work in the wastewater field require us to plug numbers into formulas and calculate the answer. There are a few basic rules that apply to solving formulas. These rules are:

1. Work from left to right.
2. Do all the multiplication and division above the line (in the numerator) and below the line (in the denominator); then do the addition and subtraction above and below the line.
3. Perform the division (divide the numerator by the denominator).

Parentheses ( ) are used in formulas to identify separate parts of a problem. A fourth rule tells us how to handle numbers within parentheses.

4. Work the arithmetic within the parentheses before working outside the parentheses. Use the same order stated in rules 1, 2, and 3: work left to right, above and below the line, then divide the top number by the bottom number.

Let's look at an example problem to see how these rules apply. This year one of the responsibilities of the operators at our plant is to paint both sides of the wooden fence across the front of the facility. The fence is 145 feet long and 9 feet high. The steel access gate, which does not need painting, measures 14 feet wide by 9 feet high. Each gallon of paint will cover 150 square feet of surface area. How many gallons of paint should be purchased?

STEP 1: Identify the correct formula.

$$\text{Paint Req, gal} = \frac{\text{Total Area, sq ft}}{\text{Coverage, sq ft/gal}}$$

or

$$\text{Paint Req, gal} = \frac{(\text{Fence L, ft x H, ft x No. Sides}) - (\text{Gate L, ft x H, ft x No. Sides})}{\text{Coverage, sq ft/gal}}$$

STEP 2: Plug numbers into the formula.

$$\text{Paint Req, gal} = \frac{(145\text{ ft x 9 ft x 2}) - (14\text{ ft x 9 ft x 2})}{150\text{ sq ft/gal}}$$

STEP 3: Work the multiplication within parentheses.

$$\text{Paint Req, gal} = \frac{(2{,}610\text{ sq ft}) - (252\text{ sq ft})}{150\text{ sq ft/gal}}$$

STEP 4: Work the subtraction above the line.

$$\text{Paint Req, gal} = \frac{2{,}358\text{ sq ft}}{150\text{ sq ft/gal}}$$

STEP 5: Divide the numerator by the denominator.

Paint Req, gal = 15.72 gal
or 16 gallons of paint will be needed

Instructions for your electronic calculator can provide you with the detailed procedures for working the practice problems below.

$$\frac{(3)(4)}{2} = 6 \qquad \frac{64}{(8)(4)} = 2$$

$$\frac{(2+3)(4)}{5} = 4 \qquad \frac{54}{(4+2)(3)} = 3$$

$$\frac{(7-2)(8)}{4} = 10 \qquad \frac{48}{(8-3)(4)} = 2.4$$

$$\frac{(15{,}000)(7.48)(24)}{(1.4)(1{,}000{,}000)} = 1.9$$

$$\frac{(225-25)(100)}{225} = 88.9$$

$$\frac{12}{(0.432)(8.34)} = 3.3$$

$$\frac{(1{,}800)(0.5)(8.34)}{(110)(2.0)(8.34)} = 4.1$$

A.15 Actual Problems

Let's look at the last four problems in the previous Section A.14, "Rules for Solving Equations," as they might be encountered by an operator.

1. A rectangular sedimentation basin treats a flow of 1.4 MGD. The volume is 15,000 cubic feet. Estimate the detention time.

Known		Unknown
Basin Volume, cu ft	= 15,000 cu ft	Detention Time, hours
Flow, MGD	= 1.4 MGD	

Calculate the detention time in hours.

$$\text{Detention Time, hr} = \frac{(\text{Basin Volume, cu ft})(7.48\ \text{gal/cu ft})(24\ \text{hr/day})}{(\text{Flow, MGD})(1{,}000{,}000/\text{M})}$$

$$= \frac{(15{,}000\ \text{cu ft})(7.48\ \text{gal/cu ft})(24\ \text{hr/day})}{(1.4\ \text{MGD})(1{,}000{,}000/\text{M})}$$

$$= \frac{(15{,}000)(7.48)(24)}{(1.4)(1{,}000{,}000)}$$

$$= 1.9\ \text{hr}$$

2. The influent BOD to an activated sludge plant is 225 mg/*L* and the effluent BOD is 25 mg/*L*. What is the BOD removal efficiency of the plant?

Known		Unknown
Influent BOD, mg/*L*	= 225 mg/*L*	Plant Efficiency, %
Effluent BOD, mg/*L*	= 25 mg/*L*	

Calculate the efficiency of the plant in removing BOD.

$$\text{Plant BOD Efficiency, \%} = \left(\frac{\text{In} - \text{Out}}{\text{In}}\right)(100\%)$$

$$= \left(\frac{225\ \text{mg}/L - 25\ \text{mg}/L}{225\ \text{mg}/L}\right)(100\%)$$

$$= \left(\frac{225-25}{225}\right)(100\%)$$

$$= 88.9\%$$

3. A chlorinator is set to feed 12 pounds of chlorine per day to a flow of 300 gallons per minute (0.432 million gallons per day). What is the chlorine dose in milligrams per liter?

Known		Unknown
Chlorinator Feed, lbs/day	= 12 lbs/day	Chlorine Dose, mg/*L*
Flow, MGD	= 0.432 MGD	

Determine the chlorine dose in milligrams per liter.

$$\text{Chlorine Dose, mg}/L = \frac{\text{Chlorinator Feed Rate, lbs/day}}{(\text{Flow, MGD})(8.34\ \text{lbs/gal})}$$

$$= \frac{12\ \text{lbs/day}}{(0.432\ \text{MGD})(8.34\ \text{lbs/gal})}$$

$$= \frac{12}{(0.432)(8.34)}$$

$$= 3.3\ \text{mg}/L$$

4. Estimate the sludge age for an activated sludge plant with a mixed liquor suspended solids (MLSS) of 1,800 mg/*L* and an aeration tank volume of 0.50 MG. The flow is 2.0 MGD and primary effluent suspended solids are 110 mg/*L*.

Known		Unknown
MLSS, mg/*L*	= 1,800 mg/*L*	Sludge Age, days
P.E. SS, mg/*L*	= 110 mg/*L*	
Tank Vol, MG	= 0.50 MG	
Flow, MGD	= 2.0 MGD	

Calculate the sludge age for the aeration tank in days.

$$\text{Sludge Age, days} = \frac{(\text{MLSS, mg}/L)(\text{Tank Vol, MG})(8.34\ \text{lbs/gal})}{(\text{P.E. SS, mg}/L)(\text{Flow, MGD})(8.34\ \text{lbs/gal})}$$

$$= \frac{(1{,}800\ \text{mg}/L)(0.50\ \text{MG})(8.34\ \text{lbs/gal})}{(110\ \text{mg}/L)(2.0\ \text{MGD})(8.34\ \text{lbs/gal})}$$

$$= \frac{(1{,}800)(0.50)(8.34)}{(110)(2.0)(8.34)}$$

$$= 4.1\ \text{days}$$

A.16 Percentage

Expressing a number in percentage is just another, and sometimes simpler, way of writing a fraction or a decimal. It can be thought of as parts per 100 parts, since the percentage is the numerator of a fraction whose denominator is always 100. Twenty-five parts per 100 parts is more easily recognized as 25/100 or 0.25. However, it is also 25%. In this case, the symbol % takes the place of the 100 in the fraction and the decimal point in the decimal fraction.

For the above example it can be seen that changing from a fraction or a decimal to percent is not a difficult procedure.

1. To change a fraction to percent, multiply by 100%.

Example: $\frac{2}{5} \times 100\% = \frac{200\%}{5} = 40\%$

Example: $\frac{5}{4} \times 100\% = \frac{500\%}{4} = 125\%$

2. To change percent to a fraction, divide by 100%.

Example: $15\% \div 100\% = 15\% \times \frac{1}{100\%} = \frac{15}{100} = \frac{3}{20}$

Example: $0.4\% \div 100\% = 0.4\% \times \frac{1}{100\%} = \frac{0.4}{100} = \frac{4}{1{,}000} = \frac{1}{250}$

In these examples note that the two percent signs cancel each other.

Following is a table comparing common fractions, decimal fractions, and percent to indicate their relationship to each other:

Common Fraction	Decimal Fraction	Percent
$\frac{285}{100}$	2.85	285%
$\frac{100}{100}$	1.0	100%
$\frac{20}{100}$	0.20	20%
$\frac{1}{100}$	0.01	1%
$\frac{1}{1,000}$	0.001	0.1%
$\frac{1}{1,000,000}$	0.000001	0.0001%

A.17 Sample Problems Involving Percent

Problems involving percent are usually not complicated since their solution consists of only one or two steps. The principal error made is usually a misplaced decimal point. The most common type of percentage problem is finding:

1. *WHAT PERCENT ONE NUMBER IS OF ANOTHER*

In this case, the problem is simply one of reading carefully to determine the correct fraction and then converting to a percentage.

Example: What percent is 20 of 25?

$$\frac{20}{25} = \frac{4}{5} = 0.8$$

$$0.8 \text{ x } 100\% = 80\%$$

Example: Four is what percent of 14?

$$\frac{4}{14} = 0.2857$$

$$0.2857 \text{ x } 100\% = 28.57\%$$

Example: The input or brake horsepower to a pump is 20 horsepower. Output or water horsepower is 15 HP. What is the efficiency of the pump?

$$\text{Efficiency, \%} = \frac{(\text{Output})(100\%)}{\text{Input}} = \frac{(\text{Water HP})(100\%)}{\text{Brake HP}}$$

$$= \frac{(15 \text{ HP})(100\%)}{20 \text{ HP}}$$

$$= 75\%$$

2. *PERCENT OF A GIVEN NUMBER*

In this case the percent is expressed as a decimal, and the two numbers are multiplied together.

Example: Find 7% of 32.

$$\frac{7\%}{100\%} \text{ x } 32 = 2.24$$

Example: Find 90% of 5.

$$\frac{90\%}{100\%} \text{ x } 5 = 4.5$$

Example: What is the weight of dry solids in a ton (2,000 lbs) of wastewater sludge containing 5% solids and 95% water?

NOTE: 5% solids means there are 5 lbs of dry solids for every 100 lbs of wet sludge.

Therefore

$$2{,}000 \text{ lbs x } \frac{5\%}{100\%} = 100 \text{ lbs of solids}$$

A variation of the preceding problem is:

3. *FINDING A NUMBER WHEN A GIVEN PERCENT OF IT IS KNOWN*

Since this problem is similar to the previous problem, the solution is to convert to a decimal and divide by the decimal.

Example: If 5% of a number is 52, what is the number?

$$\left(\frac{100\%}{5\%}\right)(52) = 1{,}040$$

A check calculation may now be performed—what is 5% of 1,040?

$$\left(\frac{5\%}{100\%}\right)(1{,}040) = 52 \text{ (Check)}$$

Example: 16 is 80% of what amount?

$$\left(\frac{100\%}{80\%}\right)(16) = 20$$

Example: An electric motor for a pump is 90% efficient. If the motor must deliver 720 watts, how many watts are required to run the motor?

$$\text{Efficiency, \%} = \frac{(\text{Output})(100\%)}{\text{Input}}$$

$$90\% = \frac{(720 \text{ watts})(100\%)}{\text{Input, watts}}$$

$$\text{Input, watts} = \frac{(720 \text{ watts})(100\%)}{90\%}$$

$$= 800 \text{ watts}$$

A.2 AREAS

A.20 Units

Areas are measured in two dimensions or in square units. In the English system of measurement the most common units are square inches, square feet, square yards, and square miles. In the metric system the units are square millimeters, square centimeters, square meters, and square kilometers.

A.21 Rectangle

The area of a rectangle is equal to its length (L) multiplied by its width (W).

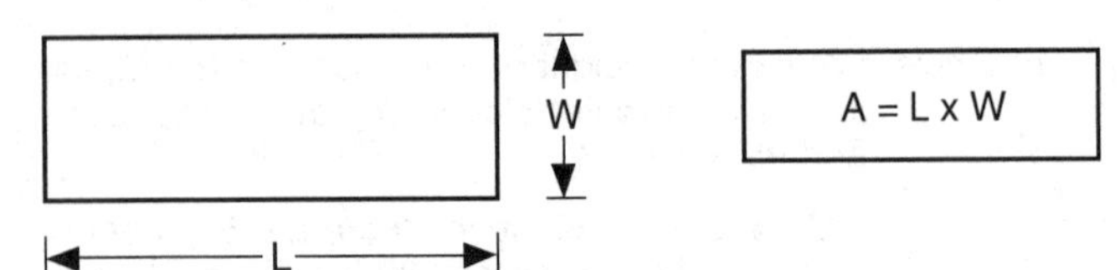

Example: Find the area of a rectangle if the length is 5 feet and the width is 3.5 feet.

Area, sq ft = Length, ft x Width, ft

= 5 ft x 3.5 ft

= 17.5 ft^2

= 17.5 sq ft

Example: The surface area of a loading dock is 330 square feet. One side measures 15 feet. How long is the other side?

A = L x W

330 sq ft = L, ft x 15 ft

$$\frac{L, ft \times 15\ ft}{15\ ft} = \frac{330\ ft^2}{15\ ft}$$ Divide both sides of equation by 15 ft.

$$L, ft = \frac{330\ ft^2}{15\ ft}$$

= 22 ft

A.22 Triangle

The area of a triangle is equal to one-half the base multiplied by the height. This is true for any triangle.

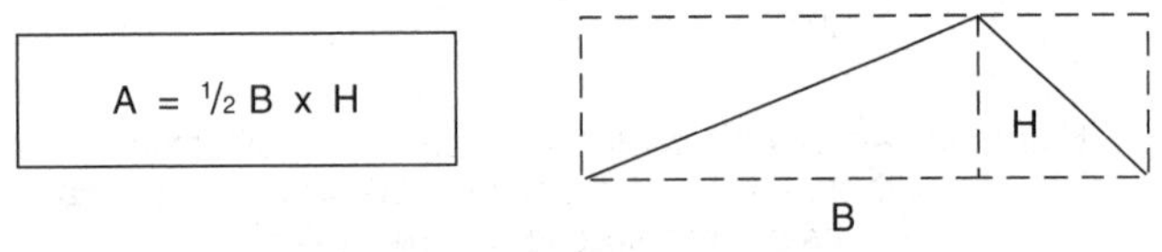

NOTE: The area of *ANY* triangle is equal to ½ the area of the rectangle that can be drawn around it. The area of the rectangle is B x H. The area of the triangle is ½ B x H.

Example: Find the area of triangle ABC.

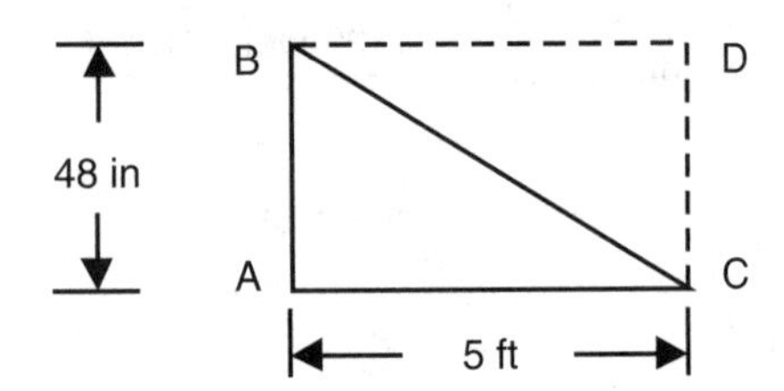

The first step in the solution is to make all the units the same. In this case, it is easier to change inches to feet.

$$48\ in = 48\ \cancel{in} \times \frac{1\ ft}{12\ \cancel{in}} = \frac{48}{12}\ ft = 4\ ft$$

NOTE: All conversions should be calculated in the above manner. Since 1 ft/12 in is equal to unity, or one, multiplying by this factor changes the form of the answer but not its value.

Area, sq ft = ½ (Base, ft)(Height, ft)

= ½ x 5 ft x 4 ft

$= \frac{20}{2}\ ft^2$

= 10 sq ft

NOTE: Triangle ABC is one-half the area of rectangle ABCD. The triangle is a special form called a *RIGHT TRIANGLE* since it contains a 90° angle at point A.

A.23 Circle

A square with sides of 2R can be drawn around a circle with a radius of R.

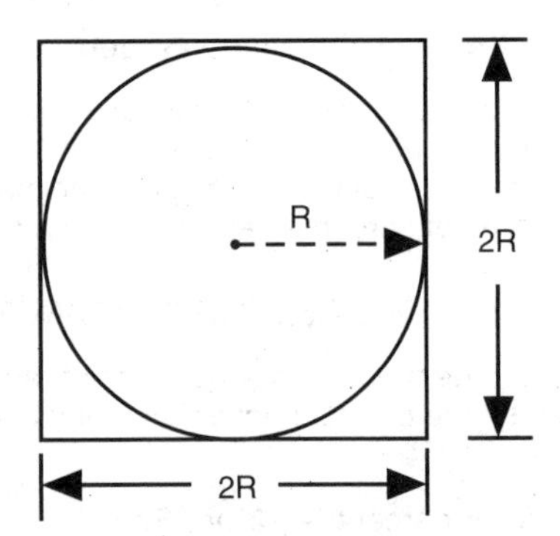

The area of the square is: $A = 2R \times 2R = 4R^2$.

It has been found that the area of any circle drawn within a square is slightly more than ¾ of the area of the square. More precisely, the area of the preceding circle is:

A circle = $3\frac{1}{7} R^2 = 3.14\ R^2$

The formula for the area of a circle is usually written:

$$A = \pi R^2$$

The Greek letter π (pronounced pie) merely substitutes for the value 3.1416.

Since the diameter of any circle is equal to twice the radius, the formula for the area of a circle can be rewritten as follows:

$$A = \pi R^2 = \pi \times R \times R = \pi \times \frac{D}{2} \times \frac{D}{2} = \frac{\pi D^2}{4} = \frac{3.14}{4} D^2 = \boxed{0.785\ D^2}$$

The type of problem and the magnitude (size) of the numbers in a problem will determine which of the two formulas will provide a simpler solution. All of these formulas will give the same results if you use the same number of digits to the right of the decimal point.

Example: What is the area of a circle with a diameter of 20 centimeters?

In this case the formula using a radius is more convenient since it takes advantage of multiplying by 10.

$$\text{Area, sq cm} = \pi (\text{R, cm})^2$$
$$= 3.14 \times 10 \text{ cm} \times 10 \text{ cm}$$
$$= 314 \text{ sq cm}$$

Example: What is the area of a gasoline storage tank with a 5-foot radius?

In this case, the formula using diameter is more convenient.

$$\text{Area, sq ft} = 0.785 (\text{Diameter, ft})^2$$
$$= 0.785 \times 10 \text{ ft} \times 10 \text{ ft}$$
$$= 78.5 \text{ sq ft}$$

Occasionally the operator may be confronted with a problem giving the area and requesting the radius or diameter. This presents the special problem of finding the square root of the number.

Example: The surface area of a garden inside a traffic circle is approximately 5,000 square feet. What is the diameter?

$A = 0.785\ D^2$, or

$$\text{Area, sq ft} = 0.785 (\text{Diameter, ft})^2$$

$5{,}000 \text{ sq ft} = 0.785\ D^2$ To solve, substitute given values in equation.

$\frac{0.785\ D^2}{0.785} = \frac{5{,}000 \text{ sq ft}}{0.785}$ Divide both sides by 0.785 to find D^2.

$$D^2 = \frac{5{,}000 \text{ sq ft}}{0.785}$$
$$= 6{,}369 \text{ sq ft. Therefore,}$$

D = square root of 6,369 sq ft, or

$$\text{Diameter, ft} = \sqrt{6{,}369 \text{ sq ft}}$$

Punch the $\sqrt{}$ sign on your calculator and get D = 79.8 ft.

Sometimes a trial and error method can be used to find square roots. Since 80 x 80 = 6,400, we know the answer is close to 80 feet.

Try 79 x 79 = 6,241

Try 79.5 x 79.5 = 6,320.25

Try 79.8 x 79.8 = 6,368.04

The diameter is 79.8 ft, or approximately 80 feet.

A.24 Cylinder

With the formulas presented thus far, it would be a simple matter to find the number of square feet in a room that was to be painted. The length of each wall would be added together and then multiplied by the height of the wall. This would give the surface area of the walls (minus any area for doors and windows). The ceiling area would be found by multiplying length times width and the result added to the wall area gives the total area.

The surface area of a circular cylinder, however, has not been discussed. If we wanted to know how many square feet of surface area are in a tank with a diameter of 60 feet and a height of 20 feet, we could start with the top and bottom.

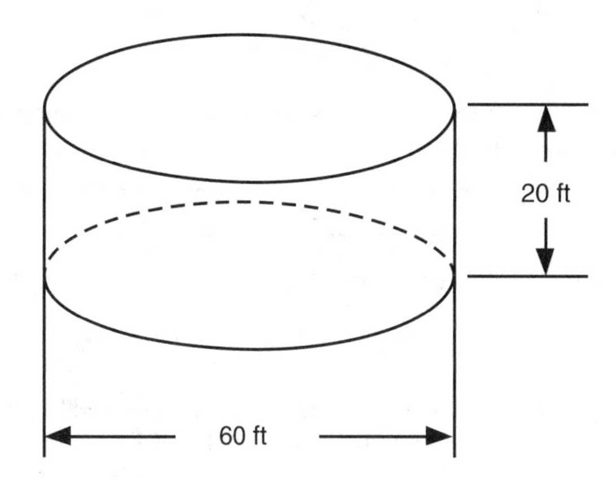

The area of the top and bottom ends are both $\pi \times R^2$.

$$\text{Area, sq ft} = (2 \text{ ends})(\pi)(\text{Radius, ft})^2$$
$$= 2 \times \pi \times (30 \text{ ft})^2$$
$$= 5{,}652 \text{ sq ft}$$

The surface area of the wall must now be calculated. If we made a vertical cut in the wall and unrolled it, the straightened wall would be the same length as the circumference of the floor and ceiling.

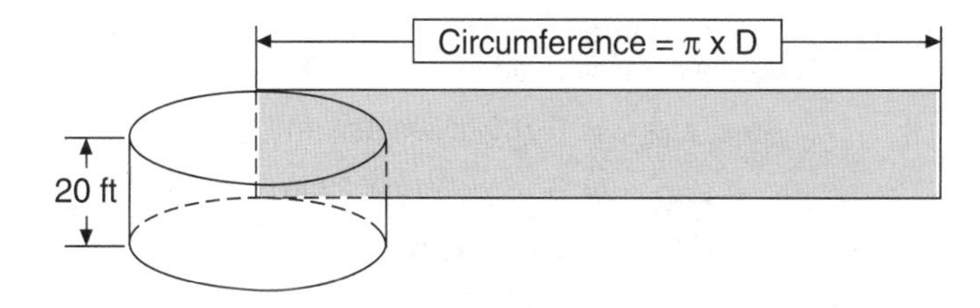

This length has been found to always be $\pi \times D$. In the case of the tank, the length of the wall would be:

$$\text{Length, ft} = (\pi)(\text{Diameter, ft})$$
$$= 3.14 \times 60 \text{ ft}$$
$$= 188.4 \text{ ft}$$

Area would be:

$$A_w\text{, sq ft} = \text{Length, ft} \times \text{Height, ft}$$
$$= 188.4 \text{ ft} \times 20 \text{ ft}$$
$$= 3{,}768 \text{ sq ft}$$

$$\text{Outside Surface Area to Paint, sq ft} = \text{Area of top and bottom, sq ft} + \text{Area of wall, sq ft}$$
$$= 5{,}652 \text{ sq ft} + 3{,}768 \text{ sq ft}$$
$$= 9{,}420 \text{ sq ft}$$

A container has inside and outside surfaces and you may need to paint both of them.

A.25 Cone

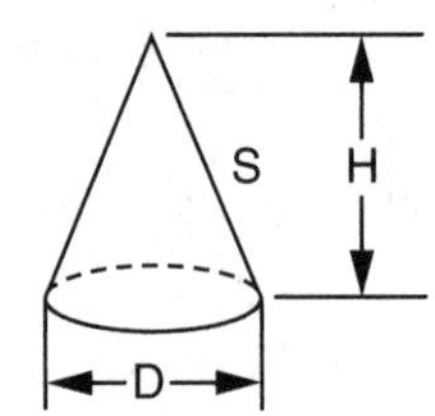

The lateral area of a cone is equal to ½ of the slant height (S) multiplied by the circumference of the base.

$$A_L = \frac{1}{2} S \times \pi \times D = \pi \times S \times R$$

In this case the slant height is not given; it may be calculated by:

$S = \sqrt{R^2 + H^2}$

Example: Find the entire outside area of a cone with a diameter of 30 inches and a height of 20 inches.

$$\text{Slant Height, in} = \sqrt{(\text{Radius, in})^2 + (\text{Height, in})^2}$$

$$= \sqrt{(15 \text{ in})^2 + (20 \text{ in})^2}$$

$$= \sqrt{225 \text{ in}^2 + 400 \text{ in}^2}$$

$$= \sqrt{625 \text{ in}^2}$$

$$= 25 \text{ in}$$

$$\text{Area of Cone, sq in} = \pi \text{ (Slant Height, in)(Radius, in)}$$

$$= 3.14 \times 25 \text{ in} \times 15 \text{ in}$$

$$= 1{,}177.5 \text{ sq in}$$

Since the entire area was asked for, the area of the base must be added.

$$\text{Area, sq in} = 0.785 \text{ (Diameter, in)}^2$$

$$= 0.785 \times 30 \text{ in} \times 30 \text{ in}$$

$$= 706.5 \text{ sq in}$$

$$= \text{Area of Cone, sq in} + \text{Area of Bottom, sq in}$$

$$\text{Total Area, sq in} = 1{,}177.5 \text{ sq in} + 706.5 \text{ sq in}$$

$$= 1{,}884 \text{ sq in}$$

A.26 Sphere

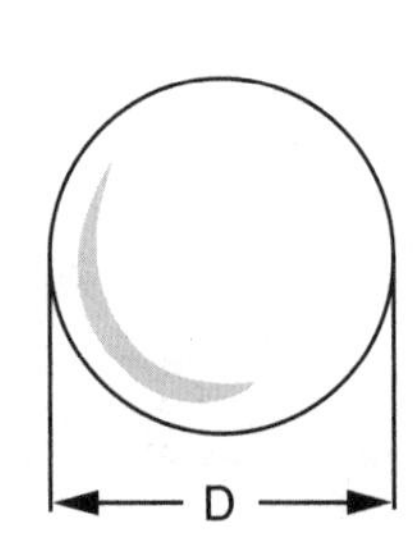

The surface area of a sphere or ball is equal to π multiplied by the diameter squared which is four times the cross-sectional area.

$$A_S = \pi D^2$$

If the radius is used, the formula becomes:

$$A_S = \pi D^2 = \pi \times 2R \times 2R = 4\pi R^2$$

Example: What is the surface area of a sphere-shaped water tank 20 feet in diameter?

$$\text{Area, sq ft} = \pi (\text{Diameter, ft})^2$$

$$= 3.14 \times 20 \text{ ft} \times 20 \text{ ft}$$

$$= 1{,}256 \text{ sq ft}$$

A.3 VOLUMES

A.30 Rectangle

Volumes are measured in three dimensions or in cubic units. To calculate the volume of a rectangle, the area of the base is calculated in square units and then multiplied by the height. The formula then becomes:

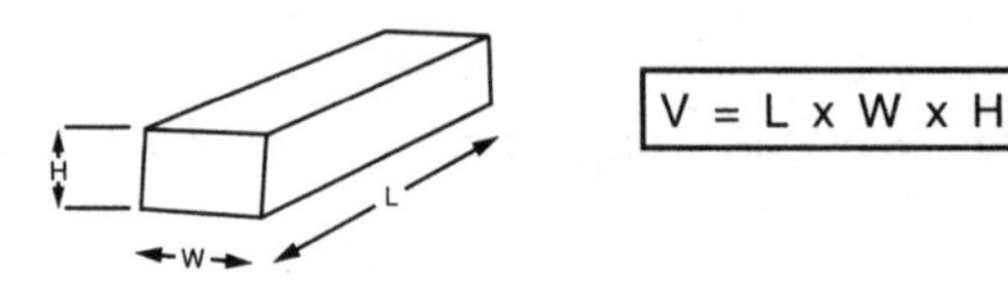

$$V = L \times W \times H$$

Example: The length of a box is 2 feet, the width is 15 inches, and the height is 18 inches. Find its volume.

$$\text{Volume, cu ft} = \text{Length, ft} \times \text{Width, ft} \times \text{Height, ft}$$

$$= 2 \text{ ft} \times \frac{15 \text{ in}}{12 \text{ in/ft}} \times \frac{18 \text{ in}}{12 \text{ in/ft}}$$

$$= 2 \text{ ft} \times 1.25 \text{ ft} \times 1.5 \text{ ft}$$

$$= 3.75 \text{ cu ft}$$

A.31 Prism

The same general rule that applies to the volume of a rectangle also applies to a prism.

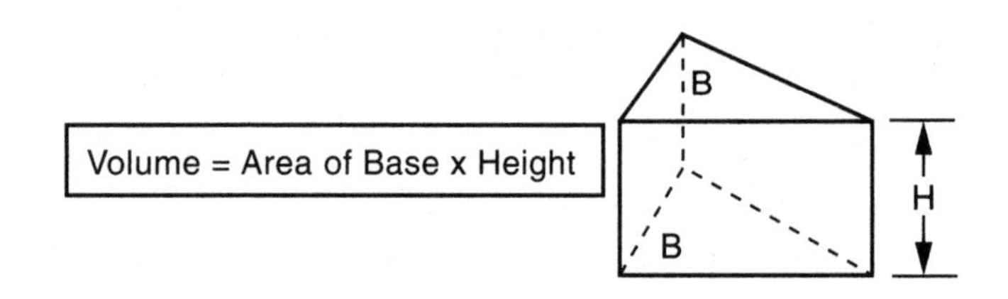

Example: Find the volume of a prism with a base area of 10 square feet and a height of 5 feet. (Note that the base of a prism is triangular in shape.)

$$\text{Volume, cu ft} = \text{Area of Base, sq ft} \times \text{Height, ft}$$

$$= 10 \text{ sq ft} \times 5 \text{ ft}$$

$$= 50 \text{ cu ft}$$

A.32 Cylinder

The volume of a cylinder is equal to the area of the base multiplied by the height.

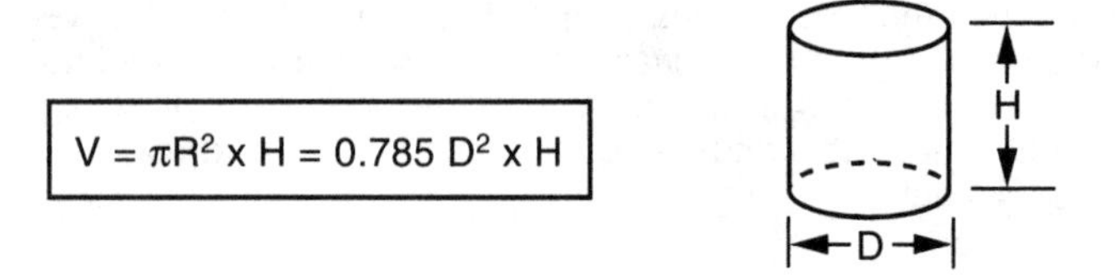

Example: A 24-inch sewer is 360 feet long. Find the volume. Use length instead of height.

$$\text{Volume, cu ft} = (0.785)(\text{Diameter, ft})^2(\text{Length, ft})$$

$$= (0.785)(2\text{ ft})^2(360\text{ ft})$$

$$= 1{,}130.4\text{ cu ft}$$

A.33 Cone

The volume of a cone is equal to 1/3 the volume of a circular cylinder of the same height and diameter.

$$V = \frac{\pi}{3} R^2 \times H$$

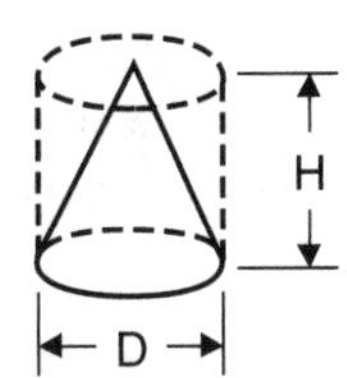

Example: Calculate the volume of a cone if the height at the center is 4 feet and the diameter is 100 feet (radius is 50 feet).

$$\text{Volume, cu ft} = \frac{\pi}{3} \times (\text{Radius})^2 \times \text{Height, ft}$$

$$= \frac{\pi}{3} \times 50\text{ ft} \times 50\text{ ft} \times 4\text{ ft}$$

$$= 10{,}500\text{ cu ft}$$

A.34 Sphere

The volume of a sphere is equal to $\pi/6$ times the diameter cubed.

$$V = \frac{\pi}{6} \times D^3$$

Example: How much gas can be stored in a sphere with a diameter of 12 feet? (Assume atmospheric pressure.)

$$\text{Volume, cu ft} = \frac{\pi}{6} \times (\text{Diameter, ft})^3$$

$$= \frac{\pi}{\not{6}} \times \overset{2}{\not{12}}\text{ ft} \times 12\text{ ft} \times 12\text{ ft}$$

$$= 904.3\text{ cubic feet}$$

A.4 METRIC SYSTEM

The two most common systems of weights and measures are the English system and the metric system (Le Système International d' Unités, SI). Of these two, the metric system is more popular with most of the nations of the world. The reason for this is that the metric system is based on a system of tens and is therefore easier to remember and easier to use than the English system. Even though the basic system in the United States is the English system, the scientific community uses the metric system almost exclusively. Many organizations have urged, for good reason, that the United States switch to the metric system. Today the metric system is gradually becoming the standard system of measurement in the United States.

As the United States changes from the English to the metric system, some confusion and controversy has developed. For example, which is the correct spelling of the following words:

1. Liter or litre?
2. Meter or metre?

The U.S. National Bureau of Standards, the Water Environment Federation, and the American Water Works Association use litre and metre. The U.S. Government uses liter and meter and accepts no deviations. Some people argue that METRE should be used to measure LENGTH and that METER should be used to measure FLOW RATES (like a water or electric meter). Liter and Meter are used in this manual because this is most consistent with spelling in the United States.

One of the most frequent arguments heard against the U.S. switching to the metric system was that the costs of switching manufacturing processes would be excessive. Pipe manufacturers have agreed upon the use of a "soft" metric conversion system during the conversion to the metric system. Past practice in the U.S. has identified some types of pipe by external (outside) diameter while other types are classified by nominal (existing only in name, not real or actual) bore. This means that a six-inch pipe does not have a six-inch inside diameter. With the strict or "hard" metric system, a six-inch pipe would be a 152.4 mm (6 in x 25.4 mm/in) pipe. In the "soft" metric system a six-inch pipe is a 150 mm (6 in x 25 mm/in) pipe. Typical customary and "soft" metric pipe-size designations are shown below:

PIPE-SIZE DESIGNATIONS

Customary, in	2	4	6	8	10	12	15	18
"Soft" Metric, mm	50	100	150	200	250	300	375	450
Customary, in	24	30	36	42	48	60	72	84
"Soft" Metric, mm	600	750	900	1050	1200	1500	1800	2100

In order to study the metric system, you must know the meanings of the terminology used. Following is a list of Greek and Latin prefixes used in the metric system.

PREFIXES USED IN THE METRIC SYSTEM

Prefix	Symbol	Meaning
Micro	μ	1/1 000 000 or 0.000 001
Milli	m	1/1000 or 0.001
Centi	c	1/100 or 0.01
Deci	d	1/10 or 0.1
Unit		1
Deka	da	10
Hecto	h	100
Kilo	k	1000
Mega	M	1 000 000

A.40 Measures of Length

The basic measure of length is the meter.

1 kilometer (km) = 1,000 meters (m)
1 meter (m) = 100 centimeters (cm)
1 centimeter (cm) = 10 millimeters (mm)

Kilometers are usually used in place of miles, meters are used in place of feet and yards, centimeters are used in place of inches and millimeters are used for inches and fractions of an inch.

LENGTH EQUIVALENTS

1 kilometer	= 0.621 mile	1 mile	= 1.61 kilometers
1 meter	= 3.28 feet	1 foot	= 0.305 meter
1 meter	= 39.37 inches	1 inch	= 0.0254 meter
1 centimeter	= 0.3937 inch	1 inch	= 2.54 centimeters
1 millimeter	= 0.0394 inch	1 inch	= 25.4 millimeters

NOTE: The above equivalents are reciprocals. If one equivalent is given, the reverse can be obtained by division. For instance, if one meter equals 3.28 feet, one foot equals 1/3.28 meter, or 0.305 meter.

A.41 Measures of Capacity or Volume

The basic measure of capacity in the metric system is the liter. For measurement of large quantities the cubic meter is sometimes used.

1 kiloliter (k*L*) = 1,000 liters (*L*) = 1 cu meter (m^3)
1 liter (*L*) = 1,000 milliliters (m*L*)

Kiloliters, or cubic meters, are used to measure capacity of large storage tanks or reservoirs in place of cubic feet or gallons. Liters are used in place of gallons or quarts. Milliliters are used in place of quarts, pints, or ounces.

CAPACITY EQUIVALENTS

1 kiloliter	= 264.2 gallons	1 gallon	= 0.003785 kiloliter
1 liter	= 1.057 quarts	1 quart	= 0.946 liter
1 liter	= 0.2642 gallon	1 gallon	= 3.785 liters
1 milliliter	= 0.0353 ounce	1 ounce	= 28.35 milliliters

A.42 Measures of Weight

The basic unit of weight in the Metric system is the gram. One cubic centimeter of water at maximum density weighs one gram, and thus there is a direct, simple relation between volume of water and weight in the Metric system.

1 kilogram (kg) = 1,000 grams (gm)
1 gram (gm) = 1,000 milligrams (mg)
1 milligram (mg) = 1,000 micrograms (μg)

Grams are usually used in place of ounces, and kilograms are used in place of pounds.

WEIGHT EQUIVALENTS

1 kilogram	= 2.205 pounds	1 pound	= 0.4536 kilogram
1 gram	= 0.0022 pound	1 pound	= 453.6 grams
1 gram	= 0.0353 ounce	1 ounce	= 28.35 grams
1 gram	= 15.43 grains	1 grain	= 0.0648 gram

A.43 Temperature

Just as the operator should become familiar with the metric system, you should also become familiar with the centigrade (Celsius) scale for measuring temperature. There is nothing magical about the centigrade scale—it is simply a different size than the Fahrenheit scale. The two scales compare as follows:

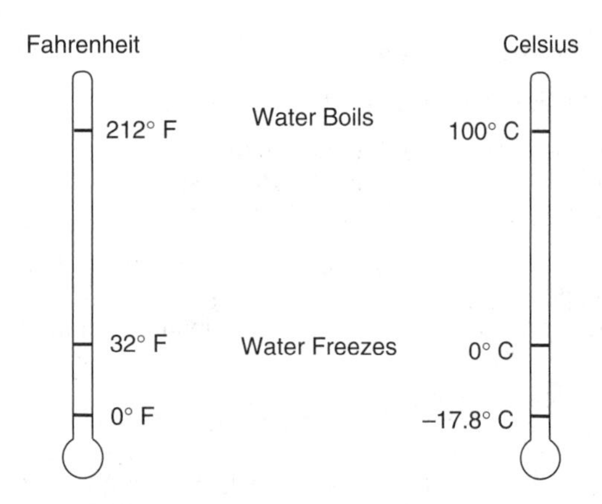

The two scales are related in the following manner:

Fahrenheit	= (°C x 9/5) + 32°
Celsius	= (°F – 32°) x 5/9

Example: Convert 20° Celsius to degrees Fahrenheit.

$$°F = (°C \times 9/5) + 32°$$
$$°F = (20° \times 9/5) + 32°$$
$$°F = \frac{180°}{5} + 32°$$
$$= 36° + 32°$$
$$= 68°F$$

Example: Convert –10°C to °F.

$$°F = (-10° \times 9/5) + 32°$$
$$°F = -90°/5 + 32°$$
$$= -18° + 32°$$
$$= 14°F$$

Example: Convert –13°F to °C.

$$°C = (°F - 32°) \times \frac{5}{9}$$
$$°C = (-13° - 32°) \times \frac{5}{9}$$
$$= -45° \times \frac{5}{9}$$
$$= -5° \times 5$$
$$= -25°C$$

A.44 Milligrams per Liter

Milligrams per liter (mg/*L*) is a unit of measurement used in laboratory and scientific work to indicate very small concentrations of dilutions. Since water contains small concentrations of dissolved substances and solids, and since small amounts of chemical compounds are sometimes used in wastewater treatment processes, the term milligrams per liter is also common in treatment plants. It is a weight/volume relationship.

As previously discussed:

1,000 liters = 1 cubic meter = 1,000,000 cubic centimeters.

Therefore

1 liter = 1,000 cubic centimeters.

Since one cubic centimeter of water weighs one gram,

1 liter of water = 1,000 grams or 1,000,000 milligrams.

$$\frac{1 \text{ milligram}}{\text{liter}} = \frac{1 \text{ milligram}}{1{,}000{,}000 \text{ milligrams}} = \frac{1 \text{ part}}{\text{million parts}} = \text{1 part per million (ppm)}$$

Milligrams per liter and parts per million (parts) may be used interchangeably as long as the liquid density is 1.0 gm/cu cm or 62.43 lb/cu ft. A concentration of 1 milligram/liter (mg/*L*) or 1 ppm means that there is 1 part of substance by weight for every 1 million parts of water. A concentration of 10 mg/*L* would mean 10 parts of substance per million parts of water.

To get an idea of how small 1 mg/*L* is, divide the numerator and denominator of the fraction by 10,000. This, of course, does not change its value since 10,000 ÷ 10,000 is equal to one.

$$1\,\frac{mg}{L} = \frac{1\ mg}{1{,}000{,}000\ mg} = \frac{1/10{,}000\ mg}{1{,}000{,}000/10{,}000\ mg} = \frac{0.0001\ mg}{100\ mg} = 0.0001\%$$

Therefore, 1 mg/*L* is equal to one ten-thousandth of a percent, or

1% is equal to 10,000 mg/*L*.

To convert mg/*L* to %, move the decimal point four places or numbers to the left.

A.5 WEIGHT-VOLUME RELATIONS

Another factor for the operator to remember, in addition to the weight of a gallon of water, is the weight of a cubic foot of water. One cubic foot of water weighs 62.4 lbs. If these two weights are divided, it is possible to determine the number of gallons in a cubic foot.

$$\frac{62.4\ \text{pounds/cu ft}}{8.34\ \text{pounds/gal}} = 7.48\ \text{gal/cu ft}$$

Thus we have another very important relationship to commit to memory.

$$8.34\ \text{lb/gal} \times 7.48\ \text{gal/cu ft} = 62.4\ \text{lb/cu ft}$$

It is only necessary to remember two of the above items since the third may be found by calculation. For most problems, 8 1/3 lbs/gal and 7 1/2 gal/cu ft will provide sufficient accuracy.

Example: Change 1,000 cu ft of water to gallons.

1,000 cu ft x 7.48 gal/cu ft = 7,480 gallons

Example: What is the weight of three cubic feet of water?

62.4 lb/cu ft x 3 cu ft = 187.2 lbs

Example: The net weight of a tank of water is 750 lbs. How many gallons does it contain?

$$\frac{750\ \text{lb}}{8.34\ \text{lb/gal}} = 90\ \text{gal}$$

A.6 FORCE, PRESSURE, AND HEAD

In order to study the forces and pressures involved in fluid flow, it is first necessary to define the terms used.

FORCE: The push exerted by water on any surface being used to confine it. Force is usually expressed in pounds, tons, grams, or kilograms.

PRESSURE: The force per unit area. Pressure can be expressed in many ways, but the most common term is pounds per square inch (psi).

HEAD: Vertical distance from the water surface to a reference point below the surface. Usually expressed in feet or meters.

An example should serve to illustrate these terms.

If water were poured into a one-foot cubical container, the *FORCE* acting on the bottom of the container would be 62.4 pounds.

The *PRESSURE* acting on the bottom would be 62.4 pounds per square foot. The area of the bottom is also 12 in x 12 in = 144 in^2.

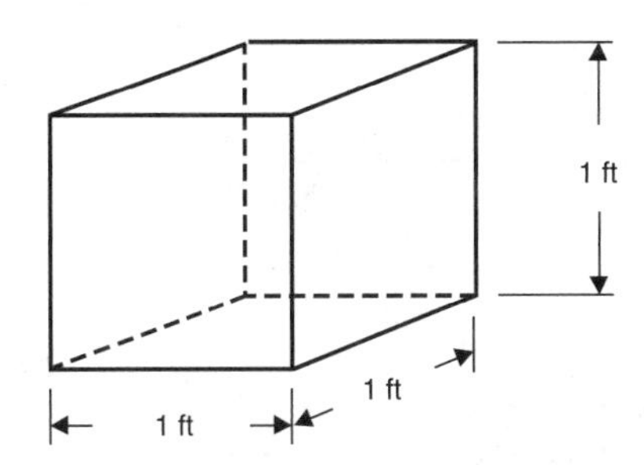

Therefore, the pressure may also be expressed as:

$$\text{Pressure, psi} = \frac{62.4\ \text{lb}}{\text{sq ft}} = \frac{62.4\ \text{lb/sq ft}}{144\ \text{sq in/sq ft}}$$

$$= 0.433\ \text{lb/sq in}$$

$$= 0.433\ \text{psi}$$

Since the height of the container is one foot, the *HEAD* would be one foot.

The pressure in any vessel at one foot of depth or one foot of head is 0.433 psi acting in any direction.

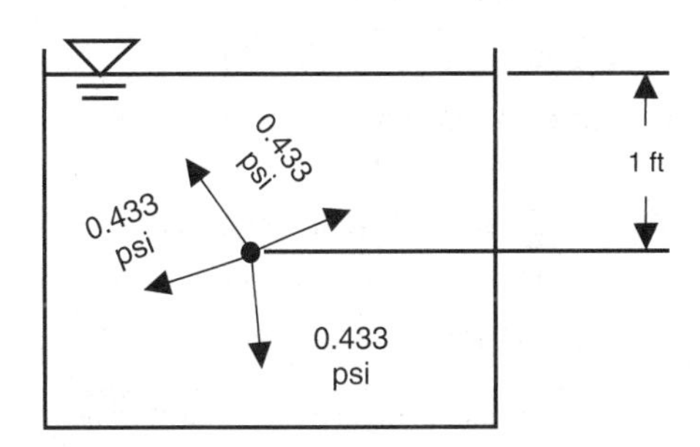

If the depth of water in the previous example were increased to two feet, the pressure would be:

$$p = \frac{2(62.4\ \text{lb})}{144\ \text{sq in}} = \frac{124.8\ \text{lb}}{144\ \text{sq in}} = 0.866\ \text{psi}$$

Therefore we can see that for every foot of head, the pressure increases by 0.433 psi. Thus, the general formula for pressure becomes:

$$p,\ \text{psi} = 0.433\ (H,\ \text{ft})$$

H = feet of head

p = pounds per square *INCH* of pressure

$$P,\ \text{lb/sq ft} = 62.4\ (H,\ \text{ft})$$

H = feet of head

P = pounds per square *FOOT* of pressure

We can now draw a diagram of the pressure acting on the side of a tank. Assume a 4-foot deep tank. The pressures shown on the tank are gage pressures. These pressures do not include the atmospheric pressure acting on the surface of the water.

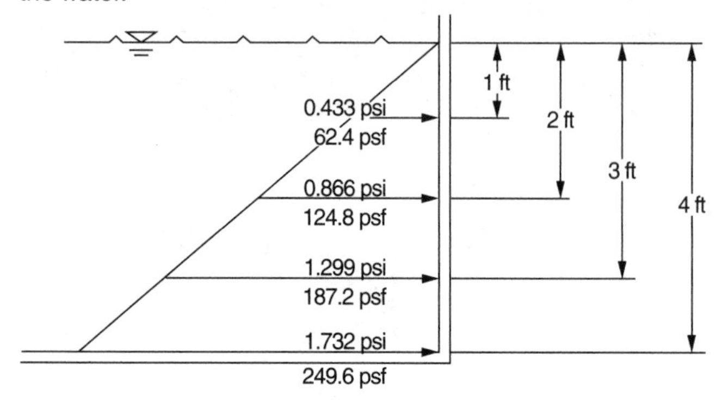

p_0 = 0.433 x 0 = 0.0 psi
p_1 = 0.433 x 1 = 0.433 psi
p_2 = 0.433 x 2 = 0.866 psi
p_3 = 0.433 x 3 = 1.299 psi
p_4 = 0.433 x 4 = 1.732 psi

P_0 = 62.4 x 0 = 0.0 lb/sq ft
P_1 = 62.4 x 1 = 62.4 lb/sq ft
P_2 = 62.4 x 2 = 124.8 lb/sq ft
P_3 = 62.4 x 3 = 187.2 lb/sq ft
P_4 = 62.4 x 4 = 249.6 lb/sq ft

The average *PRESSURE* acting on the tank wall is 1.732 psi/2 = 0.866 psi, or 249.6 psf/2 = 124.8 psf. We divided by two to obtain the average pressure because there is zero pressure at the top and 1.732 psi pressure on the bottom of the wall.

If the wall were 5 feet long, the pressure would be acting over the entire 20 square foot (5 ft x 4 ft) area of the wall. The total force acting to push the wall would be:

Force, lb = (Pressure, lb/sq ft)(Area, sq ft)

= 124.8 lb/sq ft x 20 sq ft

= 2,496 lbs

If the pressure in psi were used, the problem would be similar:

Force, lb = (Pressure, lb/sq in)(Area, sq in)

= 0.866 psi x 48 in x 60 in

= 2,494 lb*

* Difference in answer due to rounding off of decimal points.

The general formula, then, for finding the total force acting on a side wall of a tank is:

$$F = 31.2 \times H^2 \times L$$

F = force in pounds

H = head in feet

L = length of wall in feet

31.2 = constant with units of lbs/cu ft and considers the fact that the force results from H/2 or half the depth of the water which is the average depth. The force is exerted at H/3 from the bottom.

Example: Find the force acting on a 5-foot long wall in a 4-foot deep tank.

Force, lb = 31.2 (Head, ft)^2(Length, ft)

= 31.2 lb/cu ft x (4 ft)^2 x 5 ft

= 2,496 lbs

Occasionally an operator is warned: *NEVER EMPTY A TANK DURING PERIODS OF HIGH GROUNDWATER.* Why? The pressure on the bottom of the tank caused by the water surrounding the tank will tend to float the tank like a cork if the upward force of the water is greater than the weight of the tank.

$$F = 62.4 \times H \times A$$

F = upward force in pounds

H = head of water on tank bottom in feet

A = area of bottom of tank in square feet

62.4 = a constant with units of lbs/cu ft

This formula is approximately true if the tank doesn't crack, leak, or start to float.

Example: Find the upward force on the bottom of an empty tank caused by a groundwater depth of 8 feet above the tank bottom. The tank is 20 ft wide and 40 ft long.

Force, lb = 62.4 (Head, ft)(Area, sq ft)

= 62.4 lb/cu ft x 8 ft x 20 ft x 40 ft

= 399,400 lbs

A.7 VELOCITY AND FLOW RATE

A.70 Velocity

The velocity of a particle or substance is the speed at which it is moving. It is expressed by indicating the length of travel and how long it takes to cover the distance. Velocity can be expressed in almost any distance and time units. For instance, a car may be traveling at a rate of 280 miles per five hours. However, it is normal to express the distance traveled per unit time. The above example would then become:

$$\text{Velocity, mi/hr} = \frac{280 \text{ miles}}{5 \text{ hours}}$$

$$= 56 \text{ miles/hour}$$

The velocity of water in a channel, pipe, or other conduit can be expressed in the same way. If the particle of water travels 600 feet in five minutes, the velocity is:

$$\text{Velocity, ft/min} = \frac{\text{Distance, ft}}{\text{Time, minutes}}$$

$$= \frac{600 \text{ ft}}{5 \text{ min}}$$

$$= 120 \text{ ft/min}$$

If you wish to express the velocity in feet per second, multiply by 1 min/60 seconds.

NOTE: Multiplying by $\frac{1 \text{ minute}}{60 \text{ seconds}}$ is like multiplying by $\frac{1}{1}$; it does not change the relative value of the answer. It only changes the form of the answer.

Velocity, ft/sec = (Velocity, ft/min)(1 min/60 sec)

$$= \frac{120 \text{ ft}}{\cancel{\text{min}}} \times \frac{1 \cancel{\text{min}}}{60 \text{ sec}}$$

$$= \frac{120 \text{ ft}}{60 \text{ sec}}$$

$$= 2 \text{ ft/sec}$$

A.71 Flow Rate

If water in a one-foot wide channel is one foot deep, then the cross-sectional area of the channel is 1 ft x 1 ft = 1 sq ft.

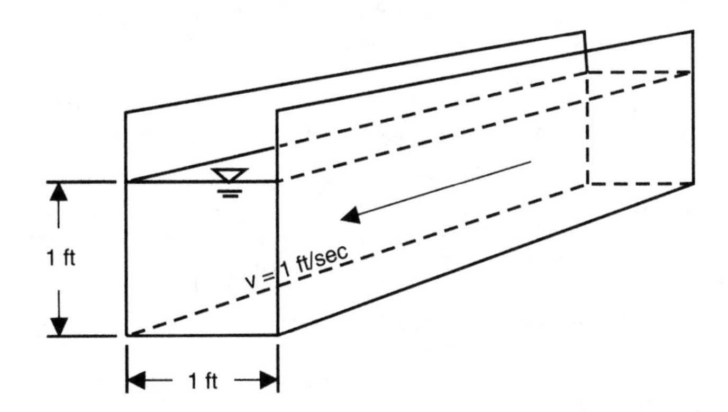

If the velocity in this channel is 1 ft per second, then each second a body of water 1 sq ft in area and 1 ft long will pass a given point. The volume of this body of water would be 1 cubic foot. Since 1 cubic foot of water would pass by every second, the flow rate would be equal to 1 cubic foot per second, or 1 CFS.

To obtain the flow rate in the above example the velocity was multiplied by the cross-sectional area. This is another important general formula.

Q = V x A

Q = flow rate, CFS or cu ft/sec

V = velocity, ft/sec

A = area, sq ft

Example: A rectangular channel 3 feet wide contains water 2 feet deep and flowing at a velocity of 1.5 feet per second. What is the flow rate in CFS?

$$Q = V \times A$$

$$\text{Flow Rate, CFS} = \text{Velocity, ft/sec} \times \text{Area, sq ft}$$

$$= 1.5 \text{ ft/sec} \times 3 \text{ ft} \times 2 \text{ ft}$$

$$= 9 \text{ cu ft/sec}$$

Example: Flow in a 2.5-foot wide channel is 1.4 ft deep and measures 11.2 CFS. What is the average velocity?

In this problem we want to find the velocity. Therefore, we must rearrange the general formula to solve for velocity.

$$V = \frac{Q}{A}$$

$$\text{Velocity, ft/sec} = \frac{\text{Flow Rate, cu ft/sec}}{\text{Area, sq ft}}$$

$$= \frac{11.2 \text{ cu ft/sec}}{2.5 \text{ ft} \times 1.4 \text{ ft}}$$

$$= \frac{11.2 \text{ cu ft/sec}}{3.5 \text{ sq ft}}$$

$$= 3.2 \text{ ft/sec}$$

Example: Flow in an 8-inch pipe is 500 GPM. What is the average velocity?

$$\text{Area, sq ft} = 0.785 (\text{Diameter, ft})^2$$

$$= 0.785 (\tfrac{8}{12} \text{ ft})^2$$

$$= 0.785 (\tfrac{2}{3} \text{ ft})^2$$

$$= 0.785 (\tfrac{2}{3} \text{ ft})(\tfrac{2}{3} \text{ ft})$$

$$= 0.785 (\tfrac{4}{9} \text{ ft}^2)$$

$$= 0.35 \text{ sq ft}$$

$$\text{Flow, CFS} = \text{Flow, gal/min} \times \frac{\text{cu ft}}{7.48 \text{ gal}} \times \frac{1 \text{ min}}{60 \text{ sec}}$$

$$= \frac{500 \text{ gal}}{\text{min}} \times \frac{\text{cu ft}}{7.48 \text{ gal}} \times \frac{1 \text{ min}}{60 \text{ sec}}$$

$$= \frac{500 \text{ cu ft}}{448.8 \text{ sec}}$$

$$= 1.114 \text{ CFS}$$

$$\text{Velocity, ft/sec} = \frac{\text{Flow, cu ft/sec}}{\text{Area, sq ft}}$$

$$= \frac{1.114 \text{ cu ft/sec}}{0.35 \text{ sq ft}}$$

$$= 3.18 \text{ ft/sec}$$

A.8 PUMPS

A.80 Pressure

Atmospheric pressure at sea level is approximately 14.7 psi. This pressure acts in all directions and on all objects. If a tube is placed upside down in a basin of water and a 1 psi partial vacuum is drawn on the tube, the water in the tube will rise 2.31 feet.

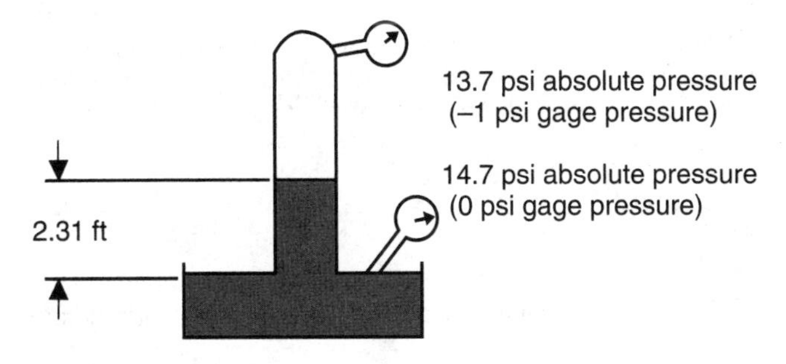

NOTE: 1 ft of water = 0.433 psi; therefore,

$$1 \text{ psi} = \frac{1}{0.433} \text{ ft} = 2.31 \text{ ft of water}$$

The action of the partial vacuum is what gets water out of a sump or well and up to a pump. It is not sucked up, but it is pushed up by atmospheric pressure on the water surface in the sump. If a complete vacuum could be drawn, the water would rise 2.31 x 14.7 = 33.9 feet; but this is impossible to achieve. The practical limit of the suction lift of a positive displacement pump is about 22 feet, and that of a centrifugal pump is 15 feet.

A.81 Work

Work can be expressed as lifting a weight a certain vertical distance. It is usually defined in terms of foot-pounds.

Example: A 165-pound man runs up a flight of stairs 20 feet high. How much work did he do?

$$\text{Work, ft-lb} = \text{Weight, lb x Height, ft}$$
$$= 165 \text{ lb x } 20 \text{ ft}$$
$$= 3{,}300 \text{ ft-lb}$$

A.82 Power

Power is a rate of doing work and is usually expressed in foot-pounds per minute.

Example: If the man in the above example runs up the stairs in three seconds, how much power has he exerted?

$$\text{Power, ft-lbs/sec} = \frac{\text{Work, ft-lb}}{\text{Time, sec}}$$
$$= \frac{3{,}300 \text{ ft-lbs}}{3 \text{ \sout{sec}}} \times \frac{60 \text{ \sout{sec}}}{\text{minute}}$$
$$= 66{,}000 \text{ ft-lb/min}$$

A.83 Horsepower

Horsepower is also a unit of power. One horsepower is defined as 33,000 ft-lbs per minute or 746 watts.

Example: How much horsepower has the man in the previous example exerted as he climbs the stairs?

$$\text{Horsepower, HP} = (\text{Power, ft-lb/min}) \left(\frac{\text{HP}}{33{,}000 \text{ ft-lb/min}} \right)$$
$$= 66{,}000 \text{ ft-lb/min} \times \frac{\text{Horsepower}}{33{,}000 \text{ ft-lb/min}}$$
$$= 2 \text{ HP}$$

Work is also done by lifting water. If the flow from a pump is converted to a weight of water and multiplied by the vertical distance it is lifted, the amount of work or power can be obtained.

$$\text{Horsepower, HP} = \frac{\text{Flow, gal}}{\text{min}} \times \text{Lift, ft} \times \frac{8.34 \text{ lb}}{\text{gal}} \times \frac{\text{Horsepower}}{33{,}000 \text{ ft-lb/min}}$$

Solving the above relation, the amount of horsepower necessary to lift the water is obtained. This is called water horsepower.

$$\text{Water, HP} = \frac{(\text{Flow, GPM})(\text{H, ft})}{3{,}960^*}$$

$$^* \frac{8.34 \text{ lb}}{\text{gal}} \times \frac{\text{HP}}{33{,}000 \text{ ft-lb/min}} = \frac{1}{3{,}960}$$

1 gallon weighs 8.34 pounds and 1 horsepower is the same as 33,000 ft-lb/min.

H or Head in feet is the same as Lift in feet.

However, since pumps are not 100% efficient (they cannot transmit all the power put into them), the horsepower supplied to a pump is greater than the water horsepower. Horsepower supplied to the pump is called brake horsepower.

$$\boxed{\text{Brake, HP} = \frac{\text{Flow, GPM x H, ft}}{3{,}960 \times E_p}}$$

E_p = Efficiency of Pump (Usual range 50-85%, depending on type and size of pump)

Motors are also not 100% efficient; therefore, the power supplied to the motor is greater than the motor transmits.

$$\boxed{\text{Motor, HP} = \frac{\text{Flow, GPM x H, ft}}{3{,}960 \times E_p \times E_m}}$$

E_m = Efficiency of Motor (Usual range 80-95%, depending on type and size of motor)

The above formulas have been developed for the pumping of water and wastewater which have a specific gravity of 1.0. If other liquids are to be pumped, the formulas must be multiplied by the specific gravity of the liquid.

Example: A flow of 500 GPM of water is to be pumped against a total head of 100 feet by a pump with an efficiency of 70%. What is the pump horsepower?

$$\text{Brake, HP} = \frac{\text{Flow, GPM x H, ft}}{3{,}960 \times E_p}$$
$$= \frac{500 \times 100}{3{,}960 \times 0.70}$$
$$= 18 \text{ HP}$$

Example: Find the horsepower required to pump gasoline (specific gravity = 0.75) in the above problem.

$$\text{Brake, HP} = \frac{500 \times 100 \times 0.75}{3{,}960 \times 0.70}$$
$$= 13.5 \text{ HP}$$ (gasoline is lighter and requires less horsepower)

A.84 Head

Basically, the head that a pump must work against is determined by measuring the vertical distance between the two water surfaces, or the distance the water must be lifted. This is called the static head. Two typical conditions for lifting water are shown below.

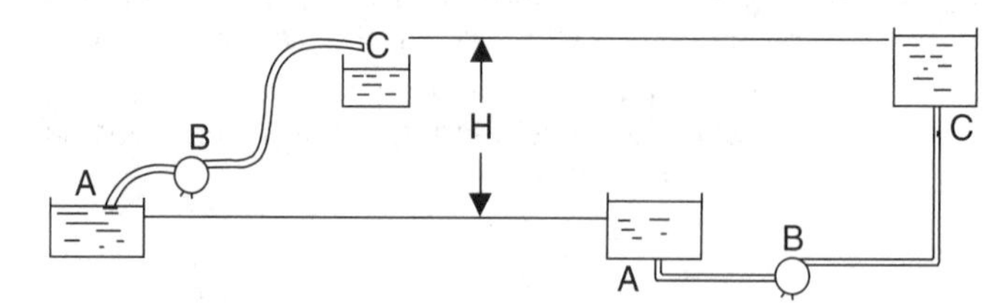

If a pump were designed in the above examples to pump only against head H, the water would never reach the intended point. The reason for this is that the water encounters friction in the pipelines. Friction depends on the roughness and length of pipe, the pipe diameter, and the flow velocity. The turbulence caused at the pipe entrance (point A); the pump (point B); the pipe exit (point C); and at each elbow, bend, or transition also adds to these friction losses. Tables and charts are available in Section A.88 for calculation of these friction losses so they may be added to the measured or static head to obtain the total head. For short runs of pipe which do not have high velocities, the friction losses are generally less than 10 percent of the static head.

Example: A pump is to be located 8 feet above a wet well and must lift 1.8 MGD another 50 feet to a storage reservoir. If the pump has an efficiency of 75% and the motor an efficiency of 90%, what is the cost of the power consumed if one kilowatt hour costs 4 cents?

Since we are not given the length or size of pipe and the number of elbows or bends, we will assume friction to be 10% of static head.

Static Head, ft = Suction Lift, ft + Discharge Head, ft

= 8 ft + 50 ft

= 58 ft

Friction Losses, ft = 0.1 (Static Head, ft)

= 0.1 (58 ft)

= 5.8 ft

Total Dynamic Head, ft = Static Head, ft + Friction Losses, ft

= 58 ft + 5.8 ft

= 63.8 ft

$$\text{Flow, GPM} = \frac{1{,}800{,}000\text{ gal}}{\cancel{\text{day}}} \times \frac{\cancel{\text{day}}}{24\ \cancel{\text{hr}}} \times \frac{1\ \cancel{\text{hr}}}{60\text{ min}}$$

= 1,250 GPM (assuming pump runs 24 hours per day)

$$\text{Motor, HP} = \frac{\text{Flow, GPM} \times \text{H, ft}}{3{,}960 \times E_p \times E_m}$$

$$= \frac{1{,}250 \times 63.8}{3{,}960 \times 0.75 \times 0.9}$$

= 30 HP

Kilowatt-hr = 30 $\cancel{\text{HP}}$ x 24 hr/day x 0.746 kW/$\cancel{\text{HP}}$*

= 537 kilowatt-hr/day

Cost = kWh x $0.04/kWh

= 537 x 0.04

= $21.48/day

* See Conversion Tables — Section A.10, "Power," page 596

A.85 Pump Characteristics

The discharge of a centrifugal pump, unlike a positive displacement pump, can be made to vary from zero to a maximum capacity which depends on the speed, head, power, and specific impeller design. The interrelation of capacity, efficiency, head, and power is known as the characteristics of the pump.

The first relation normally looked at when searching for a pump is the head vs. capacity. The head of a centrifugal pump normally rises as the capacity is reduced. If the values are plotted on a graph they appear as follows:

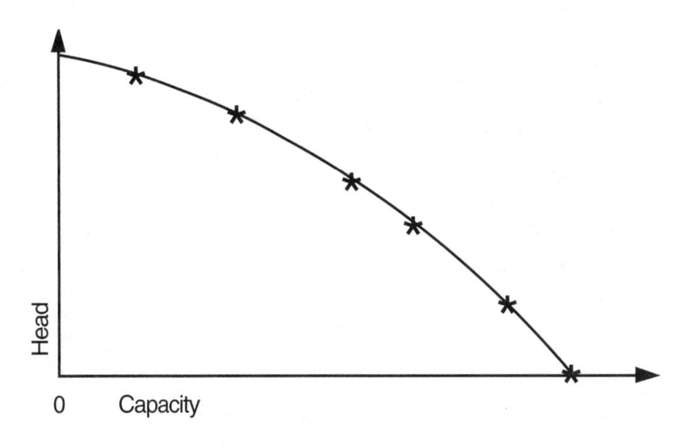

Another important characteristic is the pump efficiency. It begins from zero at no discharge, increases to a maximum, and then drops as the capacity is increased. Following is a graph of efficiency vs. capacity:

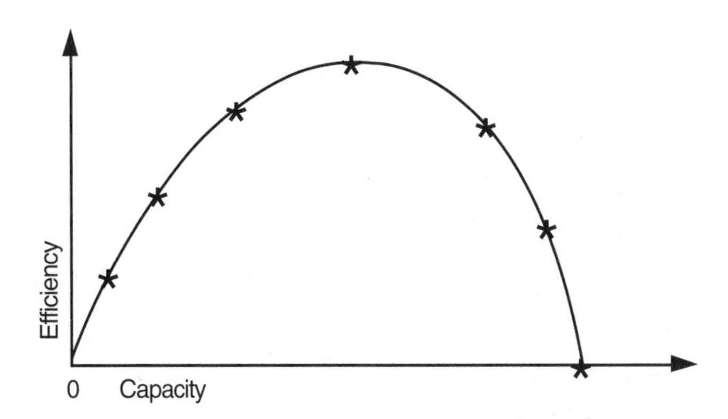

The last important characteristic is the brake horsepower or the power input to the pump. The brake horsepower usually increases with increasing capacity until it reaches a maximum, then it normally reduces slightly.

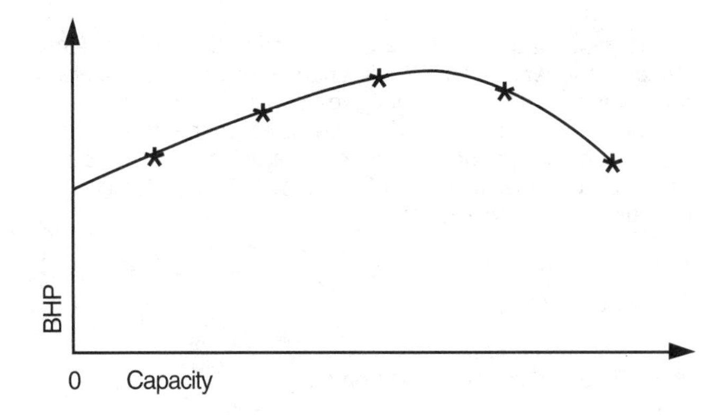

These pump characteristic curves are quite important. Pump sizes are normally picked from these curves rather than calculations. For ease of reading, the three characteristic curves are normally plotted together. A typical graph of pump characteristics is shown on page 590.

The curves show that the maximum efficiency for the particular pump in question occurs at approximately 1,475 GPM, a

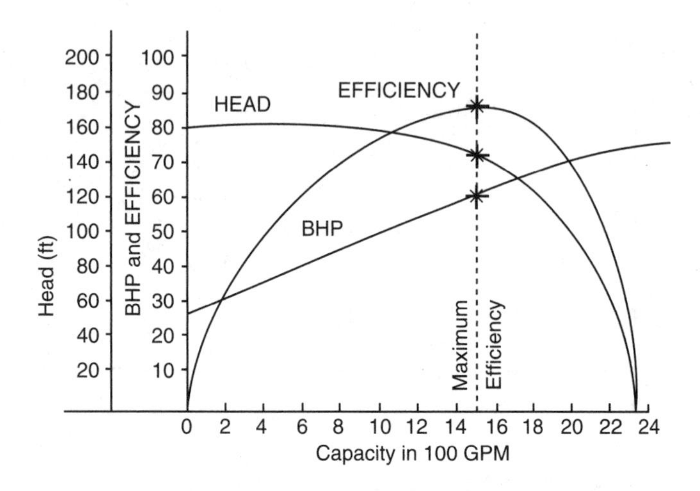

head of 132 feet, and a brake horsepower of 58. Operating at this point the pump has an efficiency of approximately 85%. This can be verified by calculation.

$$\text{BHP} = \frac{\text{Flow, GPM x H, ft}}{3{,}960 \text{ x E}}$$

As previously explained, a number can be written over one without changing its value:

$$\frac{\text{BHP}}{1} = \frac{\text{GPM x H}}{3{,}960 \text{ x E}}$$

Since the formula is now in ratio form, it can be cross multiplied.

$$\text{BHP x } 3{,}960 \text{ x E} = \text{GPM x H x } 1$$

Solving for E,

$$\text{E} = \frac{\text{GPM x H}}{3{,}960 \text{ x BHP}}$$

$$\text{E} = \frac{1{,}475 \text{ GPM x } 132 \text{ ft}}{3{,}960 \text{ x } 58 \text{ HP}}$$

$$= 0.85 \text{ or } 85\% \text{ (Check)}$$

The preceding is only a brief description of pumps to familiarize the operator with their characteristics. The operator does not normally specify the type and size of pump needed at a plant. If a pump is needed, the operator should be able to supply the information necessary for a pump supplier to provide the best possible pump for the lowest cost. Some of the information needed includes:

1. Flow range desired;
2. Head conditions:
   a. Suction head or lift,
   b. Pipe and fitting friction head, and
   c. Discharge head;
3. Type of fluid pumped and temperature; and
4. Pump location.

A.86 Evaluation of Pump Performance

1. Capacity

Sometimes it is necessary to determine the capacity of a pump. This can be accomplished by determining the time it takes a pump to fill or empty a portion of a wet well or diversion box when all inflow is blocked off.

Example:

a. Measure the size of the wet well.

Length = 10 ft

Width = 10 ft

Depth = 5 ft (We will measure the time it takes to lower the well a distance of 5 feet.)

Volume, cu ft = L, ft x W, ft x D, ft

= 10 ft x 10 ft x 5 ft

= 500 cu ft

b. Record time for water to drop 5 feet in wet well.

Time = 10 minutes 30 seconds

= 10.5 minutes

c. Calculate pumping rate or capacity.

$$\text{Pumping Rate, GPM} = \frac{\text{Volume, gallons}}{\text{Time, minutes}}$$

$$= \frac{(500 \text{ cu ft})(7.5 \text{ gal/cu ft})}{10.5 \text{ min}}$$

$$= \frac{3{,}750}{10.5}$$

$$= 357 \text{ GPM}$$

If you know the total dynamic head and have the pump's performance curves, you can determine if the pump is delivering at design capacity. If not, try to determine the cause (see Chapter 15, "Maintenance," in *OPERATION OF WASTEWATER TREATMENT PLANTS*, Volume II). After a pump overhaul, the pump's actual performance (flow, head, power, and efficiency) should be compared with the pump manufacturer's performance curves (see Volume II, Chapter 8, Section 8.25, for procedures). This method for calculating the rate of filling or emptying of a wet well or diversion box can be used to calibrate flowmeters.

2. Efficiency

To estimate the efficiency of the pump in the previous example, the total head must be known. This head may be estimated by measuring the suction and discharge pressure. Assume these were measured as follows:

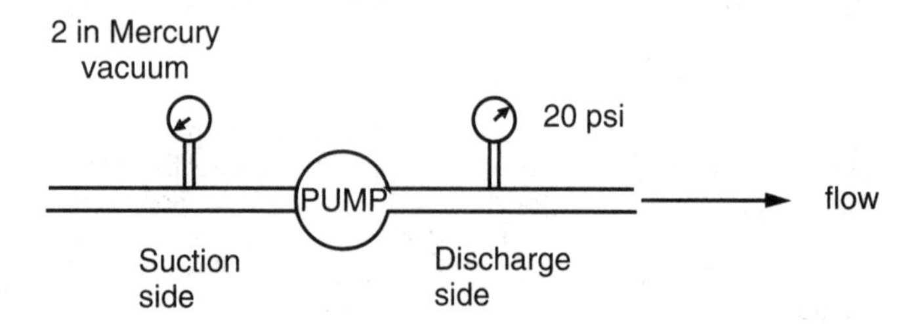

No additional information is necessary if we assume the pressure gages are at the same height and the pipe diameters are the same. Both pressure readings must be converted to feet.

$$\text{Suction Lift, ft} = 2 \text{ in Mercury} \times \frac{1.133 \text{ ft water*}}{1 \text{ in Mercury}}$$

$$= 2.27 \text{ ft}$$

$$\text{Discharge Head, ft} = 20 \text{ psi} \times 2.31 \text{ ft/psi*}$$

$$= 46.20 \text{ ft}$$

$$\text{Total Head, ft} = \text{Suction Lift, ft} + \text{Discharge Head, ft}$$

$$= 2.27 \text{ ft} + 46.20 \text{ ft}$$

$$= 48.47 \text{ ft}$$

* See Conversion Tables — Section A.10, "Pressure," page 596.

Calculate the power output of the pump or water horsepower.

$$\text{Water Horsepower, HP} = \frac{(\text{Flow GPM})(\text{Head, ft})}{3{,}960}$$

$$= \frac{(357 \text{ GPM})(48.47 \text{ ft})}{3{,}960}$$

$$= 4.4 \text{ HP}$$

To estimate the efficiency of the pump, measure the kilowatts drawn by the pump motor. Assume the meter indicates 8,000 watts or 8 kilowatts. The manufacturer claims the electric motor is 80% efficient.

$$\text{Brake Horsepower, HP} = (\text{Power to elec. motor})(\text{motor eff.})$$

$$= \frac{(8 \text{ kW})(0.80)}{0.746 \text{ kW/HP}}$$

$$= 8.6 \text{ HP}$$

$$\text{Pump Efficiency, \%} = \frac{\text{Water Horsepower, HP} \times 100\%}{\text{Brake Horsepower, HP}}$$

$$= \frac{4.4 \text{ HP} \times 100\%}{8.6 \text{ HP}}$$

$$= 51\%$$

The following diagram may clarify the above problem:

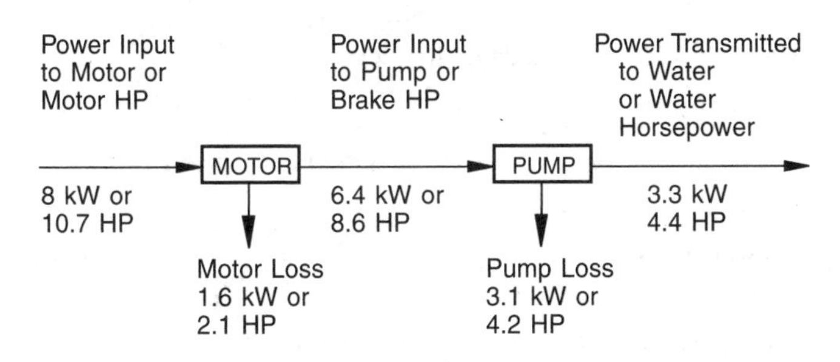

The wire-to-water efficiency is the efficiency of the power input to produce water horsepower.

$$\text{Wire-to-Water Efficiency, \%} = \frac{\text{Water Horsepower, HP}}{\text{Power Input, HP}} \times 100\%$$

$$= \frac{4.4 \text{ HP}}{10.7 \text{ HP}} \times 100\%$$

$$= 41\%$$

The wire-to-water efficiency of a pumping system (pump and electric motor) can be calculated by using the following formula:

$$\text{Efficiency, \%} = \frac{(\text{Flow, GPM})(\text{TDH, ft})(100\%)}{(\text{Voltage, volts})(\text{Current, amps})(5.308)}$$

$$= \frac{(375 \text{ GPM})(48.47 \text{ ft})(100\%)}{(220 \text{ volts})(36 \text{ amps})(5.308)}$$

$$= 41\%$$

A.87 Pump Speed-Performance Relationships

Changing the velocity of a centrifugal pump will change its operating characteristics. If the speed of a pump is changed, the flow, head developed, and power requirements will change. The operating characteristics of the pump will change with speed approximately as follows:

$$\text{Flow, } Q_n = \left[\frac{N_n}{N_r}\right] Q_r$$

$$\text{Head, } H_n = \left[\frac{N_n}{N_r}\right]^2 H_r$$

$$\text{Power, } P_n = \left[\frac{N_n}{N_r}\right]^3 P_r$$

r = rated
n = now
N = pump speed

Actually, pump efficiency does vary with speed; therefore, these formulas are not quite correct. If speeds do not vary by more than a factor of two (if the speeds are not doubled or cut in half), the results are close enough. Other factors contributing to changes in pump characteristic curves include impeller wear and roughness in pipes.

Example: To illustrate these relationships, assume a pump has a rated capacity of 600 GPM, develops 100 ft of head, and has a power requirement of 15 HP when operating at 1,500 RPM. If the efficiency remains constant, what will be the operating characteristics if the speed drops to 1,200 RPM?

Calculate new flow rate or capacity:

$$\text{Flow, } Q_n = \left[\frac{N_n}{N_r}\right] Q_r$$

$$= \left[\frac{1{,}200 \text{ RPM}}{1{,}500 \text{ RPM}}\right](600 \text{ GPM})$$

$$= \left[\frac{4}{5}\right](600 \text{ GPM})$$

$$= (4)(120 \text{ GPM})$$

$$= 480 \text{ GPM}$$

Calculate new head:

$$\text{Head, } H_n = \left[\frac{N_n}{N_r}\right]^2 H_r$$

$$= \left[\frac{1{,}200 \text{ RPM}}{1{,}500 \text{ RPM}}\right]^2 (100 \text{ ft})$$

Solution continues on next page.

$$= \left(\frac{4}{5}\right)^2 (100 \text{ ft})$$

$$= \left(\frac{16}{25}\right) (100 \text{ ft})$$

$$= (16)(4 \text{ ft})$$

$$= 64 \text{ ft}$$

Calculate new power requirement:

$$\text{Power, } P_n = \left[\frac{N_n}{N_r}\right]^3 P_r$$

$$= \left[\frac{1{,}200 \text{ RPM}}{1{,}500 \text{ RPM}}\right]^3 (15 \text{ HP})$$

$$= \left(\frac{4}{5}\right)^3 (15 \text{ HP})$$

$$= \left(\frac{64}{125}\right)(15 \text{ HP})$$

$$= \left(\frac{64}{25}\right)(3 \text{ HP})$$

$$= 7.7 \text{ HP}$$

A.88 Friction or Energy Losses

Whenever water flows through pipes, valves, and fittings, energy is lost due to pipe friction (resistance), friction in valves and fittings, and the turbulence resulting from the flowing water changing its direction. Figure A.1 can be used to convert the friction losses through valves and fittings to lengths of straight pipe that would produce the same amount of friction losses. To estimate the friction or energy losses resulting from water flowing in a pipe system, we need to know: water flow rate, pipe size or diameter and length, and number, size, and type of valve fittings.

An easy way to estimate friction or energy losses is to follow these steps:

1. Determine the flow rate,
2. Determine the diameter and length of pipe,
3. Convert all valves and fittings to equivalent lengths of straight pipe (see Figure A.1),
4. Add up total length of equivalent straight pipe, and
5. Estimate friction or energy losses by using Figure A.2. With the flow in GPM and diameter of pipe, find the friction loss per 100 feet of pipe. Multiply this value by equivalent length of straight pipe.

The procedure for using Figure A.1 is very easy. Locate the type of valve or fitting you wish to convert to an equivalent pipe length; find its diameter on the right-hand scale; and draw a straight line between these two points to locate the equivalent length of straight pipe.

Example: Estimate the friction losses in the piping system of a pump station when the flow is 1,000 GPM. The 8-inch suction line is 10 feet long and contains a 90-degree bend (long sweep elbow), a gate valve, and an 8-inch by 6-inch reducer at the inlet to the pump. The 6-inch discharge line is 30 feet long and contains a check valve, a gate valve, and three 90-degree bends (medium sweep elbows):

SUCTION LINE (8-inch diameter)

Item	Equivalent Length, ft
1. Length of pipe	10
2. 90-degree bend	14
3. Gate valve	4
4. 8-inch by 6-inch reducer	17
5. Ordinary entrance	12
Total equivalent length	57 feet

Friction loss (Figure A.2) = 1.76 ft/100 ft of pipe

DISCHARGE LINE (6-inch diameter)

Item	Equivalent Length, ft
1. Length of pipe	30
2. Check valve	38
3. Gate valve	4
4. Three 90-degree bends (3)(14)	42
Total equivalent length	114 feet

Friction loss (Figure A.2) = 7.73 ft/100 ft of pipe

Estimate the total friction losses in pumping system for a flow of 1,000 GPM.

SUCTION

Loss = (1.76 ft/100 ft)(57 ft) = 1.0 ft

DISCHARGE

Loss = (7.73 ft/100 ft)(114 ft) = 8.8 ft

Total friction losses, ft = 9.8 ft

A.9 STEPS IN SOLVING PROBLEMS

A.90 Identification of Problem

To solve any problem, you have to identify the problem, determine what kind of answer is needed, and collect the information needed to solve the problem. A good approach to this type of problem is to examine the problem and make a list of *KNOWN* and *UNKNOWN* information.

Example: Find the theoretical detention time in a rectangular sedimentation tank 8 feet deep, 30 feet wide, and 60 feet long when the flow is 1.4 MGD.

Known	Unknown
Depth, ft = 8 ft	Detention Time, hours
Width, ft = 30 ft	
Length, ft = 60 ft	
Flow, MGD = 1.4 MGD	

Sometimes a drawing or sketch will help to illustrate a problem and indicate the knowns, unknowns, and possibly additional information needed.

A.91 Selection of Formula

Most problems involving mathematics in wastewater collection systems can be solved by selecting the proper formula, inserting the known information, and calculating the unknown. In our example, we will use the basic formula for calculating detention time in a basin, tank, or wet well.

$$\text{Detention Time, hr} = \frac{(\text{Tank Volume, cu ft})(7.48 \text{ gal/cu ft})(24 \text{ hr/day})}{\text{Flow, gal/day}}$$

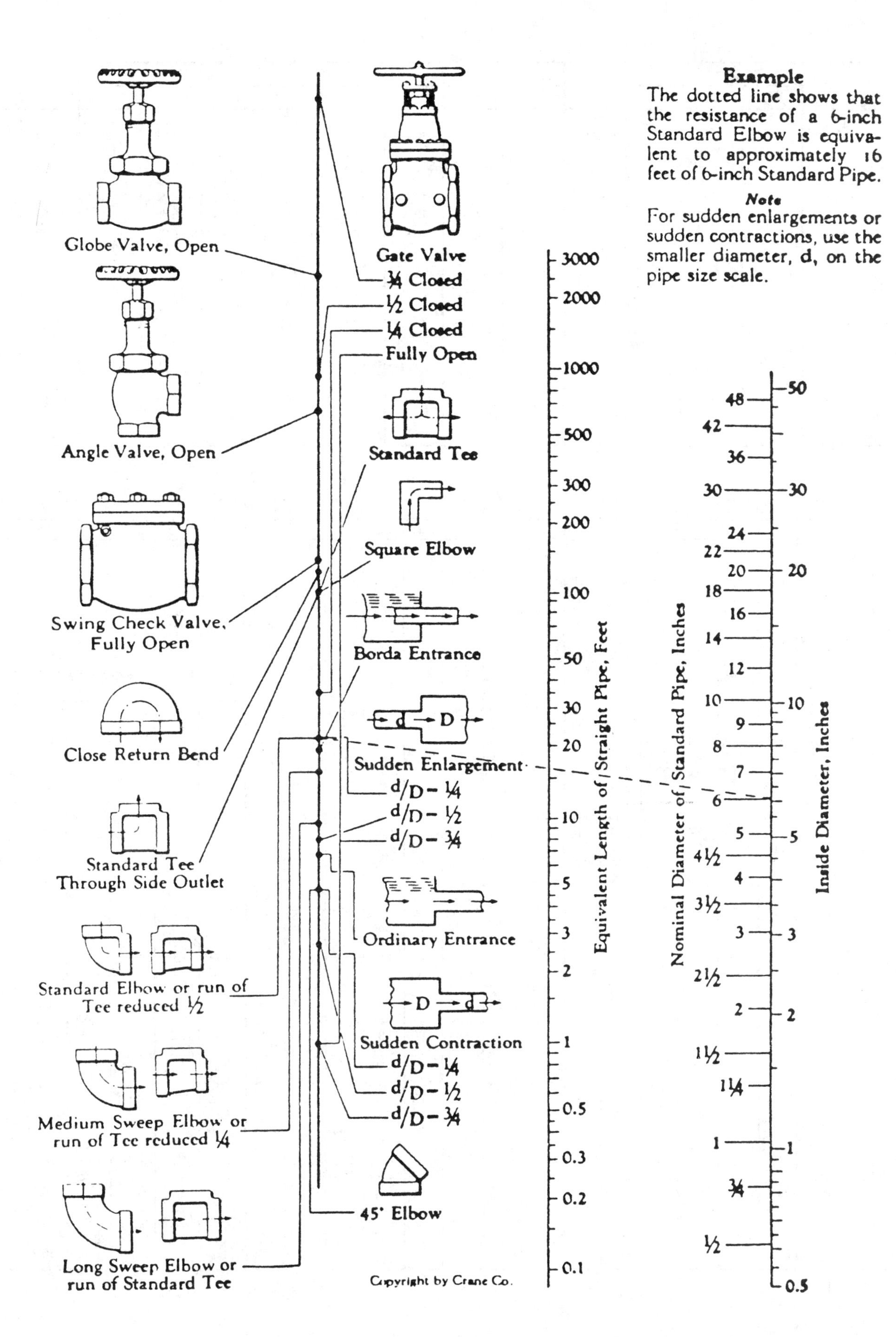

Fig. A.1 Resistance of valves and fittings to flow of water

(Reprinted by permission of Crane Co.)

U.S. GPM	0.5 in.		0.75 in.		1 in.		1.25 in.		1.5 in.		2 in.		2.5 in.	
	Vel.	Frict.	Vel.	Frict.	Vel.	Frict.	Vel.	Frict.	Vel.	Frict.	Vel.	Frict.	Vel.	Frict.
10	10.56	95.9	6.02	23.0	3.71	6.86	2.15	1.77	1.58	.83	.96	.25	.67	.11
20	....	....	12.0	86.1	7.42	25.1	4.29	6.34	3.15	2.94	1.91	.87	1.34	.36
30	....	....	....	....	11.1	54.6	6.44	13.6	4.73	6.26	2.87	1.82	2.01	.75
40	....	....	....	....	14.8	95.0	8.58	23.5	6.30	10.79	3.82	3.10	2.68	1.28
50	....	....	....	....	....	....	10.7	36.0	7.88	16.4	4.78	4.67	3.35	1.94
60	....	....	....	....	....	....	12.9	51.0	9.46	23.2	5.74	6.59	4.02	2.72
70	....	....	....	....	....	....	15.0	68.8	11.03	31.3	6.69	8.86	4.69	3.63
80	....	....	....	....	....	....	17.2	89.2	12.6	40.5	7.65	11.4	5.36	4.66
90	....	....	....	....	....	....	....	....	14.2	51.0	8.60	14.2	6.03	5.82
100	....	....	....	....	....	....	....	....	15.8	62.2	9.56	17.4	6.70	7.11
120	....	....	....	....	....	....	....	....	18.9	88.3	11.5	24.7	8.04	10.0
140	....	....	....	....	....	....	....	....		....	13.4	33.2	9.38	13.5
160	....	....	....	....	....	....	....	....		....	15.3	43.0	10.7	17.4
180	....	....	....	....	....	....	....	....		....	17.2	54.1	12.1	21.9
200	....	....	....	....	....	....	....	....		....	19.1	66.3	13.4	26.7
220	....	....	....	....	....	....	....	....		....	21.0	80.0	14.7	32.2
240	....	....	....	....	....	....	....	....		....	22.9	95.0	16.1	38.1
260	....	....	....	....	....	....	....	....		....	....	....	17.4	44.5
280	....	....	....	....	....	....	....	....		....	....	....	18.8	51.3
300	....	....	....	....	....	....	....	....		....	....	....	20.1	58.5
350	....	....	....	....	....	....	....	....		....	....	....	23.5	79.2

U.S. GPM	3 in.		4 in.		5 in.		6 in.		8 in.		10 in.		12 in.		14 in.		16 in.		18 in.		20 in.	
	Vel.	Frict.	Vel.	Frict.	Vel.	Frict.	Vel.	Frict.	Vel.	Frict.	Vel.	Frict.	Vel.	Frict.	Vel.	Frict.	Vel.	Frict.	Vel.	Frict.	Vel.	Frict.
20	.91	.15																				
40	1.82	.55	1.02	.13																		
50	2.72	1.17	1.53	.28	.96	.08																
80	3.63	2.02	2.04	.48	1.28	.14	.91	.06														
100	4.54	3.10	2.55	.73	1.60	.20	1.13	.10														
120	5.45	4.40	3.06	1.03	1.92	.29	1.36	.13														
140	6.35	5.93	3.57	1.38	2.25	.38	1.59	.18														
160	7.26	7.71	4.08	1.78	2.57	.49	1.82	.23														
180	8.17	9.73	4.60	2.24	2.89	.61	2.04	.28														
200	9.08	11.9	5.11	2.74	3.21	.74	2.27	.35														
220	9.98	14.3	5.62	3.28	3.53	.88	2.50	.42	1.40	.10												
240	10.9	17.0	6.13	3.88	3.85	1.04	2.72	.49	1.53	.12												
260	11.8	19.8	6.64	4.54	4.17	1.20	2.95	.57	1.66	.14												
280	12.7	22.8	7.15	5.25	4.49	1.38	3.18	.66	1.79	.16												
300	13.6	26.1	7.66	6.03	4.81	1.58	3.40	.75	1.91	.18												
350	....	....	8.94	8.22	5.61	2.11	3.97	1.01	2.24	.24												
400	....	....	10.20	10.7	6.41	2.72	4.54	1.30	2.55	.30												
460	....	....	11.45	13.4	7.22	3.41	5.11	1.64	2.87	.38	1.84	.12										
500	....	....	12.8	16.6	8.02	4.16	5.67	2.02	3.19	.46	2.04	.15	1.42	.06								
550	....	....	14.0	19.9	8.82	4.98	6.24	2.42	3.51	.56	2.25	.18	1.56	.07								
600	....	....	....	....	9.62	5.88	6.81	2.84	3.83	.66	2.45	.21	1.70	.08	1.25	.04						
700	....	....	....	....	11.2	7.93	7.94	3.87	4.47	.88	2.86	.29	1.99	.12	1.46	.05						
800	....	....	....	....	12.8	10.22	9.08	5.06	5.11	1.14	3.27	.37	2.27	.15	1.67	.07						
900	....	....	....	....	14.4	12.9	10.2	6.34	5.74	1.44	3.68	.46	2.55	.18	1.88	.09						
1000	....	....	....	....	....	....	11.3	7.73	6.38	1.76	4.09	.57	2.84	.22	2.08	.11						
1100	....	....	....	....	....	....	12.5	9.80	7.02	2.14	4.49	.68	3.12	.27	2.29	.13						
1200	....	....	....	....	....	....	13.6	11.2	7.66	2.53	4.90	.81	3.40	.32	2.50	.15	1.91	.08				
1300	....	....	....	....	....	....	14.7	13.0	8.30	2.94	6.31	.95	3.69	.37	2.71	.17	2.07	.09				
1400	....	....	....	....	....	....	....	....	8.93	3.40	5.72	1.09	3.97	.43	2.92	.20	2.23	.10				
1500	....	....	....	....	....	....	....	....	9.57	3.91	6.13	1.25	4.26	.49	3.13	.23	2.34	.12				
1600	....	....	....	....	....	....	....	....	10.2	4.45	6.54	1.42	4.54	.55	3.33	.25	2.55	.13	2.02	.07		
1700	....	....	....	....	....	....	....	....	10.8	5.00	6.94	1.60	4.87	.62	3.54	.29	2.71	.15	2.15	.08		
1800	....	....	....	....	....	....	....	....	11.5	5.58	7.35	1.78	5.11	.70	3.75	.32	2.87	.16	2.27	.09		
1900	....	....	....	....	....	....	....	....	12.1	6.19	7.76	1.97	5.39	.77	3.96	.35	3.03	.18	2.40	.10		
2000	....	....	....	....	....	....	....	....	12.8	6.84	8.17	2.17	5.67	.86	4.17	.39	3.19	.20	2.52	.11		
2500	....	....	....	....	....	....	....	....	....	....	10.2	3.38	7.10	1.33	5.21	.60	3.99	.31	3.15	.17		
3000	....	....	....	....	....	....	....	....	....	....	12.3	4.79	8.51	1.88	6.25	.86	4.79	.44	3.78	.24	3.06	.14
3500	....	....	....	....	....	....	....	....	....	....	14.3	6.55	9.93	2.56	7.29	1.16	5.58	.58	4.41	.32	3.57	.19
4000	....	....	....	....	....	....	....	....	....	....	....	....	11.3	3.31	8.34	1.50	6.38	.75	5.04	.42	4.08	.24
4500	....	....	....	....	....	....	....	....	....	....	....	....	12.8	4.18	9.38	1.88	7.18	.95	5.67	.53	4.59	.31
5000	....	....	....	....	....	....	....	....	....	....	....	....	14.7	5.13	10.4	2.30	7.98	1.17	6.30	.65	5.11	.38
6000	....	....	....	....	....	....	....	....	....	....	....	....	....	....	12.5	3.31	9.57	1.66	7.56	.92	6.13	.53
7000	....	....	....	....	....	....	....	....	....	....	....	....	....	....	14.6	4.50	11.2	2.26	8.83	1.24	7.15	.72
8000	....	....	....	....	....	....	....	....	....	....	....	....	....	....	....	....	12.8	2.96	10.09	1.61	8.17	.94
9000	....	....	....	....	....	....	....	....	....	....	....	....	....	....	....	....	14.4	3.73	11.3	2.02	9.19	1.18
10000	....	....	....	....	....	....	....	....	....	....	....	....	....	....	....	....	....	....	12.6	2.48	10.2	1.45

No allowance has been made for age, differences in diameter, or any other abnormal condition of interior surface. Any Factor of Safety must be estimated from the local conditions and the requirements of each particular installation. For general purposes, 15% is a responsible Factor of Safety.

Fig. A.2 Friction loss for water in feet per 100 feet of pipe

(Reprinted from the 10th Edition of the Standards of the Hydraulic Institute, 122 East 42nd Street, New York)

To convert the known information to fit the terms in a formula sometimes requires extra calculations. The next step is to find the values of any terms in the formula that are not in the list of known values.

Flow, gal/day = 1.4 MGD

= 1,400,000 gal/day

From Section A.30:

$$\text{Tank Volume, cu ft} = (\text{Length, ft})(\text{Width, ft})(\text{Height, ft})$$

$$= 60 \text{ ft} \times 30 \text{ ft} \times 8 \text{ ft}$$

$$= 14{,}400 \text{ cu ft}$$

Solution of Problem:

$$\text{Detention Time, hr} = \frac{(\text{Tank Volume, cu ft})(7.48 \text{ gal/cu ft})(24 \text{ hr/day})}{\text{Flow, gal/day}}$$

$$= \frac{(14{,}400 \text{ cu ft})(7.48 \text{ gal/cu ft})(24 \text{ hr/day})}{1{,}400{,}000 \text{ gal/day}}$$

$$= 1.85 \text{ hr}$$

The remainder of this section discusses the details that must be considered in solving this problem.

A.92 Arrangement of Formula

Once the proper formula is selected, you may have to rearrange the terms to solve for the unknown term. From Section A.71, "Flow Rate," we can develop the formula:

$$\text{Velocity, ft/sec} = \frac{\text{Flow Rate, cu ft/sec}}{\text{Cross-Sectional Area, sq ft}}$$

or $$V = \frac{Q}{A}$$

In this equation if Q and A were given, the equation could be solved for V. If V and A were known, the equation would have to be rearranged to solve for Q. To move terms from one side of an equation to another, use the following rule:

When moving a term or number from one side of an equation to the other, move the numerator (top) of one side to the denominator (bottom) of the other; or from the denominator (bottom) of one side to the numerator (top) of the other.

$$V = \frac{Q}{A} \text{ or } Q = AV \text{ or } A = \frac{Q}{V}$$

If the volume of a sedimentation tank and the desired detention time were given, the detention time formula could be rearranged to calculate the design flow.

$$\text{Detention Time, hr} = \frac{(\text{Tank Vol, cu ft})(7.48 \text{ gal/cu ft})(24 \text{ hr/day})}{\text{Flow, gal/day}}$$

By rearranging the terms

$$\text{Flow, gal/day} = \frac{(\text{Tank Vol, cu ft})(7.48 \text{ gal/cu ft})(24 \text{ hr/day})}{\text{Detention Time, hr}}$$

A.93 Unit Conversions

Each term in a formula or mathematical calculation must be of the correct units. The area of a rectangular clarifier (Area, sq ft = Length, ft x Width, ft) can't be calculated in square feet if the width is given as 246 inches or 20 feet 6 inches. The width must be converted to 20.5 feet. In the example problem, if the tank volume were given in gallons, then the 7.48 gal/cu ft would not be needed. *THE UNITS IN A FORMULA MUST ALWAYS BE CHECKED BEFORE ANY CALCULATIONS ARE PERFORMED TO AVOID TIME-CONSUMING MISTAKES.*

$$\text{Detention Time, hr} = \frac{(\text{Tank Volume, cu ft})(7.48 \text{ gal/cu ft})(24 \text{ hr/day})}{\text{Flow, gal/day}}$$

$$= \frac{\cancel{\text{cu ft}}}{} \times \frac{\cancel{\text{gal}}}{\cancel{\text{cu ft}}} \times \frac{\text{hr}}{\cancel{\text{day}}} \times \frac{\cancel{\text{day}}}{\cancel{\text{gal}}}$$

= hr (all other units cancel)

NOTE: We have hours = hr. One should note that the hour unit on both sides of the equation can be cancelled out and nothing would remain. This is one more check that we have the correct units. By rearranging the detention time formula, other unknowns could be determined.

If the design detention time and design flow were known, the required capacity of the tank could be calculated.

$$\text{Tank Volume, cu ft} = \frac{(\text{Detention Time, hr})(\text{Flow, gal/day})}{(7.48 \text{ gal/cu ft})(24 \text{ hr/day})}$$

If the tank volume and design detention time were known, the design flow could be calculated.

$$\text{Flow, gal/day} = \frac{(\text{Tank Volume, cu ft})(7.48 \text{ gal/cu ft})(24 \text{ hr/day})}{\text{Detention Time, hr}}$$

Rearrangement of the detention time formula to find other unknowns illustrates the need to always use the correct units.

A.94 Calculations

Sections A.12, "Multiplication," and A.13, "Division," outline the steps to follow in mathematical calculations. In general, do the calculations inside parentheses ( ) first and brackets [ ] next. Calculations should be done from left to right above and below the division line before dividing.

$$\text{Detention Time, hr} = \frac{[(\text{Tank, Volume, cu ft})(7.48 \text{ gal/cu ft})(24 \text{ hr/day})]}{\text{Flow, gal/day}}$$

$$= \frac{[(14{,}400 \text{ cu ft})(7.48 \text{ gal/cu ft})(24 \text{ hr/day})]}{1{,}400{,}000 \text{ gal/day}}$$

$$= \frac{2{,}585{,}088 \text{ gal-hr/day}}{1{,}400{,}000 \text{ gal/day}}$$

= 1.85, or

= 1.9 hr

A.95 Significant Figures

In calculating the detention time in the previous section, the answer is given as 1.9 hr. The answer could have been calculated:

$$\text{Detention Time, hr} = \frac{2{,}585{,}088 \text{ gal-hr/day}}{1{,}400{,}000 \text{ gal/day}}$$

= 1.846491429 . . . hours

How does one know when to stop dividing? Common sense and significant figures both help.

First, consider the meaning of detention time and the measurements that were taken to determine the knowns in the formula. Detention time in a tank is a theoretical value and assumes that all particles of water throughout the tank move through the tank at the same velocity. This assumption is not correct; therefore, detention time can only be a representative time for some of the water particles.

Will the flow of 1.4 MGD be constant throughout the 1.9 hours, and is the flow exactly 1.4 MGD, or could it be 1.35 MGD or 1.428 MGD? A carefully calibrated flowmeter may give a reading within 2% of the actual flow rate. Flows into a tank fluctuate and flowmeters do not measure flows extremely accurately; so the detention time again appears to be a representative or typical detention time.

Tank dimensions are probably satisfactory within 0.1 ft. A flowmeter reading of 1.4 MGD is less precise and it could be 1.3 or 1.5 MGD. A 0.1 MGD flowmeter error when the flow is 1.4 MGD is (0.1/1.4) x 100% = 7% error. A detention time of 1.9 hours, based on a flowmeter reading error of plus or minus 7%, also could have the same error or more, even if the flow was constant. Therefore, the detention time error could be 1.9 hours x 0.07 = ±0.13 hour.

In most of the calculations in the operation of wastewater collection systems, the operator uses measurements determined in the lab or read from charts, scales, or meters. The accuracy of every measurement depends on the sample being measured, the equipment doing the measuring, and the operator reading or measuring the results. Your estimate is no better than the least precise measurement. Do not retain more than one doubtful number.

To determine how many figures or numbers mean anything in an answer, the approach called "significant figures" is used. In the example the flow was given in two significant figures (1.4 MGD), and the tank dimensions could be considered accurate to the nearest tenth of a foot (depth = 9.0 ft) or two significant figures. Since all measurements and the constants contained two significant figures, the results should be reported as two significant figures or 1.9 hours. The calculations are normally carried out to three significant figures (1.85 hours) and rounded off to two significant figures (1.9 hours).

Decimal points require special attention when determining the number of significant figures in a measurement.

Measurement	Significant Figures
0.00325	3
11.078	5
21,000.	2

Example: The distance between two points was divided into three sections, and each section was measured by a different group. What is the distance between the two points if each group reported the distance it measured as follows:

Group	Distance, ft	Significant Figures
A	11,300.	3
B	2,438.9	5
C	87.62	4
Total Distance	13,826.52	

Group A reported the length of the section it measured to three significant figures; therefore, the distance between the two points should be reported as 13,800 feet (3 significant figures).

When adding, subtracting, multiplying, or dividing, the number of significant figures in the answer should not be more than the term in the calculations with the least number of significant figures.

A.96 Check Your Results

After having completed your calculations, you should carefully examine your calculations and answer. Does the answer seem reasonable? If possible, have another operator check your calculations before making any operational changes.

A.10 BASIC CONVERSION FACTORS

UNITS		
1,000,000	= 1 Million	1,000,000/1 Million
LENGTH		
12 in	= 1 ft	12 in/ft
3 ft	= 1 yd	3 ft/yd
5,280 ft	= 1 mi	5,280 ft/mi
AREA		
144 sq in	= 1 sq ft	144 sq in/sq ft
43,560 sq ft	= 1 acre	43,560 sq ft/ac
VOLUME		
7.48 gal	= 1 cu ft	7.48 gal/cu ft
1,000 m*L*	= 1 liter	1,000 m*L*/*L*
3.785 *L*	= 1 gal	3.785 *L*/gal
231 cu in	= 1 gal	231 cu in/gal
WEIGHT		
1,000 mg	= 1 gm	1,000 mg/gm
1,000 gm	= 1 kg	1,000 gm/kg
454 gm	= 1 lb	454 gm/lb
2.2 lbs	= 1 kg	2.2 lbs/kg
POWER		
0.746 kW	= 1 HP	0.746 kW/HP
DENSITY		
8.34 lbs	= 1 gal	8.34 lbs/gal
62.4 lbs	= 1 cu ft	62.4 lbs/cu ft
DOSAGE		
17.1 mg/*L*	= 1 grain/gal	17.1 mg/*L*/gpg
64.7 mg	= 1 grain	64.7 mg/grain
PRESSURE		
2.31 ft water	= 1 psi	2.31 ft water/psi
0.433 psi	= 1 ft water	0.433 psi/ft water
1.133 ft water	= 1 in Mercury	1.133 ft water/in Mercury
FLOW		
694 GPM	= 1 MGD	694 GPM/MGD
1.55 CFS	= 1 MGD	1.55 CFS/MGD
TIME		
60 sec	= 1 min	60 sec/min
60 min	= 1 hr	60 min/hr
24 hr	= 1 day	24 hr/day

NOTE: In our equations the values in the right-hand column may be written either as 24 hr/day or 1 day/24 hr depending on which units we wish to convert to obtain our desired results.

A.11 SUMMARY OF FORMULAS

A.110 Length

LENGTH OR CIRCUMFERENCE OF A CIRCLE:

Length, ft = 3.14 (Diameter, ft)

A.111 Area

RECTANGLE:

Area, sq ft = (Length, ft)(Width, ft)

TRIANGLE:

Area, sq ft = (1/2)(Base, ft)(Height, ft)

CIRCLE:

Area, sq ft = 0.785 (Diameter, ft)^2

CYLINDER (wall):

Area, sq ft = 3.14 (Diameter, ft)(Height, ft)

SPHERE:

Area, sq ft = 3.14 (Diameter, ft)^2

A.112 Volume

RECTANGLE:

Volume, cu ft = (Length, ft)(Width, ft)(Height, ft)

CYLINDER:

Volume, cu ft = 0.785 (Diameter, ft)^2(Height, ft)

SPHERE:

Volume, cu ft = 0.524 (Diameter, ft)^3

A.113 Velocity and Flow

$$\text{Velocity, ft/sec} = \frac{\text{Distance Traveled, ft}}{\text{Time, sec}}$$

or

$$\text{Velocity, ft/sec} = \frac{\text{Flow Rate, cu ft/sec}}{\text{Cross-Sectional Area, sq ft}}$$

$$\text{Avg. Flow, MGD} = \frac{\text{Sum of Daily Flows, MGD}}{\text{Number of Daily Flows}}$$

A.114 Slope or Grade

$$\text{Grade, ft/ft} = \frac{\text{Rise, ft}}{\text{Run, ft}}$$

A.115 Chemical Doses

Chlorine Demand, mg/*L* = Chlorine Dose, mg/*L* – Chlorine Residual, mg/*L*

Chlorine Feed Rate, lbs/day = (Dose, mg/*L*)(Flow, MGD)(8.34 lbs/gal)

Chemicals, lbs = (Concentration, mg/*L*)(Volume, MG)(8.34 lbs/gal)

A.116 Excavation and Paving

$$\text{Trench Width at Top, ft} = \text{Base, ft} + \frac{(2)(\text{Depth, ft})}{\text{Side Slope}}$$

$$\text{Where Sides Slope} = \frac{\text{Rise, ft}}{\text{Run, ft}}$$

A.117 Electricity

Ohm's Law E = IR

or

EMF, volts = (Current, amps)(Resistance, ohms)

Power, watts = (E, volts)(I, amps)

Power, watts = (I, amps)^2(R, ohms)

Motor Output, watts = (Power, watts)(Power Factor)(Motor Efficiency)

A.118 Pumps

$$\text{Water, HP} = \frac{(\text{Flow, GPM})(\text{H, ft})}{3{,}960}$$

$$\text{Brake, HP} = \frac{(\text{Flow, GPM})(\text{H, ft})}{(3{,}960)(E_p)}$$

$$\text{Motor, HP} = \frac{(\text{Flow, GPM})(\text{H, ft})}{(3{,}960)(E_p)(E_m)}$$

$$\text{Efficiency, \%} = \frac{\text{Output}}{\text{Input}} \times 100\%$$

$$\text{Pumping Rate, GPM} = \frac{\text{Volume, gallons}}{\text{Time, minutes}}$$

A.119 Storage Tanks, Wet Wells, and Storm Water Storage Basins

$$\text{Fill Time, hr} = \frac{(\text{Tank Volume, cu ft})(7.5\ \text{gal/cu ft})(24\ \text{hr/day})}{\text{Flow, gal/day}}$$

A.12 SOLUTIONS TO PROBLEMS

In this section typical math problems that have to be solved by wastewater collection system operators are presented. If a particular problem is too difficult, go to the next problem and try it. For example, if the problems on excavation become too difficult, try the problems on electricity.

A. Population Equivalent

1. A 24-inch sewer carries an average daily flow of 5 MGD. If the average daily flow per person from the area served is 100 GPCD* (gallons per capita or person per day), how many people discharge into the wastewater collection system?

Known		Unknown
Diameter, in	= 24 in	Pop. Served, persons
Flow, MGD	= 5 MGD	
	or = 5,000,000 gal/day	
Pop. Equiv., GPCD	= 100 GPCD	

How many people discharge into collection system?

$$\text{Pop. Served, persons} = \frac{\text{Flow, gal/day}}{\text{Pop. Equiv., gal/person/day}}$$

$$= \frac{5{,}000{,}000\ \text{gal/day}}{100\ \text{gal/person/day}}$$

$$= 50{,}000\ \text{persons}$$

* Usually about 70% of the water supply rate.

ANALYSIS OF RESULTS

Calculated Pop.	Actual Pop.	Comments
50,000	50,820	Results appear OK.
50,000	38,130	Since calculated population is considerably greater than actual, either there is a considerable industrial discharge or inflow/infiltration. *FIND THE CAUSE OF ADDITIONAL FLOW.*
50,000	63,617	Since calculated population is considerably less than actual population, apparently there is considerable exfiltration. *FIND THE CAUSE OR LOCATION OF EXFILTRATION.*

2. An industry discharges an average of 5,230 gallons per day. What is the population equivalent from this industry? Assume a population equivalent flow of 100 GPCD.

Known		Unknown
Average Flow, gal/day	= 5,320 gal/day	Pop. Equiv., persons
Pop. Equiv., GPCD	= 100 GPCD	

What is the population equivalent from this industry?

$$\text{Pop. Equiv., persons} = \frac{\text{Flow, gal/day}}{\text{Pop. Equiv., gal/person/day}}$$

$$= \frac{5{,}320 \text{ gal/day}}{100 \text{ gal/person/day}}$$

$$= 53.2$$

$$\text{or} = 53 \text{ persons}$$

B. Velocity and Flow

3. Estimate the velocity of wastewater flowing in a 12-inch sewer. A red dye is poured into an upstream manhole. Three minutes later the dye first appears in a manhole 400 feet downstream. The dye disappears 3 minutes and 40 seconds after the dye is poured into the upstream manhole.

Known		Unknown
Diameter, in	= 12 in	Velocity, ft/sec
Time, sec t_1	= 3 min	
Time, sec t_2	= 3 min, 40 sec	
Distance, ft	= 400 ft	

Convert travel time to seconds.

$$\text{Time, sec } t_1 = 3 \text{ min} = (3 \text{ min})(60 \text{ sec/min}) = 180 \text{ sec}$$

$$\text{Time, sec } t_2 = 3 \text{ min, } 40 \text{ sec} = (3 \text{ min})(60 \text{ sec/min}) + 40 \text{ sec} = 180 \text{ sec} + 40 \text{ sec} = 220 \text{ sec}$$

$$\text{Avg. Time, sec} = 1/2\,(t_1 + t_2)$$

$$= 1/2\,(180 \text{ sec} + 220 \text{ sec})$$

$$= 200 \text{ sec}$$

Estimate velocity in feet per second.

$$\text{Velocity, ft/sec} = \frac{\text{Distance, ft}}{\text{Avg Time, sec}}$$

$$= \frac{400 \text{ ft}}{200 \text{ sec}}$$

$$= 2.0 \text{ ft/sec}$$

ANALYSIS OF RESULTS

Velocity in the sewer is at the minimum recommended velocity of 2 FPS. Problems of solids deposition could result at low flows.

4. A float (marked stick) is dropped into a manhole to estimate the velocity of the wastewater. Two minutes and 30 seconds later the float arrives in a manhole 450 feet downstream.

Known		Unknown
Time, sec	= 2 min, 30 sec	Velocity, ft/sec
Distance, ft	= 450 ft	

Convert travel time to seconds.

$$\text{Time, sec} = 2 \text{ min, } 30 \text{ sec}$$

$$= (2 \text{ min})(60 \text{ sec/min}) + 30 \text{ sec}$$

$$= 120 \text{ sec} + 30 \text{ sec}$$

$$= 150 \text{ sec}$$

Estimate velocity in feet per second.

$$\text{Velocity, ft/sec} = \frac{\text{Distance, ft}}{\text{Time, sec}}$$

$$= \frac{450 \text{ ft}}{150 \text{ sec}}$$

$$= 3.0 \text{ ft/sec}$$

NOTE: Velocity of floats is often 10 to 15 percent higher than the average velocity.

Convert float velocity to average velocity. Assume float velocity 10 percent too high.

$$\text{Avg Vel, ft/sec} = \text{Float Vel, ft/sec} - (\text{Float Vel, ft/sec})(\text{fraction too high})$$

$$= 3.0 \text{ ft/sec} - (3.0 \text{ ft/sec})\left(\frac{10\%}{100\%}\right)$$

$$= 3.0 \text{ ft/sec} - 3.0 \text{ ft/sec } (0.1)$$

$$= 3.0 \text{ ft/sec} - 0.3 \text{ ft/sec}$$

$$= 2.7 \text{ ft/sec}$$

5. Determine the flow in a 12-inch sewer flowing full if the average velocity is 2 FPS.

Known		Unknown
Diameter, in	= 12 in	Flow, CFS
Velocity, ft/sec	= 2 ft/sec	

Calculate area of 12-inch sewer.

$$\text{Area, sq in} = \frac{\pi}{4} D^2$$

$$= \frac{3.14}{4}(12 \text{ in})^2$$

$$= (0.785)(144 \text{ sq in})$$

$$= 113 \text{ sq in}$$

or

$$\text{Area, sq ft} = \frac{\text{Area, sq in}}{144 \text{ sq in/sq ft}}$$

$$= \frac{113 \text{ sq in}}{144 \text{ sq in/sq ft}}$$

$$= 0.785 \text{ sq ft}$$

Determine the flow in cubic feet per second.

$$Q = AV$$

or

$$\text{Flow, CFS} = (\text{Area, sq ft})(\text{Velocity, ft/sec})$$

$$= (0.785 \text{ sq ft})(2 \text{ ft/sec})$$

$$= 1.57 \text{ cu ft/sec}$$

6. Convert a flow of 2 CFS to million gallons per day (MGD) and gallons per minute (GPM).

Known	Unknown
Flow, CFS = 2 CFS	Flow, MGD Flow, GPM

CONVERSION FACTORS

1 CFS = 0.646 MGD = 449 GPM

or

1.547 CFS = 1 MGD = 694 GPM

Convert 2 CFS to flow in MGD.

$$\text{Flow, MGD} = (\text{Flow, CFS})\frac{(0.646 \text{ MGD})}{(1 \text{ CFS})}$$

$$= (2 \text{ CFS})\frac{(0.646 \text{ MGD})}{(1 \text{ CFS})}$$

$$= 1.292 \text{ MGD}$$

$$\text{Flow, GPM} = (\text{Flow, CFS})\frac{(449 \text{ GPM})}{(1 \text{ CFS})}$$

$$= (2 \text{ CFS})\frac{(449 \text{ GPM})}{(1 \text{ CFS})}$$

$$= 898 \text{ GPM}$$

7. The depth of flow in a 15-inch diameter sewer is 6 inches. Determine the cross-sectional area of the flow.

Known	Unknown
D or Diameter, in = 15 in	Cross-sectional Area, sq ft
d or Depth, in = 6 in	

To determine the cross-sectional area for a sewer pipe flowing partially full use the following steps:

Find the value for the depth, d, divided by the diameter, D.

$$\frac{\text{d, in}}{\text{D, in}} = \frac{6 \text{ in}}{15 \text{ in}}$$

$$= 0.40$$

From Table 3.4 (page 38) for d/D = 0.40, Factor = 0.2934.

$$\text{Pipe Cross-sectional Area, sq ft} = \frac{(\text{Factor})(\text{Diameter, in})^2}{144 \text{ sq in/sq ft}}$$

$$= \frac{(0.2934)(15 \text{ in})^2}{144 \text{ sq in/sq ft}}$$

$$= 0.46 \text{ sq ft}$$

8. The slope of an 18-inch diameter sewer is 0.003 ft per ft, the pipe is flowing one-half full and the coefficient of pipe roughness is assumed to be 0.013. Determine the velocity and quantity (flow rate) of wastewater conveyed by the sewer.

Known	Unknown
Slope, S, ft/ft = 0.003 ft/ft	Velocity, V, ft/sec
Diameter, D, in = 15 in	Quantity, Q, CFS and MGD
Roughness, n = 0.013	

a. Calculate the cross-sectional area of the pipe when it is flowing half full.

$$\frac{\text{depth}}{\text{Diameter}} = \frac{(1/2)(15 \text{ in})}{15 \text{ in}}$$

$$= \frac{7.5 \text{ in}}{15 \text{ in}}$$

$$= 0.50$$

From Table 3.4, if d/D = 0.50, Factor = 0.3927.

b. Calculate the pipe cross-sectional flow area.

$$\text{Pipe Flow Area, sq ft} = \frac{(\text{Factor})(\text{Diameter, in})^2}{144 \text{ sq in/sq ft}}$$

$$= \frac{(0.3927)(18 \text{ in})^2}{144 \text{ sq in/sq ft}}$$

$$= 0.88 \text{ sq ft}$$

c. Calculate the wetted perimeter for a pipe flowing half full.

$$\text{Wetted Perimeter, P, ft} = \frac{(1/2)(3.14)(\text{Diameter, in})}{12 \text{ in/ft}}$$

$$= \frac{(0.5)(3.14)(18 \text{ in})}{12 \text{ in/ft}}$$

$$= 2.36 \text{ ft}$$

d. Calculate the hydraulic radius, R, in feet.

$$\text{Hydraulic Radius, R, ft} = \frac{\text{Flow Area, A, sq ft}}{\text{Wetted Perimeter, P, ft}}$$

$$= \frac{0.88 \text{ sq ft}}{2.36 \text{ ft}}$$

$$= 0.37 \text{ ft}$$

e. Calculate the flow velocity in feet per second.

$$\text{Velocity, V, ft/sec} = \frac{1.486}{0.013}(0.37\text{ ft})^{2/3}(0.003)^{1/2}$$

$$= (114.3)(0.515)(0.055)$$

$$= 3.2\text{ ft/sec}$$

f. Calculate the quantity of flow in cubic feet per second.

$$\text{Quantity, Q, CFS} = (\text{Area, sq ft})(\text{Velocity, ft/sec})$$

$$= (0.88\text{ sq ft})(3.2\text{ ft/sec})$$

$$= 2.8\text{ cu ft/sec}$$

g. Convert the quantity of flow from cubic feet per second to million gallons per day, MGD.

$$\text{Quantity, Q, MGD} = \frac{(\text{Quantity, Q, cu ft/sec})(7.48\text{ gal/cu ft})(60\text{ sec/min})(60\text{ min/hr})(24\text{ hr/day})}{1{,}000{,}000/\text{M}}$$

$$= \frac{\text{Quantity, Q, cu ft/sec}}{1.55\text{ cu ft/sec/MGD}}$$

$$= \frac{2.8\text{ cu ft/sec}}{1.55\text{ cu ft/sec/MGD}}$$

$$= 1.8\text{ MGD}$$

9. What was the average daily flow for a lift station given the following flow totalizer readings?

8:00 a.m., Monday, August 11	113,428,731 gallons
8:00 a.m., Monday, August 18	121,987,566 gallons

Known	Unknown
Totalizer Readings, gallons	Avg Flow, MGD
Time, days = 7 days	

Determine total flow for 7 days in gallons.

Final Reading, gal = 121,987,566 gal
Initial Reading, gal = 113,428,731 gal
Total Flow, gal = 8,558,835 gal
or Total Flow, MG = 8.559 MG

Calculate average flow, MGD.

$$\text{Avg Flow, MGD} = \frac{\text{Total Flow, MG}}{\text{Time, days}}$$

$$= \frac{8.559\text{ MG}}{7\text{ days}}$$

$$= 1.22\text{ MGD}$$

10. Estimate the average daily flow for a lift station given the following daily flow readings.

Day	Mon	Tues	Wed	Thurs	Fri	Sat	Sun
Flow, MG or MGD*	1.68	1.53	1.21	1.16	1.05	1.47	0.90

Known	Unknown
Daily Flows, MGD	Avg Daily Flow, MGD

**NOTE:* 1.68 MG flowed during one day so we had a flow of 1.68 MGD.

Estimate average daily flow, MGD.

$$\text{Avg Daily Flow, MGD} = \frac{\text{Sum of Daily Flows, MGD}}{\text{Number of Daily Flows}}$$

$$= \frac{1.68 + 1.53 + 1.21 + 1.16 + 1.05 + 1.47 + 0.90}{7}$$

$$= \frac{9.00\text{ MGD}}{7}$$

$$= 1.286\text{ MGD}$$

$$= 1.29\text{ MGD}$$

C. Pipes

11. A piece of pipe is 23 feet long and is to be cut into 4 equal pieces. Each piece will be ____ feet ____ inches long.

Known	Unknown
Length, ft = 23 ft	Length, ft and in
Want 4 equal pieces	

Determine length in feet and then feet and inches.

$$\text{Length, ft} = \frac{\text{Total Length, ft}}{\text{Number of Equal Pieces}}$$

$$= \frac{23\text{ ft}}{4\text{ pieces}}$$

$$= 5.75\text{ ft}$$

or

$$\text{Length, ft \& in} = 5\text{ ft} + (0.75\text{ ft})(12\text{ in/ft})$$

$$= 5\text{ ft }9\text{ in}$$

12. A sewer has failed and 127 feet of 8-inch pipe must be replaced. How many 5-foot sections of pipe will be required?

Known	Unknown
Length, ft = 127 ft	Number of Sections
Section, ft = 5 ft	

Calculate number of sections needed.

$$\text{No. of Sections} = \frac{\text{Total Length, ft}}{\text{Length, ft/section}}$$

$$= \frac{127\text{ ft}}{5\text{ ft/section}}$$

$$= 25.4\text{ sections}$$

$$= 26\text{ sections needed}$$

13. A sewer with a 12-inch diameter has how many times more carrying capacity than a 4-inch diameter sewer? Assume the velocity is the same in each sewer.

Known	Unknown
Diameter, in = 12 in	How many times more carrying capacity?
Diameter, in = 4 in	
Velocity same in each sewer	

Determine how many times more carrying capacity.

$$Q = A\,V \text{ and } A = 0.785\ D^2$$

$$\frac{12 \text{ in diameter}}{4 \text{ in diameter}} = \frac{12 \text{ inch pipe Q}}{4 \text{ inch pipe Q}}$$

$$= \frac{A\,V}{A\,V}$$

$$= \frac{\cancel{0.785}\ (12 \text{ in})^2\ \cancel{V}}{\cancel{0.785}\ (4 \text{ in})^2\ \cancel{V}}$$

$$= \frac{(12 \text{ in})^2}{(4 \text{ in})^2}$$

$$= \frac{144}{16}$$

$$= 9 \text{ times}$$

or

12-inch sewer has 9 times the capacity of the 4-inch sewer.

D. Slope or Grade

14. If the total fall in a ditch is 14 feet in 1,000 feet, what is the grade in ft/ft and in percent?

Known		Unknown
Fall, ft	= 14 ft	Grade, ft/ft
Length, ft	= 1,000	Grade, %

Calculate grade.

$$\text{Grade, ft/ft} = \frac{\text{Rise, ft}}{\text{Run, ft}}$$

$$= \frac{14 \text{ ft}}{1{,}000 \text{ ft}}$$

$$= 0.014 \text{ ft/ft}$$

or

$$\text{Grade, \%} = (0.014 \text{ ft/ft})(100\%)$$

$$= 1.4\%$$

15. How many feet of drop are in 400 feet of 8-inch sewer with a 0.045 ft/ft slope?

Known		Unknown
Diameter, in	= 8 in	Drop, ft
Length, ft	= 400 ft	
Slope, ft/ft	= 0.045 ft/ft	

Calculate drop in feet.

$$\text{Slope, ft/ft} = \frac{\text{Rise, ft}}{\text{Run, ft}}$$

Rearrange equation.

$$\text{Rise, ft} = (\text{Slope, ft/ft})(\text{Run, ft})$$

$$= (0.045 \text{ ft/ft})(400 \text{ ft})$$

$$= 18.0 \text{ ft}$$

16. What is the slope on an 8-inch sewer that is 400 feet long if the invert elevation of the upstream manhole is 428.31 feet and the invert elevation of the downstream manhole is 423.89 feet?

Known		Unknown
Diameter, in	= 8 in	Slope, ft/ft
Length, ft	= 400 ft	
Upstream MH	= 428.31 ft	
Downstream MH	= 423.89 ft	

Find difference in elevation or drop between manholes.

Upstream MH	= 428.31 ft
Downstream MH	= 423.89 ft
Drop, ft	= 4.42 ft

Calculate slope.

$$\text{Slope, ft/ft} = \frac{\text{Rise, ft}}{\text{Run, ft}}$$

$$= \frac{4.42 \text{ ft}}{400 \text{ ft}}$$

$$= 0.01105$$

$$= 0.011 \text{ ft/ft}$$

17. A 10-inch sewer has been televised and the footage meter between two manholes indicates the distance is 437 feet. An actual field measurement verifies the distance shown on the plans of 400 feet.

    a. What is the percent error in the length?

    b. If the TV camera was pulled from the upstream to the downstream manhole, what is the difference from each manhole to a broken pipe at distance 328 feet?

Known		Unknown
Meter Distance, ft	= 437 ft	a. Percent Error
Actual Distance, ft	= 400 ft	b. Location of Broken Pipe
Broken Pipe, ft	= 328 ft	

a. Calculate percent error.

$$\text{Error, \%} = \frac{(\text{Meter Dist., ft} - \text{Actual Dist., ft})(100\%)}{\text{Actual Dist., ft}}$$

$$= \frac{(437 \text{ ft} - 400 \text{ ft})(100\%)}{400 \text{ ft}}$$

$$= \frac{(37 \text{ ft})(100\%)}{400 \text{ ft}}$$

$$= 9.25\% \text{ error}$$

NOTE: Error is greater than 1 percent so all distances in log book must be corrected. Footage meter should be recalibrated.

b. Determine location of broken pipe.

$$\text{Location, ft from Upper MH} = \frac{(\text{Meter from Upper MH, ft})(\text{Actual Total, ft})}{\text{Meter Total, ft}}$$

$$= \frac{(328 \text{ ft})(400 \text{ ft})}{437 \text{ ft}}$$

$$= \frac{131{,}200 \text{ ft}}{437 \text{ ft}}$$

$$= 300 \text{ ft}$$

$$\text{Location, ft from Lower MH} = \frac{(\text{Meter from Lower MH, ft})(\text{Actual Total, ft})}{\text{Meter Total, ft}}$$

$$= \frac{(437 \text{ ft} - 328 \text{ ft})(400 \text{ ft})}{437 \text{ ft}}$$

$$= \frac{(109 \text{ ft})(400 \text{ ft})}{437 \text{ ft}}$$

$$= \frac{43{,}600 \text{ ft}}{437 \text{ ft}}$$

$$= 100 \text{ ft}$$

NOTE: Distance from upper and lower manholes to broken pipe (300 ft + 100 ft = 400 ft) is equal to the actual distance so the calculations check.

E. Leak Test

18. An 8-inch sewer 400 feet long is given a water leak test. The downstream manhole is plugged where the line enters the manhole. There are no service lines connected to the test line. At 9:00 a.m. the 36-inch upstream manhole was filled to the bottom of the cone. By 5:00 p.m. the water had dropped 4.0 feet. Calculate the leakage in terms of gallons per day per inch of sewer diameter per mile of sewer.

Known		Unknown
Diameter, in	= 8 in	Leakage in GPD/in/mi
Length, ft	= 400 ft	
Time, hr	= 8 hr	
Diameter, ft	= 3 ft	
Height, ft	= 4 ft	

Calculate the volume of water that leaked.

$$\text{Volume, cu ft} = (\text{Area, sq ft})(\text{Height, ft})$$

$$= (0.785)(3 \text{ ft})^2(4.0 \text{ ft})$$

$$= 28.27 \text{ cu ft}$$

$$\text{Volume, gal} = (28.27 \text{ cu ft})(7.48 \text{ gal/cu ft})$$

$$= 211.5 \text{ gal}$$

Calculate leakage in gallons per day.

$$\text{Leakage, GPD} = \frac{\text{Volume, gal}}{\text{Time, day}}$$

$$= \frac{(211.5 \text{ gal})(24 \text{ hr/day})}{8 \text{ hr}}$$

$$= 634.5 \text{ GPD}$$

Calculate leakage in GPD/inch of sewer.

$$\text{Leakage, GPD/in} = \frac{\text{Leakage, GPD}}{\text{Sewer Diameter, in}}$$

$$= \frac{634.5 \text{ GPD}}{8 \text{ in}}$$

$$= 79.3 \text{ GPD/in}$$

Calculate leakage in GPD/in/mile of sewer.

$$\text{Leakage, GPD/in/mi} = \frac{(79.3 \text{ GPD/in})(5{,}280 \text{ ft/mi})}{\text{Length, ft}}$$

$$= \frac{(79.3 \text{ GPD/in})(5{,}280 \text{ ft/mi})}{400 \text{ ft}}$$

$$= 1{,}047 \text{ GPD/in/mi}$$

NOTE: Maximum allowable leakage is usually 250 to 500 GPD/inch of sewer diameter/mile of sewer. Therefore sewer fails leak test. There should be very little leakage from a properly constructed sewer. Try to find all leaks and correct them.

19. Determine the force against each plug if a pipe is 18 inches in diameter and the test pressure is 5 psi.

Known		Unknown
Pipe Diameter, in	= 18 in	Force, lbs
Test Pressure, psi	= 5 psi	

a. Calculate the surface area of the plug.

$$\text{Surface Area, sq in} = (0.785)(\text{Diameter, in})^2$$

$$= (0.785)(18 \text{ in})^2$$

$$= 254 \text{ sq in}$$

b. Determine force against plug.

$$\text{Force, lbs} = (\text{Test Pressure, psi})(\text{Surface Area, sq in})$$

$$= (5 \text{ lbs/sq in})(254 \text{ sq in})$$

$$= 1{,}270 \text{ lbs}$$

How would you like to be clobbered by a force of 1,270 pounds?

> **CAUTION:** *NEVER ALLOW ANY PERSON TO ENTER A MANHOLE WHILE AIR IS BEING FORCED INTO A PIPE WITH PLUGS IN PLACE OR WHEN ANY PRESSURE REMAINS BEHIND PLUGS.*

F. Map and Blueprint Reading

20. The distance between two manholes on a map is measured as $^{15}/_{16}$ of an inch. Scale for the map is 1 inch equals 800 feet. Estimate the actual distance between the two manholes.

Known	Unknown
Distance, in = $^{15}/_{16}$ in	Actual Distance, ft
Scale, 1 in = 800 ft	

Estimate actual distance in feet.

$$\frac{\text{Actual Distance, ft}}{\text{Measured Distance, in}} = \frac{800\text{ ft}}{1\text{ in}}$$

$$\text{Actual Distance, ft} = \frac{(800\text{ ft})(^{15}/_{16}\text{ in})}{1\text{ in}}$$

$$= (800\text{ ft})\frac{(15)}{16}$$

$$= \frac{12{,}000}{16}$$

$$= 750\text{ ft}$$

G. Chemical Doses

CHLORINATION

21. Chlorine Demand

Determine the chlorine demand* of an effluent from a wastewater treatment plant if the chlorine dose is 10.0 mg/*L* and the chlorine residual is 1.1 mg/*L*.

Known	Unknown
Chlorine Dose = 10.0 mg/*L*	Chlorine Demand, mg/*L*
Chlorine Residual = 1.1 mg/*L*	

Chlorine Demand, mg/*L* = Chlor. Dose, mg/*L* – Chlor. Resid., mg/*L*
= 10.0 mg/*L* – 1.1 mg/*L*
= 8.9 mg/*L*

22. Chlorine Feed Rate

To control hydrogen sulfide and odors in an 8-inch sewer, the chlorine dose must be 10 mg/*L* when the flow is 0.37 MGD. Determine the chlorinator setting (feed rate) in pounds per day.

Known	Unknown
Dose, mg/*L* = 10 mg/*L*	Chlorinator Setting, lbs/day
Flow, mg/*L* = 0.37 MGD	

Chlorine Feed Rate, lbs/day = (Dose, mg/*L*)(Flow, MGD)(8.34 lb/gal)
= (10 mg/*L*)(0.37 MGD)(8.34 lb/gal)
= 30.9
or = 31 lbs/day

ROOT CONTROL

23. Quantity of Chemicals

A company contends their product is effective in controlling roots at a concentration of 100 mg/*L* if the contact time is 30 minutes. How many pounds of chemicals would be required assuming perfect mixing if 400 feet of 8-inch sewer were to be treated?

Known	Unknown
Concentration, mg/*L* = 100 mg/*L*	Chemicals, lbs
Contact Time, min = 30 min	
Diameter, in = 8 in	
Length, ft = 400 ft	

Determine volume of sewer.

$$\text{Volume, cu ft} = (0.785)(\text{Diam, ft})^2(\text{Length, ft})$$

$$= \frac{(0.785)(8\text{ in})^2(400\text{ ft})}{(12\text{ in/ft})^2}$$

$$= 139.63\text{ cu ft}$$

$$\text{Volume, gal} = (139.63\text{ cu ft})(7.48\text{ gal/cu ft})$$

$$= 1{,}044.4\text{ gal}$$

$$= 0.0010444\text{ MG}$$

Calculate weight of water in sewer.

Weight, lbs = (Volume, gal)(8.34 lb/gal)
= (0.001044 MG)(8.34 lb/gal)
= 0.0087 M lb

NOTE: 1 mg/*L* = 1 part/million parts = 1 lb/M lbs for water.

Calculate required pounds of chemicals.

$$100\text{ mg/}L = \frac{\text{Chemicals, lbs}}{\text{Water, M lbs}}$$

Chemicals, lbs = (100 mg/*L*)(Water, M lbs)
= (100 mg/*L*)(0.0087 M lb)
= 0.87 lbs

Order one pound of chemicals and pay the company if the chemicals work as claimed.

H. Excavation and Paving

24. A trench 3 feet wide, 8 feet deep and 70 feet long is to be filled with sand.
    a. How many cubic feet of sand required?
    b. How many cubic yards of sand?
    c. How many 5 cubic yard dump truck loads?
    d. How many tons of sand carried by each truck if sand weighs 144 lbs/cu ft?

* *STANDARD METHODS, 17th Edition, uses the term CHLORINE DEMAND when referring to stabilized water such as a domestic water supply and the term CHLORINATION REQUIREMENT when referring to wastewater. Neither term is mentioned in the 19th Edition (1995).*

Known		Unknown
Width, ft	= 3 ft	a. Sand, cu ft
Depth, ft	= 8 ft	b. Sand, cu yd
Length, ft	= 70 ft	c. Number of Trucks
Sand, lbs/cu ft	= 144 lbs/cu ft	d. Truck Load, tons

Find volume of trench and volume of sand.

$$\text{Volume, cu ft} = (\text{Length, ft})(\text{Width, ft})(\text{Depth, ft})$$
$$= (70\text{ ft})(3\text{ ft})(8\text{ ft})$$
$$= 1{,}680\text{ cu ft}$$

$$\text{Volume, cu yd} = \frac{\text{Volume, cu ft}}{27\text{ cu ft/cu yd}}$$
$$= \frac{1{,}680\text{ cu ft}}{27\text{ cu ft/cu yd}}$$
$$= 62.22\text{ cu yd}$$

Determine the number of 5 cubic yard truck loads required.

$$\text{Number of Truck Loads} = \frac{\text{Volume, cu yd}}{\text{Truck Capacity, cu yd/load}}$$
$$= \frac{62.22\text{ cu yd}}{5\text{ cu yd/load}}$$
$$= 12.44\text{ loads}$$

Thirteen truck loads will be required.

Calculate the truck load in tons.

$$\text{Truck Load, tons} = \frac{(\text{Truck Cap., cu yd})(\text{Sand, lbs/cu ft})(27\text{ cu ft/cu yd})}{2{,}000\text{ lbs/ton}}$$
$$= \frac{(5\text{ cu yd})(144\text{ lbs/cu ft})(27\text{ cu ft/cu yd})}{2{,}000\text{ lbs/ton}}$$
$$= 9.72\text{ tons}$$

25. How many cubic yards of AC paving material will be required to pave over a trench 2,400 feet long and 3 feet wide using a 3-inch deep patch?

Known		Unknown
Length, ft	= 2,400 ft	Paving, cu yards
Width, ft	= 3 ft	
Depth, in	= 3 in	

Find volume of paving in cu ft.

$$\text{Paving, cu ft} = (\text{Length, ft})(\text{Width, ft})(\text{Depth, ft})$$
$$= \frac{(2{,}400\text{ ft})(3\text{ ft})(3\text{ in})}{12\text{ in/ft}}$$
$$= 1{,}800\text{ cu ft}$$

Calculate volume of paving in cu yd.

$$\text{Paving, cu yd} = \frac{1{,}800\text{ cu ft}}{27\text{ cu ft/cu yd}}$$
$$= 66.7\text{ cu yd}$$

26. If a sewer trench is 2 feet wide at the bottom, 12 feet deep, and the walls are sloped at 3/4 horizontal to 1 vertical, how wide is the trench at the ground surface?

Known		Unknown
Trench Base, ft	= 2 ft	Trench Width, ft at top
Trench Depth, ft	= 12 ft	
Trench Slope, S	= 3/4 ft horiz. to 1 ft vert.	

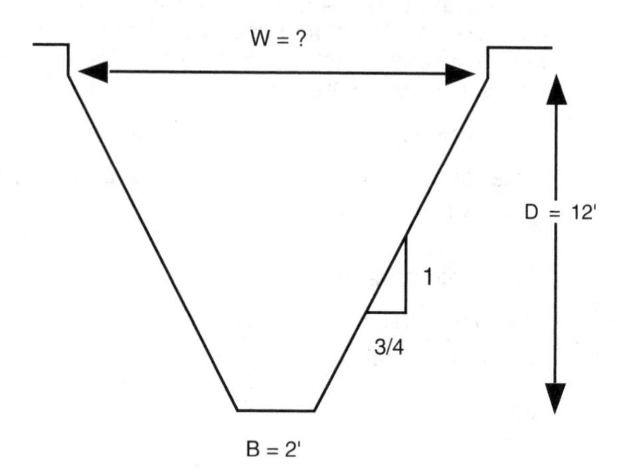

Determine trench width at top.

$$\text{Slope, S} = \frac{\text{Rise}}{\text{Run}}$$
$$= \frac{1}{3/4}$$
$$= 4/3$$

$$\text{Width, ft} = \text{Base, ft} + (2\text{ sides})\frac{(\text{Depth, ft})}{\text{Slope}}$$
$$= 2\text{ ft} + (2)\frac{(12\text{ ft})}{4/3}$$
$$= 2\text{ ft} + (2)(12\text{ ft})\frac{(3)}{4}$$
$$= 2\text{ ft} + 18\text{ ft}$$
$$= 20\text{ ft}$$

27. Prepare a sewer grade (cut) sheet for a sewer laid on a 0.5% grade with the given stake elevations and an invert grade of 100.00 feet for station 0+00.

Station	Stake Elevation	Invert Grade	Cut
0+00	105.60	100.00	______
0+50	106.12	______	______
1+00	106.75	______	______
1+50	107.37	______	______
2+00	108.02	______	______
2+50	108.41	______	______
3+00	108.98	______	______
3+50	109.22	______	______
3+91	108.69	______	______

To solve this problem, complete the above cut sheet by first calculating the invert grades.

$$\text{Grade, ft/ft} = \frac{\text{Grade, \%}}{100\%}$$

$$= \frac{0.5\%}{100\%}$$

$$= 0.005 \text{ ft/ft}$$

or $$= (0.005 \text{ ft/ft})\left(\frac{100 \text{ ft}}{100 \text{ ft}}\right)$$

$$= 0.5 \text{ ft/100 ft}$$

Since the grade is 0.5 ft per 100 ft, increase the invert grade by 0.25 ft for each 50 ft, or multiply the distance times the grade.

$$\text{Invert Grade, ft} = 100.00 \text{ ft} + (\text{Distance, ft})(\text{Grade, ft/ft})$$

$$= 100.00 \text{ ft} + (50 \text{ ft})(0.005 \text{ ft/ft})$$

$$= 100.00 \text{ ft} + 0.25 \text{ ft} = 100.25 \text{ ft}$$

Station

1+00 = 100.00 ft + (100)(0.005) = 100.50

1+50 = 100.00 ft + (150)(0.005) = 100.75

3+91 = 100.00 ft + (391)(0.005) = 101.96

To determine cut, subtract invert from stake elevation.

Station	Stake		Invert		Cut
0+00	105.60	–	100.00	=	5.60
0+50	106.12	–	100.25	=	5.87
1+00	106.75	–	100.50	=	6.25
3+91	108.69	–	101.96	=	6.73

Completed sewer cut sheet.

Station	Stake Elevation	Invert Grade	Cut
0+00	105.60	100.00	5.60
0+50	106.12	100.25	5.87
1+00	106.75	100.50	6.25
1+50	107.37	100.75	6.62
2+00	108.02	101.00	7.02
2+50	108.41	101.25	7.16
3+00	108.98	101.50	7.48
3+50	109.22	101.75	7.47
3+91	108.69	101.96	6.73

I. Electricity

28. A 12-volt battery operates a radio with a 4-ohm resistance. What amp fuse is necessary?

Known		Unknown
EMF, volts	= 12 volts	Current, amps
Resist., ohms	= 4 ohms	

Calculate the current, amps.

$$\text{Current, amps} = \frac{\text{EMF, volts}}{\text{Resistance, ohms}}$$

$$= \frac{12 \text{ volts}}{4 \text{ ohms}}$$

$$= 3 \text{ amps}$$

A three-amp fuse is necessary.

29. A 30-ohm resistor requires a current of 8 amps. How many watts will the resistor draw?

Known		Unknown
Resist., ohms	= 30 ohms	Power, watts
Current, amps	= 8 amps	

Calculate the power in watts.

$$\text{Power, watts} = (\text{I, amps})^2(\text{R, ohms})$$

$$= (8 \text{ amps})^2(30 \text{ ohms})$$

$$= 1{,}920 \text{ watts}$$

30. A pump motor is supplied with 220 volts and draws 15 amps. What is the horsepower output of the motor if the power factor is 0.8 and the motor efficiency is 90%?

Known		Unknown
EMF, volts	= 220	Power, watts Output, horsepower
Current, amps	= 15 amps	
Power Factor	= 0.8	
Efficiency, %	= 90%	

Calculate the power in watts.

$$\text{P, watts} = (\text{E, volts})(\text{I, amps})$$

$$= (220 \text{ volts})(15 \text{ amps})$$

$$= 3{,}300 \text{ watts}$$

Calculate the output horsepower.

$$\text{Output, HP} = \frac{(\text{P, watts})(\text{Power Factor})(\text{Efficiency})}{746 \text{ watts/HP}}$$

$$= \frac{(3{,}000 \text{ watts})(0.8)(0.9)}{746 \text{ watts/HP}}$$

$$= 3.18 \text{ HP}$$

Use a 3.5 horsepower motor.

J. Pumps

31. Determine the water horsepower delivered by a pump if the flow is 100 gallons per minute and the total dynamic head is 70 feet.

Known		Unknown
Flow, GPM	= 100 GPM	Water HP, HP
TDH, ft	= 70 ft	

Calculate water horsepower.

$$\text{Water HP} = \frac{(Q, \text{GPM})(\text{TDH, ft})}{3{,}960} = \frac{(Q, \text{CFS})(62.4 \text{ lb/cu ft})(H, \text{ft})}{550 \text{ ft-lb/sec-HP}}$$

$$= \frac{(100 \text{ GPM})(70 \text{ ft})}{3{,}960}$$

$$= 1.77 \text{ HP}$$

32. How efficient is a pump if the water horsepower is 1.77 HP and the brake horsepower is 2.50 HP?

Known	Unknown
Water HP = 1.77 HP	Pump Efficiency, %
Brake HP = 2.50 HP	

Calculate the pump efficiency.

$$\text{Pump Efficiency, \%} = \frac{(\text{Water HP})(100\%)}{\text{Brake HP}}$$

$$= \frac{(1.77 \text{ HP})(100\%)}{2.50 \text{ HP}}$$

$$= 71\%$$

33. If a pump motor draws 2,000 watts and the water horsepower from the pump is 1.77 HP, what is the overall efficiency of the system?

Known	Unknown
Pump Motor, watts = 2,000 watts	Overall Eff., %
Water HP = 1.77 HP	

Find input horsepower.

$$\text{Input HP} = \frac{\text{Power Drawn, watts}}{746 \text{ watts/HP}}$$

$$= \frac{2{,}000}{746 \text{ watts/HP}}$$

$$= 2.68 \text{ HP}$$

Calculate overall efficiency.

$$\text{Overall Eff., \%} = \frac{\text{Output, HP}}{\text{Input, HP}}$$

$$= \frac{(1.77 \text{ HP})\ 100\%}{2.68 \text{ HP}}$$

$$= 66\%$$

34. A pump has a rated capacity of 500 GPM, develops a head of 120 feet, and has a power requirement of 20 HP when operating at 900 RPM. If the efficiency remains constant, what will be the operating characteristics if the speed increases to 1,200 RPM?

Known	Unknown
Flow, GPM = 500 GPM	Q_n, GPM
Head, ft = 120 ft	H_n, GPM
Power, HP = 20 HP	P_n, GPM
N_r, RPM = 900 RPM	
N_n, RPM = 1,200 RPM	

Calculate the new flow rate or capacity.

$$Q_n, \text{GPM} = \left(\frac{N_n}{N_r}\right)Q_r$$

$$= \left(\frac{1{,}200 \text{ RPM}}{900 \text{ RPM}}\right)(500 \text{ GPM})$$

$$= \left(\frac{4}{3}\right)(500 \text{ GPM})$$

$$= 667 \text{ GPM}$$

Calculate the new head.

$$H_n, \text{ft} = \left(\frac{N_n}{N_r}\right)^2 H_r$$

$$= \left(\frac{1{,}200 \text{ RPM}}{900 \text{ RPM}}\right)^2 (120 \text{ ft})$$

$$= \left(\frac{4}{3}\right)^2 (120 \text{ ft})$$

$$= 213.3 \text{ ft}$$

Calculate the new power requirement.

$$P_n, \text{HP} = \left(\frac{N_n}{N_r}\right)^3 P_r$$

$$= \left(\frac{1{,}200 \text{ RPM}}{900 \text{ RPM}}\right)^3 (20 \text{ HP})$$

$$= \left(\frac{4}{3}\right)^3 (20 \text{ HP})$$

$$= 47.4 \text{ HP}$$

35. The rate at which a wet well fills can be used to calibrate a flowmeter. The same procedure can be used to determine the capacity of a pump. Determine the pump capacity in GPM if an 8-foot long and 6-foot wide wet well is lowered 4 feet in 5 minutes. Assume no additional water flows into the wet well.

Known	Unknown
Length, ft = 8 ft	Flow, GPM
Width, ft = 6 ft	
Depth, ft = 4 ft	
Time, min = 5 min	

Determine volume of water pumped.

$$\text{Volume, cu ft} = (\text{Length, ft})(\text{Width, ft})(\text{Depth, ft})$$

$$= (8 \text{ ft})(6 \text{ ft})(4 \text{ ft})$$

$$= 192 \text{ cu ft}$$

$$\text{Volume, gal} = (192 \text{ cu ft})(7.48 \text{ gal/cu ft})$$

$$= 1{,}436 \text{ gal}$$

Calculate pumping rate.

$$\text{Flow, GPM} = \frac{\text{Volume, gal}}{\text{Time, min}}$$

$$= \frac{1{,}436 \text{ gal}}{5 \text{ min}}$$

$$= 287 \text{ GPM}$$

36. Water is flowing into a wet well at a rate of 342 GPM. The level of the wet well is rising at a rate of 28 GPM. What is the pump capacity under these conditions?

Known	Unknown
Inflow, GPM = 342 GPM	Pump, GPM
Rise, GPM = 28 GPM	

Determine pump capacity.

Inflow = Outflow + Storage

Inflow, GPM = Pump, GPM + Rise, GPM

or

Pump, GPM = Inflow, GPM – Rise, GPM

= 342 GPM – 28 GPM

= 314 GPM

37. A lift station receives wastewater from a 24-inch intercepting sewer with a capacity of 5 MGD. The elevation of the sewer invert entering the wet well is 51.80 feet. The elevation of the lowest manhole lid on the intercepting sewer is 75.20 feet and the elevation of the lowest home plumbing fixture that could overflow and cause damage is 76.50 feet. The slope of the intercepting sewer is 1.5 ft/1,000 ft. If the lift station pumps fail or there is a power outage, how much time is available to get the lift station back on the line if the inflow is 2.5 MGD?

Known		Unknown
Diameter, in	= 24 in	Time, hour
Sewer Capacity, MGD	= 5 MGD	
Sewer Flow, MGD	= 2.5 MGD	
Elev. of Invert to L.S., ft	= 51.80 ft	
Elev. of MH Lid, ft	= 75.20 ft	
Elev. of Fixture, ft	= 76.50 ft	
Sewer Slope, ft/1,000 ft	= 1.5 ft/1,000 ft	

PROCEDURE

1. Calculate the storage capacity in the intercepting sewer when it is flowing at 2.5 MGD. Use only the capacity in the sewer between the wet well and the lowest manhole lid. A safety factor will be provided by any connecting sewers and a portion of the sewer whose crown elevation is above the lowest manhole lid but whose invert is below the elevation of the lid.

2. Calculate the time to fill the intercepting sewer by dividing the flow into the storage capacity.

Determine elevation at crown of sewer entering wet well.

Elevation of Sewer Invert, ft	= 51.80 ft
Diameter of Sewer, ft	= 2.00 ft
Elevation of Crown of Sewer, ft	= 53.80 ft

Calculate the elevation of vertical distance the water can rise before overflowing from the lowest manhole lid. Use crown of sewer rather than water surface at well because the empty portion of the sewer is the space or storage capacity that will be filled with water.

Elevation of Manhole Lid, ft	= 75.20 ft
Elevation of Crown of Sewer, ft	= 53.80 ft
Difference in Elevation, ft	= 21.40 ft

Use 21 feet with a 0.4 foot additional safety factor.

Estimate length of sewer that could be filled before the lowest manhole starts overflowing.

$$\text{Length of Sewer, ft} = \frac{\text{Diff. in Elev., ft}}{\text{Sewer Slope, ft/ft}}$$

$$= \frac{21.00 \text{ ft}}{1.5 \text{ ft}/1{,}000 \text{ ft}}$$

$$= (14 \text{ ft})(1{,}000)$$

$$= 14{,}000 \text{ ft}$$

NOTE: If slope is not 1.5 ft/1,000 ft over entire 14,000-foot section, length will have to be recalculated on basis of actual slopes and lengths until difference in elevation is reached.

Calculate the cross-sectional area of the sewer.

$$\text{Area, sq ft} = (0.785)(\text{Diameter, ft})^2$$

$$= (0.785)(2 \text{ ft})^2$$

$$= 3.14 \text{ sq ft}$$

Determine volume of 24-inch sewer 14,000 feet long.

$$\text{Volume, cu ft} = (\text{Area, sq ft})(\text{Length, ft})$$

$$= (3.14 \text{ sq ft})(14{,}000 \text{ ft})$$

$$= 43{,}982 \text{ cu ft}$$

NOTE: If 24-inch sewer is not 14,000 feet long, volume will have to be calculated on basis of lengths and diameters of sewer in 14,000-foot section.

What portion of volume is available for storage? If flow is 2.5 MGD and capacity is 5.0 MGD, then sewer is flowing half full.

$$\frac{\text{Actual Flow}}{\text{Capacity}} = \frac{2.5 \text{ MGD}}{5.0 \text{ MGD}}$$

$$= 1/2 \text{ full}$$

Therefore, 1/2 volume of sewer is available for storage.

$$\text{Storage Volume, cu ft} = (1/2)(\text{Volume, cu ft})$$

$$= (1/2)(43{,}982 \text{ cu ft})$$

$$= 21{,}991 \text{ cu ft}$$

Estimate time to fill sewer before manhole overflows.

$$\text{Time, hr} = \frac{(\text{Storage Volume, cu ft})(7.5 \text{ gal/cu ft})(24 \text{ hr/day})}{\text{Flow, gal/day}}$$

$$= \frac{(21{,}991 \text{ cu ft})(7.5 \text{ gal/cu ft})(24 \text{ hr/day})}{2{,}500{,}000 \text{ gal/day}}$$

$$= 1.58 \text{ hr or } 1 \text{ hr} + 0.58 \text{ hr } (60 \text{ min/hr})$$

$$= 1 \text{ hr } 35 \text{ min}$$

38. In the previous problem (number 37) we estimated a time of 1 hour and 35 minutes would be available before the lowest manhole would start overflowing. In actual practice we would watch the rate of rise of the water backing up to estimate how much time is available before the manhole would overflow. Estimate the expected rate of rise and calculate the actual rate of rise if water is rising 1.0 foot in 5 minutes.

Known	Unknown
From problem 37:	Estimated Rise, ft/min
Distance, ft = 21.00 ft	Actual Rise, ft/min
Time, hr = 1.58 hr	
= 1 hr 35 min	
This problem:	
Distance, ft = 1.0 ft	
Time, min = 5 min	

Calculate estimated rise.

$$\text{Est. Rise, ft/min} = \frac{\text{Distance, ft}}{\text{Time, min}}$$

$$= \frac{21.0 \text{ ft}}{(1.58 \text{ hr})(60 \text{ min/hr})}$$

$$= 0.22 \text{ ft/min}$$

How long would we estimate it takes for the water to rise 1 foot?

$$\text{Est. Rise, min/ft} = \frac{\text{Distance, ft}}{\text{Est. Rise, ft/min}}$$

$$= \frac{1 \text{ ft}}{0.22 \text{ ft/min}}$$

$$= 4.54 \text{ min to rise 1 ft}$$

What is the actual rate of rise?

$$\text{Actual Rise, ft/min} = \frac{\text{Distance, ft}}{\text{Time, min}}$$

$$= \frac{1.0 \text{ ft}}{5 \text{ min}}$$

$$= 0.2 \text{ ft/min}$$

NOTE: Since the actual time for the water to rise 1 foot is 5 minutes and we estimated 4.5 minutes, we'll have a little longer than our estimated time of 1 hour and 35 minutes to get the lift station back on the line before the lowest manhole starts to overflow. Continue to watch the actual rate of rise for any changes.

K. Ventilation

39. What capacity blower (CFM) is required to ventilate a manhole 48 inches in diameter and 17 feet deep with 15 air changes per hour or one air change every 4 minutes?

Known	Unknown
MH Diameter, in = 48 in	Blower Capacity, CFM
MH Depth, ft = 17 ft	
Air Change, min = 4 min	

Determine volume of manhole. Area = 0.785 D^2

$$\text{MH Volume, cu ft} = (\text{Area, sq ft})(\text{Depth, ft})$$

$$= (0.785)(4 \text{ ft})^2(17 \text{ ft})$$

$$= 213.6 \text{ cu ft}$$

Calculate blower capacity.

$$\text{Blower Capacity, CFM} = \frac{\text{MH Volume, cu ft}}{\text{Air Change, min}}$$

$$= \frac{213.6 \text{ cu ft}}{4 \text{ min}}$$

$$= 53.4 \text{ cu ft/min}$$

or at least 55 CFM

L. Costs

40. Estimate the total cost and cost per linear foot of a sanitary sewer construction project. The project consists of 1,550 lineal feet of 10-inch VCP with four manholes equally spaced. The average depth of the trench is 9 feet and the average width is 3 feet (including excavation for manholes). Estimated costs are as follows:

Excavation and Backfill	$ 30.00 per linear foot
Pipe Cost	$ 5.00 per linear foot
AC Paving	$ 4.00 per square foot
Manholes, completed	$ 1,500.00 each

Known	Unknown
Costs given above	Total Cost, $
Length, ft = 1,550 ft	Cost/ft, $/ft
Diameter, in = 10 in	
Depth, ft = 9 ft	
Width, ft = 3 ft	

Determine excavation and backfill cost.

$$\text{Cost, \$} = (\text{Cost, \$/ft})(\text{Length, ft})$$

$$= (\$30.00/\text{ft})(1{,}550 \text{ ft})$$

$$= \$46{,}500$$

Determine pipe cost.

Cost, \$ = (Pipe, cost, \$/ft)(Length, ft)

= (\$5.00/ft)(1,550 ft)

= \$7,750.00

Determine paving cost.

Cost, \$ = (Paving Cost, \$/sq ft)(Length, ft)(Width, ft)

= (\$4.00/sq ft)(1,550 ft)(3 ft)

= \$18,600.00

Determine manhole cost.

Cost, \$ = (MH Cost, \$/MH)(Number of MH)

= (\$1,500/MH)(4 MH)

= \$6,000.00

Determine total cost of project.

Exc. and Backfill	=	\$46,500.00
Pipe	=	7,750.00
Paving	=	18,600.00
MH	=	6,000.00
Total Cost	=	\$78,850.00

Determine cost per linear foot.

$$\text{Cost, \$/ft} = \frac{\text{Total Cost, \$}}{\text{Length, ft}}$$

$$= \frac{\$78,850.00}{1,550 \text{ ft}}$$

= \$50.87/ft

or = \$51/ft

41. A review of previous records indicated that a certain repair job could be completed by a crew of two workers in 20 hours. How long would it take a crew of five workers to complete a similar job and what would be the labor costs at \$10.00 per hour?

Known	Unknown
Two workers can do job in 20 hours.	Time for 5 Workers
Labor = \$10.00/hour	Labor Costs

Determine worker-hours to do the job.

Time, worker-hours = (Number of Workers)(Time, hours)

= (2 workers)(20 hours)

= 40 worker-hours

Calculate time for 5 workers to do job.

$$\text{Time, hours} = \frac{\text{Time, worker-hours}}{\text{Number of workers}}$$

$$= \frac{40 \text{ worker-hours}}{5 \text{ workers}}$$

= 8 hours

Estimate labor costs.

Labor Costs, \$ = (Time, worker-hours)(Labor, \$/worker-hour)

= (40 worker-hours)(\$10/worker-hour)

= \$400

M. Performance Indicators

42. A collection system manager wishes to determine performance indicators and performance ratings for both the city's sewer system and lift stations. Annual costs for operating and maintaining the sewer system are \$150,000 per year and \$50,000 per year for the lift stations. The city serves a population of 50,000 people with 150 miles of sewers and 20 lift stations. Last year there were 15 sewer line stoppages which caused flooding and one lift station failure which caused flooding. The lift station consumed 2.4 million kilowatt hours last year. Assume a sewer line standard of 90 and a lift station standard of 1,000.

SEWERS

Known		Unknown
Length of Sewers, mi	= 150 mi	1. Performance Indicator
Annual Cost, \$/yr	= \$15,000 yr	2. Performance Rating
Number of Stoppages	= 15/yr	
Assumed Standard	= 90	

Calculate the Performance Indicator for the sewer lines.

Performance Indicator = (Costs, \$/mi)(Stoppages, number/mi)

$$= \frac{(\$150,000)(15 \text{ stoppages})}{(150 \text{ mi})(150 \text{ mi})}$$

= 100

Calculate the Performance Rating for the sewer lines.

$$\text{Performance Rating} = \frac{(\text{Standard})(100\%)}{\text{Performance Indicator}}$$

$$= \frac{(90)(100\%)}{100}$$

= 90%

LIFT STATIONS

Known		Unknown
Number of Lift Stations	= 20	1. Performance Indicator
Annual Costs, $/yr	= $50,000/yr	2. Performance Rating
Number of Failures	= 1/yr	
Energy Consumed, mkWh/yr	= 2.4 mkWh/yr	
Assumed Standard	= 1,000	

Calculate the Performance Indicator for the lift stations.

$$\text{Performance Indicator} = \frac{(\text{Costs, \$/station})(\text{Failures, number/station})}{\text{Energy Consumed, mkWh/station}}$$

$$= \frac{(\$50{,}000/20)(1/20)}{(2.4/20)}$$

$$= \frac{(2{,}500)(0.05)}{0.12}$$

$$= 1{,}042$$

$$\text{Performance Rating} = \frac{(\text{Standard})(100\%)}{\text{Performance Indicator}}$$

$$= \frac{(1{,}000)(100\%)}{1{,}042}$$

$$= 96\%$$

A.13 ADDITIONAL READING

1. *BASIC MATH CONCEPTS FOR WATER AND WASTEWATER PLANT OPERATORS* and *APPLIED MATH FOR WASTEWATER PLANT OPERATORS* by Joanne Kirkpatrick Price. Obtain from CRC Press LLC, Attn: Order Entry, 2000 Corporate Boulevard, NW, Boca Raton, FL 33431-9868. Prices: *BASIC MATH* (TX8084) $49.95; *APPLIED MATH* (TX69892) $67.95.
2. *COMPUTATION PROCEDURES FOR WASTEWATER TREATMENT OPERATIONS* by Thurlow M. Morrow. Obtain from Thurlow M. Morrow, PO Box 372, Tacna, AZ 85352. Price, $21.95.

A.14 PRACTICE PROBLEMS

1. Estimate the number of people contributing to a flow of 400,000 gallons per day in an 8-inch sewer. Assume a population flow equivalent of
   a. 100 GPCD
   b. 85 GPCD.
2. A large cannery discharges 8 MGD during the peak canning season. What is the typical population equivalent using 100 GPCD?
3. A subdivision serves 1,500 people. What is the expected average daily flow in MGD assuming 100 GPCD?
4. A colored dye is dumped into a manhole. The dye first appears 4 minutes later in a manhole 750 feet downstream and disappears 6 minutes after first dumped into manhole. Estimate the velocity of flow in the sewer in
   a. ft/min
   b. ft/sec.
5. A float required 3 minutes and 28 seconds to travel 428 feet between two manholes. Estimate the velocity of the wastewater in the sewer assuming the surface velocity (float velocity) is 12 percent higher than the average velocity.
6. Estimate the flow in a 15-inch sewer flowing full if the average velocity is 2.6 ft/sec.
   a. CFS
   b. MGD
   c. GPM
7. Calculate the flow in a 6-inch sewer flowing half full when the average velocity is 2.3 ft/sec.
   a. CFS
   b. MGD
8. An 8-inch sewer carries a flow of 0.5 MGD when flowing full. What is the average velocity?
9. A sewer must carry a maximum flow of 1 MGD at a velocity of 2 ft/sec. What is the diameter? Select the next smaller nominal diameter so the actual flow will exceed 2 ft/sec and assume adequate grade is available to achieve the necessary velocity.
10. What is the average daily flow during August for a lift station given the following flow totalizer readings:

8:00 a.m. August 1	2,023,761,405 gallons
8:00 a.m. September 1	2,232,109,478 gallons

    There are 31 days in August.
11. Estimate the average daily flow on Mondays for a lift station with the following readings:

Date	Flow, MG
August 4	1.75
August 11	1.78
August 18	1.63
August 25	1.72

12. A piece of pipe is 17 feet long and is to be cut into 5 equal pieces. Each piece will be ______ feet ______ inches long.
13. A piece of pipe is 11 feet, 6 inches long and is to be cut into 4 equal pieces. Each piece will be ______ feet ______ inches long.
14. One side of the grounds for a lift station is 100 feet long. How many trees will be required if the trees are to be 25 feet apart with a tree at each end? HINT: Draw a sketch.
15. How many feet will a 12-inch sewer rise in 420 feet when laid on a grade of 0.009?
16. A trench drops 13 feet in 828 feet. What is the grade in ft/ft and in percent?
17. How many feet will a 6-inch sewer drop in 385 feet when laid on a 0.7% grade?
18. Determine the slope on a 10-inch sewer that is 367 feet long if the invert elevation at the downstream manhole is 77.23 feet and upstream invert elevation is 81.39 feet.
19. An 8-inch sewer has been televised and the footage meter between two manholes indicates the distance is 381 feet. An actual field measurement verifies the distance shown on the plans as 392 feet.
    a. What is the percent error in the length?
    b. If the TV camera was pulled from the upstream to the downstream manhole, what is the distance from each manhole to a protruding tap at a distance of 238 feet?

20. A 6-inch sewer 383 feet long is given a water leak test. The downstream manhole is plugged where the line enters the manhole and there are no service lines connected to the test line. At 9:00 a.m. on Thursday, the 36-inch upstream manhole was filled to the bottom of the cone. At 9:00 a.m. on Friday, the water had dropped 3 feet 8 inches. Calculate the leakage from the sewer in terms of gallons per day per inch of sewer diameter per mile of sewer.

21. A section of sewer in a swampy area is going to be televised in an attempt to locate and identify the causes of excessive infiltration. The distance to be televised measures $2\,^{13}/_{16}$ inches and the scale is 1 inch equals 800 feet. How long (in feet) is the line to be televised?

22. How many square feet of surface area are in a maintenance corporation yard that measures $2\,^{3}/_{4}$ inches by $5\,^{3}/_{8}$ inches on a drawing whose scale is 1 inch equals 40 feet?

23. A new manhole has been installed 254 feet from an existing manhole. How far would this new manhole be located from the old manhole on a map with a scale of 1 inch equals 40 feet?

24. Estimate the chlorine demand of a wastewater if the chlorinator is dosing at a rate of 20 mg/*L* and a residual of 0.4 mg/*L* of chlorine remains in the wastewater 30 minutes later.

25. If a wastewater is to be dosed at a rate of 10 mg/*L* and the flow is 2 MGD, what should be the chlorinator setting in pounds of chlorine per day?

26. A chlorinator is set to dose at a rate of 100 pounds of chlorine per day. What is the dose in mg/*L* when the flow is 0.8 MGD?

27. How many pounds of chemicals are required for root control if a concentration of 300 mg/*L* is recommended by the manufacturer to dose a 6-inch sewer 326 feet long?

28. Ten pounds of chemicals could produce what concentration (mg/*L*) if perfectly mixed in a 10-inch sewer 407 feet long?

29. A trench 2.5 feet wide, 6.5 feet deep and 65 feet long is to be filled with select backfill.
   a. How many cubic feet of backfill required?
   b. How many cubic yards of backfill?
   c. How many 5-cubic yard dump truck loads?
   d. How many tons of backfill if the backfill weighs 130 lbs/cu ft?

30. How many cubic yards of AC paving material will be required to pave a maintenance yard 100 feet wide and 200 feet long if the paving material is to be 4 inches thick?

31. How wide is a sewer trench at the top if the trench is 3 feet wide at the bottom, 9 feet deep, and the walls are sloped at 1 foot horizontal to 1 foot vertical?

32. Prepare a sewer grade (cut) sheet for a sewer laid on a 0.7% grade with the given stake elevations and invert grade.

Station	Stake Elevation	Invert Grade	Cut
0+00	57.28	50.00	______
0+50	57.91	______	______
1+00	58.13	______	______
1+50	58.55	______	______
2+00	59.67	______	______
2+48	60.03	______	______

33. A 12-volt battery operates a radio with a 3.5-ohm resistance. What amp fuse is necessary?

34. A pump motor is supplied with 440 volts and draws 35 amps. What is the horsepower output of the motor if the power factor is 0.85 and the motor efficiency is 89%?

35. A 40-ohm resistor requires a current of 11 amps. How many watts will the resistor draw?

36. Determine the water horsepower delivered by a pump if the flow is 500 GPM and the total dynamic head is 58 feet.

37. What is the total dynamic head that a pump can overcome when the flow is 2 CFS (cubic feet per second) and the delivered water horsepower is 10 horsepower?

38. What flow (GPM) could a 6 horsepower pump deliver against a total dynamic head (TDH) of 48 feet under these flow conditions? *NOTE:* TDH depends on flow and increases with flow. Usually a flow is known, the TDH is determined, and a pump to meet these requirements is selected.

39. How efficient is a pump if the brake horsepower is 8 HP and the water horsepower is 5.8 HP?

40. If a pump motor draws 3,800 watts and the water horsepower from the pump is 3.5 HP, what is the overall efficiency of the system?

41. A variable-speed pump delivers 500 GPM at a speed of 1,200 RPM. What should be the speed to deliver 550 GPM?

42. Estimate the flow into a wet well 5 feet wide and 8 feet long if the level rises 3.5 feet in 6 minutes.

43. Estimate the pumping rate of a pump when water is flowing into a wet well at 500 GPM and the wet well water level is dropping 0.8 feet in 3 minutes. The wet well is 5 feet by 5 feet.

44. Estimate the total cost and cost per linear foot of a sanitary sewer construction project. The project consists of 378 feet of 6-inch sewer and one manhole. Costs are estimated as follows:

Excavation and backfill	$ 12.00	per linear foot
Pipe cost	$ 2.00	per linear foot
Manhole, complete	$650.00	each

45. Estimate the time for a four-operator crew to complete a particular job. The labor costs are $6 per hour. Previous records indicate that the job could be completed by a crew of three operators in 36 hours. Also estimate the total cost of labor.

46. The temperature of wastewater in a sewer is reported as 12°C. What is the temperature in degrees Fahrenheit?

47. The temperature of wastewater is measured as 75°F. What is the temperature in degrees Celsius (centigrade)?

48. Determine the performance indicators and performance ratings for both the sewer lines and lift stations described in this problem. Annual costs for operating and maintaining the 176 mile sewer system are $176,392 per year and $39,186 per year for the 17 lift stations. Last year there were 23 sewer line stoppages which caused flooding and two lift station failures which caused flooding. The lift station consumed 0.572 million kilowatt hours last year. Assume a sewer line standard of 120 and a lift station standard of 6,000.

A.15 ANSWERS TO PRACTICE PROBLEMS

1. a. 4,000 persons
   b. 4,706 persons
2. 80,000 persons
3. 0.150 MGD
4. a. 150 ft/min
   b. 2.5 ft/sec
5. 1.81 ft/sec
6. a. 3.19 CFS
   b. 2.06 MGD
   c. 1,433 GPM
7. a. 0.23 CFS
   b. 0.15 MGD
8. 2.2 ft/sec
9. 12 in
10. 6.72 MGD
11. 1.72 MGD
12. 3 ft $4\frac{13}{16}$ in
13. 2 ft $10\frac{1}{2}$ in
14. 5 trees
15. 3.78 ft
16. 0.0157 ft/ft or 1.57%
17. 2.70 ft
18. 0.0113 ft/ft or 1.13%
19. a. 2.8%
    b. 245 ft from upper
       147 ft from lower
20. 446 GPD/in/mi
21. 2,250 ft
22. 23,650 sq ft
23. $6\frac{3}{8}$ in
24. 19.6 mg/*L*
25. 170 lbs/day
26. 15 mg/*L*
27. 1.2 lbs
28. 722 mg/*L*
29. a. 1,056 cu ft
    b. 39 cu yd
    c. 8 truck loads
    d. 68.6 tons
30. 247 cu yds
31. 21 ft
32.

Station	Invert Grade	Cut
0+00	50.00	7.28
0+50	50.35	7.56
1+00	50.70	7.43
1+50	51.05	7.50
2+00	51.40	8.27
2+48	51.74	8.29

33. 3.5 amps
34. 15.6 HP
35. 4,840 watts
36. 7.32 HP
37. 44 ft
38. 495 GPM
39. 72.5%
40. 68.7%
41. 1,320 RPM
42. 175 GPM
43. 550 GPM
44. $5,942
    $15.72/ft
45. 27 hours
    $648
46. 53.6°C
47. 23.9°C
48. Sewers
    PI = 131
    PR = 92%

    Lift Stations
    PI = 8,095
    PR = 74%

SUBJECT INDEX

G

H

I

J

K

L

Q

R

S

NOTES

NOTES

NOTES

NOTES

NOTES

NOTES

NOTES

NOTES